International Conference on Nuclear Technologies and Sciences (ICoNETS 2017)

Journal of Physics: Conference Series Volume 962

Makassar, Indonesia
12 October 2017

Editors:

Geni Rina Sunaryo
Julwan Hendry Purba

ISBN: 978-1-5108-5987-6
ISSN: 1742-6588

Printed from e-media with permission by:

Curran Associates, Inc.
57 Morehouse Lane
Red Hook, NY 12571

Some format issues inherent in the e-media version may also appear in this print version.

Printed by Curran Associates, Inc. (2018)

For permission requests, please contact the Institute of Physics at the address below.

Institute of Physics
Dirac House, Temple Back
Bristol BS1 6BE UK

Phone: 44 1 17 929 7481
Fax: 44 1 17 920 0979

techtracking@iop.org

Additional copies of this publication are available from:

Curran Associates, Inc.
57 Morehouse Lane
Red Hook, NY 12571 USA
Phone: 845-758-0400
Fax: 845-758-2633
Email: curran@proceedings.com
Web: www.proceedings.com

Table of contents

Volume 962

International Conference on Nuclear Technologies and Sciences (ICoNETS 2017)
12 October 2017, Makassar, Indonesia

Accepted papers received: 19 January 2018
Published online: 16 March 2018

Preface

Papers

IOP Conf. Series: Journal of Physics: Conf. Series **962** (2018) 011001 doi:10.1088/1742-6596/962/1/011001

PREFACE

ICoNETS, an acronym for International Conference on Nuclear Energy Technologies and Sciences, has been launched in line with the National Nuclear Energy Agency of Indonesia (BATAN) objectives to give researchers and professionals from universities, research institutions and nuclear related industries the opportunities to present their latest studies, excellent ideas and thoughts related to the development of nuclear energy technologies and it's social culture aspects. It also provides exciting opportunities to all possible stakeholders to amplify their active involvements in future nuclear power plant constructions in Indonesia. ICoNETS2017 is the second in a series of conference on *nuclear technologies and sciences*. It follows the successful of ICoNETS2015 in Denpasar. ICoNETS2017 in Makassar, once again aims at covering state of the art research and development in all aspects related to *nuclear technologies and sciences*.

ICoNETS2017 in conjunction with the Forth-International Symposium on Smart Material and Mechatronics (ISSMM2017) was held in Makassar and organized by Deputy of Nuclear Energy Technology of National Nuclear Energy Agency (BATAN) and Engineering Faculty of Hasanuddin University. It was also co-sponsored by the Association of Indonesian Nuclear Society (HIMNI), the Indonesian Researcher Union BATAN Branch (HIMPENINDO-BATAN), and the local government of the South Sulawesi Province. The theme of ICoNETS2017 is *Nuclear Energy for a Realistic Approach of the National Energy Security and Long Term Sustainability*.

The ICoNETS2017 proceeding consists of a series of invited lectures by distinguished professors and individual presentations. Papers submitted to ICoNETS2017 were peer - reviewed by competent Editors and Reviewers for their intellectual content without regard to race, gender, sexual orientation, religious belief, ethnic origin, citizenship, or political philosophy of the authors. For confidentiality reason, the editor and any editorial staff must not disclose any information about a submitted manuscript to anyone other than the corresponding authors, reviewers, potential reviewers, other editorial advisers, and the publisher, as appropriate. In a total of 65 papers from various institutions from Republic of Indonesia, Japan and Republic of Korea has been accepted to be presented in the conference and included in the proceeding.

Thanks are due to all contributors, referees, sessions' chairs, and program committee members of ICoNETS2017 for their kind co-operation and enthusiasm for ICoNETS2017; to Chairman of National Nuclear Energy Agency (BATAN), Rector of Hasanuddin University, and Chairman of Nuclear Energy Regulatory Agency (BAPETEN) for their roles as honorary speakers; and to Frederik Reitsma (Division of Nuclear Power IAEA), Prof. Keigo WATANABE (Okayama University, Japan), and Dr. Ir. Trenggono Sutioso, MM. for their roles as invited speakers.

Dr. Syaiful Bakhri
ICoNETS2017 General Chair
Makassar, October 2017

IOP Conf. Series: Journal of Physics: Conf. Series **962** (2018) 011001 doi:10.1088/1742-6596/962/1/011001

EDITORS

Dr. Geni Rina Sunaryo, M.Sc. (BATAN)
Dr. Julwan Hendry Purba, ST., M. App.IT. (BATAN)

REVIEWERS

Prof. Dr. Djarot S. Wisnubroto (BATAN)
Prof. Dr. Nesimi Ertugrul (UoA, Australia)
Prof. Dr. M. Wihardi Tjaronge (UNHAS Makassar)
Prof. Dr. Akio Gofuku (Okayama Univ., Japan)
Prof. Dr. Nguyen Trung Tinh (TIC-VARNS, Vietnam)
Prof. Dr.-Ing. Nandy Putra (UI, Depok)
Prof. Dr. Jung Jae-cheon (KINGS, Korea)
Prof. Dr. Ridwan (BATAN)
Prof. Dr. Ir. Dedi Priadi, DEA. (UI, Depok)
Dr. Hadid Subkhi (IAEA)
Mr. Frederik Reitsma (IAEA)
Dr. Jim Kuijper (NRG, Netherlands)
Dr. Mark Mitchell (PBMR, South Africa)
Dr. Mike Davies (AFW, UK)
Dr. Kunihiko Nabeshimaa(JAEA-Japan)
Dr. Sun Jun (Tsinghua University, China)
Dr. Phongpaeth Pengvanich (CU,Thailand)
Dr. Sidik Permana (ITB Bandung)
Dr. Alexander Agung (UGM, Yogyakarta)
Dr. Deendarlianto (UGM, Yogyakarta)
Rafiuddin Syam, Ph.D. (UNHAS, Makassar)
Dr. Wayan Nata Septiadi (UNUD, Bali)

IOP Conf. Series: Journal of Physics: Conf. Series **962** (2018) 011001 doi:10.1088/1742-6596/962/1/011001

IOP Conf. Series: Journal of Physics: Conf. Series **962** (2018) 011002 doi:10.1088/1742-6596/962/1/011002

Peer review statement

All papers published in this volume of *Journal of Physics: Conference Series* have been peer reviewed through processes administered by the proceedings Editors. Reviews were conducted by expert referees to the professional and scientific standards expected of a proceedings journal published by IOP Publishing.

Counter Action Procedure Generation in an Emergency Situation of Nuclear Power Plants

A Gofuku

Graduate School of Natural Science and Technology, Okayama University, 3-1-1 Tsushima-naka, Kita-ku, Okayama, 700-8530, Japan

gofuku-a@cc.okayama-u.ac.jp

Abstract. Lessons learned from the Fukushima Daiichi accident revealed various weak points in the design and operation of nuclear power plants at the time although there were many resilient activities made by the plant staff under difficult work environment. In order to reinforce the measures to make nuclear power plants more resilient, improvement of hardware and improvement of education and training of nuclear personnel are considered. In addition, considering the advancement of computer technology and artificial intelligence, it is a promising way to develop software tools to support the activities of plant staff.This paper focuses on the software tools to support the operations by human operators and introduces a concept of an intelligent operator support system that is called as co-operator. This paper also describes a counter operation generation technique the authors are studying as a core component of the co-operator.

1. Introduction

Lessons learned from the Fukushima Daiichi accident revealed various weak points in the design and operation of nuclear power plants at the time [1] although there were many resilient activities made by the plant staff under difficult work environment. The factors that obstructed suitable counter activities can be summarized as weak power source and service systems against tsunami, excessive dependence on remote control, insufficient communication means for information sharing, weakness of lightning and monitoring outside the building, and insufficient tools to remove a heap of rubbles. These mean that the measures necessary for the fourth layer in the concept of defence in depth [2] were insufficient.

In order to reinforce the measures in the fourth layer to make nuclear power plants more resilient, improvement of hardware and improvement of education and training of nuclear personnel are considered. In addition, considering the advancement of computer technology and artificial intelligence, it is a promising way to develop software tools to support the activities of plant staff.

This paper focuses on the software tools to support the operations by human operators and introduces a concept of an intelligent operator support system that is called as co-operator [3, 4]. This paper also describes a counter operation generation technique the authors are studying as a core component of the co-operator.

2. Concept of co-operator

2.1. Roles of human operators

One of the expected roles of human operators in plant operation is to carry out situation awareness [5]. The situation awareness is an abstract conception that includes detecting an anomaly, identifying the anomaly cause, and predicting future plant behaviour.In addition, human operators are asked to take suitable counter operations based on counter operation procedures inan abnormal plant situation.

The newly constructed nuclear power plants equip with a large screen to share important information among human operators and VDU (Visual Display Unit) based operation panels to give flexible ways of monitoring and operating a plant. In this new control room, some autonomous intelligent systems support human operators. In this style of plant operation, information sharing among human operators and autonomous intelligent systems becomes important.

2.2. Co-operator

Considering the technological advancement, especially, in computer technology, artificial intelligence, and robotics, the concepts of system configuration of a future operation support system will include remote operation support, co-operator, augmented sense, and online system maintenance [6].

The co-operator is a concept of intelligent system to support human operators as shown in Fig. 1 [3, 4]. It monitors plant condition through plant instrumentation system and interacts mutually with human operators. Its expected roles are a subordinate to accurately execute the tasks requested by human operators, a partner to share operation tasks, and an adviser to give useful knowledge and information.

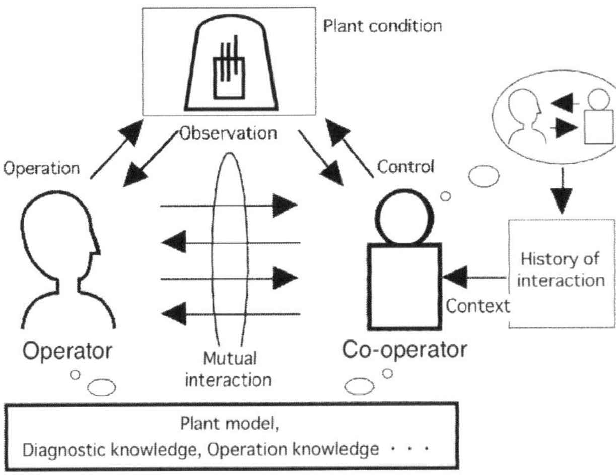

Fig. 1 Co-operator

The topics of developing a co-operator include monitoring plant condition by different viewpoints from those of human operators as well as the same viewpoints of them, moderate backup of the errors made by human operators, easy understandable explanation generation of processes and conditions of reasoning made by the co-operator, and understanding the intentions of human operators.

3. Counter operation procedure generation

3.1. Functional information

The authors are studying a technique to generate plausible counter operation procedures in an emergency plant situation as a core component of the co-operator [7, 8]. The technique utilizes a qualitative reasoning based on a functional model constructed in the framework of Multilevel Flow

Modelling (MFM) [9, 10, 11], although quantitative information can be generated by a combination of numerical simulation package [12].

Functional information expresses the role and objective of a component. The features of a functional model are a) correlation of role and purpose with system behavior, b) implicit representation of causal relations, c) hierarchical modeling capability, d) involvement of linguistic representation, and so on.

We utilize MFM as a functional modeling framework. An MFM model expresses an engineering system from the standpoint of means and goals. The characteristic feature of the MFM is to use a set of primitive function concepts to represent goals, functions, and their relations. An MFM model expresses the functional information of a system in a graphical way using objective and functional nodes and relation arcs.

3.2. Technique to generate counter operation procedures

In the counter operation procedure generation, the influence of an operation on plant future behavior is inferred along the causality relations implicitly expressed in an MFM model. For the inference, inference propagation rules [7, 11, 13] are derived in advance for each pattern of relation among function nodes and objective nodes. An algorithm to generate plausible counter operation procedures is developed.

The applicability of the algorithm is examined in several small break LOCA (Loss-Of-Coolant Accident) cases of a PWR (Pressurized Water Reactor) plant [7]. Three severe accidental situations of a PWR plant are considered. All of them are LOCA (Loss Of Coolant Accident) cases after the detection of partial core damage with the conditions such that (A) the ECCS (Emergency Core Cooling System) and main steam relief valve fail to initiate, (B) the residual heat removal system does not work, and (C) internal spray system fails to initiate. As shown in Table 1, the developed system successfully generates counter operation procedures including the procedures called AM (Accident Management) prepared for Japanese PWR plants.

Table 1. Generated operation procedures

Case		Generated procedure	Accident management
A	a	1. Open MSIV (Main Steam Isolation Valve) 2. Open TBV (Turbine Bypass Valve)	(Usage of turbine bypass system) 1. Open MSIV 2. Open TBV
	b	1. Open pressurizer relief valve	
B	a	1. Open MSIV 2. Open TBV	
	b	1. Open tie-line valve 2. Open residual heat removal valve	(Alternative recirculation) 1. Open tie-line valve 2. Open residual heat removal valve
	c	1. Open pressurizer relief valve	
	d	1. Open main steam relief valve	
C	a	1. Open pressurizer relief valve	
	b	1. Open main steam relief valve	
	c	1. Open MSIV 2. Open TBV	
	d	1. Open residual heat removal valve	
	e	1. Open fire protection system valve 2. Open containment spray valve	(Water injection into reactor containment) 1. Open fire protection system valve 2. Open containment spray valve

4. Conclusions

This paper deals with software tools to support the operations by human operators to make nuclear power plant more resilient andintroduces a concept of an intelligent operator support system that is called as co-operator. This paper also describes the outline of a counter operation generation technique as a core component of the co-operator.

Although there are many technological developments are necessary to implement an integrated intelligent system like co-operator, the author hopes that this kind of software tools will improve the safety of nuclear power plants to reinforce the fourth layer of defence in depth.

Acknowledgements
A part of the results of this study is obtained by the support of Japan Society for the Promotion of Science (JSPS) [KAKENHI grant number 16H03136].

References

[1] Investigation committee on the accident at the Fukushima nuclear power stations of Tokyo Electric Power Company, *Final report*, 2012.
[2] IAEA, "Defence in Depth in Nuclear Safety", *INSAG-10*, 1996.
[3] A. Gofuku, T. Ohi, K. Ito, Y. Niwa,"Dynamic Operation Permission System Based on Operation Procedure as a Co-operator", *The Transactions of Human Interface Society*, Vol. 6, No. 1, pp. 29-39, 2004. (in Japanese)
[4] A. Gofuku, "Support Systems of Plant Operators and Designersby Function-based Inference Techniques Based on MFM Models",*International Journal of Nuclear Safety and Simulation*,Vol. 2, No. 4, pp. 327-338 (2011).
[5] M. R. Endsley, "Toward a Theory of Situation Awareness in Dynamic Systems", *Human Factors*, Vol. 37, No. 1, pp. 32-64, 1995.
[6] A. Gofuku, "Necessary elemental technologies and image of future plant operation", *Instrumentation Control Engineering*, Vol. 49, No. 9, pp. 17-23, 2006. (in Japanese).
[7] A. Gofuku, T. Inoue, T. Sugihara, "A technique to generate plausible counter- operation procedures for an emergency situation based on a model expressing functions of components",*Journal of Nuclear Science and Technology*, http://dx.doi.org/10.1080/00223131.2017.1292966, 2017.
[8] T. Inoue, A. Gofuku, "A technique to prioritize plausible counter operation procedures in an accidental situation of plants", *International Journal of Nuclear Safety and Simulation*, Vol. 7, No. 2, pp. 144-152, 2016.
[9] M. Lind, "Modeling goals and functions of complex plant," *Applied Artificial Intelligence*, vol. 8, no. 2, pp. 259–283, 1994.
[10] M. Lind, "An introduction to multilevel flow modeling," *International Journal of Nuclear Safety and Simulation*, vol. 2, no. 1, pp. 22–32, 2011.
[11] M. Lind, X. Zhang, "Functional modeling for fault diagnosis and its application for NPP", *Nuclear Engineering and Technology*, vol. 46, no. 6, pp. 753-772, 2014.
[12] A. Gofuku, Y. Condo, "Quantitative Effect Indication of a Counter Action in an Abnormal Plant Situation",*International Journal of Nuclear Safety and Simulation*, Vol. 2, No. 3, pp. 255-264, 2011.
[13] A. Gofuku, K. Adachi, Y. Tanaka, "Finding out counter actions in an anomalous plant situation based on functions and behavior",*Transactions of The Institute of Systems, Control and Information Engineers*, Vol. 11, No. 8, pp. 458-465, 1998 (in Japanese).

Reliability Analysis of RSG-GAS Primary Cooling System to Support Aging Management Program

Deswandri, M.Subekti, Geni Rina Sunaryo

Center for Nuclear Reactor Technology and Safety – BATAN
Kawasan Puspiptek Gd. 80, Setu, Tangerang Selatan

wandri@batan.go.id

Abstract. Multipurpose Research Reactor G.A. Siwabessy (RSG-GAS) which has been operating since 1987 is one of the main facilities on supporting research, development and application of nuclear energy programs in BATAN. Until now, the RSG-GAS research reactor has been successfully operated safely and securely. However, because it has been operating for nearly 30 years, the structures, systems and components (SSCs) from the reactor would have started experiencing an aging phase. The process of aging certainly causes a decrease in reliability and safe performances of the reactor, therefore the aging management program is needed to resolve the issues. One of the programs in the aging management is to evaluate the safety and reliability of the system and also screening the critical components to be managed. One method that can be used for such purposes is the Fault Tree Analysis (FTA). In this papers FTA method is used to screening the critical components in the RSG-GAS Primary Cooling System. The evaluation results showed that the primary isolation valves are the basic events which are dominant against the system failure.

1. Introduction

In order to support the national development program, BATAN has been conducting the research activities, development and application of nuclear energy, isotope and radiation. These activities are conducted with the aim of accelerating the welfare of the nation. To implement such tasks, it is needed several facilities that can be used to carry out various activities. One of the main facilities of BATAN is Multipurpose Research Reactor G.A. Siwabessy (RSG-GAS), a 30 MW research reactor located in PUSPIPTEK, Setu, South Tangerang.

BATAN has been operating RSG-GAS since 1987. During that period, RSG-GAS reactor was successfully operated safely. However, since it has been in operation for almost 30 years, the Structures, Systems and Components (SSC) of the reactor are already beginning to experience an aging phase. The aging process is a natural process characterized by the declining ability of the SSC to perform its functions following the function of age or time [1]. There are two categories of aging: non-physical aging and physical aging. Non-physical aging is associated with obsolete equipment, which is already experiencing difficulties in terms of spare parts support or has not met the latest regulatory standards. Physical aging occurs due to the influence of various operating environment factors on the accumulated SSC over a long period of time, resulting in degradation in terms of functional ability of the SSC.

The aging process certainly leads to a decrease in reliability and safety performance of the reactor operation. RSG-GAS is part of BATAN's means to provide nuclear science and technology utilization services for stakeholders or the community. Decreasing reliability and safety performance will decreases the quality of service to stakeholders. To maintain the mission of BATAN in terms of providing excellent service in the utilization of nuclear science and technology, it is needed to take a systematic activity to manage the aging process that occurs in RSG-GAS. This systematic activity is embedded in an activity called aging management, and as regulated in BAPETEN rules, aging activity is one of the requirements for the process of extending the operating license and regulation.

Aging management is an integrated technical, operational and maintenance measures to control aging of SSC within acceptable safety limits [1]. The systematic program for aging management in research reactors includes elements, such as: screening the essential SSCs for aging management, identifying and understanding aging of SSCs, minimizing aging of SSCs, detecting, monitoring and mitigating SSC aging degradation, as well as several other programs [2].

The SSC screening program is an early stage in aging management activities. There are a large number of components required in the operation of the reactor. In selecting an important component for aging management, a systematic method is needed to determine which components are important for safety and the difficulty of replacement. There are various methods that can be used in screening the SSCs, including with expert judgment, operational experience analysis or by probabilistic safety analysis method (PSA) [3,4,5].

As a center assigned to operate the RSG-GAS reactor, Center for Multi-Purpose Research Reactor (PRSG) has already processed SSCs screening for aging management programs based on safety function and ease of replacement [6]. These SSCs screening program has not been systematically conducted using a particular method. PRSG's SSCs screening program is only a result of the assessment and evaluation of PRSG personnels based on daily operational experience and has not used the methods referred to in References 3, 4 and 5.

In this paper, one of the most frequently used elements in the probabilistic safety analysis method (PSA), was used to screen components of RSG-GAS Primary Cooling System. This element, the Fault Tree Analysis (FTA) method, is used to screen SSCs through the development of RSG-GAS Primary Cooling System model in the form of fault tree diagram. The development of RSG-GAS Primary Cooling System model in the form of fault tree diagram had been done in previous research activities [7] and also for the Pool Venting System and Low Pressure Ventilation System of RSG-GAS [8]. In contrast to the current model developed, the development of the reliability model in Reference 7 is intended to see the reliability level of the RSG-GAS Primary Cooling System while it is operating and the results of its analysis provide an estimated probability value of the loss of primary coolant flow initiating event in the RSG-GAS due to failure of Primary Cooling System. The reliability analysis of the Primary Cooling System by the FTA method carried out in the present activity is based on the real operating conditions of the current reactor, i.e. continuous operation for 45 days (average operating time), operating at 50% power of design power and estimated from the time of system started until completion of 45 day mission time. The purpose of analysis is to get information on which key components fail to have an impact on the function of the RSG-GAS Primary Cooling System and should receive more attention in day-to-day operations. Accordingly, the results of this research activity can be used to verify and validate the screening process of RSG-GAS Primary Cooling System components as provided in Reference 6.

2. Primary Cooling System of RSG-GAS

When the reactor is in operation, the reactor core will produce continuous amounts of heat as a result of fission reactions occurring in the fuel. In order for the temperature in the core to remain constant in accordance with the design temperature, this heat must be transferred to the environment. The Primary Cooling System serves to extract heat directly from within the reactor core by draining free mineral water to the fuel grille and transferring it to the secondary cooling system through heat exchanger components.

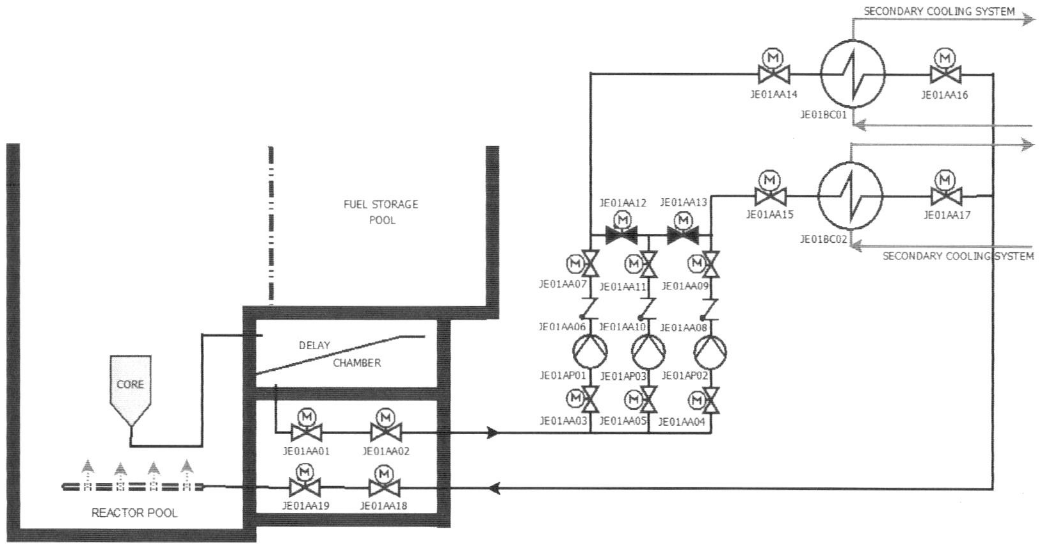

Figure 1. Diagram of RSG-GAS Primary Cooling System [9]

Figure 1 shows the schematical diagram of RSG-GAS Primary Cooling System. The components of the primary cooling system consist of the stainless steel pipes, three centrifugal pumps, motor operated valves and two heat exchangers. The pumps serve to drain cooling water to heat exchangers in a closed loop, where under normal circumstances two pumps operate and one is in standby state. The primary cooling water flows from the top to the bottom, through the porch and the chamber delay tank (to decrease the concentration of radioactive material N16) [9].

The valves in the primary cooling system have different functions. Some valves serve as a cooling water flow regulator when the system is working normally. Some valves serve to isolate the cooling water in the reactor pool when the system is not working. Some valves serve to isolate the spare pump, so that the cooling water flow does not enter the backup pump while the system is working. Some valves also serve to isolate one pump channel at the time the pump is in maintenance. The heat exchanger is isolated by two valves (at the entrance and exit lines) when the system is not working or the heat exchanger is undergoing maintenance.

The heat exchanger component serves to move heat from the primary system to the secondary system. There are two components of heat exchanger used, each having a capacity of 50% and installed in the piping line after the pump. The horizontally-mounted, vertical-shell and tube multy-pass heat exchanger component with the inlet and outlet side of the primary and secondary cooling water. The primary cooling water flows on the shell and the secondary cooling water flows in the center of the tube [9].

The function of the Primary Cooling System is to cool the core and reactor pool during normal operation. The success criteria of the system is the success of the heat transfer system generated from the core when the reactor operates normally to the secondary cooling system, so that the temperature of the core and pool is maintained according to the design temperature. In order for the primary cooling system to successfully perform its functions, then:

1. JE01AA01, AA02, AA14, AA15, AA16, AA17, AA 18 and AA19 should always be open when the system is in operation.
2. Two of the three Primary Cooling System Pumps (JE01AP01, JE01AP02 and JE01AP03) should be able to function while the system is working. In normal operation, two pumps operate and one is left in standby [6].

3. Each of the above mentioned pumps should be able to operate continuously while in service. When in standby, every pump must be immediately operated in the event that one pump in operation is suddenly dead (due to failure).
4. The JE01AA03, AA06 and AA07 valves must always be open when the JE01AP01 pump is operating. In contrast, the JE01AA03 and AA07 valves must be shut down when the JE01 AP01 pump is not operating.
5. The JE01AA04, AA08 and AA09 valves must always be open when the JE01AP02 pump is operating. In contrast, the JE01AA04 and AA09 valves must be shut down when the JE01 AP02 pump is not operating.
6. The JE01AA05, AA10 and AA11 valves must always be open when the JE01AP03 pump is operating. Conversely, the JE01AA05 and AA11 valves must be shut down when the JE01 AP03 pump is not operating.
7. The JE01AA12 valve should be able to open when the required JE01AP03 pump operates to replace the JE01AP01 pump.
8. The JE01AA13 valve must be able to open when the required JE01AP03 pump operates to replace the JE01AP02 pump.
9. JE01BC01 and BC02 Heat Exchangers should be able to function to transfer heat from the reactor core to the secondary cooling system.

3. Methodology

3.1. *Fault Tree Analysis (FTA).*
The FTA method is a graphical method that models how the failure of a component evolves into a system failure [10]. This method is deductive and presented in the form of directed acyclic graph consisting of two types of nodes, event and gate. The event consists of a basic event which is a spontaneous occurrence and an intermediate event which is an intermediate event caused by one or more other events. The gate illustrates how failures spread and combine resulting in system failure. Gates commonly used in the FTA method is AND gate, OR gate and VOTING gate. In the AND gate, the output gate occurs only if all input gates occur. In the OR gate, the output gate occurs if one or more input gates occur. While on VOTING gate, output gate occurs if a certain amount of input gate occurs.

The development of the model in the FTA method begins by defining a general event (Top Event), proceeding to an intermediate event that causes the general event. The analysis continues to the next event that causes the intermediate event. The analysis ends up to the most basic event level that cannot be elaborated, such as component failure events or human failure. Each event is related to the cause of the event through a gate. The use of gate (AND, OR or VOTING) depends on how the relation between input occurrences and output events.

3.2. Analytical Techniques.
The analytical techniques in the FTA method can be divided into two categories: qualitative techniques and quantitative techniques [10]. Qualitative techniques are used to detect weaknesses based on minimal cutset parameters, minimal path set and common cause failure. Quantitative techniques calculate the numerical value of Fault Tree diagrams in the form of Importance Measure parameters that show how critical (important) a component to the system and Stochastic Measure that gives probability value of system failure.

3.2.1. Qualitative Analysis Technique. In qualitative analysis, mathematical reduction follows the laws of the Boolean Algebraic equation to the failure tree diagram representing the system reliability model. The mathematical reduction results of the Boolean Algebra result in the minimal cutset parameter, which represents the smallest combination of basic events that can cause top events. The less of amount basic events in the minimal cutset the more critical the basic events in the system.

3.2.2. Quantitative Analysis Technique. In quantitative analysis, the probability calculation of top event occurrences is based on the probability of basic event events. To calculate the probability of basic event events, the reliability of basic events or components needs to be modeled. There are several component reliability models, such as fixed model, rate model, MTTF model, dormant model, standby model, and others [11]. In this paper, components are modeled based on the rate model. The mathematical equation of the rate model can be written as in Equation 1.

$$Q(t) = \frac{\lambda}{\lambda+\mu}\left[1 - e^{-(\lambda+\mu)t}\right] \tag{1}$$

where:

Q(t) = component unavailability
λ = component failure rate
μ = component repair rate
t = component mission time

The probability of a top event occurrence can be calculated by the rare event method or Esary-Proschan [11]. The Esary-Proschan method produces a more accurate calculation and is used in this paper. The mathematical equation for calculating the probability of a top event occurrence based on the Esary-Proschan method is given in the form as in Equations 2 and 3.

$$Q_{Sys} = \prod_{i=1}^{m} Q_i\left[1 - \prod_{j=1}^{n}\left(1 - Q_{Cutset\ j}\right)\right] \tag{2}$$

$$F_{Sys}(t) = 1 - e^{-\left(1-Q_{Sys}(t)\right)} \tag{3}$$

where:

F_{Sys} = system unreliability
Q_{Sys} = system unavailability
Qi = unavailability of i_{th} common event
$Q_{Cutset\ j}$ = cutset j_{th} unavailability with exclude of common event
m = number of common event in all of cutsets
n = cutset number
t = system mission time.

In addition to the calculation of the probability of a top event occurrence, quantitative analysis techniques in the FTA method also calculate the Importance Measure parameters. Importance Measure parameters calculated in this paper are F-Vesely Importance, Risk Achievement Worth (RAW) and Risk Reduction Worth (RRW) parameters. The Importance parameter F-Vesely is a parameter that shows how much the contribution of the basic event to the probability of top event occurrence. The RAW Importance parameter is a parameter that shows how much the probability of a top event increase if the related basic events always occur (components always fail). The RRW Importance parameter is a parameter that shows how much the probability of the top event decreases if the related basic event never occurs (the component is always successful). The mathematical equations for these three Importance Measures are given in Equation 4, 5 and 6 below [12]:

$$IMP_{FV} = \frac{Q(base)-Q(x=0)}{Q(base)} \tag{4}$$

$$IMP_{RAW} = \frac{Q(x=1)}{Q(base)} \tag{5}$$

$$IMP_{RRW} = \frac{Q(base)}{Q(x=0)} \tag{6}$$

where:
IMP_{FV}= F-Vesely importance
IMP_{RAW} = Risk Achievement Worth (RAW) importance
IMP_{RRW} = Risk Reduction Worth (RRW) importance
$Q_{(base)}$ = system unavailability
$Q_{(x=0)}$ = system unavailability with the asumption of the related basic events are full reliable (never fail)
$Q_{(x=1)}$ = system unavailability with the asumption of the related basic events are full unreliable (always fail)

3.2.3. Flow Chart of FTA Method.

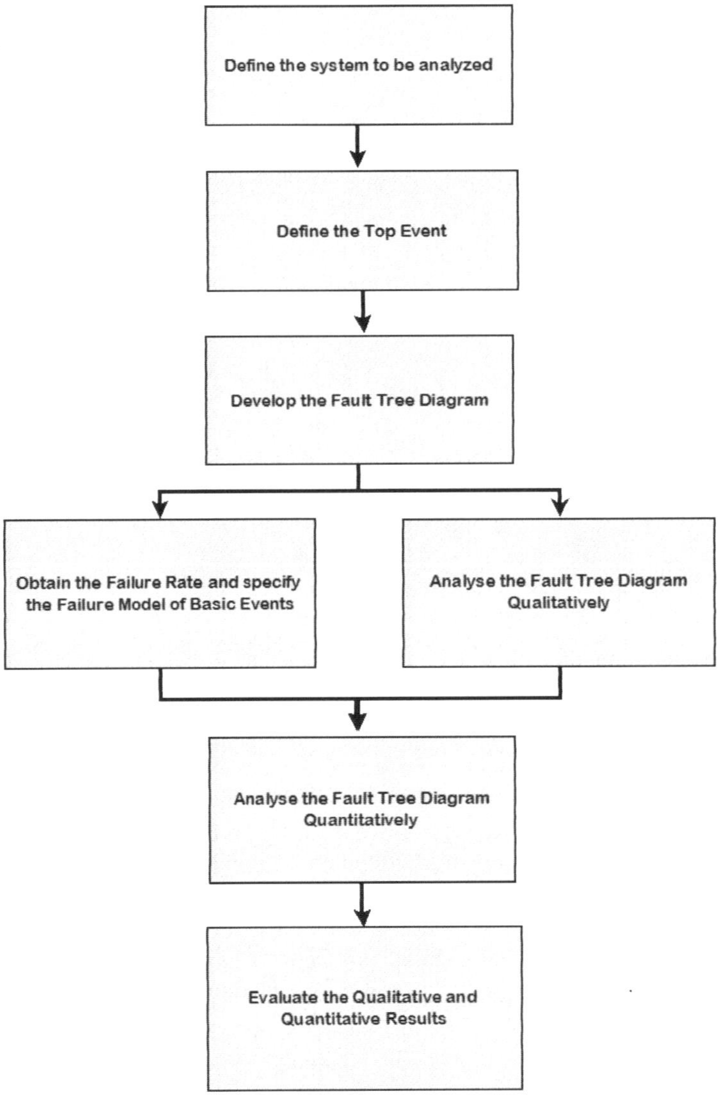

Figure 2. FTA Flow Chart Diagram

Figure 2 shows the flow chart diagram on developing a fault tree diagram of the system and analyzing process for the fault tree diagram that had been developed. The tool for analyzing the fault tree diagram on this paper is used the ITEM TOOLKIT software.

4. Result and Discussion

4.1. Development of Fault Tree Diagram.
In general, there are three factors needed to construct a fault tree diagram model: knowledge of the components in the system and their functions, the relationships between the components in the system and the logic and numerical evaluation of those relationships [13]. Failure Mode and Effect Analysis (FMEA) methods, Functional Block Diagram, and several other methods are commonly used for the initial step in constructing the fault tree diagram [14]. In this paper, Functional Block Diagram was developed to help understanding the system and the relationships among the components in the system, as shown in Figure 3 below.

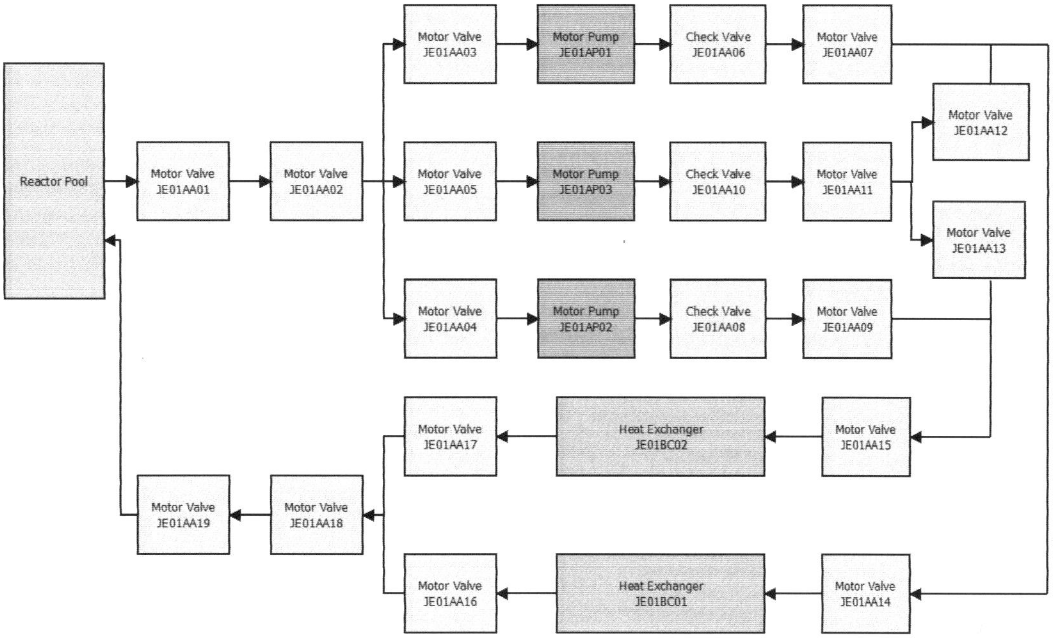

Figure 3. Functional Block Diagram of the RSG-GAS Primary Cooling System.

Based on the Functional Block Diagram of RSG-GAS Primary Cooling System in Figure 3, then we developed the fault tree diagram, as shown below. It should be noted that although in design the RSG-GAS primary cooling system is configured 2 of 3 (pump lines) [9], but since RSG-GAS reactor is operated only at 15 MWt (50% full power), the assumptions taken in the development the fault tree diagram are 1 out of 3 configuration for the pump lines and the heat exchanger used depend on the pump line operated.

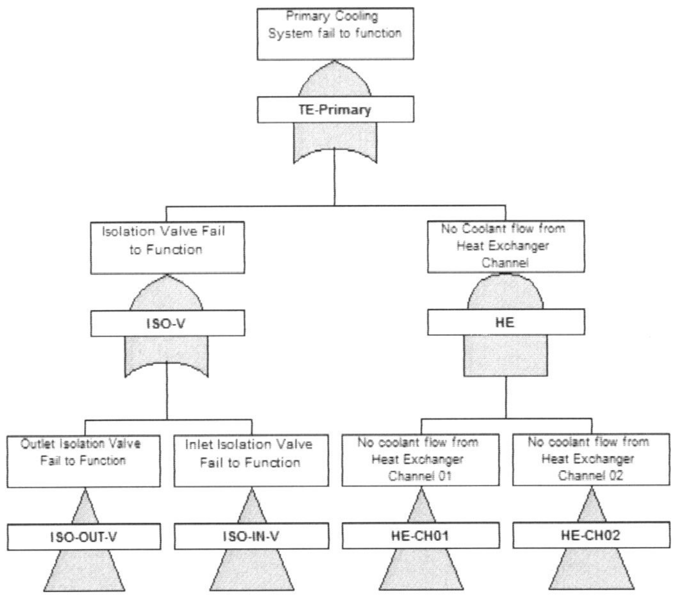

Figure 4. Fault Tree Diagram of RSG-GAS Primary Cooling System; Top Event.

In Figure 4 it is shown that the failure of the RSA-GAS Primary Cooling System is caused by the failure of the primary system isolation valves to function (open) or the failure of the coolant flow on the heat exchanger lines. The failure of the primary isolation valves occur when one of the JE01AA01 AA02, AA18 and AA19 fails to open or fails to continue to open (spurious closed). The primary isolation valves fault tree diagrams in more detail are given in Figures 5 and 6 below.

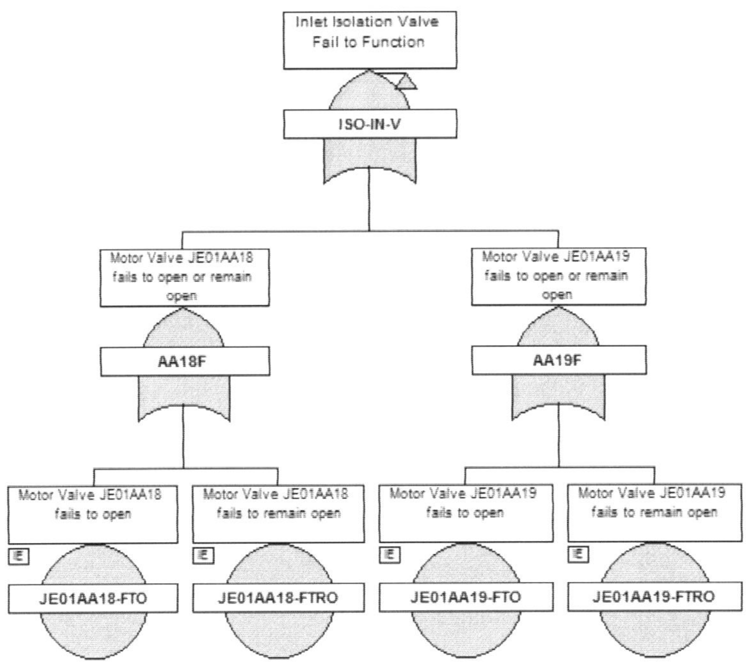

Figure 5. The fault tree diagram of the RSG-GAS primary isolation valves; Inlet part

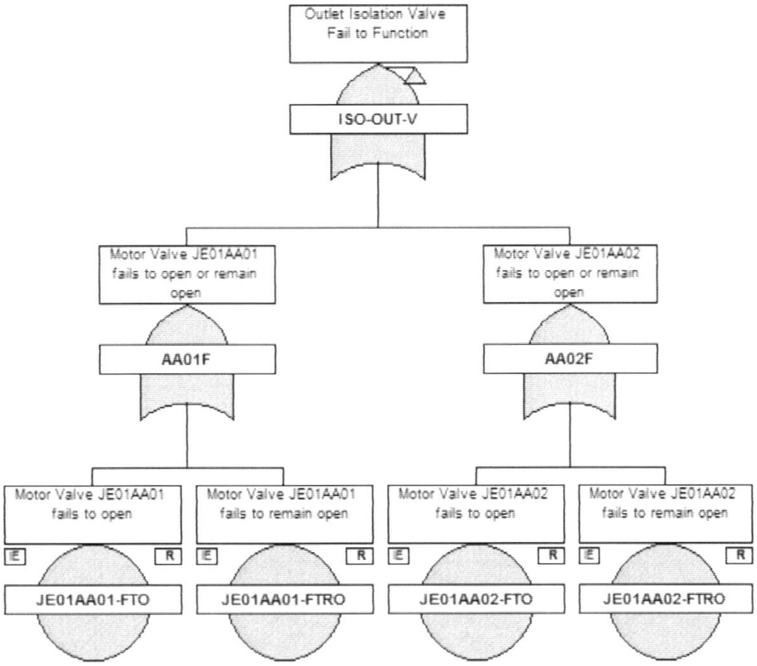

Figure 6. The fault tree diagram of the RSG-GAS primary isolation valves; Outlet part

The failure of no coolant flow in the heat exchanger line occurs if one of the heat exchanger line 1 or 2 fails to provide a coolant flow. The fault tree diagram of the absence of coolant flow in the RSG-GAS Heat Exchanger channel 1 is given in Figure 7. In the diagram, the failure of the absence of coolant flow in the heat exchanger channel 1 occurs due to the clogged in the heat exchanger, the isolation valves (JE01AA14 and AA16) not opening (failing to open or failing to keep open) or due to the failure of absence of coolant flow from the pump line 1. The failure model of the absence of coolant flow in the RSG-GAS Heat Exchanger line 2 is logically the same as line 1 and it is not shown in this paper.

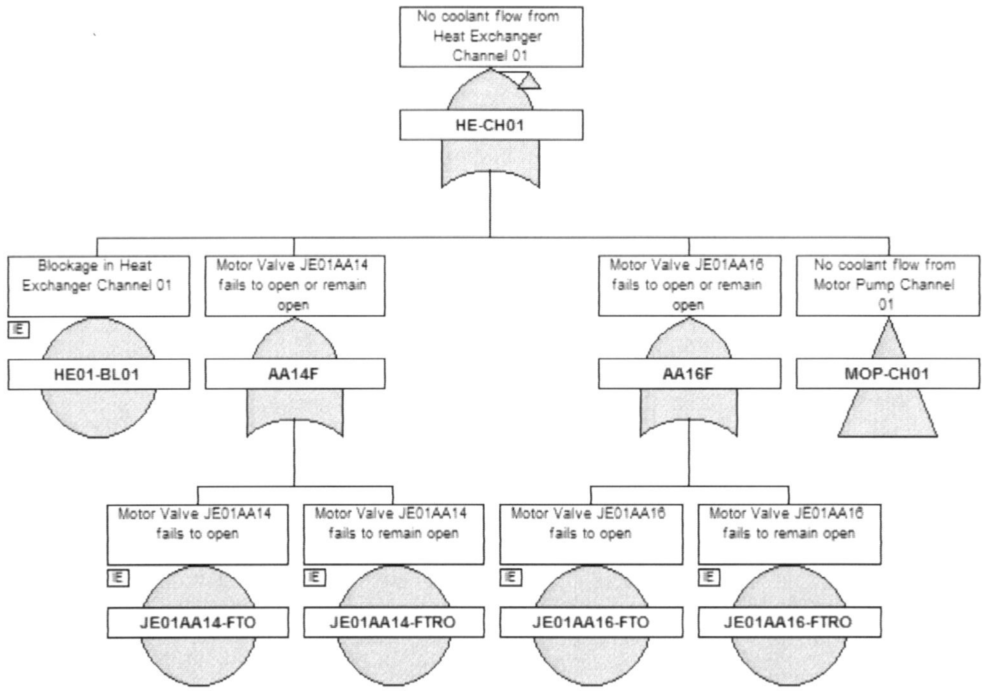

Figure 7. The Fault Tree Diagram of the failure of no coolant flow in the RSG-GAS Heat Exchanger line 1

The fault tree diagram of the absence of coolant flow from the pump line 1 is shown in Figure 8. In the figure, it is seen that the absence of coolant flow in the Pump line 1 is due to the blocked in the check valve JE01AA06, the pump isolation valves (JE01AA03 and AA07) not opening or fail to continue opening, Pump channel 1 (JE01AP01) fails to start or fails to remain running after a successful start, or failure of coolant flow from the backup pump line 1. The fault tree diagram of the pump line 2 is logically same as the pump line 1, therefore it is not shown in this paper.

The fault tree diagram of no coolant flow from the pump line 3, as a backup for the pump line 1, is shown in Figure 9. It should be noted that the functions of the pump line 3 are for the backups of the pump line 1 or the pump line 2. Interface between the pump line 3 to line 1 is a valve JE01AA12 and interface between the pump line 3 to line 2 is valve JE01AA13. In Figure 9, it is shown that the failure of no coolant flow from the pump line 3 is caused by the valve JE01AA12 fails to open or fails to continue opening, the check valve JE01AA10 blocked, the pump isolation valves (JE01AA05 and AA11) not opening or fail to continue opening, Pump channel 3 (JE01AP03) fails to start or fails to keep operating after a successful start.

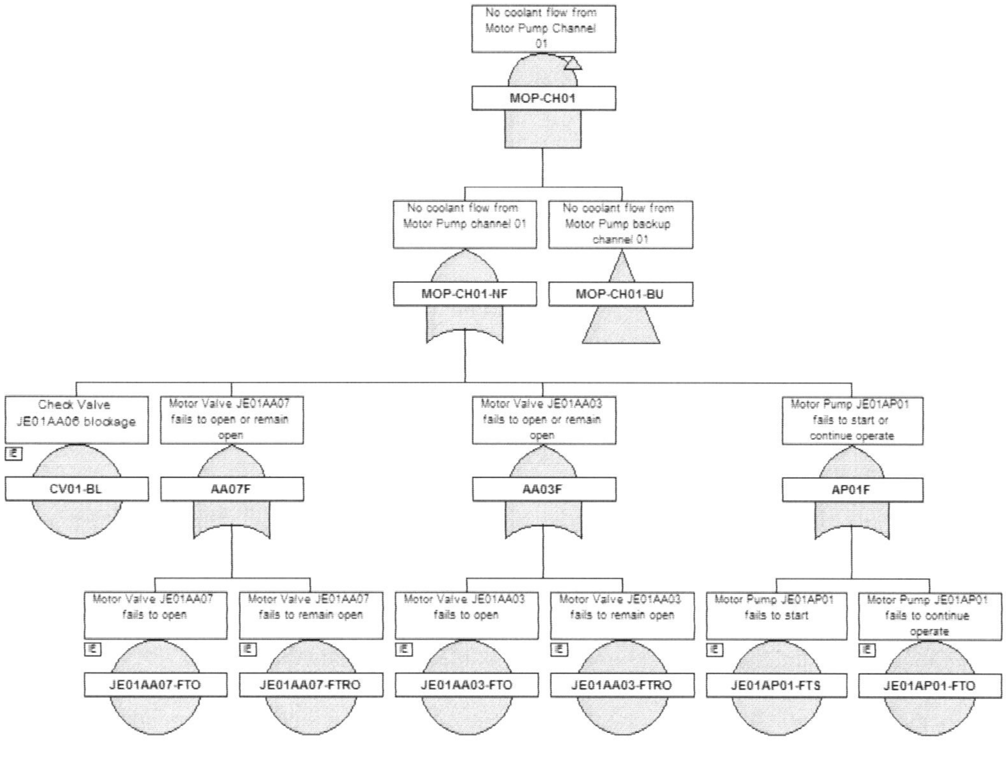

Figure 8. The Fault Tree Diagram of the failure of no coolant flow in the pump line 1.

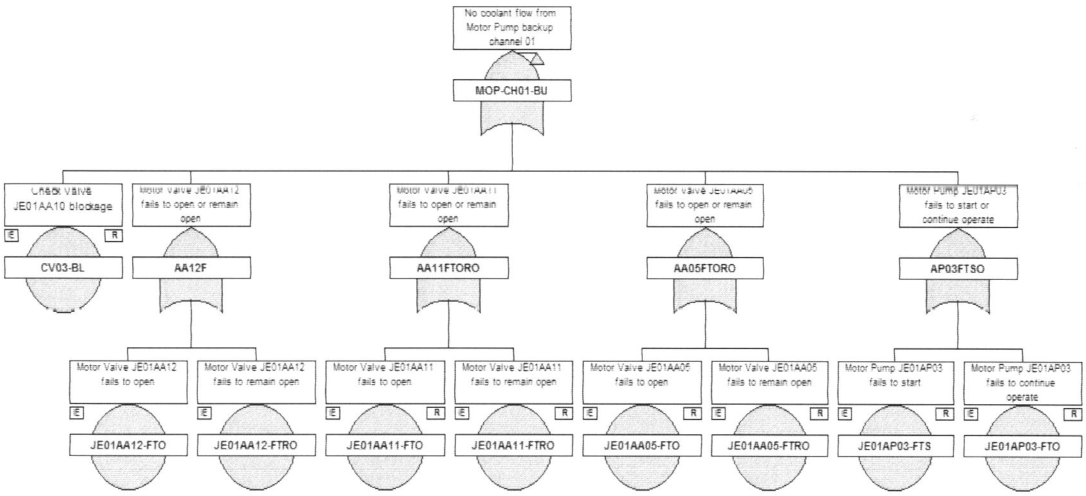

Figure 9. The Fault Tree Diagram of the failure of no coolant flow in the pump line 3 (as a backup for the pump line 1).

4.2. Quantification Result and Discussion.

The fault tree diagram of the RSG-GAS primary cooling system was quantified using the reliability analysis software of ITEM TOOLKIT. The purpose of quantification is to get a minimal cutsets which are the smallest combination of basic event failures that lead to the system failure. In addition, quantification of fault tree diagram is also intended to obtain the probability of occurrence of top event and the importance of each basic event. To obtain the value of probability, it is needed the failure data of each basic event (component). The failure rates of basic events in this paper were obtained from the Reference 15, as shown in Table 1 below.

Table 1. Failure Rate Data of Components

No	Component	Failure Mode	Failure Rate
1.	Motor Operated Valve	- Fail to open	4.1181E-04/demand [15]
		- Fail to remain open	3.3089E-08/h [15]
2.	Check Valve	Fail to open (blocked)	4.8600E-06/demand [15]
3.	Motor Operated Pump	- Fail to Start	1.3104E-03/demand [15]
		- Fail to continue running	3.2495E-06/h [15]
4.	Heat Exchanger	Blocked	3.6846E-07/h [15]

Quantification results are done with several assumptions, i.e.:

1. The reactor operates for 45 days, according to the data (average value) provided in the RSG-GAS component maintenance report.
2. As long as the reactor operates, the power supply is assumed to be always successful (without a trip).
3. Valve and pump instrumentation system and operator action to activate valves and pumps are assumed to be always successful.

Based on the failure rate data and the above assumptions, calculations are done by using ITEM TOOLKIT software. The result of the calculation giving the probability of system unavailability is shown in Table 2 and the Minimal Cutset list and the probability of occurrence are given in Table 3.

Table 2. System Unavailability dan Availability

Unavailability Q	0.00179
Availability	0.99821
No of Cut Sets	1202

Table 3 only shows 55 minimal cuts of 1202 minimal cuts available. The combination of basic events from all the minimal cuts produced varies between combinations of 1 basic event to 3 basic events. 8 minimal cutsets consist of single basic event. Table 3 shows that there are 4 minimal cuts that have the greatest unavailability value (0.00041) and consists of a combination of 1 basic event, namely: failure of the isolating valve of the primary system to open (JE01AA01, AA02, AA18 and AA19).

Table 3. Minimal Cutset

No.	Unavailability	Minimal Cutset
1	0.00041	JE01AA18-FTO
2	0.00041	JE01AA02-FTO
3	0.00041	JE01AA19-FTO
4	0.00041	JE01AA01-FTO
5	3.57355E-5	JE01AA19-FTRO
6	3.57355E-5	JE01AA18-FTRO
7	3.57355E-5	JE01AA02-FTRO
8	3.57355E-5	JE01AA01-FTRO
9	1.69587E-7	JE01AA16-FTO ::JE01AA15-FTO

IOP Conf. Series: Journal of Physics: Conf. Series **962** (2018) 012002 doi:10.1088/1742-6596/962/1/012002

10	1.69587E-7	JE01AA14-FTO ::JE01AA17-FTO
11	1.69587E-7	JE01AA16-FTO ::JE01AA17-FTO
12	1.69587E-7	JE01AA14-FTO ::JE01AA15-FTO
13	1.63842E-7	JE01AA16-FTO ::HE01-BL02
14	1.63842E-7	JE01AA14-FTO ::HE01-BL02
15	1.63842E-7	HE01-BL01 ::JE01AA15-FTO
16	1.63842E-7	HE01-BL01 ::JE01AA17-FTO
17	1.58291E-7	HE01-BL01 ::HE01-BL02
18	4.29967E-8	JE01AP01-FTO ::JE01AP03-FTO ::JE01AP02-FTO
19	1.60828E-8	JE01AP01-FTO ::JE01AP03-FTO ::JE01AP02-FTS
20	1.60828E-8	JE01AP01-FTO ::JE01AP03-FTS ::JE01AP02-FTO
21	1.60828E-8	JE01AP01-FTS ::JE01AP03-FTO ::JE01AP02-FTO
22	1.47162E-8	JE01AA16-FTRO ::JE01AA17-FTO
23	1.47162E-8	JE01AA16-FTO ::JE01AA15-FTRO
24	1.47162E-8	JE01AA14-FTO ::JE01AA15-FTRO
25	1.47162E-8	JE01AA14-FTRO ::JE01AA17-FTO
26	1.47162E-8	JE01AA16-FTO ::JE01AA17-FTRO
27	1.47162E-8	JE01AA14-FTRO ::JE01AA15-FTO
28	1.47162E-8	JE01AA16-FTRO ::JE01AA15-FTO
29	1.47162E-8	JE01AA14-FTO ::JE01AA17-FTRO
30	1.42176E-8	HE01-BL01 ::JE01AA15-FTRO
31	1.42176E-8	JE01AA16-FTRO ::HE01-BL02
32	1.42176E-8	HE01-BL01 ::JE01AA17-FTRO
33	1.42176E-8	JE01AA14-FTRO ::HE01-BL02
34	6.01570E-9	JE01AP01-FTO ::JE01AP03-FTS ::JE01AP02-FTS
35	6.01570E-9	JE01AP01-FTS ::JE01AP03-FTS ::JE01AP02-FTO
36	6.01570E-9	JE01AP01-FTS ::JE01AP03-FTO ::JE01AP02-FTS
37	5.05422E-9	JE01AA14-FTO ::JE01AP03-FTO ::JE01AP02-FTO
38	5.05422E-9	JE01AP01-FTO ::JE01AP03-FTO ::JE01AA17-FTO
39	5.05422E-9	JE01AP01-FTO ::JE01AP03-FTO ::JE01AA15-FTO
40	5.05422E-9	JE01AA07-FTO ::JE01AP03-FTO ::JE01AP02-FTO
41	5.05422E-9	JE01AA16-FTO ::JE01AP03-FTO ::JE01AP02-FTO
42	5.05422E-9	JE01AP01-FTO ::JE01AA11-FTO ::JE01AP02-FTO
43	5.05422E-9	JE01AP01-FTO ::JE01AP03-FTO ::JE01AA09-FTO
44	5.05422E-9	JE01AP01-FTO ::JE01AA05-FTO ::JE01AP02-FTO
45	5.05422E-9	JE01AA03-FTO ::JE01AP03-FTO ::JE01AP02-FTO
46	5.05422E-9	JE01AP01-FTO ::JE01AP03-FTO ::JE01AA04-FTO
47	4.88298E-9	JE01AP01-FTO ::JE01AP03-FTO ::HE01-BL02
48	4.88292E-9	JE01AP03-FTO ::HE01-BL01 ::JE01AP02-FTO
49	2.25015E-9	JE01AP01-FTS ::JE01AP03-FTS ::JE01AP02-FTS
50	1.89051E-9	JE01AA16-FTO ::JE01AP03-FTS ::JE01AP02-FTO
51	1.89051E-9	JE01AP01-FTO ::JE01AA11-FTO ::JE01AP02-FTS
52	1.89051E-9	JE01AP01-FTO ::JE01AA05-FTO ::JE01AP02-FTS

53	1.89051E-9	JE01AP01-FTS ::JE01AP03-FTO ::JE01AA09-FTO
54	1.89051E-9	JE01AA07-FTO ::JE01AP03-FTS ::JE01AP02-FTO
55	1.89051E-9	JE01AP01-FTS ::JE01AP03-FTO ::JE01AA04-FTO
...	....	....

Quantification results also provide the importance parameters for each basic event, as shown in Table 4. The importance parameters calculated in this paper are the importance of F-Vesely, Risk Achievement Worth (RAW) and Risk Reduction Worth (RRW). From Table 4 it can be seen that the greatest contribution to system failure comes from the basic events of the isolation valves fail to open (JE01AA01-FTO, JE01AA02-FTO, JE01AA018-FTO and JE01AA019-FTO), respectively contributing 22.96%. Based on the quantification of RAW values, failure of each of the primary isolation valves (JE01AA01, JE01AA02, JE01AA018 and JE01AA019) in either the fail to open or fail to remain open will result in an increase in the probability of failure of the primary cooling system in a factor of 558.39 times. Based on the quantification of the RRW value, the events of the primary isolation valves fail to open (JE01AA01-FTO, JE01AA02-FTO) decreases the probability of failure of the primary cooling system by 1.298 times, while for other components almost the same.

Table 4. Importance Measures of Basic Events

No.	Event	F-Vesely	RAW	RRW
1	JE01AA01-FTO	0.2296332	558.38965	1.2980829
2	JE01AA18-FTO	0.2296332	558.38965	1.2980829
3	JE01AA19-FTO	0.2296332	558.38965	1.2980829
4	JE01AA02-FTO	0.2296332	558.38965	1.2980829
5	JE01AA19-FTRO	0.019919299	558.38965	1.0203241
6	JE01AA01-FTRO	0.019919299	558.38965	1.0203241
7	JE01AA02-FTRO	0.019919299	558.38965	1.0203241
8	JE01AA18-FTRO	0.019919299	558.38965	1.0203241
9	JE01AA15-FTO	0.00030436823	1.7387943	1.0003045
10	JE01AA17-FTO	0.00030436823	1.7387943	1.0003045
11	JE01AA16-FTO	0.00030436823	1.7387943	1.0003045
12	JE01AA14-FTO	0.00030436823	1.7387943	1.0003045
13	HE01-BL02	0.00029405195	1.7387943	1.0002941
14	HE01-BL01	0.00029405195	1.7387943	1.0002941
15	JE01AP03-FTO	9.1918229E-5	1.0261456	1.0000919
16	JE01AP01-FTO	7.8706708E-5	1.0223877	1.0000787
17	JE01AP02-FTO	7.8706708E-5	1.0223877	1.0000787
18	JE01AP03-FTS	3.4306184E-5	1.0261456	1.0000343
19	JE01AP01-FTS	2.9375314E-5	1.0223877	1.0000294
20	JE01AP02-FTS	2.9375314E-5	1.0223877	1.0000294
21	JE01AA14-FTRO	2.6402114E-5	1.7387943	1.0000264
22	JE01AA16-FTRO	2.6402114E-5	1.7387943	1.0000264
23	JE01AA17-FTRO	2.6402114E-5	1.7387943	1.0000264
24	JE01AA15-FTRO	2.6402114E-5	1.7387943	1.0000264
25	JE01AA11-FTO	1.0771466E-5	1.0261456	1.0000108

26	JE01AA05-FTO	1.0771466E-5	1.0261456	1.0000108
27	JE01AA07-FTO	9.2232702E-6	1.0223877	1.0000092
28	JE01AA03-FTO	9.2232702E-6	1.0223877	1.0000092
29	JE01AA09-FTO	9.2232702E-6	1.0223877	1.0000092
30	JE01AA04-FTO	9.2232702E-6	1.0223877	1.0000092
31	JE01AA12-FTO	1.6836569E-6	1.0040867	1.0000017
32	JE01AA13-FTO	1.6836569E-6	1.0040867	1.0000017
33	JE01AA11-FTRO	9.3435995E-7	1.0261456	1.0000009
34	JE01AA05-FTRO	9.3435995E-7	1.0261456	1.0000009
35	JE01AA07-FTRO	8.0006324E-7	1.0223877	1.0000008
36	JE01AA03-FTRO	8.0006324E-7	1.0223877	1.0000008
37	JE01AA09-FTRO	8.0006324E-7	1.0223877	1.0000008
38	JE01AA04-FTRO	8.0006324E-7	1.0223877	1.0000008
39	JE01AA12-FTRO	1.4604711E-7	1.0040867	1.0000001
40	JE01AA13-FTRO	1.4604711E-7	1.0040867	1.0000001
41	CV03-BL	1.2706836E-7	1.0261456	1.0000001
42	CV01-BL	1.0880467E-7	1.0223877	1.0000001
43	CV02-BL	1.0880467E-7	1.0223877	1.0000001

Thus, the primary system isolation valves (JE01AA01, AA02, AA018 and AA19) are the most important components in the system. These valves are dual-functioned, i.e. as safety-related components as well as its success can increases availabilities of the reactor. In the event of an incident or accident in the reactor pool, the success of one of the inlet valves (one of JE01AA18 and AA19) and one of the outlet valves (one of JE01AA01 and AA02) to close so that the reactor pool is isolated from the other system will make the reactor safe. In contrast to normal operation, the success of all primary system isolation valves (JE01AA01, AA02, AA018 and AA19) to open and remain open will increase reactor availability to provide service, because if one valve closes then the reactor will shut down automatically. Therefore, these valves become important object for aging management and needs to be tested frequently for its opening and closing functions, as well as other components of the primary cooling system.

5. Conclusion

Reliability analysis using FTA method can be used to screen the components which are important for aging management program at RSG-GAS Reactor, which in this paper is implemented in Primary Cooling System. The results of this paper may also verify and validate the results of the SSCs screening of the primary cooling system for aging management as listed in the SOP of the RSG-GAS Aging Management Program. Based on the results of qualitative and quantitative analysis of the RSG-GAS primary cooling system, it can be concluded that the basic events of the failure of the primary isolation valve (JE01AA01, JE01AA02, JE01AA018 and JE01AA019) to open or remain open are the most critical basic events in the system. Compared with the results of SSCs screening of the Primary Cooling System for aging management as listed in the SOP of the RSG-GAS Aging Management Program; JE01AA01, JE01AA02, JE01AA018 and JE01AA019, are all included in the list. Therefore, these all valves maintenance and testing program should be given top priority in the RSG-GAS aging management program so that the availability and safety level of the reactor remains high.

References

[1]. IAEA 2007, "IAEA safety glossary : terminology used in nuclear safety and radiation protection", International Atomic Energy Agency, Vienna.

[2]. IAEA 2010, "Ageing management for research reactors: specific safety guide" No. SSG-10, International Atomic Energy Agency, Vienna.
[3]. Nitoi, M., and M. Pavelescu 2009, "Identification of ageing signs." Romanian Journal Of Physics 54, no. 9-10, pg. 869-885.
[4]. Poghosyan, S., Amirjanyan, A., and Malkhasyan, A. 2009, "Approach on Component Selection for Ageing-Trend Analysis Within PSA Models." 17th International Conference on Nuclear Engineering, pp. 307-311. American Society of Mechanical Engineers.
[5]. Simola, K.1999, "Reliability methods in nuclear power plant ageing management", Technical Research Centre of Finland.
[6]. PRSG 2014, Standard Operating Procedure of RSG-GAS Reactor Aging Management Program, No. SOP 002.001/RN 00 02/RSG 3, PRSG – BATAN. In Indonesian.
[7]. Deswandri and Situmorang, J. 2012, "Reliability Modeling of Primary Cooling System for Estimating Probability of Loss of Cooling Flow Accident in the RSG-GAS Research Reactor", Proceeding of 18th National Seminar on Technology and Safety of Nuclear Power Plant and Nuclear Facilities, page 267-282, Bandung, 29 September 2012. ISSN: 0854-2910. In Indonesian.
[8]. Deswandri and Johnny Situmorang 2012, "Reliability Evaluation of RSG-GAS Pool Venting System and Low Pressure Ventilation System Using Fault Tree Analysis", Scientific Magazine Sigma Epsilon, Vol. 16, No. 2, pg. 32-47. ISSN: 0853-9103. In Indonesian.
[9]. Anonim 2011, "Safety Analysis Report of RSG-GAS Research Reactor Rev. 10.1", PRSG BATAN. In Indonesian.
[10]. Ruijters, E. and Stoelinga, M.2015, "Fault Tree Analysis: A Survey Of The State-Of-The-Art In Modeling, Analysis And Tools", Computer Science Review, Vol.15-16, pg. 29-62, Elsevier.
[11]. Anonim 2005, "ITEM TOOLKIT; User Manual Fault Tree Module", Item Software, Inc.
[12]. Borst, M. V. and Schoonakker, H.2001, "An Overview of PSA Importance Measures", Reliability Engineering and System Safety, Vol.72, pg. 241-245.
[13]. Wolfe, W.A.1978, "Fault Tree Analysis", AECL-6172, Chalk River Nuclear Laboratories, Ontario-Canada.
[14]. Rausand, M. and Hoyland, A.2004, System Reliability Theory: Models, Statistical Methods, and Applications, Wiley.
[15]. Nuclear Regulatory Commission 2012, Component Reliability Data Sheets Update 2010.

Acknowledgement
This work was funded through the annual budget of PTKRN-BATAN. The authors thank the PRSG-BATAN staffs who had provided the information on the systems analyzed in this work.

IOP Conf. Series: Journal of Physics: Conf. Series **962** (2018) 012003 doi:10.1088/1742-6596/962/1/012003

Sensor Failure Detection of FASSIP System using Principal Component Analysis

Sudarno, Mulya Juarsa, Kussigit Santosa, Deswandri, Geni Rina Sunaryo

Center for Nuclear Reactor Technology and Safety, Puspiptek Area Building 80, Serpong, South Tangerang City, 15310, INDONESIA

sudarno@batan.go.id

Abstract. In the nuclear reactor accident of Fukushima Daiichi in Japan, the damages of core and pressure vessel were caused by the failure of its active cooling system (diesel generator was inundated by tsunami). Thus researches on passive cooling system for Nuclear Power Plant are performed to improve the safety aspects of nuclear reactors. The FASSIP system (Passive System Simulation Facility) is an installation used to study the characteristics of passive cooling systems at nuclear power plants. The accuracy of sensor measurement of FASSIP system is essential, because as the basis for determining the characteristics of a passive cooling system. In this research, a sensor failure detection method for FASSIP system is developed, so the indication of sensor failures can be detected early. The method used is Principal Component Analysis (PCA) to reduce the dimension of the sensor, with the Squarred Prediction Error (SPE) and statistic Hotteling criteria for detecting sensor failure indication. The results shows that PCA method is capable to detect the occurrence of a failure at any sensor.

Keywords : Sensor, fault detection, FASSIP, Principal Component Analysis

1. Introduction

Safety is a condition that must always be achieved in the management of a nuclear reactor during the construction, the operation, until the completion of the decommissioning process. The operation reactor safety is closely linked to the reliability of structures, systems and components (SSC) of the nuclear reactor. All SSC will be affected by aging process and its functionality will be degraded, then the level of SSC reliability will be reduced.

The background of this research is the Fukushima Daiichi nuclear reactor accident in Japan in 2011, where the core damage was caused by failure of the active cooling system (diesel generator submerged in water). Then, passive cooling system researches (without pump driven force) at the nuclear power plant are performed in many countries in order to improve the safety aspects of nuclear reactors. An understanding of the passive cooling system is a cooling system that operates without the need of power supply, cooling flow occurs due to the phenomenon of natural circulation. Such phenomenon can be seen in the form of a closed loop thermosyphon heat, its water mass flow rate as an indicator of a natural circulation is obtained based on the difference in temperature, altitude and pressure between the hot section and the cold section of the loop [1, 2].

Sensor measurement accuracy at FASSIP system is essential, as the basis for determining the characteristics of a passive cooling system. In general, sensor failure can be very disadvantageous, because it can affect the safety and quality of the output of the system. Failure of the sensors can be categorized into four types, namely: (1) signal output is constant, does not represent the value changes of the measured signal, (2) the output signal change in a short time, (3) the signal measurement contains noise, so that its values change randomly. Also with increasing operating time, (4) the sensor output signal may experience a drift [3].

Therefore, monitoring and analysis of sensor measurement should be done to detect if there are sensor failures. The purpose of this research is to develop a sensor fault detection of FASSIP system, so that an occurrence of sensor failure can be detected quickly. Many sensors have been installed at the FASSIP system for detecting the phenomena that occur in the passive cooling system. Thus in this research, the analytical data-driven method is appropriate for sensor failure detection. The method used is the Principal Component Analysis (PCA) to reduce the dimension of the sensor, and the criteria Squared Prediction Error (SPE) and statistics Hotteling value for detecting sensor failure indication. PCA is a multivariate statistical method that can be used to perform online monitoring [4, 5, 6, 7].PCA method has advantages compared with other methods, such as PCA model transform high dimension original data into a representation of significantly reduced dimension. Therefore it will reduce the diagnosis time of sensor failure. The technology of online monitoring in a variety of industries has grown to the point where many cases of equipment failure and/or maintenance needs can be predicted quickly in a matter of days, weeks, or even months before the occurrence of equipment or system failures [8].

2. Theory

2.1. Sensor Fault Detection

Validation of the sensor is very important in safety-critical processes such as Nuclear Power Plant (NPP). The conventional method for sensor validation is to check and recalibrate the sensors at fixed time intervals [9]. Although this method has been widely applied in industry to detect a sensor failure, it is unable to detect a sensor failure when the system is being operated. In addition, because of the increasing number of sensors, it will be too expensive and even impossible to inspect all of the sensors regularly one by one. Therefore, efforts have been done to develop a more systematic method, which can generally be categorized into hardware redundancy and analytical redundancy approaches [10].

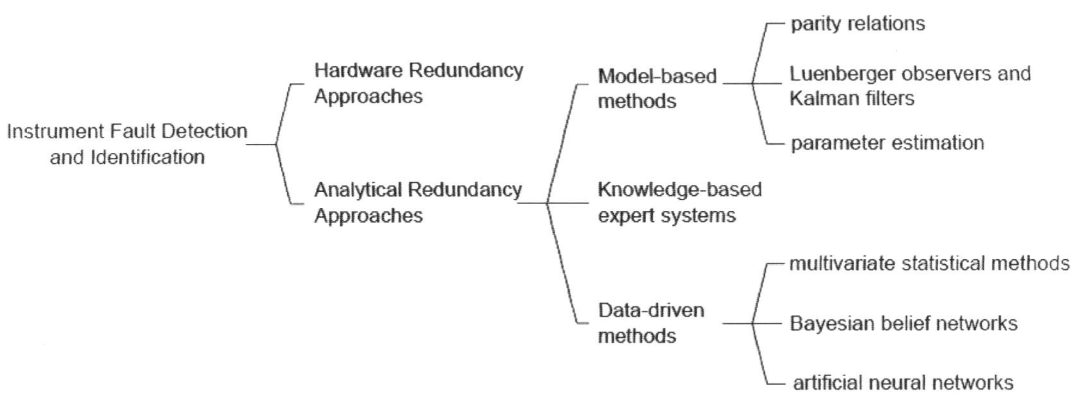

Figure 1. Method classification of detection and identification of instrumentation failure [10].

The concept of sensor failure detection method using hardware redundancy is by measuring each critical variable using two or more sensors. The faulty sensors are then detected and isolated by checking the consistency and voting logic. This approach has been widely used in safety-critical systems due to its simplicity and reliability.

The analytical redundancy approach does not use any additional sensor. It identifies the functional relationship between the measured variables by a mathematical model that can be developed based on the relation of physical phenomena underlying the system, or is derived directly from the measured data. The residual difference between the sensor measurement and output model permits to detect and isolate the faulty sensor. As illustrated in Figure 1, the analytical redundancy approach can be further classified into model based methods, knowledge-based expert systems, and data-driven methods.

Increasing the number of sensors will provide data that can be collected from multiplying. It is very suitable for development of various methods of data-driven, for examples: multivariate statistics, Bayesian Belief Network and Principal Component Analysis (PCA).

2.2. Principal Component Analysis (PCA) Method

PCA is a linear transformation that is commonly used in data compression. PCA is also a common technique used to draw the features of the data on a high-dimensional scale. PCA can reduce the dimensionality of the data without losing important information from these data.

PCA model is constructed by changing the high dimensional data matrix into less dimensional representation. PCA model is build based on normal operating condition data, then it is used as the basis for process monitoring and fault sensor diagnostic. PCA-based failure detection focuses on a small number of PCs that represent a large number of proses variables. Model PCA was established by the decomposition of eigenvalues of the covariance matrix of the original data. Here is the PCA algorithm used in this research [4,5] :

1. Collect data into a matrix. Suppose X is a data matrix where rows and columns represent data sample and variables consecutively.
2. Make a normal data by subtracting each data with the mean value.

$$\hat{X} = X - \bar{X} \tag{1}$$

where $\hat{X}$ is the normalized vector, X is the column vector and is the average of X.

3. Calculate the covariance matrix (C) from the normalilzed data.

$$C = \frac{\hat{X}^T \hat{X}}{m - 1} \tag{2}$$

4. Calculate the eigen value and eigen vectors of the covariance matrix.
5. Construct principal components (PC).

   The eigenvalues are arranged sequentially decreased, then the eigenvectors are arranged according to their eigenvalues. PCA model divides the monitoring data into two orthogonal parts, namely the principal component (PC) and the Residual Space (RS). Thus, the PCA model divides X into a number of products of n loading vectors (P_i) and score vectors (T_i), plus residual matrix (E), using the equation :

$$X = \sum T_i P_i^T + E \tag{3}$$

Fault detection

In the PCA, the latent variables (Pi and Ti) summarize the majority of the variation in the process. Therefore, monitoring of these two variables by using various statistical measures may provide significant information about the characteristics of the process. The new observations can be projected on a plane defined by the loading vector (Pi) to get a score value (Ti). Hotelling statistic τ2 is the sum of the squares of normalization, defined as follows:

$$\tau^2{}_i = T_i \lambda_k^{-1} T_i^T \tag{4}$$

where λ_k are eigenvalues associated to the principal components. In general, the higher the value of τ2, the further measurement distance from the mean. The matrix λ^{-1} is a diagonal matrix containing the inverse of the eigenvalues associated to the kth eigenvector.

Squarred Prediction Error (SPE) is a part of the measurement that is not represented by the PCA model and is defined as the sum of squared errors.

$$SPE = e_i e_i^T \tag{5}$$

where $e_i = (I - P_k P_k^T)$.

When a sensor error occurs, the faulty sensor will change the correlation with other sensors. If SPE value exceeds the limit value, then the variables with greater fault contribution is considered as a variable that is likely to fail.

2.3. FASSIP System

FASSIP system (Passive System Simulation Facility) is a passive cooling systems research facility. The design of FASSIP system is shown in Figure 2.

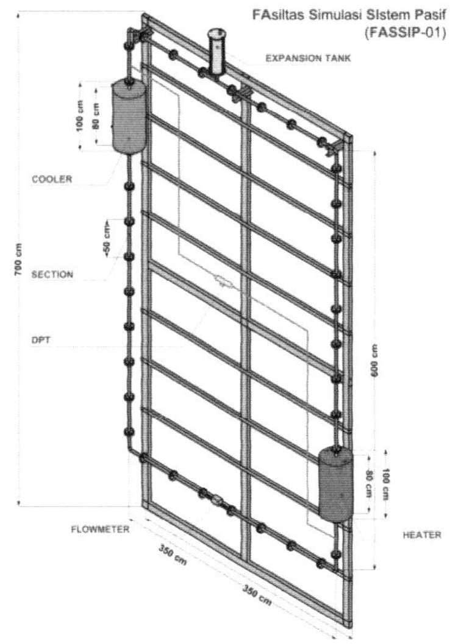

Figure 2. Passive System Simulation Facility (FASSIP) [courtesy by M. Juarsa]

In order to validate models of the passive cooling system, the experiment data on FASSIP system are recorded using a Digital Data Acquisition System. It is important to ensure a valid value of sensor output signal, so that the generated parameter measurements are also accurate. There are three sensor types in FASSIP system:

- 68 Thermocouples,
- 1 Flowmeter,
- 2 Pressure Transducers (PTs).

The number of sensors and their placement are determined in order to acquire optimum information of the natural circulation phenomena in FASSIP system.

3. Methodology

The data used for PCA fault detection testing is obtained from FASSIP experimental data. The first experiment is conducted with all sensors to function normally, this data is used for data learning of PCA model. The next experiment is performed with failure simulation on one sensor with drift and noise error signals. This data is used to test the PCA sensor failure detection system. After that the failure is simulated on different sensors to test the overall system failure detection capabilities.

4. Results and Discussion

PCA model is constructed by transforming the high dimension data matrix into fewer dimension representation space. PCA models are based on the data of normal conditions of FASSIP operation, which is then used as the basis for process monitoring and sensors diagnostic. The data used for the PCA learning process are FASSIP commissioning, with the heater voltage of 180V, and the cooler pump is operated at frequencies of 10 to 50 Hz.

In this testing, the sampled sensor failure detection is a temperature sensor which is located below the cooler tube (see Fig. 2). It consists of four temperature sensors (four variables), the measurement result is shown in Figure 3.

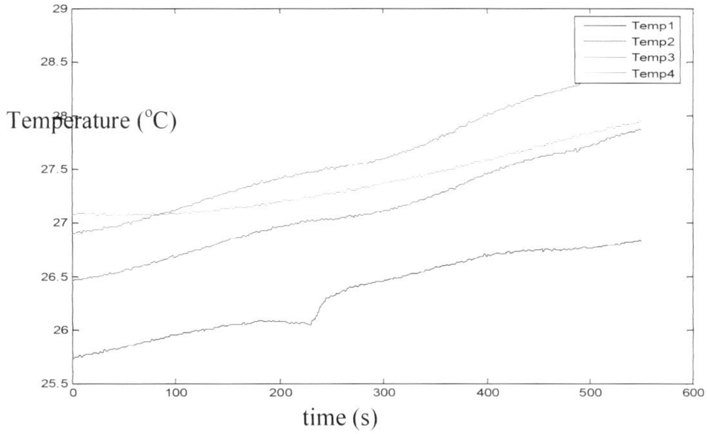

Figure 3. FASSIP temperature measurement in normal operation.

After the process of data normalization for each variable, we calculate the eigenvalues (*lambda*) and eigenvectors (*eigvec*) of the covarian matrix.

$$lambda = [3.9263, 0.0545, 0.0190, 0.0001] \quad (6)$$

$$eigvec = \begin{bmatrix} 0.4947 & 0.8364 & 0.2358 & -0.0085 \\ 0.5029 & -0.1306 & -0.5688 & 0.6376 \\ 0.5035 & -0.2043 & -0.3588 & -0.7590 \\ 0.4990 & -0.4915 & 0.7015 & 0.1316 \end{bmatrix} \quad (7)$$

According to the eigenvalues, if we choose one principal component, we will get a degree of certainty of 98.16%, while two principal components will result 99.5%. In this research we use the degree of certainty of 95%, with one principal component. The acceptance criteria are determined by the Hotteling method (T2) or Squared Prediction Error (SPE). If the value T2 or SPE is higher than the limit value, it is considered that the measurement data is not normal (fault occurrence). Based to the data test, the degree of certainty of 95% corresponded to :

$$T2\_lim(0.95) = 6.0353 \quad (8)$$

$$SPE\_lim(0.95) = 0.0715 \quad (9)$$

To test the ability of this method to detect a sensor failure, we simulated the erroneous input variable sensor. The fault sensor signal is generated following a mathematical equation:

$$x(t) = \hat{x}(t) + w(t) \quad (10)$$

where $\hat{x}(t)$ is the target signal value and w (t) is the signal errors, a random value is used here. Figure 4 displays the input signals used to test the PCA, which is a continuation of the previous input signals (Fig. 3).

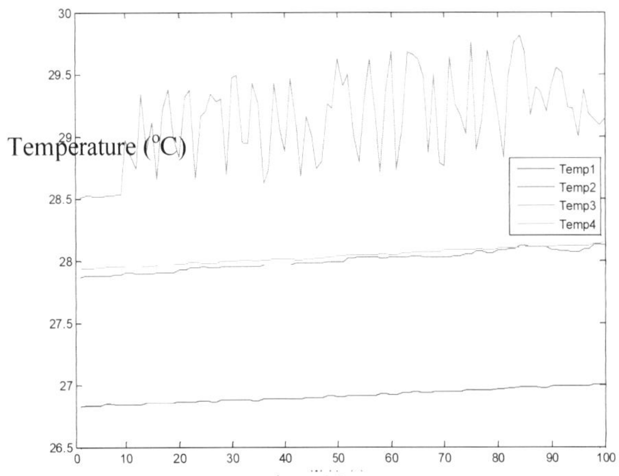

Figure 4.The input signal to the sensor failure detection.

The calculation result of Hotteling T2 value indicates a sensor error, if it exceeds the limit value (T2_lim). The Hotteling T2 value is shown in Figure 5.

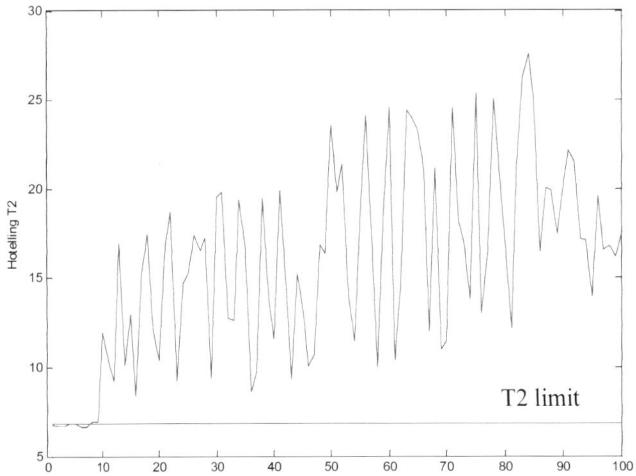

Figure 5. The value T^2

Figure 5 shows that the sensor error appears when *t*> 10 seconds, as the value T2 exceeds the fault limit value.

The identification of the faulty sensors is determined by its Fault Contribution value. Figure 6 shows the Contribution value, the higher the value indicates the higher probability of sensor error. In this test, the sensor error occurs at the temperature sensor number 3 (see Figure 4). From Figure 6 we can state that according to the PCA method developed in this research, the major fault contribution comes from sensor number 3.

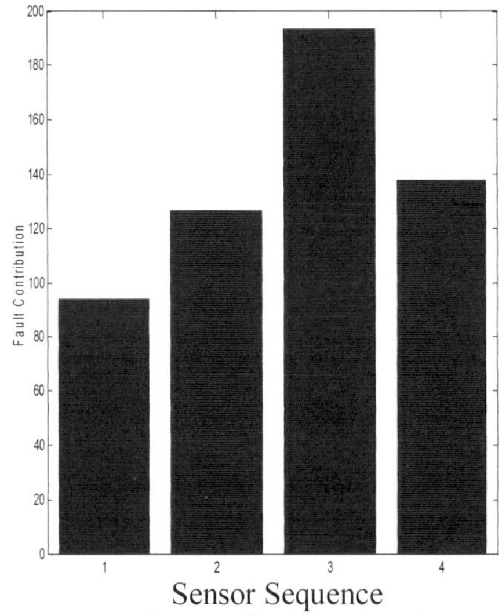

Figure 6. Fault Contribution of each sensor.

From the above test result, the PCA method is capable to detect sensor faults. To determine whether the PCA method can detect any failure of the sensor, the test is done by simulating each sensor failure. Figure 7 shows a bar graph of each test.

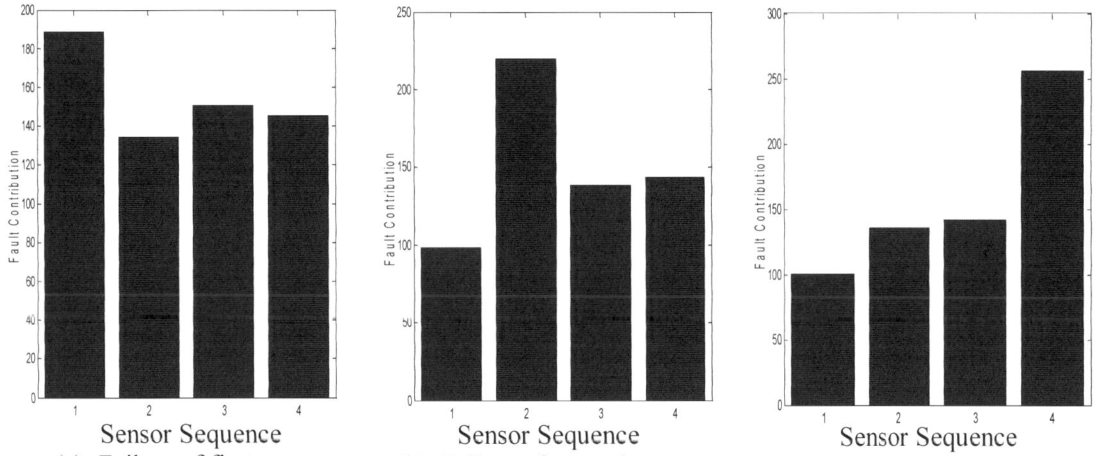

(c) Failure of first sensor (b) Failure of second sensor (a) Failure of third sensor

Figure 7.Testing the detection of each sensor fault.

The failure of the first sensor (temperature 5) detected on the bar graph Figure 7(a) where the contribution of the first sensor failure is the highest value. Similarly, the test results in Figure 7(b) and 7(c) in accordance with each simulation used as an input sensor failure of PCA method.

5. Conclusion

In this research a detection method of FASSIP system sensor failure has been developed, using Principal Component Analysis (PCA). The learning process of PCA model is based on the data FASSIP operation under normal conditions. Testing of the sensor failure detection is done by simulating fault on one of the sensors. The results showed that PCA method is able to detect the occurrence of a failure in any sensor failure based on their fault contribution value.

Acknowledgment
The research was funded by DIPA PTKRN-BATAN fiscal year 2016. Also, the authors would like to thank BPFKR colleagues who have helped in collecting FASSIP experiment data.

References

[1] Juarsa M 2011Experimental Study of Water Mass Flow Rate of Natural Circulation Case Based on Changes of Tilt Angle Using USSA-FT01 Natural Circulation Simulation*J. Material danEnergi Indonesia*vol**1**pp 22–30 Indonesian
[2] Suhendra 2012Characterization of Heat Exchanger Based on Changes of Flow Rate on the Primary Side of BETA Test Loop *Proc.National Conf. Mech. Dept. XI* (SNTTM XI) Thermofluid IV Gajah Mada Univ.pp 16–7 Indonesian
[3] Sharma AB, Golubchik L andGovindan R 2010 Sensor Faults: Detection Methods and Prevalence in Real-World Datasets*ACM Trans. Sens. Networks*vol**6**pp 1–39
[4] Munesawa Y, Minowa H and Suzuki K 2013 Development of Fault Diagnosis System Using Principal Component Analysis for Intelligent Operation Support System *Chem. Eng. Trans.*vol**31**
[5] Sharifi M 2009 Sensor Fault Diagnosis Using Principal Component Analysis *Dissertation of Doctor of Philosophy* Texas A&M Univ.
[6] Benaicha A, Mourot G, Ragot J andBenothman K. 2013Fault detection and isolation with Interval Principal Component Analysis*Proc. Eng. Technol.- Int. Conf. Control Eng. Inf. Technol.*(CEIT'13), pp. 162–7.
[7] Qiu T and Bai X 2012 Comparison of Different Fault Detection Statistics Detectability in PCA *Proc. 2nd Int. Conf. Comput. Appl. Syst. Model.* (Paris: Atlantis Press)
[8] Anonymous 2010 On-Line Monitoring of Accuracy and Reliability of Instrumentation and Health of Nuclear Power Plants *Analysis and Measurement Services Corp. Final Project Report, DOE Grant No. DE-FG02-06ER84626*
[9] Pike P andPennycook K. 1992 Commissioning of Building Energy Management System - a code of practice *BSRIATechnical Report*
[10] Ibarguengoytia PH, Sucar LE andVadera S 2001 Real time intelligent sensor validation*IEEE Trans. Power Syst.*vol **16**(4) pp 770–5

IOP Conf. Series: Journal of Physics: Conf. Series **962** (2018) 012004 doi:10.1088/1742-6596/962/1/012004

Preliminary Analysis of High-Flux RSG-GAS to Transmute Am-241 of PWR's Spent Fuel in Asian Region

M Budi Setiawan and S Kuntjoro

Center for Nuclear Reactor Technology and Safety, Puspiptek Complex, building no. 80, Serpong, Tangerang Selatan 15410 Indonesia

Email: setiawan@batan.go.id

Abstract. A preliminary study of minor actinides (MA) transmutation in the high flux profile RSG-GAS research reactor was performed, aiming at an optimal transmutation loading for present nuclear energy development. The MA selected in the analysis includes Am-241 discharged from pressurized water reactors (PWRs) in Asian region. Until recently, studies have been undertaken in various methods to reduce radiotoxicity from actinides in high-level waste. From the cell calculation using computer code SRAC2006, it is obtained that the target Am-241 which has a cross section of the thermal energy absorption in the region (group 8) is relatively large; it will be easily burned in the RSG-GAS reactor. Minor actinides of Am-241 which can be inserted in the fuel (B/T fuel) is 2.5 kg which is equivalent to Am-241 resulted from the partition of spent fuel from 2 units power reactors PWR with power 1000MW(th) operated for one year.

1. Introduction

Generation of electricity by nuclear power plant (NPP) involves the production of radioactive waste. It is reported that the total discharged spent fuel derived from nuclear power plants in the world, with a capacity of approximately 400GWe, will accumulate more than 450,000 tonnes in 2020 [1], which should be managed in a geological disposal.

Spent fuel contains about 3,000 tonnes of Pu, about 140 tonnes of Np-237 (which will be increased to approximately 500 tonnes due to the decay of Pu-241 and Am-241) and about 120 tonnes of the amount of americium. Long life fission products (LLFP) will also accumulate in spent fuel, such as the I-29 (about 60 tonnes), and Tc-99 (about 250 tonnes) [2]. On the other hand, activities of decommissioning nuclear power plants also produce significant quantities of waste for disposal in geological disposal.

To support the geological disposal in the back-end technology, the transmutation of actinides and LLFP contained in high level waste (HLW) have been studied extensively. To minimize the long term of risk radiotoxicity, theoretical and experimental studies, especially in Europe and Japan have been underway to transmute LLRN (long-lived radioactive nuclides): actinides, particularly Pu, minor actinides (MA) such as Np, Am and Cm; and fission products (FP) to short-lived fission products (SLFP) or stable species.

Minor actinides in HLW is a difficult problem since they have a very long half-life. The development of nuclear technology and the geological disposal of HLW including MA is currently regarded as the most realistic method. However, there is a problem of uncertainty after about 1,000 years, since the immobilization technology can only presume the scenario for about 10^3 years [3,4].

Potential risks of HLW can largely be calculated until 10^3 years after the time out, as defined for engineering barrier, due to the technology of glass or ceramics to accommodate HLW known only capable for around those mentioned years. The most potential risk in the HLW is the isotope with a half-length, i.e Tc-99, I-129, and MA isotopes, namely Np-237, Am-241, Am-243, except FP isotopes (Sr-90, Cs-137) and MA from Cm-244 which has a half-life until about 10^3 years.

To solve this problem, the concept of geological disposal of HLW supported by partitioning and transmutation technologies is proposed [4]. It is primarily focused on the transmutation of long-lived radionuclides into shorter-lived species by utilizing neutron fission. This method is very attractive to eliminate MA (Am-241, Am-243 and Np-237) and LLFP (Tc-99, I-129) while producing energy[5,6]. Among MA isotopes which regarded as a long-term radiological risk is Am-241

RSG-GAS is a research reactor with a maximum power of 30MWt. The fuel is a plate type fuel. RSG-GAS core configuration is arranged to yield high neutron flux of $1x10^{14}n/cm^2s$. This high flux profile is very effective to burn/transmute MA nuclide obtained from PWR's spent fuel.

A preliminary study of Am-241 transmutation in the high flux profile RSG-GAS research reactor was performed, aiming at an optimal transmutation loading for present nuclear energy development. The analysis assumed the Am-241 nuclide is obtained from partitioning process discharged from pressurized water reactors (PWR) in Asian region.

2. Theory

In this study, the concept of transmutation was evaluated to develop the transmutation system of a high-flux RSG-GAS research reactor for significantly transmute MA nuclides to reduce the potential risk of geologic disposal. Figure 1 shows the conceptual design of recycle transmutation loop, which is connecting transmutation reactor with reprocessing and partitioning process.

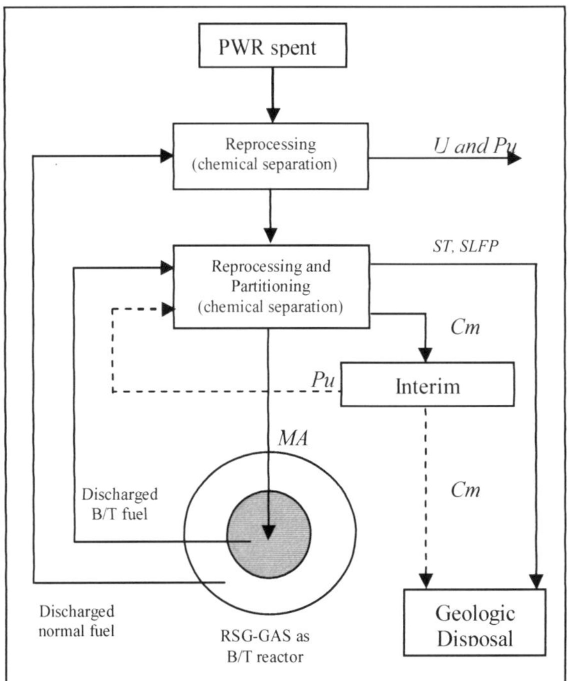

Figure 1. Conceptual design utilization of RSG-GAS as the high-flux reactor to transmute minor actinides derived from the processing, separation and fabrication of PWR spent fuel in Asia.

As shown in Figure 1, the spent fuel from PWR power reactors in Asia were reprocessed to gain the uranium and plutonium to be reprocessed and used in the MOX fuel reactors. The next step after reprocessing was to separate and take out the stable and short-life FP which will be stored in geologic disposal facility. Based on their half-life Cm nuclides was assumed to be stored in an interim storage, and eventually stored in a geologic disposal facility. Plutonium nuclides as a result storing Cm nuclides in an interim storage was return back to the reprocessing and partitioning process.

The MA nuclides (in this study is Am-241 selected due to its longest half-life time), which was reprocessed and partitioned from the facility, was manufactured as B/T (burning and transmutation) fuel for RSG-GAS reactor. This study is aimed to preliminary calculate the amount of Am-241 can be transmuted in the RSG-GAS.

3. Calculation

Flowchart of computation used in this study is shown in Figure 2. Code SRAC2006 executes calculations cells with the collision probability method (P_{ij} module) to cover the entire range of neutron energy calculation. This Code solves linear equations which describe the behaviour of neutrons in a steady state condition is described by following the following equations, where the system under consideration is divided into N regions and neutron energy range is divided into group G. Some other physical quantities are defined as follows[7]:

Volume region i:

$$V_i = \int \mathrm{dV} \tag{1}$$

Flux for the whole region i of energy group g:

$$\varphi_{\mathrm{ig}} = \int \mathrm{dV} \int \mathrm{dE}\varphi(\mathrm{r}, \mathrm{E}) \tag{2}$$

Collision probability within the region i to region j for the group g:

$$P_{\mathrm{ijg}} \tag{3}$$

Scattering cross section which is used to evaluate the neutron equilibrium

$$\sum\nolimits_{a,mg} = \sum_{g' \not\subset \mathrm{G}} \Sigma_{\mathrm{smg} \to g'} + \sum_{g' \not\subset \mathrm{G}} \chi_{\mathrm{mg}'} \upsilon \Sigma_{\mathrm{f,mg}} \tag{4}$$

where:

Σ_{mg} = total cross section of nuclide m of group g

$\Sigma_{\mathrm{f,mg}}$ = fission cross section of nuclide m of group g

Σ_{amg} = absorption cross section of nuclide m of group g

$\Sigma_{\mathrm{smg} \to g'}$ = scattering cross section of energy group g to g'

The mass balance of burning and transmutation (B/T) fuel was calculated from the cell depletion calculation. The time required to achieve the equilibrium state of the core can be estimated by performing a series of cell depletion calculations starting with a typical cell composition.

In each successive calculation, the same isotopic composition with that used in the original cell calculation, were added to the remains of the B/T fuel at the end of cycle (EOC). The amount of new fissile materials and the fuel density were adjusted accordingly to keep the fuel cycle length and burn up in the same conditions. The atomic density of isotope-i in each region, N_i, can be expressed by

$$N_i = \frac{F_i \rho_F N_A}{M_F} \tag{5}$$

where

F_i : weight fraction of nuclide-i in fuel
ρ_F : material density g/cm^3
N_A : Avogadro's number, 6.022045 x10^{23}mol^{-1}
M_F : average molecular weight of fuel, g/g-mol

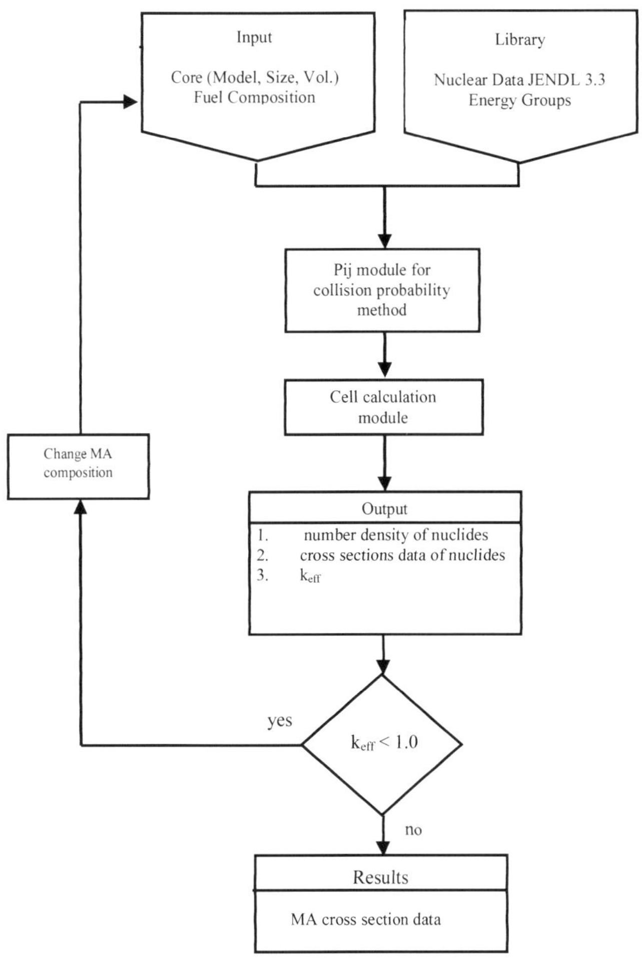

Figure 2. Flowchart of calculation using computer code SRAC2006

4. Results and Discussion

The literature retrieval was undertaken to obtain the data of high level waste (HLW) generated from spent fuel of PWR nuclear power plants in the region. The overall spent fuel assumed to be processed and separated in the processing and separation facilities in the Asian region as well. Similarly, the fabrication of fuel for burning and transmutation (B/T-fuel) is performed at a fuel fabrication facilities in Asia.

Data on nuclear power plants in Asia can be seen in Table 1, included in the table also is data regarding the processing facilities, separation and fabrication.

Table 1. Data of nuclear power plants (NPP) operating in the region and fuel recycling capabilities possessed by some Asian countries [8].

Country	NPP in operation	NPP under construction	NPP planned	Research Reactor	*Fuel Cycle* Capability
Australia	0	0	0	1	UM
Bangladesh	0	0	2	1	
China	30	24	40	16	UM, C, E, FF
India	21	6	22	4	UM, FF, R, WM
Indonesia	0	0	1	3	FF
Japan	43	3	9	14	C, E, FF, R, WM
S. Korea	25	3	8	2	C, FF
Malaysia	0	0	0	1	
Pakistan	3	2	2	1	UM, E, FF
Philippines	0	0	0	1	
Taiwan	6	2	0	1	
Thailand	0	0	0	2	
Vietnam	0	0	4	1	
Total Asia	128	40	88	48	

Note:
UM : Uranium Mining
C : Conversion
E : Enrichment
FF : Fuel fabrication
R : Reprocessing
WM : Waste management on Geologic Disposal

HLW handling scenarios (Figure 1) of PWR spent fuel in the Asian region prepared by RSG-GAS research reactor as a burning and transmutation reactor; using radionuclide transmutation into short-lived fission products or unstable nuclides.

The study aimed to evaluate the available technologies of RSG-GAS research reactor to operate with fuel transmutation (B/T-fuel). The evaluation includes the analyzing of the effectiveness of the transmutation of MA into fission products by utilizing the high flux of RSG-GAS reactor.

Nuclide selected for transmutation is Am-241 separated / partitioned in the spent fuel PWR in Asian countries, in this case selected third countries that intensively use the nuclear power plants in the region, which are Japan, China and South Korea.

Table 2. Data of PWR operating in Asia (Japan, China and South Korea)[9]

Country		**Power (MW)**	**Burn-up (MWd/THM)**	**U enrichment (%)**
Japan				
	Genkai	3478	60000	4
	Ikata	2022	48000	4-5
	Mihama	1666	37000	4
	Ohi	4710	70000	4

Sendai	1780	37000	4
Takahama	3304	55000	4
Tomari	2070	48000	4-5
Tsuruga	1160	35000	4
China			
Daya Bay	1888	40000	4
Ling Ao	3876	55000	4-5
Hongyanhe	1080	35000	4-5
Ningde	1080	35000	4
Tianwan	2120	48000	4-5
Yangjiang	1086	35000	4-5
South Korea			
Ulchin/Hanul	5881	80000	4-5
Yeonggwang /Hanbit	5875	80000	4-5
South Korea			
Ulchin/Hanul	5881	80000	4-5
Yeonggwang /Hanbit	5875	80000	4-5
Kori	4414	55000	4-5

Based on these data and the composition of the spent fuel generated at end of cycle (EOC) of PWR, and then calculated the composition of transuranic (especially the minor actinides) which can be separated from the fuel. A typical composition of actinides from LWR spent fuel 1 unit 1 GWe enrichment 3.3% U-235, 150 days after the EOC and the burn-up = 33000 MWD / Mg (HM) in Table 3.

Table 3. Actinide of 1 GWe LWR spent fuel with 3.3% U-235, 150 days after the EOC, burn-up = 33 GWD / Ton (HM) [10].

Radionuclide	Half-life (y or d)	Mass (kg/y)
^{234}U	2.47×10^5 y	3.14
^{235}U	7.1×10^8 y	2.15×10^2
^{236}U	2.39×10^7 y	1.14×10^2
^{237}U	6.75 d	9.15×10^{-7}
^{238}U	4.51×10^9 y	2.57×10^4
Total U		2.60×10^4
^{237}Np	2.14×10^6 y	2.04×10^1
^{239}Np	2.35 d	2.05×10^{-6}
Total Np		2.04×10^1
^{236}Pu	2.85 y	2.51×10^{-4}
^{238}Pu	86 y	5.99
^{239}Pu	24,400 y	1.44×10^2
^{240}Pu	6,580 y	5.91×10^1
^{241}Pu	13.2 y	2.77×10^1
^{242}Pu	3.79×0^5 y	9.65
Total Pu		2.46×10^2
^{241}Am	458 y	1.32
^{243}Am	7,950 y	2.48
Total Am		3.81

It is assumed target of Am-241 of 2.5 kg is inserted in the reactor core RSG-GAS. Calculation of homogenized fuel cells are consists of Am-241 with AlMg2 cladding and water moderator. Calculations were used 8 neutron energy group. The results of calculation of the macroscopic cross section of Am-241 can be seen in Figure 3.

From the graph of the fission and absorption cross-section (Figure 3) it can be concluded that the Am-241 which has a cross section of the thermal energy absorption in the region (Group 8) is relatively large, it will be easily burned (transmuted) in the RSG-GAS reactor.

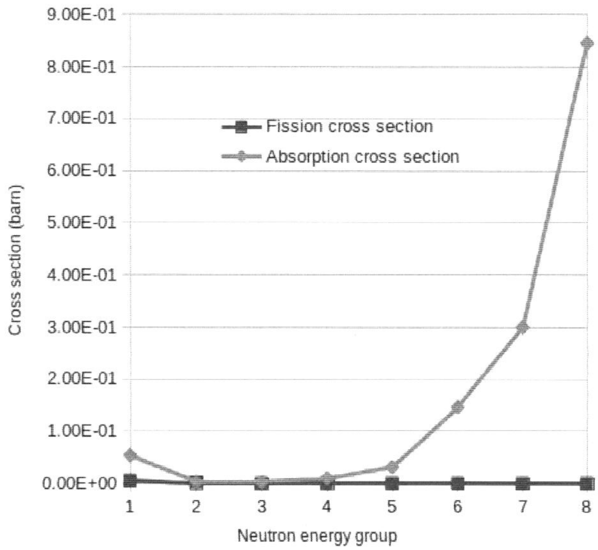

Figure 3. Macroscopic cross section of Am-241

5. Conclusions

Minor actinide Am-241 which has a cross section of the thermal energy absorption in the area (group 8) is relatively large; it will be effectively transmuted in the RSG-GAS reactor. Am-241 which can be inserted in the fuel transmutation (B/T fuel) of 2.5 kg which is equivalent to Am-241 resulting from the partition of spent fuel from 2 units of 1000MWt PWR power reactors operated for one year. For the spent fuel from power reactors PWR in Asia (Japan, China and South Korea), which amounted to 17 units, then the minor actinides from spent fuel will be burnt in the RSG-GAS for around ~ 9 years.

Acknowledgments
Authors wishing to acknowledge Jati Susilo for his continuing assistance on SRAC2006 code, and financial support from PTKRN Batan DIPA budget for fiscal year 2016.

References

[1] Yoshihara A et al 2000 Estimation of spent fuel compositions from Light Water Reactors *Journal of Nuclear Science and* Technology **37 10** 924-933
[2] Charles M 2000 Overview of the Hydrometallurgical and Pyro-metallurgical Processes Studied Worldwide for the Partitioning of High Active Nuclear Wastes, *Proceeding of NEA/OECD 6th Information Exchange Meeting on Actinide and Fission Product Partitioning and Transmutation*, 53-64
[3] Kitamoto A and Setiawan M B 2002 Recycle transmutation of MA and LLFP using BWR for sustaining geologic disposal *Progress in Nuclear Energy* **40 3** 465-72

[4] Setiawan M B and Kitamoto A 2001 Evaluation on transmutation performance of minor actinides with high-flux BWR *Annals of Nuclear Energy* **28** 443-56
[5] Liu K et al 2012 Application of fine depletion calculation for LLFPs in-core transmutation evaluation in PWRs *20th International Conference on Nuclear Engineering and the ASME 2012 Power Conference* (American Society of Mechanical Engineers) 5 137-44
[6] Liu K et al 2013 A code development for LLFP transmutation analysis based on the whole pin-wise calculation in PWRs *Nuclear Engineering and Design* **256** 56-66
[7] Okumura K et al 2007 SRAC2006: A Comprehensive neutronics calculation code system *Japan Atomic Energy Agency* JAEA-Data/Code 2007-004
[8] Yapici H 2003 Study on transmutation of minor actinides discharged from high burn-up PWR-MOX spent fuel in the force-free helical reactor *Annals of Nuclear Energy* **30 4** 413-36
[9] Übeyli M 2004 Transmutation of minor actinides discharged from LMFBR spent fuel in a high power density fusion reactor *Energy Conversion and Management* **45 20** 3219-38
[10] Kora K et al 2016 A study on transmutation of LLFPs using various types of HTGRs *Nuclear Engineering and Design* **300** 330-38

IOP Conf. Series: Journal of Physics: Conf. Series **962** (2018) 012005 doi:10.1088/1742-6596/962/1/012005

Kr-85m activity as burnup measurement indicator in a pebble bed reactor based on ORIGEN2.1 Computer Simulation

I Husnayani, P M Udiyani, S Bakhri, G R Sunaryo

Center for Nuclear Reactor Technology and Safety, Puspiptek Complex, building no. 80, Serpong, Tangerang Selatan 15314 Indonesia

Abstract. Pebble Bed Reactor (PBR) is a high temperature gas-cooled reactor which employs graphite as a moderator and helium as a coolant. In a multi-pass PBR, burnup of the fuel pebble must be measured in each cycle by online measurement in order to determine whether the fuel pebble should be reloaded into the core for another cycle or moved out of the core into spent fuel storage. One of the well-known methods for measuring burnup is based on the activity of radionuclide decay inside the fuel pebble. In this work, the activity and gamma emission of Kr-85m were studied in order to investigate the feasibility of Kr-85m as burnup measurement indicator in a PBR. The activity and gamma emission of Kr-85 were estimated using ORIGEN2.1 computer code. The parameters of HTR-10 were taken as a case study in performing ORIGEN2.1 simulation. The results show that the activity revolution of Kr-85m has a good relationship with the burnup of the pebble fuel in each cycle. The Kr-85m activity reduction in each burnup step,in the range of 12% to 4%, is considered sufficient to show the burnup level in each cycle. The gamma emission of Kr-85m is also sufficiently high which is in the order of 1010 photon/second. From these results, it can be concluded that Kr-85m is suitable to be used as burnup measurement indicator in a pebble bed reactor.

1. Introduction

Pebble Bed Reactor (PBR) is a high temperature gas-cooled reactor which uses graphite as a moderator and helium as a coolant. PBR is categorized as a Generation IV reactor because it offers many advances compared to conventional reactor such as Light Water Reactor (LWR). The advantage of PBR over LWR is mainly due to its fuel best performance and reliability to operate in high temperature condition. The fuel of PBR is a 6-cm in diameter graphite sphere (pebble) containing thousands of UO2 micro-kernels coated by Tri-structural Isotropic (TRISO) carbon layers; Inner Pyrolitic Carbon (IPyC), Silicon Carbide (SiC), and Outer Pyrolitic Carbon (OPyC) [1]. Typically, there are two schemes of fuel loading in PBR which are once-through-then-out (OTTO), where the fuel pebbles circulate only one time in the core, and multi-pass scheme, where the fuel pebbles circulate for several cycles in the core before being discharged out of the reactor core [2].

In a multi-pass Pebble Bed Reactor, when the fuel pebble completes each cycle, its burnup level must be measured in order to determine whether the burn-up of the pebble is still below the burn-up target so that it must be reloaded into the core for another cycle or it has reached the burn-up target so that it must be discharged out of the core. Burnup measurement in PBR is performed by online measurement without shutting down the reactor. Burnup is a variable that shows the amount of energy produced by the fuels from fission reaction of the fissile materials. Performing online burnup measurement is usually conducted by means of passive system, which is by detecting signal from the fuels and then relates it with the burnup. Those signals include neutron and gamma rays.

The feasibility of burnup measurement using passive neutron counting technique has been studied by Su, et al. [3] and gamma emission technique has been studied by Yan et al. [4]. Regarding the burnup measurement using gamma emission spectroscopy, type of radionuclide that will be used as burnup measurement indicator must fulfill some criteria. For online burnup measurement, the radionuclide must have a short half-life, simple mode of production and decay, and low neutron absorption cross section [5]. In this work, the feasibility of Krypton-85m (Kr-85m) as a burnup indicator in a Pebble Bed Reactor was studied. The activity and gamma emission of Kr-85m was analysed in relation with the burnup of the PBR fuel pebble in each cycle. The activity and gamma emission of Kr-85m is calculated by using ORIGEN2.1 computer code. ORIGEN2.1 is a well-known and worldwide-used fuel depletion code developed by Oak Ridge National Laboratory [6]. Parameters of Chinese 10-MWth PBR (HTR-10) were taken as a case study and in constructing the input file for ORIGEN2.1 simulation.

2. Theory

Kr-85m is an isomer of krypton produced from fission reaction in nuclear reactor fuels. Some characteristics of Kr-85m are shown in the Table 1. These characteristics mainly fulfill the criteria for burn-up measurement indicator. The half-life of Kr-85m is relatively short, which is less than 24 hours, so that it is suitable for online burn-up measurement. The highest probability of gamma emission is in the energy of 151.195 KeV. The fission yield of Kr-85m is relatively high so that the concentration of Kr-85m in the fuel as well as its activity will be sufficient to be detected. The thermal absorption cross section of Kr-85m is far less than one barn so that neutron activation reaction of Kr-85m can be ignored and the concentration of Kr-85m in the fuel pebble will be mainly related to the burnup level of that fuel pebble.

Table1. Characteristics of Kr-85m [5]

Characteristic	Value
Half life	268.8 min
Emission probability	
151.195 KeV	75%
129.81 KeV	0.3%
451 KeV	0.01%
304.87 KeV	14%
U-235 fission yield	5.89E-05
Thermal absorption cross section	2.35E-02 barn
Mode of production	1

2.1. ORIGEN2.1 Computer Code

ORIGEN2.1 is a versatile depletion code used for calculating the isotopic composition and some decay characteristics of the reactor fuel. The nuclide density rate calculation in the ORIGEN code is done using formula as follows [7]:

$$\frac{dN_i}{dt} = \sum_{j=1}^{N} l_{ij}\lambda_j X_j + \varphi \sum_{k=1}^{N} f_{ik}\sigma_k X_k - (\lambda_i + \varphi\sigma_i + r_i)X_i + F_i, i = 1, \ldots\ldots, N \quad (1)$$

N_i = density of nuclide i
N = number of nuclides
l_{ij} = fraction of radioactive disintegration by other nuclide, which lead to formation of species I
φ = position- and energy-averaged neutron flux (n cm^{-2} s^{-1})
f_{ik} = fraction of neutron absorption by other nuclides, which lead to formation of species i
σ_k = spectrum-averaged neutron absorption cross section of nuclide k (barn)

r_i = continuous removal rate of nuclide i from the system
F_i = continuous feed rate of nuclide *i*

The photon release rates (photon/second) of the nuclide *i* can be calculated from the equation as below:

$$P_i = g_i \lambda_i N_i \tag{2}$$

where P_i is the photon release rate, g_i is the photon released per decay, which is given in ORIGEN2.1 photon library, $\lambda_i N_i$ are described above.

3. Methodology

HTR-10 was taken as a case study for PBR fuel handling system as well as reactor parameters used in this study. In HTR-10, the fuel pebbles circulates in the reactor core for five cycles. The fuel loading scheme of HTR-10 is shown in the Figure 1. SG01 transports the fuel pebbles one by one to BP01 where the burnup of the pebble is measured. If it has not reached its maximum burnup, the pebble is transported back to the core through the lighting pipe 2. If the fuel pebble has reached its maximum burnup, then it is transported to the FS02 through the lifting pipe 3. If some fresh fuel pebbles need to be added into the core, the fuel pebbles will be loaded into the core from FS01 through the lifting pipe 1.

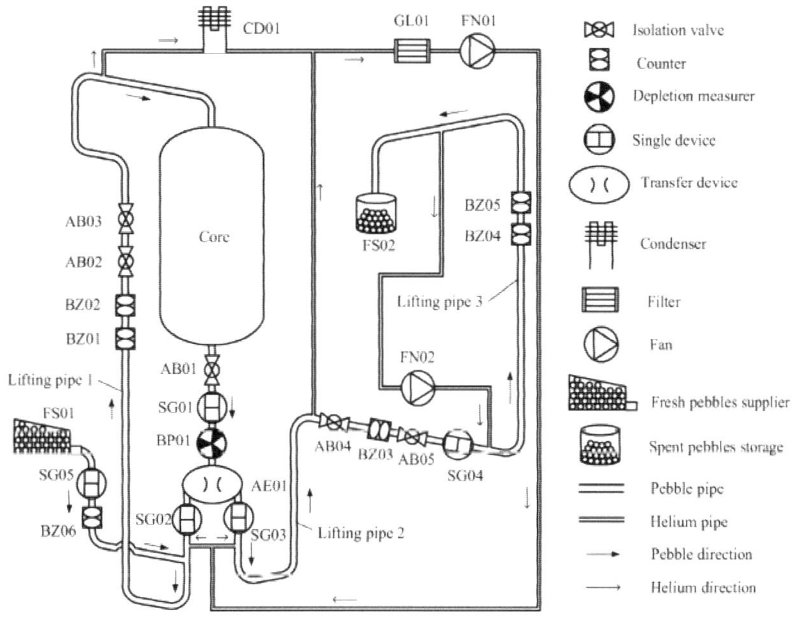

Figure 1. Fuel handling system of the HTR-10 [9]

In HTR-10, the average target burnup of the fuel is designed for 80.000 MWD/MTU. In the equilibrium state, the number of fresh pebbles and recycled pebbles needed per EFPD is 25 and 100, respectively [8]. It means that there are 125 pebbles performed burn-up measurement per day. It was assumed that one pebble takes about 11 minutes from leaving the reactor core, performing burn-up measurement, and then recycling into the core or moving out into the discharged tube. During this 11 minutes, because of its position is not in the core, there is no fission reaction in the fuel, it is only the radionuclide decay process that occurs in the pebble fuel.

The irradiation process of the pebble fuels during their residence in the reactor core was simulated by using ORIGEN2.1 computer code, from cycle 1 through cycle 5. The reactor parameters used in creating input data for ORIGEN2.1 simulation are shown in Table 2. The material composition of the fuel pebbles, which is shown in Table 3, was calculated based on the heavy metal loading and enrichment of the fuel. Since the thermal power of the reactor is 10 MW and designed maximum burnup of the pebble is 80.000 MWD/MTU, it is assumed that the pebble resides in the core for about 216 days per cycle or 1080 days for full cycle. After each cycle, the fuel pebble was simulated to undergo decay for about 11 minutes. The activity and photon intensity of Kr-85m were acquired during this decay period.

Table 2. Parameters used as input in ORIGEN2.1 simulation [10]

No	Parameter	Nominal	Unit
1	Thermal power	10	MW
2	Number of pebble	27000	pebbles
3	Heavy metal loading	5	grams
4	Enrichment	17	%

Table 3. Material composition in the fuel pebbles

No	Material	Nominal	Unit
1	U-235	22937.32	grams
2	U-234	41.15	grams
3	U-238	111946.9	grams
4	Oxygen	18180.19	grams
5	Carbon	5283195	grams
6	Silicon	34818.25	grams

4. Results and Discussion

The results for Kr-85m activity as a function of burnup are shown in Figure 2. Around the first 20 days of irradiation, the activity of Kr-85m increases significantly until it reaches 3.83 Ci. After that, the activity of Kr-85m decreases consistently when the burn-up of the fuel pebble increases. Several sharp activity reduction emerged at each cycle indicate the decay process of Kr-85m when it leaves the reactor core and moves to the burnup measurement position. This rapid reduction is due to the short half-life of Kr-85m. The concentration and activity of Kr-85 reduce rapidly since there is no Kr-85m produced from fission reaction during this decay period. From this Kr-85m activity revolution trend, it can be implied that it has a good relationship with the fuel burnup increment. The percentages of the activity reduction of Kr-85m at several burnup steps are shown in Table 4. This reduction percentage is sufficient to determine the burnup level of the fuel at each cycle. These reduction percentages are considered acceptable since the best range of activity reduction for burnup measurement is between 5% and 35% [11].

This result also indicates that the use of Kr-85m as burnup indicator, from the point of view of its activity reduction in relation with the burnup increment, is suitable for online measurement. Compared to other radionuclide such as Cs-137 whose half-life is about 37 years, the activity of Cs-37 will be steady and not be significantly different at each burnup step. Thus, a short half-life radionuclide, which in this case is Kr-85m, is the potential option for use in the online burnup measurement in pebble bed reactor. In order to obtain a better measurement results, the instrument used for burnup measurement must be quite sensitive with high accuracy and with detection error of less than 4%.

The activity of Kr-85m when performing decay at each cycle is shown in Figure 3. It has been stated before that Kr-85m is assumed to undergo decay for 11 minutes after each cycle. If the burnup measurement is performed within 2 minutes after the pebble leaving the reactor core, the activity of Kr-85m, in approximation, at cycle 1 is 3.38 Ci, at cycle 2 is 3.09 Ci, at cycle 3 is 2.91 Ci, at cycle 4 is 2.76 Ci, and at cycle 5 is 2.65 Ci. Since the half-life of Kr-85 in quite long (268 minutes) compared to the decay time (11 minutes), the activity of Kr-85m does not alter very significantly in the process of the decay within this 11 minutes period. This makes the measurement at each cycle unaffected by the alteration of the activity due to reasonable comparison between the half-life of Kr-85m and the decay period. Again, since the activity reduction is varied between 12% at cycle 1 to 4% at cycle 5, this condition requires the use of instrument that has small detection error in order to obtain the more accurate measurement data.

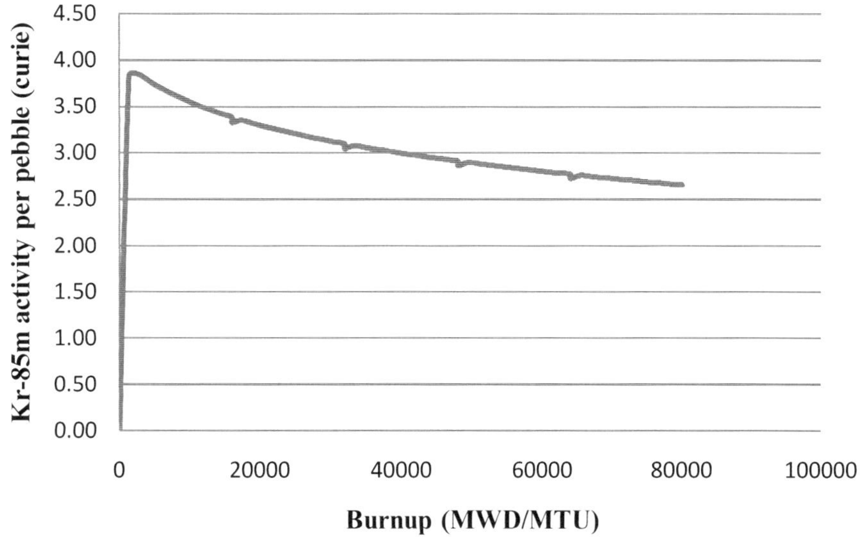

Figure 2. The activity of Kr-85m as a function pebble fuel burnup

Table 4. Percentage of Kr-85m activity reduction at several burnup steps

No	Cycle	Burnup increment	Activity reduction
1	Cycle 1	0 – 16000 MWD/MTU	12%
2	Cycle 2	16000 – 32000 MWD/MTU	9%
3	Cycle 3	32000 – 48000 MWD/MTU	6%
4	Cycle 4	48000 – 64000 MWD/MTU	5%
5	Cycle 5	64000 – 80000 MWD/MTU	4%

The results for photon intensity from Kr-85 decay are shown in Figure 4. This photon is 0.151 MeV gamma emitted from beta decay reaction of Kr-85m. The emission probability of this gamma is 75% which is the most dominant compared to the other gamma rays emitted by Kr-85m which are 0.3%, 0.01% and 14% for gamma energy 0.129 MeV, 0.451 MeV, and 0.304 MeV, respectively. This photo intensity, which is in the order of 10^{10} photon per second, is considered more than sufficient to be detected since it can be categorized as a strong signal produced from a radionuclide. The instrument that will be used to detect this gamma emission should have a good resolution, which can differentiate the gamma rays emitted by Kr-85m from another gamma emission emitted by other radionuclides in the fuel pebble. The challenge is that there is another radionuclide which emits almost similar gamma energy as Kr-85 which is Sr-85m. However, the yield of Sr-85m in U235 fission reaction is much lower than Kr-85m yield which is merely 1.47×10^{-13}. The yield of Kr-85m is about 8.5×10^{-5}. It can be implied that the concentration of Sr-85m will be much lower than Kr-85m and the gamma rays with the energy of 0.151 MeV detected by the instrument can be considered mainly emitted by Kr-85m and the contribution of Sr-85m can be ignored.

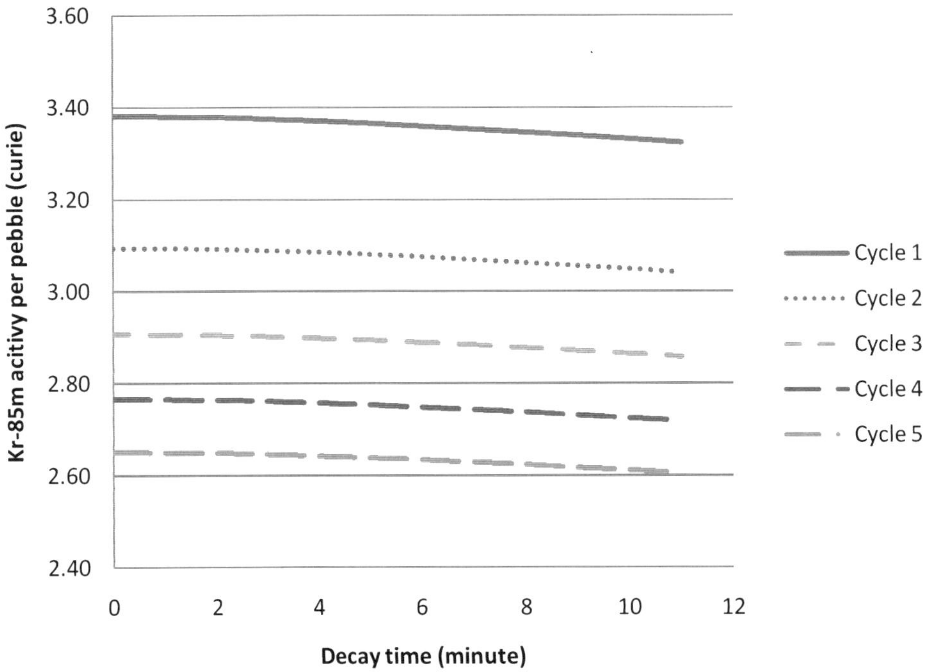

Figure 3. The activity of Kr-85m performing decay in each cycle

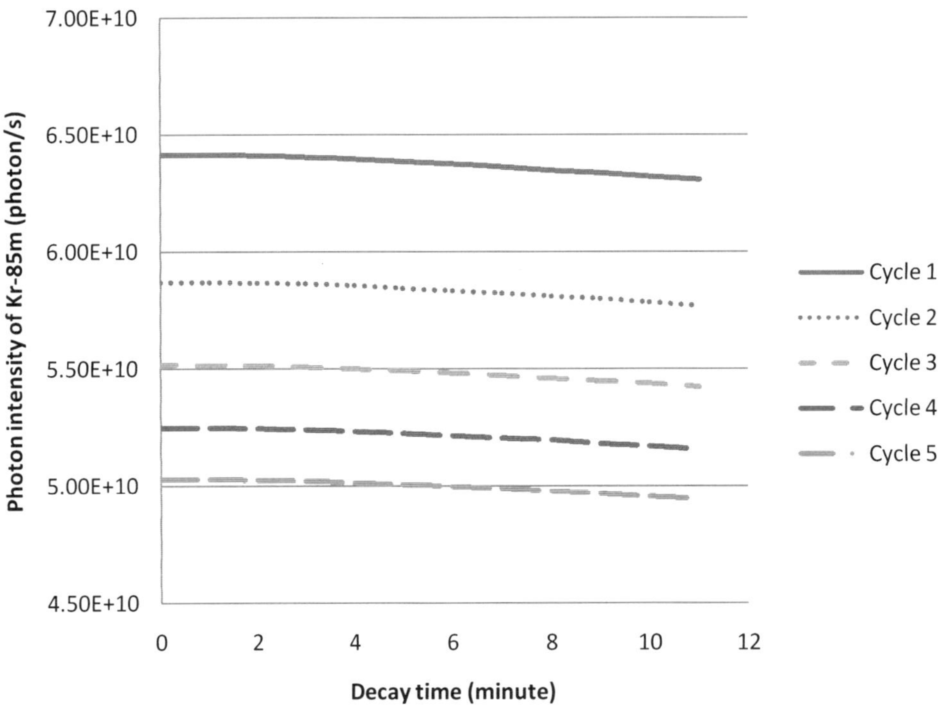

Figure 4. Photon intensity of Kr-85m at each cycle

5. Conclusion

The activity and gamma emission of Kr-85m have been studied in order to investigate the feasibility of Kr-85m to be used as an online burnup measurement indicator in pebble bed reactor. The calculation of the activity and photon intensity of Kr-85 were performed by using ORIGEN2.1. The results show that Kr-85m activity has a good relationship with the burnup of the PBR fuel pebble. The trend of Kr-85m activity revolution is consistent with the fuel burnup increment. The activities of Kr-85m reduce for about 12%, 9%, 6%, 5%, and 4% at cycle 1, cycle 2, cycle 3, cycle 4, and cycle 5, respectively. This reduction percentage is acceptable for determining the burnup level of the fuel at each cycle. The results for gamma emission of Kr-85m show that the emission intensity is in the order of 10^{10} photon/second. This intensity is sufficiently high to be detected by the instrument of fuel burnup measurement. From these results, it can be concluded that Kr-85m isfulfilled the criteria as an online burnup indicator in a pebble bed reactor.

Acknowledgement

The author would like to acknowledge the Head of Center for Nuclear Reactor Safety and Technology (PTKRN) BATAN for supporting this work by granting the government research fund (DIPA-PTKRN-BATAN) of the year 2017.

References

[1] Futterer M A, Fu L, Sink C et al 2014 Status of the very high temperature reactor system*Progress in Nuclear Energy***77** 266 – 281
[2] IAEA TECDOC 2010 High Temperature Gas-Cooled Fuels and Materials *International Atomic Energy Agency*
[3] Bingjing Su,Zhongxiong Zhao, Jianwei Chen et al 2006 Assessment of on-line burnup monitoring of pebble bed reactor fuel by passive neutron counting Progress in Nuclear Energy **48**686 – 702
[4] Yan Wei-Hua et al 2014 Prototype studies on the nondestructive online burnup determination for the modular pebble bed reactors*Nuclear Engineering and Design***267** 172 – 179
[5] Akyurek T, Tucker, L P and Usman S 2014 Review and characterization of best candidate isotopes for burnup analysis and monitoring of irradiated fuel*Annals of Nuclear Energy***69** 278 – 291
[6] Benkharfia H, Zidi T and Belgaid M2016 Lumped pseudo fission products during burnup step in MCNP5 – ORIGEN coupling system*Progress in Nuclear Energy***88**277 – 284
[7] I. Husnayani 2016 Calculation of radionuclide content of nuclear materials by using ORIGEN2.1 computer code*Sigma Epsilon***19** 20 – 25
[8] Yongwei Y, Zhengpei L, Xingqing J andZongxinW2002 Fuel management of the HTR-10 including the equilibrium state and the running-in phase*Nuclear Engineering and Design***218** 33 – 41
[9] Liu H, Du D, Han Z and Chang B 2017Pneumatic Transportation Pattern of Fuel Pebbles in a Pebble Bed Reactor*Annals of Nuclear Energy***99**434 – 443
[10] Scari M E, Costa A L and Pereira C2016 HTR steady state and transient thermal analyses**41** 7192 - 96
[11] JianweiChien2004 Online Interrogation of Pebble Bed Reactor Fuel Using Passive Gamma Ray SpectroscopyPhD Dissertation *University of Cincinnati*

IOP Conf. Series: Journal of Physics: Conf. Series **962** (2018) 012006 doi:10.1088/1742-6596/962/1/012006

Operator Support System Design forthe Operation of RSG-GAS Research Reactor

S Santoso, J Situmorang, S Bakhri, M Subekti, G.R Sunaryo

Center for Nuclear Reactor Technology and Safety-BATAN, PuspiptekComplex, Building no.80, Serpong,Tangerang Selatan 15310, Indonesia

sigitsan@batan.go.id

Abstract. The components of RSG-GAS main control room are facing the problem of material ageing and technology obsolescence as well, and therefore the need for modernization and refurbishment are essential. The modernization in control room can be applied on the operator support system which bears the function in providing information for assisting the operator in conducting diagnosis and actions. The research purpose is to design an operator support system for RSG-GAS control room. The design was developed based on the operator requirement in conducting task operation scenarios and the reactor operation characteristics. These scenarios include power operation, low power operation and shutdown/scram reactor. The operator support system design is presented in a single computer display which contains structure and support system elements e.g. operation procedure, status of safety related components and operational requirements, operation limit condition of parameters, alarm information, and prognosis function. The prototype was developed using LabView software and consisted of components structure and features of the operator support system. Information of each component in the operator support system need to be completed before it can be applied and integrated in the RSG-GAS main control room.
Keywords: operator support system, RSG-GAS, control room

1. Introduction

RSG-GAS reactor has been operating for about 30 years, so it is certain that many components will experience the problem of ageing or obsolescence technology. To anticipate these problems, the program for refurbishment or modernization was established. According to this program and challenge, an integrated computer-based operator support system(OSS) in the control room is developed. The operator support system is aimed at improving operator effectiveness and performance by providing information for diagnosis of system condition, detection system, operation procedure, and component information presented at control room. Refurbishment of the main control room requires good planning and comprehensive consideration on operation characteristics and design features. Moreover, the design process must follow the requirements and principles of human factor engineering in order to obtain a safe, efficient and reliable control room[1-3].

The control program of RSG-GAS is distinguished as control of reactor operation safety and target irradiation control [4]. Most of the drive system is regulated and controlled from RKU and a small portion through local panels. Reactor control is performed by the reactor operator at the reactor control desk in the main control room. There are two modes in the reactor control performing which are manual mode and automatic mode. The status of the control rods used as a main device in power generation settings is shown by an indicator located on the reactor control desk [4,5].

In advanced reactors, auxiliary systems are designed by applying the modern technology in instrumentation, information and computation [1,5,6]. Additional function applied on the operator support system includes a high-level information display that allows the operator to know the condition of the reactor quickly, early fault detection system, and computer-based procedures [5-7]. Those functions are expected to reduce the operation failure due to human error (operator error). In addition to technological approaches, consideration of human aspect in the design of modern support system is also implemented, such as consideration on cognitive activities of the operator [5,8-10]. To achieve a suitable operator support system, the design of support system and its features should be tested and evaluated using modeling and simulators before being applied to the real control room [3,11-14].

The objective of this research is to obtain a model/prototype of computer based operator support system design that can be further implemented in RSG-GAS control room. The design adopts the progress of technology especially for information processing and visualization. The model obtained in this study can then be further developed by adding operational scenarios as well as other relevant information. The research methodology includes the study of operation scenario of RSG GAS, identification of component at control room based on design document and observation, and discussion with reactor personnel. Based on the information obtained in above stage, structure and components of the operator support system prototype was designed.

2. Theory

2.1. Components of RSG-GAS control room

The design of main control room as the center of interaction and control should be continously improved to be more suitable with operator needs and task functions [1,2]. The Main Control Room of RSG GAS consists of information related with the main functions, auxiliary systems and supporting installations, managing the basic requirements for the operation of the reactor and other important systems. Figure1.shows the components and layout of the main control room of RSG GAS. The reactor protection panel is installed in the RKU and is a part of the reactor protection system. In the reactor protection system (RPS), testing instruments and switches for conducting signal testing including the signal characteristics of engineered safety features is installed.

The reactor power regulation system is carried out by maneuvering the control rod (8 pieces, JDA01 to JDA08) inside the core. This reactor operation can be done manually or automatically. Reactor power regulation and controls are performed from the CWA01 reactor control desk, control rod status information which consists of control rod group (7 bank control rod) and a regulatory control rod (1 piece) are provided in control desk. Control line from the operator will be stopped if the reactor protection system generates a scram signal and control rod will be rapidly dropped and cause the reactor to shut-down. Indicators shown for the operation of each control rod show the position/status of the control rods, that are: up position, down position, 100% absorber down position, armature drop position, armature coupled position and overload insert position. The control rod position setting is done via a push button located on the control desk CWA01, that also displays the control rod position indicator. The signal from the control desk is connected to the electronic cabinet in the Marshalling box before it is connected to the drive system in the reactor [4].

For reactor operation, the display provided at the RKU includes the status of the primary cooling system, secondary cooling, auxiliary system, ventilation system, and control rod position. Furthermore, display for safety-related system variables are installed on control equipment and RKU control desk. The information is grouped into three groups: controlled component state information from the Engineered Safety Feature System (ESFS), Status information on the reactor protection panel and other variables groups. The MCR safety panel also displays the status of reactor protection system and the amount of the output signal, as well as information from the system variables that are important for safety. Moreover the reactor protection panel is also provided with the buttons/switch used to monitor the output signal and test back from ESFS. Hence, general condition of Reactor Instrumentation and Control System (SIK), Reactor Protection System (SPR), and mechanical facilities connected to SPR can be monitored through the Reactor Protection Panel. The reactor protection panel consists of 15 columns and several rows that are grouped into three groups, namely group A for alarms, group B includes the information of the neutron flux density, unbalanced load, gamma dose rate, while group C consists the test button, the triggering criterion of the scram signal and instrument for testing information [4,14].

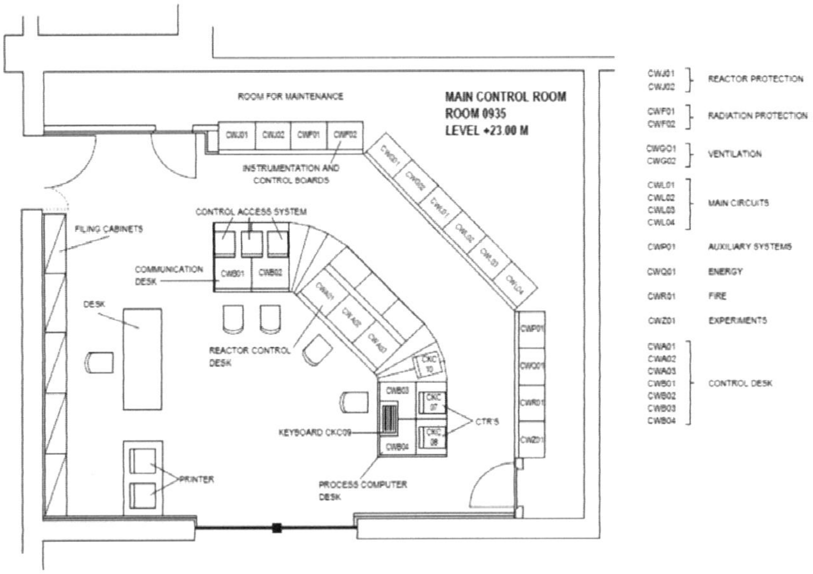

Figure 1. Components and layout of RSG GAS Main Control Room [4]

2.2. RSG-GAS operation

Operation of RSG-GAS is categorized into four types: start-up, low power reactor operation, high power operation, power change/maneuvering, shutdown and reactor scram. The requirements and characteristics of the operation as well as the indicators to be considered for each mode of operation may vary. One of them is the use of indicators JKT 01 to JKT 04, each have functions in different ranges. In addition to the information content, communication in the control room is required to convey the conditions and requirements of operating parameters, information on limit conditions (scram reactor period, reactivity, etc.) and to determine actions in operation. Operating characteristics that need to be considered include the reactor start-up operation to the high power during the decreasing of Xenon. The parameters that should be noticed by operator in conducting the settings/operations include reactor periods and reactivity. The terms and limitations of reactivity and power settings are taken into account to avoid reactor scram [4,15].

To maintain the operatorperformance in conducting the task, the shift work system is applied on RSG-GAS operation. In one day operation is divided into 3 shifts, each shift for 8 hours. One shift operation consists of at least one supervisor, two operators, one radiation protection officer and one maintenance/maintenance worker. To support the reactor operation there are 11 supervisors, 21 operators, 20 person for maintenance technicians, and 9 person for maintenance supervisors. Qualifications and training for personnel working in the reactor facility is established and stated in BAPETEN Chairman Regulation No.6 of 2013 [16].

3. Methodology

The main objective of this study is to develop a prototype design for the operator support system based on the reactor operation characteristics and system features. The research methodology included the study of the RSG GAS operating system based on documents/ procedures, control room observations, and conducting discussions with reactor personnel. Identification is focused on the availability of data and information systems on the existing control room, operator needs and requirements, input-output components, operational scenario, and operator tasks in some operational scenarios. The following step is determining the features and structure of the operator support system based on the data collected data. The features and structures obtained are subsequently used in the design of support system prototype/model using LabView software.

4. Result and discussion

4.1. *Reactor operation scenario applied in the OSS design*

Most of the reactor operation action at RSG-GAS is performed by the operator from the control desk in control room in manual or automatic mode. Prior to reactor operation all required conditions and limits for operation must be satisfied, i.e., all safety-related systems must be able to operate or perform their function. The design of the support system should contain information for the preparation of the operation. In the event of deviation during operation, the reactor is allowed to continue operating until the end of the cycle if the limits and operating requirements are not exceeded. The simulated operating scenario of the reactor is divided into three categories: Power operation (Mode 1) that is start-up operation up to power operation > 3% full power; Low power operation (Mode 2) which is distinguished into low power operation with natural convection cooling and Low power operation with forced convection cooling; and Reactor shutdown(Mode 3) as shown in Table 1.

The parameters regarding operating requirements and safety constraints are shown in the information display of support system, such as in forced convection is covering the maximum power parameters of the reactor, maximum power factor of inlet pool temperature, reactor water level, and minimum flow rate. While in natural convection parameters that are considered include the thermal power of the reactor, the minimum period of reactor and the height of the pool water. In addition, the requirements for activating the reactor protection system to be fulfilled prior to the operation are also presented, e.g. for the operation mode 2 reactor shall not be turned on if non-functioning channel exist. In principle, the information system in the support system contains all parameters related to the safety system that must be met as a condition for operation. In addition, the support system also displays deviations occur during operation, and information according to the limitations/operating requirements that should not be exceeded.

The operating characteristics and parameters shown in the information system design and procedures in the simulation/model of operator support systems are:

- Characteristics of core loading and burning fractions: the cycle in using of fork absorber or operation time (MWD), fluens boundary, maximum drop test of control rods, burning fraction for spent fuels.
- Shutdown reactivity and control rod configuration: shutdown reactivity margin , control rod reactivity, maximum deviation of bank control rod position, control rod position indicator status, maximum drop time for 80% drop and average time drop of all absorber (Mode 1 and Mode 2) .
- Reactor cooling parameters: minimum primary system flow rate, total mass flow rate to cool irradiated insertion at core and reflector, grid position, nuclear heat channel factor for each configuration of steady-state core and specific core (axial maximum 1.6 and radial max 2.6).

Table 1. Reactor operation scenarios applied in the OSS prototype/model

Reactor operation mode	Mode 1 (power operation)	Mode 2 (low power operation)	Mode 3 (shutdown reactor)
Operation characteristics	Reactor power > 3% of full power	*Mode 2a:* -primary cooling stopped -Reactor power < 1% of full power *Mode 2b :* -Primary cooling operated -Reactor power < 3% of full power	Control rod position at the lowest level Meter indicator : zero Position of 6 contact system switch : open Start-up key position : off

4.2. Structure and features of the OSS prototype

The prototype of the operator support system contains three main information groups, namely the preparation of operations, the implementation/execution of the operation and the database system procedure. Operating sequences in support system designs/model are based on RSG-GAS operation scenarios. The OSS design would be useful in performing appropriate actions. For example in power operation (Mode 1) and low power operation (Mode 2), if required condition is not yet fulfilled, the OSS will displayed the task/action that should be performed by operator such as related to core cooling, eg: keep reactor remain in shut down condition, control flap position of the inlet isolation valve at heat exchanger in the primary system done flap position positioning from valve Inlet of heat transfer isolation of the primary system, or check another factor initiated the flow increase /decrease. Furthermore, reduction of irradiated target number or increase of hydraulic resistance, and changing of core configuration are carried out. The structure and sample picture of the designed operator support system prototype/model is presented in figure 2.

Other features embeded in the prototype were designed to assist the operator in conducting diagnosis of operation condition, such as in a scram condition. Reactor scram mode is generally performed automatically by the reactor protection system. Therefore the information due to parameter that triggers the occurrence of reactor shutdown/scram is displayed in the designed operator support system prototype prototype. The scram parameters shown in the display includes start-up flux > 105, start-up reactivity accident (periods of interim flux <5 second), reactivity accident at power range, beam tube flooding, cooling channel closed, and primary pump fails.

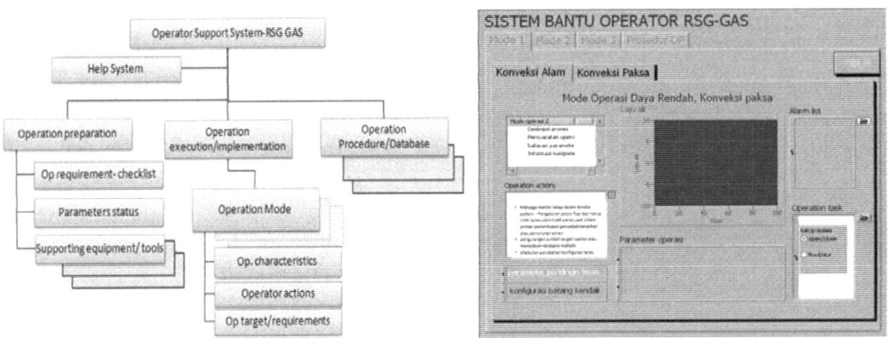

Figure 2. Component structure and sample mimic of the OSS ptototype

5. Conclusion

Refurbishment and modernization of RSG GAS control room has been planed and partly implemented. The approach for modernization was conducted in this study through the provision of operator support system at main control room. The prototype of operator support system in this study was developed based on reactor operation scenario, system features, and operator needs. Operation scenarios applied in the OSS prototype are low power operating scenario, power operation and reactor scram. The designed OSS consist of main components which related to the operation preparation, operation execution, and operation procedure. The prototype design provides essential information that would be useful for operator in performing appropriate actions and in conducting diagnosis especially during abnormal conditions. Information of each component in the operator support system structure should be completed before it can be applied and integrated in the RSG-GAS main control room.

Acknowledgements

This research was a part of research project supported and funded by the Nuclear Research & Development Program (DIPA 2016) at the Center for Nuclear Reactor Technology and Safety of the National Nuclear Energy Agency. Information from the discussion with reactor personell, especially Mr. Sudjarwono were very helpful in improving the author perspective on reactor operation. The authors is grateful to all of them.

References

[1] Mishima T, Nishi H, Tamura Y, and Nakagawa Y, 2010 *Int. J. of Nucl. Safety and Simulation.* **1(2)** 166-169
[2] O'Hara JM, Higgins JC, Fleger SA, Pieringer PA. 2012 Human factors engineering program review model. NUREG-0711, Rev.3
[3] Lee M S, Hong J H, Suh J K, Lee S H, and Hwang D H 2009 *J. Nucl. Sci. Technol.* **46(1)** 90–101
[4] RSG GAS 2011 Sistem Instrumentasi dan Kendali, LAK RSG - GAS, Rev. 10.1
[5] Yoshikawa H, 2013 *J. Nucl. Eng and Technol* **45(2)** 165-178
[6] Gofuku A, Inoue T 2016 *J. Nucl. Safety and Simulation*, **7(1)** 35-41
[7] Boring RL, Kenneth D 2015 *Procedia Manufacturing* 3, pp 5261-5268
[8] Xu S, Song F, Li Z, Zhao Q, Luo W 2008*J. Reliability Engineering & System Safety* **93(10)** 1500–11
[9] Lee S J, Kim M C, and Seong P H 2008 *J. Reliability Eng. Syst. Saf.* **93(4)** 567–77
[10] Kim J H, Seong P H 2007 *J. Reliaility Eng. Syst. Saf.* **92,** 171
[11] Lind M., and Zhang X. 2014 *J, Nucl. Engineering and Technology*. **46 (6)** 753-772
[12] Lee S J et al 2009 *J. Nucl. Sci. Technol.* **46(5)** 424–35
[13] Tanaka T, Ueyama I, Tamura Y and Nakagawa K 2011 *J. of Nucl. Safety and Simulation* **2(1)** 77-82.
[14] IEEE Standards1786-2011, IEEE Guide for Human Factors Applications of Computerized Operating Procedure Systems (COPS) at Nuclear Power Generating Stations and Other Nuclear Facilities, Newyork
[15] Villim R B et al, 2013 Monitoring and Diagnosis of System Faults (*Nucl. Engineering International*), November p 24-27
[16] Peraturan Kepala BAPETEN No.6 2013 Tentang ijin bekerja petugas instalasi dan bahan nuklir, Jakarta

V&V Plan for FPGA-based ESF-CCS Using System Engineering Approach.

Restu Maerani, Joyce Mayaka, Mohamed El Akrat, Jung Jae Cheon

Department of Nuclear Power Plant Engineering, KINGS Instrumentation and Control Lab, KEPCO International Nuclear Graduate School, Ulsan, 45014 Republic of Korea.

maerani@email.kings.co.kr

Abstract. Instrumentation and Control (I&C) systems play an important role in maintaining the safety of Nuclear Power Plant (NPP) operation. However, most current I&C safety systems are based on Programmable Logic Controller (PLC) hardware, which is difficult to verify and validate, and is susceptible to software common cause failure. Therefore, a plan for the replacement of the PLC-based safety systems, such as the Engineered Safety Feature – Component Control System (ESF-CCS), with Field Programmable Gate Arrays (FPGA) is needed. By using a systems engineering approach, which ensures traceability in every phase of the life cycle, from system requirements, design implementation to verification and validation, the system development is guaranteed to be in line with the regulatory requirements. The Verification process will ensure that the customer and stakeholder's needs are satisfied in a high quality, trustworthy, cost efficient and schedule compliant manner throughout a system's entire life cycle. The benefit of the V&V plan is to ensure that the FPGA based ESF-CCS is correctly built, and to ensure that the measurement of performance indicators has positive feedback that "do we do the right thing" during the re-engineering process of the FPGA based ESF-CCS.

Keywords: *ESF-CCS, FPGA, Verification, Validation*

1. Introduction

Advanced Power Reactor 1400 (APR1400) is equipped with multi-loop controllers based on microprocessors for the safety and non-safety control systems. Principles such as redundancy, diversity, and segmentation have been incorporated in order to achieve both the desired availability and reliability of these systems [1]. The purpose of the Verification and Validation (V&V) plan is to meet the reliability requirements for components for the Instrumentation and Control (I&C) systems in the Nuclear Power Plant (NPP) [2]. Programmable Logic Controllers (PLCs) have previously been used to implement real-time controllers in the Reactor Protection Systems (RPS). However, the increase in the complexity of I&C systems as well as high maintenance cost, require a more reliable application development platform that still meets the measures of effectiveness (MOE's) of the I&C system. One of the technologies currently being looked at is the Field-Programmable Gate Array (FPGA) [3].

The design and implementation of an FPGA-based ESF-CCS must pass through several verification stages. Because of the FPGA's are dual nature, being both hardware and software, and its inherent complexity, we need a specific regulatory document for FPGA that would address issues such as system safety assessment, design life cycle, verification and validation (V&V) and related documentation requirements for NPP safety system[4]. Since the specific standard of FPGA is not found on United States Nuclear Regulatory Commission (USNRC) so we can use the alternative document from International Electrotechnical Commission IEC 62566:2012 about "Nuclear Power Plants – Instrumentation and control important to safety- Development of HDL-programmed integrated circuits for systems performing category A functions". While carrying out reverse engineering, V&V activities are required to ensure that the system is redesigned in accordance with the original design basis requirements and system design requirements [5]. Then for the re-engineering process the verification process will be conducted after completing the design and implementation process; which includes system, hardware, and software verification. Based on the review of IEEE Std. 15288-2008 [6], R Halligan [7] said that "a verification system is a system like any other system, but is also a subsystem of the bigger system, of which verification is a part of the solution".

V&V is a means by which the product is checked, and by which its performance is demonstrated and assured to be a correct interpretation of the requirements. The V&V processes provides assurance that the specified functional and reliability requirements of the system are met [8]. V&V process can prove the success of a design, but cannot prove that the model is true and accurate for all possible conditions and applications. It can only provide evidence that the model is accurate enough and its output can be utilized. Therefore, the V&V process is complete when sufficiency is achieved [9].

To make certain that the V&V process is carried out correctly, it needs to be implemented using System Engineering approach to ensure the realization of a successful system. The problem statement should begin by determining the top level function, which the system will do, as well as explaining the mission and concepts to be performed. In addition, the possibility of deficiencies or barriers should be defined, as well as how to overcome them and trace the requirements document. Explain how the process of implementation and design simulation. All these formulations must also be in accordance with the client's request as they have the ease of operation in the future [10].

The use of life cycle model is also recommended by the IAEA in "IAEA safety standards on Design of Instrumentation and Control Systems for Nuclear Power Plants" [5]. For safety critical I&C systems, it is very common to use the life cycle model for system that related to electronic system and instrumentation. In NPPs this practice has been applied for the documentation of the development process of I&C system to facilitate the process of verification of each activity and review of the final product by validation process, to ensure that the final product is suitable for use. It is important to control a process with the necessary management systems to achieve the objectives, provide the means to meet all requirements and provide the organization product to be identified, and its development should be planned, implemented, assessed and continuously improved.

This paper outlines how to verify and validate an FPGA-based ESF-CCS system for the APR1400. The paper does not discuss the overall development process of an FPGA-based ESF CCS or give detailed information about ESF and FPGA, neither does it discuss the results of the V&V activities. These shall be developed in later work.

2. Definition

FPGA based EFS- CCS Operation

The digital Instrumentation and Control System (I&C) technology of nuclear power plants, is expected to have the ability to meet the safety criteria of nuclear reactors in order to avoid three problems that can occur: software common-cause failure, failure interaction between operator and Man Machine Interface System (MMIS) and the non-detectability of software failure [11].

In the recent past, FPGA technology has gained attention in the nuclear industry. A field-programmable gate array (FPGA) is an integrated circuit designed to be configured by a customer or a designer after manufacturing – hence "field-programmable". FPGAs contain an array of programmable logic blocks, and a hierarchy of reconfigurable interconnects that allow the blocks to be "wired together", like many logic gates that can be inter-wired in different configurations.

The ESF-CCS is composed of the electrical and mechanical devices and circuitry, from sensors to actuation device input terminals that are involved in generating those signals, which actuate the required ESF system components. This system consists of four independent channels (Channels A, B, C, and D) which are electrically and physically independent. The ESF-CCS configuration is currently based on a PLC platform. The ESF-CCS performs the functions of actuating Engineered Safety Features Actuation System (ESFAS) and executing component control by interfacing with ESFAS portion of Plant Protection System (PPS) as shown in Figure 1.

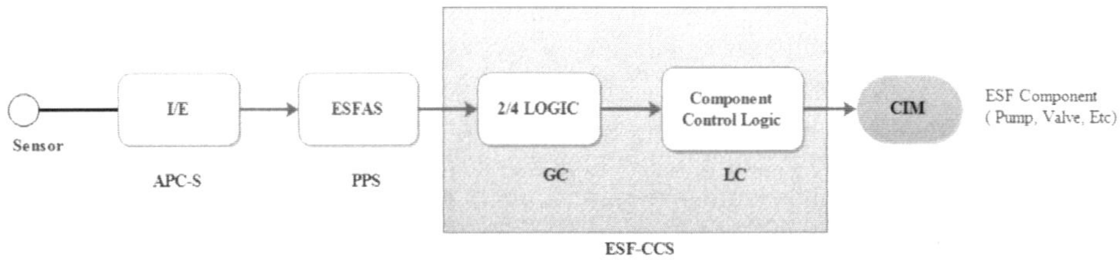

Figure 1: EFS-CCS Operation

Design Life Cycle

Based on the IAEA recommendations that I&C all the activities associated with the development, implementation and operation of the overall I&C architecture, individual I&C systems and I&C components should be carried out in the framework of a documented development life cycle [5]. Because this design is related to hardware for the FPGA Board also VHDL code as a software tools to generated signal for simulation of ESF CCS design, we also need to refer to IEEE 1074-2006: Standard for Developing a Software Project Life Cycle Process [12]. There are many life cycle models to use for the system engineering process of developing large-scale complex systems. One famous model is the V-Model [13]. What follows is a brief description of the lifecycle detailed in EPRI 1019181, which is also referenced by IAEA [14] and is consistent with the one suggested by IEC 62566 [15]. The V-model lifestyle is depicted in Figure 2.

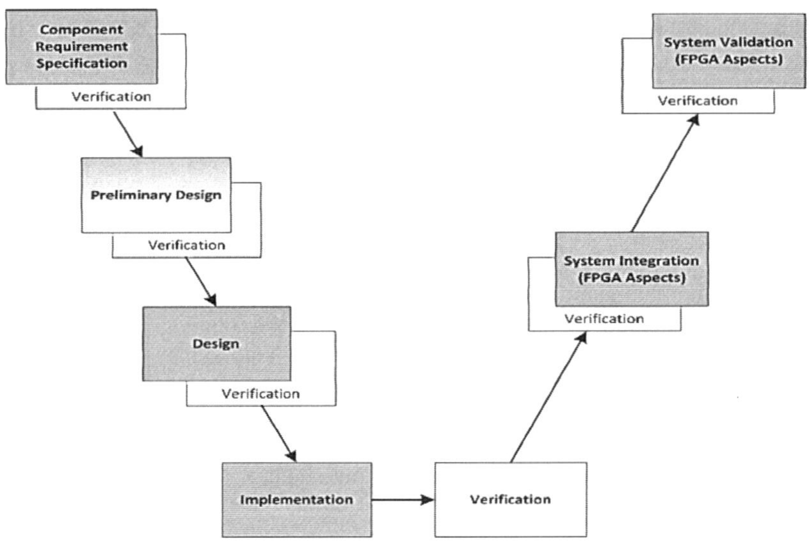

Figure 2: V Model for FPGA [14]

Verification and Validation (V&V)

Verification and validation is a process of checking the results of a design implementation to ensure the reliability of its performance and has been ensured in accordance with applicable requirements standards document [8]. The benefit of the V&V process here is to ensure the reliability of the I&C system developed using FPGA, appropriate design steps may require reiterating the design until the verification requirements are met. And verification steps can also had revision of the initial requirements so it need to update all the previous required design steps [4].

3. Methodology

NUREG /CR-7006 [4] recommends the life cycle illustrated in Figure 3. Based on this recommendation, the Y model in Figure 4 was developed with the aim of minimizing the system complexity.

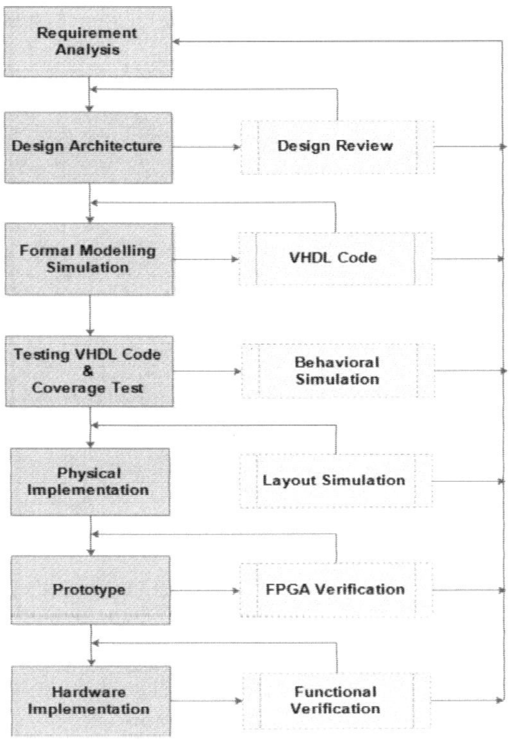

Figure 3: FPGA Design Flow [4].

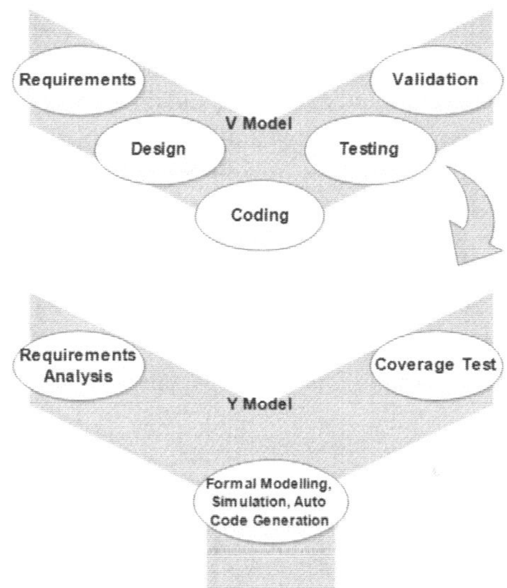

Figure 4: Modification of Development Process from V Model to Y Model [16].

During development process for the FPGA-based ESF-CCS design, there are several activities that should be verified, referenced based on the design flow in Figure 3.

The following tasks should be performed:

- Analyze the requirements document for the ESF CCS by referring:
  - 10 CRF 50
  - Design Document
  - Korea Utility Requirements Document
  - Standards
  - Regulatory Guides
  - Branch Technical Position
  - Standard Review Plant
- Develop Design Architecture of ESF-CCF based FPGA using Basys 3 board

- Create the formal modelling for the operating scenario of ESF CCS, make the code for generate the signal to FPGA board.
- Develop test case to verify the VHDL code
- Develop coverage test to verify the successful of the system
- Validate the design by reporting the technical performance.

4. Discussion

4.1 Requirement Analysis

This is a description of the system to be implemented. It should lay out the required functionality, interfaces and operational environment of the system. Additional requirements, such as critical timing, power, and board size, may be added to the requirements [4].

When we start to design a system, in this case I&C system, we must first submit the design that related to design requirement from the regulator to be verified. The regulator will determine whether the I&C system will meet the requirements of standard regulation document [17].

I&C systems are described in Chapter 7 of the Standard Review Plant (SRP), specifically section 7.3 deals with the Engineered Safety Features Actuation System (ESFAS). In the SRP, reference is found to the Regulatory Guides (RG) which must be implemented in the design for the licensee to be in compliance with the established law of 10 Code Federal Regulations (CFR). These RGs in turn, reference standards which give a more detailed guide on the implementation of a given RG. The hierarchy of the requirements document is as shown in Figure 5.

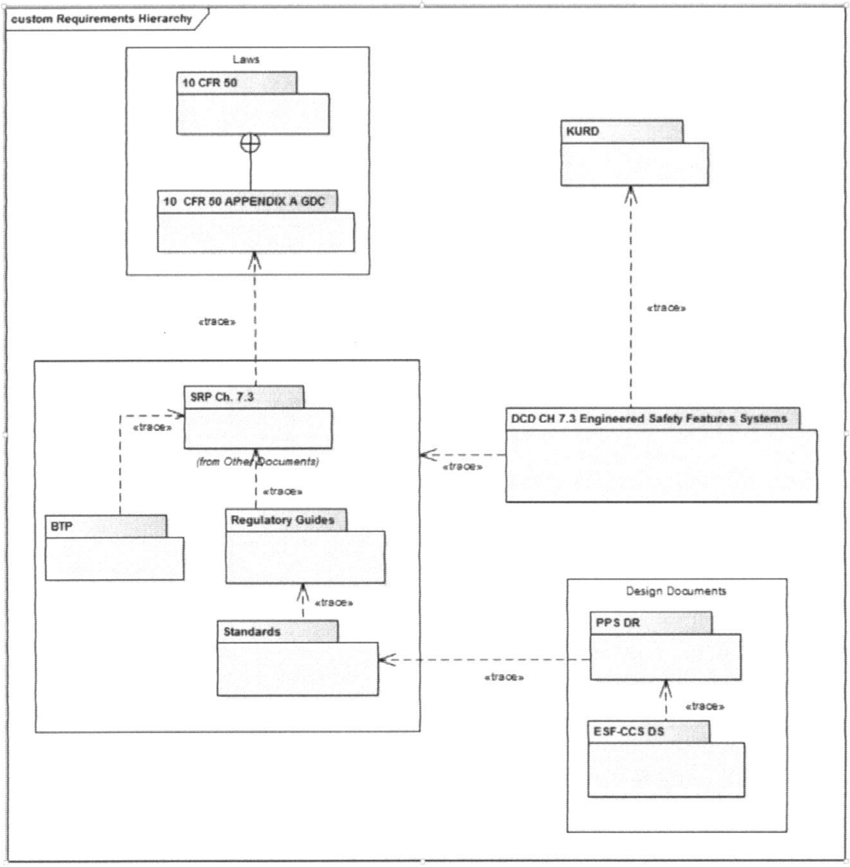

Figure 5: Requirements Hierarchy and Inter-relationship for ESF-CCS

4.2 Formal Modelling

Elisabeth Hull [18] mentioned that "Those methods that use diagrammatic representations are usually referred to as 'structural methods', those that use object-orientation are referred to as 'object-oriented method' and those that use mathematics are referred to as 'formal methods'. The benefit of using formal modelling for I&C system because the formal method relies on the use of unambiguous formalisms for specifying systems. With that methods we can develop a larger system with complex software systems for re-engineering program because formal modelling are easy to understand [16]. Therefore, the correct checking can be done perfectly to avoid human error, because formal modelling describes in detail about design requirements [18]. During the implementation process, the requirements allocated to the system elements are used to design, fabricate, code, or build each individual element using specified materials, processes, physical or logical arrangements, standards, technologies, and/or information flows outlined in detailed drawings or other design documentation [19].

4.3 Testing and Coverage Test

FPGA verification is performed by giving input which is will be simulated based on software. According to FPGA Basys3 board hardware features it has limited digital inputs/output ports (24 Input/Output/digital ports). According to system design there are more than 25 Input/output signals.

Every VHDL implementation goes through extensive verification. In this case VHDL code will be verified using Xilinx Vivado Software where the logic design is created to be simulated with FPGA Board.

In this process, the following steps should be done to verify the code:

1. Use generated VHDL to simulate the VHDL code for verification of all logic gates and hardware functions.
2. Run the simulation under all possible scenarios
3. Check the output waveform and compare it with the designed truth table
4. Verify the possibility to generate bit stream file that can be deployed on FPGA
5. Simulation.

Figure 6 shows the steps to verify the code.

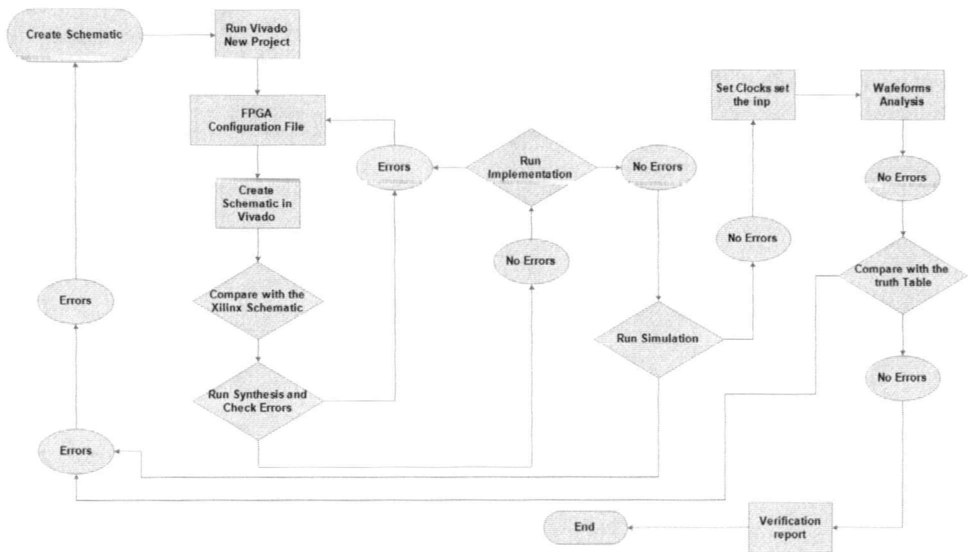

Figure 6: VHDL Code Verification Flow Chart.

4.3.1 Develop Test Case for VHDL Code

Test cases for this system were developed, based on[20] as follows:

1. Requirements based test case selection,
2. Test coverage analysis.

The test cases were developed according to [8]. This guide outlines different test that ensure 100 percent coverage of a gate or decision point. Figure 7 was the test case from VHDL code from Vivado which is will be use for generate coverage test in Aldec software for Modified Condition/Decision Coverage (MC/DC) test.

```
--COMPONENT PROCESSING MODULE TEST CASES--
==========================================================================
-- Test Case TC01--
==========================================================================
MCR_Start <= '1'; MCR_Enable<='0'; MCR_Conf<='1'; RSR_Transfer   <='0';
wait for Clock_period*10;

MCR_Start <= '1'; MCR_Enable<='1'; MCR_Conf<='0'; RSR_Transfer   <='0';
wait for Clock_period*10;

MCR_Start <= '1'; MCR_Enable<='1'; MCR_Conf<='1'; RSR_Transfer   <='1';
wait for Clock_period*10;

MCR_Start <= '1'; MCR_Enable<='1'; MCR_Conf<='1'; RSR_Transfer   <='0';
wait for Clock_period*10;

MCR_Start <= '0'; MCR_Enable<='1'; MCR_Conf<='1'; RSR_Transfer   <='0';
wait for Clock_period*10;

--Reset signals
MCR_Start <= '0'; MCR_Enable<='0'; MCR_Conf<='0'; RSR_Transfer   <='0';
wait for Clock_period*10;

==========================================================================
```

Figure 7: Applied Test Case in Test Bench Vivado

Table 1: Sample of Test Case

Test Case ID	**TC01**
Test Case Summary	The objective is to observe if the component start command is generated only when there are valid conditions at the MCR.
Related Requirements	2.1.2.1.2 2.1.2.1, 2.1.2.3, 2.1.2.5, 2.1.2.6
Test Procedure	Set the inputs of to the defined states and observe the component output.
Test Data	see below

Test No.	MCR ESCM START	ESCM ENABLE	CONF. SWITCH	RSR TRANSFER
1	1	0	1	0
2	1	1	0	0
3	1	1	1	1
4	1	1	1	0
5	0	1	1	0

Expected Result

Test No.	Component Start
1	0
2	0
3	0
4	1
5	0

Actual Result

Test No.	Component Start
1	0
2	0
3	0
4	1
5	0

Status	Passed
Created By	Joyce Mayaka
Executed By	Mohamed Elakrat
Date	5/12/17
Test Environment	Vivado

4.3.2 Coverage Test

Coverage is a measure and not a method of testing. Definition of the coverage required for testing gives a measure of the adequacy of the verification. There are two measures of coverage depicted in Figure 8, requirements coverage and software structure coverage. Requirements coverage analysis determines how well the requirements based testing verified the implementation of the software requirements, and establishes traceability between the software requirements and the test cases.

Structural coverage analysis determines how much of the code structure was executed by the requirements-based tests and establishes traceability between the code structure and the test cases. For type of structural coverage metric used was the Modified Condition/Decision Coverage (MC/DC).

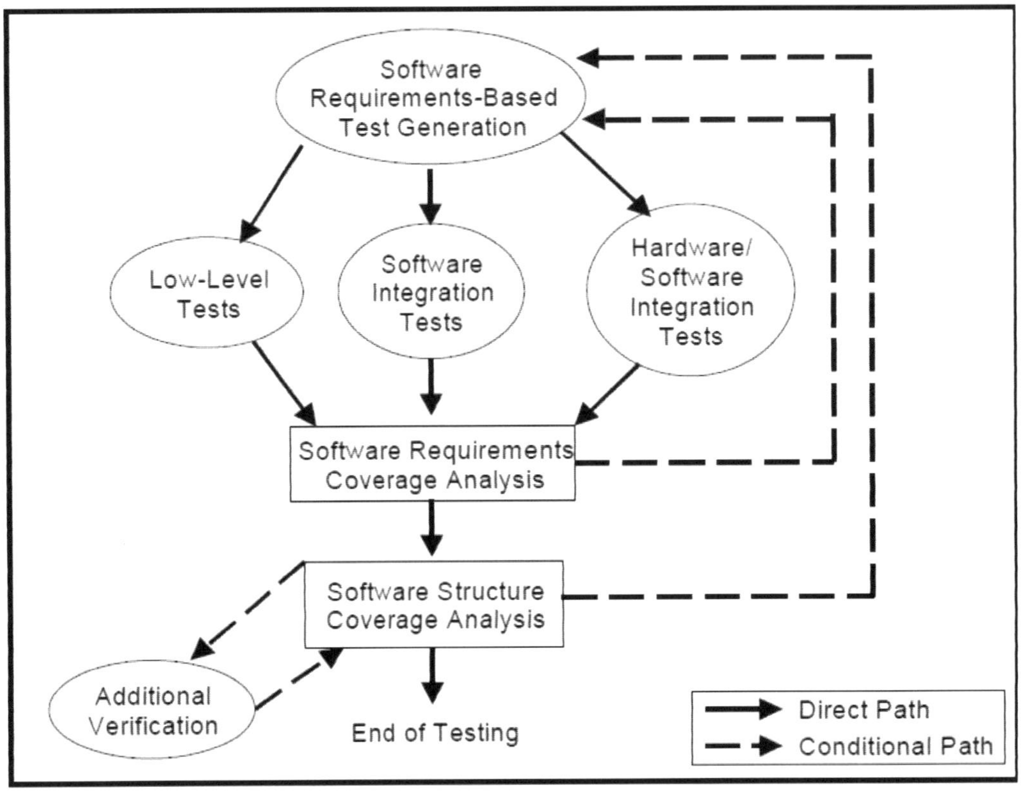

Figure 8: Coverage Measures[21]

5. Validation

After completing all verification activities, the last activity is to validate the system. Validation checks compliance of the whole system with the initial system requirements specification. The FPGA function validation method is different from general software. We will measure the performance by define the Measures of Performance (MOP) and Technical Performance Measures (TPMs).

6. Conclusion

The V&V activity plan are includes verification at the system level, component level and software level of the FPGA-based ESF-CCS. In this V&V plan, the stages of the V&V process are structured by applying a systems engineering approach. It aims to produce a system that can be justified both from the design concept and the design implementation before it is applied significantly to ESF-CCS based FPGA.

The goal of this V&V plan is to verify the development and implementation for the stage of reverse engineering program which it applied to elicit the requirements of the system as well as gain an understanding of the current system architecture, and the re-engineering program with the aim of improving the performance of ESF-CCS.

The result should prove whether FPGA has been able to surpass the performance of PLC which is already applied in I&C system. The benefit of V&V activity that it allows for the early identification of problems, any design failure or system malfunction during the design process, so that errors and modifications can be made as soon as possible. The hope that from the good structure of V&V plan it will produce a capable design and ready to apply in the I&C system of NPP.

This research was supported by the 2017 Research Fund of the KEPCO International Nuclear Graduate School(KINGS), Republic of Korea.

7. References

[1] H. Kim, "The Design Characteristics of Advanced Power Reactor 1400," vol. IAEA-CN-16.

[2] Y. Wu, X. Shui, Y. Cai, J. Zhou, Z. Wu, and J. Zheng, "Development, verification and validation of an FPGA-based core heat removal protection system for a PWR," *Nucl. Eng. Des.*, 2016.

[3] J. Yoo, J. H. Lee, and J. S. Lee, "A research on seamless platform change of reactor protection system from PLC to FPGA," *Nucl. Eng. Technol.*, vol. 45, no. 4, pp. 477–488, 2013.

[4] U.S.NRC, "NUREG/CR-7006, Review Guidelines for Field-Programmable Gate Arrays in Nuclear Power Plant Safety Systems," 2009.

[5] IAEA, "Design of Instrumentation and Control Systems for Nuclear Power Plants," in *IAEA Safety Standards*, no. SSG-39 Specific Safety Guide, 2016.

[6] IEEE, "Systems and software engineering - System Life Cycle Processes," in *IEEE Std 15288-2008*, 2008.

[7] R. Halligan, "Application Guidance for : ISO / IEC 15288 (IEEE Std 15288 2008) Second Edition 2008 02 01 Systems and software engineering – System Life Cycle Processes by Robert Halligan , FIE Aust," vol. 15288, pp. 1–14, 2009.

[8] IAEA Technical Report Series 384, "Verification and Validation of Software Related to Nuclear Power Plant Instrumentation and Control," 1999.

[9] B. Thacker, S. Doebling, F. Hemez, M. Anderson, J. Pepin, and A. Edward, "Concepts of Model Verification and Validation," *Natl. Nucl. Secur. Agency (NNSA), Los Alamos Natl. Laboratotry*, vol. LA-14167-M, no. October 2004, 2004.

[10] "What is Systems Engineering." [Online]. Available: http://www.incose.org/AboutSE/WhatIsSE. [Accessed: 27-Jun-2017].

[11] H.-W. Huang, C. Shih, S. Yih, and M.-H. Chen, "Integrated software safety analysis method for digital I&C systems," *Ann. Nucl. Energy*, vol. 35, no. 8, pp. 1471–1483, 2008.

[12] IEEE, *IEEE 1074-2006: IEEE Standard for Developing a Software Project Life Cycle Process* , vol. 2006, no. July. 2006.

[13] A. Sanders and J. Klein, "Systems engineering framework for integrated product and industrial design including trade study optimization," *Procedia Comput. Sci.*, vol. 8, pp. 413–419, 2012.

[14] IAEA, "Application of Field Programmable Gate Arrays in Instrumentation and Control Systems of Nuclear Power Plants," *Iaea Nucl. Energy Ser. Publ. NP-T-3.17*, vol. 1995–7807, no. Viena, 2016.

[15] EPRI, "Guidelines on the Use of Field Programmable Gate Arrays (FPGAs) in Nuclear Power Plant I&C Systems," 2009.

[16] J. Jung, H.-S. Chang, and H.-B. Kim, "'3+3 process' for safety critical software for I&C system in nuclear power plants," *Nucl. Eng. Technol.*, vol. 41, no. 1, pp. 1–8, 2009.

[17] S. Yih and C.-F. Fan, "Analyzing the decision making process of certifying digital control systems of nuclear power plants," *Nucl. Eng. Des.*, vol. 242, pp. 379–388, 2012.

[18] E. Hull, *Requirements Engineering*, vol. 41, no. 4. 2009.

[19] IEEE, "System and Software Engineering-System Life Cycle Process," in *IEEE Std 15288-2008*, vol. Second edi, 2012.

[20] IEEE, "Software life cycle processes," in *IEEE Std 12207-2008*, 2008, p. Software & Systems Engineering Standards Committe.

[21] K. J. Hayhurst, D. S. Veerhusen, J. J. Chilenski, and L. K. Rierson, "A Practical Tutorial Decision Coverage on Modified Condition," 2001.

Preliminary Investigation of Time Remaining Display on the Computer-based Emergency Operating Procedure

T J Suryono and A Gofuku

Graduate School of Natural Science and Technology, Okayama University, 3-1-1 Tsushimanaka, Kita-ku, Okayama, 700-8530, Japan

Email: jojok.s.mif@s.okayama-u.ac.jp

Abstract. One of the important thing in the mitigation of accidents in nuclear power plant accidents is time management. The accidents should be resolved as soon as possible in order to prevent the core melting and the release of radioactive material to the environment. In this case, operators should follow the emergency operating procedure related with the accident, in step by step order and in allowable time. Nowadays, the advanced main control rooms are equipped with computer-based procedures (CBPs) which is make it easier for operators to do their tasks of monitoring and controlling the reactor. However, most of the CBPs do not include the time remaining display feature which informs operators of time available for them to execute procedure steps and warns them if the they reach the time limit. Furthermore, the feature will increase the awareness of operators about their current situation in the procedure. This paper investigates this issue. The simplified of emergency operating procedure (EOP) of steam generator tube rupture (SGTR) accident of PWR plant is applied. In addition, the sequence of actions on each step of the procedure is modelled using multilevel flow modelling (MFM) and influenced propagation rule. The prediction of action time on each step is acquired based on similar case accidents and the Support Vector Regression. The derived time will be processed and then displayed on a CBP user interface.

Keywords: computer-based procedure, time remaining, multilevel flow modelling, support vector regression, PWR type reactor, steam generator tube rupture

1. Introduction

In an emergency situation, operators receive a lot of information related with the plant condition, such as alarm and warning messages. In this condition, they have to investigate the initiating event based on the alarm signals and then mitigate the accident in order to prevent the core melting and the release of radioactive material to the environment. However, they only have short time to do the actions. In some cases, the overload information and cognitive workload will make them confused and may cause wrong decision and actions. Therefore, the proper investigation of initiating event is important in order to reduce the worst case.

The response of operators in mitigating the accidents can be divided into three phases: initiating event; diagnose the event; and perform safety-related operator action. Figure 1 shows the phases [1]. In the initiating event, which is indicated by T_0, operators investigate the initiating event based on the alarms and other information. Then, in the period time between T_a (even alarm) and T_{dc}(end of event diagnose), operators start to diagnose the event and find the adequate operating procedure to mitigate the accident. Finally, safety related actions are conducted in time interval T_t (operator first significant response triggered by the procedure step), T_i (first significant action initiation) an T_c (action completion) following the operating procedure and accident management.

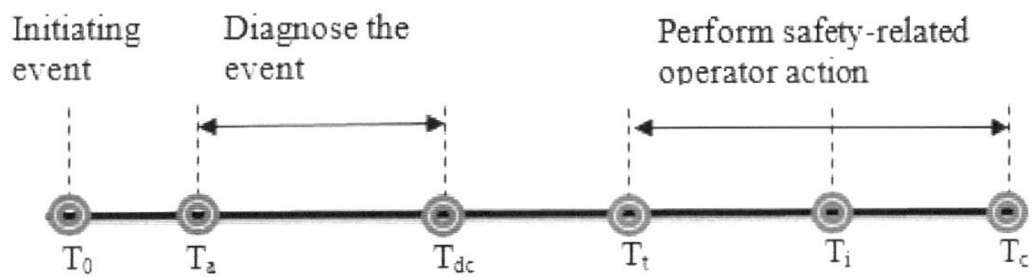

Figure 1. Response time of accident management[1]

This paper discusses the third phase of figure 1, which is the time available for operators to take the action on each step of emergency operating procedure. Information about the time available for operators to take actions on each step is very important. However, the information should be properly presented on the computer-based procedure (CBP) and does not cause more cognitive workload to the operators. In this paper, the sequence and impact of operators' action on each procedure step are modeled using Multilevel Flow Modeling (MFM). MFM [2] [3] is a methodology to model the industrial plants in term of objectives, function and structures which represent the physical structure, and also relation among them. MFM implements influence propagation rule and cause-effect relations, which are useful for determining the flow of the counter actions. In addition, the operators' actions are represented by a control function in MFM model.

The data for estimating the time available for operators to take the action on each procedure step is taken from historical data of time needed to execute counter actions from similar accidents. A support vector regression method is used to predict the time. Then the time is presented on a CBP user interface as remaining time information. Operators should do an action before the time is exceeded, otherwise they will get the warning message. In addition, the presentation of the information will consider the emergency situation and human factor engineering. This paper presents the result of a preliminary investigation to implement a necessary feature proposed in [4].

2. Safety of Nuclear Power Plant

Safety is the most important thing in a nuclear power plant. A nuclear power plant, including main control rooms have been developed to increase the safety and minimize the consequences of accidents. The main objective of safety measure of a nuclear power plant is to prevent the spreading the accident by automatic shutdown and preventing the release of radioactive to the environment by cooling the reactor and the containment. In addition, based on the literature [5] there are four main objectives of accident management: 1) Prevention of the accident from leading to core damage; 2) Core damage termination; 3) Maintaining the containment integrity; and 4) minimizing the release of radioactive on-site and off-site. In order to achieve the objectives, the mechanism of reactor accidents, including controlling safety and non-safety related system should be understood by operators.

When an anomaly happens in the reactor, which is indicated by changing parameter levels shown on the visual display unit, the reactor safety system will trip the reactor automatically to prevent the spreading the accidents. During this time, operators should diagnose the cause of accidents and try to find the suitable counter actions and emergency operating procedure to mitigate the accident.

In the main control rooms of nuclear power plant, the action managements of operators, especially in mitigating accidents are related with the control and time. Control means that they have to take action by controlling or changing the parameter levels of components following the operating procedure and accident management. On the other hand, the term "time" means that their actions are limited by time. They have to take the action during the time provided, otherwise the worst condition will happen. Therefore, the dependency between operators performance and time to execute the control actions should be considered in order to get an adequate performance of control system [6].

3. Computer Based Procedure

Most of advanced and modern of nuclear power plants are equipped with computerized procedures (CPs). According to EPRI 1015313 [7] there are three types of computerized procedures: 1) Electronic procedures (EP), which is the duplicate of PBPs which are displayed on the VDU in text or graphical format; 2) Computer-based procedures (CBPs) 3) CBPs with procedure-based automation (PBA), which have the capability to perform multiple procedure steps and take the control actions as mentioned in the procedure steps.

This paper only discusses the CBP. A CBP is not just the electronic version of paper based procedure (PBP) but provides some features which can overcome the drawbacks of PBPs, for example, the ease of finding, retrieving and displaying information related with a procedure step; linking with other procedures easily; and keeping track of operator' position on the procedure. In addition, a CBP is also intended to increase the usability of emergency operating procedure (EOP); to minimize operators workload in while finding the correct procedure and controlling the reactor in emergency condition; and to reduce the time needed for making decision and taking the actions to mitigate the accident.

CBPs are already used in several nuclear power plants around the world such as COMPRO in Switzerland, N4 Computerized Procedure System in France, Plant safety monitoring and assessment system (PLASMA) in Hungary, Computerized Procedure System in Korea and On-Line Procedure System (OLPS) in Taiwan [7]. Because the purposes of the development and implementation of CBPs are to reduce operators workload and increase safety, the design should be based on human factor engineering.

4. Desirable Features

As mentioned in section 3 that CBPs offer some benefits compare with PBPs. However, some problems will be faced by operators when using the CBPs. A literature [4] proposed the desirable features that will be added on the CBP to overcome the problems. This section discusses the CBP problems and the proposed features.

Firstly, reading the information on the visual display unit (VDU) is not as easy as on papers because of the limited information displayed on the VDU [8] that can cause operators lose a sense of their current position in completing the procedure steps. In addition, reading on VDU is slower and more fatiguing which can reduce human reliability and impair operator response [9]. The functional information feature is proposed to overcome these problems. The feature provides information about components influenced and future plant behaviour caused by the actions of operators on each procedure step. This information will help operators to understand the procedure step and make it easier for them to make decision and take action. The investigation of this feature has been conducted by first author.

Secondly, operators tend to spend much time on a specific step of the procedure before making decision and taking action related with the plant condition. This behaviour will inhibit the mitigation process of the plant to the safe operation condition. It can be solved by adding the remaining time display feature on the CBP which gives information and warning to operators to take a specific before reaching the provided time. This paper will discuss the preliminary investigation of techniques to generate and display the time remaining display.

Finally, some operators do not aware of the importance of all steps in the procedure or because of they do not understand the purpose of the procedure steps, then they try to skip the steps. This action is kind of commission error which may cause other impacts on the plant condition. The dynamic operation permission feature which based on work developed by [10] [11] is included on the CBP and used to reduce commission error by operators. Warning messages will be displayed if operators skip or do not follow one or more steps.

5. MFM as a Base Technique to Model a Plant for Generating Information

5.1. Overview of MFM

Multilevel flow modelling (MFM) was developed by Lind [2] to represent the complex system in graphical format and in terms of functions, structures and objectives and also interconnected relation among them. MFM is one of the functional modelling methods, which represents the plant in several levels of abstraction. The MFM models a system in means (tools) to achieve the ends (objectives) and part-whole expression (description of aggregation in different levels). MFM has been widely used to model the large and complex systems such as nuclear power plant, chemical plant and other industrial processes. In addition, MFM is also applied for operator support system [12] [13] and dynamic operation permission [10]; alarm and root cause analysis [14]; and supervisory control [15].

As mentioned before that MFM is a representation of system in graphical format, some graphical symbols are needed. Figure 2 shows the MFM symbols including functions and relations. The function primitives (source, transport, storage, etc) are connected each other (interconnected) using influence relations to build mass flow or energy flow structure. The function primitives are used to represent physical systems. For example, a tank can be represented as a storage function and a pump as a transport function in an MFM model.

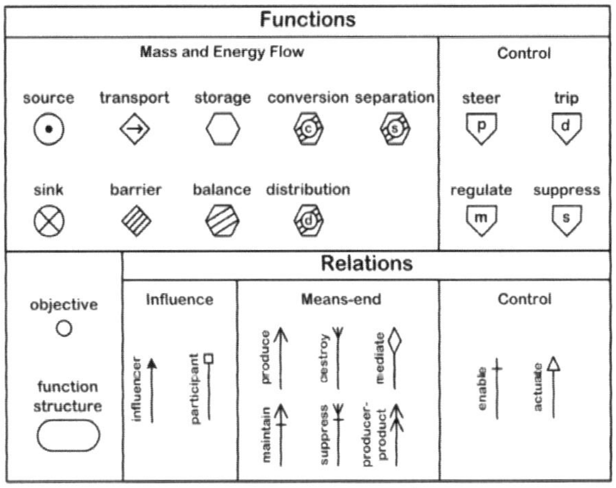

Figure 2. MFM symbols [3]

Figure 3 shows the example of MFM model of a PWR plant. It can be seen that there are some mass flow (mfs) and energy flow structures (efs). The primary system is represented by mfs1 which contains sto3 (reactor vessel), sto10 (pressurizer), bal7 (steam generator) and tra5 (reactor coolant pump). On the other hand, the secondary system (mfs2) includes sto4 (steam generator), tra16 (main steam isolation valve), sto6 (turbine) and sto7 (condenser). The MFM model also include a model of safety systems such as efs9 (ECCS), mfs4 (PORV). In this model, the control rod is represented by mfs3 to control the fission reaction to produce heat in reactor vessel. The heat generated in reactor vessel (efs1) is then delivered from primary to secondary system in efs7 through steam generator. In efs6 (turbine) the heat energy is converted to mechanical energy. Subsequently, the mechanical energy is converted by generator in efs8 to produce the electricity which is delivered to the grid. In this case, the objective obj1 to produce the electricity is achieved.

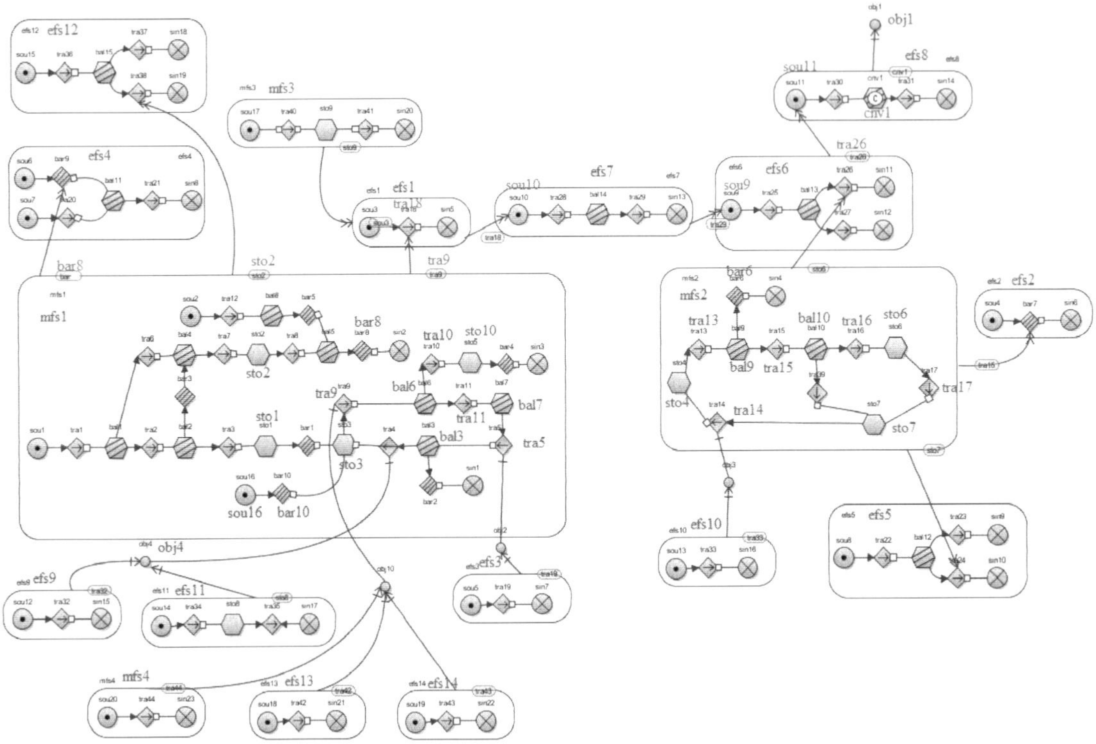

Figure 3. MFM model of PWR plant in normal condition

5.2. MFM for generating functional information

The mitigation process of accident depends on controls and time. The controls are conducted by operators following the EOP. The controls and their impacts to the plant can be modelled and investigated using MFM. In this case, operators' actions are represented by the control structure in MFM model as shown in figure 4. The control structure consists of control function primitive (pco1) and parameter to be observed (obj1). As can be seen in figure 4, the control function pco1 is actuated to control the transport function tra1 in mfs1 if the level of storage function sto1 is met the requirement in the operating procedure or accident management. For example, the low level in sto1 will actuate the control function pco1 to start the transport function tra1 to flow, which is the objective of the control action (obj2).

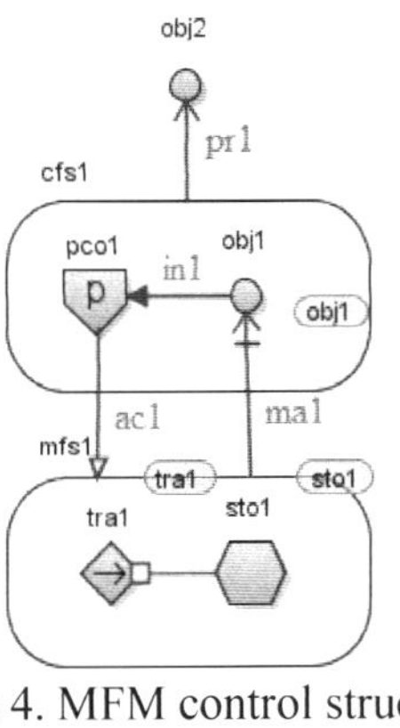

Figure 4. MFM control structure

In addition, MFM is based on cause-effect relation. It means that the action which change the state of a function primitive will impact the connected function primitive. The impact then will propagate to other function primitives, function structure, objective and whole system. This is the concept of influence propagation rule. Table 1 shows the definition of states in MFM. The definition will be used to investigate the cause-effect and influence propagation in MFM model.

Table 1. Definition states of MFM symbols

Symbols	States
source	normal, high output flow potential, low output flow potential, no output flow potential
sink	normal, high input flow, low input flow, no input flow
transport	normal, high flow, low flow, no flow
storage	normal, high volume, low volume, no volume
barrier	normal, leak
balance	normal, unbalance (fill or leak)
threat	exist (high), exist (low), non-exist
objective	true (high), true (low), false

The control structure, the definition of state and the influence propagation rule together can be used to investigate the components influenced and future plant behaviour because of the operator actions on procedure steps. These information is useful for operators and can be used as consideration before taking the action on each procedure step. The study of investigating the techniques to derive additional information (components influenced and future plant behaviour) has been conducted by first author.

6. Time remaining display

Information of time available for operators to take the action is displayed on the CBP user interface. The time is started when operators select the procedure steps. The time is counting down. When the counting reach the limit, warning message is given to the operators which order operators to take the actions. The information is presented by considering the emergency condition and human factor engineering in order to prevent the worst condition and reduce cognitive work load. In addition, the information is presented clearly and easy to watch. The time information is intended to increase the awareness of operators about the safety and conduct the proper action to mitigate the accident.

The process of deriving the remaining time is provided in figure 5. From the figure, it can be seen that when operators open the CBP user interface, it will initialize the MFM model subsystem and operating procedure subsystem. Furthermore, if the procedure step is selected, it will initiate the MFM model of the procedure step. From this point, the information about the components influenced and future plant behaviour is displayed on the functional information field of the CBP. The next step is, on selected procedure step, the system tries to find the time to take the action in the time list. Once it is found, the system starts to count the time and displayed on the CBP user interface as the time remaining display.

Figure 6 shows the proposed of CBP user interface with the time remaining display feature. If operators click on the procedure step, the time is started to count and displayed in the warning field on the CBP user interface. Operators should take actions during the available time, otherwise the warning message will be given to operators and order them to take the actions as soon as possible

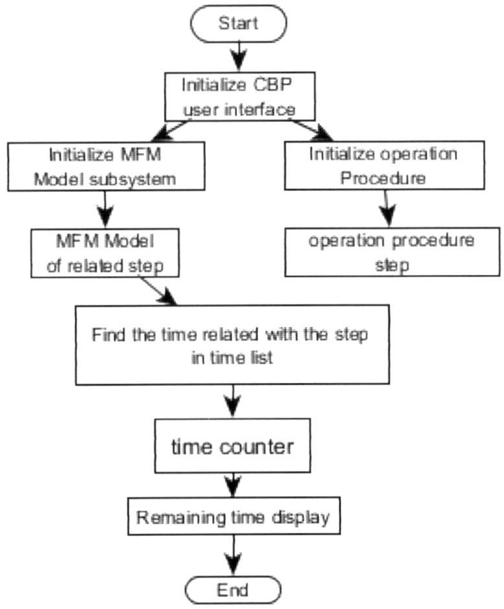

Figure 5. Process of time remaining display

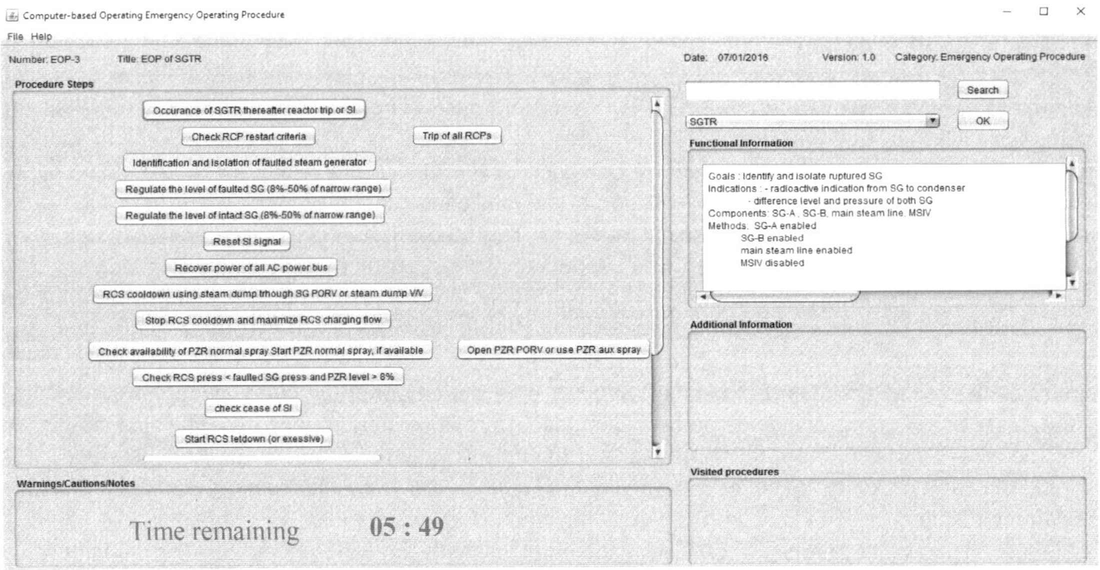

Figure 6. Proposed CBP user interface with time remaining display feature

7. SVR as a Promising Technique for Time Remaining Display

7.1. Overview of SVR

This section discusses the overview of support vector machine (SVM) and support vector regression (SVR) which will be used to predict the time provided for operators to take the actions on each procedure step. Support vector machine (SVM) was developed by Vapnik [16]. Initially was used for classification and then expanded to solve regression problems. Compared with other methods, SVM is relatively new and has better performance. SVM and SVR have been widely implemented for several purposes such as for forecasting [17], travel time prediction [18] and nuclear engineering [19][20]. SVM can be applied for linearly separable data and non-linearly separable data.

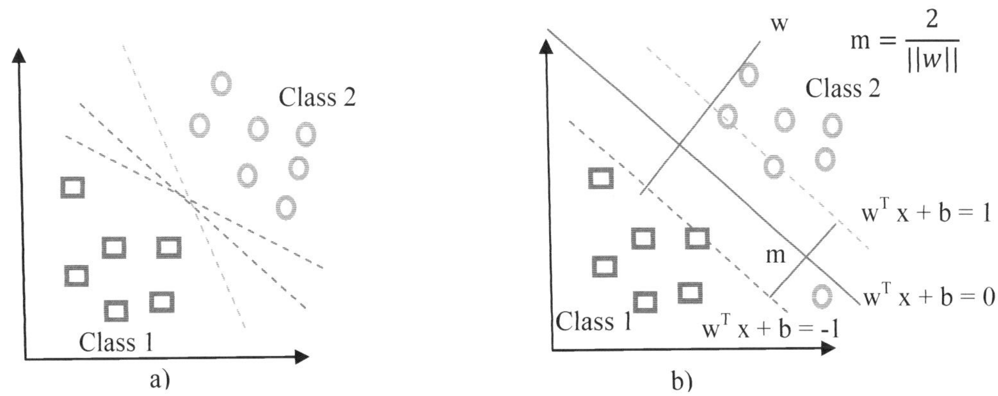

Figure 7. SVM for linearly separable data [21]

Figure 7 shows the example of SVM for linearly separable data. Given the dataset $\{x_1,..,x_n\}$ and $y_1 \in \{1, -1\}$ is the class of data x_i. In this case, data are classified into two classes (class 1 and class 2) that are separated by decision boundary. The boundary can be defined as follows:

$$w^T x + b = 0 \tag{1}$$

where w is the weight vector and b is the bias.

In figure 7a there are many decision boundaries but the better boundaries are unidentified. Therefore, the optimum decision boundary should be derived which can classify the data and has highest margin. As can be seen in figure 7b, the idea of support vector regression is to find the equation (1) with the highest margin m by using the following expression:

$$m = \frac{2}{||w||} \tag{2}$$

In addition, data that located on the boundary is called the support vector. The detailed explanation of the support vector regression is provided in literature [16].
Therefore, the linear SVR is

$$y = \sum_{i=1}^{N} (\propto_i - \propto_i^*) . \langle x_i, x \rangle + b \tag{3}$$

For non-linear separable data, SVM implements kernel trick. The idea is the input which low dimensional space is taken and converted into a higher dimensional space. The SVR function for non-linear is shown in equation (4)

$$y = \sum_{i=1}^{N} (\alpha_i - \alpha_i^*) . \langle \varphi(x_i), \varphi(x) \rangle + b \tag{4}$$

$$y = \sum_{i=1}^{N} (\alpha_i - \alpha_i^*) . K(x_i, x) + b \tag{5}$$

with $K(x_i, x)$ is the kernel function. For Gaussian Radial Basis Function the kernel function is

$$k(x_i, x_j) = exp\left[-\frac{||x_i - x_j||^2}{2\sigma^2}\right] \quad (6)$$

Based on the previous explanation, the SVM gives better classification approach for linear and non-linear separable data and optimum generalization bound for regression. Therefore, it can be used to predict the time required by operator to execute the procedure steps.

7.2. A technique of generating time remaining display based on SVR

Time taken by operators to take the action on each procedure step for similar case can be varied for different nuclear power plant. The data can be used in this paper for calculating the remaining time. Based on the data, the time available for operators to take the action on each procedure step is predicted using support vector regression.

This paper and future works will use steam generator tube ruptured (SGTR) accident of PWR accident as a case study. The consideration of choosing SGTR accident due to it is one of common accident in PWR plant. The accident is caused by the degradation of SG due to long time operation or chemical/mechanical thermal stress and pressure shock. The degradation will cause leak of cooling water from primary to secondary side and then to turbine and condenser.

There are many cases of SGTR accident in nuclear power plant in the world, for example in Mihama Unit 2 in Japan, Prairie Island NPP unit 1, and etc. The data of time to take the action on each step of emergency operating procedure of SGTR is collected and then used for predicting the time available for operators to take the action on the proposed CBP.

The prediction is made by using Support Vector Regression. The result of predicted time is saved in the time list which the format as follow:

"Step no "+ "Step description" + "time to take the action"

The flow of deriving the prediction time is shown in figure 8.

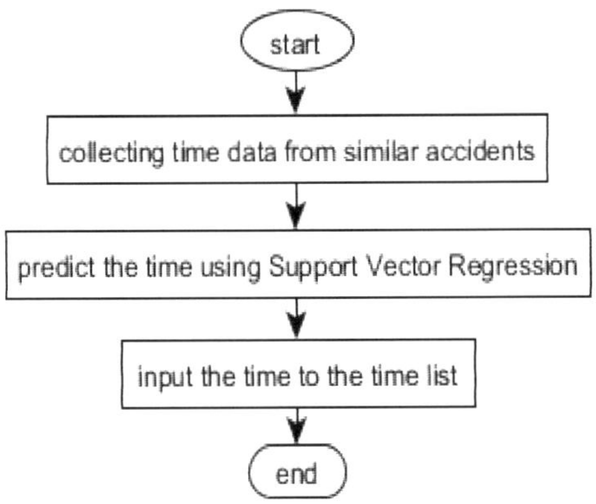

Figure 8. Process of time prediction

8. Conclusion

This paper discussed the preliminary investigation of technique to generate the information of time available for operators to the action on each procedure step. The method includes determining the accident sequence, components influenced and future plant behaviour using MFM. Moreover, the time is predicted using support vector regression based on the data of time taken by operators to mitigate accidents following operating procedure of similar accident case. The information of the remaining time then properly displayed on the CBP user interface.

Future work is implementing the support vector regression to predict the time for operators to take the action on each procedure step.

Acknowledgements
This study was supported by the scholarship Program for Research and Innovation in Science and Technologies (RISET-Pro) World Bank Loan under Grant number 8245-ID. In addition, a part of the results of this study is obtained by the support of Japan Society for the Promotion of Science (JSPS) [KAKENHI grant number 16H03136].

References

[1] Lee M, Lee S, Park M, Lee G, and Kim S 2014 A Study of Time Response for Safety-Related Operator Actions in Non-LOCA Safety Analysis *Korea Trans. Korean Nucl. Soc. Autmn.* Meet. 30–31.
[2] Lind M 1994 Modeling goals and functions of complex plant, *Appl. Artif. Intell. an Int. J.*, **8**, pp. 259–283.
[3] Lind M 2011 An introduction to multilevel flow modeling *Nucl. Saf. Simul.*, vol. 2, no. 1, pp. 22–32.
[4] Suryono T and Gofuku A 2016 The Desirable Features of Computer Based Emergency Operating Procedure for Nuclear Power Operation *IFAC-PapersOnLine* **49**, no. 19, pp. 403–407
[5] IAEA 2004 Implementation of Accident Management Programmes in Nuclear Power Plants *Safety Reports Series No. 32* **32**, pp. 129.
[6] Carvalho P, de Oliveira M, and Luquetti dos Santos I 2009 A computerized tool to evaluate the cognitive compatibility of the emergency operational procedures task flow *Prog. Nucl. Energy*, **51**, no. 3, pp. 409–419.
[7] EPRI. 1015313 2007 Computerized Procedures Design and Implementation Guidance for Procedures, Associated Automation and Soft Controls, *Electric Power Research Institute,* Palo Alto, USA.
[8] Lee Y, Hwang S, and Wang E 2005 Reducing cognitive workload of a computer-based procedure system *Int. J. Hum. Comput. Stud.*, 63 (6), pp. 587–606.
[9] O'Hara J, Higgins J, Stubler W, and Kramer J 2000 Computer-based procedure systems: technical basis and human factors review guidance *Nucl. Eng. Des.*
[10] Gofuku A and Sato T 2007 Development of a Dynamic Operation Permission Agent for Preventing Commission Errors of Operators *Proc. 2nd Int. Conf. Innov. Compt. Inf. Cont (ICICIC 2007)* Kumamoto, Japan pp. 108
[11] Gofuku A and Sato T 2009 Dynamic operation permission system for oil refinery plants, *Proc. IEEE Int. Conf. Networking, Sens. Control. ICNSC 2009*, 14(2), pp. 746–751.
[12] Gofuku A 2011 Applications of MFM to intelligent systems for supporting plant operators and designers : function-based inference techniques *Int. J. Nucl. Saft. Sim.* 2(3).
[13] Gofuku A, Inoue T and Sugihara T 2017 A technique to generate plausible operating procedure for an emergency situation based on a functional model *J. Nuc. Sci. Tech.* **3131**, pp. 578-588
[14] Larsson J, Öhman B, Nihlwing C, Jokstad H, Iren L, Kvalem J, and Lind M 2005 Alarm reduction and root cause analysis for nuclear power plant control rooms *Norway Proc. Enlarged Halden Prog. Group Mtg*
[15] Lind M 1999 Plant modelling for human supervisory control *Trans. Inst. Meas. Control*, **21**(4–5), pp. 171–180

[16] Smola A and Schölkopf B 2004 A Tutorial on Support Vector Regression *Stat. Comput.* **14**(3), pp. 199–222.
[17] Chen K 2007 Forecasting systems reliability based on support vector regression with genetic algorithms *Reliab. Eng. Syst. Saf.* **92**(4), pp. 423–432
[18] Wu C, Ho J, and Lee D 2004 Travel-time prediction with support vector regression *IEEE Trans. Intell. Transp. Syst.* **5**(4), pp. 276–281.
[19] Kim S, No Y, and Seong P 2015 Prediction of severe accident occurrence time using support vector machines *Nucl. Eng. Technol.* **47**(1), pp. 74–84.
[20] Na M, Park W, and Lim D 2008 Detection and diagnostics of loss of coolant accidents using support vector machines *IEEE Trans. Nucl. Sci.* **55**(1), pp. 628–636
[21] Law M 2005 A simple introduction to support vector machines *Lect. CSE*, vol. 802.

Design of Fishing Boat for Pelabuhanratu Fishermen as One of Effort to Increase Production of Capture Fisheries

Iswadi Nur[1], Purwo Joko Suranto[1]
Study Program of Naval Architecture[1], Engineering Faculty of UPN "Veteran" Jakarta
Email : iswadinur62@gmail.com

Abstract Design of fishing boat for Pelabuhanratu fisherman as one of effort to increase production of capture fisheries. The fishing boat should be proper for the characteristic of its service area, as ;capacity of fishing boat up to 60 GT, the fishing boat has minimum 6 fish holds and location of fish hold in the middle body, the fishing boat hull has the bilge keel plate, and the material of hull fishing boat to be made of wooden, steel, aluminium, or fiberglass. Main dimesion of fishing boat is Length Over All = 25.436 m, Breadth = 4.55 m, Draft = 1.6 m, Speed = 12.5 knots. The research had been known every thing that will be supporting the production of capture fisheries like ; amount of fish production = 25.030 ton per day, the fishing port capacity approximately 268.957GT per day, the area of fishing port < 30 hectares, the zone of fish processing industry had not completed, therefore all data research result less than standard of Oceanic Fising Port. So Pelabuhanratu National Fishing Port can not be changed to Oceanic Fishing Port.

Keyword : fishing boat, national fishing port, oceanic fishing port

1. Introduction

Design of fishing boat for Pelabuhanratu fisherman as one of the effort to increase production of capture fisheries. The research have to know every thing that will be supporting the production of capture fisheries like; type of fishing boat fleet, total fishing boat capacity, service area , quay wall fasility, port quay dimension, deep of sea water, total fish production, fishing port capacity, quality assurance facility of fish production, area of fishing port, and zone of fish processing industry, that all based on standard of Oceanic Fishing Port.Because there were several issue of news before that Pelabuhanratu National Fishing Port can be changed to Pelabuhanratu Oceanic Fishing Port, but it should be proper to the class characteristic regulation of **KKP** number 16, 2006. Related to the fishing boat that should be proper to the characteristic of its service area,socapacity of fishing boat, minimum speed, minimum fish holds, location of fish hold, stability of fishing boat, and the material of hull fishing boat like wooden, steel, aluminium, or fiberglass. Type of general arrangement should be suitable for fishermen culture, especially type of stem, location of deck house. Based on 4 class characteristic regulation of **KKP** number 16, 2006, therefore Indonesia has 6 Oceanic Fishing Port, 15 National Fishing Port, 44 Beach Fishing Port, and least 937 Fish Landing Quay

1.1. Research Objective

Design of fishing boat for Pelabuhanratu fishermen as one of the effort to increase production of capture fisheries.

2. Literature Study

Indonesia is archipelago country and Indonesia has many extraordinary resources as; 17480 islands, 95.181 km beach line, 5.8 million km^2 sea water area, that is consist of 3.1 million km^2 territorial ocean, and 2.7 million km^2 ZEE ocean, **Lukito** detiknews.com, 2009. In the ZEE ocean has many

important recources like various of fish,coral, minning, maritime tourism destination, etc, thereof many important recources will come up foreign exchange.
Pelabuhanratu Natioanl Fishing Port has been landed up to 100 GT fishing boat, and production of capture fisheries come from tuna, yellow fin tuna, big eye tuna, albacora tuna, southern blue fin tuna, longtail tuna, eastern litle tuna, bullet tuna, frigate tuna, cakalang, shrimp, etc, thereof of many recources of fish will come up foreign exchange from foreign countries like Korea, China, Japan, etc.

2.1. Based on Regulation and Research Before

2.1.1. Abbas Salim, 2005, Many factors should be attention and consideration to development the fishing port. as;

a. Growth of local economic of local government (hinterland) come from its port.
b. Growt of industry sector come from local government (hinterland)
c. Forcasting and Analysis data of commodities circulation.
d. Type and capacity of fishing boat that will come out to the port
e. Networking of train track and outer ring road to the port
f. Impact of port savety and environment
g. Analysis of economic and finance

2.1.2. The class characteristic regulation of KKP, number 16, 2006 for the oceanic fishing port. It has several condition, as ; the fishing vessel fleet have service area in the ZEE territorial and the Internastional Ocean, the quay wall fasility > 60 GT, the port quay dimension has length > 300 m and deep of sea water > 3 m, amount of fish production 60 ton per day, the fishing port capacity > 6000 GT per day, it has the quality assurance facility of fish production, the area of fishing port > 30 hectares, it has the zone of fish processing industry

3. Research Method

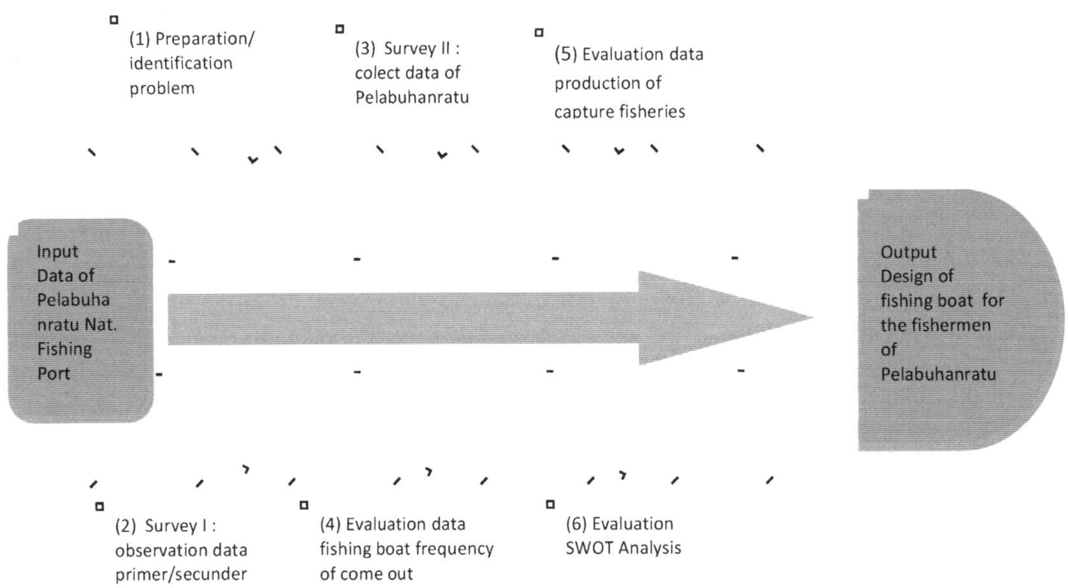

Figure 1. Fishbone of Research Method

4. Calculation and Conclusion

Total Outboard Boat & Inboard boat frequency of Come Out

Table 4.1 : Total Boat Boat Frequency of Come Out (0 – 50 GT)

No	Type of Boat	Boat Frequency of Come Out (Boat/Year)	Capacity (GT/Year)	Capacity (GT/day)
1	Outboard Boat 2.5 GT	20,962	52,405	143,575
2	Inboard Boat 5 – 10 GT	3,548	26,610	72,904
3	Inboard Boat 10 – 20 GT	47	705	2
4	Inboard Boat 20 – 30 GT	198	4,950	13,561
5	Inboard Boat 30 – 50 GT	50	2,000	5,479
Total		24,805	86,670	237,519

Table 4.2 : Total Boat Frequency of Come Out (50 – 200) GT

No	Kind of Boat	Boat Frequency of Come Out (Boat/Year)	Capacity (GT/Year)	Capacity (GT/day)
1	Inboard Boat 50 – 100 GT	61	4,575	12,534
2	Inboard Boat 100 – 200 GT	46	6,900	18,904
Total		107	11,475	31,438

Based on calculation data on table 4.2, thereby total inboard boat of 50 – 200 GT frequency of come out is very little, so local govermance should be help Pelabuhanratu fisherman in order they have 60 GT fishing boat, its mean that the local government should be support to Pelabuhanratu fishermen to increase the fish production of capture fisheries. One of Pelabuhanratu fishermen said that ZEE area have not been exploited by Pelabuhanratu fisherman, but it have been exploited by Kalibaru fishermen come from Noth Jakarta, because they have 60 GT fishing boat

Table 4.3. Average production of capture fisheries for 3 years

no	Month	2013	2014	2015
1	Januari	637,257	941,617	1,048,216
2	Februari	439,013	619,133	514,008
3	March	329,656	582,889	436,176
4	April	469,653	701,496	572,952
5	Mei	853,987	1,074,079	914,047
6	Jun	945,444	1,240,977	1,168,243
7	July	696,582	1,027,744	1,312,412
8	August	395,329	410,135	621,668
9	September	600,024	717,588	703,195
10	October	976,085	985,647	785,104
11	November	753,449	873,037	551,916
12	December	832,594	1,182,681	494,383
	Total	7,929,073	10,357,023	9,122,320

Data source : The Annual Statistical Report Capture Fisheries Year 2015

Average production of capture fisheries = $\{\frac{Prod.2013+Prod.2014+Prod.2015}{3}\}$kg/year

Average production of capture fisheries = $\{\frac{7.929.073+10.357.023+9.122.320}{3}\}$ kg/year

Average production of capture fisheries = $\{\frac{27.408.416}{3}\}$ = 9,136,138.667 kg/year

Average production of capture fisheries = 250,030.516 kg/day

This calculation result = 250,030.516 kg/day is less than standard of oceanic fishing port = 60 ton per day

Table 4.4. The consept design of fishing boat for the Pelabuhanratu fisherman, it should be suitable from the charracteristic of Pelabuhanratu ZEE ocean, as:

No.	Subject Research	Ship Design Consideration
1.	Fishing Boat Capacity ≥ 60 GT	The fishing boat capacity ≥ 60 GT, it means the fishing noat should be operated in ZEE sea area, where it has many fish resources will come up foreign exchange.
2.	The main dimension of fishing boat ≥ 60 GT	This fishing boat main dimension for 60 GT capacity, should be many correction like stability, trim, manouvering, and main dimesion ratio. If L/B ratio is expand, therefore the fishing boat has bad manouvering, on the other side if L/B ratio is small, the fishing boat has good manuvering If L/H ratio is expand, therefore the fishing boat has bad hull strength, on the other side if L/H is small, the fishing boat has good hull strength. If H/T ratio is expand, therefore the reserve displacement should be increased, on the other side if H/T ratio is small, the reserve dislacement should be decreased
3.	The minimum speed = 12.5 knots	This minimum speed is dicided in order to speed up the operational voyage to ZEE area.
4.	The fish hold is in the middle ship, and it has 6 numbers of fish hold.	The fish hold is in the middle ship, it means that the draft immersion of fishing boat should be even keel. The fishing boat has 6 numbers of fish hold, therefore the fishing boat has many bulkhead of fish hold. This bulkhead will protect the fish from ship motion.
5.	Amount of fishermen on boat = 20 – 30 persons	Amount of fishermen on boat should be suitable of general arrangement layout or boat capacity
6.	Hull of boat should be installed the bilge keel	The function of bilge keel will retard the rolling period of fishing boat, therefore the fish in cold storage will save from rolling motion. The fish will not shake from rolling motion
7.	Material of boat made of steel, wooded, or fiberglass.	Selection material of fishing boat should be depended many matters as; the fishermans culture for maintaining hull boat, cost consideration, boat material life time consideration, and boat material strength
8.	Type of general arrangement based on result of observation in many local government of Indonesia.	General arrangment should be suitable form fisherman culture, especially type of stem. The super structure or deck house of fishing boat to be on after deck, the store hold to be in fore hull of fishing boat, and the engine room to be in after hull of fishing boat.

9	The General Arrangement of 60 GT fishing boat, see figure 2	Loa = 25.436 m, B = 4.55 m, H = 2.60 m, T = 1.60 m
10	The Lines Plan of 60 GT fishing boat, see figure 3	Loa = 25.436 m, B = 4.55 m, H = 2.60 m, T = 1.60 m

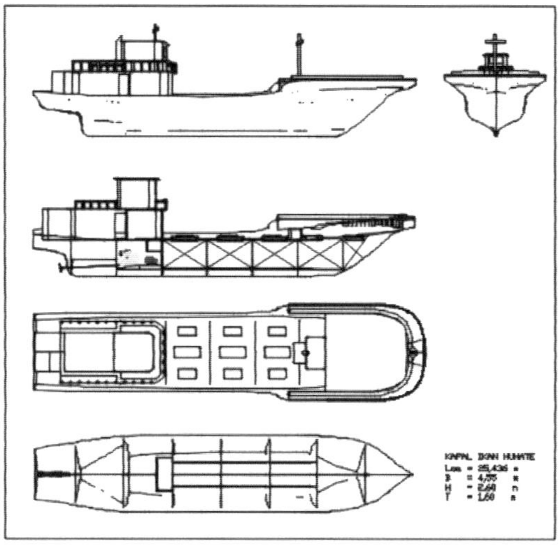

Figure 2. General Arrangement of Fishing Boat

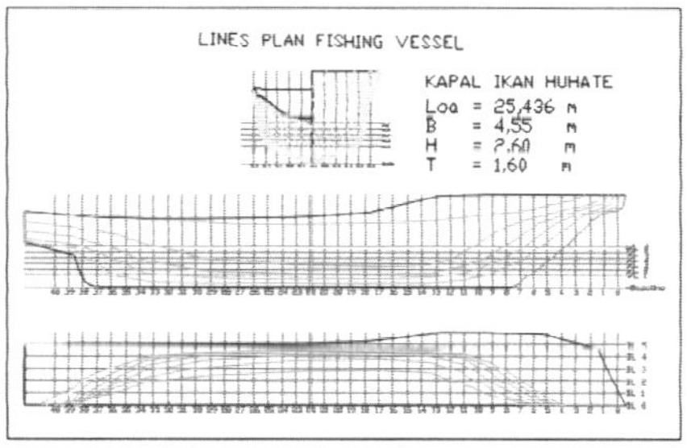

Figure 3. Lines Plan of Fishing Boat

4.1 Conclusion

a. The result of research had many factors that had not been proper to the KKP regulation, as; amount of fish production = 25.030 ton per day, the fishing port capacity approximately 268.957 GT per day, the area of fishing port < 30 hectares, the zone of fish processing industry had not completed, so Pelabuhanratu National Fishing Port can not be changed to Pelabuhanratu Oceanic Fishing Port

b. Fishermen Group of Pelabuhanratu Nasional Fishing Port have the fishing boat fleet, but generally the capacity of fishing boat is ≤ 30 GT. The various of fish that have been captured as ; yellow fin tuna, big eye tuna, albacora tuna, southern blue fin tuna, longtail tuna, eastern litle tuna, bullet tuna, frigate tuna, cakalang, shrimp.

c. The area of Pelabuhanratu fishing boat to be ZEE ocean that more than 50 sea mile from the beach line. Pelabuhanratu ZEE has many important recources of fish. ZEE area have not been exploited by Pelabuhanratu fisherman, but it have been exploited by Kalibaru fishermen, they come from Noth Jakarta, because they have 60 GT fising boat.

d. The local govermance should be help Pelabuhanratu fishermen in order they have 60 GT fishing boat, it means that the local government should be support to increase the production of capture fisheries.

e. The main dimension of fishing boat of Pelabuhanratu fishermen is: Fishing boat capacity ≥ 60 GT, Length Over All = 25.436 m, Breath = 4.55 m, Draft = 1.6 m, Speed = 12.5 knots, Inboard Machineries System, Handline Fish Capture Applinaces, 6 Cold Storage in the middle area, The Fishing Boat has Bilge Keel, Amont of fisherman = 20 – 30 persons.

5. References :

[1]. Aep Saepurahman, Pengembangan Pelabuhan Perikanan Di Pelabuhan Ratu, 2012, Hasil Penelitian. (Aep Saepurahman, Development of Fishing Port of Pelabuhanratu), Research Report (unpublished)

[2]. Baheramsyah, Kementerian Kelautan Dan Perikanan Soroti Perkembangan Perikanan, 2012, www.infopublik.kominfo.go.id (Baheramsyah, Ministry of Marine and Fisheries Attention to Fisheries Development), 2012, www.infopublik.kominfo.go.id

[3]. Daniel dkk, Model Pengembangan Industri Perikanan Berbasis Pelabuhan Perikanan di Makassar Sulawesi Selatan, 2011, Jurnal Penelitian UMI Makassar. (Daniel et all, Development Model of Fishery Industry Base on Fishing Port in Makasar South Celebes), 2011, Researh Jurnal of UMI Makasar.

[4]. Helmi Yusuf dkk, Dampak Pengembangan Pelabuhan Perikanan Terhadap Penyerapan Tenaga Kerja Dan Pendapatan Masyarakat, 2005, www.journal.ipb.ac.id (Helmi Yusuf et all, Impact of Fishing Port Development for Absorption Labor and People Income), 2005, www.journal.ipb.ac.id

[5]. Kementerian Kelautan dan Perikanan, Pusat Informasi Pelabuhan Perikanan (PPIP), April 2015 (Ministry of Marine and Fisheries Fishing, Port Information Centre), April 2015

[6]. Kementerian Kelautan dan Perikanan Profil Pelabuhan Perikanan Nusantara Pelabuhanratu, 2009, www.kkp.go.idb(Ministry of Marine and Fisheries, Profile of Pelabuhanratu National Fishing Port), 2009, www.kkp.go.id.

[7]. Kementerian Kelautan dan Perikanan, Profil Pelabuhan Perikanan Samudera Nizam Zachman Jakarta, 2012, www.kkp.go.id (Ministry of Marine and Fisheries Profile of Nizam Zachman Jakarta Oceanic Fishing Port), 2012, www.kkp.go.id.

[8]. Kementerian Kelautan dan Perikanan, Peraturan Menteri Kelautan Dan Perikanan nomor: Per.16/Men/2006, Tentang Pelabuhan Perikanan (Ministry of Marine and Fisheries, Regulation of Minister of Marine and Fisheries Number: 16/Men/2006, about Fishing Port)

[9]. Keputusan Menteri Perhubungan nomor: KM 53 Tahun 2002, Tentang Pelabuhan Nasional (Ministry of Transportation Number : 53, Year 2002, About National Port)

[10].Kementerian Kelautan dan Perikanan, Peraturan Menteri Kelautan dan Perikanan nomor 2 tahun 2015, tentang Larangan Penggunaan Alat Penangkap Ikan (API) Pukat Hela (Trawl) dan Pulat Tarik (Seine Nets) di Wilayah Pengelolaan Perikanan Negara Republik Indonesia. (Ministry of Marine and Fisheries, Regulation of Minister of Marine and Fisheries Number 2 Year 2015, About Warning to Fishing Gears for The Trawl and The Seine Nets in Ocean of Indonesia).

[11].Peraturan Menteri Kelautan dan Perikanan nomor 56 tahun 2014, tentang Pelarangan Berlayar Bagi Kapal Eks Asing (Ministry of Marine and Fisheries, Regulation of Minister of Marine and Fisheries Number 56 year 2014, About Warning to Voyage of Secondhand Fishing Boat From Abroad)

[12].Pelahunan Perikanan Nusantara Pelabuhanratu, Buku Laporan Tahunan Statistik Perikanan Tangkap Tahun 2015 (Pelabuhanratu National Fishing Port, The Annual Statitistical Report Capture Fisheries Year 2015)

[13]. Pramudya Edy dkk, Analisis Efisiensi Pelabuhan Perikanan Dan Strategi Pengembangan Pelabuhan Perikanan Samudera Cilacap, 2006, Universitas Diponegoro, Semarang (Pramudya Edy et all, Effisiency Analisys of Fishing Port and Development Strategy of Cilacap Oceanic Fishing Port), Diponegoro University, Semarang (unpublished)
[14]. Rizald Max Rompas, Garis Pantai RI Terpanjang Keempat Di Dunia, 2009, www.antaranews.com dan www.goblue.or.id (Rizald Max Rompas, Indonesia Has Beach Line 4th Longest in The World, 2009, www.antaranews.com and www.goblue.or.id
[15]. Salim H.A. Abbas Manajemen Pelayaran Niaga Dan Pelabuhan, 2005, Cetakan Kedua, Dunia Pustaka, Jakarta. (Salim H.A. Abbas Management of Trading Shipping and The Ports), 2005, Second Edition, Dunia Pustaka, Indonesia.
[16]. Kementerian Kelautan dan Perikanan, UPT Dirjen Perikanan Tangkap, 2012, www.kkp.go.id. (Technical Implementation Unit Directorate General of Capture Fisheries, List of Indonesia Fishing Port). 2012, www.kkp.go.id.

IOP Conf. Series: Journal of Physics: Conf. Series **962** (2018) 012009 doi:10.1088/1742-6596/962/1/012009

CURRICULUM VITAE

Personal

Name : Iswadi Nur, born in Langkat North Sumatera, 24 February 1962

Educational;

Bachelor 3 years of Naval Architecture of Maritime Academic of UPN "Veteran" Jakarta
Bachelor 4 years of Naval Architecture of Darma Persada University Jakarta.
Magister of Sea Transportation of Institute of Sepuluh Nopember Surabaya

Lecturer Experiences:

Lecturer of Naval Architecture Study Program of Engineering Faculty UPN "Veteran" Jakarta since 1993. Subject lecture ; Ship Design, Ship Vibration, Production & Operation Management.

Job Experiences

Secretary of Naval Architecture Departmen, Head of Naval Architecture Department, Vice Dean Of Academic Affair

Optimization of Parameters for Manufacture Nanopowder Bioceramics at Machine Pulverisette 6 by Taguchi and ANOVA Method

Hendri Van Hoten[a], Gunawarman[b], Ismet Hari Mulyadi[b], Afdhal Kurniawan Mainil[a] dan Putra Bismantolo[a]

[a)]Mechanical Engineering, Bengkulu University, Indonesia
[b)]Mechanical Engineering, Andalas University, Indonesia

Corresponding author's: hendri_m00@yahoo.com

Abstract. This research is about manufacture nanopowder Bioceramics from local materials used Ball Milling for biomedical applications. Source materials for the manufacture of medicines are plants, animal tissues, microbial structures and engineering biomaterial. The form of raw material medicines is a powder before mixed. In the case of medicines, research is to find sources of biomedical materials that will be in the nanoscale powders can be used as raw material for medicine. One of the biomedical materials that can be used as raw material for medicine is of the type of bioceramics is chicken eggshells. This research will develop methods for manufacture nanopowder material from chicken eggshells with Ball Milling using the Taguchi method and ANOVA. Eggshell milled using a variation of Milling rate on 150, 200 and 250 rpm, the time variation of 1, 2 and 3 hours and variations the grinding balls to eggshell powder weight ratio (BPR) 1: 6, 1: 8, 1: 10. Before milled with Ball Milling crushed eggshells in advance and calcinate to a temperature of 900°C. After the milled material characterization of the fine powder of eggshell using SEM to see its size. The result of this research is optimum parameter of Taguchi Design analysis that is 250 rpm milling rate, 3 hours milling time and BPR is 1: 6 with the average eggshell powder size is 1.305 μm. Milling speed, milling time and ball to powder weight of ratio have contribution successively equal to 60.82%, 30.76% and 6.64% by error equal to 1.78%.

Keywords: Ball Milling, nanopowder, bioceramics, Eggshell, Taguchi and ANOVA

1. Introduction

The disease that happened at human being generally cured use the medicines consists of the chemical compound substance called medical medicine (synthetic medicine). Medical medicine is the modern medicine made of a substance of synthetic or nature substance processed modernly. Source for the making of the medicine generally come from plants, animals, microbe structure and biomaterial engineering. The form of the raw material medicine is powder before next processed. In science Farmaceutical is multiform medicine, for example, powder, capsule, pill and emulsion [1].

In biomaterial engineering, all researcher develop the biomedical technology representing an application of technique medical. This field combines the design and problem-solving ability of an engineer with knowledge of the medical and biological sciences in the medical field, such as medicine, diagnosis, monitoring, and therapy. One application is the biomedical disease prevention, which includes aspects of medicine and related equipment diagnosis and treatment of disease [2].

In the medicine, research is directed to find sources of biomedical materials that will be in the nanopowders. It can be used as raw material for medicine. One of the biomedical materials that can be used as raw material for medicine is bioceramics. One local bioceramics materials are widely available and unutilized which according to researchers could be used as raw material for the medicine were eggshells. Eggshell is widely found in nature and thrown away unused. Eggshells also take a long time to decompose in nature [3]. The compounds found in the eggshells can be used as natural medicines. Among the chemical compounds contained in eggs, one of which is a substance that has the properties can neutralize the acid that is calcium carbonate (CaCO3) [4]. So that these compounds utilized in the pharmaceutical world as an antacid because of its ability to neutralize these acids. Besides, the CaCO3 is the most compound in the chicken egg shells reach 94-97% [5]. CaCO3 addition to other compounds that are beneficial to the medical world to be used as a supplement or as raw material for bone construction. To obtain the compounds contained in the eggs, then have to turn it into a powder and analyzed nanoscale characterization. So it can be any compounds that can be used and can not be utilized.

One of the tools to manufacture nanomaterial is a Ball Mill. Ball mill manufacturing is one method of making nanomaterials on techniques MM-PM (mechanical alloying-powder metallurgy). MM-PM use the energy of the crashed between balls and the vial wall. At the present, the researchers attempted to produce nanomaterials in large quantities and very short time range [6].

Therefore, in this researchers will be optimization of the manufacturing process of nanomaterials from chicken eggshells used Ball Milling with the Taguchi method and ANOVA. Then, a checked of nanopowder characterization using SEM.

2. Methodology

2.1. Preparation of samples

The chicken eggshells were collected from household waste in Padang city and washed with water. The eggshells were then washed with aquades and place in the sun for 3 days. The eggshells were milled in Blender and the eggshell coarse powder was calcined at 900°C for 10 hours. Calcination uses the furnace of mark Nabertherm P320. Then, the eggshell powder was milled in a Pulverisette 6 Classic Line Fritsch Planetary Mono Mill.

Scanning electron microscope (SEM Hitachi S 3400) was used to determine the morphology and microstructures of the eggshells. Methodology of the research are show in figure 1.

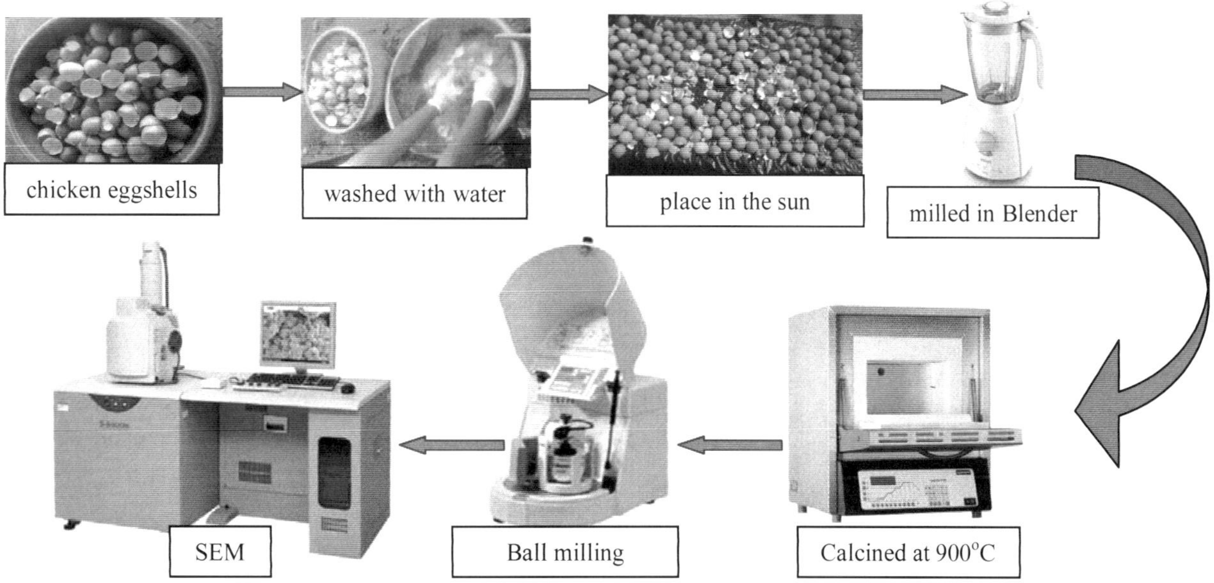

Figure 1. Methodology of the research.

2.2. Experimental conditions by Taguchi method for Ball Milling

To look at the effect of three factors and their impacts in three different levels, 27 samples are required to be milled, but the experimental stages can be reduced by using the Taguchi method [7]. For automatic design and analysis of Taguchi experiments, Minitab16 software was used. Minitab16 software is a windows software for Taguchi experimental design based on selected factors and levels. This program also evaluates the main effects of each factor in each level, analysis of variance and optimum conditions. In this research, in order to optimize the ball mill process and to find the effective factors having the most influence on the particle size of the eggshell powder, three factors with three levels were chosen and studied based on the Taguchi method which is shown in table 1. An orthogonal array of $L_9(3^3)$ designed by the Taguchi method which presents the experimental conditions is shown in table 2. The samples were prepared according to the orthogonal array of $L_9(3^3)$ conditions.

Table 1. Factors and levels used in this experimental.

No.	Factors	Levels		
1	Milling rate (rpm)	150	200	250
2	Milling time (hour)	1	2	3
3	Ball to Powder Ratio	1 : 6	1 : 8	1 : 10

Table 2. Taguchi orthogonal array table of $L_9(3^3)$ for experimental conditions.

No.	Milling rate (rpm)	Milling time (hour)	BPR
1	150	1	1 : 6
2	150	2	1 : 8
3	150	3	1 : 10
4	200	1	1 : 8
5	200	2	1 : 10
6	200	3	1 : 6
7	250	1	1 : 10
8	250	2	1 : 6
9	250	3	1 : 8

3. Results and discussion

Results from this research are the distribution of the size of the eggshell powder. The results obtained after the eggshell is processed using a Ball Mill machine by setting parameters that have been determined as can be seen in Table 2 is based on the analysis of the Taguchi method using Software minitab16. The parameter is adjusted by Ball Milling tool's ability Pulverisette 6.

Results of the eggshell powder size distribution check using SEM can be seen in figure 2 and table 3. Also, can be seen in table 3 process parameter gives the average powder size smallest is 250 rpm of milling rate, milling time 3 hours and BPR of 1: 8. Then, the examination results analyzed using Taguchi method.

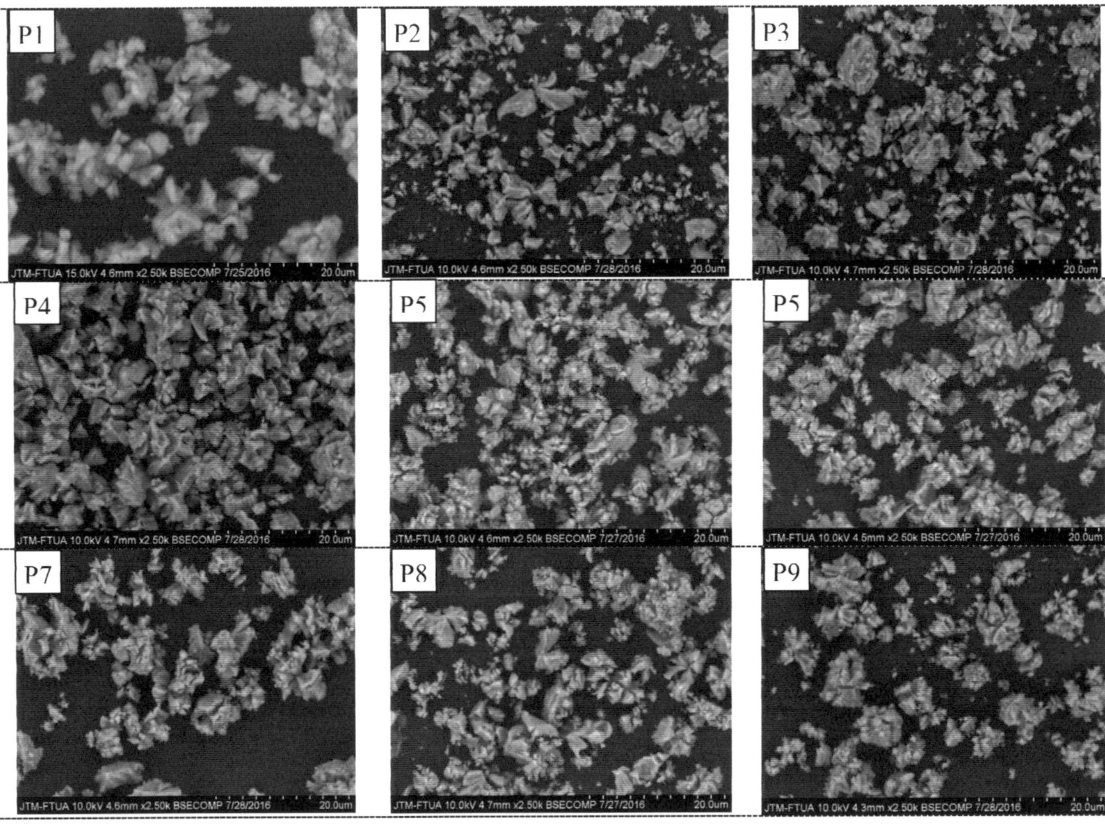

Figure 2. The Result of SEM images of the 9 samples by Taguchi L-9 orthogonal array.

Table 3. The results of average powder size eggshells after the check using SEM

Experiment no.	Milling rate (rpm)	Milling time (h)	BPR	Average powder size (μm)
P1	150	1	1 : 6	3
P2	150	2	1 : 8	4.2
P3	150	3	1 : 10	2.4
P4	200	1	1 : 8	6.7
P5	200	2	1 : 10	6.05
P6	200	3	1 : 6	3.8
P7	250	1	1 : 10	3.5
P8	250	2	1 : 6	3.4
P9	250	3	1 : 8	2.05

From the analysis of Taguchi design will be obtained factors influence the level of process parameters on the size of the eggshell powder. It serves to determine the average value of each individual factor and level in order to know the difference between the average value of each individual level and rank. The results of this analysis as shown in table 4. From the table, it appears that the milling rate has the most influence to get smallest size of an eggshell powder compared to two other factors. Milling time second ranks then followed by the BPR. The mean effect were calculated by statistical software minitab16 are show in Figure 3, indicating that the size of the eggshell powder was at its minimum, with milling rate, milling time and BPR respectively, 250 rpm, 3 hour and 1:6. The result from the analysis of Taguchi design are the optimum condition parameter. To determine the percentage contribution of the process parameters, then do ANOVA.

Table 4. The average of response for each level of factors and the main factor effect

Level	Milling rate (rpm)	Milling time (hour)	BPR
L1	3.2	4.4	3.4
L2	5.52	4.55	4.317
L3	2.98	2.75	3.983
Lmax - Lmin	2.53	1.8	0.917
Ranking	1	2	3

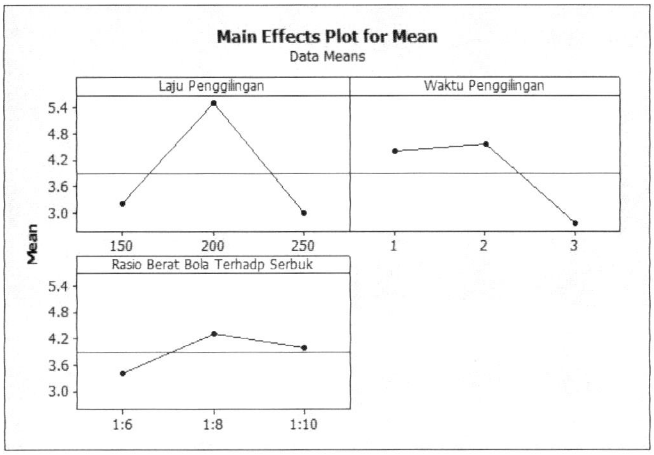

Figure 3. The main factor effect plots.

ANOVA results can be seen in figure 4. From figure 4 it can be seen that the result of the eggshell powder size is influenced by 60.82% of the milling rate. Milling time influenced by 30.76% and BPR at 6.64%, the possibility of error of 1.78%. This indicates that the setting value milling rate affects the size of a chicken egg shell powder.

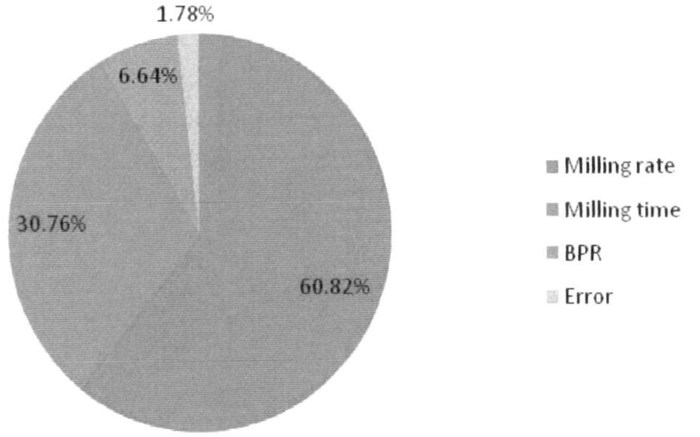

Figure 4. Percent contributions of each parameters milling..

The Eggshell Powder Size analyses at optimal condition was shown in figure 5. Figure 5 shows the result of SEM image from twice test the Ball Milling with the same parameters.
Process 1 dan 2 shows the average Eggshell Powder Size value of respectively, 1.36 μm dan 1.25 μm. So, the average Eggshell Powder Size value from 2 samples are 1.305 μm.

Figure 5. The result of SEM image of the samples by optimal parameter.

4. Conclusion

In the analysis of Taguchi design obtained table level influence of factors on the size of the powder, which rank 1, 2 and 3 are respectively the rate of milling, grinding time and BPR. Knowing how much influence the level of these factors to the size of the powder is to create a table or ANOVA analysis of variance. Results obtained using this method is in the form of a percentage contribution of factors which influence the rate of milling, grinding time and BPR respectively as follows 62.83%, 30.76% and 6.64% to 1.78% error. The result of this research is optimum parameter of Taguchi Design analysis that is 250 rpm milling rate, 3 hours milling time and BPR is 1: 6 with the average eggshell powder size is 1.305 μm.

References

[1]. http://citra-dewi-ff13.web.unair.ac.id/artikel_detail-86643-newbie%20pharmacist-bentuk%20sediaan%20obat.html, accessed 01 april 2015.
[2]. http://www.prospects.ac.uk/biomedical_engineer_job_description.htm, 01 April 2015.
[3]. repository.usu.ac.id/bitstream/123456789/25090/4/Chapter%20II, accessed 01 April 2015.
[4]. repository.usu.ac.id/bitstream/123456789/16191/4/Chapter%20II, accessed 01 April 2015.
[5]. A.Nurlaela, S.U. Dewi, K. Dahlan, D.S. Soejoko, The Use of Hen's and Duck's Eggshell as Calcium Source to Synthesis Bone Mineral, Jurnal Pendidikan Fisika Indonesia 10, hal. 81-85, UIN Syarif Hidayatullah, Jakarta, 2014.

IOP Conf. Series: Journal of Physics: Conf. Series **962** (2018) 012010 doi:10.1088/1742-6596/962/1/012010

[6]. Singh, Kaushal, Kishore and Sudipto, Bhattacharjee, Study On The Effect Of High Energy Ball Milling (A Nano Material Process) On The Microstructure And Mechanical Properties Of A (Al-Si-Fe-Cu) ALLOY, thesis for the degree of bachelor, Department of Metallurgical and Materials Engineering National Institute of Technology, Rourkela, 2007.

[7]. Farbod, Mansoor, Rafati, Zahra, and Shoushtari, Morteza Zargar. "Optimization of parameters for the synthesis of Y2Cu2O5 nanoparticles by Taguchi method and comparison of their magnetic and optical properties with their bulk counterpart", Journal of Magnetism and Magnetic Materials, 2016.

Study on Gold and base metal occurrence in Uluwai Prospect, Western Latimojong Mountain, South Sulawesi

Adi Maulana[1], Asri Jaya[1], Akira Imai[2]

[1]Department of Geology, Faculty of Engineering, Hasanuddin University
[2]Department of Earth Resources Engineering, Akita University

adi-maulana@unhas.ac.id

Abstract. Uluwai Prospect is located in the northern part of South Arm of Sulawesi along the eastern part of the Kalosi Fold Belt and Latimojong Mountain. The area is generally characterized by moderate to rugged topography area with elevation in the range of 700 to 1400 m above sea level in the mountainous complex called Latimojong Mountain Complex. The mineralization is characterized by a relatively simple sulphide ore mineral assemblage consists of pyrite, sphalerite and chalcopyrite. Samples were collected in areas showing abundant sulphide minerals where younger faults cut the bedding and foliation of country rocks. A number of silicified zones have been observed, as well as float material containing disseminated pyrite, chalcopyrite, and sphalerite with hematite, goethite and limonite. Some alteration types have been observed including sericitization, albitization, carbonatization and silisification. The samples collected indicated that the mineralisation is contained within metasedimentary (sandstone to mudstone) and greenschist. Geochemical analyses from 16 samples including 5 stream sediment samples indicated that the most promising mineralization occur in the prospect area are copper (Cu) and zinc (Zn). This is also supported by the abundance of chalcopyrite and sphalerite in some highly altered samples. Assaying of the collected samples revealed most of samples contain relatively low gold (Au) concentration. However, two samples contain 0.007 and 0.01 ppm of Au. In the mineralized area, Zn concentrations are up to 134 ppm, Cu up to 120 ppm and Pb up to 18 ppm and As up to 70 ppm. There is no clear relationship that exists between Au and the base metals except that one of the samples with highest Au values tend to have high Zn and As. This unclear pattern also shown by Cu, Pb and Zn. Base metal concentration in stream sediment samples show a relatively stable pattern than in rock samples.Arsenic tends to be elevated in base metal rich samples. Sb and Mo are relatively low in all sample type. However, Mo values will be high in the samples which contain highest Cu and Zn.

Keywords: Gold, Base metal, Uluwai, Latimojong, South Sulawesi

1. Background

The Uluwai polymetallic prospect area is located at the border between Toraja and Enrekang Regency, South Sulawesi Province. It is centered in the southern tip of Toraja Regency and northeastern tip of Enrekang Regency, at the geographic center of South Arm of Sulawesi, approximately 340 kilometres North of Makassar, the capital city of South Sulawesi Province (Fig.1). It is three kilometres northeast of Curio village (Enrekang Regency) in a sparsely populated agricultural area. The closest city to the

IOP Conf. Series: Journal of Physics: Conf. Series **962** (2018) 012011 doi:10.1088/1742-6596/962/1/012011

prospect is Enrekang City, approximately 60 kilometres to the SW, and the nearest town is Kalosi, 20 kilometres to the west.

The project area is generally characterized by moderate to rugged topography area with elevation in the range of 700 to 1400 m above sea level in the mountainous complex called Latimojong Mountain[1-2]. Slope gradients are steepest within the northern and eastern portions grading to moderate toward the western and southern sections which were controlled by the complex magmatic events and regional structure [2-4].History of ore mineral exploration in this prospect has been dated since colonialization era according to formal report. [5] reported that [6] had mentioned the occurrence of so called "Koper-Lai" (Copper-Slate) Formation, now knowns as Toraja Formation, in the foothill of the Latimojong Mountain as small pocket of native copper and malachite in a shally sequence. In 1980`s a group of local people tried to dig some holes and find some quartz veins contain some gold. Local government office staff also reported the placer gold occurrence in Malua River which originated from Uluwai area. World class gold deposit, Awak Mas, is located just about 15 km to the southeast within the same geological formation. However, serious exploration for mineralization in Uluwai Area has never been conducted despite its high prospect of economic metal.

The data contained in this study was sourced from the result of stream sediment sampling, field observation and geological mapping. The purposes of this project are:1) to study alteration and mineralization process, and 2) to do a sampling program to evaluate the concentration of Au and other base metalin the studied area.

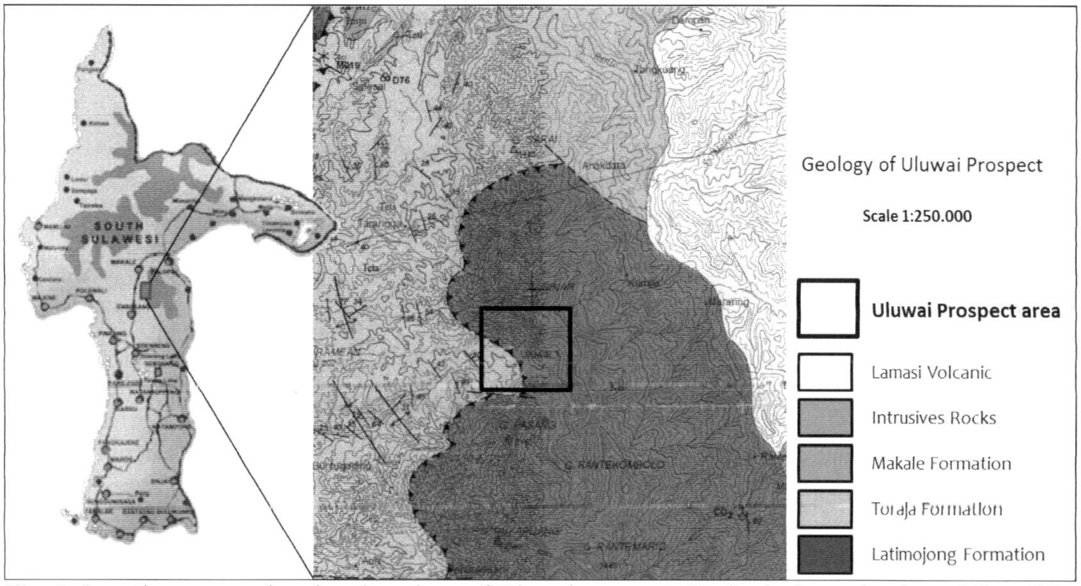

Fig. 1. Location map and regional geology of Uluwai prospect. Geological map is adopted from [2].

2. Material and Method

Forty (60) samples including 40 fresh rock samples and 20 stream sediment samples were collected during field work in Uluwai area. Of these, 10 thin sections were prepared from fresh rock samples and studied petrographically to determine the rock types, mineral assemblages, fabric and textural relations. In addition, 5 polished sections were prepared and studied under reflected light in Optical Mineralogy Laboratory, Hasanuddin University to study the ore mineral type.

For stream sediment, each sample of about 2 kg constituted composite from at least four sites across the stream bed, 10-15m on either side of the sampling point. Fine silt sediments were collected as much as possible, avoiding the organic material. Care was also taken to collect samples from up a tributary above the point of influence of water flow from the main stream.The samples were kept in a well labelled polythene sample bags for easy and safe transportation. The sample sites were immediately indicated on the field map and the site on the field flagged with a red flag bearing the sample number. Garmin GPSMap 60CSx was used to assist in locating the sample points. Prior to dispatch to the laboratory, all samples were checked to ensure that labels and seals are intact to avoid contamination and tampering. Care was also taken during sampling and handling to avoid the introduction of any foreign material that could give bias results.

All the samples were sent to Intertek laboratory at Jakarta for geochemical analyses. Au was analysed using Fire Assay (FA) method, whereas Cu, Pb, Zn and Ag method were analysed using Atomic Absorption Spectrometry (AAS) method and As, Sb and Mowere analysed using X-ray fluorescence (XRF) method.

Sample location map, geological map, mineralization/vein map and alteration map are shown in appendix.

3. Results

3.1. Mineralization

The Uluwai prospect falls into the broad category of a metasedimentary-hosted gold deposit. Mineralization at Uluwai rarely attains true massive sulphide levels; however horizons containing 15-20% sulphides most commonly define areas of significant base and precious metal mineralization. The main body of the deposit appears to be stratiform in nature. The highest grade zones are very likely to be controlled by strike-slip faults mineralization hosted in sedimentary to metasedimentary rocks.

These faults zones can be interpreted as related to feeder zones developed after the currently visible mineralization was deposited. The typical Uluwai polymetallic deposit contains pyrite, black sulphide, sphalerite and chalcopyrite with oxide mineral occurs as hematite, magnetit and goethite. These minerals are typically associated with ubiquitous 10-20% disseminated pyrite (Fig. 2.a). Pyrite is the dominant sulphide mineral and often replaced by chalcopyrite and sphalerite. When observed under microscope, the sulphide crystal are mainly euhedral to subhedral and it appears that they have been corroded. Fig. 2.b shows a photomicrograph of a polished section for sample ST-16 and Fig. 2.c shows quartz vein in thin section from sample no RC-12. Pyrite grains occur as a cluster or disseminated in the silicate or carbonate gangue. Chalcopyrite and sphalerite inclusions are commonly observed within pyrite. Chalcopyrite is the second dominant sulphide phase. Most of the grains are coarse (up to 2 mm) and display subhedral to irregularly shaped crystal form. Sphalerite occurs in the form of up to 3 mm wide continuous to discontinuous monomínerallic veinlets. It may contain abundant chalcopyrite blebs and/or inclusion of pyrite and chalcopyrite.

The mineralization in metamorphic rocks occurs in stop 68 in the southeastern part of Uluwai Village. The greenschist shows brown to greenish color, highly foliated (175°/40°) and has been altered into potassic and silicic alteration. In stop 64, which is located approximately 1 km southwest of Pekaparan village, two abandoned tunnels were found in greenschist rocks. The host rocks show brown to yellow in colour, composed of mostly sericite and clay mineral. Quartz vein found in the tunnel with the length ranges from 1 to 2 meters. Sandstones often contain mineralization in the form of vein cross cut the bedding as shown in Fig. 2.d. The sandstone show grey to brown in colour, slightly oxidized as shown by reddish surface colour (limonite), interbedded with silt and mudstone. Mineral composition composed of quartz, feldspar and clay. The quartz vein thickness range from 3 to 10 cm and associated with fault shear zone.

Fig. 2. a. Mineralization occurrence in metasedimentary rocks showing disseminated pyrite black sulphide and chalcopyrite.b. Polished section of ST-16. Pyrite (Py) has been partly replaced by sphalerite (Sph) and Chalcopyrite (Cp).c. Quartz vein in thin section showing some mineralization.d. Mineralization in greenschist rocks showing boulder of quartz vein.

3.2. Alteration

The rocks in the studied area locally suffered pervasive sericitization, albitization, carbonatization and silisification alteration. Quartz and calcite occur also as continuous and discontinuous veins.

Based on field observation study, hydrothermal alteration style in Uluwai prospect consist of (1) quartz – sericite – pyrite (sericitization); (2) albite – quartz – chlorite (albitization); (3) carbonate – quartz – pyrite (carbonatization); (4) quartz – pyrite (silicification).

Mineralization is typically hosted within a flysch sequence that generally dips 20-55°. Ore is usually associated with abundant quartz veining and silica-albite-pyrite alteration (alibitization) and quartz-pyrite alteration (silicification). Field observation indicates that ore mineralization sulphide mineral (black and yellow sulphide) and oxide mineral (hematite, goethite, limonite). Gold occurrence will be possibly associated with albitization and silicification alteration.

3.2.1 Quartz – sericite – pyrite (sericitization). Sericitization occur in light to dark grey mudstone, associated with hematite within quartz vein in Stop-61 near Pekaparan Village. The alteration was indicated by the occurrence of sericite and pyrite within quartz vein. Quartz vein ranges from 10 to 25 cm, showing cockade texture (Fig. 3.a).

3.2.2 Albite – quartz– chlorite (albitization). Albitization commonly occur in metasandstone with disseminated and associated with quartz vein in the fracture (Fig.3.b).

3.2.3 Carbonate – quartz– pyrite (carbonatization). Carbonatization present in dark to light grey metaclaystone which has been metamorphosed into greenschist facies condition in stop-43 (a creek to the south of Parombean). The alteration process was indicated by the occurrence of carbonate vein and disseminated pyrite associated quartz in vein or fracture (Fig. 3.c).

3.2.4 Quartz– pyrite (silicification). The silicification zone is widely distributed in Uluwai Village (stop-76) and developed in sandstone to siltstone (Fig. 3.d). It is indicated by the occurrence of quartz vein cutting the bedding of the sandstone and massive silicified block.

Fig. 3. a. Quartz vein showing cockade texture in sericitization alteration zone. b. Hand specimen of albite-quartz-chlorite alteration samples showing a thin quartz vein within sandstone. c. Carbonatization alteration in Parombean Village. D. Quartz-pyrite (silicification) alteration in Uluwai Village.

3.3 Geochemistry

Sixteen (16) samples consists of five (5) stream sediment samples, eleven (11) rock samples (rock chip and rock float) were analysed. The assay results of the samples collected are shown in Table 1.The geochemical result show that the concentration of Au in stream sediment (SS) samples is very low and only found in rock chip (RC) sample (RC 005). Ag also show very limited concentration (<1) whereas other elements show variation. In some samples, base metal (Cu, Pb and Zn) yield high concentration.

Table 1. Assay result of the analysed samples

IDENT	Au1	Au2	Au3	Au4	Cu	Pb	Zn	Ag	As	Sb	Mo
UNITS	PPM	PPM	PPM	PPM	PPM	PPM	PPM	PPM	PPM	PPM	PPM
DET.LIM	0.005	0.005	0.005	0.005	2	4	2	1	1	1	1
SCHEME	FA50	FA50	FA50	FA50	GA02	GA02	GA02	GA02	XR01	XR01	XR01
SS 33 #40	<0.005				34	16	69	<1	12	1	<1
SS 35 #40	<0.005				39	18	81	<1	10	1	1
SS 37 #40	<0.005				35	10	80	<1	11	2	1
SS 38 #40	<0.005				44	18	88	<1	19	2	1
SS 39 #40	<0.005				46	19	75	<1	6	3	<1
VC 01	<0.005				38	13	63	<1	7	2	2
RC 005	0.007				24	17	120	<1	70	2	<1
RC 006	<0.005	<0.005			34	6	71	<1	<1	2	1
RC 010	<0.005				21	11	28	<1	3	2	<1
RC 011	<0.005				120	10	134	<1	8	3	6
RC 012	<0.005				9	<4	57	<1	1	3	<1
RC 013	<0.005				22	12	61	<1	2	2	<1
RC 014	<0.005				74	7	28	<1	<1	2	<1
RC 015	<0.005				48	5	26	<1	<1	2	<1
RF 002	<0.005			<0.005	30	<4	17	<1	3	2	<1
HS 02	0.01				41	5	68	<1	2	1	1

4. Discussion and Conclusion

The assay results from sampling campaign and geological mapping indicate that the mineralisation occur in Uluwai Project area. The most significant mineralization is Cu and Zn mineralization as shown by geochemical analyses result with low Au content. The copper mineralisation is associated with disseminated pyrite, chalcopyrite and sphalerite both in metasedimentary (sandstone to mudstone) and greenschist host rocks.

The alteration and mineralisation of the sampled rocks indicate that the area has been subjected to hydrothermal alteration. Confidence in the assay results is strengthened by the outcrops (intensity of sulphide-bearing quartz vein) and polished section scale mineralization where significant sulphide mineralization were observed. In addition, widespread alteration zone within the project area is also support the promising mineralization occurrence.

The mineralization in Uluwai Project appears to be related to calcite ⊥ quartz veining event.Now that some anomalies have been identified on the project area, it will be necessary to conduct a detailed mapping and soil sampling program to characterise the nature of the geology. Effort should be made to determine whether the spatial location of the alteration is typical of sedimentary-hosted gold deposit and therefore can be used as a guide to exploration.

Assaying of collected samples revealed most of samples contain relatively low gold concentration. However, two samples contain 0.007 and 0.01 ppm. Highest values were detected from metasedimentary rock (sandstone). In the mineralized area, Zn concentrations are up to 134 ppm, Cu up to 120 ppm and Pb up to 18 ppm and As up to 70 ppm. There is no clear relationship that exists between Au and the base metals except that one of the samples with highest Au values tend to have high Zn and As. This unclear pattern also shown by Cu, Pb and Zn. Base metal concentration in in stream sediment samples show a relatively stable pattern than in rock samples.

Arsenic tends to be elevated in base metal rich samples. Sb and Mo are relatively low in all sample type. However, Mo values will be high in the samples which contain highest Cu and Zn. The most significant of these include RC-005, RC-011, RC-014, HS-02 and RC-006. RC-011 was taken in stop-62 near Pekaparan Village within a sericitized sandstone that contains chalcopyrite, pyrite and sphalerite and it returned grades of 120 ppm Cu, 134 ppm Zn, 8 ppm As and 6 ppm Mo. The sample at location RC-011 is shown in Figure 20.RC-005 was sampled from a slate in stop-41, with the predominant opaque mineral being pyrite, followed by chalcopyrite and sphalerite. The results at this location included a zinc grade of 120 ppm and relatively low Cu (24ppm) but high As content (70 ppm). It is interesting to note that this sample contain high As which is a good indicator of sedimentary-host gold deposit in other deposit in the world.RC-014 was taken from schist on top of the Hill near POS-2 with 74 ppm of Cu and 28 ppm of Zn. No significant gold were detected from all analysed samples so far.HS-02 was samples near RC-006 from metasedimentary rocks outcrops in stop-71. This sample shows an Au anomaly content compared to other sample (0.01 ppm), 41 ppm Cu and 68 ppm Zn.

References

[1] BergmanSC, Coffield DQ, Talbot JP, and GarrardRA 1996Tectonic Evolution of Southeast Asia, ed R Hall and D J Blundell (Geological Society of London Special Publication106) p 391.
[2] Djuri and Sudjatmiko 1998 Geology of Majene Quadrangle 1:250.000 in scale. Geology Research and Development Centre, Bandung
[3] Elburg MA and FodenJ 1999 *Chem. Geology* **156** 67-93
[4] Maulana A, Imai A, van Leeuwen T, Watanabe K, Yonezu K, Nakano T, Boyce A, Page L, Schersten A 2016*J. of Asian Earth Sciences* **102** 104-125.
[5] Van Leeuwen T M and Pieters P E 2011*Proc. of the Sulawesi Mineral Resources*, MGEI-IAGI, 1-110.
[6] Van Bemmelen R W 1949 *The geology of Indonesia* Vol. 1A(Government Printing Office, The Hague)

Appendix:

1. Sample location map.
2. Geological map
3. Mineralization/vein map
4. Alteration map

IOP Conf. Series: Journal of Physics: Conf. Series **962** (2018) 012011 doi:10.1088/1742-6596/962/1/012011

App. 1

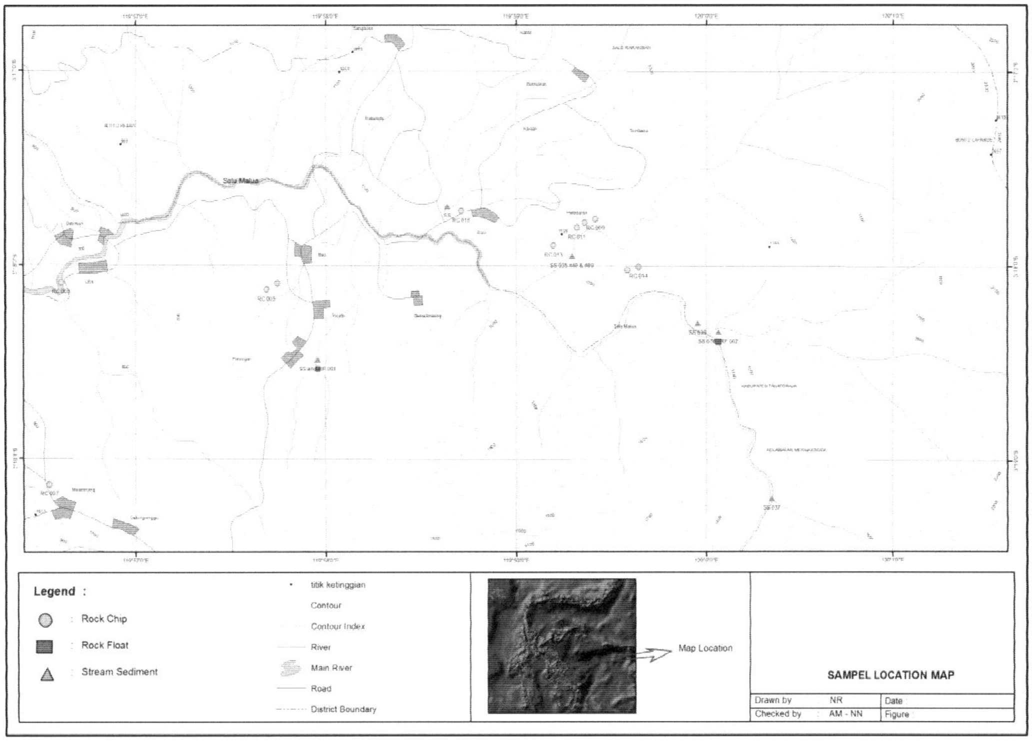

App. 2

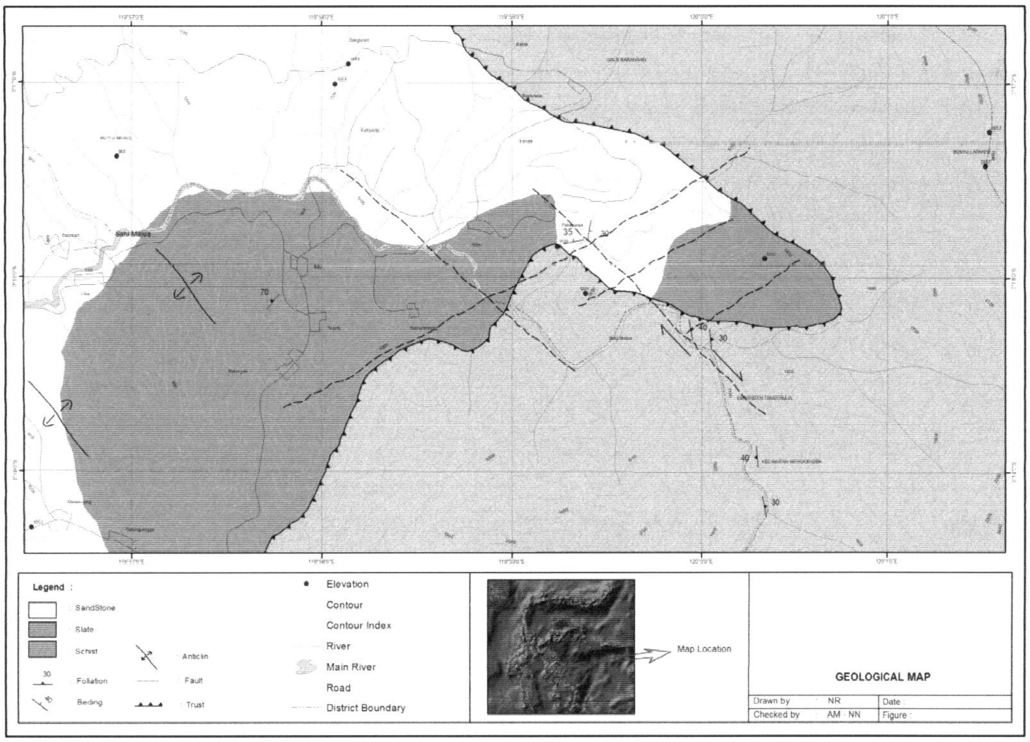

App. 3

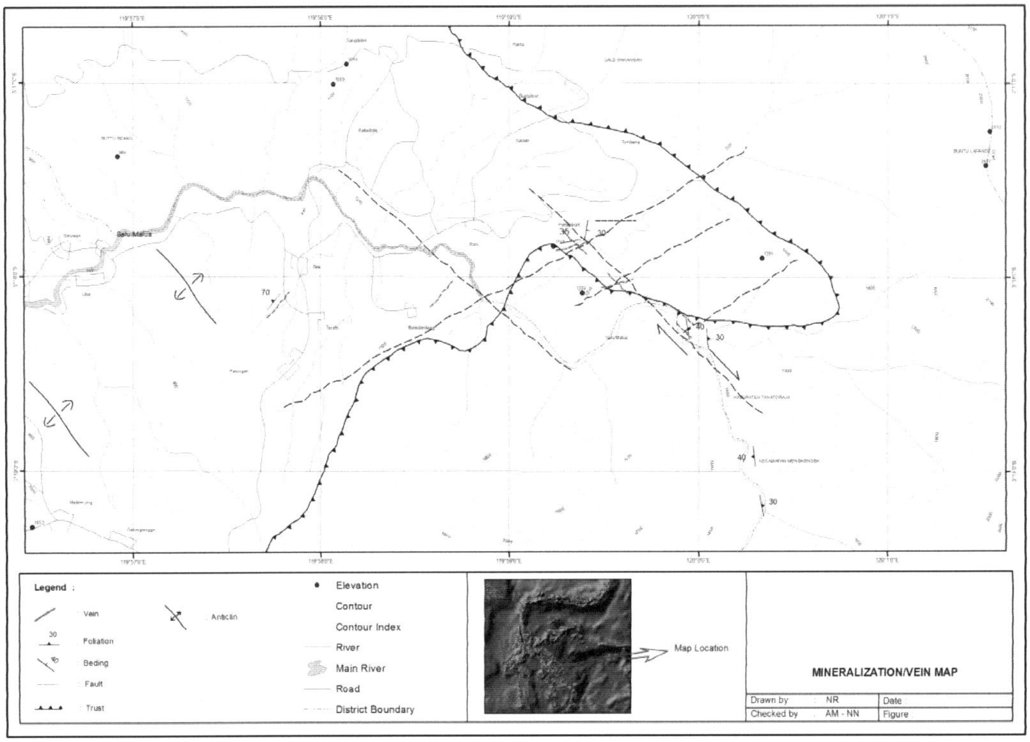
Legend :
Vein
Anticlin
Foliation
Beding
Fault
Trust
Elevation
Contour
Contour Index
River
Main River
Road
District Boundary
Map Location
MINERALIZATION/VEIN MAP
Drawn by NR Date
Checked by AM - NN Figure

App. 4

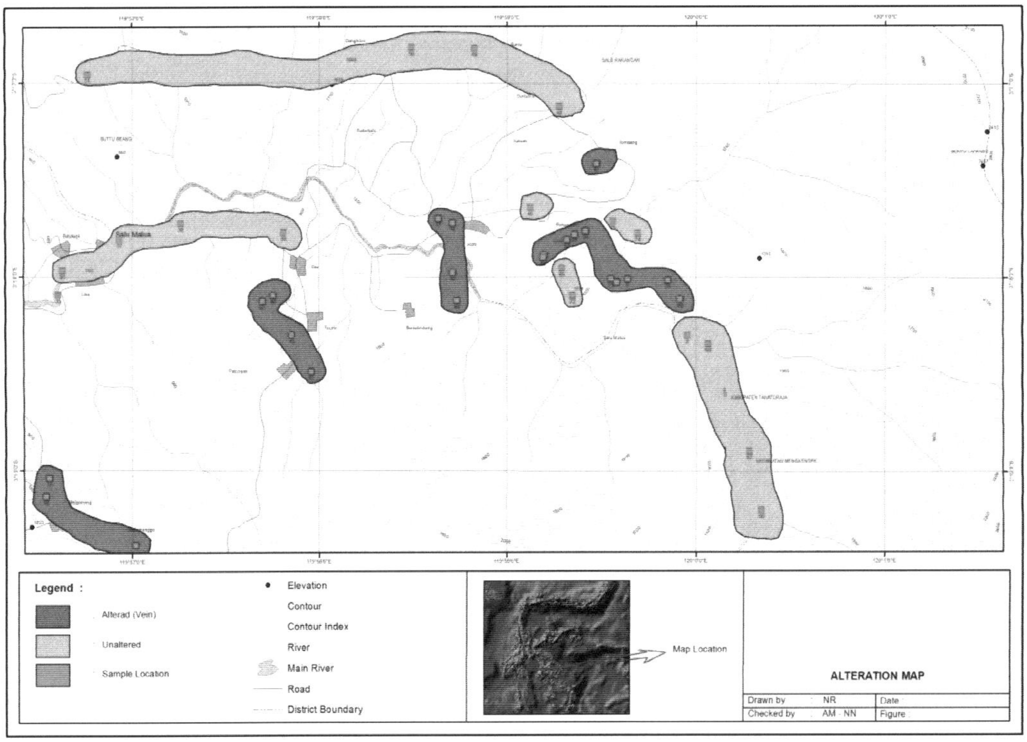
Legend :
Alterad (Vein)
Unaltered
Sample Location
Elevation
Contour
Contour Index
River
Main River
Road
District Boundary
Map Location
ALTERATION MAP
Drawn by NR Date
Checked by AM - NN Figure

IOP Conf. Series: Journal of Physics: Conf. Series **962** (2018) 012012 doi:10.1088/1742-6596/962/1/012012

A sub-target approach to the kinodynamic motion control of a wheeled mobile robot

Kimiko Motonaka*, Keigo Watanabe, and Shoichi Maeyama****

*Kansai University
**Okayama University

motonaka@kansai-u.ac.jp

Abstract. A mobile robot with two independently driven wheels is popular, but it is difficult to stabilize it by a continuous controller with a constant gain, due to its nonholonomic property.It is guaranteed that a nonholonomic controlled object can always be converged to an arbitrary point using a switching control method or a quasi-continuous control method based on an invariant manifold in a chained form. From this, the authors already proposed a kinodynamic controller to converge the states of such a two-wheeled mobile robot to the arbitrary target position while avoiding obstacles, by combining the control based on the invariant manifold and the harmonic potential field (HPF).On the other hand, it was confirmed in the previous research that there is a case that the robot cannot avoid the obstacle because there is no enough space to converge the current state to the target state.In this paper, we propose a method that divides the final target position into some sub-target positions and moves the robot step by step, and it is confirmed by the simulation that the robot can converge to the target position while avoiding obstacles using the proposed method.

1. Introduction

In these days, many mobile robots with two independently driven wheels are developed for transportation or giving service.These robots need an autonomous locomotion, and there are many researches about an obstacle avoidance and a path planning for the two-wheeled mobile robot.For the smooth moving of the robot in an environment including obstacles, it is desirable to consider about the kinematic constraint and the dynamic constraints.Therefore, the method called "kinodynamic motion planning" is proposed[1].The kinodynamic motion planning aims to design the control input so that the kinematic constraints generated from the environment and the dynamic constraints generated from the dynamic characteristics of the controlled object are both considered simultaneously.Masoud realized the kinodynamic motion planning for a point mass by combining two control inputs: one is generated by considering the dynamics of the controlled object and the other is based on the gradient information of the potential field called harmonic potential field (HPF) which is generated from the boundary information of the environment[2].In this research, it is aimed at applying the kinodynamic motion planning to the two-wheeled mobile robot by extending the method proposed by Masoud.

The kinodynamic motion planning using the HPF is achieved by adding the gradient information of an HPF to the control input that has considered the dynamic characteristics of the controlled object[3].To apply the kinodynamic motion planning based on an HPF, at first, it needs the control input that considers the dynamic model of the controlled object.Here, the stabilizing control to the origin using an invariant manifold is given as one of the methods to control the two-wheeled mobile robot.By using the control based on the invariant manifold, the controller can reliably stabilize the controlled object to the origin kinematically and dynamically[4].In the previous research, the authors confirmed that the proposed method was able to guide the robot to the target position while avoiding obstacles to a certain degree.However, it was also confirmed that there was the case that the robot was not able to avoid the obstacle because there was no enough space to converge the state to the target state.In this paper, we propose the method that dividing the target position into some pieces and moving step by step, and it is confirmed by the simulation that the robot can converge to the target position while avoiding obstacles using the proposed method.

2. A model of two-wheeled mobile robot

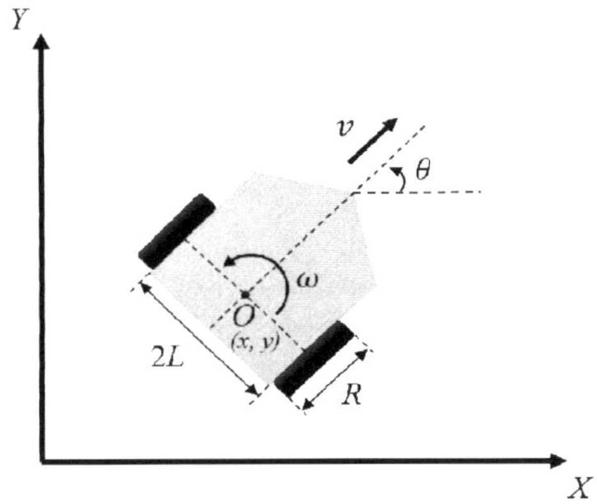

Figure 1. The two-wheeled mobile robot

Figure 1 shows the schematic structure of the two-wheeled drive mobile robot when assuming that the world coordinate system is the X-Y coordinate system. Note that, $2L$, R, and θ in Fig. 1 mean the tread, the radius of the wheel, and the direction angle from the X-axis respectively. The controlled point of the two-wheeled drive mobile robot, $O(x, y)$exists at the center of the tread. v means the forward velocity, and omega means the angular velocity around the point O. Hereafter, the authors call the two-wheeled drive mobile robot "the robot." The dynamic model of the robot with two-inputs and five-states is shown as:

$$\begin{bmatrix} \dot{x} \\ \dot{y} \\ \dot{\theta} \\ \dot{v} \\ \dot{\omega} \end{bmatrix} = \begin{bmatrix} v\cos x \\ v\sin x \\ \omega \\ 0 \\ 0 \end{bmatrix} + \begin{bmatrix} 0 \\ 0 \\ 0 \\ \frac{1}{M} \\ 0 \end{bmatrix} F + \begin{bmatrix} 0 \\ 0 \\ 0 \\ 0 \\ \frac{1}{I} \end{bmatrix} \tau \tag{1}$$

Here, M, I, F and τ mean the mass of the robot, the moment of inertia, the force and the torque, respectively.

3. Control using the invariant manifold

The invariant manifold means a manifold constructed by the scalar functions whose differential value is 0 with respect to time. There are two types of control using the invariant manifold. One is the switching control that is able to divide the first-step (until on the invariant manifold) and the second step (on the invariant manifold), and the other is the quasi-continuous exponential stabilization control that excludes the switching control by combining the first and second steps. It is known that the robot can converge to the origin by using these control methods. Generally, a nonholonomic system includes several complex nonlinear terms, and it is difficult to design the controller from the original model. On the other hand, by using the canonical model, such as the chained form, it becomes easy to make a control system, which assures a globally asymptotic stability.

It needs to consider the error from the origin to an arbitrary point to converge the robot state value to the arbitrary point. The chained form transformation, which takes account of the error, is called an "error model." The error model is shown as:

$$\dot{\boldsymbol{e}} = \begin{bmatrix} \dot{e}_1 \\ \dot{e}_2 \\ \dot{e}_3 \\ \dot{e}_4 \\ \dot{e}_5 \end{bmatrix} = \begin{bmatrix} e_4 \\ e_5 \\ e_2 e_4 \\ u_1 \\ u_2 \end{bmatrix} \tag{2}$$

Here, u_1 and u_2 denote the control inputs on the chained form.

$$\begin{aligned} u_1 &= \frac{\tau}{I} \\ u_2 &= -z_4^2 z_2 - \frac{\tau}{I} z_3 + \frac{F}{M} \end{aligned} \tag{3}$$

Then, the error state $[e_1 e_2 e_3 e_4 e_5]^T$ is represented by

$$\begin{bmatrix} e_1 \\ e_2 \\ e_3 \\ e_4 \\ e_5 \end{bmatrix} = \begin{bmatrix} z_1 - z_{1d} \\ z_2 - z_{2d} \\ z_2 - z_{2d} - \frac{1}{2} e_2^2 z_{2d} - z_{2d} e_1 \\ z_4 - z_{4d} \\ z_5 - z_{5d} \end{bmatrix} \tag{4}$$

Note that, the $[z_1 z_2 z_3 z_4 z_5]^T$ is

$$\begin{bmatrix} z_1 \\ z_2 \\ z_3 \\ z_4 \\ z_5 \end{bmatrix} = \begin{bmatrix} \theta \\ x\cos\theta + y\sin\theta \\ x\sin\theta - y\cos\theta \\ \omega \\ v - (x\sin\theta - y\cos\theta)\omega \end{bmatrix} \tag{5}$$

4. The quasi-continuous exponential stabilization control

The quasi-continuous exponential stabilization control is the method that is realized in one unified manner, instead of conducting the control one by one in the first and second steps described previously. This control excludes the switching, and it can be controlled without a sudden change of the input. The control input of the quasi-continuous exponential stabilization control is given by

$$\begin{aligned} u_1 &= -2ke_4 - k^2 e_1 \\ u_2 &= \frac{fS(e)}{e_1} - k^2 e_2 - 2ke_5 \end{aligned} \tag{6}$$

Here, the time response of $S(e)$ becomes

$$S(t) = S\big(\boldsymbol{e}(0)\big)e^{-\frac{f}{4k}t} \tag{7}$$

5. Combining of the control using invariant manifold with the HPF

The aim of this method is to achieve the kinodynamic motion planning for a two-wheeled drive mobile robot by combining the quasi-continuous exponential stabilization control with the HPF. Concretely, the kinodynamic motion planning is achieved by adding the directional force in gradient to the control inputs, u_1 and u_2 in the chained form.However, if the control input with the information from the HPF is used in the whole area of the environment, the differential value of the invariant manifold does not become 0, then it is considered that the state does not reach the target point. In other words, if always adding the directional force from the HPF to the controller, the stability of the manifold is broken. Therefore, in this chapter, basically the quasi-continuous exponential stabilization control is used, and added the information of the HPF to its control input locally. In this method, the convergence to the target position can be achieved finally, because only the quasi-continuous exponential stabilization control is used in the area without the obstacles and then the stability of the manifold is not broken. In what follows, it is described a force in the gradient of the HPF and how to design the input torque for the robot by introducing its information as a force, and state the control that combines the quasi-continuous exponential stabilization control with the HPF.

From the HPF, the x-directional and y-directional gradient vector of each position of the environment can be calculated. These gradient vectors show the direction that the robot should move. By using the x-directional gradient $\Delta x$and the y-directional gradient Δy, the gradient direction θ_k, which is the direction that the robot should move, can be written as follows:

$$\theta_k = \tan^{-1}\frac{\Delta y}{\Delta x} \tag{8}$$

Assuming that the input torque with the gradient directional force is $\boldsymbol{\tau}'$, it can be written as the following equation:

$$\tau' = \frac{L}{R}(\tau_1 - \tau_2) + k_1\theta_k \tag{9}$$

with the gain $\boldsymbol{k_1} > 0$.Therefore, the kinodynamic motion planning should be able to be achieved by adding the gradient directional force of the HPF $\boldsymbol{\theta_k}$ to $\boldsymbol{u_1}$. Assuming the gain $\boldsymbol{k_1} > 0$, the control inputs of the quasi-continuous exponential stabilization control can be written by

$$\begin{aligned} u_1 &= -2ke_4 - k^2e_1 + k_1\theta_k \\ u_2 &= \frac{fS(e)}{e_1} - k^2e_2 - 2ke_5 \end{aligned} \tag{10}$$

When guiding the robot from the initial position to the target position by the proposed controller, the robot sometimes ignores the obstacles because the controller prioritizes the convergence of the orientation.Therefore, we propose a method based on using the potential of the environment which has been already calculated, to calculate the ideal trajectory in the assumed environment and approach to the target state step by step. Concretely, the ideal trajectory is divided into an arbitral number, whose divided points are used as the temporary target positions, i.e., sub-target positions.The controlled object moves to the nearest sub-target position, and removes to the next sub-target position if the current state would be converged.

6. Simulations

In this section, some simulations are conducted to verify the effectiveness of the kinodynamic control for the two-wheeled mobile robot. For the simulations, the MATLAB and PDE Toolbox (one of the toolbox for MATLAB) is used as the calculation tool. The conditions and the results of simulations are described below.

6.1. Conditions

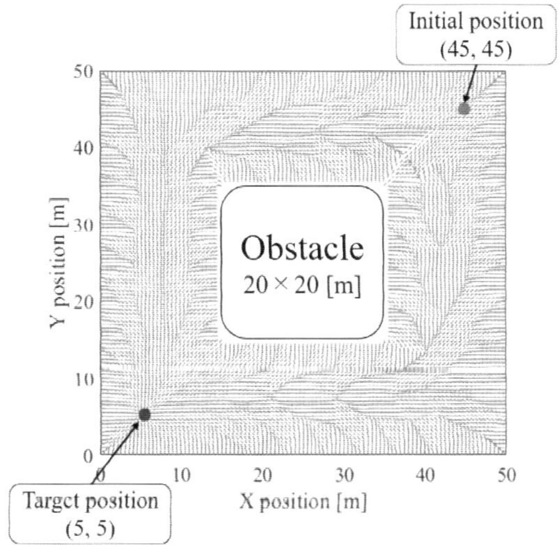

Figure 2. An assumed environment

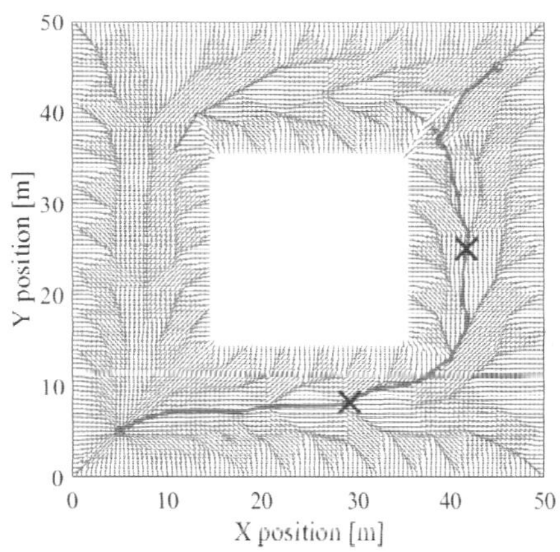

Figure 3. An ideal trajectory

In this simulation, it is confirmed that the initial state of the robot $\boldsymbol{x_0} = [x_0 y_0 \theta_0 v_0 \omega_0]^T$ can be converges to the target value $\boldsymbol{x_d} = [x_d y_d \theta_d v_d \omega_d]^T$ by using the quasi-continuous exponential stabilization control and the HPF in the environment shonw in Fig. 2.The physical parameters of the robot are set as M= 5 [kg], I = 1 [kg/m^2], $R = 0.03$ [m], and $L = 0.06$ [m], whereas the sampling interval and the threshold value are set to $\Delta t = 0.01$ [s] and $\epsilon = 0.01$, respectively.The feedback gains are set to k = 1 and f = 9, and Eq. (10) is used as the control input.In the assumed environment, there are one rectangle obstacle shown in Fig. 2.Here, initial state $\boldsymbol{x_0} = [45\ 45\ 1.5\ 0\ 0]^T$ and the target value $\boldsymbol{x_d} = [5\ 5\ 0\ 0\ 0]^T$. Note from the Eq. (10) that, the gain k_1 is set to 1.The red line in Fig. 3 shows the ideal trajectory which is calculated from the gradient of the HPF. The dividing number is

3, and the divided points are shown by blue crosses in Fig. 3.

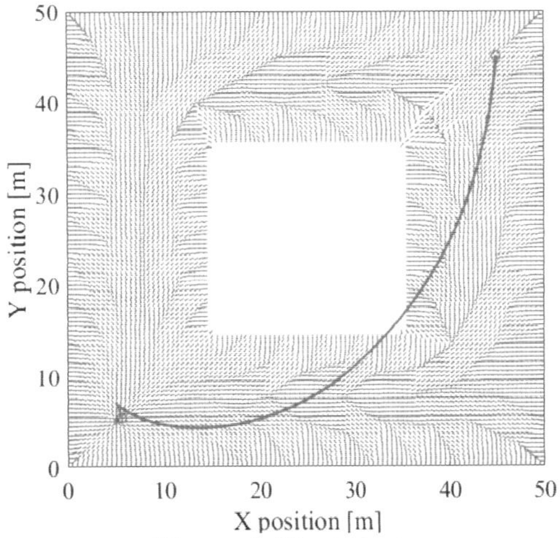

Figure 4. The trajectory with the conventional method

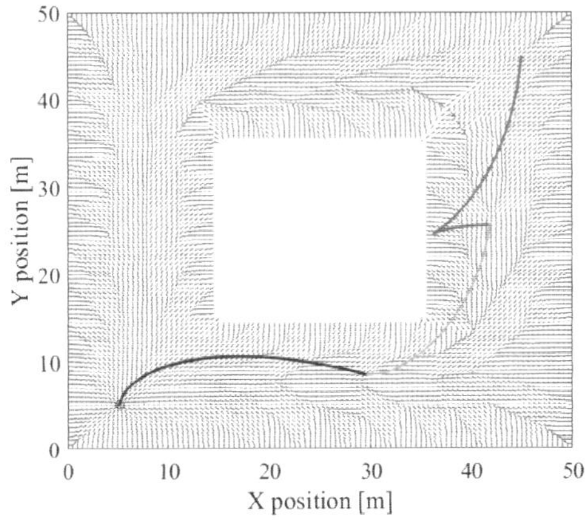

Figure 5. The trajectory with the proposed method

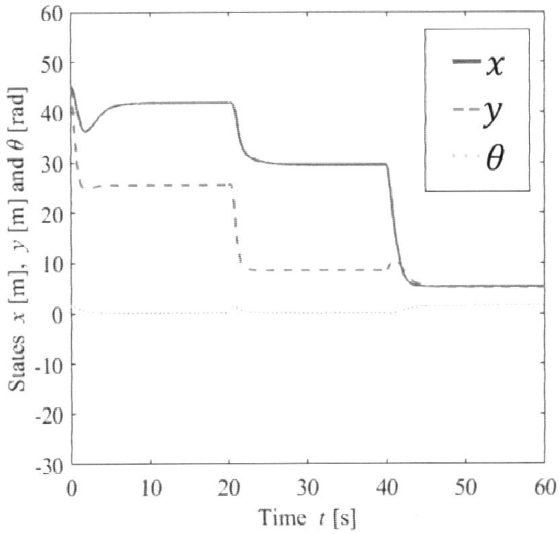

Figure 6. The trajectory with the conventional method

6.2.Results

Figures 4 and 5 show the trajectories of the robot in the X-Y plane with the conventional method and the proposed method, respectively. In Fig. 5, the red solid line shows the trajectory connected from the initial state to the point where the robot reaches to the first sub-target position. The yellow solid line is similarly the trajectory connected from the first sub-target position to the second sub-target position, and the blue line is one connected from the second sub-target position to the final target position.Figure 6 shows the time response of the states x, y, and θ in the robot when using the proposed method.

6.3.Discussions

As shown in Fig. 4, the robot was not able to avoid the obstacle in the assumed situation when using the conventional method.On the other hand, it is confirmed from Fig. 5 that the robot can be converged to the target state while avoiding the obstacle using the proposed method, which is of having introduced some sub-target positions generated by dividing the ideal trajectory.It is found in Fig. 5 that the robot takes also a switchback route to adjust its orientation. Moreover, as shown in Fig. 6, it is confirmed that the position states x and y in the robot are finally converged to the target state (5, 5) [m].Thus, the usability of the proposed method is confirmed in this simulation.However, the number of divisions for the ideal trajectory, i.e., the number of sub-targets is decided using the empirical rule by a human, so that it needs to formulate it using any rule.

7. Conclusions

In this paper, a control method using the quasi-continuous exponential stabilization control and the HPF has been proposed for the two-wheeled drive mobile robot. In the proposed method, the control input was designed to reach the target point while avoiding obstacles by combining the quasi-continuous exponential stabilization control with the invariant manifold and the directional force in gradient of the HPF. It was confirmed in the simulation that the robot was able to reach the target position by approaching step by step to the final target position.

References

[1] Donald, B., Xavier, P., Canny, J., & Reif, J. (1993). Kinodynamic motion planning: *Journal of the ACM, 40(5)*, 1048-1066.

[2] Masoud, A., (2010). Kinodynamic motion planning: *IEEE Robotics & AutomationMagazine, 17(1)*, 85-99.

[3] Motonaka, K., Watanabe, K., & Maeyama, S., (2013). Kinodynamic motion planning and control using anisotropic damping forces: In *Proceedings ofInternational Symposium on Artificial Life and Robotics*, (pp. 57–60). Oita, Japan.

[4] Tsiotras, P., Corless, M., & Longuski, J., M. (1993). Invariant manifold techniquesfor attitude control of symmetric spacecraft: In *Proceedings of 32th Conference on Decision and Control* (pp. 1470–1475).

[5] Goto, T., Watanabe, K., Maeyama, S., & Motonaka, K. (2014). Stabilization control to the arbitrary target point for a two-wheeled drive mobile robot considering the dynamical model: In *Proceedings of SICE System and Information Division Annual Conference* (pp. 599-604). (*in Japanese)*.

[6] Watanabe, K., Yamamoto, T., Izumi, K., & Maeyama, S. (2010) Underactuatedcontrol for nonholonomic mobile robots by using double integrator model and invariant manifold theory: In *Proceedings of IEEE/RSJ International Conference on Intelligent Robots and Systems (IROS)* (pp. 2862-2867).

[7] Motonaka, K., Goto, T., Watanabe, K., & Maeyama, S. (2016). Kinodynamic motion planning for a two-drive mobile robot using a harmonic potential field: In Proceedings of Annual Conference of the IEEE Industrial Electronics Society.

Kinematics Analysis of End Effector for Carrier Robot of Feeding Broiler Chicken System

Rafiuddin Syam[1], Hairul Arsyad[1], Ruslan Bauna[1], Ilyas Renreng[1], Syaiful Bakhri[2]

[1]Mechanical Engineering Department, Faculty of Engineering, Hasanuddin University, Indonesia
[2]Center for Nuclear Reactor Technology and Safety – National Nuclear Energy Agency- BATAN, Indonesia

email: rafiuddinsyam@gmail.com

Abstract. The demand for commodities, especially Broiler chicken farms are increasing, the volume of feed requirements Broiler chickens increased with age up to the age of 30-57 days required feed 3,829 grams /day/head, so if the chicken population is 3,000 needed transporting feed 11 487 kg/day, This research aims to produce a robot capable of transporting feed in the top of the cage by using a control system so as to make efficient use of manpower. Design robot performed using software design three-dimensional Solidworks2010, process of making the robot is started with the design manufacture three (3) units of mechanical systems (mechanical system for holder feed, mechanical systems for lifter feed and mechanical systems for transporting feed), then do the design process framework as a component buffer so that the mechanical system will work properly and safely when the robot operates. Furthermore, the manufacture of electronic circuits and control are using Arduino Mega microcontroller. After assembling all components mechanical systems and installation of electronic systems and control, then experiments to evaluate the performance of the robot have been made. The results of experiments showed that all components work well according to plan, in particular the speed and acceleration of end effector motion so it can hold and release the feed well. This strongly supports the robots perform tasks in accordance with the intent, i.e., holding, lifting and moving feed.

Keywords: End effector, Mechanical System for Broilers System Arduino, control design

1. Introduction

Demand for livestock commodities tends to increase along with increasing income, population, and public awareness to improve nutrition. Livestock commodities containing high protein nutrients in 1993 were set to be consumed 10 grams per day, 4 grams of livestock origin and 6 grams of fish origin. Nevertheless, the survey results show that the new Indonesian population can consume as much as 7.60 grams.

The Broiler is a fast-growing race chicken that can produce meat in a relatively short time (5-7 weeks). Broiler has an important role as a source of animal protein from livestock. To maintain the health of cattle is needed vaccine at a certain age. Vaccination is an effective way to prevent humans and animals from some infectious diseases [1].

Body weight of broiler chickens is determined at 7 to 28 days and feeding is recorded periodically to calculate the ratio of feed conversion (FCR) (Liu et al., 2016). Feed quantity is classified into four age groups are: week 5 (age 30-36 days) 111 grams / day / cage, week 6 (age 37-43 days) 129 grams /day/ cage, week 7 (age 44-50 days) 146 grams / day /chicken and week 8 (age 51-57 days) 161 grams /day/ chicken. So the total amount of feed per tail at age 30-57 days is 3.829 gram /days/chicken [2].

Ammonia concentrations of ambient air ambient lows are achieved in enclosures mixed with husk, shell charcoal, zeolite, and soil. In the cage, this place looks drier and not damp base. Probably the ability of litter to absorb dirt, liquids, and ammonia is better than other treatments. In addition to the presence of soil, it is thought that some ammonia has been converted to nitrates by nitrifying bacteria, as seen in the following reaction [3].

Observational data of the study found that the chicken population capacity of each cage ranged from 3,000 to 5,000 heads. For a population of 3,000 chickens at the age of 30-57 days requires the transportation of feed above the cage 11487 kg. The amount of feed that must be transported above the cage during the cultivation process is still done manually by using human power. Then, the process of transporting this feed will be very helpful when using a mechanical system and automatic control.

2. Research Method

Our research begins with a design process that is divided into two main parts, namely the design of mechanical systems which then proceed with the process of designing the frame as a place of support when the mechanical system works.

2.1. Mechanical System Design

The manipulator as part of a mechanical system is a parallel platform, and the end effector is connected with several linear mechanisms [4]. Robot manipulators are designed for home services, such as for handling multiple household objects or serving dishes with mobile platforms [5]. Acceleration of manipulator arises from four sources 1. motion link, 2. Grounding motion, 3. Wheel rotation and 4. Rotation of flexible robotic manipulator rotor [6]. The design of mechanical system that we made in the manufacture of robot consists of 3 (three) units, namely: mechanical feed holder system, mechanical lifting system and feed mechanical transfer system. The manufacturing process of each can be seen as follows:

The design of an End Effector Mechanical System:

We cannot generally assure that a fixed connection unit will produce the desired end-effector velocity as desired [7]. The feed holder is a series of connecting rods that will move at relative speeds of each other whose magnitude can be determined through a kinematic speed or differential polygon system. The kinematic differential is the study of joint and stems velocity, especially the speed of the effector end. The connecting rod will form a new configuration of the manipulator according to the minimal spacing configuration between the end effector [8]. The design of the feeding mechanical system is based on the form of feed media which will be held by the robot. The feed that will be held by the robot to move is the feed still in the sack, so the design of the feed holder mechanism that is able to hold the feed in the form of a sack with dimension approach 25 cm x 55 cm x 75 cm. So as to enable to hold the feed properly it is made to design the feeding mechanism by using four (4) fingers as in figure 1 (a, b).

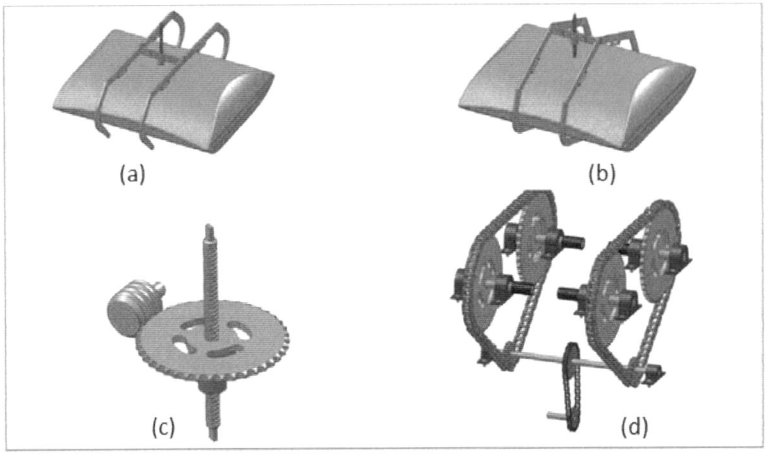

Figure 1. a) and b) Design of removing the mechanical system and feeding holder, c) Mechanical lifting system, d) Mechanical mechanical system of feed. [9]

The feed lifting process will be done by lifting the feed in a sack weighing 50 kg with vertical lift upward with a lift height of 25 cm. Based on the type of lifting process to be carried out and the amount of load to be lifted, the design of the feed lifting mechanism using bolts and nuts is selected. The position of the lifter gear is mounted horizontally so that the gear axle where mounting screw lift becomes vertical which form angle 90° to side gear lift as in picture 2c.

2.1.1. Design of Mechanic System of Feeder Transportation

To maintain the load balance during feed transfer process, the mechanical desiccant system of feed transfer uses 4 (four) wheels, each of the two wheels on the left side and two wheels on the right side. Each wheel is driven by a gear (4th gear) each of which has 40 teeth per gear. The pair of gears on each side (left and right) are driven by a moving gear (Gear 3) through a chain. The 3-valve gear mechanism of this mechanism is 2 (two) pieces driven by a driving gear (tooth 2) through a shaft. Teeth 2 is driven by a driving gear (tooth 1) through a chain. For more details can be seen in Figure 2d.

2.1.2. Frame Design

To support the mechanical system capable of moving the power optimally at the time of operation, then made a frame design for each mechanism. The basic consideration of the framework design is done based on the type of loading that will be handled and the shape of the mechanical system to be sustained.

The design of the frames we make is, of course suitable to handle loads of the three mechanical systems used. For more details can be seen in Figure 3.

2.2. Control System Design

The movement of robotic components occurs due to the process of transmitting power from the motor to a mechanical system in the form of a finite kinematic circuit. The process of power transmission that occurs causes the relative movement with a certain speed on the kinematic stem circuit that aims to give movement to the robot in accordance with its function through the control system. The variable to be controlled is the angular velocity of the component end while the control system considers the link speed angle of the manipulator. In order for the reference command in the form of point velocity to be changed to the angular velocity parameter, it must use intercellular kinematic interfaces [10].

2.3. Experimental Works

Robotic mechanical system testing procedure is divided into two groups, namely testing on each unit of a mechanical system, and testing when all mechanical systems are installed in their respective positions on the robot.

2.3.1. Testing of each mechanical system

This testing process is carried out separately to determine the performance of mechanical systems that have been made, the purpose is to find the possible deficiencies in the mechanism so that the deficiencies can be resolved wherever possible before the assembly robot intact. The testing of each mechanical system are:

In the feeding system, test the end effector capability of opening and closing at the time the motor is operating so that it can hold the feed well.

In the feed lifting system, testing the ability of whether to lift the weight of the feed or not, in this case the minimum feed mass to be lifted is 50 kg. The task of this mechanism is the most severe compared to other system mechanisms because it is tasked to change the position of feed vertically.

In the feeding transfer system, the ability of the system is going well. It is be able to move to the target well. In this experiment there are several things that must be considered include the process of transmitting the force from the teeth 1 to the 2nd and 3rd gear to the 4th gear using the chain, whether the chain rotates well and not caught when the motor is operated, so does the wheel axle position does not shift At the time the driving force is run.

2.3.2. Testing complete mechanical system on a robot.

This test is performed when the entire system of mechanisms is mounted in position, including the wheels already installed on the tracks that have been provided. The mechanical system testing is done in the robot working order during the load removal process. That is the first is to test the feed holder system to do the job well is to process the clamping (holding) the load in the sack that has been provided, after the process of holding the feed done then continued the process of feeding appointment, this process to test the mechanical system lifter installed on the robot Can lift the target load well. After the feed is lifted, then the next testing process exists.

3. Results of Experiments

From the results of the research, the design of robot feed robot Broiler chicken capable of holding, lifting and transporting feed from one place to another place that has been determined.

3.1. Direction of Motion Point on End Effector

The end effector is one part of the feeding robot that is supported by a driving mechanism. End effector in performing its function is driven a DC motor that is connected directly with screw shaft feed holder. When the motor is operating, the base end of the end effector moves up and down following the direction of motor rotation. The movement resulted in other parts of the end effector also moving in a certain direction..

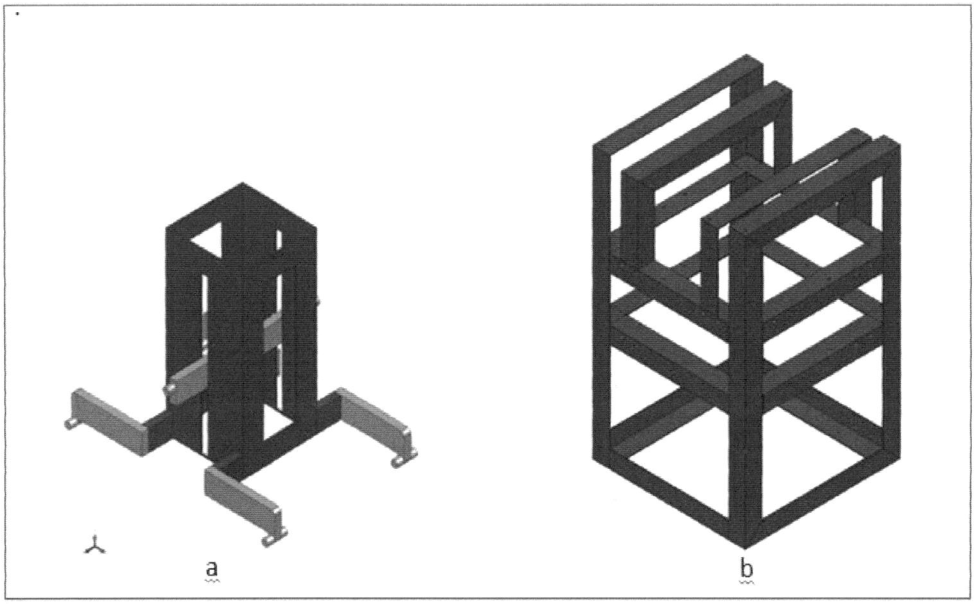

Figure 2. (a, b) Frame End Effector b) Lifting frame and feed carrier. (Source: RuslanBauna, Thesis Robot Feeder Broiler Chickens, 2016)

Figure 2 (a, b) shows the starting position as the motor rotates causing the lift nest (A1) to move upwards relative to the right position of the pin point 1, this causes the point to the left of pin 1 (B1, C1 and D1) to move Form a counter clockwise pattern with the center point of point 1, while the point centered at pin 2 (B2, C2 and D2) moves in a clockwise direction, consequently the points are on the outside of the end effector (B1-B2, C1-C2 , And D1-D2) move closer together so that both ends of the end effector move clamped, this movement is used as a feeding process as shown in figure 24, there is only one wheel in contact with the ground, and the other wheels rotate on the wheel axis In contact with the runway due to rotation

3.2. Speed of Motion Point on End Effector

DC motor data and end effector used:

Torque (Tm) = 100 kg.cm
Round (n) = 500 rpm
Pic thread (p) = 1.6 mm
A large change in the velocity of the base end effector in the path ($\Delta\omega$)= -2°.
Distance point 1-A = 138.3 mm
Distance point 1-B = 208.1 mm
Distance point 1-C = 132 mm
Distance point 1-D = 205.3 mm
Linear velocity point A (V_A) :
$V_A = n \, . \, p$ (1.6 mm)
$V_A = 500x1.6$

V_A = 800 mm/menit
From the VA value, then another point velocity value can be obtained by using the speed polygon as in figure 5c.
Corner rod speed (ω_1) :
1 rpm = 0.10467 radian/sec
V_A = 1A. 6.28.ω_1
6.28 . ω_1 = V_A /1A
ω_1 = 800 /(138.3 x 6.28)= 0.92 rpm ≈ 1 rpm
Stem angle acceleration (α_1)
The length of the end of the end effector (s) = 111.8 mm, the end effect effect (t) = 0.14 minutes = 8.4 seconds, then
$\alpha_1 = \frac{\Delta\omega}{t\,x\,\pi} = \frac{-2}{8.4\,x\,6.24}$ = -0.04 rpm (angular deceleration)
Acceleration of point A (A_A) :
$A_A = A_A{}^n \mathrel{+\!\!\rightarrow} A_A{}^t$
A_A = (1-A). $\omega_1{}^2 \mathrel{+\!\!\rightarrow}$ (1-A). α_1
A_A = 138.3x1$^2 \rightarrow$ 138.3 x (-0.04)
A_A = 138.3 $\mathrel{+\!\!\rightarrow}$ (-5.532)
A_A = 138.32 mm/min^2

Transient acceleration (AAT) is very small compared to normal acceleration (AAN). From the results of normal acceleration calculation (AAN) and tangential acceleration (AAT) can be obtained acceleration polygon as in figure 5d.

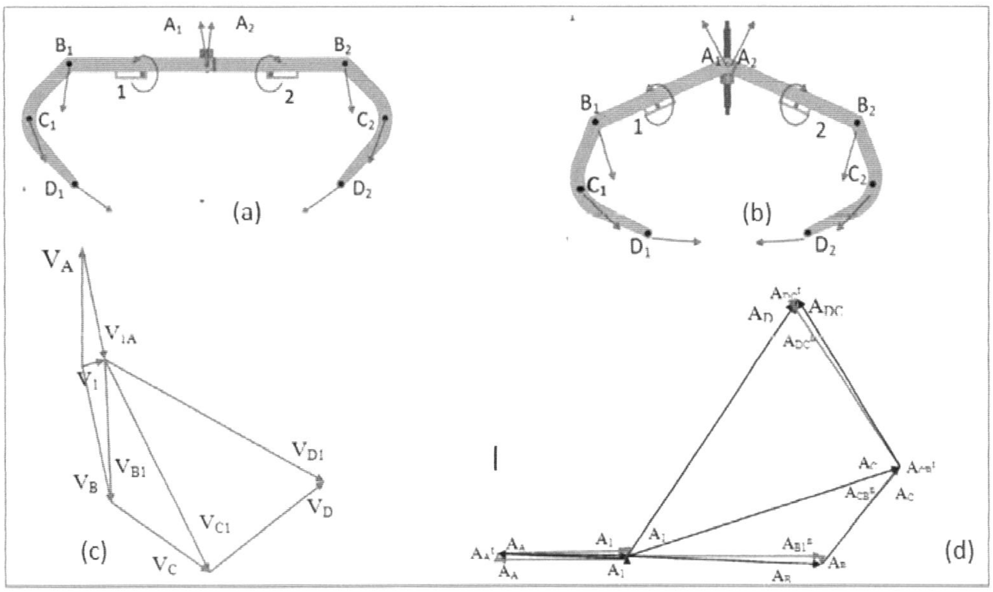

Figure 3. a) Position end effector release, b) Position end effector holding, c) Polygon speed, d) Polygon acceleration. (Source: RuslanBauna, Thesis Robot Feeder Broiler Chickens, 2016)

4. Discussion and Conclusion

From the results of the end-effector motion testing it is found that, although point 1, point A, point B, point C, point D is located on the same rod, but from polygon image perception seen no difference in velocity value. This occurs because the points have different distances to the momentary center of the stem during a movement. The farther the point from the center of the moment will result in greater speed. While the angular velocity of point 1, point A, point B, point C, and point D are the same, this occurs because all of those points lie on a rigid stem with a momentary center point (point 1).

The propeller end shaft drive source used is a DC motor with a spin (n) of 500 rpm, this means that the shaft rotation of the end-effector shaft is constant. The screw rotation of the end effector will be converted into translational motion by the nut end effector with a constant velocity, this is what causes point An on the end effector moving up and down which produces the holding motion and release the feed on the end effector pair (end effector 1 and 2). But constant point A motion is not constant with constant angular velocity (ω), where from the test results it is seen that when point A moves upwards it will cause a smaller angular velocity, and vice versa when point A moves down the angular velocity will Increase at a constant linear velocity of point A. This happens because when point An end effector move upwards will cause point A to grow away from the center of momentary point 1 so that radius (r) rotation is bigger. Because the linear velocity of point A remains while the radius of rotation is greater, so to keep up with constant velocity at the time of the increase of radius than the effect of turning angle decreases. Similarly vice versa when point A moves down, the radius of rotation is smaller so that the time required for one spin will be shorter so that the effect of angular velocity is greater.

Because at the point An end effector moves upward there is a decrease in angular velocity (ω) due to the increase of rotary radius, meaning that at that moment angular acceleration (α) is negative (slowing). While at point An on the end effector moves downward the increment of angular velocity (ω), in this case the angular acceleration (α) is positive (accelerated).

From the above acceleration polygon image shows that the tangential acceleration at each point on the end effector is very small compared to the normal acceleration. This happens because the angular acceleration (α) that occurs at the very small end of the effector end which causes the tangential acceleration also becomes very small compared to the tangential acceleration.

4.1. Conclusions and Recommendations

From the results of the above analysis and discussion can be drawn some conclusions include:

Rotational motion of the motor can be transformed into linear motion by the end-effector mechanism system. Motion from point An end effector causes rotating motion at point B, C and D which is still part of the end effector with non-constant angular velocity. Instantaneous revolving motion alternately point B, C and D on the end effector is used as the motion of holding and removing the feed by the feeding robot

To achieve more optimal results, it is advisable in planning to choose the type of frame material more suited to the type of loading task experienced, allowing the construction of robots to become lighter and stronger. For operational toughness, it is possible to select a stronger DC motor with a smaller current but still be able to operate the robot properly. This final task can still be developed again in subsequent studies.

References

[1] S. A. S. et.al, "Toward theckontrol of necrotic enteritis in broiler chickens with in-feed antibiotics

phasing-out worldwide.," *International Journal of Animal Nutrition,* vol. 1, pp. 1-11, 2016.

[2] Anonymous, "Raising of Broiler (in Bahasa) (Original: Budidaya Ayam Ras Pedaging)," BPP Teknologi, Jakarta., 2000.

[3] M. Alexander, Introduction to Soil Microbiology. Second Edition., New York: John Wiley and Sons, Inc., 1961.

[4] e. Shukla A, "Modelling Simulation & Control of6-DOF Parallel Manipulator using PID Control and Compensator.," in *Third International Conference on Advances in Control and Optimalization of Dynamical Sistems.*, New Delhi, India, 2014.

[5] H. Lim, "Design Optimization of the Robot ManipulatorBased on Global Performance Indices Using the Grey-based Taguchi Method," in *5th IFAC Symposium on Mechatronic Sistems*, Boston Cambridge., 2010.

[6] M. Korayem, A. Shafei and H. Shafei, "Dynamic modeling of nonholonomic wheel mobile manipulators with elastic joint using recursive Gibbs-Appell formulation," *International Journal of Scientia Iranica,* vol. 19, no. 4, pp. 1092-1104, 2012.

[7] P. C.D and e. al., "An Analitycal approach to Operational Space Control of Robotic Manipulators with Kinematik Constrain. Proceeding of," in *The 19th World Congress The International Federation of Automatic Control*, South Africa, 2014.

[8] A. Hentout and e. al, "Multi-agent control Approach for Autonomous Mobile Manipulatotrs: Simulation Results on RobuTER/ULM. Proceeding of the 19th World Congress," in *The International Federation of Automatic Control.*, South Africa., 2014.

[9] R. Banua, "Thesis Robot Feeder Broiler Chickens," Hasanuddin University, Makassar, Indonesa, 2016.

[10] O. Bulen, "Guidance and Kontrol of a Planar Robot Manipulator Used in a Mounting Line. Proceeding of," in *the 18th World Congress The International Federation of Automatic Control*, Milano, Italy, 2011.

The Ultimate Strength of Double Hull Oil Tanker Due to Grounding and Collision

Samuel Izaak Latumahina, Muhammad Zubair Muis Alie, Ganding Sitepu

Department of Naval Engineering, Faculty of Engineering, Hasanuddin University Makassar, South Sulawesi, Indonesia

samuellatumahina@student.unhas.ac.id

Abstract. The damaged tanker by grounding and collision may totally collapse if loss its buoyancy, stability and suffer structural failure. The objective of the present study is to investigate the ultimate strength of double hull oil tanker under vertical bending moments due to grounding and collision. The damages are modelled by removing the elements consist of stiffened and unstiffened plates from the damages part. One-frame space of the double hull oil tanker is taken to be analysed. Two damages cases are considered in the analyses those are grounding and collision. The transversal damage extent for grounding are 10%, 25%, 40% and 55%. The groundings are placed at symmetric position on the outer bottom part. For the case of collision, the vertical damage extent are taken as 10%, 20%, 40% and 60%. The transversal damages extent is taken to be B/16 and it is constant for all collision damages. The investigation of the ultimate strength is performed by the Non-Linear Finite Element Analysis method under moment control. The boundary condition is applied with fully constrained on all nodes at the aft-end, while the rigid linked on all nodes is attached at the fore-end with respect to the reference point on the neutral axis. The initial imperfection, welding residual stress and crack are not considered in the analyses. The results obtained by Non-Linear Finite Element Analyses for the ultimate strength are compared with the in-house program using Smith's method implemented in HULLST. The stress distribution and deformation for every case of damages including intact are also discussed in the present study.

1. Introduction

The grounding and collisions experienced by the ships are the two biggest contributing factors of the 12591 maritime incidents occurring in 2011-2015 [1]. The ships damaged may experience the total collapse if lose stability and buoyancy and constructional failure. Based on constructional failure, it is important to investigate the ultimate strength of bending moment of the ship. The Non-Linear Finite Element Analysis (NLFEA) method is very accurately and recommended by the classification and many researchers to calculate the ultimate strength of ships were complex structures.

Many researches have been done regarding to the ultimate strength analysis of ships damaged by grounding and collision. Jiang *et al* [2]performed finite element analysis to investigate the residual strength of ship hulls due to collisions and grounding, with comparing two models of damage. The first form of damage is a hole with adjacent plastic deformation and the second by removing the damaged frames. Luis *et al* [3] analyzed the reliability of double hull Suezmax tankers damaged by grounding. The damage is assumed to occur in the middle of the keel and simulated by removing damaged elements. Various measures of damage were analyzed and linked to reliability indices related to bilge ultimate strength. Muis-Alie [6] analyzed the residual strength of unsymmetrical damaged bilge hulls in the effect of longitudinal bending using a beam method adopted for testing of residual strength of two bulk carriers (Ship B1 and Ship B4) and a three-hold-model of a bulk carrier with a single hull Panamax type under hogging and sagging conditions. Muis-Alie *et al* [8] used Smith's method to analyze the residual strength of bulk carrier and tankers who suffered collision damage. Paik *et al* [9] developed a procedure for identifying the possibility of hull girder failure after collision and grounding based on the closed-form formulae of hull girder strength and cross-sectional modulus after damage. Muis-Alie [7] analyzed the effect of asymmetric damage on a bulk carrier type Panamax using finite-element analysis of DYTRAN.

Based on the background, as a follow-up of the Muis-Alie [8] research, this study aims to investigate the ultimate strength of damaged tanker by grounding and collisions using the NLFEA method. The results from this study will be compared with the Smith's method.

2. Method of Analysis

In the present study, The Non-Linear Finite Element Analysis is used coded by ANSYS [10]. The double hull oil tanker [8] which is 234 m in length, 44 m of breadth and 22.1 m in depth. The cross sectional and material properties of double hull oil tanker as shown in Figure 1.

Unit : N, mm

Main Dimensions
Breadth : 44000 mm
Depth : 21200 mm

Materials
Density : 7.8E-05 N/mm3
Elastic Modulus : 210000 N/mm2
Poisson Ratio : 0.3

Frame Space
Bottom : 890 mm
Hull : 890 mm
Deck : 890.9 mm

Yield Stresses
* : 290
** : 370

Stiffener No	Description	Type	Mark
1	300 x 11 + 90 x 16	Angle Bar	**
2	350 x 12 + 100 x 17	Angle Bar	*
3	400 x 11.5 + 100 x 16	Angle Bar	*
4	489 x 11 + 150 x 19	Angle Bar	**
5	619 x 12 + 150 x 19	Angle Bar	**
6	966 x 16 + 200 x 16	Angle Bar	*
7	1016 x 16 + 200 x 16	Angle Bar	**
8	1116 x 16 + 200 x 16	Angle Bar	*
9	1116 x 16 + 200 x 16	Angle Bar	**
10	1145 x 16 + 200 x 16	Angle Bar	**
11	200 x 13	Flat Bar	**
12	200 x 14	Flat Bar	**
13	306 x 11 + 125 x 16	Tee Bar	*
14	324 x 12 + 125 x 19	Tee Bar	**
15	324 x 11 + 125 x 19	Tee Bar	*
16	344 x 11 + 125 x 19	Tee Bar	**
17	367 x 11 + 150 x 22	Tee Bar	**
18	379 x 11 + 125 x 19	Tee Bar	**
19	[illegible]	Tee Bar	**
20	386 x 11 + 125 x 16	Tee Bar	*
21	412 x 11 + 150 x 22	Tee Bar	**
22	429 x 11 + 125 x 19	Tee Bar	*
23	431 x 11 + 125 x 16	Tee Bar	**
24	431 x 12 + 125 x 16	Tee Bar	**
25	452 x 11 + 150 x 22	Tee Bar	**
26	454 x 11 + 150 x 19	Tee Bar	*
27	459 x 11 + 125 x 19	Tee Bar	*
28	469 x 11 + 150 x 19	Tee Bar	*
29	472 x 11 + 150 x 22	Tee Bar	**
30	472 x 11 + 150 x 22	Tee Bar	*
31	474 x 11 + 150 x 19	Tee Bar	*
32	475 x 11 + 150 x 25	Tee Bar	**
33	485 x 11 + 150 x 30	Tee Bar	*
34	487 x 11 + 180 x 32	Tee Bar	**
35	487 x 11 + 180 x 32	Tee Bar	*
36	489 x 11 + 125 x 19	Tee Bar	*
37	492 x 12 + 125 x 25	Tee Bar	**
38	507 x 11 + 200 x 32	Tee Bar	**
39	507 x 11 + 200 x 32	Tee Bar	*
40	522 x 11 + 150 x 22	Tee Bar	**
41	722 x 14 + 150 x 22	Tee Bar	**

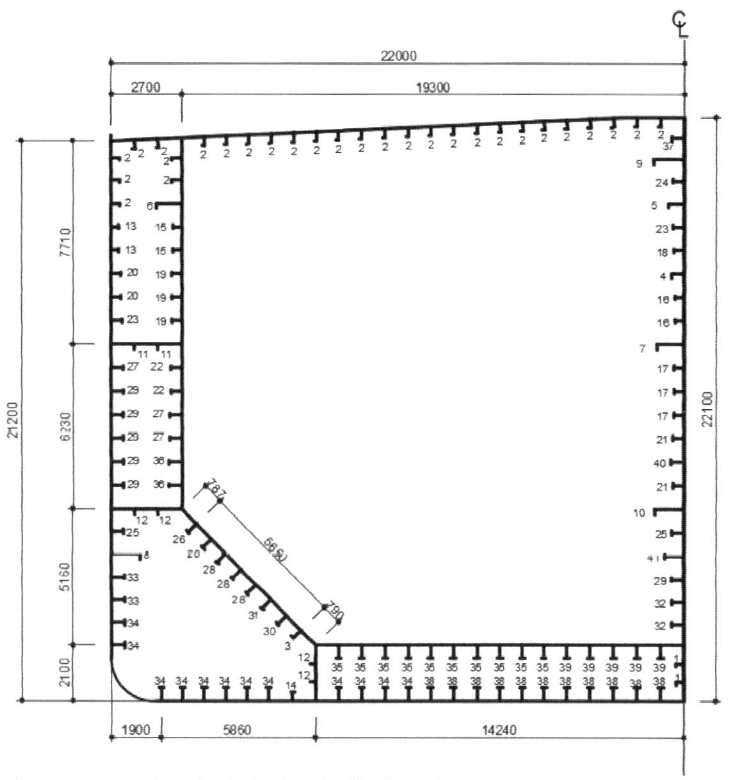

Figure 1. The cross sectional of double hull oil tanker.

The grounding and collision damages are assessed for the ultimate strength of bending moment using ANSYS [3].

The grounding and collision damages are simulated based on DNV [5]. The percentage of collision damage are taken as 10%, 20%, 40% and 60%. The transversal damage extent is considered to be B/16 and it is constant for collision damages. The element composed of plates and stiffened plates for the damage part are removed as illustrated in Figure 2.

IOP Conf. Series: Journal of Physics: Conf. Series **962** (2018) 012014 doi:10.1088/1742-6596/962/1/012014

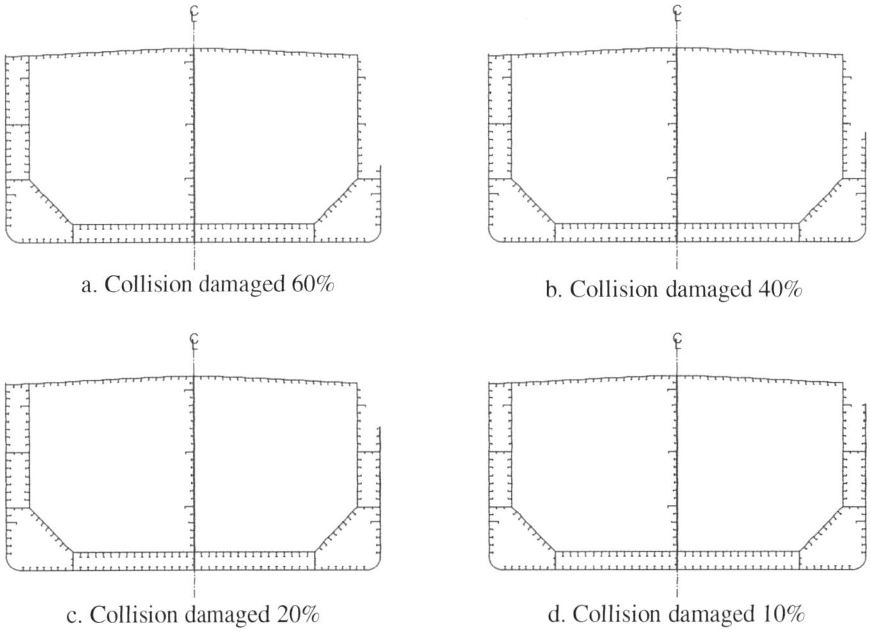

a. Collision damaged 60% b. Collision damaged 40%

c. Collision damaged 20% d. Collision damaged 10%

Figure 2. Tanker's cross sectional configuration due to collision.

For the case of grounding, the damages are taken as 10%, 25%, 40% and 55%. The damages are place at symmetric position at the outer bottom part. The vertical damage extent is 1/20 of the width of the vessel or two meters (whichever is smaller) and damage breadth by 10% of the breadth of the ship. Similarly, all the elements consists of plates and stiffened plates of the damage part are also removed as shown in Figure 3.

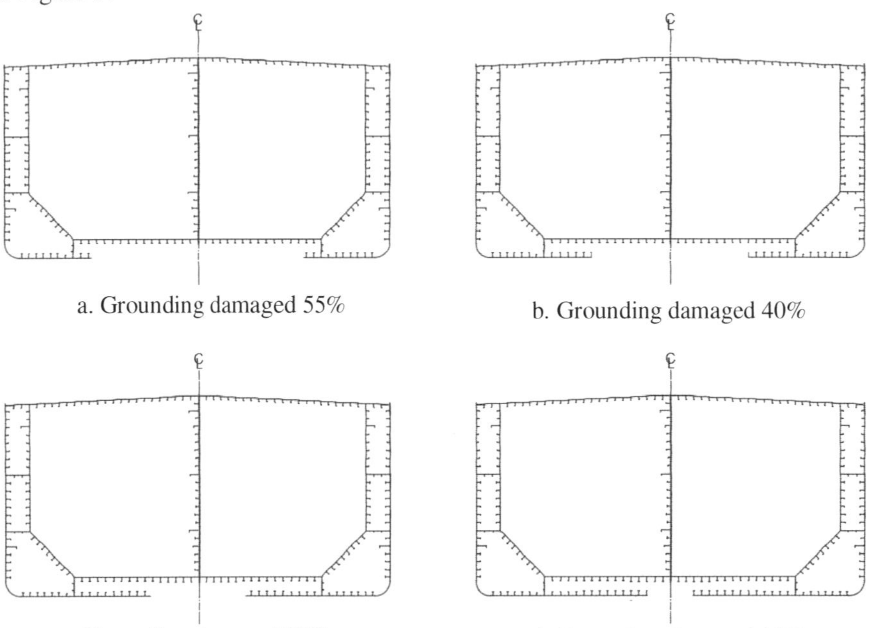

a. Grounding damaged 55% b. Grounding damaged 40%

c. Grounding damaged 25% d. Grounding damaged 10%

Figure 3. Tanker's cross sectional configuration due to grounding.

The 3D of the finite element model is created by using shell element type 181 with configuration and material specifications for each element as shown in Figure 4. After that, the material properties value for each element is included in Table 1. The size of mesh is 500 mm and it is applied for whole model.

Table 1. Material properties.

Material Properties	Type of Materials	
	AH27	AH36
Density (kg/m^3)	7850	7850
Modulus Young (N/mm^2)	210000	210000
Poisson's Ratio	0.3	0.3
Yield Strength	290	370
Tangent Modulus	625	675

In the present study, the ultimate strength is analysed under moment control process in order to know the progressive collapse of double hull oil tanker before and after damage. The boundary condition is applied to the overall end point of the back model is a full constraint. There is no loading applied at the reference point. The fore end of section model at the neutral axis reference point, the bending moment of 1.6 x 10^{13} Nmm is given. Then, all nodes at the fore-end are rigidly linked to the neutral axis reference point. The illustration of boundary condition and loading as shown in Figure 4.

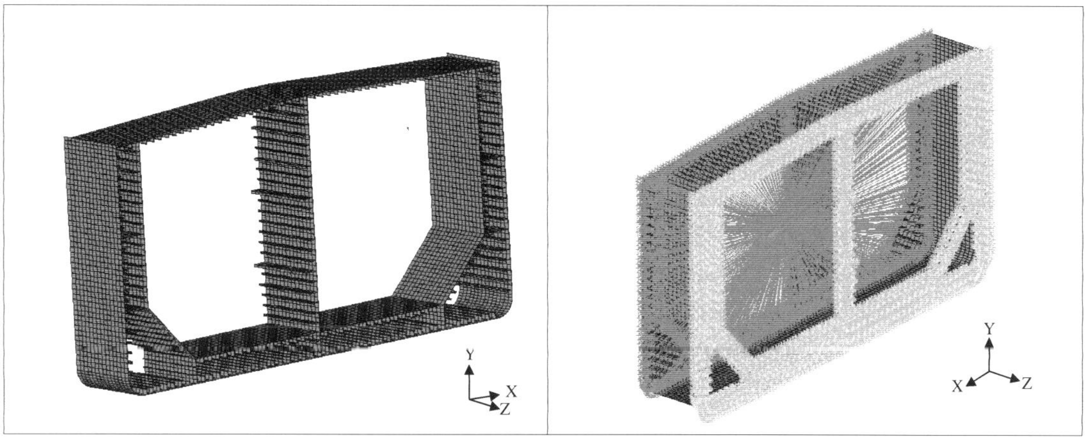

Figure 4. Modelling and boundary condition of double hull oil tanker hull girder by NLFEA.

The Non-Linear Finite Element Analysis (NLFEA) used static analysis by arc length method to get the ultimate buckling and post buckling value from the structure. The ultimate bending strength then evaluated and compared using previous research conducted by Muis-Alie *et al* (2012).

3. Results And Discussion

Table 2 shows result of the ultimate bending moment strength calculated at each percentage of damage in both sagging and hogging conditions using NLFEA. Prior to the damage, the ultimate strength of double hull oil tanker's obtained as 1.38983x10^{13} Nmm under hogging conditions and -1.17019x10^{13} Nmm under sagging conditions.

Table 2. The ultimate bending strength of double hull oil tanker (in Nmm).

Condition of tanker		Ultimate bending strength x 10^{13} Smith's Method	NLFEA	Ratio
Intact	$M_{u\text{ - Hogging}}$	1.3632	1.38983	1.0195
	$M_{u\text{ - Sagging}}$	-1.07898	-1.17019	1.0845
Collision 10%	$M_{u\text{ - Hogging}}$	1.3181	1.3879	1.053
	$M_{u\text{ - Sagging}}$	-1.0212	-1.0513	1.029
Collision 20%	$M_{u\text{ - Hogging}}$	1.2956	1.3243	1.022
	$M_{u\text{ - Sagging}}$	-0.99470	-1.04204	1.048
Collision 40%	$M_{u\text{ - Hogging}}$	1.2789	1.3214	1.033
	$M_{u\text{ - Sagging}}$	-0.97931	-0.98724	1.008
Collision 60%	$M_{u\text{ - Hogging}}$	1.2701	1.2723	1.002
	$M_{u\text{ - Sagging}}$	- 0.97285	- 0.98190	1.009
Grounding 10%	$M_{u\text{ - Hogging}}$	1.3181	1.3864	1.052
	$M_{u\text{ - Sagging}}$	-1.0672	-1.1521	1.079
Grounding 25%	$M_{u\text{ - Hogging}}$	1.2524	1.3532	1.080
	$M_{u\text{ - Sagging}}$	-1.0486	-1.1256	1.073
Grounding 40%	$M_{u\text{ - Hogging}}$	1.2025	1.2561	1.045
	$M_{u\text{ - Sagging}}$	-1.0319	-1.0635	1.031
Grounding 55%	$M_{u\text{ - Hogging}}$	1.1309	1.1563	1.022
	$M_{u\text{ - Sagging}}$	-1.0065	-1.0123	1.006

In hogging conditions, the ultimate strength of damages due to grounding is significantly decreased. This occurs because the breakdown of the element in the bottom causes the compression force on the bottom not to be massively supported by the bottom structure. At grounding with 60% of damage, the ultimate strength decreases from 16.8% to 1.1563 x 10^{13} Nmm, whereas at 55% collision damage, the ultimate bending strength decreases from 8.46% with 1.2723 x 10^{13} Nmm.

In the sagging condition, the greatest the ultimate strength occurs under the collision damage condition. The ship wall damages due to collision resulted neutral axis position becomes not symmetrical. Unsymmetrical neutral axis makes the distance of the point of damage to the neutral axis becomes greater. This causes the working stress on the damaged part becomes greater. At grounding 60%, there was a decrease of the ultimate strength value of 13.49% to -1.0123x10^{13} Nmm whereas at collision 55%, there was a decrease of the ultimate bending strength value of 16.09% to 0.9819x10^{13} Nmm.

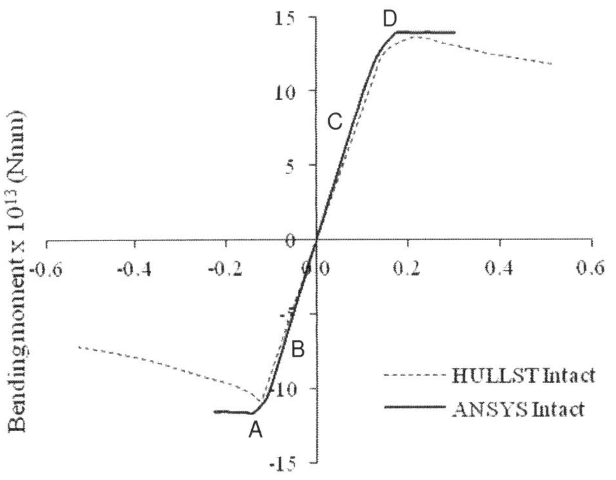

Figure 5. Moment curvature in intact condition

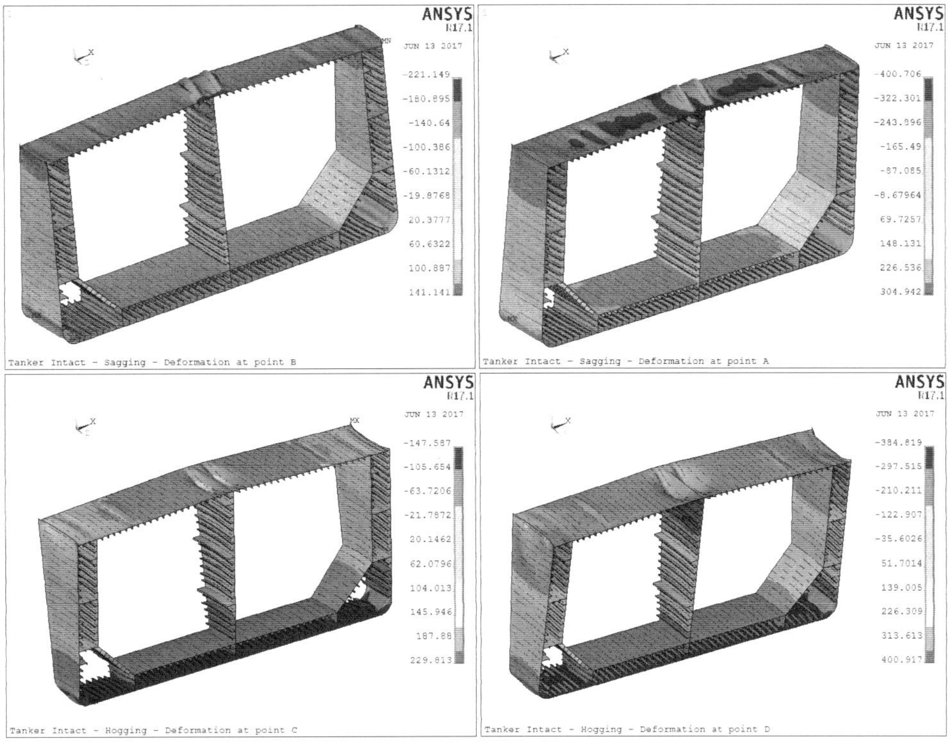

Figure 6. The behaviour and stress distribution of intact tanker structure at the ultimate bending strength in sagging and hogging condition.

Figure 5 shows the comparison of the ultimate strength obtained by Smith's method (HULLST) and NLFEA methods in intact condition. The ultimate strength obtained by NLFEA is greater as 1.95% for hogging and 8.45% in sagging conditions rather than the Smith's method. The stress distribution at the ultimate bending strength of double hull oil tanker for intact is showed in Figure 6.

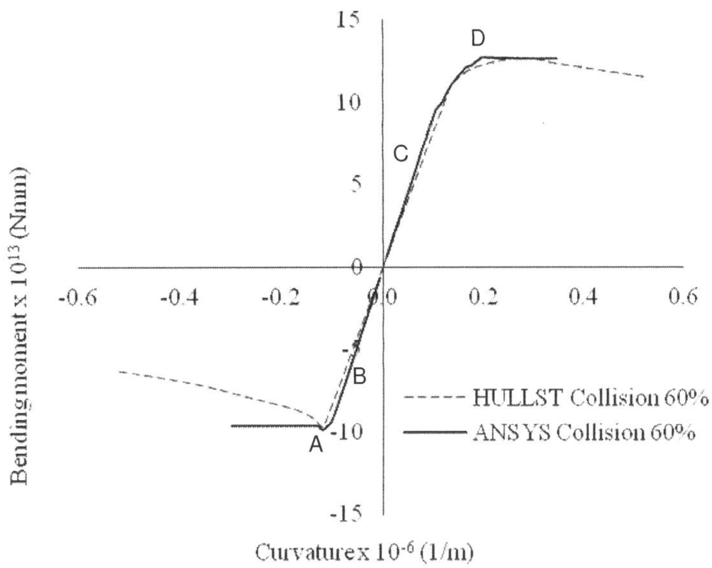

Figure 7. Moment curvature in collision 60% condition.

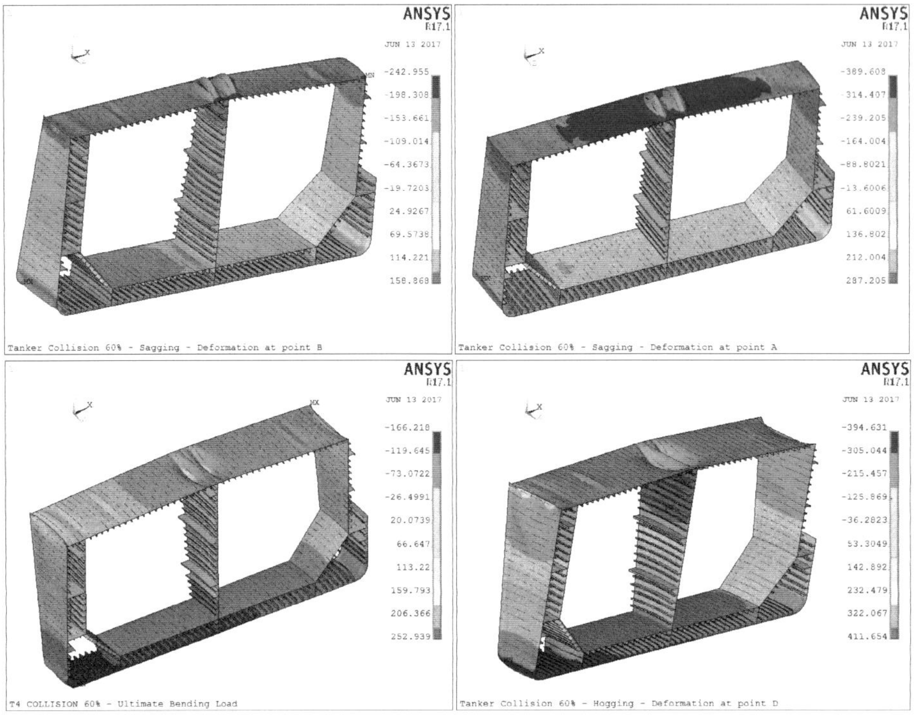

Figure 8. The behaviour and stress distribution of collision tanker structure at the ultimate bending strength in sagging and hogging condition.

Figure 7 shows the comparison of the ultimate strength obtained by Smith's method and NLFEA methods at 60% collision damage condition, the ultimate strength obtained by NLFEA is greater 0.2% in hogging condition and 0.9% in sagging condition than Smith's method. The stress distribution at the ultimate bending strength of double hull oil tanker for collision tanker is shown in Figure 8.

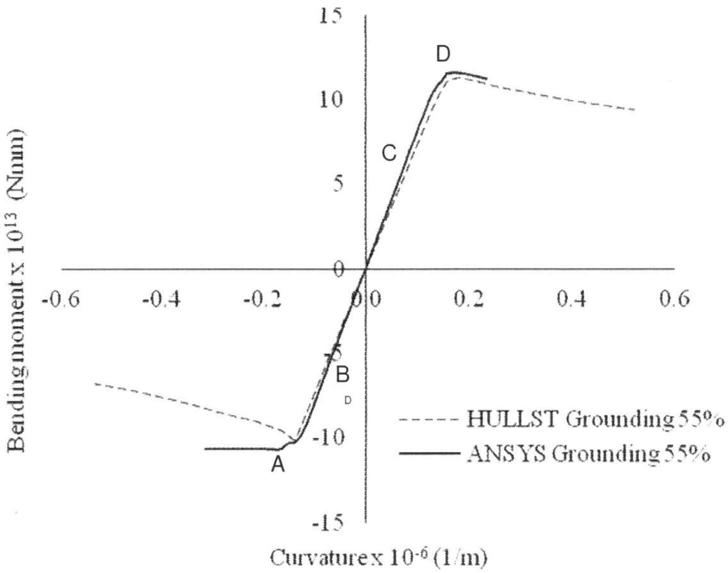

Figure 9. Moment curvature in grounding 55% condition.

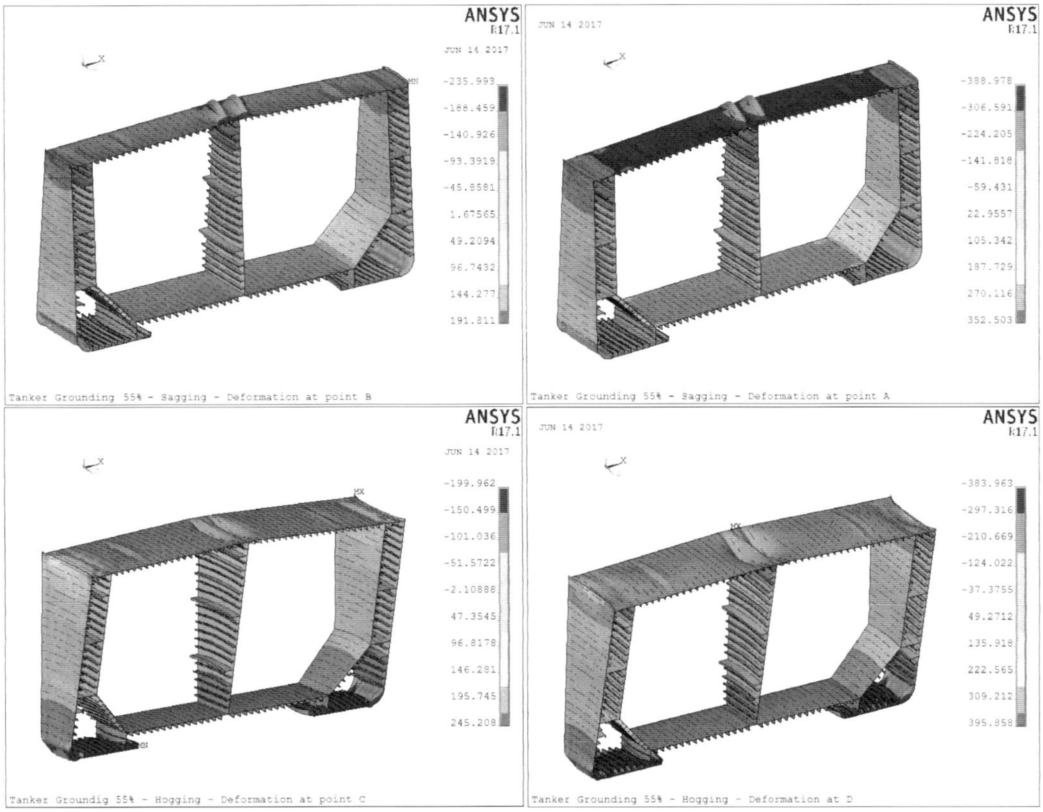

Figure 10. The behaviour and stress distribution of grounding tanker structure at the ultimate bending strength in sagging and hogging condition.

Figure 9 shows the comparison of the ultimate strength obtained by Smith's method and NLFEA methods for 55% damage conditions, the ultimate strength obtained by NLFEA is greater 2.2% in hogging conditions and 0.6% in sagging conditions than in the Smith's method. The stress distribution at the ultimate bending strength of double hull oil tanker for grounding is illustrated in Figure 10.

Figures 11 and 12 show the accumulation of the ultimate strength of bending moment-curvature relationship for damages conditions due to grounding and collision for hogging and sagging. The ultimate strength of double hull oil tanker decreased by the increasing the percentage of damage. The ultimate strength of double hull oil tanker is larger in hogging conditions because the bottom has a massive structural configuration that can withstand compressive forces better than structural configurations in the deck part. After the structure reaches the peak of the ultimate strength, the trend of the moment-curvature relationship tends to decrease until it reaches the point of collapse because the structure has become plastic, so that when buckling takes place the structure cannot return to its original shape.

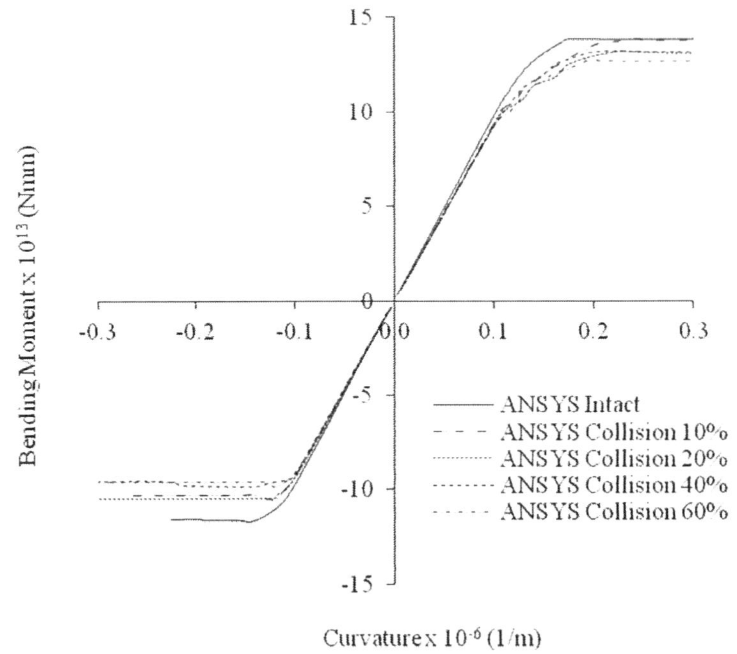

Figure 11. Moment curvature in intact and collision tanker.

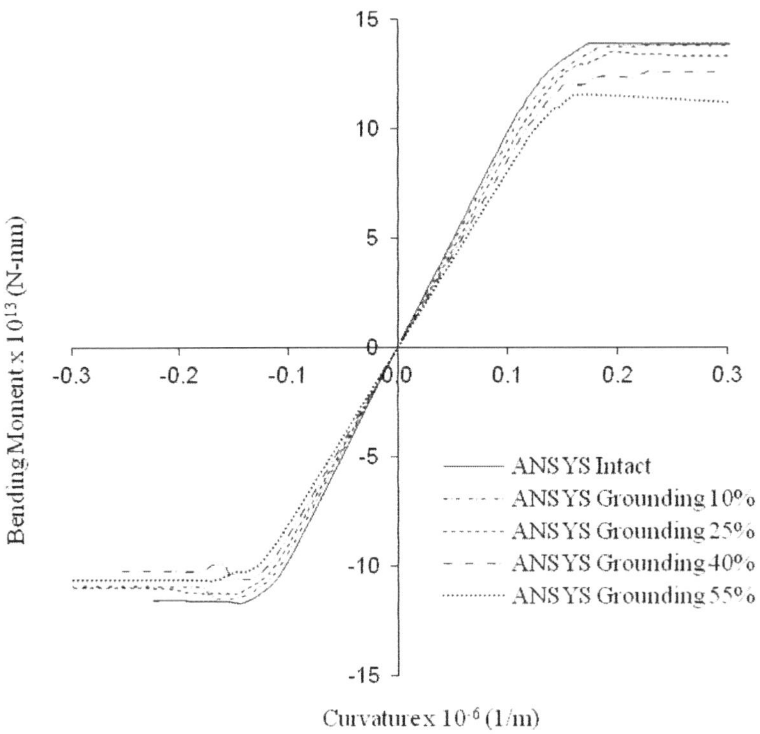

Figure 12. Moment curvature in intact and grounding tanker.

4. Conclusions

The Nonlinear Finite Element Analysis (NLFEA) method to investigate the ultimate bending strength of a double hull oil tanker due to grounding and collision under vertical bending moment is conducted in the present study. The following conclusions can be drawn:

a. The increased of damaged percentage of the hull girder effects on the increase of working stress and decreased for the ultimate bending strength of the hull girder.
b. Damage caused by grounding and collision indicates that the structure still has a bending moment strength greater than the external bending moment. Based on this, the ship is totally collapsed on 55% and 60% of grounding and collision, respectively.
c. The result obtained by NLFEA for the ultimate strength is in good agreement with the Smith's method implemented in HULLST.

The behaviour of double hull oil tanker obtained for this research can be developed for future works such as: ship strength analysis due to weld cracking at critical part in ship structure, analysis of ship's residual strength due to grounding and collision and ship strength analysis due to torsion load and load combination.

References

[1]. EMSA 2016 Annual Overview of Marine Casualties and Incidents 2016.
[2]. Jiang X H Yu and M L Kaminski 2014 *Assessment Of Residual Ultimate Hull Girder Strength Of Damaged Ships* Proc. of the International Conference on Offshore Mechanics and Arctic Engineering – OMAE
[3]. Luis R M *et al* 2009 *Longitudinal Strength Reliability of a Tanker Hull Accidentally Grounded* Journal Marine Structures Vol 31 p 224-233
[4]. Hughes O F dan Paik J K 2010 *Ship Structural Analysis and Design* The Society of Naval Architects and Marine Engineers – SNAME (New Jersey)
[5]. DNV GL 2017 Rules for Classification: Ships (Norway)
[6]. Muis Alie M Z 2016 *Residual Strength Analysis of Asymmetrically Damaged Ship Hull Girder using Beam Finite Element Method* Makara J. Technol. 20/1 p 7-12
[7]. Muis Alie M Z 2016 *Finite Element Analysis On The Hull Girder Ultimate Strength Of Asymmetrically Damages Ships* Proc. of the ASME 2016 35th International Conference on Ocean and Artic Engineering, OMAE.
[8]. Muis Alie M Z *et al* 2012 *Residual Longitudinal Strength Analysis of Ship's Hull Girder with Damages* Proc. 22nd International Offshore and Polar Engineering Conference ISOPE (Rhodes, Greece) Vol 4. p 831-838
[9]. Paik J K *et al* 1998 *Residual Strength Assessment of Ships after Collision and Grounding* Journal Marine Technology Vol 35 p 38-54
[10]. Suman Kar D G Sarangdhar and G S Chopra 2008 *Analysis Of Ship Structures Using ANSYS* SeaTech Solutions International (S) Pte Ltd

IOP Conf. Series: Journal of Physics: Conf. Series **962** (2018) 012015 doi:10.1088/1742-6596/962/1/012015

Development of autonomous controller system of high speed UAV from simulation to ready to fly condition

Herma Yudhi Irwanto
Indonesian National Institute of Aeronautics and Space
Jl. Pemuda, No. 1, Jakarta Timur, DKI Jakarta, Indonesia
Email: hyirwant@gmail.com

Abstract. The development of autonomous controller system that is specially used in our high speed UAV, it's call RKX-200EDF/TJ controlled vehicle needs to be continued as a step to mastery and to developt control system of LAPAN's satellite launching rocket. The weakness of the existing control system in this high speed UAV needs to be repaired and replaced using the autonomous controller system. Conversion steps for ready-to-fly system involved controlling X tail fin, adjusting auto take off procedure by adding X axis sensor, procedure of way points reading and process of measuring distance and heading to the nearest way point, developing user-friendly ground station, and adding tools for safety landing. The development of this autonomous controller system also covered a real flying test in Pandanwangi, Lumajang in November 2016. Unfortunately, the flying test was not successful because the booster rocket was blown right after burning. However, the system could record the event and demonstrated that the controller system had worked according to plan.

1. Introduction

1.1. Background
The development of high speed UAV since 2013 by LAPAN, called RKX-200TJ, developed from the laboratory phase, flight tested using pneumatic launcher and using EDF (electric ducted fan) as engine, and also fly manually by the pilot at the start of the flight using a rocket booster, has been well implemented. And the next is how to seek the function of auto take off system as a replacement pilot for the initial condition of flying. So hopefully with 1 button, RKX-200TJ can fly directly to auto pilot phase by way points. It's just that there are less references related to high speed UAV, booster rocket and auto take off [1].

It is not an easy task to develop an autonomous controller system that consists of hardware and software for high speed UAV, a vehicle designed to fly with high speed of 250 km/h, launched by a booster rocket, equipped with auto take off system and auto pilot by way points. Although there are a bunch of systems produced in the market, it would be more appropriate if a development is done specifically for high speed UAV, which has different characteristics than other vehicles have. Moreover, it can be done as a step to mastery and further development of control system of LAPAN's satellite launching rocket.

1.2. Problem Identification
The controller system like ArduPilot [2], which had been considered successfully implemented in high speed UAV, in its test was reset within speed of 280 km/h, so a development of autonomous controller system was urgently needed. In the long term road map of rocket controller system development, speed will be improved step by step until 600 – 700 km/h. Thus, a research related to mastery in controlling technology for the high sub sonic vehicle needs to be done. The development step of this research, starts using hardware in the loop simulation (HILS) and referring to the standard flight procedures of the available systems on the market, needs major adjustment to make the real flying test system, especially for the purpose of flight safety since take off until safe landing. This research will

discuss more specifically the things related to migration from the simulation system to the system ready for flight test of high speed UAV.

1.3. Research scope and significance
Technology mastery of autonomous controller system is a complex research because the system needs to solve the problems of a vehicle's flight since it takes off until it lands. Because of that, this research is done as:

- Mastery in technology of control for high speed vehicles.
- Mastery in development of controller system in details, starting from simulation until readiness to fly.
- Embryo of control system for the LAPAN's satellite launching rocket.

2. Methodology

This research is a continuation after the simulation stage is considered complete and a system for flight test needs to be realized.

2.1. Ready system of simulation stage
For the moment, flight simulation for high speed UAV uses HILS, until the auto take off and auto pilot by way points stage is considered done [1,3]. In this simulation, even a real IMU is successfully added and located on top of hexapod and controlled by close loop PID replacing data-data attitude coming from flight simulator XPlane.

2.2. ArduPilot as a reference for procedure and setting of a vehicle's flight
The development of autonomous controllers system for high speed UAV adopts much ArduPilot system that has been successfully implemented as auto take off and auto pilot by way points system in a real flight test.

2.3. Implementation of autonomous controller system
The development of the system is done by combining several literatures and adjustment to specification and performance of the hardware micro controller in use, so the system can handle all flying process from take-off to the last way point target.

3. Results and Discussion

3.1. Last status of the simulation system
The development of the HILS version autonomous controller system has completed several steps, which mark the base before being used in the real flight test. The steps are:

3.1.1. Designing high speed UAV vehicle that is used in the simulation. As has been discussed in previous papers [3,4], the design of high speed UAV frame refers to the design of F18. For this simulation, the frame of the vehicle is created using assistance from XPlane Maker, as seen in figure 1. All specifications of this vehicle are included in the setting of dimensions fuselage, wing, control geometry, load and balance, engine specification and landing gear. Even, the software can also create JATO (Jet Assisted Take Off) setting as a replacement for booster simulation.

Figure 1. Design frame high speed UAV

3.1.2. Networking all sub computers involved in HILS (hardware in the loop simulation). To simulate the entire flight system of high speed UAV, it needs 3 computers and their own main micro controllers. They consist of dynamic computer for controlling XPlane flight simulator, monitoring computer for controlling hexapod and functions as ground station, and controller computer as connection to main micro controller [5]. The three computers are connected with each other using UDP, while they connect to the micro controller using shared memory, because it uses the same software, labview. HILS diagram block is seen in figure 2.

3.1.3. Implementation of MyRIO as the main micro controller of the autonomous controller system.

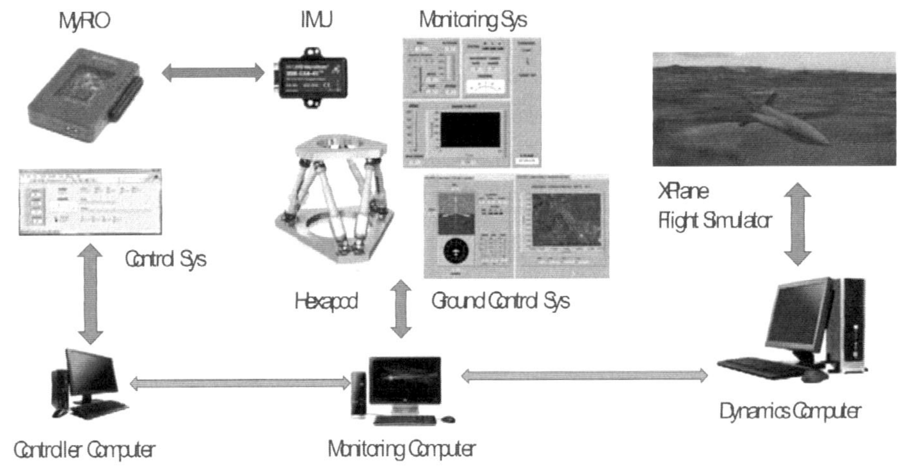

Figure 2. Hardware in the loop simulation

The selected main micro controller of high speed UAV is NI MyRIO [6,7], regarding its ability in providing all input and output as standard peripheral of flying vehicle in the form of simulation in auto take off system and auto pilot by way points. Figure 3 shows a ready-to-fly system that is used both in simulation and flight test.

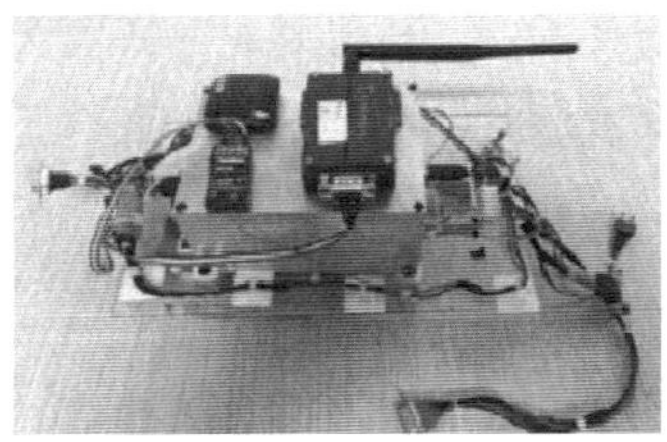

Figure 3. Main controller RKX-200TJ

3.1.4 Integration of IMU (inertial measurement unit) and hexapod in switching attitude data from XPlane. The next step that has been completed in the simulation is making IMU placed on top of hexapod as the replacement of data attitude, which had been supported by XPlane. This ensures that IMU has been feasible and ready to fly (Figure 4).

3.2. Ready system for flight test

After all simulation system of auto take off by booster rocket using JATO and auto pilot by way points is considered sufficient, it is time to prepare a system for the real flight test. The realization of transfer module from simulation to the flight system is not simple. There are many components to prepare.

3.2.1. Control X Tail Fin Actuator. It needs a function that can translate the motion field condition of aileron and elevator into the movement of X Tail Fin Actuator. In the flight simulation using bank to turn control method, the function of rudder is almost negligible (too minor). For that reason, it needs only data of aileron and elevator to couple. Thus, the condition of control in flight simulator XPlane must be changed to number for those 4 fin tails, as seen in figure 5, so it needs a formula script that has to convert the mentioned conditions [8]. To make it simple, 2 input aileron and elevator will create a value for each of those fin actuators in the form of PWM (pulse width modulator) number, which is easy to translate to the hardware.

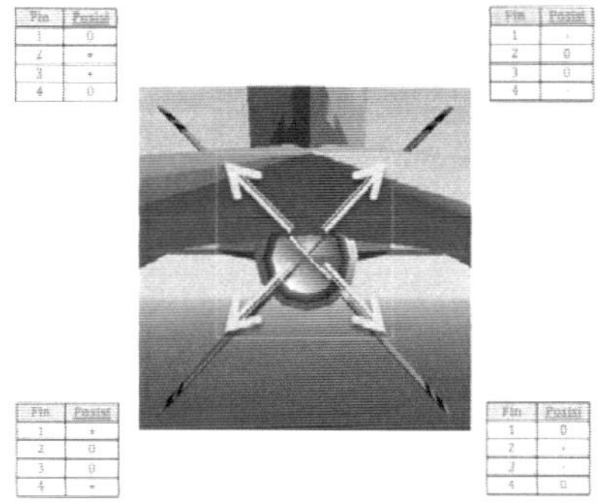

Figure 5. Actuator simulation

3.2.2. Auto Take Off System. Realization of flight procedure using JATO (jet assisted take off) in the simulation system must be replaced with the real booster rocket when take-off. In the real condition, the setting of throttle EDF before and after booster burning out significantly influences the success of auto take off system. Figure 6 shows a vehicle in simulation that is launched using booster rocket and auto take off system being developed. The ignition of EDF must be automatic, so in reality it needs additional sensor (accelerometer) to detect when the vehicle starts launching and when the burning out happens. The procedure of auto take off is very crucial, defining whether the vehicle can fly successfully of not. If it successful, the next step will be handed over to auto pilot system. Compared with ArduPilot related to the auto take off system, ArduPilot only gives option on altitude to reach in the system, while this autonomous controller system can give option on altitude and speed to reach before heading to its first way point. It means that if either altitude or speed is reached first, the vehicle will maneuver to the first way point. In auto take off

Figure 6. Auto take-off simulation

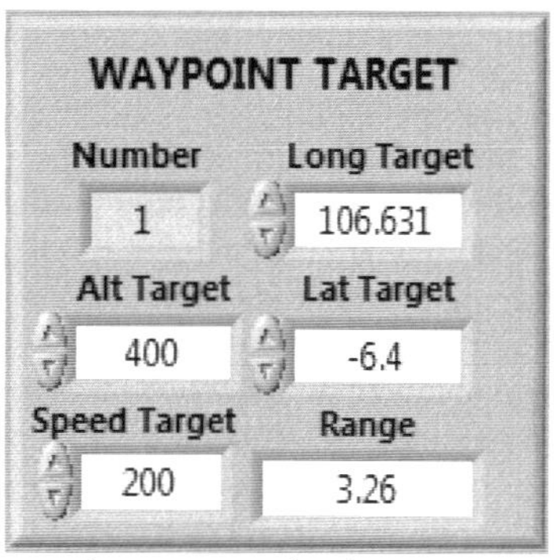

Figure 7. Monitoring target

condition, the system is maintained in anti-roll condition, meaning that the flying vehicle is kept stable without roll until it reaches either option mentioned before.

3.2.3. Loading way points. In the simulation, there is only 1 way point is set. If the vehicle can successfully fly to the way point, it is assumed that the vehicle will make through any other way points. However, the procedure of reading way points by the vehicle must be set when all system is ready to fly. The setting of way points can be done in the ground station and saved in a flash disk. From the flash disk, the way points will be loaded to the system memory. Then, the system has to detect and give report to the ground station that the vehicle is directing to way point X, with distance and heading that is automatically calculated, as seen in figure 7. The calculation of one target way point to the next way point is done based on PID control. If the vehicle has reached 100 meter to the way point target, the system will immediately calculate the next way point target. For now, the vehicle can maneuver bank to turn maximum 30 deg [9], and can detect target to reach the nearest heading.

3.2.4. Communication with Ground Station. It is not easy to communicate all activities of a flying vehicle to the ground station [10], as seen in figure 8 below. For the moment, communication of data works only for one way, i.e. from the vehicle to the ground station. Data in the ground station is supposed to demonstrate all flight condition of the vehicle, starting from attitude, position, way point target etc, in a user friendly display to make it easy in interpreting the condition and position of the vehicle within a split second. Moreover, all data coming to the ground station is saved for the purpose of flight analysis and performance improvement. The autonomous controller system is not only able to send real time data through telemetry (frequency 20Hz), but also able to save data directly into a memory flash disk at the speed of 10 times frequency of telemetry, i.e. 200Hz. So if the vehicle can land safely, the received data will be more detailed for further analysis.

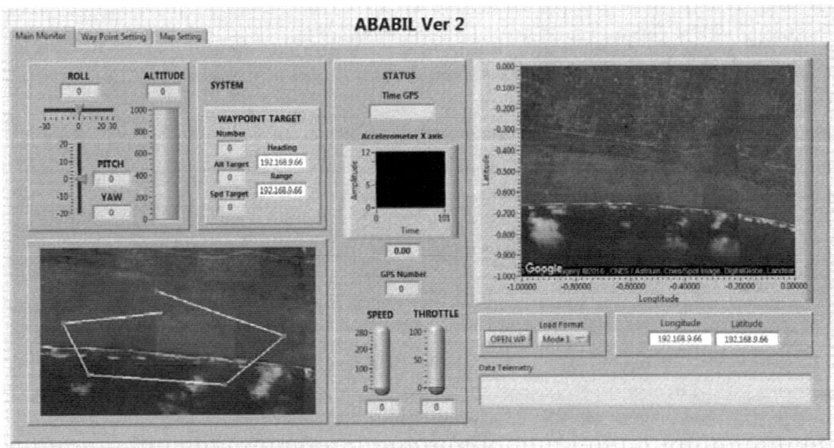

Figure 8. Ground station

3.2.5. Safety landing. One important thing in the process of preparing the autonomous controller system into a ready to fly system is safety issue. In the auto take off system, there are many procedures in engine ignition, related to the safety of booster rocket launcher. The highlight on safety landing [11] needs to be emphasized, because the autonomous controller system does not have 'manual' mode at all for a human pilot to control and the vehicle flies entirely in auto take off and auto pilot until the last set way point. Thus, automatic and manual safety is inserted in the system using remote control. The safety system has several functions, namely:

- Engine will automatically shut down when reach distance of 200 m from the last way point

This is supposed to reduce velocity when landing, as in the flight scenario, the last way point is set at altitude 20 m.

- Manual engine shut down can also be done anytime whenever the system is reached by a remote control
  One of switches in the remote control has been successfully functioned to shut down engine. The manual shut down can be used if unexpected things happen when take-off preparation or if the vehicle flies heading to dangerous condition when landing.

3.3. Flight test

Pre-flight test preparations had been done, including its real monitoring through the ground station using radio telemetry. Included in the preparation were durability test of system against G shock and distance test to make sure that plot and tracking of vehicle in the map is accurate. After that, a real flight test was done in Pandanwangi, Lumajang, East Java on November 6, 2016 (figure 9), along with series of other tests for high speed UAV. Unfortunately, the autonomous controller system did not make it in the test because its booster rocket launcher was blown (figure10). However, the procedure of take-off system went according to the planned program.

Figure 9. RKX-200EDF ready to fly

Figure 10. Booster rocket blown

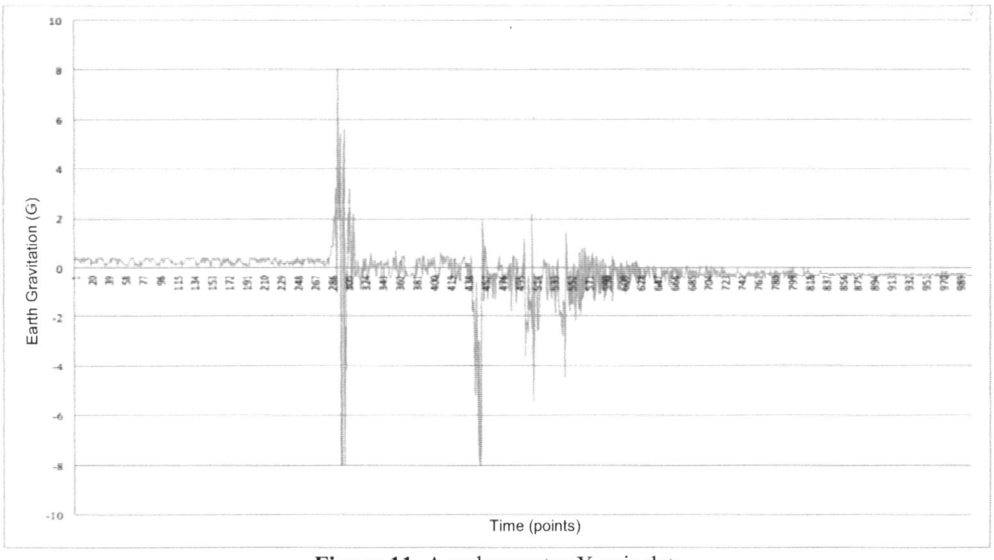

Figure 11. Accelerometer X axis data

Based on recorded data, though not too complete, there were several encouraging conditions. Firstly, the accelerometer X axis detected the initial G shock of 8G (figure 11) and it was seen on IMU pitch data that showed pitch up condition 'for a moment' and pitch down, and ended with a collision with the ground (figure 12). Moreover, throttle data also shows that it gave response to initial G shock by increasing throttle to 70% and continuing it to 90% before collision with the ground (figure 13). This proves that the procedure of EDF engine ignition went according to the plan. However, the blown booster put a stop to the entire agenda in testing this autonomous controller system.

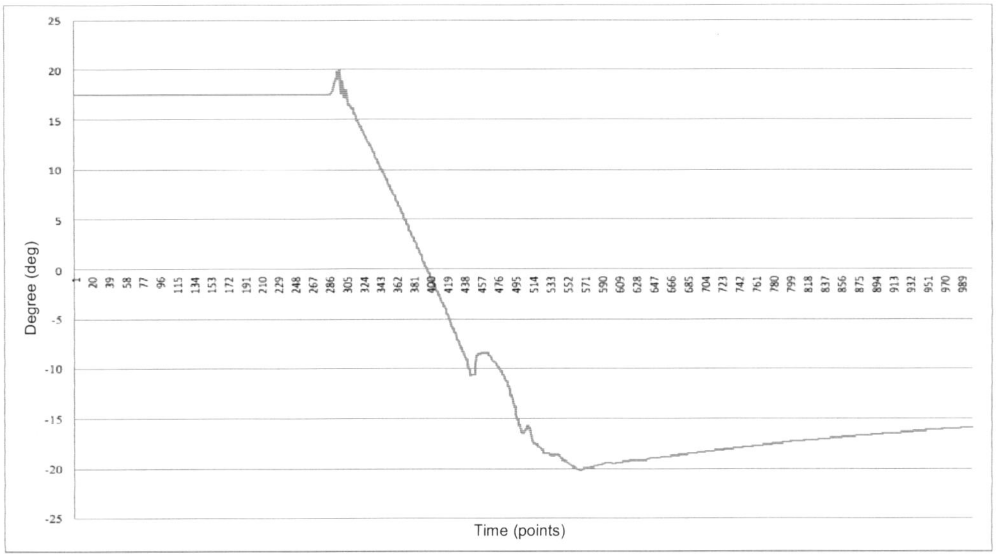

Figure 12. Pitch data

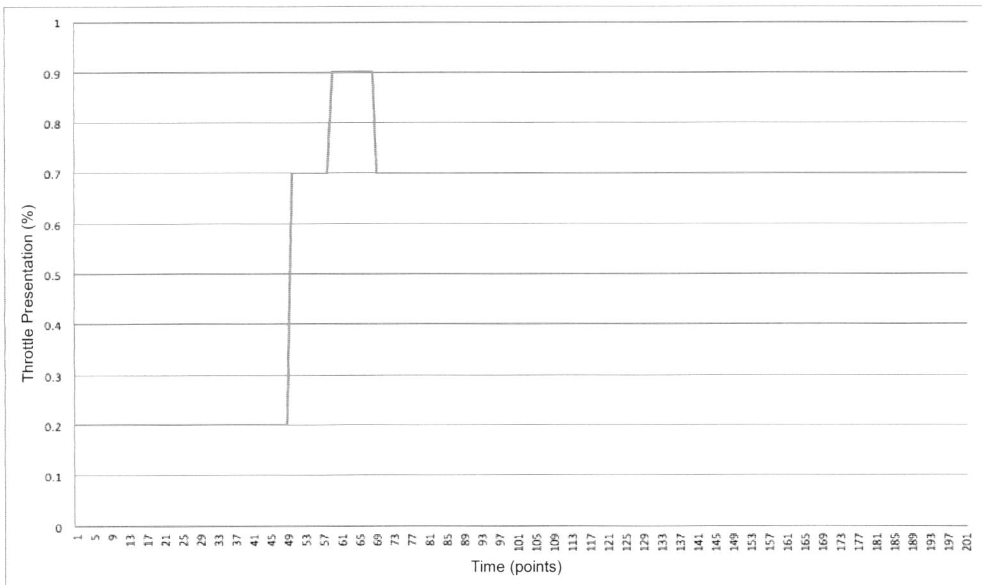

Figure 13. Throttle data

4. Conclusion and recomendation

At this time, conversion of result from simulation system needs more activities to be done in real condition, including adding sensor or new module to be used in the real flight test. This autonomous controller system is developed exclusively for our high speed UAV series. Thus, the development program may become simpler. Moreover, this autonomous controller system is easier to explore for further developments by adding new module or sensor for the purpose of performance or speed improvement. The result of high speed UAV flight test, although not completely successful because of the blown booster rocket, shows that the autonomous controller system has proper recorded data according to the plan and flight procedure implanted in the system. This should be completed in 2017 flight test.

Acknowledgements
Appreciation is addressed to Deputy of Aerospace Technology and Head of Rocket Technology Centre for facilities and support in conducting this research. Thanks to colleagues in 2016 Controller System Program for the assistance both directly and indirectly.

References

[1] Herma Yudhi Irwanto 2016 HILS of Auto Take Off System for High Speed UAV using Booster Rocket In *ISITIA 2016 Recent Trend in Intelligent Computational Technologies for Sustainable Energy* Institute Technology Sepuluh Nopember (ITS), Lombok 2016, 373-380, http://doi.org/10.1109/ISITIA.2016.7828689
[2] Bin, H., & Justice, A 2009 The design of an unmanned aerial vehicle based on the ArduPilot *Indian Journal of Science and Technology*, *2*(4), 12–15. http://doi.org/10.17485/ijst/2009/v2i4/29423
[3] Herma Yudhi Irwanto 2015 Rocket Booster Effect on RKX-200TJ Flight Stability In *Proceedings of 2015 4th International Conference on Instrumentation, Communications, Information Technology and Biomedical Engineering* Institut Technologi Bandung 2015, 130-133, http://doi.org/10.1109/ICICI-BME.2015.7401349
[4] Herma Yudhi Irwanto 2015 Integrated Simulation System for High Speed UAV In *Bunga Rampai Hasil Litbangyasa: Teknologi pada Pesawat Terbang, Roket dan Satelit*, LAPAN.
[5] Wang, L., Liu, N., Wang, J., Hu, J., & Liang, F. 2015 Design of an intelligent vehicle control system based on LabVIEW *Metallurgical and Mining Industry*, *7*(7), 21–27. http://doi.org/10.1109/iCECE.2010.1241
[6] Herma Yudhi Irwanto 2016 Development of New Autonomous Controller for Auto Take Off System of High Speed UAV *Advance in Aerospace Science and Technology in Indonesia 2016"*, 1-7, Indonesia Book Project
[7] Ribeiro, L. R., & Oliveira, N. M. F. 2010 UAV autopilot controllers test platform using Matlab/Simulink and X-Plane. In *Proceedings - Frontiers in Education Conference, FIE.* http://doi.org/10.1109/FIE.2010.5673378
[8] Jung, D., &Tsiotras, P. 2008 Bank-to-turn control for a small UAV using backstepping and parameter adaptation In *IFAC Proceedings Volumes (IFAC-PapersOnline)* (Vol. 17). http://doi.org/10.3182/20080706-5-KR-1001.1869
[9] Herma Yudhi Irwanto 2015 Complete Monitoring of Ground Control System for High Speed UAV. In *Advance Research in Material Sciences, Manufacturing, Mechanical and Mechatronic Engineering Technology International Conference.* AR4MET 2015
[10] Yanhui, L., Shuguang, Z., & Lei, G. 2011 A safety strategy for high-speed UAV landing taxiing control. In *Procedia Engineering* (Vol. 17, pp. 197–211). http://doi.org/10.1016/j.proeng.2011.10.022

A Method for Calculating the Amount of Movements to Estimate the Self-position of Manta Robots

Takuya Imahama, Keigo Watanabe, Kota Mikuriya and Isaku Nagai

Graduate School of Natural Science and Technology
Okayama University
Okayama, Japan

imahama-t@usmm.sys.okayama-u.ac.jp

Abstract. In recent years, the demand of underwater investigation is increasing in the circumference of a dam, the environmental research of the shallow where approach by ship is difficult, etc. It is known, however, that for man, all over the sea, danger exists mostly, and prolonged diving has a bad influence to a human body. Then, the development of underwater exploration robots that investigate underwater instead of humans is expected. Among underwater exploration robots, it is known that robots imitating aquatic organisms have little influence on underwater environment. Therefore, at this laboratory, a Manta robot using propulsive mechanisms with pectoral fins was developed, imitating the pectoral fin of Manta. Although underwater environmental research needs a function for estimating the self-position, it is not mounted in this Manta robot. This paper explains the amount estimation of movements using optical flows. Especially, a gimbal mechanism is introduced to reduce the influence on the optical flow calculation by pitch motion of the Manta robot. Several experiments are conducted to demonstrate the usefulness of the proposed method.

Keywords: Manta robot, Self-position estimation, Optical flow

1. Introduction

In recent years, the demand of underwater investigation is increasing in a dam, the environmental research of a shallow, etc. However, in the inside of the sea, there is much danger for man. Moreover, it is known that a bad influence will appear to a human body due to a prolonged diving. Therefore, an underwater exploration robot is expected to investigate underwater instead of man [1]. Among underwater exploration robots, robots that imitate aquatic organisms don't have a rotor as a propulsive mechanism. Therefore, they are less likely to involve animals and seaweeds in the water, and they also have the feature that there are little underwater radiation noises. The load on the environment is small, so that it is thought that they are effective as underwater exploration robots [2].

Our laboratory has developed a Manta robot that has a propulsion mechanism with pectoral fins, which are constructed by mimicking the pectoral fin of the manta ray [3]. The Manta robot generates the propulsive force by undulating a propulsor composed of the right and left pectoral fins, and it is possible to make a forward and backward motion, a turning motion, and a diving motion. Since the Manta robot swims like actual aquatic lives without making a sound, it is possible to investigate aquatic lives without affecting seriously for them and its environment. In order for underwater robots to investigate an environment to be surveyed all over, it needs to have a function of estimating the self-position. However, such function is not mounted on this Manta robot. Since any radio wave doesn't reach underwater environments, it is impossible to acquire the positional information from GPS etc.

Therefore, in this research, we aim to implement a function of estimating the self-position using the optical flow, which expresses the motion of the object as flow vectors from the image taken by a camera. In this paper, a method for calculating the amount of movements is described by using the Lucas-Kanade method, which is one method of detecting optical flows. In addition, this manta robot causes a pitch motion during propulsion, so that it is a troublesome problem that the accuracy in the calculation of optical flows is deteriorated. Therefore, a technique of preventing the calculation accuracy of an optical flow from being degraded is proposed by introducing a gimbal mechanism, which controls a camera posture to follow a desired value, regardless of a pitch motion of the Manta robot. This validity is verified through some experimental results with the ground setup consisting of a camera experimental device mounted on a gimbal mechanism, which is attached to the finger-tip of a manipulator.

2. Manta robot

A Manta robot is an underwater exploration robot, which has a propulsion mechanism that imitates the pectoral fin of a Manta ray. Fig. 1 shows the appearance of the Manta robot developed in our laboratory, Okayama University.

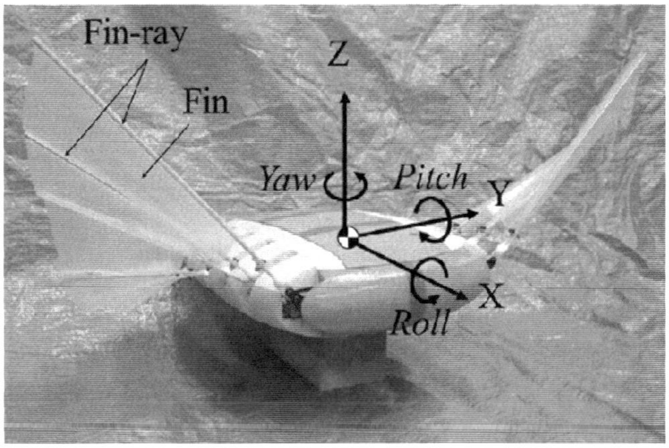

Fig. 1. Manta robot

The size of this manta robot is 420 mm in length, 620 mm in width, 90 mm in height and 5.3 kg in weight. The propulsive mechanism with a pectoral fin has six "fin-rays," one of which is equivalent to the skeleton in the fin of Manta ray, on each side of the fuselage, where the fins are attached between fin-rays. Each fin-ray is fitted with one servo motor housed in a waterproof box, and a sine wave is formed to obtain a thrust by operating the fin-ray up and down. In addition, it is possible to periodically transmit the photographed data and to instruct an action by human in emergency by communicating with the ground via the wireless communication function. Coral reef survey in shallow water is considered as an application of manta robot.

In this research, we are aimed at implementing the function of estimating the self-position, by incorporating a visual sensor into the Manta robot shown in Fig. 1. The amount of movements of the Manta robot is calculated using optical flows, which are constructed from the moving images acquired by the camera. The function of estimating the self-position is realized by accumulating this.

3. Method of estimating self-position

3.1. Optical flow

An optical flow is described by a vector, which represents the movement of an object between consecutive images. The optical flow can be calculated by analyzing images obtained from a visual sensor to be typified by a camera. After acquiring stationary images such as landscape etc. from the moving Manta robot, it performs the detection and integration of optical flows. Thus, the relative amount of movements of Manta robot with respect to the background obtained from the image is realized.

In this research, the Lucas-Kanade method (i.e., LK method) is adopted as a method for detecting optical flows [4]. In the LK method, it is assumed that optical flows are uniform in a certain small region after detecting pixels, which have a feature suitable for tracking on the image as feature points. Thereafter, the least squares solution of flow vectors is obtained after imposing a constraint, i.e., the condition that the luminance is not changed for each pixel in the small region, on optical flows. As an advantage of LK method, the calculation is performed only on a feature point, so that the calculation load is light. The LK method is suitable for detecting optical flows in real-time due to such an advantage.

3.2. Gimbal mechanism

The gimbal mechanism is a mechanism that keeps the direction of a sensor constant by utilizing motors to control the posture, or a suspended weight is oriented in a vertical direction. In this research, a stable calculation of amount of movements is realized even if the posture of the manta robot changes, by acquiring optical flows, keeping the posture of the camera constant with a gimbal mechanism.

Fig. 2 is a graph showing the posture of the manta robot swimming in water. A periodic oscillation about 6 deg has occurred in the pitch angle, so that there may appear a certain influence on the acquisition of optical flows. Therefore, in this paper, one-axis gimbal is used to reduce the influence of this pitch motion.

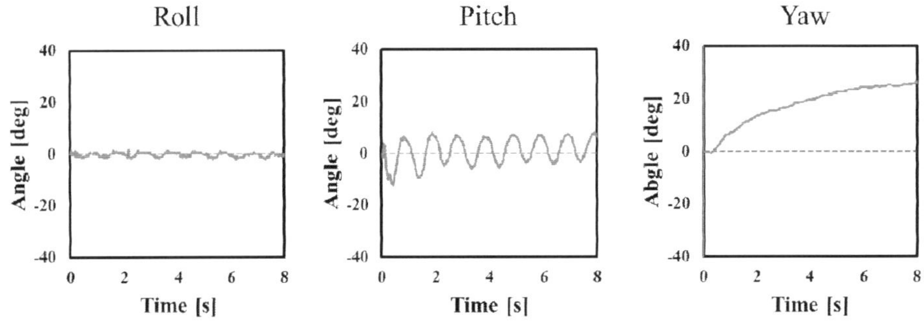

Fig. 2. Attitude of swimming Manta robot [5]

Fig. 3 shows the gimbal mechanism used in this research. A brushless motor with an excellent response is adopted as an actuator for controlling the attitude. It is expected that the attitude of the camera can be controlled by following the change of posture of the Manta robot without any delay. In addition, as sensors for detecting the changes in the posture of the Manta robot, an inertial measurement unit (IMU: Inertial Measurement Unit) that detects the posture is mounted on the main body side and camera side of the Manta robot, respectively.

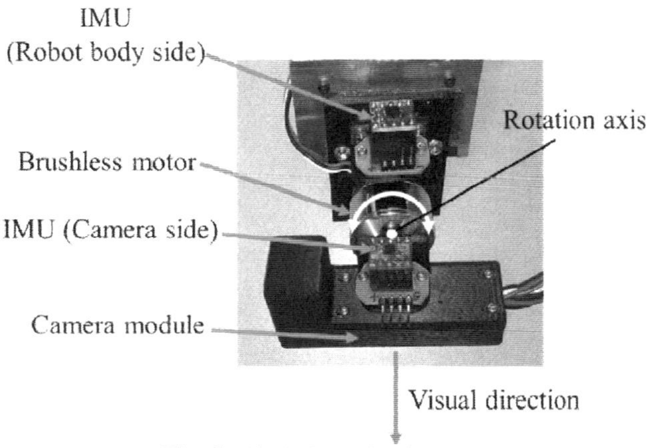

Fig. 3. Gimbal mechanism

Fig. 4 shows the block diagram of controlling a gimbal used in this study. The posture control of the gimbal consists of combining feedforward (FF) control, which is based on using the attitude angle of the main body side, and feedback (FB) control, which is based on using the attitude angle of the camera side. Properly speaking, in the block part of FF control, we should create the inverse dynamical model of the gimbal when the Manta robot performs the pitch motion. In this way, if the controlled variable of the gimbal is known, then the manipulated input can be determined for the minimum-time control of the gimbal by using the FF control. However, since the target value of the gimbal is known, but constant, the output of an inverse system is also reduced to a constant, and therefore we implemented FF control to calculate the manipulated input of the gimbal from the difference between the attitude angle of the Manta robot and the target angle of the gimbal. In addition, in order to improve the stability of the gimbal against mechanical errors and disturbances, PID control using the attitude angle of the camera was implemented in the FB block. Hereinafter, the combination of FF control and PID control is expressed as FF+PID control.

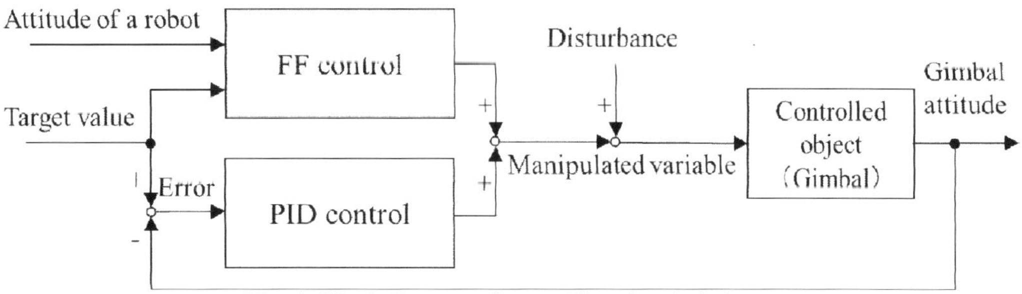

Fig. 4. Block diagram of controlling the gimbal

Fig. 5 shows each manipulated variable due to FF, PID and FF+PID control in gimbal control. The attitude control of the gimbal is realized by sending these signals as the manipulated variables to the brushless motor. The manipulated variable from the FF control was represented by a regular sine wave, whereas the manipulated variable from the PID control or the FF + PID control was represented by an irregular sine wave. As a reason why the manipulated variable from the PID control or the FF + PID control is represented by an irregular sine wave, it is considered that any noise is included in the measurement of the controlled variable because the manipulated variable is determined by calculating the deviation between the target value and the current controlled variable in the FB control.

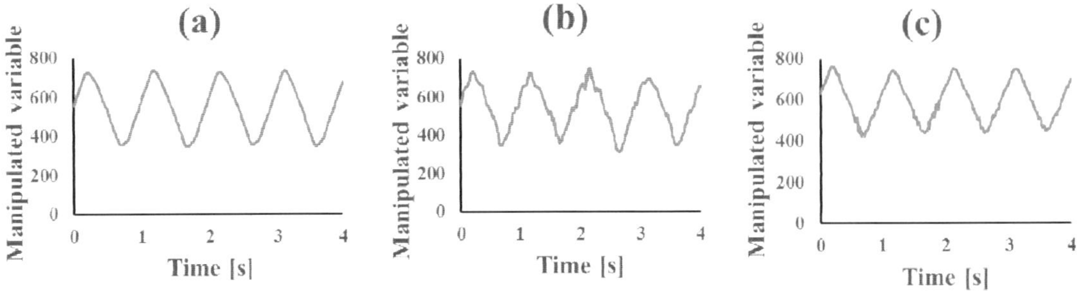

Fig. 5. The manipulated variables due to FF, PID and FF+PID control:
(a)FF control (b) PID control (c) FF+PID control

Fig. 6 shows the graph of controlling the attitude of the gimbal, under a situation in which the manipulated variable by each control method shown in Fig. 5 is given to the brushless motor to generate a certain pitch motion. An input signal whose amplitude is 6.0 deg and period is 1.0 s was set to a servomotor so as to generate a pitch motion. This pitch motion was reproduced by adopting the periodical oscillation acquired when the Manta robot was swimming, referring to Fig. 2. In the case of only FF control, it doesn't converge to the target value due to the mechanical error of the brushless motor, though the responsiveness is high. Originally, the FF control is a control method being capable of obtaining a controlled variable in a shortest time, under an environment where there exist no disturbances etc. However, as one of causes that the attitude of gimbal does not converge to the target value like this time, it is thought that the validity of a model for the FF control created is not high. When comparing the performance of PID control and FF + PID control, both converged to the target value, and there was no significant difference. Therefore, we conduct in this paper some experiments on the position detection for the Manta robot, using PID control and FF+PID control respectively to control the gimbal and select what is a suitable control method.

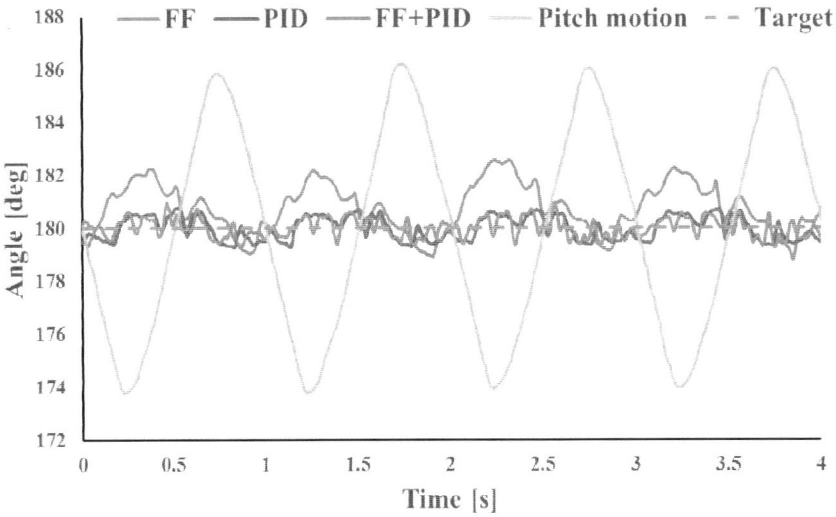

Fig. 6. The attitude control of the gimbal for a certain pitch motion

4. Experiments for estimating the self-position

4.1. Estimating the self-position without oscillations

It is confirmed that the LK method, which is one method of detecting optical flows, is useful in estimating the self-position of the Manta robot. The camera (BSW13K08H) was attached to the robot arm (RV-M2) and was moved 300 mm in the horizontal direction. It was confirmed whether the self-position was able to be accurately estimated by using optical flows detected from the images obtained during movements. The distance from the camera to the landscape on a seafloor was set to 300 mm. Considering an actual environment to be measured, the paper which printed a submarine landscape was installed in the photography side. Since the maximum speed of the Manta robot is about 300 mm/s, the experiment was conducted with the moving speed changed in the range from 50 mm/s to 300 mm/s. When estimating the self-position, it causes errors, if any shadows appear in the photography side. In laboratory experiments, the shadow of the robot arm and the gimbal mechanism may be reflected on the photograph side due to the ceiling illumination. Therefore, the ceiling lighting was turned off during the experiment, and the photography surface was illuminated with a desk light from the side-direction. In addition, a stabilized power supply was used as the power source for the brushless motor in the gimbal mechanism, where the voltage applied to the brushless motor was 5.0 V. Using a notebook PC with Core i5 2.4~Hz and RAM 4~GB for the measurement, a 320 × 240 pixel image acquired from the camera at 30 fps was calculated using Open CV 2.4.11. The measurement was conducted five times under all conditions.

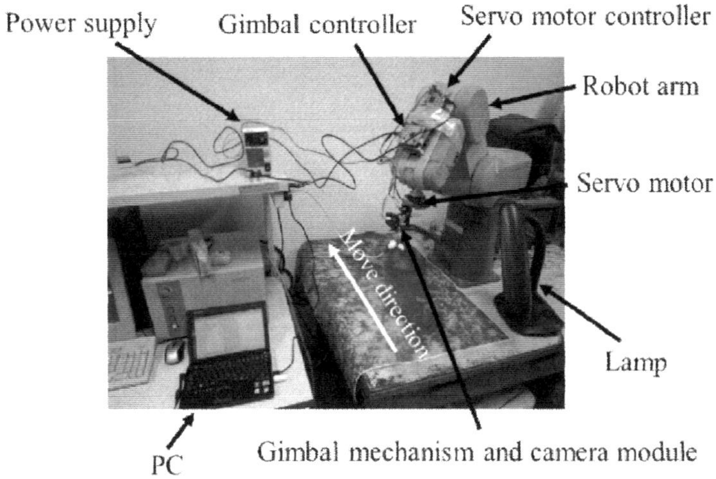

Fig. 7. Experimental environment

4.2. Estimating the self-position with oscillations

It is confirmed that the attitude control of a camera using a gimbal mechanism is useful for reducing periodic oscillations generated when the Manta robot is propelled. In addition to the conditions given in the previous section, some experiments were conducted with a pitch motion generated by adding an input of amplitude 6.0 deg and period 1.0 s to a servomotor. On the case with gimbal control, and the case of being nothing, the experiments were conducted, respectively. Especially, on the case with gimbal control, experiments and comparisons were carried out for the case of only PID control and for the case of FF+PID control, respectively.

5. Experimental results and considerations

5.1. Estimating the self-position without oscillations

Fig. 8 shows the experimental result on estimating the self-position without pitch motion. The self-position is estimated stably at the moving speed less than 300 mm/s, so that it is concluded that calculating the amount of movements by optical flows is effective as a method for estimating the self-position. In addition, a reason why the amount of movements calculated by optical flows is smaller than the actual one may be attributed to the fact that there is a processing dropout in program. It is thought that an improvement in accuracy can be expected by taking an image at a frame rate lower than 30 fps, and reducing the resultant amount of data.

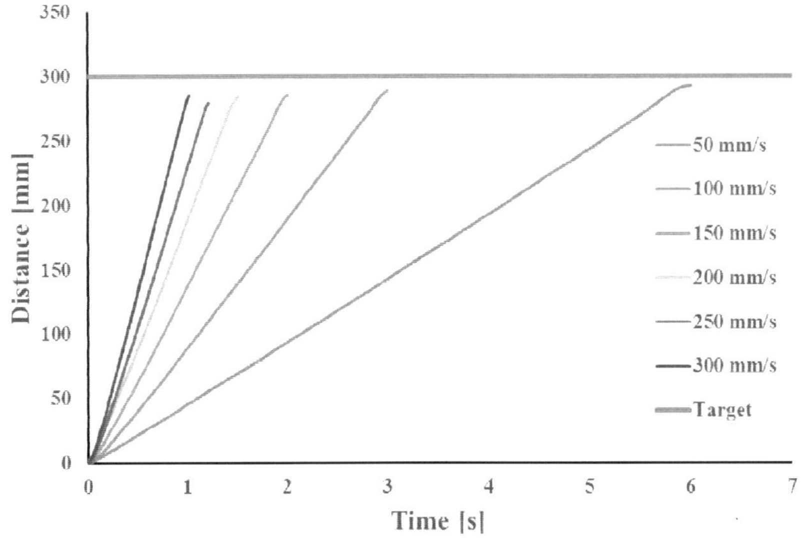

Fig. 8. Experimental result of estimating the self-position without pitch motion

5.2. Estimating the self-position with oscillations

Fig. 9 shows the experimental result on estimating the self-position accompanied by pitch motion, but without gimbal control. Any pitch motion has affected on the calculation accuracy in the amount of movements, so that it was impossible to estimate the self-position accurately. Also, there was a remarkable deterioration in the calculation accuracy on the amount of movements, at the moving speed more than 150 mm/s. As this cause, it is attributed to the fact that the acquisition of a flow vector becomes inaccurate by the existence of oscillations of a camera, and that if especially a certain fixed forward speed is exceeded, then the estimated speed with a very large variation is calculated through an optical flow method. From the above, it was shown that the influence of a pitch motion is so large that it can't be ignored, and any countermeasures are indispensable.

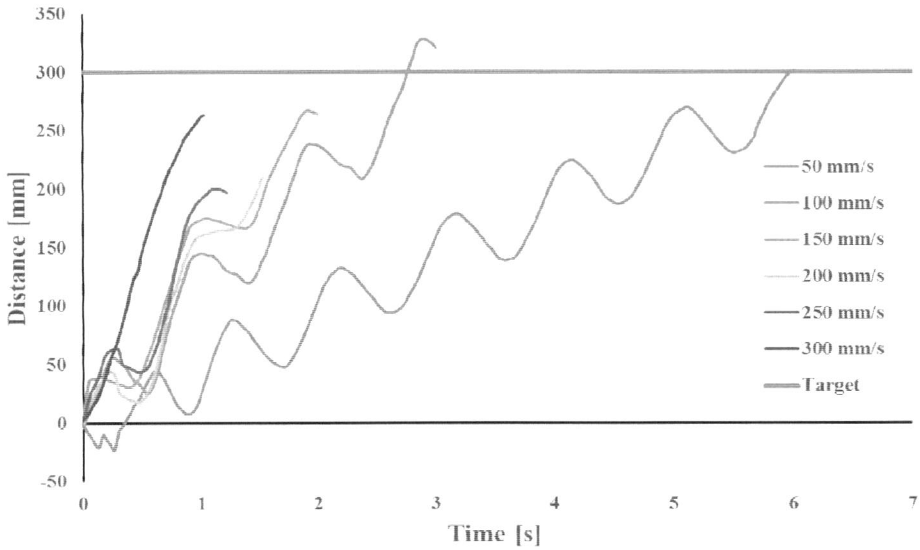

Fig. 9. Experimental result of estimating the self-position accompanied by pitch motion (no gimbal control)

Fig. 10 shows the experimental result on estimating the self-position accompanied by pitch motion, when the gimbal is controlled only by the PID control. The influence of a pitch motion was suppressed, and the result close to a target value of 300 mm was obtained in all speed ranges. In addition, it was confirmed that the accuracy in estimating the self-position during movement was improved, compared to the previous gimbal control. The variation in position information adversely affects the performance in position control of the Manta robot. Therefore, the attitude control of a camera by a gimbal mechanism is considered to be effective for improving the performance in position control of the Manta robot. However, it was observed that the accuracy in estimating the self-position was deteriorated immediately after the start of movement, due to the influence of the rising time in PID control.

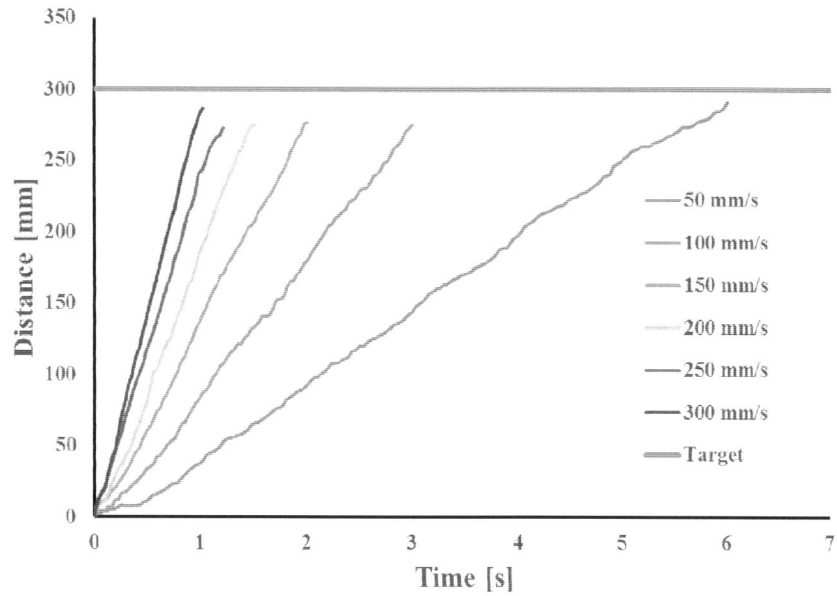

Fig. 10. Experimental result of estimating the self-position accompanied by pitch motion (with PID control)

Fig. 11 shows the experimental result on estimating the self-position accompanied by pitch motion, when the gimbal is controlled by the FF+PID control. An improvement was found in the accuracy of estimating the self-position immediately after the start of movement, compared to the case where the gimbal was controlled only by the PID control. In particular, the influence was remarkable in the speed range less than 100 mm/s. It is attributed to the fact that since the rising time in the gimbal control was improved by the FF control, the accuracy in estimating the self-position was improved immediately after the start of the movement. From this result, the FF+PID control is used for gimbal control in future research.

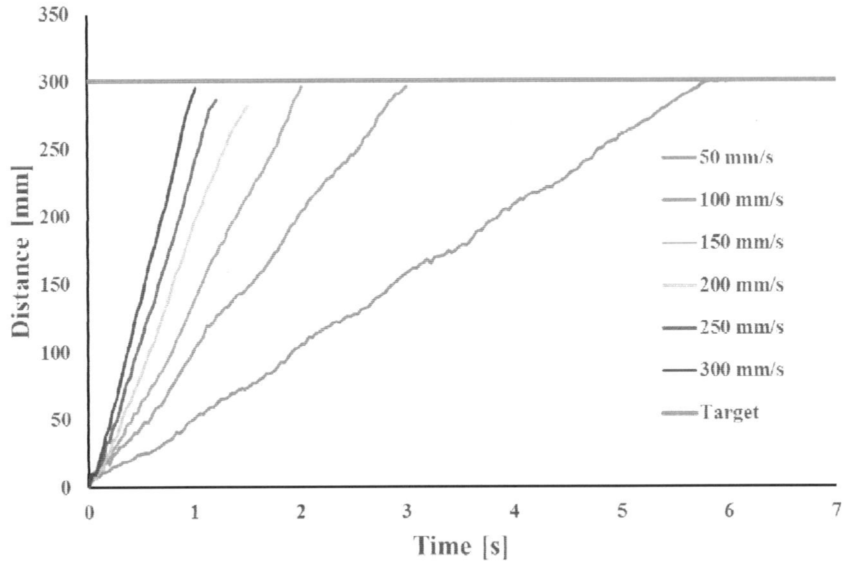

Fig. 11. Experimental result of estimating the self-position accompanied by pitch motion (with FF+PID control)

6. Conclusion

In this paper, we have been aimed at implementing a function of estimating the self-position by optical flows into a Manta robot, and demonstrated the usefulness of Lucas-Kanade method which is one of detection methods of optical flows. As a result, it was found that a method of calculating the amount of movements by optical flows is effective as the method for estimating the self-position. In addition, we conducted some experiments to confirm an improvement in stability for estimating the self-position by a gimbal mechanism, taking account of the influence of a pitch motion occurring in the Manta robot when acquiring optical flows. It was proved that the attitude control of a camera by a gimbal mechanism was effective for improving the performance in position control of the Manta robot.

References

[1] Ura T 2004 Robots for underwater world *Journal of the Robotics Society of Japan* vol 22 (Tokyo: The Daily Industrial News) pp 692–6

[2] Kato M 2001 Application of Study on Aqua Bio-Mechanisms to Ocean Engineering *The Society of Naval Architects of Japan 16th Ocean Engineering Symp.* (Tokyo: The Society of Naval Architects of Japan) pp 132–8

[3] Mikuriya K, Watanabe K, Ikeda M and Nagai I 2015 Development of a New Manta Robot Considering the Form Drag *Proc. of the SICE Annual Conference 2015* (Tokyo: The Society of Instrument and Control Engineers) pp 1397–1400

[4] Lucas B D and Kanade T 1981 An Iterative Image Registration Technique With an Application to Stereo Vision *Proc. 7th Int. Joint Conf. Artif. Intell.* (Vancouver: Morgan Kaufmann Publishers Inc.) pp 674–9

[5] Deng L, Wang Q, Mikuriya K and Watanabe K 2016 The Integration of Some Sensors for Measuring the Attitudes of a Manta Robot *Proc. of 2016 IEEE International Conference on Mechatronics and Automation* (New York: IEEE) pp 519–24

IOP Conf. Series: Journal of Physics: Conf. Series **962** (2018) 012017 doi:10.1088/1742-6596/962/1/012017

Design of Omni Directional Remotely Operated Vehicle (ROV)

Rahimuddin, Hasnawiya Hasan, Haryanti A Rivai,Yanu Iskandar, Claudio P

Marine Engineering Department, Hasanuddin University

E-mail: rahimnav@gmail.com, rahimnav@unhas.ac.id

Abstract Nowadays, underwater activities are increased with the increase of oil resources finding. The gap between demand and supply of oil and gas cause engineers to find oil and gas resources in deep water. In other side, high risk of working in deep underwater environment can cause a dangerous situation for human. Therefore, many research activities are developing an underwater vehicle to replace the human's work such as ROV or Remotely Operated Vehicles. The vehicle operated using tether to transport the signals and electric power from the surface vehicle. Arrangements of weight, buoyancy, and the propeller placements are significant aspect in designing the vehicle's performance. This paper presents design concept of ROV for survey and observation the underwater objects with interaction vectored propellers used for vehicle's motions.

1. Introduction

1.1. *Background*

As a maritime country, Indonesia has vast areas of the oceans 70% of the total area that stretches from Sabang to Merauke. The region of Indonesia has abundant marine wealth which has the potential to be managed for the welfare of the Indonesian people. The ocean is being in between the Pacific Ocean, India ocean and South China Ocean. The region is very strategic for many marine biotas live. The Climatic influenced by the climatic in the south (Australia) and in the north (South China Ocean) cause the current stream change regularly in every year. In another side, Indonesia is on track of tectonic plates, which stretches from west to east and from north to south. This causes the region is very rich in minerals, gas, and oil. The seabed layers are quite active with many basins especially in eastern ocean of Indonesia.

In 2010, the diversity of marine ecosystems Indonesia has been known throughout the world where researcher from Indonesia and America working together in an expedition called INDEX/SATAL. The expedition explored the deep ocean around Sangihe Talaud, founding many new ecosystems in deep water and the contour of seabed explains that there are many untouched areas that give information for human such as; the information of organism which can help the human lives, such earthquakes and tsunamis prediction, ancient artefacts, underwater maintenance and seabed basins patterns that can be used to predict the distribution of oil resources in the seabed.

In INDEX/SATAL 2010 event, many researchers explored the underwater ecosystem in varies depth even in underwater where the human cannot work because of the high-water pressure. For this condition, research activities required an underwater vehicle that can operate remotely or autonomously in a risk environmental conditions, such as: ROV (remoted Operated Vehicle), AUV (Autonomous Underwater Vehicle), or UUV (Unmanned Underwater Vehicle). These vehicles can perform some commercial underwater operations such as survey and observation, pipe work installation of submarine cables, instrument maintenance under the sea, and another oceanographic research that can be operated directly from the ships on the surface of the sea (Mother Ship). Using either wired or using acoustical communication works automatically with artificial intelligence control system, [1,2,3].

1.2. *Underwater Vehicle*

Generally, underwater vehicle can be divided in two types; Autonomous Underwater Vehicle (AUV) and Remotely Operated Vehicle (ROV). An autonomous underwater vehicle **(AUV)** is a robot which travels underwater without requiring input from an operator. AUVs constitute part of a larger group of undersea systems known as unmanned underwater vehicles, and vehicles that can be driven from the operator which is at sea level is a classification that includes non-autonomous or remotely operated underwater vehicles. Communication data and source power supplied using through tether. The vehicle controlled and powered from the surface by an operator using remote control.

ROV is a tethered underwater vehicle, commonly used in deep water industries such as offshore hydrocarbon extraction. The vehicle is linked to a surface ship by a buoyant tether or, often when working in rough conditions or in deeper water, a load-carrying umbilical cable is used along with a tether management system (TMS). The TMS is separate assembly which sits on top of the ROV. The purpose of the TMS is to lengthen and shorten the tether so the effect of cable drags where there are underwater currents is minimized. The umbilical cable is an armoured cable that contains a group of electrical conductors and fibre optics that carry electric power, video, and data signals between the operator and the TMS. Once at the ROV, the electric power is distributed between the components of the ROV. However, in high-power applications, most of the electric power drives a high-power electric motor which drives a hydraulic pump. The pump is then used for propulsion and to power equipment such as torque tools and manipulator arms where electric motors would be too difficult to implement in subsea. Most ROVs are equipped with at least a video camera and lights. Additional equipment is commonly added to expand the vehicle's capabilities. These may include sonars, magnetometers, a still camera, a manipulator or cutting arm, water samplers, and instruments that measure water clarity, water temperature, water density, sound velocity, light penetration, and temperature.

1.3. *Brief history of ROV Design*

British and US navies developed the vehicles during the 1950s and '60s. The early British ROV was named Cutlet, used in the 1950s by the Royal Navy to retrieve practice torpedoes and mines. These developments led to the ability to perform deep-sea rescue and recovery operation; as early example, the ROV used to retrieve the atomic bomb lost in the Mediterranean in 1966 when a B-52 crashed during a refuelling operation. This operation employed an early generation ROV, the CURV-I (Cable-controlled Undersea Recovery Vehicle) and in 1973, CURVIII performed the deepest underwater rescue in history when it rescued two men who were stranded 1,575 feet (480 m) below the ocean surface. CURVIII was equipped with an array of TV cameras and lights and a 35-mm camera with a 500-frame color film capacity. It could operate at depths of up to 10,000 feet (~3,000 m), weighed approximately 5,400 pounds (2,400 kg) and could be equipped with a range of mechanical tools and sensors. The CURV family of ROVs are still in use and the latest generation product, the CURV-21, weighs 6,400 lbs (2,909 kg), operates at a depth of 20,000 feet (~6,000 m) and is fitted with high resolution digital cameras, sonar and a fibre optic multiplexing system which can combine up to eight channels of video, sonar, USBL (ultra-short baseline) positioning data, RS232/422/485 data communications and navigation data on a single fibre optic. The vehicle using 400 MHz digital communications network in control navigation.

These military developed the technological groundwork that provided the offshore oil and gas industry with so-called "work-class" ROVs. These became essential tools in the 1970s and '80s when much of the new offshore oil and gas development was at a depth that exceeded the reach of human divers. Today's modern ROVs are used extensively in the construction of sub-sea oil and gas developments and in their subsequent inspection, repair and maintenance and vary hugely is terms of depth rating, size, complexity and capabilities. A typical commercial ROV comprises a large flotation pack above an aluminium chassis to provide the necessary buoyancy. In addition to powering the various electrical systems, electrical power can be used to drive a hydraulic pump that is used for propulsion as an alternative to the more normal electrically powered thrusters and to actuate high power equipment such as torque tools, cutters and manipulator arms. Most ROVs are equipped with a video camera and lights and often include sonar systems and in the case of ROVs aimed at scientific research or environmental monitoring, sensors for monitoring water quality, sampling systems and high-resolution cameras are also deployed, [4].

Therefore, the system in ROV is a highly interrelated group of subsystems that, when functioning synergistically, provides an impressive subsea capability. Since the characteristics of the vehicles are multi-variant, with no unique solution, the design process is necessarily iterative. Because of this highly interdependent relationship, ROV system performance is a delicate balance of design and operational characteristic. Commonly, the ROV design system can be divided into a number of major subsystems, these would include; ROV Vehicles, Tools and Sensors, Control / display console, Electrical power distribution, Umbilical and tether cables, handling system

2. Design of ROV, Propulsion System, Communication Data, and Power System

2.1. *Design of ROV*

The concept of the developed ROV is a vehicle for underwater survey and observation with omni directional movement. With this ability, the vehicle can conduct survey and observation activities where the vehicle does not need to maneuver when it moves to sideways, back, front, and others.

Two important parts considered in the design of the vehicle to be stable; point of the center of gravity or center of weight and center of the buoyancy force. These determine the vehicle's stability characteristic, where the vehicle's stability is an ability of the vehicle to return to the equilibrium position when there any external force or moment disturbance. Good stability indicates the center of gravity is underneath the center of buoyancy force and the longer distance between center of gravity and center of buoyancy the higher moment stability and stiffness of motion. Therefore, the vehicle design place buoyant structure at the top of the vehicle and place more weight at the bottom.

In the design of propeller, the vehicle uses ducted propeller where the ducted propeller has advantages than the conventional propeller. The ducted propeller can increase the efficiency of the thrust propeller, has lower diameter compare to non-ducted propeller, and the ducted propeller can protect the propeller from other objects. Ducted propeller installed on the main structure with fixed mounting angles. However, besides the fixed angle mounted on the structure, rotatable propeller also able to be installed that can decrease the number of propeller installed for application to multi-direction movement. In this paper, the fixed propeller angle chooses as it has a simpler in design and installation.

In another side, the vehicle designed to have a capability to keep the vehicle's position in stable position even under force disturbance from current. The vehicle also designed to have same weight to the buoyancy force that ease the vehicle move in vertically. To adjust the weight, pumping system run the pump to fill the ballast system that can pump water in and out to the ballast tank. In construction, the vehicle construction is made using galvanized pipe with the space inside the pipe allowed for ballast tank. Design of the ballast system also consider the total weight required and centre of the weight or gravity. Therefore, in design of the vehicle's weight distribution and the buoyancy of every part are designed properly so it will not decrease the vehicle's stability. The higher stability can make the vehicle very stiff and decrease the ability of the vehicle to rotate in pitch and roll motion. The concept design of the ROV as shown in figure1.

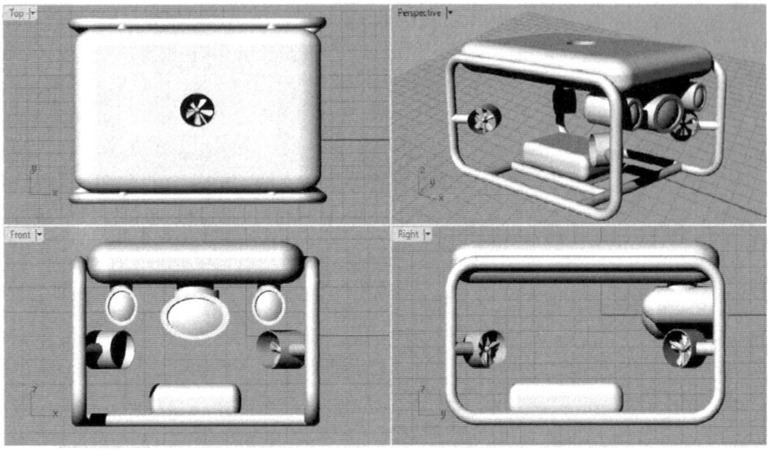

Figure 1. Proposed Developed Design of ROV

In this research, the ROV designed to operate in 100m under the water surface. The dimension of the developed design has 58.60cm length, 42.21cm breadth and 32.60cm high. The model equipped five propellers with 7.0cm diameter, one buoyant tank at top, four dry batteries placed at bottom to increase the vehicle stability and as additional source power. Four propellers are installed at each corner with 45 degrees angle to the axis and one propeller placed vertically at the buoyant tank. The propellers used to control the vehicle motions in six degrees. The design speed of vehicle is 0.5m/s.

2.2. *ROV System*

The ROV system designed to control its motion automatically. It is equipped five brushless motors with electronic speed control (ESC) system, motion sensors, water leak sensor, temperature sensor, current flow sensor, and pressure sensors. The ROV also equipped two cameras and lights as remote eyes for operators and one joystick used as operator command. The speed and rotation of motor are controlled simultaneously. The motion sensor consists of accelerometer, gyrometer, and compass. Four dry batteries placed at lower ROV with total capacity about 28Amph. The electronic systems be placed in box with water tight compartments to protect them from the water leaks and protect being crushed by the extreme pressure exerted on the ROV.

ROV has two systems; remotely operation system and onboard vehicle system. The remote operation system consists of console for display data and video captured, calculate the joystick signals and convert it to the data require for the system on the vehicle. The onboard vehicle system uses a microcontroller with network module to do calculation, control and communication between the sensors and actuator, and linked to remote operator. The diagram for the whole system as shown in figure 2.

2.3. *Propulsion System*

Propulsion system of ROV derived from the advanced of omni-wheel robot moves in omni-directional movement. Propulsion systems designed using five propellers. Four propellers are used to drive the vehicle in horizontal and longitudinal direction. One propeller is placed on the upper side, in line with the centre of gravity, used to drive the vehicle moves vertically. The vehicle uses duct propeller with six blades with some advantages as follows; The ducted propeller has more efficient in producing thrust than a conventional propeller. For the same static thrust, a ducted propeller has a smaller diameter than conventional propeller. The water velocity through the ducted can be adjusted to allow it operate more efficiency. Ducted propeller offers enhanced safety from contact with others object, as shown in figure3.

The four propellers installed at each corner and crossing each other at 45 and 135 degrees to its directional axis. The vehicle's movement resulted by vectored forces of the propeller's thrust. The vehicle's movement is designed to have a motion like omni-wheel robot where it can move in omni-directional or move at certain degree. The omni-four wheels robot and ROV's movement concept are shown in the figure 4.

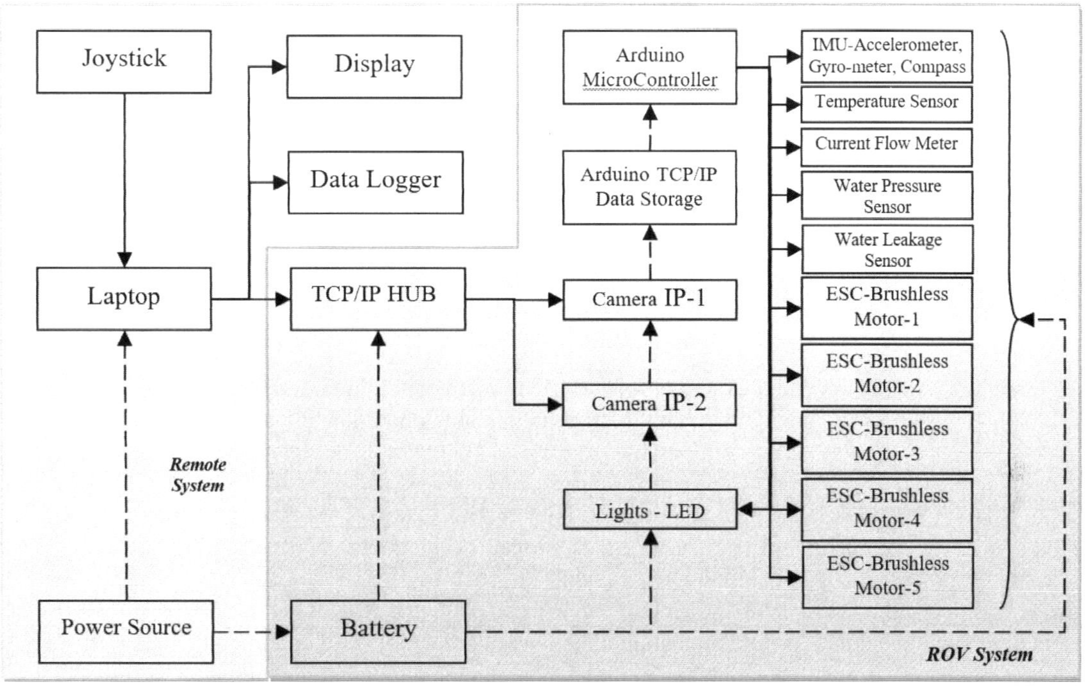

Figure 2. Diagram of ROV System, on-vehicle and remote system

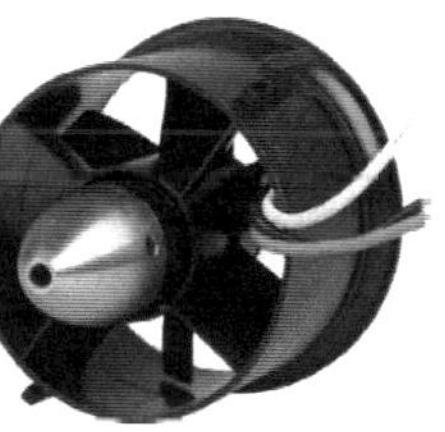

Figure 3. Duct Propeller with Six blades

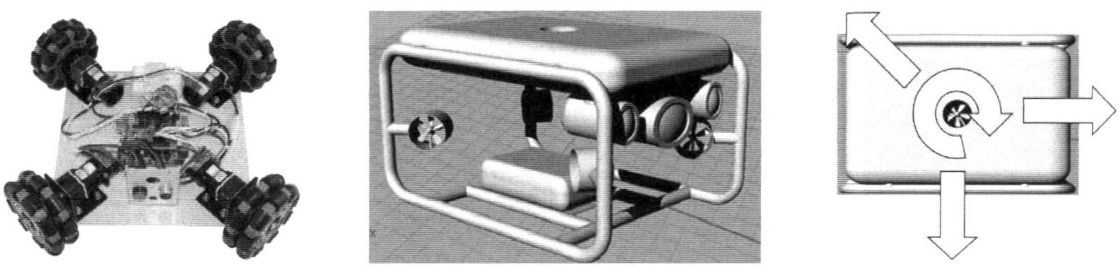

(a) (b) (c)

Figure 4. (a) Omni four-wheel robot with 45-degree wheel to origin axis, (b) design of ROV, (c) Direction movement of ROV

2.4. *Navigation System*

Navigation the vehicle using four propellers for transversal and longitudinal movement while one propeller for vertical movement. Combinations of the propeller thrust direction in navigation are shown in figure 5.

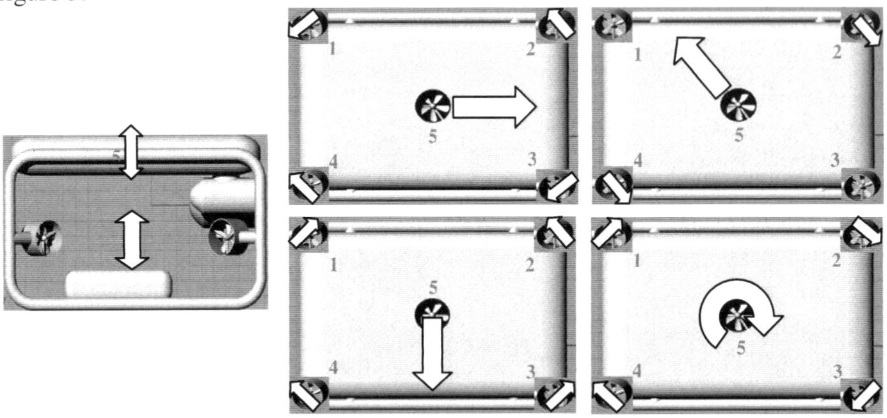

Figure 5. Navigation concept of ROV, yellow arrow is propeller thrust direction and white arrow is vehicle movement directions.

Installation of propeller at each corner of the vehicle must consider the centre of vehicle's movement. The propeller position installed at proper position that will not result a contra force between propeller's thrust. Positioning the propeller in such can make the effectivity of the propeller's forces are decreased. Based on the omni-four-wheel robot, the kinematic model of transformation propeller thrust to the directional force applied are written as follows, [5,6];

$$\begin{bmatrix} F_x \\ F_y \\ F_z \\ M_x \\ M_y \\ M_z \end{bmatrix} = \begin{bmatrix} c\theta & -c\beta & 0 & 0 & 0 \\ s\theta & -s\theta & 0 & 0 & 0 \\ 0 & 0 & 0 & 0 & 1 \\ d\,s\theta & d\,s\theta & 0 & 0 & -l \\ 0 & -d\,c\beta & d\,c\theta & 0 & -l \\ d\,c\theta & -d\,c\beta & -d\,c\theta & -d\,c\beta & 0 \end{bmatrix} \begin{bmatrix} T_1 \\ T_2 \\ T_3 \\ T_4 \\ T_5 \end{bmatrix} \tag{1}$$

$$T_i = K_T \times D_i^4 \times n_i^2 \tag{2}$$

Where *F, M, T* are force, moment and propeller thrust respectively, *c* is cosines, *s* is sinus, *d* is horizontal propeller arm to centre of rotation, *l* is vertical propeller arm to centre of rotation, *D* is propeller diameter, *n* is propeller rotation and K_T is propulsion coefficient of thrust. θ, β are angle of propeller where θ is angle for propeller 1 and 3, and β is angle for propeller 2 and 4.
In order to drive the propellers in two rotational direction, brushless motor with Electronic Speed Controller are applied. The motor has high torque and high rpm to high provide power for high thrust propeller.

2.5. *Communication Data*

Communication data between remote operator and vehicle is through optic fibre. The advance of optic fibre compare to cable data communication are the optic fibre that can ignore the signal noise, high speed data, and high capacity data. The optic fibre likely has no limited length to transfer the signal and data. However, in this early research, the optic fibre is replaced using network cable data TCP/IP. This communication data length can go about 100m length.

2.6. *Power System*

The ROV has two power source system, main power from the remote station or operator station and additional power placed at the vehicle. The source of power supplied from the surface ship where the electrical power uses 12-13.6 Volt DC. The vehicle uses four dry batteries placed at the bottom as additional power source for the vehicle system. In order to maintain the voltage value for sensors and microcontroller, some voltage regulators required. The voltage regulator used to supply constant voltage for the whole sensors in the vehicle.

3. Weight, Buoyancy, and Moment Stability Calculation

3.1. *Weight and Buoyancy Calculation*

The ROV is designed to have a high stability and can fly underneath water surface. In designing the ROV, the stability is an important factor that ROV must meet the stability requirement where the center of floatation should be at upper than center of gravity. To increase the stability of the ROV can be done by placing the buoyant tank at upper position and the place the weight at lower position.

In the design of the ROV, the main construction used are galvanized pipes. There fourteen galvanized pipes and eight elbow connectors used. While the buoyant tank made from the fiberglass material. Calculation of the ROV weight, displacement, center of buoyancy and center of gravity provide using the following equations.

The weight of the galvanized pipe obtained by calculating the pipe's material volume than multiply with the specific material, as follows;

$$W_{material} = \pi \times \rho_{pipe} \times L \times \frac{(D_{out} - D_{in})}{4} \quad (3)$$

Where D_{out}, D_{in} are diameter outside and inside of pipe. L is the length of pipe, and ρ is the weight specific of the pipe materials.

The added weight by water ballast inside the pipe is obtained by calculating the inside volume of the pipe using the following equation;

$$W_{ballast} = \pi \times \rho_{pipe} \times L \times \frac{D_{in}}{4} \times h \quad (4)$$

where h is the water high inside the pipe.

The buoyancy of the vehicle could be obtained by calculating the displacement volume of the vehicle and the volume multiply with the water weight specific of the vehicle as follows;

$$\Delta_{Bouyancy} = \pi \times \rho_{water} \times \frac{D^2{}_{out}}{4} \times L \quad (5)$$

The design of the origin axis of the vehicle placed at the middle of x-y axis and at bottom for z-axis. The center of gravity to the origin axis provided from the following equations;

$$M_x = \sum_{i=0}^{i=n} W_i \, x \, X_i + W_{ai} \, x \, X_{ai}$$

$$M_y = \sum_{i=0}^{i=n} W_i \, x \, Y_i + W_{ai} \, x \, Y_{ai}$$

$$M_z = \sum_{i=0}^{i=n} W_i \, x \, Z_i + W_{ai} \, x \, Z_{ai} \quad (6)$$

W_i, W_{ai}, and X_i, Y_i, Z_i, and X_{ai}, Y_{ai}, Z_{ai} are the weight component, added weight by ballast water, the offset of center of weight component, and added weight component respectively. In general, the buoyancy force of the vehicle's component are calculated by calculating the weight displacement of the component equal to the buoyancy force. Then, the moment buoyancy provided by the following equations;

$$M_{bx} = \sum_{i=0}^{i=n} B_i \, x \, X_i$$

$$M_{by} = \sum_{i=0}^{i=n} B_i \, x \, Y_i$$

$$M_z = \sum_{i=0}^{i=n} B_i \, x \, Z_i \tag{7}$$

According to the equations above, calculations above and origin of axis located at the baseline then found the weight of the ROV is 12,17kg, the center gravity of the vehicle and the center of buoyancy are;

G_x = 0.0cm G_y = 0.0cm G_z = 13.32cm

B_x = 0.0cm B_y = 0.0cm B_z = 20.60cm

3.2. *Moment Stability Calculation*

The ability of the vehicle to return to its original state when trim is a measure of the underwater vehicle stability. The vehicles can return to the original state when moment between gravity and buoyancy forces are positive. As mentioned before, the way to increase the stability is to set the distance between center of gravity and the center of buoyancy is high where the center of buoyancy is upper then center of gravity, [2,7]. The moment stability of the vehicle at certain heal angle obtained using the following equation, and the moment stability is shown in the figure 6.

$$M_S = w \times (B_z - G_z) \times \sin(\frac{\theta}{180} \times \pi) \tag{8}$$

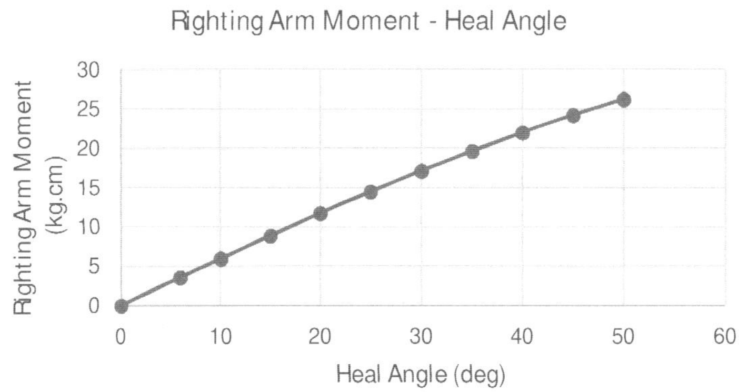

Figure 6. Righting moment arm required to keep the vehicle in stable condition at every heal angle.

3.3. *Drag Calculation*

Drag force is the resistance force caused by the motion of a body through a fluid, such as in water or in air. A drag force acts opposite to the direction of the oncoming flow velocity. This is the relative velocity between the body and the fluid. The drag force calculated using the following equation;

$$D = \frac{1}{2} \, C_\rho \, A \, V^2 \tag{9}$$

Where C_ρ is the drag coefficient, which can vary along with the speed of the body. Typical values range from 0.4 to 1.0 for different fluids (such as air and water), ρ is the density of the fluid through which the body is moving, V is the speed of the body relative to the fluid, A is the projected cross-sectional area of the body perpendicular to the flow direction (perpendicular to V), [8]. The drag coefficients are obtained from the data and the drag force calculated with assumtion no heal angle during the variation of the speed as shown in figure 7.

4. Discussion

Underwater vehicle has various types and systems in accordance with the purpose of vehicle design. ROV is a vehicle operated from the surface of the sea as a work vehicle where underwater environment can disturb vehicle motion performance such as velocity of ocean currents, high pressure water depth. This vehicle has four propellers so that the vehicle can move surge, sway, heave, yaw, pitch and roll and with a propeller placed vertically. The vehicle's design takes the concept of omni-wheel robot where the robot can move in omni directions, [6, 9]

The design of the ROV construction has high stability where the centre buoyancy at 17.28cm above the centre of gravity. From the stability calculations, the righting moment showed an increased graph trend moment up to 90 degrees. The graph showed the stability of the vehicle is always positive so that the vehicle will not run into stability problems when the vehicle in underwater.

The calculation shows an increase of drag force within increase of vehicle speed and the gradient of the drag curve also increase. The ration of drag force to vehicle's weight at 0.5m/s speed is 0.47 and required a minimum resultant propeller force about 57N.

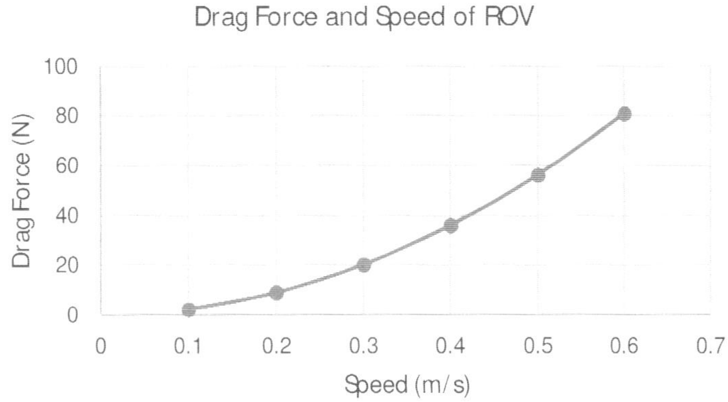

Figure 7. Drag force of ROV for certain speed

5. Conclusion

According to the design of the ROV and the static stability and drag calculations, the ROV design was concluded as following;

- The ROV has a positive stability where the centre of buoyancy at 17.28cm above the centre of gravity.
- The static drag force calculation showed the force required to propel the vehicle still required more analysis where the non-linearity aspect is should be considered

6. Future work

As the future work, this research will continue on developing a numerical time domain simulation program to explore the nonlinearity of the vehicle's response and develop the control method for vehicle's keeping position under current disturbance.

Acknowledgements

This research was funded by Kementerian Riset dan Pendidikan Tinggi of Indonesia. Thanks to all support team in this research, the students of marine engineering, particularly the student in the laboratory of control and electrical of marine engineering department of Hasanuddin University.

References

[1]. Tadahiro Hyakudome 2011 *Conceptual Design of Navigation of An AUV For Monitoring CCS Site at Deep Sea Bottom* Proceedings of the 30th International Conference On Ocean, Offshore and Arctic Engineering

[2]. Tadahiro Hyakudome 2011 *Design of Autonomous Underwater Vehicle. International Journal of Advanced Robotic Systems* INTECH pp 131-139

[3]. Thomas B. Curtin 2005 *Autonomous Underwater vehicles, Trends and Transformation* Marine Technology Society Journal pp 65-75

[4]. Robert Bogue 2015 *Underwater Robots: a review of technologies and applications* Industrial Robot Journal, Vol. 42 pp186-191

[5]. Helder P. Oliveira, Armando J, Sousa, A. Paulo Moreira, Paulo J. Costa 2009 *Modelling and assessing of Omni-Directional Robots with Three and Four Wheels* INTECH Open Access book publisher.

[6]. Benjamin Meyer, Kristian Ehlers, Christoph Osterloh, Erik Maehle 2013 *An Autonomous Omnidirectional Underwater Robot* PALADYN Journal of Behavioural Robotics vol 4 pp 204-210

[7]. Tehrani Nima Harsamizadeh, Mahdi Heidari, Yadollah Zakeri, & Jafar Ghaisari 2010 *Development, depth control and stability analysis of an underwater Remotely Operated Vehicle (ROV)* Control and Automation (ICCA) 2010 8th IEEE Conference

[8]. Rui M.F.Gomes, Alexandre Sousa, Sergio Loureiro Fraga, Alfredo Martins, Joao Borges Sausa, Fernando Lobo Pereira 2005 *A new ROV Design: Issues on Low Drag and Mechanical Symmetry*, IEEE Conference

[9]. Ranjit Barua, Sajal Mandal, Samiran Mandal 2015 *Motion Analysis of A Mobile Robot With Three Omni-Directional Wheels* IJISET - International Journal of Innovative Science, Engineering & Technology, vol 2 pp 644-648

IOP Conf. Series: Journal of Physics: Conf. Series **962** (2018) 012018 doi:10.1088/1742-6596/962/1/012018

Map generation in unknown environments by AUKF-SLAM using line segment-type and point-type landmarks

Sho Nishihta, Shoichi Maeyama, and Keigo Watanebe,

Graduate School of Natural Science and Technology, Okayama University,
3-1-1 Tsushima-naka, Kita-ku, Okayama 700-8350 Japan

nishihata-s@usmm.sys.okayama-u.ac.jp, {maeyama, watanabe}@sys.okayama-u.ac.jp

Abstract. Recently, autonomous mobile robots that collect information at disaster sites are being developed. Since it is difficult to obtain maps in advance in disaster sites, the robots being capable of autonomous movement under unknown environments are required. For this objective, the robots have to build maps, as well as the estimation of self-location. This is called a SLAM problem. In particular, AUKF-SLAM which uses corners in the environment as point-type landmarks has been developed as a solution method so far. However, when the robots move in an environment like a corridor consisting of few point-type features, the accuracy of self-location estimated by the landmark is decreased and it causes some distortions in the map. In this research, we propose AUKF-SLAM which uses walls in the environment as a line segment-type landmark. We demonstrate that the robot can generate maps in unknown environment by AUKF-SLAM, using line segment-type and point-type landmarks.

1. INTRODUCTION

In recent years, the development of an autonomous mobile robot that gathers information at disaster sites has been actively developed. It is difficult to give the map in advance in disaster sites where people cannot enter. Therefore, the robot needs to realize autonomous behavior not only in known environments with maps but also in unknown environments without maps. In order to enable robots to move under unknown environments, it needs to estimate the self-location, as well as to generate the map. This is called a Simultaneous Localization and Mapping (SLAM) problem[1][2]. As a method to solve the SLAM problem, an Extended Kalman Filter (EKF) and an Unscented Kalman Filter (UKF) which adopt stochastic concept, and an Augmented UKF (AUKF) was also proposed so as to treat systematic errors included in odometry, such as measurement errors in a wheel diameter and a tread are proposed[3][4]. In these methods, features in the environment such as corners where walls vertically intersect are registered as landmarks. Whenever the robot moves, it acquires a landmark and corrects the self-location from the positional relationship between the robot and the landmark. However, when the robot moves in an environment like a corridor consisting of few point-type features, the robot decreases the accuracy in the estimation of self-location and creates a distorted map.

There are several researches using line segments in SLAM research[5][6][7][8]. Therefore, the robot registers the wall as a landmark, which is the feature of the line segment in the environment, and uses it for the estimation of self-location. In this research, the present SLAM is assumed to use an AUKF indoors, and adopts walls as landmarks, as well as corners. Experiments are conducted using an autonomous mobile robot, and it is verified that the line segment-type landmark is useful. In this paper, a SLAM problem is first described using an AUKF. A conventional point-type landmark and the proposed line segment-type landmark are next explained. Finally, experiments using an actual mobile robot are conducted to check the usefulness of the line segment-type landmark.

2. SLAM solution method using AUKF

Figure 1 shows the proposed AUKF-SLAM model. In figure 1, (x_t, y_t, θ_t) is the position and posture of the robot at time t, (R_R, R_L, T, S) are the right wheel radius, the left wheel radius, the tread, and the sensor mounting offset. Let the ith landmark information of the corner estimated by the robot be (l_{xi}, l_{yi}, l_{si}). Also, let the kth landmark information of the wall estimated by the robot be (l_{x1k}, l_{y1k}, l_{x2k}, l_{y2k}). Then, using the position, orientation and kinematic parameters of the robot and the landmark information, the state $\boldsymbol{x}_t$ of the robot is expressed by

$$\boldsymbol{x}_t = (x_t, y_t, \theta_t, R_R, R_L, T, S, \ldots, l_{xi}, l_{yi}, l_{si}, \ldots, l_{x1k}, l_{y1k}, l_{x2k}, l_{y2k}, \ldots) \quad (1)$$

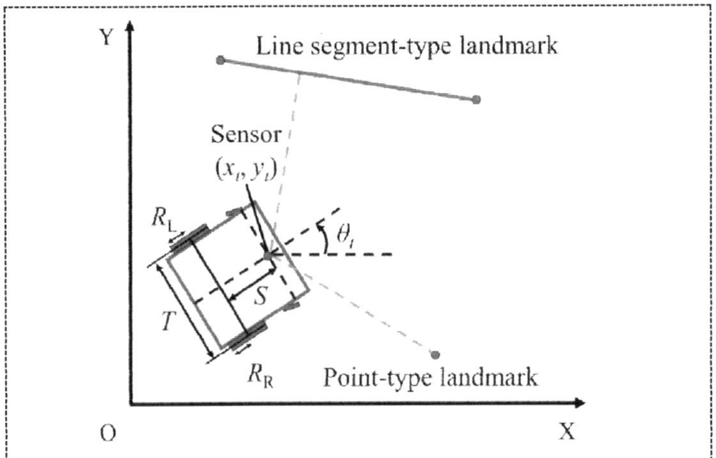

Figure 1. Model of AUKF-SLAM

The state $\boldsymbol{x}_t$ is represented by the mean and covariance of a normal distribution to reduce the influence by noises. In the AUKF, those values are corrected by prediction and updating, and the appropriate estimated values are calculated. The following equations are defined as the motion model and the measurement model, respectively, which are necessary for estimation.

$$\boldsymbol{x}_t = g(\boldsymbol{u}_t, \boldsymbol{x}_{t-1}) + \boldsymbol{\delta}_t \quad (2)$$

$$z_t = h(\boldsymbol{x}_t) + \varepsilon_t \quad (3)$$

g and h in each model are nonlinear functions. z_t is the measured value and $\boldsymbol{u}_t$ is the control input. $\boldsymbol{\delta}_t$ is a motion noise vector, and ε_t is a measurement noise vector. It is assumed that the noise is Gaussian with zero-mean and additively added.

3. Line segment-type and point-type landmarks

The landmark is a feature used by the robot to estimate the self-location and create a map. From the sensor data that measured the environment, the wall is used as a line segment-type landmark, and the corner at which the walls intersect is used as a point-type landmark. Increasing the types of landmarks eliminates the shortage of landmarks to be used by robots for estimation, so that it is thought to be able to cope with any environment. Here, we describe the acquisition method and its measurement model of each landmark.

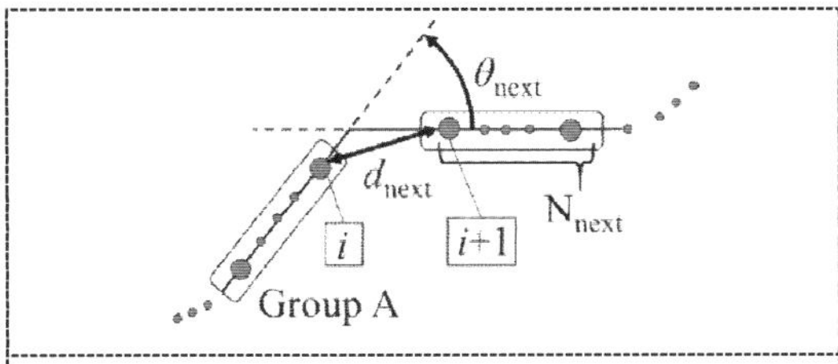

Figure 2. Grouping sensor data

3.1. Line segment-type landmark

In the case of using line segment-type landmarks, it needs to group the sensor data into each wall. The grouping method of points is explained with figure 2. Assume that the group A, whose *i*th point is a terminal, is determined. Compare the group A and a point group, which is composed of from the (*i*+1)th point to N_{next} ahead. If the angle θ_{next} formed by two straight lines is less than a threshold value, and the distance d_{next} between the *i*th point and the (*i*+1) th point is less than a threshold value, unify them as a one group. On the other hand, when either θ_{next} or d_{next} exceeds the threshold value, the group from the (*i*+1)th point is set as a group of different straight lines. After grouping is completed, the position at both ends of the obtained straight line (l_{x1}, l_{y1}, l_{x2}, l_{y2}) is registered as a line segment-type landmark.

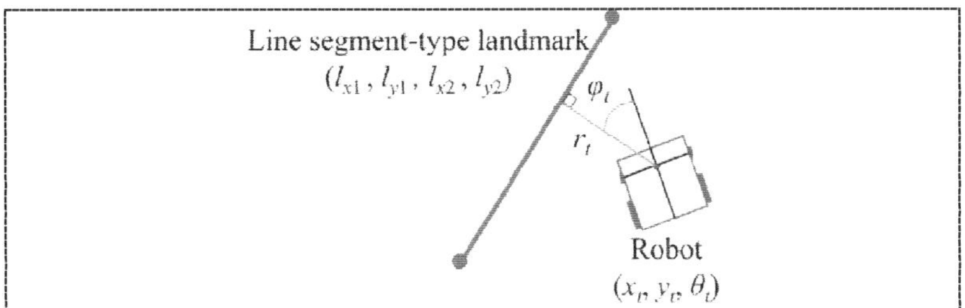

Figure 3. Measurement model of the line segment-type landmark

The relationship between a robot and a line segment-type landmark is shown in figure 3. The measured value z_t obtained from the line segment-type landmark is the distance r_t from the robot to the landmark, and the relative angle φ_t. Therefore, the measured value is expressed by the following equation.

$$\mathbf{z}_t = \begin{bmatrix} r_t \\ \varphi_t \end{bmatrix} \tag{4}$$

Using the position (x_t, y_t), and attitude θ_t of the robot, and also using the linear parameters, i.e., (a, b, c) related to the line segment-type landmark, the measurement function h in equation (2) is expressed by the following equation:

$$h = \begin{bmatrix} \frac{|\mathrm{a}x_t + \mathrm{b}y_t + \mathrm{c}|}{\sqrt{\mathrm{a}^2+\mathrm{b}^2}} \\ \tan^{-1}\frac{\mathrm{b}}{\mathrm{a}} - \theta_t \end{bmatrix} \tag{5}$$

Here, the linear parameters a, b, c of the line segment-type landmark used in equation (5) are expressed, using the position at both ends of the obtained straight line (l_{x1}, l_{y1}, l_{x2}, l_{y2}), by the following equations:

$$a = l_{y2} - l_{y1} \tag{6}$$
$$b = -(l_{x2} - l_{x1}) \tag{7}$$
$$c = -(al_{x1} + bl_{y1}) \tag{8}$$

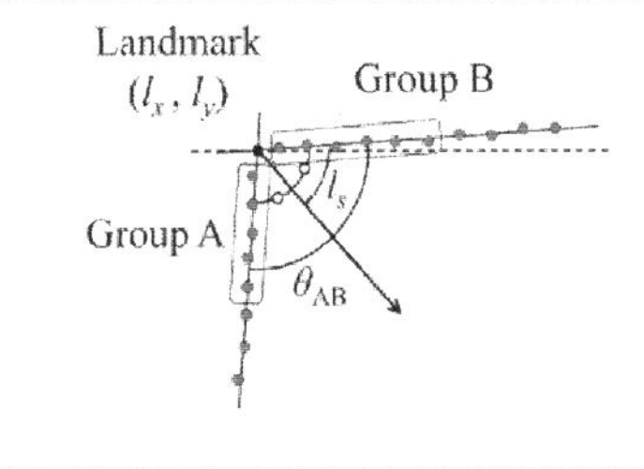

Figure 4. Position detection of the point-type landmark

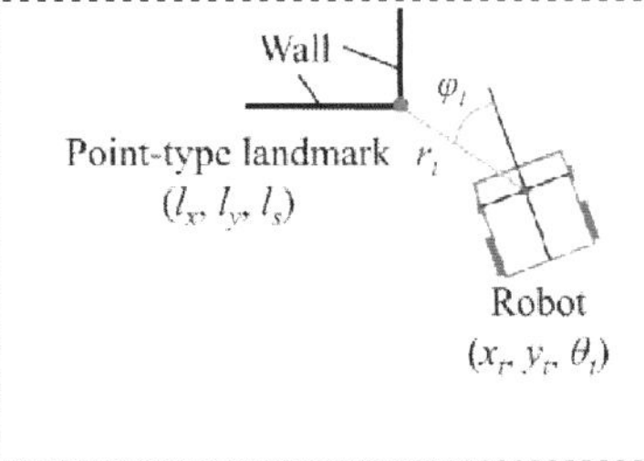

Figure 5. Measurement model of the point-type landmark

3.2. Point-type landmark

The position detection of point-type landmarks is explained with figure 4. First, compare the information on the wall acquired when the line segment-type landmark was measured. Next, the angle θ_{AB} formed between the walls is calculated using the linear parameters of the wall. When the angle θ_{AB} is around 90 deg, the intersection of the walls is estimated as a point-type landmark. Assuming that a straight line of group A is $a_A x + b_A y + c_A = 0$ and a straight line of group B is $a_B x + b_B y + c_B = 0$, the estimated point-type landmark position (l_x, l_y) is expressed as follows:

$$\begin{bmatrix} l_x \\ l_y \end{bmatrix} = \frac{1}{(a_A b_B - a_B b_A)} \begin{bmatrix} b_A c_B - b_B c_A \\ a_B c_A - a_A c_B \end{bmatrix} \tag{9}$$

The relationship between the robot and a point type landmark is shown in figure 5. The measured value z_t obtained from the point-type landmark is the distance r_t from the robot to the landmark, the relative angle φ_t and its direction s_t. Therefore, the measured value is expressed by the following equation:

$$\mathbf{z}_t = \begin{bmatrix} r_t \\ \varphi_t \\ s_t \end{bmatrix} \tag{10}$$

The measurement function h is expressed, using the position (x_t, y_t), the attitude θ_t of the robot and the position (l_x, l_y) and the direction l_s of the point type landmark, by the following equation:

$$h = \begin{bmatrix} \sqrt{(l_x - x_t)^2 + (l_y - y_t)^2} \\ \tan^{-1}\left(\frac{l_y - y_t}{l_x - x_t}\right) - \theta_t \\ l_s \end{bmatrix} \tag{11}$$

4. Experiments

Figure 6 shows the robot used in this research. The experimental robot is a front wheel steering type mobile robot driven by a rear wheel. An encoder is attached to the driving wheels, and an odometry is calculated based on the information. The size of this robot is 85 cm in length, 60 cm in width and 65 cm in height. The mounted sensor has a 30 m range and 0.25 deg angular resolution from -135 deg to 135 deg.

Figure 6. A mobile robot

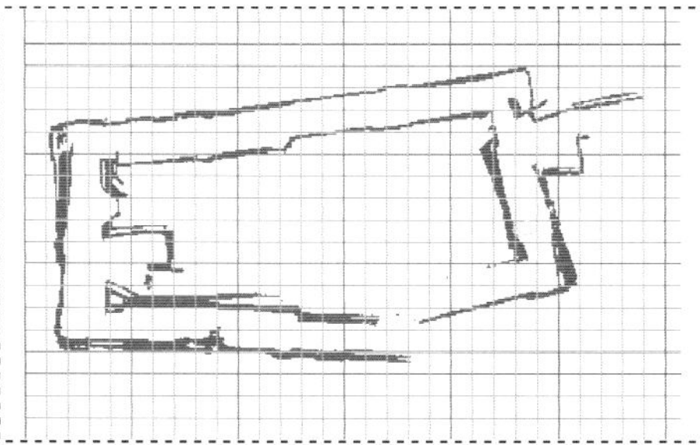

Figure 7. A map created by AUKF-SLAM using only point-type landmarks

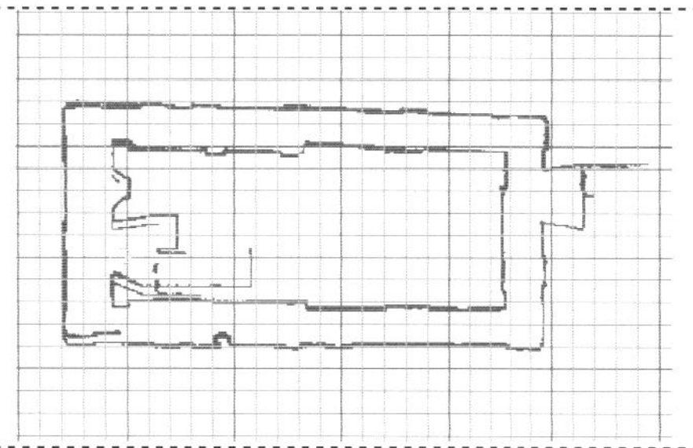

Figure 8. A map created by AUKF-SLAM using both line segment-type and point-type landmarks

The mobile robot autonomously moved in the environment composed of few point-type features. The robot does not possess the map information in advance, and automatically follows the route generated by using the DT method. Then, the robot carries out SLAM using only point-type landmarks, and similarly does SLAM using line segment-type and point-type landmarks, and then, the performances are compared each other. The robot corrects the self-location and updates a map every 250 ms. Figure 7 shows a search result using only point-type landmarks. Since there were few point-type features, the robot was not able to correct the self-location when moving through the corridor. As a result, the map was distorted. The distortion was accumulated, so that the robot was not able to create a closed map. Consequently, the route was not able to be generated due to the gap between the map and the sensor data, and the robot was left in the unknown environment. On the other hand, figure 8 shows a map generated using both line segment-type and point-type landmarks. The self-location of the robot was able to be corrected from the wall by using a line segment-type landmark. For this reason, the distortion that occurred during the turning motion and the translational motion was corrected. As a result, the robot succeeded in searching for an unknown environment while generating a map without blocking the route.

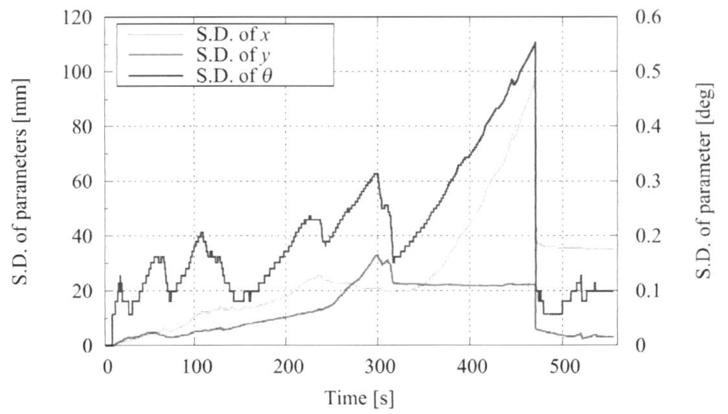

Figure 9. The standard deviations of estimated parameters

The standard deviation of self-location is shown in figure 9. The standard deviations of x, y, and θ are increasing with lapse of time. Comparing the standard deviations of x and y, it is found that the former is larger than the latter. In this environment, the robot is always moving along the wall. When using line segment-type landmarks, the robot estimates them in the direction perpendicular to the wall. Therefore, it was impossible to estimate them sufficiently in the x direction where few walls exist. On the other hand, since the wall always existed in the y direction, it is considered that the standard deviation of y became smaller than that of x. There were many changes in the standard deviation of θ because the orientation of the robot changed frequently due to the turning motion and the adjustment for following the route. However, it was confirmed that the increase in the standard deviation of θ was suppressed by the correction using the line segment-type landmark. Furthermore, in near 470 s, the self-location was greatly corrected and the standard deviations became small by remeasuring the highly reliable landmark detected at the start. It was confirmed by this experiment that the line segment-type landmark was very useful in the environment where it was difficult to obtain the point-type features.

5. Conclusion

In this paper, we have described AUKF-SLAM, line segment-type and point-type landmarks. Next, we conducted experiments using an actual robot and verified that line segment-type landmarks were useful. In the future, we will conduct an experiment to verify the estimated kinematic parameters of the robot.

References

[1] Masahiro T 2011 Probabilistic Approaches to Localization and Mapping for Mobile Robots *Journal of the Robotics Society of Japan* **29** 423 –6
[2] Thrun S, Burgard W, and Fox D 2006 *Probabilistic Robotics* (Cambridge: The MIT Press)
[3] Kurt-Yavuz Z and Yavuz S 2012 A Comparison of EKF, UKF, FastSLAM2.0, and UKF-based FastSLAM Algorithms *IEEE 16th International Conference on Intelligent Engineering Systems* 37-43
[4] Maeyama S, Takahashi Y and Watanabe K 2015 A Solution to SLAM Problems by Simultaneous Estimation of Kinematic Parameters Including Sensor Mounting Offset with an Augmented UKF *Advanced Robotics* **29** 1137-49
[5] Andrew P G and Walterio M 2006 Real-Time Model-Based SLAM Using Line Segments *Advances in Visual Computing* 354-63
[6] Hirose K and Saito H 2012 Fast Line Description for Line-based SLAM *British Machine Vision Conference* **83** 1-11

[7] Eduardo P, Luis M L and Jose M C 2013 LineSLAM: Visual Real Time Localization Using Lines and UKF *ROBOT2013: First Iberian Robotics Conference* **252** 663-78

[8] Sunhyo K and Se-Young O 2008 SLAM in Indoor Environments using Omnidirectional Vertical and Horizontal Line Features *Journal of Intelligent and Robotic Systems* **51** 31-43

The study of production performance of water heater manufacturing by using simulation method

M Iqbal[1], OAA Bamatraf[2] and M Tadjuddin[3]

[1,3]Department of Mechanical Engineering, Faculty of Engneering, Syiah Kuala University, Banda Aceh, Indonesia
[2]Department of Manufacturing and Material Engineering, Kulliyah of Engineering, International Islamic University Malaysia

email: mohd.iqbal@unsyiah.ac.id

Abstract. In industrial companies, as demand increases, decision-making to increase production becomes difficult due to the complexity of the model systems. Companies are trying to find the optimum methods to tackle such problems so that resources are utilized and production is increased. One line system of a manufacturing company in Malaysia was considered in this research. The Company produces several types of water heater and each type went into many processes, which was divided into twenty six sections. Each section has several operations. The main type of the product was 10G water heater which is produced most compare to other types, hence it was taken under consideration to be studied in this research. It was difficult to find the critical section that could improve the productions of the company. This research paper employed Delmia Quest software, Distribution Analyser software and Design of Experiment (DOE software) to simulate one model system taken from the company to be studied and to find the critical section that will improve the production system. As a result, assembly of inner and outer tank section were found to be the bottleneck section. Adding one section to the bottleneck increases the production rate by four products a day. The buffer size is determined by the experiment was six items.

Keywords: production parameters, production performances, utilizations, production rate, waiting, simulation method.

Introduction

A production system is defined as the arrangement that includes all the required processes, mechanisms, machines and so on to produce a specific output. This system consists of three main categories. The first one is input which includes all raw materials, labours, machines, components or parts. The second category is conversion process which converts the raw materials from one form to another whether manually, chemically or mechanically. Also, it consists of supporting activities such as purchase of raw materials, production planning and control, storing raw materials and finished goods and many more which can help the conversion process. Final category is the output which includes customer service and the final products (finished goods) [1, 2].

Moreover, one of the most important tool in industries all over the world is scheduling [3]. Scheduling is a matter of organizing, controlling and enhancing the work to increase the production efficiently so that it meets the customers' demands. This tool is used to appoint equipment and means, design the production process, strategize the purchase materials and plan the human resources. One other aspect which is important in the industries is decision-making to improve the production performance. Long system and big industries face a problem in finding the processes that could make a difference in the production system, which is mostly complicated and tricky to get the optimum processes. Many techniques used by hand can solve such problems, however, complex long systems need computer [4, 5, 6] programs to be study and analyse the situation. That is why the need of computer literate work-force is essential and important to meet the massive growing of customer demands [1, 2].

One of the most efficient software is Delmia Quest, which is used to simulate a real-life of everyday process [7, 8]. Delmia stands for Digital Enterprise Lean Manufacturing Interactive Application. It is a very powerful tool to visualize the flow of the products from raw materials to finished products by setting number of sources, machines, buffers and sinks, cycle times, capacity and other related parameters. The software enable the user to conduct experiment provide optimum study under simulation model. The software also shows accelerations/decelerations of the production performances, and plan layout. It becomes easier to evaluate the effects of production parameters to production performances without interrupting the real processes in the production line.

Objectives of this study were to study the effect of the production parameters (inter arrival time, number of machine and buffer size of bottle neck sections) to the production performances (utilization, throughput and average waiting time). Furthermore, the related mathematical models would be developed to be used to predict the performance at any production condition.

Manufacturing of The Water Heater

The study focussed on the production process of the 10 G water heater. Figure 1 showed the components of the water heater and Figure 2 showed its process sequence. The production line consisted of 26 sections. All of the sections were operated manually. The production line worked 8 hours per shift, one shift a day and five days a week. The queuing rule was first-in-first-out (FIFO) and the raw materials were always available.

The system started with two lines which produced inner tank and outer tank. Inner tank went through 13 sections and outer tank went through 6 sections. Then, those two lines intersected in one section to be assembled through 7 sections.

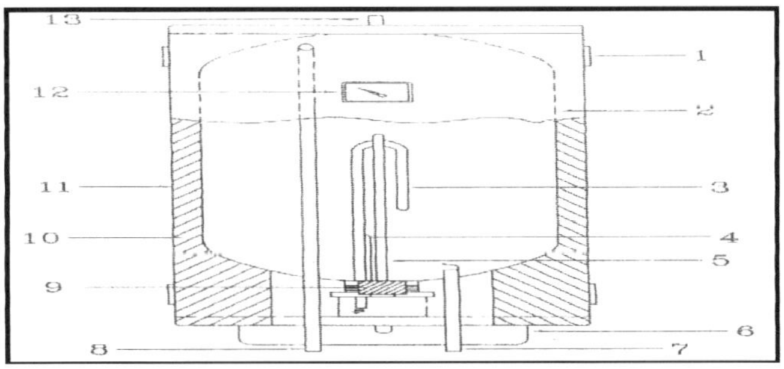

1) Bracket, 2) inner body, 3) heating element, 4) thermal fuse pocket, 5) thermostat pocket, 6) inspection cover, 7) inlet pipe, 8) outlet pipe, 9) thermal cut out device, 10) insulation, 11) outer body, 12) thermometerand 13) vent pipe

Figure 1. Components of the heater

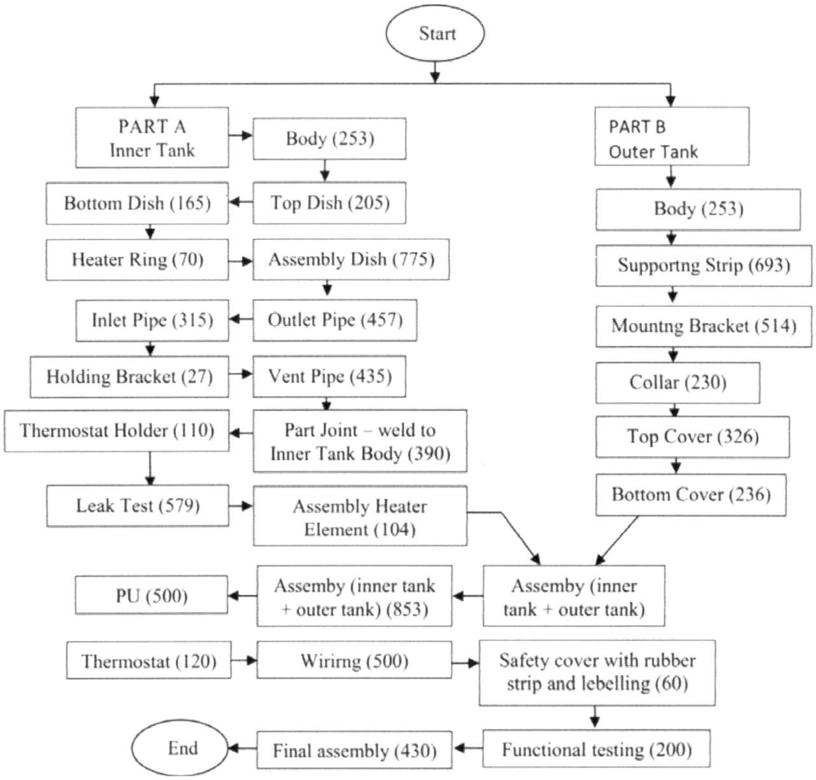

Figure 2. Process flow water heater production lines.

Methodology

The research started with the development of simulation model by using Delmia Quest simulation software. All of the stations, buffers, material generators and finish product storage were generated in the model to represent the real production activities of the production line, as shown in Figure 3.

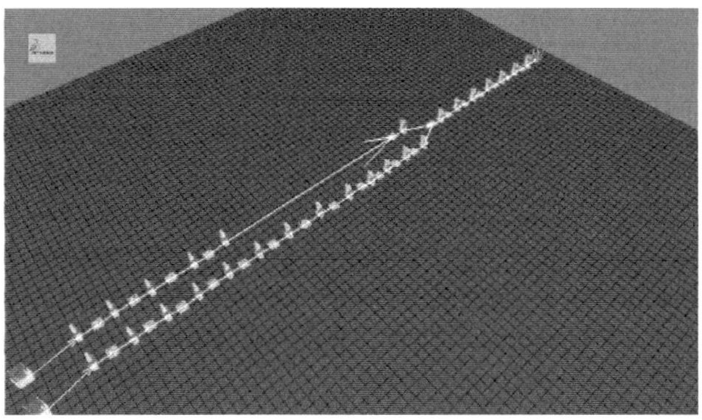

Figure 3. Simulation model in Delmia Quest software

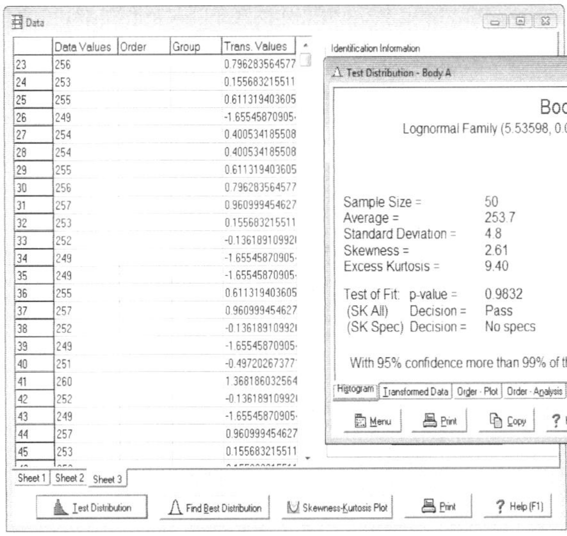

Figure 4. Result of Type of Distribution for Body A

Table 1. The cycle time of the stations

#	Machine Name	Distribution Type	Cycle time
1	Body A	Lognormal	M=253.7 SD=4.8
2	Top Dish A	Uniform	Min=200 Max=210
3	Bottom Dish A	Uniform	Min=180 Max=193
4	Heater Ring A	Exponential	M=72.1
5	Assembly Dish A	Uniform	Max=770 Min=780
6	Outlet Pipe A	Normal	M=454.3 SD=3.1
7	Inlet Pipe A	Uniform	Min=310 Max=320
8	Vent Pipe A	Uniform	Min=430 Max=440
9	Holding Bracket A	Uniform	Min=23 Max=30
10	Thermostat Holder A	Uniform	Min=105 Max=115
11	Part Joint- Weld to inner tank body A	Uniform	Min=388 Max=395
12	Assembly-Heater element A	Uniform	Min=100 Max= 107
13	Leak Test A	Uniform	Min=577 Max=585
14	Body B	Uniform	Min=157 Max=165
15	Supporting Strip B	Uniform	Min=688 Max=697
16	Mounting Bracket B	Uniform	Min=510 Max=518
17	Collar B	Uniform	Min=227 Max=235
18	Top Cover B	Normal	M=326.2 SD=2.1
19	Bottom Cover B	Uniform	Min=230 Max=244
20	Assembly (Inner tank + Outer Tank) C	Normal	M= 853.4 SD= 2.4
21	PU C	Uniform	Min=498 Max=505
22	Thermostat C	Uniform	Min=117 Max=123
23	Wiring C	Uniform	Min=498 Max=505
24	Safety cover with rubber strip and labeling C	Uniform	Min=57 Max=63
25	Functional testing C	Uniform	Min=197 Max=203
26	Final assembly C	Uniform	Min=427 Max=433

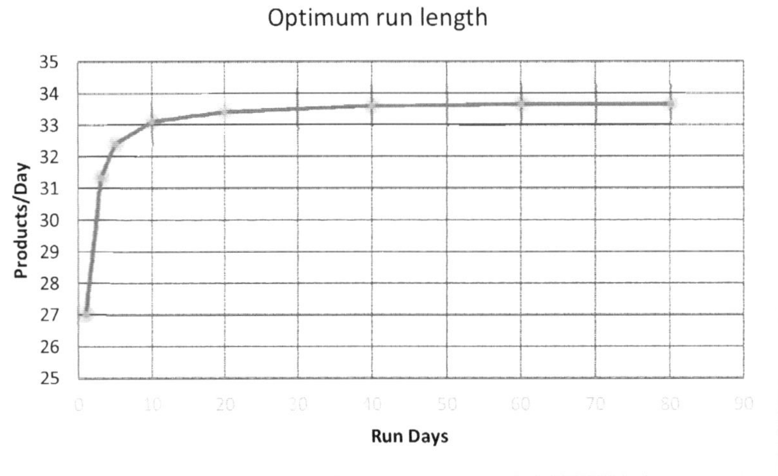

Figure 5. Optimum Run Length

The cycle time (operation time and handling time) of each section was collected from both previous company record and from direct measure in the production line, then formulated into the simulation model. Fifty cycle time data was selected for each of the station. The group of data was analysed by using Distribution Analyser Software in order to find out the type of statistical distribution of the cycle time and its properties. One of the results of Distribution Analyzer software are shown in Fig 4. It will be used as an input data into Delmia Quest software in the next step. The cycle times of all of the sections are shown in Table 1.

Three parameters were selected as independent variables and their limit were determined. They were inter arrival time to the line (1 to 3600 seconds), buffer size (1 to 5) and number of machine of the bottle neck section (1 to 3). Three performances were selected as dependent variables or responses. They were throughput of the line, average waiting time and utilization of the whole line. The Design Of the Experiment (DEO) was developed by using Box-Behnken technique of Response Surface Methodology (RSM) [9, 10]. With the help of Design Expert software, the DOE method created 17 experiment conditions as shown in Table 2.

The simulation study started with steady-state study to find the optimum run length where the production did not change much. The simulation started with run length of 10 days and repeated by adding each 10 days and the throughput was recorded. Figure 5 showed that the throughput per day increases as the number of days the simulation was run till the number of day reaches 60, where the change in production per day is very small. Hence, the optimum run length is found to be 60 days.

Validation of the simulation model was conducted through the way the products movement from one section to another in the model and compare it to the real production line. The other way was through the actual daily production. The real production line produced about 33 heaters per day (Taken from the cycle time sheet provided by the Company). The result of the simulation run showed that the production per day was about 33 heaters. Hence, this system was valid and the analysis could be based on it.

The model was run for 60 days which was the optimum run length and then the bottleneck section was determined based on the highest utilization. The results showed that there were four bottleneck sections in the production line. A further investigation must be made to determine a real bottleneck section. Therefore, one identical machine was added to each bottle neck station and then the simulation is run to examine the effects of adding the machine to the utilization of the section. The result showed that the first section of the assembly line (Assembly Inner + Outer tank) has a great effect on the throughput by 37.033 heaters/day. Therefore, Assembly Inner + Outer Tank section was considered as the bottleneck section.

Table 2. Results of the experiment

Std	Run	Faktor 1 A: Number of Machine	Faktor 2 B: Buffer Size	Faktor 3 C: Inter Arrival Time	Response 1 Utilazazion (%)	Response 2 Througput (pieces)	Response 3 Average Waiting Time (hour)
12	1	2	15	3600.00	10.8235	598	1.62368
16	2	2	10	1800.50	19.4795	1077	1.7586
15	3	2	10	1800.50	19.4795	1077	1.7586
11	4	2	5	3600.00	10.824	598	1.62348
17	5	2	10	1800.50	19.4795	1076	1.7056
2	6	3	5	1800.50	18.8038	1077	1.75148
10	7	2	15	1.00	50.5314	2222	29.6986
8	8	3	10	3600.00	10.4411	597	1.61448
4	9	3	15	1800.50	18.8017	1076	1.74966
5	10	1	10	1.00	52.41	2222	29.6992
7	11	1	10	3600.00	11.2244	598	1.6364
6	12	3	10	1.00	48.8004	2222	29.683
9	13	2	5	1.00	50.5403	2222	29.6752
13	14	2	10	1800.50	19.4795	1077	1.76
14	15	2	10	1800.50	19.4795	1077	1.76
1	16	1	5	1800.50	20.1994	1077	1.76628
3	17	1	15	1800.50	20.1988	1077	1.7672

Results and Discussion

The simulation experiment was conducted based on the DOE. The responses were collected and recorded to the Design Expert software for further statistical analysis as sown in Table 2. The result of Analysis of Variance (ANOVA) showed that the model term was significant with F is < 0.05 as shown in Table 3. Furthermore, Factor A (number of machine) and Factor C (inter arrival time) were significant since their $F < 0.0001$. Factor B (buffer size) was not significant, and there was not any significant interaction between the factors.

The experimental result was plotted into 3D graph to show the effect of the production parameters to the production performances. From figure 6, it can be seen clearly that as the inter arrival time decreases the throughput of the system increases. However, the effect of number of machine to the throughput could not be seen in the graph. Figure 7 showed that buffer size did not give any effect to the throughput. At the other hand, it could be seen clearly that the throughput was affected significantly by the number of machine. Adding 1 machine to the bottleneck section will increase the throughput. However, adding 1 more machine did not give any change to the throughput. It meant that the optimum number of the machine in the bottle neck section was 2.

Figure 8 showed that inter arrival time affects the utilization significantly. Less inter arrival time gave higher utilization. Higher number of machine made lower average utilization. It was obvious that the more machines added to the system, the less load imparted to the system which results in lower average utilization of the whole system.

IOP Conf. Series: Journal of Physics: Conf. Series **962** (2018) 012019 doi:10.1088/1742-6596/962/1/012019

Table 3. Analysis of Variance

Transform | Fit Summary | Model | ANOVA | Diagnostics | Model Graphs

Use your mouse to right click on individual cells for definitions.

Response: Utilization

ANOVA for Response Surface Reduced Quadratic Model

Analysis of variance table [Partial sum of squares]

Source	Sum of Squares	DF	Mean Square	F Value	Prob > F	
Model	3699.60	4	924.90	34040.66	< 0.0001	significant
A	*6.45*	*1*	*6.45*	*237.54*	*< 0.0001*	
C	*3158.89*	*1*	*3158.89*	*1.163E+005*	*< 0.0001*	
C²	*532.26*	*1*	*532.26*	*19589.60*	*< 0.0001*	
AC	*2.00*	*1*	*2.00*	*73.49*	*< 0.0001*	
Residual	0.33	12	0.027			
Lack of Fit	*0.33*	*8*	*0.041*			
Pure Error	*0.000*	*4*	*0.000*			
Cor Total	3699.93	16				

The Model F-value of 34040.66 implies the model is significant. There is only
a 0.01% chance that a "Model F-Value" this large could occur due to noise.

Values of "Prob > F" less than 0.0500 indicate model terms are significant.
In this case A, C, C^2, AC are significant model terms.
Values greater than 0.1000 indicate the model terms are not significant.
If there are many insignificant model terms (not counting those required to support hierarchy),
model reduction may improve your model.

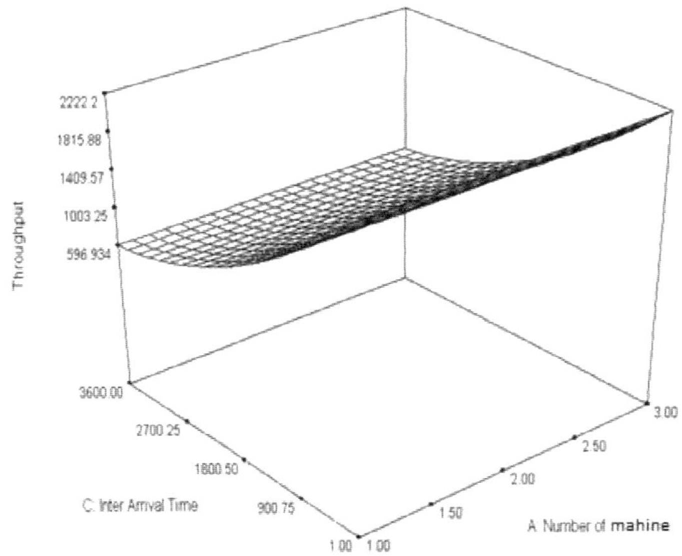

Figure 6. Throughput VS inter arrival time and number of machine

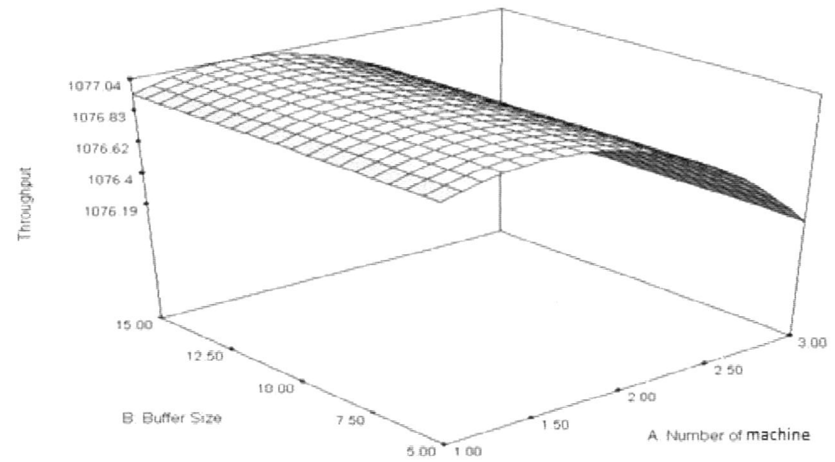

Figure 7. Throughput VS buffer size and number of machine

Figure 9 showed the relation between the numbers of machine and buffer size to the the average waiting time of the part to be processed by the sections. From the 3D graph shown in figure 9, as number of machine added to the bottleneck section increased, the average waiting time decreased. Similar trend happened to the buffer size. As the buffer size increased, the average waiting time decreased.

Mathematical model for each performance was developed by the help of Design Expert Software using the same method of the DOE. All those formulas were based on the bottleneck section found earlier, which is the assembly of inner and outer tank of the water heater. Therefore, number of machine in the formula refers to the sections added to the bottleneck section and buffer size also refer to the work in progress for the bottleneck section as well. Thus, using those formula will save time for the company to calculate and examine the average utilization of the manufacturing line, throughput of the whole manufacturing line and average waiting of the products in the line at any given parameters.

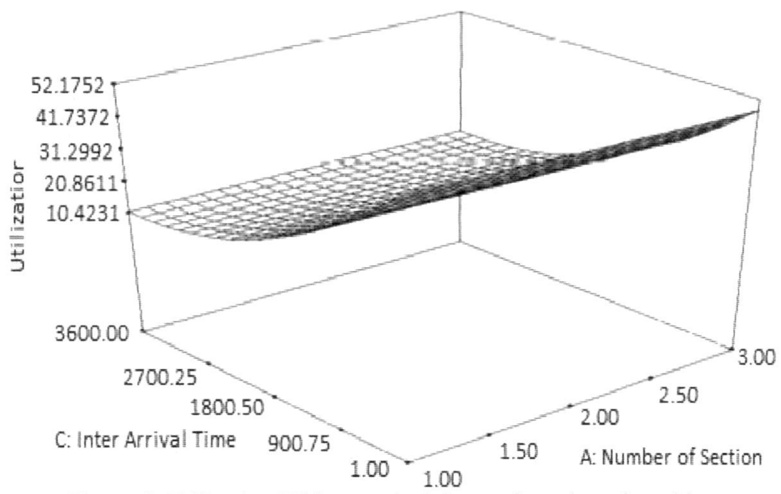

Figure 8. Utilization VS inter arrival time and number of machine

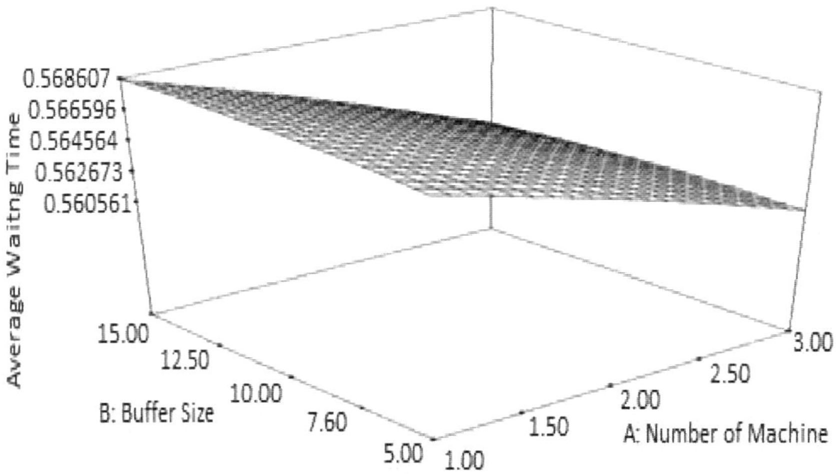

Figure 9. Average waiting time VS number of machine and buffer size vs

The mathematical models are as follow:

Utilization = 53.80428 – 1.60514 * Number of sections – 0.024294 * Inter arrival time + 3.4619E-006 * Inter arrival time^2 + 3.92633E-004 * Number of machine * Inter arrival time.

Throughput = 2221.73584 + 1.40145 * Number of machine – 0.82147 * Inter arrival time-0.38158 * Number of machine^2 + 1.02871E-004 * Inter arrival time^2 -1.38927E-004 * Number of machine * Inter arrival time

Average waiting time = 3.39178 + 4.86547E-004 * Number of machine + 3.93494E-004 * buffer size – 2.33108E-003 * Inter arrival time – 2.44034E-004 * Number of machine^2 -1.20593E-005 * Buffer size^2 + 4.23974E-007 * Inter arrival time^2 -5.58647E-005 * Number of machine * Buffer size - 1.47816E-006 * Number of machine * Inter arrival time -1.84785E-008 * Buffer size * Inter arrival time

Conclusion

The assembly system of the 10 G water heater has been study by using simulation method with the help of Delmia Quest software. Result of the experiment successfully showed the effect of productions parameters to the production performances. The best solution that chosen based on the maximum throughput of the whole system would be achieved by 1 second of inter arrival time, 2 machine in the bottleneck section and 6 part of the buffer size. The production performances at different level of the prpduction parameters could be predicted by using mathematical models developed in the study.

References

[1] Bhadur, R. (2008). Production and operation management. Jaipur, India: Book Enclave.

[2] Groover, M. P. (2008). Automation, Production Systems, and Computer-Integrated Manufacturing. Third Edition. New Jersey: Pearson Education Inc, pp144.

[3] Taj, S., Nedeltcheva, G. N., Pfeil, G., & Roumaya, M. (2012). A spread-sheet model for efficient production and scheduling of a manufacturing line/cell. International Journal of Production Research, 50(4), 1141-1154.

[4] Penker, A. (2007). Bottleneck analysis in MDF-production by means of discrete event simulation. International Journal of Simulation Modelling, 6(1), 49-57.

[5] Um, I., Cheon, H., & Lee, H. (2009). The simulation design and analysis of a Flexible Manufacturing System with Automated Guided Vehicle System. Journal of Manufacturing Systems, 28(4), 115-122.

[6] Mak, L. C., Wong, W. K., & Leung, Y. S. (2014). A simulation analysis of the impact of production lot size and its interaction with operator competence on manufacturing system performance. Simulation Modelling Practice and Theory, 49, 203-214.

[7] Bzymek, Z. M., Nunez, M., Li, M., & Powers, S. (2008). Simulation of a Machining Sequence Using Delmia/Quest Software. Computer-Aided Design and Applications, 5(1-4), 401-411.

[8] Lu, H., Liu, X., Pang, W., Ye, W. H., & Wei, B. S. (2012). Modeling and Simulation of Aircraft Assembly Line Based on Quest. AMR, 569, 666-669.

[9] Montevechi, J. A., De Pinho, A. F., Leal, F., & Marins, F. A. (2007). Application of design of experiments on the simulation of a process in automotive industry. 2007 Winter Simulation Conference.

[10] Antony, J., Coleman, S., Montgomery, D. C., Anderson, M. J., & Silvestrini, R. T. (2011). Design of experiments for non-manufacturing processes: benefits, challenges and some examples. Proceedings of the Institution of Mechanical Engineers, Part B: Journal of Engineering Manufacture, 225(11), 2078-2087

IOP Conf. Series: Journal of Physics: Conf. Series **962** (2018) 012020 doi:10.1088/1742-6596/962/1/012020

Design Multi-Sides System Unmanned Surface Vehicle (USV) Rocket

Rafiudin Syam, Onny Sutresman, Abdullah Mappaita, Amiruddin, Ardi Wiranata

Mechanical Engineering Department, Engineering Faculty, Hasanuddin University, Indonesia

Corresponding author's: rafiuddinsyam@gmail.com

Abstract. This study aims to design and test USV multislide forms. This system is excellent for maneuvering on the x-y-z coordinates. The disadvantage of a single side USV is that it is very difficult to maneuver to achieve very dynamic targets. While for multi sides system easily maneuvered though x-y-z coordinates. In addition to security defense purposes, multi-side system is also good for maritime intelligence, surveillance. In this case, electric deducted fan with Multi-Side system so that the vehicle can still operate even in reverse condition. Multiple-side USV experiments have done with good results. In a USV study designed to use two propulsions.

1. Introduction

The state of Indonesia has consists mainly of the sea and has separated islands extending from Sabang city to Merauke city. Therefore, the government, especially the Navy has difficulties and limitations for monitoring the security of each island. Along with changes in emphasis, new capabilities will require navies in the field of maritime intelligence, surveillance, reconnaissance and assault.

Unmanned surface vehicle (USV) is an unmanned vehicle that is operated on the surface (surface) for a particular purpose. USV is also known as autonomous surface vehicle (ASV) or an automatic surface vehicle because it uses a global positioning system (GPS) in determining the direction of the purpose of the movement of the vehicle [1] - [2].

Many researchers have done on USV topics. [3] and [4] researchers presents an automatic method to acquire, identify, and track obstacles from an Unmanned Surface Vehicle (USV) location in marine environments using 2D Commercial of The Shelf (COTS) video sensors, and analyzing video streams as input. The researcher also shows the guiding line of this research is to develop real-time automatic identification and tracking abilities in marine environment with COTS sensors.

Another researcher [5] states a USV with a trajectory planning and tracking approach for following a differentially constrained target vehicle operating in an obstacle field. The developed approach includes a novel algorithm for computing a desired pose and surge speed of the target boat, jointly defined as a motion goal, and tightly integrates it with trajectory planning and tracking components of the entire system. The trajectory planner generates a dynamically feasible, collision-free trajectory to allow the USV safely reach the computed motion goal.

However, in this study, the authors propose new design of USV rocket. The USV consist of four sides with two propellers as electric deducted fan (EDV).

Unmanned surface vehicle (USV) is indispensable in everyday life. Especially in the field of surveillance or observation in places that are harmful to humans. System Unnamed to reduce the risk to the manned forces. Research on USV has been widely practiced in various countries of the world in both the hull model and the USV drive system, but there less study of USV for multi-sides systems that use rockets as propulsion.

2. Literature Review

2.1 Rocket

A rocket is a spacecraft, a missile, or a flying vehicle that gets a boost through a rocket reaction to the rapid discharge of the fluid material from the engine output of the rocket. The action of the output in the combustion chamber and the developer nozzle, capable of making the gas flow at hypersonic speed so as to generate a large reactive impulse for the rocket is proportional to the counter reaction, in accordance with the Law of Newton III.

The movement of a rocket is determined by the angle of movement of the three rocket fins. In the presence of elevator control fins, rudder control fins and aileron control fins. The elevator control fins are the controls that adjust the motion of a rocket up and down with a certain degree (pitch angle change). Full rudder fins are controls that can deflect the rocket nose to the left and right or so-called yaw movements. While the aileron control is a surface control that can control the motion of rocket roll [6]

USV rocket system works more or less the same as a rocket in general, but rocket work on the water surface is much easier. The system moves in control by adjusting the speed of both USV rocket boosters, turn by temporarily halting or reducing the speed of one of the driving forces.

Mechanical made in the form of ship with catamaran concept. The ship type has 4 symmetrical main hulls (Hulls) on the right and left sides and has two sides of the upper and lower hull that allow the ship to operate even in reverse and can. Move (Maneuver) more balanced.

2.2. Control System

Control issues are closely related to strategies that allow a microcontroler to direct the movements of a robot through the sensor and deliver the robot's response to the microcontroller. The control system has three elements: input, process and output [7].

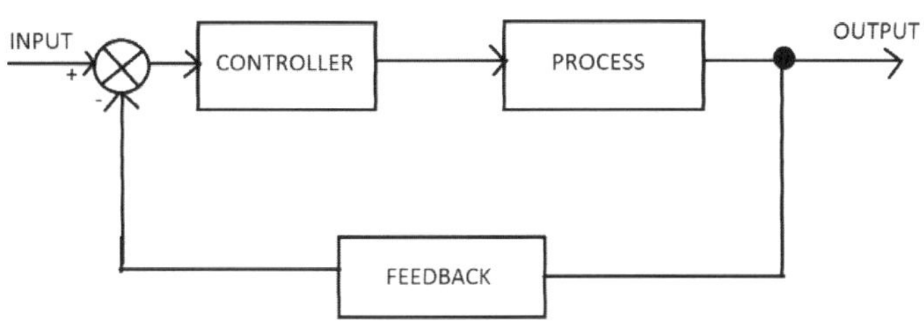

Figure 1. Control System

Input is generally a signal and a transducer is a tool that can change the physical quantities into electrical quantities, such as buttons, limit switches, thermostat and others. The transducer provides a quantifiable quantity, then this quantity can be processed by the process part, may be a control circuit using electrically assembled equipment or a programmable control system such as a microprocessor and microcontroller system.

IOP Conf. Series: Journal of Physics: Conf. Series **962** (2018) 012020 doi:10.1088/1742-6596/962/1/012020

Information processing called input signal, generates an output signal, which is then used to activate the actuator as output equipment, which can be an electric motor, contactor, solenoid valve, lamp etc. With output equipment, the quantity of electricity converted into physical quantities.

3. Research Mehod

3.1 Work Design

Mechanics made in the form of USV rocket with catamaran concept. The USV rocket has four hulls. Two symmetrical hulls on the right and left sides at the bottom of the USV rocket, while the other two hulls are at the top of the USV Rocket in the same position. Such gastric construction allows the USV rocket to remain in operation even in reverse.

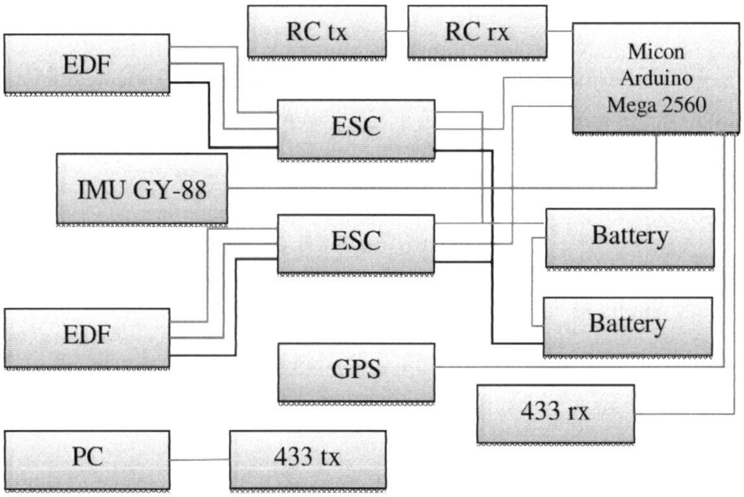

Figure 2. Electronic Circuit Scheme

3.2 Electronic Device System

Electronic devices on USV rockets that will be made include Electric Ducted Fan (EDF), IMU GY-88 Sensor, Electronic Speed Controler (ESC), arduino mega 2560 and Remote Control. The circuit schematic shown in Figure 3.

Figure 3. Model of Multiside System unmanned surface vehicle

3.3 User Interface with Google Earth

The user interface display can be seen in figure 19. The displayed data consists of roll, pitch, yaw, compass and current direction. The data shown is the current position data of the rides and the trajectory.

The selection of Google Earth to show the position of the rides is due to its easy use and user friendly or easy to use interface. Google Earth is also available with unpaid maps making it easy to access the locations used.

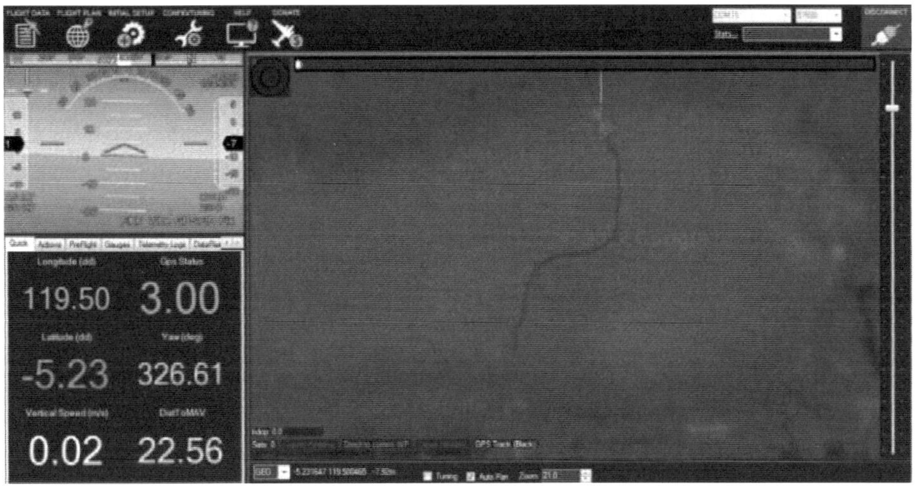

Figure 4. Google Earth User Interface

4. Results and Discussion

4.1. Experiment Results

4.1.1. Measurement of thrust force.

The thrust force in the multimotor is due to the propeller-generated lift force. Since the pitch angle of the propeller is fixed, then to increase or decrease the lift force is by increasing or decreasing the speed of the propeller angle. This can be found in figure 4

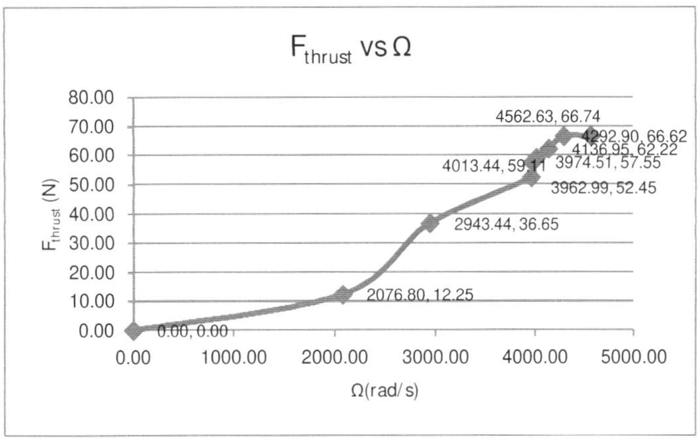

Figure 5. Relationship between thrust force Fthrust (N) with rotation speed Ω (rad / s).

Because the propeller is rotated has a dimension and mass large enough. To be able to provide high angular speed, the motor requires considerable power (watts). This can be seen in Figure 6.

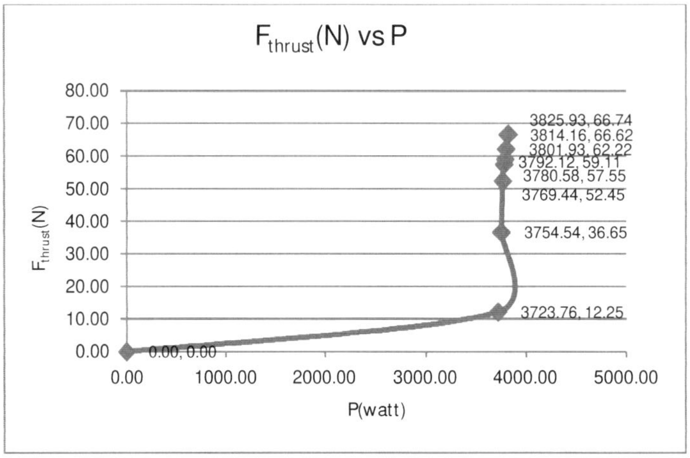

Figure 6. The relationship between the thrust force Fdorong (N) and power P (watts).

Because the measurement is done with tools that are less precision, such as tachometer that still uses the touch system. So there is an inaccuracy of data retrieval. This can be seen in the measurement of the thrust force constant (Ns2).

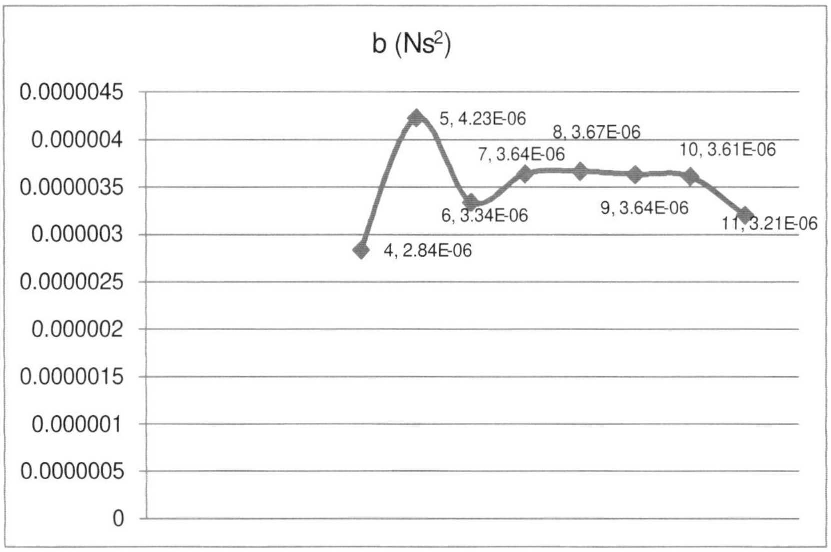

Figure 7. The result of the measurement of the thrust force constant (Ns2).

4.1.2. Measurement of moments.

Then the motor will produce a moment that direction in the opposite direction with the direction of propeller rotation, because the motor rotates propellers that have mass. In the USV rocket, this moment is used to rotate the USV rocket body on the z axis. The greater the rotation speed of the propeller, the greater the moment produced. This can be seen in figure 8.

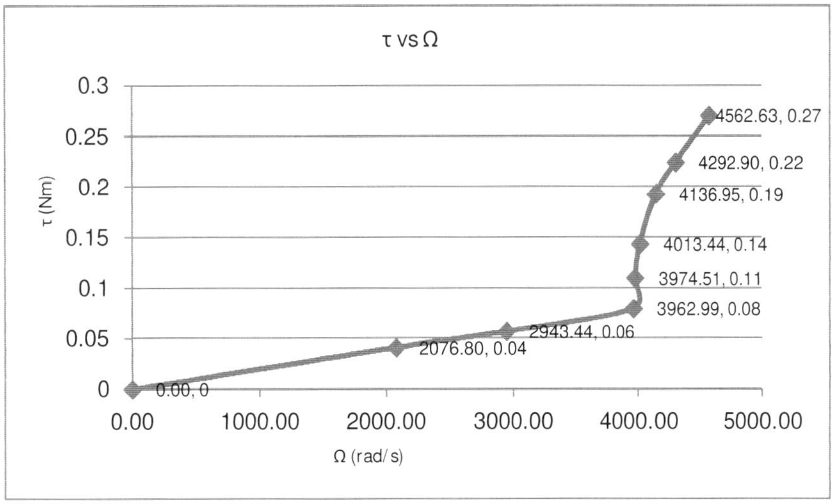

Figure 8. The relationship between the moments τ (Nm) with the angular velocity of propeller Ω (rad / s).

Because the size of the power affect the speed of rotation, the higher the power given the higher the greater the moment generated. This can be seen in Figure 9.

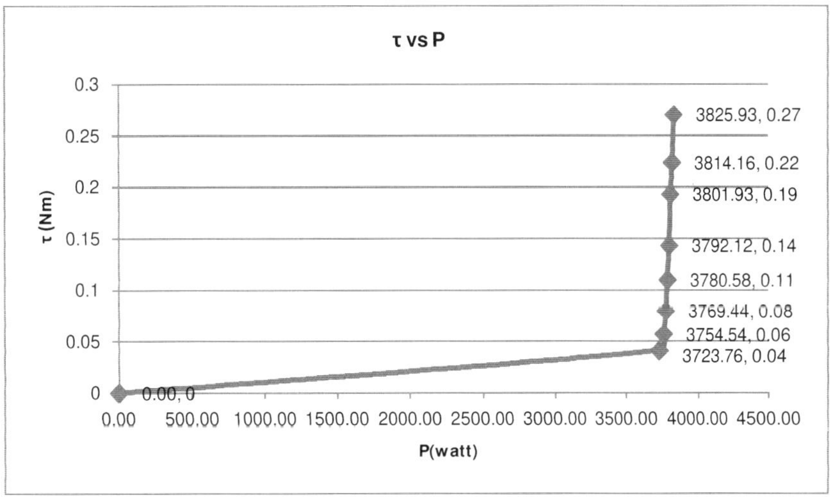

Figure 9. Relationship between moment τ (Nm) with power P (watts).

Just like the thrust gauge, the moment measurement also uses a tachometer with the touch system. Thus the propeller rotation readings become less accurate. This results in less accurate data on the measurement of drag factor d (Nms2).

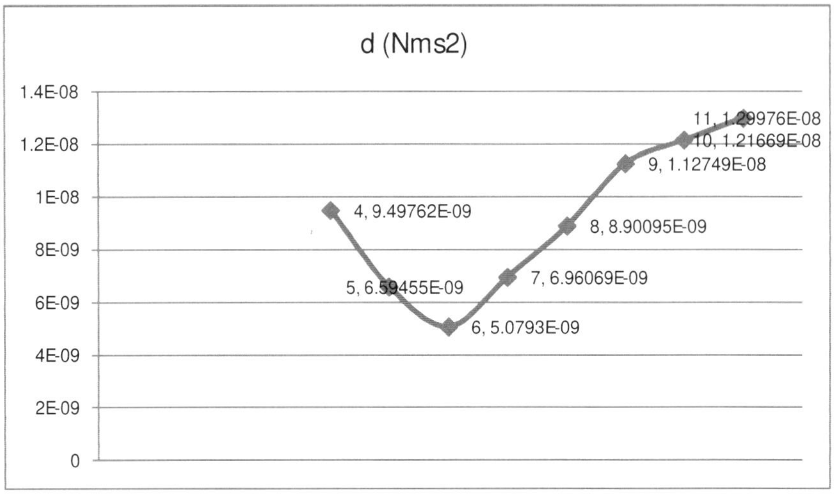

Figure 10. The drag factor measurement result d (Nm2)

5. Conclusion

Unmanned surface vehicle rockets have been made in accordance with the design. The USV rocket has a length of 720 mm, a width of 475 mm, a height of 305 mm, and a total weight of 5579 grams. This rocket is driven by two electric motor ducted fan 90 mm 1600 kv, two ESC 160 A, FlySky i6 remote controller, and Li-Po 18.5 V 4000 mAh battery power source as much as two pieces.

Using a 4000 5S 18.5 Volt battery composed in series, the amount of power the motor needs to reach its maximum speed is 3825.93 watts and the maximum thrust force the motor can generate at a maximum rotation of 4562,629 rad / s is 66,378 N. The moment that the motor can generate at maximum rotation is 0.270578 Nm. This is due to the Voltage of used batteries and the load of the rides.

Based on the observation of trajectory USV rocket when moving through variation of the path. Total error of 1896 cm and the largest deviation is 200 cm at the time of USV rocket cornering. This is because USV rockets need a big turn for cornering.

Acknowledgments
We would to thank to LPPM of Unhas and Directorate General Research of Technology and Higher Education, Ministry of National Education Republic of Indonesia (Ristekdikti) for PUPT funding.

References

[1] J. E. Manley, "Unmanned Surface Vehicles, 15 Years of Development," pp. 1-4, 2008.

[2] R. Syam and Mustari, " Simulation and Experimental Works of Quadcopter Model for Simple Maneuver," *International Journal on Smart Material and Mechatronics,* vol. 2, no. 1, pp. 29-33 , 2015.

[3] O. Sutresman, R. Syam and S. Asmal, "Controlling Unmanned Surface Vehicle Rocket using GPS Tracking Method.," *International Journal of Technology,* vol. 8, no. 4, pp. 709-718, 2017.

[4] O. Gal, "Automatic Obstacle Detection for USV's Navigation Using Vision Sensors,," in *Robotic Sailing: Proceedings of the 4th International Robotic Sailing Conference pp 127-140,*, Norway, 2011.

[5] Š. Petr and et.al, "Target following with motion prediction for unmanned surface vehicle operating in clustttuered environments," *Autonomous Robots,* vol. 36, no. 4, pp. 383-405, 2014.

[6] S. P. I. E. P. Putra Setya Bagus, "Sistem Kendali Roket RKX-200 LAPAN dengan Pengendali PID, in Bahasa (.(in Bahasa) (Control System of Rocket RKX-2000 Lapan using PID method))," ITS, Surabaya, Indonesia, 2010.

[7] K. Ogata, Modern Control Engineering, New Jersey: Prentice Hall, 2002.

Flow rate and temperature characteristics in steady state condition on *FASSIP-01 loop* during commissioning

M Juarsa, Giarno, A. N. Rohman, G.B. Heru K., J.P. Witoko, D.T. Sony Tjahyani

Center for Nuclear Reactor Technology and Safety, National Nuclear Energy Agency of Indonesia (BATAN), Kawasan PUSPIPTEK Tangerang Selatan 15340 Banten, Indonesia

juars@batan.go.id

Abstract. The need for large-scale experimental facilities to investigate the phenomenon of natural circulation flow rate becomes a necessity in the development of nuclear reactor safety management. The FASSIP-01 loop has been built to determine the natural circulation flow rate performance in the large-scale media and aimed to reduce errors in the results for its application in the design of new generation reactors. The commissioning needs to be done to define the capability of the FASSIP-01 loop and to prescribe the experiment limitations. On this commissioning, two scenarios experimental method has been used. The first scenario is a static condition test which was conducted to verify measurement system response during 24 hours without electrical load in heater and cooler, there is water and no water inside the rectangular loop. Second scenario is a dynamics condition that aims to understand the flow rate, a dynamic test was conducted using heater power of 5627 watts and coolant flow rate in the HSS loop of 9.35 LPM. The result of this test shows that the temperature characterization on static test provide a recommendation, that the experiments should be done at night because has a better environmental temperature stability compared to afternoon, with stable temperature around 1°C - 3°C. While on the dynamic test, the water temperature difference between the inlet-outlets in the heater area is quite large, about 7 times the temperature difference in the cooler area. The magnitude of the natural circulation flow rate calculated is much larger at about 300 times compared to the measured flow rate with different flow rate profiles.

Keywords: natural circulation, flow rate, temperature, commissioning, large-scale, FASSIP-01

1. Introduction

The effectiveness of natural circulation depends on various thermal-hydraulic parameters. Natural circulation in the reactor for core heat removal takes place due to the density difference between the cold regions coming down in the down comer with the heated and/or two-phase flow going upward in the core, referred to the buoyancy. The change of density due to the increase of temperature will affect the buoyancy force. Many researches on passive cooling system with focusing on the investigations on natural circulation phenomena have been carried out. Since 1967, Welender has investigated the effects of buoyancy force, pressure drop and friction loss in the pipes against a fluid flow driving force [1]. Several years later, K. Vijayan et al. [2] and Zvirin [3] studied the effect of geometrical variation on the natural circulation flow. The research was conducted using a small rectangular loop by 2.6 meter in length and 0.7 meter in width using 1 inch in diameter stainless-steel pipe. The research focused on analyzing the stability of steady and transient flow using heater and cooler control. The problem in natural circulation phenomenon is the stability flow, which was affected by boundary condition (BC) changes, by geometrical changes such as diameter, length and height. Thermal hydraulics parameter has given an influence also to the changes of BC. The differences of thermal boundary conditions, geometrical variations, a heat transfer effect and the mass flow rate parameters during natural circulation have been studied [4-9]. However, the research was done in a steady state condition using a small scale experimental apparatus. Then, the development of test facilities and thermal-hydraulic system codes have significant role in the assessment of passive safety system design, evaluation and certification to the design of advanced reactors and safety features in light water reactor (LWR) in the present.

A large-scale facility has been constructed to study natural circulation phenomenon of passive cooling system in LWR and SMRs. KAERI (Korea) made a large-scale experimental facility (ATLAS, advanced thermal-hydraulics test loop for accident simulation). The ATLAS condition refers to APR-1400 (advanced power reactor-1400MWe) reactors type [10] and using 3 inch size for natural circulation experiment. Object of experiments focused on identifying characteristics of the loop of the reactor coolant system (RCS) and compare the results with computer code. The PANDA facility (Switzerland) was constructed to simulate in RPV's natural circulation BWR reactor type. The LINX facility is a medium-scale experimental facility constructed to study passive cooling system [11]. In Indonesia [12-14], several researches were conducted to investigate natural circulation phenomenon as a preliminary experimental research in passive cooling system development. The research was done using the USSA-FT01 Loop as a small-scale rectangular loop (1.5 x 1.0 meters) with 1 inch size pipe diameter and a NC-QUEEN loop with rectangular geometry is 2.7 x 0.5 meters and 1 inch size tube diameter. However, it is very difficult to obtain information and comparison data such as geometric data, primary data from those researches, the variation of working fluid and also power magnitude which was used as well as experimental results from these facilities. To extend the research goal in the field of passive cooling system development, from 2015 to 2016 in BATAN, the new experimental facility called FASSIP-01 (Indonesian: *FAsilitas Simulasi SIstem Pasif*) or Passive System Simulation Test Facility with medium-scale rectangular loop (3.5 m x 6 m) has been build. The first step of research activity is conduct a commissioning of FASSIL loop to understand and to collect technical data, performance and characteristics of FASSIP-01 loop before experiment scenario (base on experimental matrix) will be conducted.

2. Methodology

2.1. Experiment apparatus

From 2015 until 2016, FASSIP-01 loop has been constructed as a research facility used to investigate the flow of natural circulation on the rectangular loop in medium size. FASSIP-01 loop consists of two main parts, the HSS (heat sink system) loop and the rectangular loop, where the research object is a rectangular loop with geometrical size are 3.5 m wide and 6 m high, which was placed in a vertical position. The experiment setup of FASSIP-01 loop is shown in Figure 1.

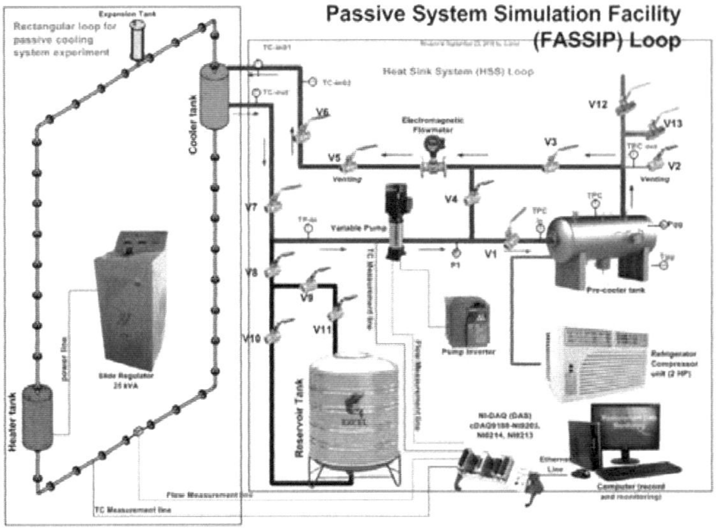

Figure 1. Experiment setup of FASSIP-01 loop

As shown in Figure 1, the rectangular loop consists of a heater tank as a heat source (non-direct, water to water, hot water is heated-up by submerged heater connected to the slide regulator 25 kVA), cooler tank (non-direct, water to water, cold water from HSS loop) and an expansion tank to compensate a sudden pressure rise during experiment. The piping system of rectangular loop composed of 32 sections and 4 elbows made by 1 inch diameter SS304 pipe with length of each section is 50 cm. Each section was connected by flanges. Around 50 thermocouples of type K (error average: 4.2%) were installed along the rectangular loop (including 6 thermocouples in every tank) and 1 flow meter FLR1009ST-I type for low flow with maximum flow 0.5 LPM (error: 6.43%). The HSS loop consists of pre-cooler tank as a cold water source which is cooling by coil connected to the refrigerant. High speed pump, a Grunforce Rotation Varity model was installed in HSS loop with maximum flow of 45.09 LPM and it connected to an inverter to vary the flow rates. A reservoir tank is connected to the HSS loop as a water supply during operation. The explanation of FASSIP-01 loop including error analysis, construction and other reports were explained in previous paper [15-18].

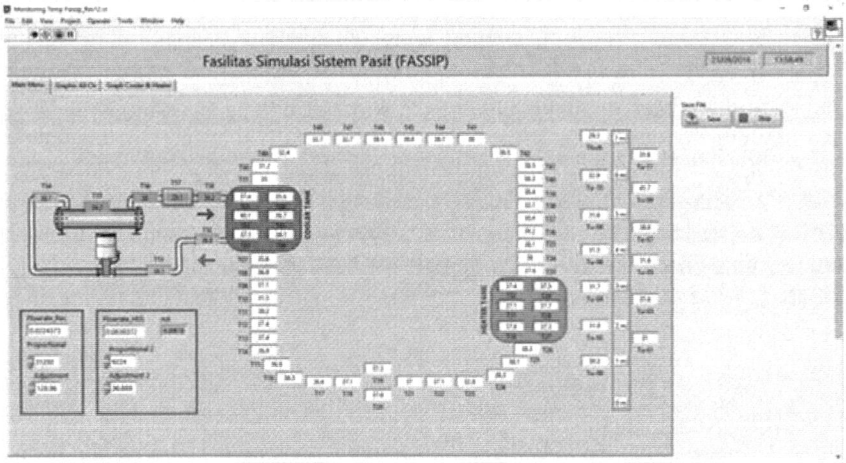

Figure 2. FASSIP-01 loop VI display

Figure 2 shows the indication of temperature measurement in cooler tank with blue colour and temperature in heater tank with red colour including natural circulation flow rate in rectangular loop and water flow rate in HSS loop. As measurement system, 14 thermocouples of type K and 1 electro-magnetic flow meter (error: 2.5%) were installed in HSS loop. All data from measurement points was collected in PC through data acquisition system (DAS) consisting of a few of temperature module (NI-9213 and 9214) and a universal module for flow and pressure (NI-9203). The modules were plugged in cDAQ-9188 using USB cable connecting to PC. In PC monitor, a virtual instrumentation (VI) was made using LabView 2011 software to evaluate parameter changes during experiment. Figure 2 shows the VI display in PC monitor.

2.2. *Experiment procedure for commissioning*

Experiment scenario was made to conduct the FASSIP-01 Loop commissioning which consists of two scenarios i.e.: a static test and a dynamics test. The static test was done for about 24 hours, with two steps of testing. The first step is done without water in the inside of rectangular loop, heater tank and cooler tank. All measurement temperature was done for about 24 hours as well. This test is intended to understand the characteristics and response of thermocouples reading to the air around the laboratory for 24 hours. The second step is done also by filling water into the inside of rectangular loop pipe, heater tank and cooler tank for 24 hours experiment. The dynamic test is done by filling the water into the FASSIP-01 loop and heating water inside heater tank and cooling water inside cooler tank. The dynamics test was conducted by the water flow rate in HSS loop is 9.35 LPM and using heater with electric power of 5627 watt for heating the water inside heater tank for 7 hours.

2.3. *Analysis*

The analysis was done by fitting all temperature histories data into several graphs for static and dynamics test. Natural circulation flow rate from the temperature changes in rectangular pipes inside heater tank and cooler tank is also calculated. The result calculation of the flow rate compared to flow rate measurement. Calculation was done using equation which was derivate from natural circulation law inside rectangular loop [19]. Natural circulation flow rate is a function of total loop length, height differences between heater and cooler areas, pipe diameter, temperature and constantan K have influenced into flow rate.

$$Q = A\frac{-64\mu L + \sqrt{(64\mu L)^2 + 8gHK\rho(\rho_h - \rho_c)D^4}}{2D^2\rho K} \tag{1}$$

Where Q is flow rate (m^3/s; LPM), D is pipe diameter (m), L is total length of rectangular pipes, A is pipe cross-section area (m^2), ρ is water density (m^3/kg) which is the value depends on the temperature change, μ is dynamics viscosity (N.s/m^2), K is frictional loss of piping components, and g is gravitational acceleration (9.89 m/s^2).

3. Result and discussion

3.1. *Temperature characteristics*

The temperature history curve for statics test in 24 hours both for no-water and there is water inside pipe is shown on Figure 3a and Figure 4a. Ambient temperature in the air around laboratory also was measured for 24 hours, shown in Figure 3b and Figure 4b. Figure 3a shows a temperature profile during 24 hours from 13.00 o'clock on July 19, 2016 to 13.00 o'clock on July 20, 2016. As shown in Figure 3a and Figure 3b, since the beginning recording data was started at 13.00 until the end of recording data during 24 hours, the air temperatures which were measured for all points of thermocouples have a same profile. At 13.00, ambient temperature around 32°C and higher temperature in the FASSIP-01 loop was 40.5^{o}C. All temperatures data indicating begin to drop at 15.00 due to rain, but all temperature data shows a stable condition start at 17.30 on July 19, 2017 until 06.30 on July 20, 2016 with only a temperature change of about 2.2^{o}C.

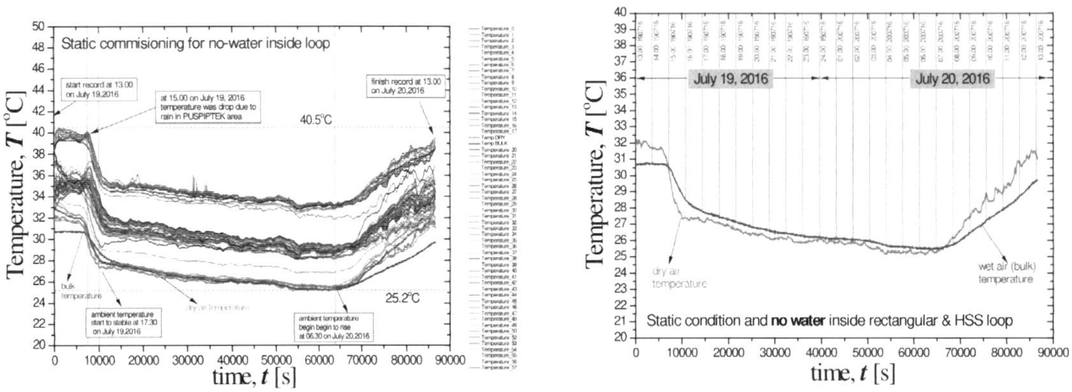

(a) All temperatures measurement for static no water inside loop scenario

(b) Ambient temperature measurement for static no water inside loop scenario

Figure 3. Characteristics of temperature measurement for static with-out water inside FASSIP-01 Loop

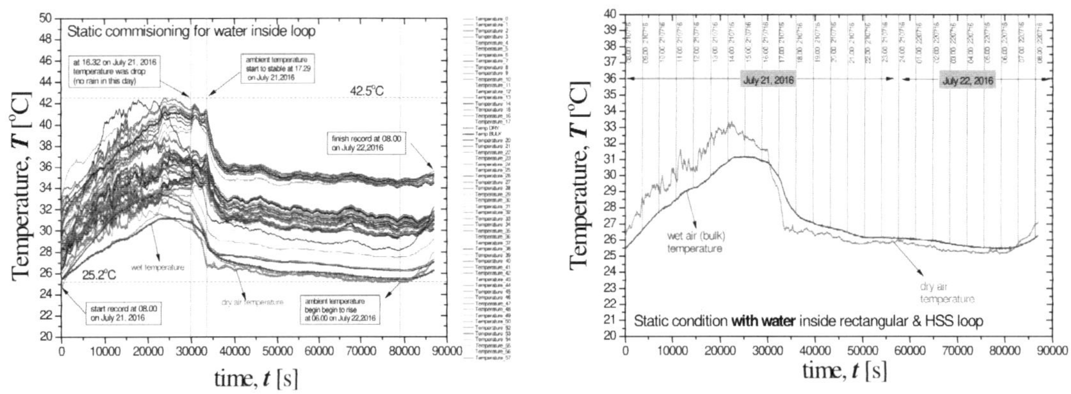

(a) All temperatures measurement for static with water inside loop scenario

(b) Ambient temperature measurement for static with water inside loop scenario

Figure 4. Characteristics of temperature measurement for static with water inside FASSIP-01 Loop

The same conditions also occur for static tests with water inside pipe of the FASSIP-01 loop, shown in Figure 4a and Figure 4b. But, there is no rain at July 21, 2016 in the afternoon. Ambient temperature become drop at 16.32 the become stable since 17.29 on July 21, 2016 until 06.00 on July 22, 2016 with only a temperature change of about 1.5°C. Based on these conditions, the experiment should be done from 18:00 to 06:00 (12 hours) with temperature differences between 1°C - 3°C. But, if the experiment must be conducted in the morning the temperatures differences from the morning until afternoon is around 7°C – 8°C. Figure 4 shows temperature histories for several measurement points inside pipe in rectangular loop of FASSIP-01.

Measurement points are consists of a heater area and a cooler area. In heater area there are 4 points, inlet temperature in pipe into heater area, outlet temperature in pipe from heater area, and 2 points in pipe inside heater tank. The same number of measurement points for cooler area. Inside curve of Figure 4, there is 1 point for bulk temperature to measure the ambient temperature. As shown in Figure 5, water temperature inside pipe in heater area has a stable condition at 100°C and 103°C. It is seems there is boiling inside pipe due to temperatures reached saturation temperature of water. But it is believed that there is no boiling in the heater area due to an increase of pressure inside pipe of rectangular loop. Then, the differences of water temperature inlet and outlet between heater areas are around of 49°C. In the cooler area, the differences of water temperature inlet and outlet between cooler areas are around of 7°C. It can be said that there is heat loss when the natural circulation flow was flowing and bring heat from the heater are to the cooler area and vice versa. Thus, the consideration to coat pipe surface of the FASSIP-01 loop with a thermal insulator becomes a necessity in the further experiments.

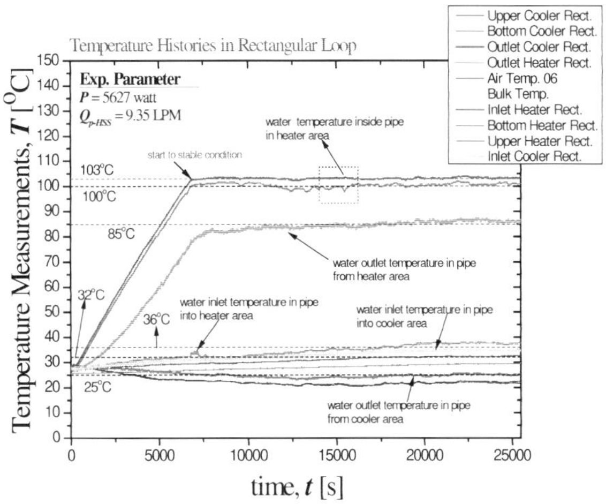

Figure 5. Characteristics of all temperatures measurement in FASSIP-01-01 Loop during dynamic experiment

As shown in Figure 5, there is an interesting condition about this dynamic test is that there is an unstable condition during temperature rise from 0 second to 6935 seconds. Then, from 6936 until the end of experiment at 25000 seconds, water temperatures inside pipe and in inlet-outlet pipe at the heater area are stable. This condition indicates that there is an average of natural circulation flow rate of water inside rectangular pipe of FASSIP-01 loop.

3.2. Flow rate characteristics

Natural circulation flow rate has been obtained through the calculation result and the measurement result, as shown in Figure 6.

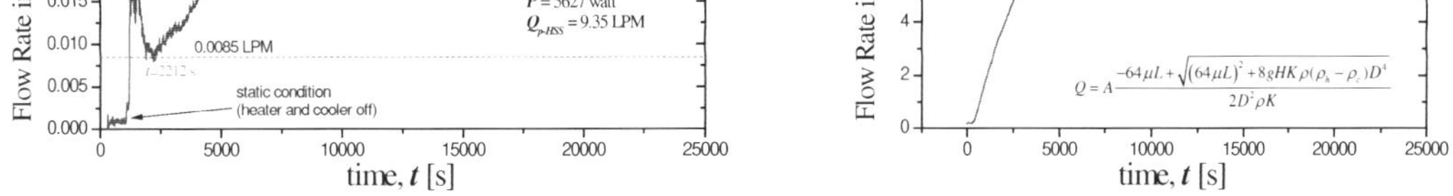

(a) Flow rate measurement (b) Flow rate calculation

Figure 6. Characteristics of flow rate in rectangular loop of FASSIP-01-01 during dynamic experiment

Figure 6 shows characteristic of natural circulation flow rate both for calculation and measurement. Figure 6a explains natural circulation flow rate profile based on measurement result using flow meter FLR1009ST-I type. And Figure 6b explains the result of calculation for natural circulation flow rate profile using equation (1). Based on Figure 6a, the measured flow rate profile is very interesting, since there is 2 times the increase in flow rate. The flow rate was increased starting from of 0.0085 LPM to 0.0325 LPM in 5229 seconds (from 2212 seconds up to 7441 seconds). Then, starting from 7441 seconds, the flow rate was increased from 0.0325 LPM to 0.0415 LPM. The different results are shown from flow rate calculations, where there is no increase in flow rate as much as 2 times as the result of measurement. Since the beginning of heating at 0 second until 6994 seconds, the flow rate increased from 0 LPM to 9.75 LPM. Then since that time, the natural circulation flow rate becomes stable until the end. Meanwhile, a particular concern is the difference between the calculation results with the measurement results, where the calculation results are approximately 300 times greater than the measurement results. In the meantime, the analysis believes more on measurement results. Thus, it is necessary to modify and improve on equation (1) the calculation result with the dimensionless number. Various data with variations of parameters in the experimental matrix forward must be made to obtain the result of modification of the equation which results close to the measurement data. In the meantime, the analysis believes more on measurement results. Thus, it is necessary to modify and improve on equation (1) by applying a dimensionless number. Various data with variations of parameters in the experimental matrix in the future must be made to obtain the result of modification of the equation which has the data of the calculation results close to the measurement data. Meanwhile, in addition there is a different flow rate profiles. Where, the flow rate profile of the calculation results almost similar to the water temperature change curve in the heater region, especially in the outlet pipe of the heater and in the heater area. The condition seems reasonable, since the flow rate equation includes the average temperature parameters in the heater area and in the cooler area as the input data into calculation.

4. Conclusion

The commissioning of FASSIP-01 loop has been done to characterized temperature histories of water inside pipe and natural circulation flow rate using measurement data. Also a calculation to define natural circulation flow rate was occurred using temperature data from experiment. The result of temperature characterization on static test shows that, experiments should be done at night with better environmental temperature stability compared to noon, about $1^{o}C$ - $3^{o}C$. While on the dynamic test, the water temperature difference between the inlet-outlets in the heater area is quite large, about 7 times the temperature difference in the cooler area. The magnitude of the natural circulation flow rate calculated is much larger at about 300 times compared to the measured flow rate with different flow rate profiles. This test indicates that the natural circulation flow rate has been formed within the FASSIP-1 strand rectangular pipes and the flow rate measurements need to be validated by visualization.

Acknowledgement

This research project was funded by DIPA PTKRN BATAN with No. SP DIPA-**080.01.1.450310/2016**. This work is a part from the research program activity (KAK) entitled Assessment of Safety and Technical Design of Nuclear Power Plant of PTKRN-BATAN, year 2016. Authors would like to thank to the Head of Center for Nuclear Reactor Technology and Safety (PTKRN), Dr. Geni Rina Sunaryo, MSc. for granting permission to use the research budget. Authors would like to express deeply thanks to the Thermal Hydraulics Experimental Laboratory of PTKRN BATAN members for their support during construction, installation and commissioning. The travel expense funded by INSINAS Program from the Ministry of Research, Technology and Higher Education (Kemenristekdikti) of Indonesia with contract number **02/INS-2/PPK/E/E4/2017**.

References

[1] IAEA, Design Safety Considerations for Water Cooled Small Modular Reactors Incorporating Lessons Learned from The Fukushima Daiichi Accident, IAEA-TECDOC 1785, Vienna, 2016.

[2] P. Welander, On the oscillatory instability of a differentially heated fluid loop, Journal of Fluid Mechanics, Vol.29, No.1, pp.17-30 (1967).

[3] P.K.Vijayan, H. Austregesilo, V. Teschendorff, Simulation of unstable oscillatory behaviour of single-phase natural circulation with repetitive flow reversals in a rectangular loop using computer code ATHLET, Nuclear Engineering and Design, Vol.155, pp.623-641 (1995).

[4] Y. Zvirin, A review of N. C. loops in PWR and other systems, Nuclear Engineering and Design 67 (1981) p.203–225.

[5] M. Misale, F. Devia, P. Garibaldi, Some considerations on the interaction between the fluid and wall tube during experiments in a single-phase natural circulation loops, Proceedings of the 3rd IASME/WSEAS Int. Conf. on Heat Transfer, Thermal Engineering and Environment, Corfu, Greece, August 20-22, p p.128-133 (2005).

[6] Y.Y. Jiang, M. Shoji, Flow stability in a natural circulation loop: influence of wall thermal conductivity, Nuclear Engineering Design, Vol. 222, No.1, pp.6–28 (2003).

[7] M. Misale et al., Experiments in a single-phase natural circulation mini-loop, Experiment Thermal Fluid Science, Vol. 31, pp.1111-1120 (2007).

[8] F. D'Auria, M. Frogheri, Use of a natural circulation map for assessing PWR performance, Nuclear Engineering Design, Vol. 215, pp.111-126 (2002).

[9] F. D'Auria et al., Scaling of natural circulation in PWR Systems, Nuclear Engineering and Design, Vol.132, pp.187-205 (1991).

[10] P.K. Vijayan, Experimental observations on the general trends of the steady state and stability behavior of single-phase natural circulation loops, Nuclear Engineering and Design, Vol. 215, pp.139–152 (2002). Experimental thermal-hydraulics group, http://www.psi.ch/teg/facilities, accessed in January 8th, 2015, time 15.00.

[11] J. Vikas et al., Experimental investigation on the flow instability behavior of a multi-channel boiling natural circulation loop at low-pressures, Experimental Thermal and Fluid Science, Vol. 34, pp.776–787 (2010).

[12] M. Juarsa, dkk., *Analysis on Natural Circulation Flow with Reynolds Number Based on Slope Angle Variation in Natural Circulation Simulation Loop*, , Prosiding Seminar Nasional Thermofluid V dan Pameran Teknologi Gas Tahun 2011, Jurusan JTMI, FT UGM, Yogyakarta, 4-5 Oktober 2011, (in Indonesian).

[13] Y.S. Gaos, dkk., *The effect on Slope Angle Revolutiom to Heat Transfer and Water Flow Rate in Natural Circulation Loop*, Tri Dasa Mega, Vol.14, No.1, (2012) pp.39-53.

[14] M. Juarsa, et al., Preliminary Study on Mass Flow Rate in Passive Cooling Experimental Simulation During Transient Using NC-Queen Apparatus, Atom Indonesia Vol.40, No.3, (2014) pp.141-147.

[15] Giarno, dkk., *Characterization on the Pre-Cooler as the Heat Sink System of the Loop FASSIP-01*, Prosiding Seminar Nasional Teknologi Energi Nuklir 2016, Batam, 4-5 Agustus 2016, p.425-432 (in Indonesian).

[16] Giarno, dkk., *Designing the Heat-Sink System of Fassip-01 Loop by Software Cycle-Tempo*, Prosiding Seminar Nasional Teknologi Energi Nuklir 2015, Bali, 15-16 Oktober 2015, p.489-195 (in Indonesian).

[17] G. Bambang Heru K, dkk., *Design on Calculation Program of Water Mass Flow Rate Based on Temperature Changes Using LabView*, Prosiding Seminar Nasional Teknologi Energi Nuklir 2015, Bali, 15-16 Oktober 2015, p.481-488 (in Indonesian).
[18] Khairina Natsir, dkk., *Evaluation on Uncertainty of Flowmeter Measurement Results Using Data Acquisition System*, Prosiding Seminar Nasional Teknologi Energi Nuklir 2015, Bali, 15-16 Oktober 2015,p.534-541 (in Indonesian).
[19] Mulya Juarsa , et al., Passive System Simulation Facility (FASSIP) Loop for Natural Circulation Study, Prosiding Seminar Nasional Teknologi Energi Nuklir 2016, Batam, 4-5 Agustus 2016, p.673-680.

The Pneumatic Actuators As Vertical Dynamic Load Simulators On Medium Weighted Wheel Suspension Mechanism

Simon Ka'ka[1], Syukri Himran[2], Ilyas Renreng[2] and Onny Sutresman[2]

[1] Mechanical Engineering Department, State Polytechnic of Ujung Pandang, 90245, SouthSulawesi, Indonesia, kakasimon@ymail.com
[2] Mechanical Engineering Department, Engineering Faculty, Hasanuddin University, 90245, South Sulawesi, Indonesia

The corresponding e-mail address: kakasimon@ymail.com

Abstract. Almost all of road damage can be caused by dynamic loads of vehicles that fluctuate according to the type of vehicle that passes through. This study aims to calculate the vertical dynamic load of the vehicle actually occurs on road construction by the mechanism of vehicle wheel suspension. Pneumatic cylinders driven by pressurized air directly load the spring and shock absorber installed on the wheels of the vehicle. The load fluctuations of the medium weight categorized vehicles are determined by the regulation of the amount of pressurized air that enters into the pneumatic cylinder chamber, pushing the piston and connecting rods. The displacement that occurs during compression on the spring and shock absorber, is substituted into the equation of vehicle dynamic load while taking into account the spring stiffness constant, and the fluid or damper gas coefficient. The results show that the magnitude of the displacement when the compression force works has significant influences to the amount of vertical dynamic load of the vehicle that overlies the road construction. The presence of dynamic load of vehicles that fluctuates and repeats, also affects on the reduction of road ability to receive the load. Experimental results using pneumatic actuators instead of real dynamic vehicle loads illustrate the characteristics of the relationship between work pressure and dynamic load. If the working pressure of P_2 (bar) is greater, the vertical dynamic load F_t (N) that overloads the road structure is also greater. The associate graphs show that the shock absorber has a greater ability to reduce dynamic load vertically that burden the road structure when compared with the ability of screw spring.

Keywords: Pneumatic Cylinder, Dynamic load, Pressurized air, Suspension, Road

1 Introduction

A number of four or more wheel type vehicles that pass over the surface of the highway at any time result in a number of dynamic loads that fluctuate. In addition to being obtained from dynamic load variation, DLV, this fluctuating loading can also be caused by overload, OL, and repetition loads (RL) [1]. The effects of transfer the fluctuating load (FL), OL and RL will decrease the capability/stability of the road structure in the form of hollow or bumpy surface deformations. The ability of the road structure to receive the vehicle's vertical dynamic load is dependent on the type of surface layer, the elastic foundation layer and the sliding foundation layer.

Variations in the weight of vehicles: light, medium and heavy, that pass over the road surface will simultaneously affect much the ability of the elastic foundation layer of the road in receiving such loads. Based on these problems, it is necessary to study the some influencing variables that both naturally direct and indirect. Fluctuated loading is influenced by the variables of weight of wheel axis and vehicle. The conditions of repetitive loading, RL, are also strongly influenced by the repetitive flow of traffic, the average volume of traffic flow (amount of vehicles/hour) generated from each vehicle and the length of the vehicle passing through the road. An attempt was made to examine the dynamic load of the medium weight vehicles that cross the road. This needs an experiment on the loading of a quarter part of a vehicle centered on a front wheel of a vehicle. This innovative study is based on a basic knowledge of the mechanism of vertical load transfer through a wheel suspension system consisting of spring work and shock absorber.

A study of dynamic loads transformed through the mechanism of suspension work on the vehicle wheels towards the road aims to know the form of formula and the magnitude of the vertical dynamic load of each type of vehicle weight passing the road, knowing the characteristics of the capability of the road in receiving repetitive loads, RL, and over loads, OL, and resulting in a relationship/comparison between vehicle vertical dynamic load and the capability of the road in the form of dimensionless parameters.A recentness element targeted in this study is to utilize pneumatic actuators on test equipment/ experiments as a replacement for dynamic loads of vehicles when loading the road surface. Pressurized air from the compressor provides compression forces variedly to the spring and shock absorber contained in the suspension system.

The passive suspension mechanism and the condition of a quarter vehicle loading structure are shown in Figure 1.1, [2]. The weight of the vehicle (*mass sprung*), m_2, with the stiffness of the spring, k_2, and the damping coefficient, c, will overload the axis of vehicle wheel (*mass un sprung*), m_1, furtherly gives an action force to the contour of the road surface[3].For light categorized vehicles : have axis load m_1 = 400 kg, and vehicle weight, m_2 = 835 to 1394 kg. Medium categorized vehicles : have axis load, m_1 = 480 to 600 kg and vehicle weight, m_2 = 1185 to 1990 kg. While heavy weight vehicles: have axis load, m_1 = 525 to 850 kg, and vehicle weight, m_2 = 5200 to 8730 kg (source: P.T. Astra International, 2014. The over loading conditions, OL, depend on the ***k***and axis load $\boldsymbol{m_1}$ (kg, ton) for each vehicle type. The value of ***k*** for a single axis : k = 1, the double axis : k = 0.086 and the triple axis : k = 0.031.

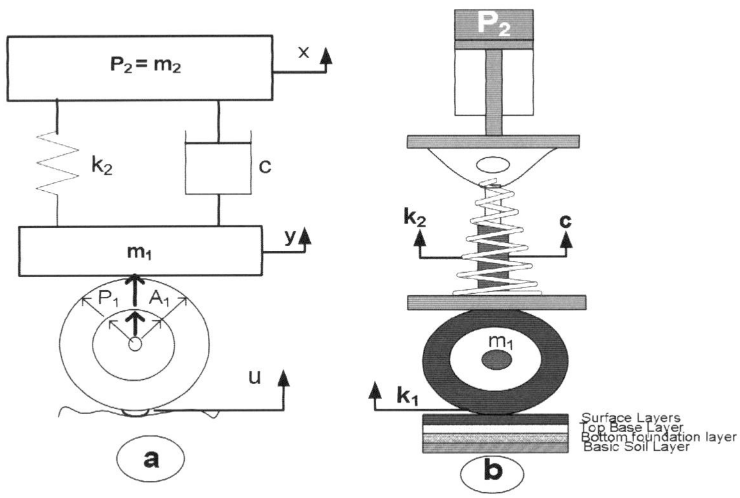

Figure 1.1. (a) Model of passive suspension. (b) Suspension unit, shock absorber, axis and a wheel on the structure of road layer.

The reaction force of the road contour to the tire will be distributed in the direction of u through the tire elasticity constant, $\boldsymbol{k_1}$. The magnitude of the vertical dynamic force by the equivalent pneumatic actuator in the weight force of the vehicle body (2) is described in equation (1.1). Compression displacement as far as x_2 (cm) will vary according to the magnitude of the working pressure P_2 (bar) that is varied from 1 bar to 8 bar.

$$\left.\begin{aligned} F_{p2} &= F_{ef} = F_k - R_f = F_k - 0.1F_k = 0.9F_k \\ &\sum F_{v=0} \\ &F_{ef} - (F_{k2} + F_{c2}) = 0 \\ &0.9\,F_k - (k_2 + c)x_2 = 0 \\ F_k &= \frac{(k_2+c)x_2}{0.9} \\ F_{p2} &= F_k = 1.1(k_2 + c)x_2 \end{aligned}\right\}(\mathrm{N}) \qquad (1.1)$$

If the force F_{p2} works together with the axis force on the wheel (1) then the relationship is obtained as :

$$\left.\begin{aligned} F_{t1} &= F_{p2} + F_{r1} \\ F_{t1} &= 0.9F_k + (m_1.g) \end{aligned}\right\}(\mathrm{N}) \qquad (1.2)$$

If the piston diameter, D = 100 mm (0.100 m), k_2 = spring constant, c = damping coefficient, m_1 = weight of wheel axis, gravity acceleration g = 10 m/s^2 and push force of the pneumatic cylinder piston is $F_k = \frac{\pi}{4}D^2P_2$, then equation (1.2) becomes :

$$F_{t1} = 785\,P_2 + 10m_1 \text{ (N)} \qquad (1.3)$$

The effective force, $\mathbf{F_{ef}}$ (N) of the piston cylinder in the intake stroke is the difference between the theoretical force, F_k (N) and frictional force $\mathbf{R_f}$ (N) [4, 5]. If the friction force $\mathbf{R_f}$ is set to $10\,\%\,x\,F_k$ (N) then the overall load transfer mechanism on the asphalt road structure refers to the following Free Body Diagram (FBD) as shown in Figure 1.2.

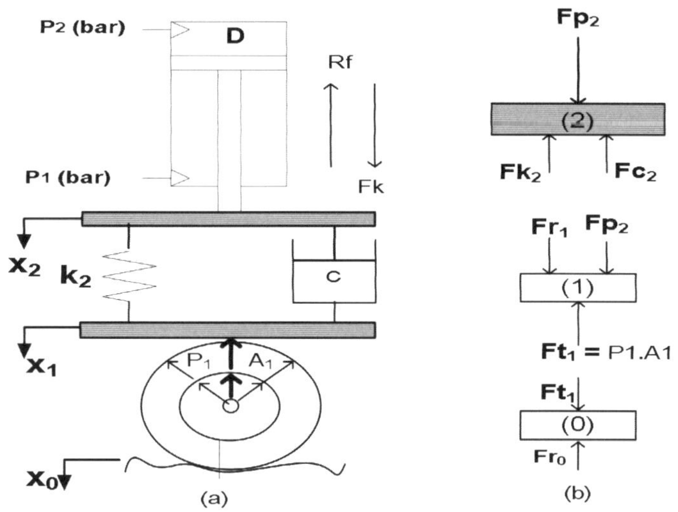

Figure 1.2. The mechanism of experimental loading of pneumatic system.

If the spring and shock absorber are subjected to varying loads of pressure P = 1 to 8 bar, there will be a deviation toward the compression as far as x (mm)[6]. Based on the action-reaction law, the load applied to the spring is proportional to the magnitude of the deflection multiplied by the spring constant.The magnitude of average displacement on the optimized suspension system x = 0.009264 m with maximum vertical acceleration is 15.571 m/s^2. In this condition it is obtained the coefficient of stiffness of the suspension spring $\mathbf{k_s}$ = 41821 N/m, and the suspension damper coefficient $\mathbf{c_s}$ = 68574 Ns/m [5] and $\mathbf{c_s}$ = 1920 J/kg.K . The suspension prototype tested by [7] has spring constant $\mathbf{k_s}$ = 10581.292 N/m, shock damping coefficient, $\mathbf{c_s}$ = 96.073 Ns/m and tire elastic constants, $\mathbf{k_e}$ = 98041.246 N / m. A preliminary study of the relationship between piston diameter D and hole diameter $\mathbf{d_i}$ has been carried out [5, 8] in the form of equation (1.4).

$$d_i = 0.065713\, D \tag{1.4}$$

The dimension of air duct to be used is adjusted to the dimension of the selected pneumatic cylinder piston.Based on the specifications of tires/wheels for light and medium vehicles it is used standard wind pressure within tires of **2 bar**, and **5.86 bar** for heavy vehicles. If the size of the tire/wheel used has dimension of 215/60/16 which means, width of tire S = 215 mm, height T = 60% x S (mm) = 129 mm = 0.129 m, and wheel diameter, D_1 = 16" (inches) = 16 x 25 mm = 400 mm = 0.4 m, and the air pressure within the tire is P_1 = 2 bar = 200000 N/m^2, and the tire area $A_1 = \pi[r_o^2 - r_i^2] = 3.14[0.102^2 - 0.020^2] = 0.031412\ m^2$,then the inner force of the tire/wheel in holding the vehicle's vertical load is $F_{t1} = P_1 \times A_1$ = 6282.4 N.The magnitude of the vehicle dynamic load coefficient is formulated [9] as a comparison between the average dynamic load and the static load as follows :

$$DLC = \frac{Average\ of\ dynamic\ loads}{Static\ load} = \frac{FD_{rms}}{F_s} \tag{1.5}$$

For vehicles with hydro-pneumatic suspension, the dynamic load can be reduced by 20%. Under this condition Dynamic Load Coefficient (DLC) becomes

$$DLC = \frac{(100-20)\,\%}{100\ \%} = 0.80. \tag{1.6}$$

When the wheel of quarter vehicle is rolling over the road surface, then the vertical dynamic load of the vehicle is formulated as follows :

$$\left.\begin{aligned} F_{DV} &= 0.8\ x\ F_{t1} \\ &= 628P_2 + 8m_1 \le F_{A0} \end{aligned}\right\}\ \text{(N)} \tag{1.7}$$

2 Research Method

2.1 Experiments

Experimentally the loading to the road foundation structure is carried out by equivalenting the total weight of the vehicle body m_1 (kg) and the weight of wheel axis m_2 (kg), with the compressive force of pneumatic cylinder piston, $\mathbf{F_{ef}}$. (N) as an actuator at a working pressure P_2 = 1 to 8 (bar). The research that is oriented to simulations of pneumatic cylinder piston force identified by [10]as real dynamic vehicle loads is shown in Figure 2.1. If the compressive force of the pneumatic cylinder piston presses the spring and the shock absorber then there will be a compression process indicated by the designation of the pointer x at the displacement scale (mm).

2.2 Collecting Data

Based on the loading simulation, the experimental data collection is done by reading the scale (11) pointer reveal/dial indicator of the displacement, x (mm) a while after compression on the spring (10). The magnitude of the change in displacement is determined by the amount of air pressure, P_2 (bar) set from 1 (bar) to 8 (bar). In order that the measurement results $x\ (mm)$ is closer to the real state then it is necessary to measure x (mm) 5 times at every working pressure setting by calculating the average value of x (mm)[11, 12].

2.3 Data Analysis

The magnitude of the mean displacement x (mm) obtained is substituted into equations (1.1) and (1.3) as vehicle weight force, Fp_2 (N) and the vehicle's real vertical dynamic force Ft_1 (N) to the road construction. With the help of Mat-Lab/Simulink program analysis, the magnitude Fp_2 (N) and Ft_1 (N) can be obtained. The loading characteristics done to the road construction can be shown through graphic curves.

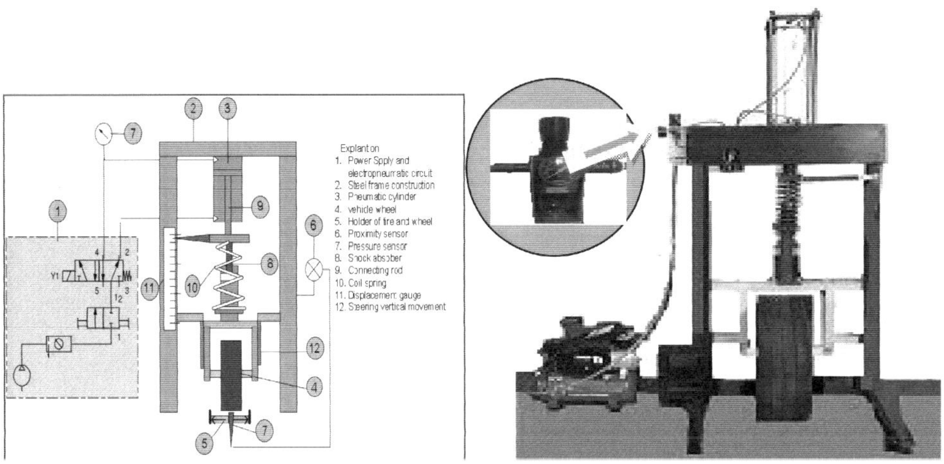

(a) (b)

Figure 2.1. (a) Design of motion control simulation with pneumatic actuators. (b) Test equipment of loading on pneumatic system wheel suspension.

3. Results And Discussion

3.1. Experiment of Pneumatic System

The dynamic load simulation of vehicles played by pneumatic cylinders/actuators results in the magnitude of the dynamic load of vehicles, loading the road structure with refers to equation (1.3). If the working pressure, P_2 instead of the load of m_2 is varied from 1 bar to 8 bar, then the working characteristic of the dynamic load to the working pressure is shown in Figure 3.1.

The graph of the relationship between the working pressure of P_2 (bar) and the vertical dynamic that loads road structure represents a significant increase in dynamic load with the total vehicle load increase. This significance condition is shown in the graphic figure as a straight line.

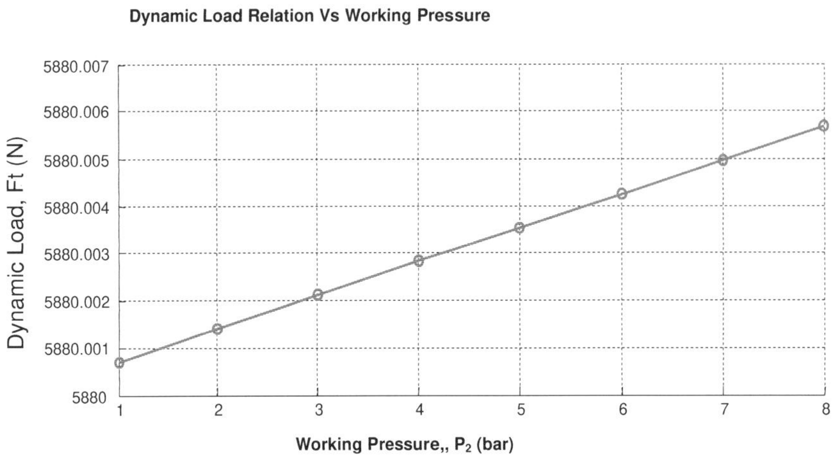

Figure 3.1. Dynamic load characteristic of pneumatic system

The dynamic load characteristics re obtained from the relationship between the operating pressure and the spring constant k_2. Pressing force of pneumatic cylinder piston, F_{p2}, shock absorber damping, c_2 and the magnitude of the vertical dynamic load that burdens the road structure, F_{t1},are shown in Figure 3.2 (a). The effect of attenuation by the shock absorber gives a huge influence on the dynamic load burdening the road structure. Such a condition causes the dynamic load of F_{t1} (N) according to Figure 3.2 (b) to be around 10000 (N) or equivalent to 1000 kg.

Hence, if the medium weight categorized vehicle being tested has the weight of wheel axis m_1 = 600 kg, body weight m_2 = 1665 kg and total/net weight m_t = 2265 kg according to the data in Table 3.1 then it is obtained the reduced dynamic vehicle load that is $m_r = m_t$ - 1000 Kg = 2265-1000 = 1265 kg.

The percentage of dynamic load drop burdening the road structure due to the attenuation of the shock absorber is $\frac{2265}{1265}x100\% = 55.85\ \%$.

3.2. Vertical Dynamic Load of Vehicle

Figure 3.3 (a) shows a fairly fantastic picture that the attenuation of the shock absorber attached to the vehicle suspension mechanism provides a damping value of **c (Ns/cm)** to a very large dynamic load that is between 4523.729 Ns/cm and 48702.041 Ns/cm.

The smallest effect of attenuation that is also used in the suspension system of the test vehicles is the contribution of the attenuation of threaded spring **k_2 (N/cm)** whose values range between 1330.508 N/cm and 6408.163 N/cm.

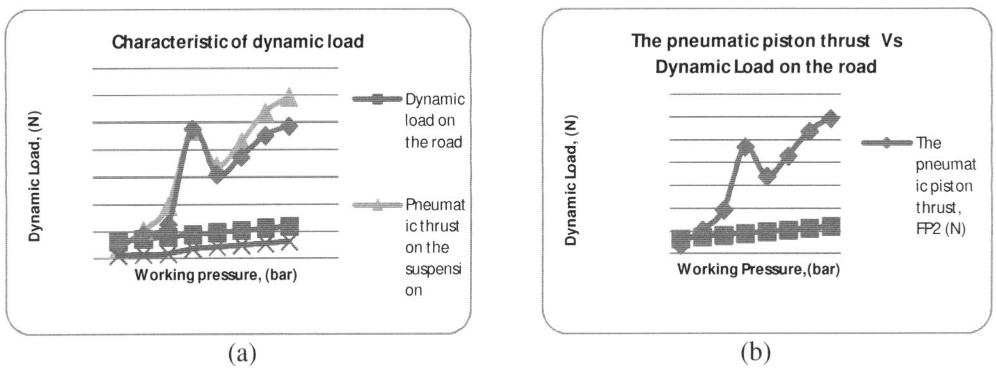

(a) (b)

Figure 3.2. (a) Dynamic load characteristics by piston force, F_{P2}, to spring and shock absorber. (b) Relationship between piston force, F_{P2}, and dynamic load, F_{t1}, burdening the road structure .

The gap value which occurring between the shock absorber and the threaded spring, indicates that the shock absorber plays an enormous role in reducing the vibrations generated by the vehicle's vertical dynamic load.Figure 3.3 (b) describes the relationship characteristics between the dynamic load of the experimental results and the actual dynamic load that occurs. By using the Dynamic Load Coefficient (DLC) value as 20% acts for vehicles with hydro-pneumatic suspension then the actual vertical vehicle dynamic load transformation follows the equation (1.5) with a value between 5332 (N) and 9728 (N) or less than the weight value of 10000 (N).

A number of important data information obtained from several formulas for experimental test vehicles are shown in Table 4.1, which can also present the results of dynamic load analysis using equations (1.2), (1.4) and equations (1.6). The magnitude of the vertical dynamic load, F_{Dv} (N) as contained in Table 4.1 can be identified as the normal force, F_N (N) acting in a perpendicular direction to the horizontal force, F_H (N) and the tangential force, F_T (N).

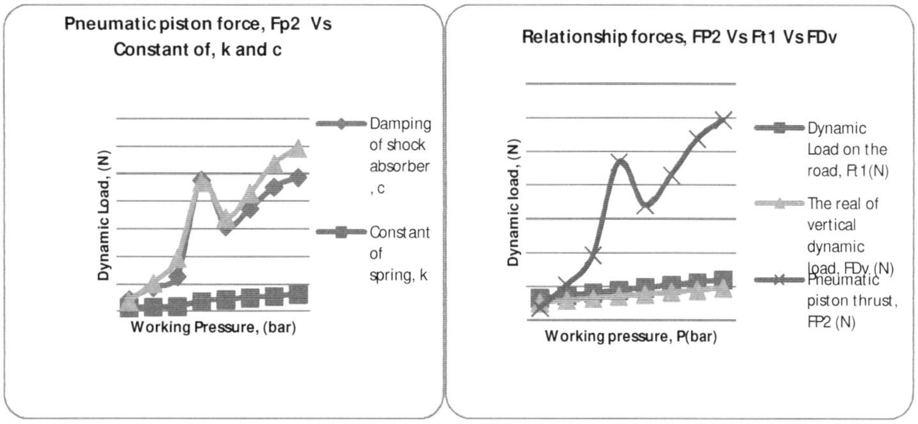

(a) (b)

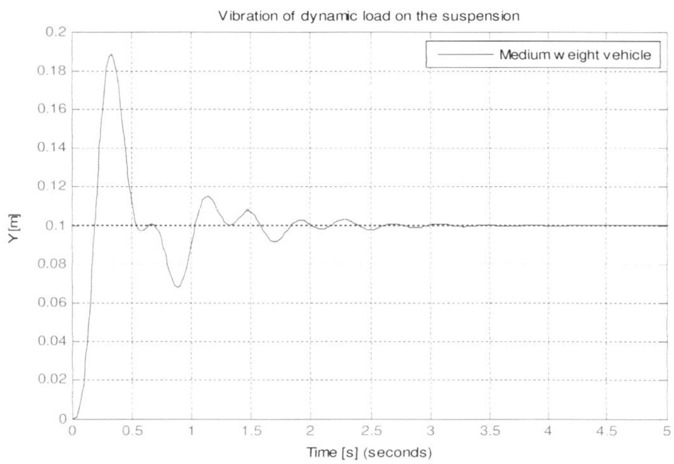

(c)

Figure 3.3. (a) Relationship of piston force, F_{P2} (N), with shock absorber and threaded spring.(b) Relationship of pneumatic cylinder piston force, F_{P2} (N), with dynamic load of test vehicle, F_t (N), and the real vertical dynamic load, F_{Dv} (N). (c)Vibration dynamic load on the suspension mechanism

Table 3.1 Analysis results of vertical dynamic load of medium weight categorized vehicle.

Working pressure P_2(bar)	Wheel Axis weight $W1=m_1.g$	Vertical dynamic load $Ft_1=Fr_0$ (N)	Real Vertical dynamic load F_{Dv} (N)	Thrust force of pneumatic cylinder piston Fp_2 (N)
1	5880	6665	5332	3799.400
2	5880	7450	5960	10534.700
3	5880	8235	6588	19428.750
4	5880	9020	7216	46974.400
5	5880	9805	7844	34108.250
6	5880	10590	8472	43002.300
7	5880	11375	9100	53796.050
8	5880	12160	9728	59408.800

Acknowledgement
The authors would like to thank to the Department of Mechanical Engineering, Hasanuddin University and also to the Department of Mechanical Engineering, State Polytechnic of Ujung Pandang for providing the facilities in Workshop and Laboratory.

4. Conclusion

Characteristics of the suspension mechanism on the quarter vehicle wheel of medium weight vehicle type experiences longest displacement of 0.18 m with a duration of 0.35 seconds. In an interval of 2.5 seconds later, the vibrations begin to shrink and eventually disappear in the last 3 seconds.

Medium weight categorized vehicle with a total weight : 2265 kg > 1942.8 kg obtains a dynamic vertical load that is greater than 658.8 kg. The percentage of the dynamic load decrease that burdens the road structure due to the attenuation of the shock absorber is 55.85%.

References

[1] A Rehnberg, "Vehicle dynamic analysis of wheel loaders with suspended axles," *TRITA-AVE,* vol. 15, 2008.
[2] D J Brembeck, "LPV Control of Full-Vehicle Vertical Dynamics using Semi-Active Dampers," *Science Direct,* vol. 49-11, p. 8, 2016.
[3] E A Finney, "Dynamic Aspect Of Vehicle Size and Weight," p. 23, 2010.
[4] T A Ilchmann*, "Pneumatic cylinders: modelling and feedback force-control," *International Journal of Control,* vol. 79, , 2006.
[5] Simon Ka'ka, M.M. "A method to Compute the Dimension of An Air Distribution Channel of A Pneumatic Cylinder," presented at the IGCESH Conference, Johor Bahru, Malaysia, 2013.
[6] M.TongyiXu[a],ChuanLi[b,*],ShuaiYang[a], "Design and analysis of a shock absorber with variable moment of inertia for passive vehicle suspensions," *Journal of Sound and Vibration,* vol. 355, p. 20, 2015.
[7] C S Shrikant S Dodamani[#1,] Nagaraj D[#3], "Design and Fatigue Analysis of McPherson Strut Assembly Coil Spring," *International Journal of Innovative Research in Science, Engineering and Technology,* vol. 5, p. 11, 7, July 2016 2016.
[8] V G Vijaya, "Design and Fabrication of Sequencing Circuit with Single Double Acting Cylinder," *International Journal of Innovative Research in Science, Engineering and Technology (An ISO 3297: 2007 Certified Organization),* vol. 2, p. 8, 2013.
[9] M A Mohd Nizam Sudin1, Shamsul Anuar Shamsuddin1, Faiz Redza Ramli1, Musthafah Mohd Tahir1, "Review of Research on Vehicles Aerodynamic Drag Reduction Methods,"

IOP Conf. Series: Journal of Physics: Conf. Series **962** (2018) 012022 doi:10.1088/1742-6596/962/1/012022

Review of Research on Vehicles Aerodynamic Drag Reduction Methods, vol. Vol.14, p. 35, 2014.

[10] L D Y A Asst. prof. Munaf F. Badr[1,], Engineer Ahmed Kadhiam Jaliel3, "Position Control of the Pneumatic Actuator Employing ON/OFF Solenoids Valve," *International Journal of Mechanical & Mechatronics Engineering IJMME-IJENS,* vol. Vol.17 No.02, p. 9, 2016.

[11] P B M Lalanne, J.Der Hagopian. (2007). *Mechanical Vibrations for Engineers.*

Geological Study and Regional Development of Mamberamo Raya Disctrict of Papua Province, Indonesia

Adi Tonggiroh,Asri Jaya HS,Ulva Ria Irfan

Geological Engineering Department, Hasanuddin University, Jl. Poros Malino KM.6, Gowa 92171, South Sulawesi, Indonesia

Email address: adi_unhas@yahoo.com

Abstract. The goverment of Mamberamo Raya district was established through Act No. 19 of 2007 dated 15 March 2007 as part of the administrative area of Papua Province. The administrative age of this district is relatively young requires hard work of all components in facing development challenges so that necessary strategic steps of vision and mission of regional development to achieve ideal conditions of spatial which as direction of the desired embodiment in the future. Regional development covers all technical aspects including the geological aspect that the area is located on the morphology of the mountains and Mamberamo watershed. Strategic steps require policy as an action to achieve the goal with the elaboration of operational steps to realize the welfare of peoples equally and sustainably according to the potential physiogeography of Mamberamo watershed. The geological aspect as the consideration of technical that this region belongs to the regional tectonic which is divided into the difference of fault in the north there is Yapen fault and in the south is Mamberamo-Gauttier Fault and also a consideration on the stratigraphic structure of various rock types including the dominance of sedimentary rocks. This study examines geological aspects as an element of earth science in spatial planning in Mamberamo district, especially Kasonaweja and Burmeso. The analysis is presented based on field data, in the form of geographical map data of geological structure, geological map, and earthquake data described by cluster pattern indicating regional motion relationship and rock characteristics that make up Mamberamo watershed. It finds land characteristics controlled by geological structures, rock arrangements and landforms in response to landslide, flood and seismic changes.
Keywords: geological aspect,Gauttier Faut, Kasonaweja,Burmeso,Mamberamo

1. Introduction

Mamberamo River is a main river with a length of 670 km, as the largest and longest river in Papua province that flows from south to north to the Pacific Ocean, the river flows between Burmeso and Kasonaweja areas.

Mamberamo Raya district, Papua Province was established under Act No. 19, 2007 (15 March 2007) with an area of 23.813.91 km^2 (BPS, 2013) or 16.852.16 km^2 or 31.136.85 km^2 (in various sources). The boundaries of Mamberamo Raya as follows: North - Pacific Ocean, South - Puncak Jaya and Tolikara, West - Waropen and Yapen, East - Sarmi.

Although there are differences in the size of region, this district is very broad for a district government size with a population of 18.365 people [1] especially since the geology of the district connects mountain to coastal morphologies. Generally, land use in the area of Mamberamo Raya district consists of primary forests, secondary forests, primary mangroves, secondary mangroves, primary swamp forests, secondary swamp forests, community gardens, dry land farming, shrubs, and open land.

Certainly, to build a substantial district with uneven distribution of population required hard work for the government by involving stakeholders and various disciplines to build for the advancement of society. Multidisciplinary involvement is a necessity to provide much input to the government as a consideration of policies and decisions to build the physical area; this can be seen in the view of geological discipline regionally and locally to the location of the district in Mamberamo watershed and its relation to land carrying capacity, fault structures and morphological types.

2. Regional geology

Regionally, the location of Mamberamo Raya is influenced by active dynamic motion of Gautthers fault in the central part of Papua Island that impact can be seen with the emergence of several local faults that change the carrying capacity of rock. The territory of Mamberamo entered on the geology of the central mountains and the northern island of Papua with lithology in northern part composed by sediments (Formations of Unk, Mamberamo and Makats), it called North Irian basin [2] or Mamberamo basin [3]. Lithology of sediment spread following the morphology of mountains to the plains and also encountered metamorphic rock exposure in the southern part that spreads on the mountainous topography.

The condition of rocks distribution is closely related to the tectonic structure and the regional geological structure of New Guinea, which pose a sinistral shear fault structure. Northern part is fault zone of Mamberamo Raya towards East-West and in the Southern is Tarera-Aiduna (figure 1).

This paper is based on the results of research on geological mapping, mineral resource mapping and geological structure pattern mapping in the district of Burmeso and Kasonaweja in 2016. The benefits of this research can be used as a reference in regional development planning of Mamberamo district and geological disaster mitigation.

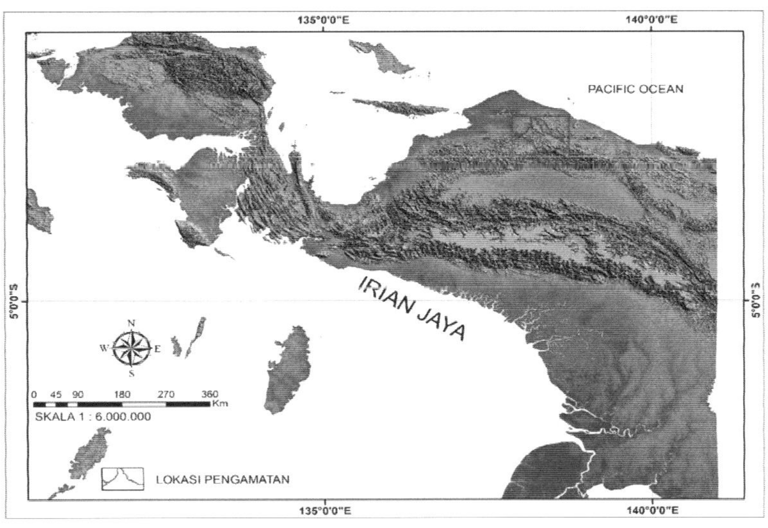

Figure 1. Regional geological map of Papua (*after* [4]) and the research site.

3. Methods

A quantitative research is done directly in the field by descriptive and measurement which include (a) morphological data collection, flow pattern recorded in the map and then followed by observation at the representative descriptive point (b) completeness of morphological data also combined with the symptoms of mass movement such as landslides, creep or fallout (c) geological structure data measurement is done on rocks that record the type of structure, and generally occur in sandstone sedimentary rocks.

Furthermore, field data is processed through making slope map through a method of landform interpretation, land use interpretation, recording on slope table, making digital slope map with *Arcgis 10.0* and 3D Analyst tool.

The data obtained in the field, also equipped with earthquake data recording from the Agencies of Meteorology, Climatology and Geophysics and then processed using statistica v.10 software to obtain a model form that represents the condition of field.

4. Results and discussions

4.1. Mamberamo watershed

Mamberamo River has a catchment area of 138.877 km². The depth of river ranges from 8 to 33 m. According to research in 1983, the water debit was able to reach 5.500 m^3/sec. The landscape around this river varies.

Generally, watersheds condition in the southern part are upstream of bermeander river such as S. Apauwar, S. Tor, S. Baedasifu, S. Roufaer, S.Verkam, S. Tariku, S. Burmeso and S. Kasonaweja towards S. Mamberamo which flows to the northern whose estuary is in contact with the Pacific Ocean. A number of rivers are upstream from fault structures and downstream along the Mamberamo River. The geological structure control of Mamberamo watershed is characterized by the presence of some well-known rapids among geologists; Marina Vallen, Marvalen and Batavia. The Mamberamo River downstream into the Pacific Ocean and Yapen where it is also used as water traffic by public and goverment shipping.

4.2. Geomorphology

The morphology of plains is located in the north and south of this district that separated by the mountains of Foja and Rouffaer. In northern part, the plain lies between the coastline and the mountains that extends to west-east in the central part of the district with the highest peak of 2.164 m above sea level. In southern part, the plain lies in the sedimentary basin situated above a basin flanked by the mountains of Foya and Nassau to Jayawijaya.

The southern plains lie between Mamberamo watersheds and flowed by large rivers, such as Tariku River (Rouffaer River) flowing from west to east and Taritatu River (Idenburg River) flowing from east to west. The two rivers then merge into one and become Memberamo River flowing to north through the Foja-Rouffaer Mountains.

Generally, the morphology of study area is controlled by the geological structures of fault and regional folds that characterized by a denritic flow pattern.

4.2.1. Morphology unit of lowland. The morphology of lowland enters residential areas of Kasonaweja and Burmeso with a slope of 10% to 12%. These morphological constituent rocks consist of strongly laid sandstones and alluvial sediment deposits (figure 2).

4.2.2. Morphology unit of hillside. This morphology enters residential areas of Anggreso Baru and Trimuris with a slope of 15% to 20%. The constituent rocks of this region are high-decaying in contact with the mud deposits.

4.2.3. Morphological unit of steep hill. This morphology spreads in Saromaja and Pisano areas, Van Rees, Gauttier and Siduarsi mountains were characterized by a slope of 22° to 45^0 degrees. Although the constituent rocks are generally old geological ages, they are compact and hard but in some places exhibit their brittle properties especially on fracture compiled by siltstone, mudstone, sandstone, and conglomerate, elevation 50 m to 75 m above sea level, the slope between 10% to 12%. Fracture morphology control is spreading in the Middle East Mamberamo, Mamberamo Hulu and Roufaer, west and southeast Mamberamo, Burmeso, Kasonaweja with the general direction North-Northwest-South-Southeast and Southwest-Northeast.

Figure 2. Lowland morphology.

4.3. Structure

4.3.1. Folds. The structure of major and minor folds as anticline and syncline, the slope of wings 30° to 50°, the general direction of axis is irregular northwest-southeast to east-west, steep corner. This structure is found in sedimentary rock of Unk Formation that make up Burmeso and Kasonaweja districts.

4.3.2. Fault. Regional fault pattern of Roufaer-guinea-Gauttier mountain-and Van Ress affecting local faults spreads in Kasonaweja-Burmeso and the middle parts-benuki-bira lake-marvalen rapids-batavia rapids-marina valen until downstream Mamberamo.

4.4. Lithology

Distribution of sedimentary rocks consisting of claystone, sandstone, siltstone that dominate the residential area. Claystone shows the fresh color of gray-blackish, fragments: clay, matrix: clay, cement: clay, N3250E/350 in position. Sandstone, grayish-white, fragments: very coarse sand, matrix: medium sand, cement: claystone and coal insertion (lignite). Siltstone shows fresh gray to brown, fragment: silt, matrix: silt, cement: silt, strong decay.

4.5. Potential of mineral resources

Result of inventory of metallic and non-metallic mineral resources by the Department of Energy and Mineral Resources of Mamberamo Raya district that mining in 2014 found some potential of excavation materials as follows:

- Fine sand, sirtu, coal, clay, gold, copper and iron, stibnit, black lead, zinc and energy.
- Coal, fine sand and clay are found in Roufaer District.
- Coal is found in Middle-East Mamberamo district.
- Sirtu (sand and stone), gems, gold, copper and iron, stibnit, lead and zinc are found in upstream Mamberamo district.

- Sirtu (sand and stone), gold, gemstone and coal are found in Central Mamberamo district.
- Indications of energy found in downstream Mamberamo and Sawai districts.

Especially the coal in the area of Mamberamo Raya district, where based on the results of laboratory analysis has a calorific value with a range of 5000 cal/gr, it is suitable as a raw material of Steam Power Plant. In addition to Steam Power Plant, the coal in Mamberamo Raya is also suitable to be made as coal briquettes which have the potential and distribution of coal in this region has a wide distribution.

Based on the potential data of metallic and nonmetallic resource above that related to the size of region, geologic location of rock variation as mineral source is estimated there are still some potency of other excavation material that has not been disclosed.

4.6. Slope

The slope has a variable angle that is affected by rock layer resistance and the location of region within the regional fault zone. This factor plays an important role in the formation of soil movement so it should be used in planning the construction of roads and bridges.

Based on the slope calculation is known that there are three slope classifications, namely: (1) steep (30% to 40%), (2) slight steep (15% to 30%), (3) ramps (8% to 15%). Generally, the steep slope has a greater gravity effect than the slight steep and ramps, which is due to the property of gravity is proportional to the sloping surface of the horizontal plane. This gravity is an absolute requirement of the process of detachment, transportation, and sedimentation (Wiradisastra, 1999).

Slope analysis of Kasonaweja-Burmeso region using morphometric parameters, quantitative assessment of landform as supporting aspects of morphographic and morphogenetic. These closely parameters are characterized by the properties of contour lines. The density of the contour lines on the map indicates that the slope is steeper and vice versa if the contour line is looser then the slope will be more ramps.

Data collection and interpretation approach of topographic map are used as morphometric data using slope formula (α):

$$\text{slope } (\alpha) = \text{Arc Tan (high difference-scaled distance)}$$

Based on the calculation using the above formula, it is known that there is a difference of slope in the district area which is described in the form of slope map (figure 3). Generally, the slopes distribution of flat slope, almost flat is located in the north or downstream area of Mamberamo River and in the northeast which borders Sarmi district. While, the slope of 9° to 15° spreads in Kasonaweja - Burmeso, where the spreads is limited by slight steep to steep.

4.7. Geological disaster mitigation

Mitigation in disaster terminology is defined as human efforts with the main objective of reducing the risks or impacts of natural disasters caused by natural or human activities (Sudibyakto, 2011). If an event that has a potential hazard occurs in an area with a vulnerable condition, then the area can be categorized as at risk of disaster [5].

The general potential of geological disasters is tectonic movement, tropical climate, soil movement and human activities that tend to change topography. In terms of regional geology, Mamberamo Raya district is flanked by regional fault of Gauttier located in the south and Yapen fault in the north and some local fault. Both of these regional faults are relatively active. Based on geological history, the dynamic tectonic motion of the past can be seen in the traces of local fault structures and folds recorded on sedimentary rocks. The impact of this structure, although it is local but can evolve to form a labile zone that characterized by the formation of a steep slope to a very steep, reduced rock compaction gradually change the size (fragmentation) from coarse to fine and even the size of clay and the formation of the slip due to layer changes and rock slope.

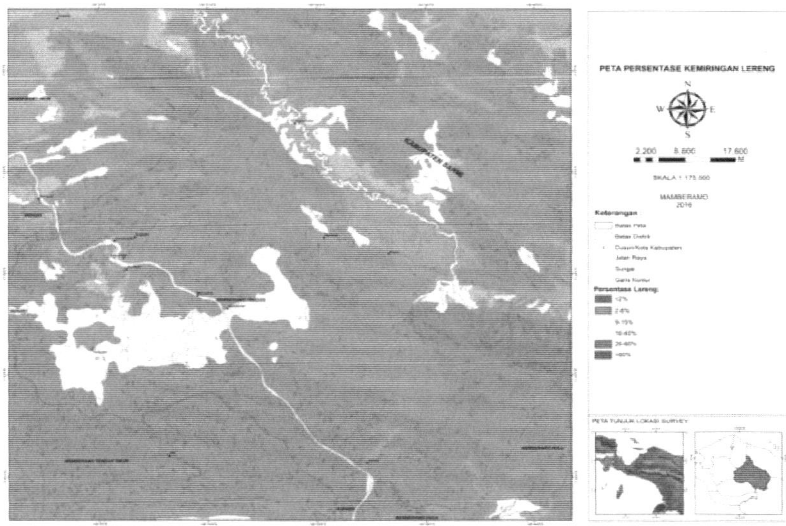

Figure 3. Map of slope percentage of Mamberamo Raya district. (1) 0°-2° = flat to almost flat, green (2) 2°-8° = slightly angled, light green (3) 9°-15 ° = angled with higher magnitude, yellow (4) 16°-40° = slightly steep, orange (5) 26°-60 ° = steep, pink (6) 35°-55° = very steep, red (7) > 60° = very steep, purple.

The data of soil movement potential is seen from the slope difference and seems to have a different kind of material contribution as well. Tilt of slope 0° to 15° although the topography is relatively flat but enough potency to cause sheet and groove erosions. Then, groove erosion that develops into landslides that also occur on the slope of slightly steep. The denudation process that occurs in rocks is quite active on steep conditions so it can increase the intensity of land motion. The condition of the soil movement develops on a very steep until very-very steep, which is composed by rock sediment (colluvial).

Data (1950-1990) average rainfall Mamberamo region in one year for each month is about 262.5 mm. Average rainfall above 200 mm every month and above 250 mm occurred in December, January, February, March and April [6]. Based on the accumulation of rainfall data, it quite high and influential as a trigger for the emergence of symptoms of mass movement, this is correlated with field conditions where usually after the rainfall activity found real erosion. Another trigger is the change in the compactness structure of rocks that occur in silt and sand that causes the development of real erosion into large-scale erosion or gully.

The high rainfall also affects several changes in Mamberamo watershed, namely the increase in surface water volume, the rising water levels of the river branches and Mamberamo River and the erosion of river walls. The above change factor can be used as an agenda in preparing the concept of regional development, another consideration is the field data of floods in 2015 which is an unusual event by the overflowing of Mamberamo River and inundating several houses located a few meters from the riverbank.

Although the district is included in watershed and the morphology of the plains, it cannot be said that the abundance of water can meet the needs of clean water. It is well-founded that the results of some field observation points on the outcrop of sedimentary rocks with the layered structure indicate *impermeable* physical properties. Because the distribution of sedimentary rock is wide enough it is necessary to note the availability of aquifer layer as a consideration material in preparing regional development planning. Given that the availability of groundwater as a need for clean water is a major and increasing need.

4.8. Synergy

Efforts to integrate spatial plans and regional development are undertaken by the Government of Mamberamo Raya district of Papua Province, but these efforts require maximum implementation due to constraints, technical and non-technical, and budget availability.

4.8.1. The role of stakeholder. The success of spatial planning will be determined by how much society can be involved in planning activities, spatial use, and spatial use controls facilitated by the Government. As the first stage of spatial planning, participatory planning provides greater opportunities for the creation of integrated and synergic space utilization, as well as effective and efficient spatial controls.

4.8.2. The role of central and provincial governments. Planning and development target of Mamberamo district need to synergize with various disciplines including the target of how the geology database. (First): the coordination, integration and communication of the Government of Mamberamo Raya district of Papua Province, (Second): Principal target of Metallic and Non- Metallic Resources Management, (Third): Mapping vulnerable zone of geological disasters, (Fourth): mapping ground water fluctuations, (Fifth): performing mapping of river characteristic, chemical properties and flow patterns of Mamberamo watershed.

4.8.3. General planning. (1) Government efforts to open roads, should be continued by prioritizing road openings, hardening connecting the districts to the villages that allow the passage of four-wheeled vehicles (2) plan a wider Airport from now, or look for a wider airport location alternative (3) availability of economic facilities for the community in utilizing river and estuary potentials.

4.8.4. Watershed management. The function of Mamberamo River for communities and government is as a source of livelihood of fisheries and transportation routes that connect the economy between villages, districts and other districts. Along with the development of uncontrolled land use in the upper river, it also disrupted the mechanism of river function and river hydrological behavior in Mamberamo watershed. Initially influenced by nature properties as a watershed system but gradually influenced also by humans. Therefore, a planned and integrated participatory watershed management should be undertaken in order to maintain and improve the continuity and quality of river water resources by maintaining upstream areas and their zones including: (1) plan to remove wood residual that embedded in Mamberamo river from Kasonaweja to downstream (2) plan to making dam along the riverbank in Kasonaweja - Burmeso to prevent erosion of river walls.

4.8.5. Agricultural planning. The planning of mapping program is an early step for realizing the characteristics of the land for the determination of the type of plant, mapping of agricultural land and its designation and establishing the type of mainstay.

4.8.6. Planning of estuary and coastal areas. Large coastal areas contain natural resources such as fisheries and mangrove forests to oil and gas so that guidance is required for the management of estuarine and coastal areas development. Several alternatives that can be conducted such as making industrial development centers, ports and shipping, tourism, agribusiness, human settlements and activities, both within the coastal and outer (upper land and high seas) systems. The development of tourism aspect is quite prospective and promising for example mangrove ecosystem tours in the form of views of various types of mangrove trees and biota supported by the thick atmosphere and Rombebai Lake tourism inhabited by various types of freshwater fish, endemic and marine biota.

4.8.7. Mineral resource management. The management of metallic and non-metallic mineral resources including rocks involves the local community which leads to the original income of the Mamberamo Raya district of Papua Province. Despite the enactment of Act No. 23 of 2014 regarding re-withdrawing decentralized mining management authority to the Central and Provincial Governments. However, it is necessary to review and involve the local community given the geographical conditions and hierarchy of land ownership.

5. Conclusions

The study concludes that (1) the application of geological-based regional development is fully undertaken (2) the provision of thematic maps containing technical information of a rock, soil, or land; information on the prevention and prevention of potential geological hazards (3) implementation of community-based mineral center development as an optimization step (4) availability of plain morphology as land that can grow agriculture and involve transmigration in developing area (5) necessary strong spatial development to geological disasters by arranging and mapping geomorphology of regional zoning according to zoning and constituent rocks (6) integrated-planning of coastal and estuarine areas requiring multi-disciplines and stakeholder integrity.

Acknowledgments

The author thanks to Mr. Dorinus Dasinapa, James Britay, and Alfons Sesa for their assistances on the implementation of the research.

References

[1] BPS 2013 Biro pusat statistik Papua
[2] Mc Adoo R L and Haebig J C 1999 *Tectonic elements of The North Irian Basin* IPA Proceedings 27th
[3] Baldwin S L, Fitzgerald P G and Webb L E 2012 Tectonics of the New Guinea Region *Earth Planet Sci.*
[4] Dow D B and Sukamto R 1984 Later Tertiary to Quaternary tectonics of Irian Jaya *Episodes* **7**
[5] Sadisun I A 2007 *Peta rawan bencana: suatu informasi fundamental dalam program pengurangan resiko bencana* Fakultas Ilmu dan Teknologi Kebumian ITB Bandung Indonesia
[6] Conservation International 2007 *Integrating biodiversity and hydrological processes into conservation planning at the landscape scale,* Workshop on Integrating Hydrological Biodiversity and Socio-economic Analysis for Conservation
[7] Wiradisastra, U. S., B. Tjahjono, K. Gandasasmita, B. Barus, Khursatul Munibah.1999. Geomorfologi dan Analisis Lansekap. Institut Pertanian Bogor. Bogor.

Analyze Experiment For Vigas and Pertamax to Performance and Exhaust Gas Emission for Gasoline Motor 2000cc

Muhamad As'adi, Diachirta Chrisna Ayu Dwiharpini Tupan

Universitas Pembangunan Nasional "Veteran" Jakarta
Jl. RS Fatmawati No 1., Pondok Labu, Jakarta Selatan

adi_shiddiq@yahoo.com, diachirta.harpini@gmail.com

Abstract. The purpose and target for this analyze experiment is we get the performance variabel from gasoline motor which used LGV for fuel and Pertamax, so can give knowledge to community if LGV can be using LGV for fuel to transportation industry and more economic. We used experiment method of engine gasoline motor with 2000 cc which is LGV and Pertamax for fuel. The experiment with static experiment tes above Dyno Test. The result is engine perform to subscribe Torque, power, fuel consumption. Beside the static test we did the Exhaust Steam Emission. The result is the used LGV with the commercial brand Vigas can increase the maximum Engine Power 20.86% and Average Power 14.1%, the maximum torque for Motor which is use LGV as fuel is smaller than Motor with Pertamax, the decrease is 0.94%.Using Vigas in Motor can increase the mileage until 6.9% compare with the Motor with pertamax.Air Fuel Ratio (AFR) for both of the fuels still below the standard, so still happen waste of fuel, specially in low compression.Using Vigas can reduce the Exhaust Steam Emission especially CO_2

Keywords: Gasoline motor, Vigas, Pertamax, Engine Perform, Emission.

1. Introduction

The vehicle which is used fuel in Indonesian always increase every year. In Traffic Indonesian Police Data Station noted, the total Vehicle in operation at 2013 is 104,211 million units, increased 11% from 2012 which is 94,299 million units. With the increased vehicle in operation, so fuel consumption or BBM that we called, follow being increase too. With big fuel consumption so we needed the fuel production more than now, but in the reality the oil in Indonesian since 2000 have been reduce, so in 2005, Indonesian be an Oil Importir. The high cost in crude oil, is making Governor should give the subsidy BBM.

In 2013. As a planned, Head Governor would increase the point of subsidy BBM in APBN-2013 Changed 46 million kilo litre. So if the vehicle in operation will increase so the subsidy BBM will increase too which is the Governor should alocated every year. Beside energy source and that subsidy so another problem arises like environmental pollution from result Fuel combustion at Vehicle, such as CO_2, PM10, and Pb. Data from Department of Transportation, The CO_2 pollution in 2003 from Transportation system is 168 million ton, if linear with the incremental the vehicle operation so in 2007 will be approximately 324 million ton.

Growth the environment pollution will give the bad affect for the human healthy. The big sector is power plants and transportation, so need the special handling. The president 'Susilo Bambang Yudhoyono' in his speech at the meeting of Growing Country (*Group of 77*), promised will reduce the emission 26% at 2020. For the action from he speech need clearly divide contribute reduce emission in every sector.

From transportation sector the main problem from the fuel, the used fuel give the big pollution. So need the change of fuel which is more adaptable in environment, one of is Liquid Petrolium Gas. Gas in here is means Liquid Gas for Vehicle (LGV). LGV had been choosen because still have more rest same with words in MP3EI (Masterplan Percepatan dan Perluasan Pembangunan Ekonomi Indonesia – " Masterplan quick and widely indonesian economic growth) with the rest natural gas is 165 TCF.

See from the big rest and the low price, resut pollution from LGV more small compare with Gasoline. So alternate to use LGV for change BBM is one of the choose which is should do by Governor.

Total Vehicle use Gas for fuel still increase since 1987 is approximately 300 units and in 2000 is 6633 units, but in 2001 had been decrease until 500 units at 2004, because the infrastructure is not ready enough. So the Governor at 2007 start socialize again Fuel Convert to LGV in Taxi, Bus and Bajaj with socialize and give the converter kit especially in DKI Jakarta but the result is not same with the willing so at 2012 Governor start socialize again Fuel Convert to LGV for 4 wheels vehicle.

2. Literature Review

2.1 Combustion Engine

Combustion Engine is one of calor engine, engine which is change from thermal energy to mechanism energy. Before being mechanism power the chemistry energy fuel had been changed to heat energy by fuel combustion with air. This combustion

What Happen inside calor engine itself and outside calor engine. This 4 steps combustion engine include in inside combustion engine category where the engine movement from the result combustion in cylinder. Engine 4 steps is called spark ignition engine too, is means the engine with the combine fuel – air ignition use spark fire from spark plugs.

2.2 Liquid Gas for Vehicle (LGV)

Liquid Gas for Vehicle (LGV) is gas fuel for vehicle use the spark ignition engine, with the combine propane C3 and butane C4. Some of literarure tell is LGV more compatible for vehicle with small cylinder volume like Bajaj, Taxi, public transportation, office operational car and personal car, because have big capacity tank for the same milleage compare with Oil Fuel (BBM). And have low pressure plus minus 15 kg/cm^2.

The benefit LGV compare with CNG such as :

- We have a lot of LGV in our country ,
- Quick help for Program using Gas in Transportation Sector,
- Decrease subsidy Premium BBM which is will change by LGV,
- Growth of SPBG LGV (Fuel Station LGV) more easy than SPBG CNG/BBG (Fuel Station CNG/BBG)
- More applicable at Areas that are not or have not been reached by gas pipelines,
- Low price for conversion kit LGV than CNG or BBG,
- The power of pressure in car and SPB LGV more low (maximum 15 bar than CNG or BBG (200 bar))

LGV already used since a long time ago for vehicle fuel in some country like United Stated of America, Mexico, Russia, Dutch, Germany, Ireland, Sweden, Finland, Italy, Indian, Turkey, Japan, China, Filipina, Thailand, Korea, Australia, and New Zealand. But in Indonesia LGV more fast develop being subtitute energy BBM for Quick help Program blue sky and energy diversification beside CNG / BBG.

2.3 Combustion Theory
Combustion is Fast reaction between Fuel and Oxygen which is the heat and light is the result. One of the requirement can happen the perfect fuel combustion is the right oxygen for the source. The total Oxygen (O_2) in free air until 20.9% from air. Solid Fuel or Liquid need to change to Gas first before combustion in combustion chamber. Usually need heat for change liquid or solid to be Gas. BBG will be fired in normal state if there is enough air.

The purpose from the good combustion is take out all the heat in combustion chamber. This is need to do with the 3 T Controlling combustion like :
a) High temperature for starter and maintain lightning fuel,
b) Turbulence or mixed between oxygen and good fuel, and
c) enough Time for the perfect combustion.

The used common fuel like natural gas and propan which is contain between carbon and hydrogen. The water vapour is the side product hydrogen combustion, which is can take heat from exhaust gas, and maybe can use for heat transfer continued. Natural gas contain more hydrogen and less carbon per kg compare oil fuel, so will more produce water vapour. And the affect, more heat will bring to waste when natural gas combustion. More or less fuel in some air combustion, will affect didn't fire the fuel and will make Carbon monoxide. The quantity of specific oxygen need for perfect combustion with need the add some air for make sure the perfect combustion. Even that, more air will affect loose heat and efficiency.

2.4 Engine Performamce
Engine Performance is the dimensi for how big the efficiency form that engine, with the performance parameter contains as :
a) Torque,
b) Power, and
c) Spesific Fuel Consumption.
For count the Torque need test bed at dynometer and shown as force times with length in Newton Meter (N.m). Combustion Type as shown in Figure 1.

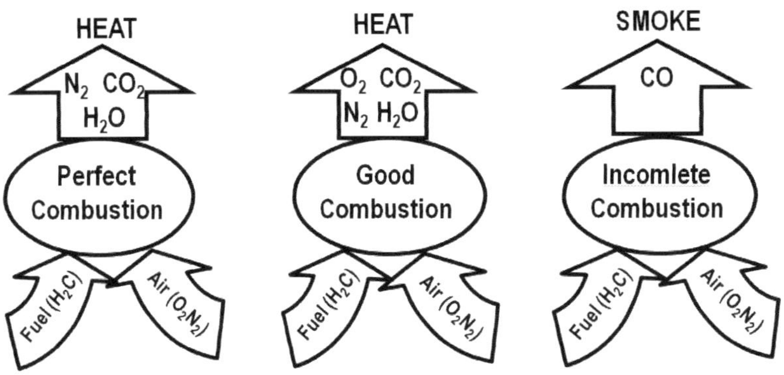

(a) (b) (c)

Figure 1. (a) Perfect combustion, (b) Good combustion, (c) Bad Combustion (unperfect combustion)

Power is the big work for system result per times in Watt (SI) and horse power, horse-power; british-system. Rate of fuel consumption in engine per output power is called specific fuel consumption (sfc) and tell in fuel mass per power or kg/kWh, and this is the indicator too for economic rate from engine operational in use the fuel, more low the point sfc so need low cost and otherwise

2.5 Exhaust gas emissions

Exhaust gas emissions in conventional engine need more attention from all around the world. Because the affect form Exhaust Gas can destroy the environment. The affect from this can make the glass house which is not human wish.

In Conventional engine the output Exhaust gas emissions is HC, CO, CO_2, O_2, NOx and other particulars. Many experiment had been doing for reduce the content Exhaust gas emissions conventional engine itself. Output Exhaust gas emissions from the unperfect combustion in combustion chamber, where is only half fuel being reacting with the oxygen specially near the wall of cylinder between Torak and cylinder, commonly the reason is low fire and low combustion temperature. If the combustion temperature is low and spark ignition is low too and surrounding is wide with low temperature combustion chamber, this condition will meet at the new motorcycle starting or in idle, natural condition the motorcycle will give output Exhaust gas emissions more than ussually and can give the bad affect for human healthy.

Some parameter from Exhaust gas emissions motorcycle like this:

- Hydrocarbon (HC) is Exhaust Gas from the fuel where didn't fire and counting in part per million (ppm), light mass molecul, unseen. Because have the characteristic can Binds hemoglobin so dangerous for health. If the Hydrocarbon is low so will be better.
- Carbon Dioxide (CO_2) is the element which is indicate combustion heat and count in presentage, The ideal around (11 – 16) %. Carbon Dioxide have unseen light mass and not dangerous, but main potential to Glass House, floral and Sea community As well as peat plants
- Carbon Monoxide (CO) is Gas from Reaction the unperfect combustion, light mass, unseen, dangerous for healthy like Acute Respiratory Infections – "Infeksi Saluran Pernapasan Akut (ISPA)" , the cancer, and will reduce the Intelligence. The ideal (0.3 – 3) %.
- Oxygen (O_2) is indicator from Combustion Quality, The ideal less from 2%, if smaller so the combustion was perfect, and didn't dangerous for healthy.
- Nitrogen Oxide (NOx) is Exhaust gas emissions from Oxidized nitrogen because the pressure and heat when compression. This Gas is toxic so dangerous fro healthy and environment.
- Particulate
  This Particulate will Wreathed in the lining cells of the lungs so the function will broken and will be black.

Method for reduce the bad affect from Exhaust gas emissions to combustion quality, healthy, and environment is :

- Revise the design and combustion system
- Increase the fuel quality.

One of the method can be use is use the Gas Fuel like LGV.

2.6 Previous Experiment

Using LGV in gasoline motorcycle without doing adjustment in injection pressure and injection timing give Carbon Monoxide, Hydro Carbon, More oxygen than Premium Fuel, but lamda levels and Carbon Dioxide more high [1].

Using the dual fuel (CNG-Solar) give the engine power same with solar engine. Beside that with dual fuel (CNG-Solar) give reduce smoke, but the hydrocarbon (HC) and NOx being increase than engine with 100% Diesel. [2], others experiment tell using CNG direct injection system with Combustion controlling, reduce the orifis nozzle dimentional and injection corner will help quality of fuel mixed. This will give the combustion efficiency more high even knocking because there is rich mixed air – fuel in some areas in Combustion Chamber [3].

3. Experimental Method

The experiment of this research is engine with the cylinder capacity is 2000cc with carburator. After that the engine installed converter kit so this car have dual fuel system its means can use Liquid Fuel or Elpiji Fuel. Do the Engine Tune Up when the optimum condition for both of fuel, so while changing form one fuel to other fuel can run smooth.

The next step is Performance Trial and Exhaust gas emissions, Performance Trial above Static Dyno, and trial fuel consumption with the road test.
The experiment flow as shown Figure 2.

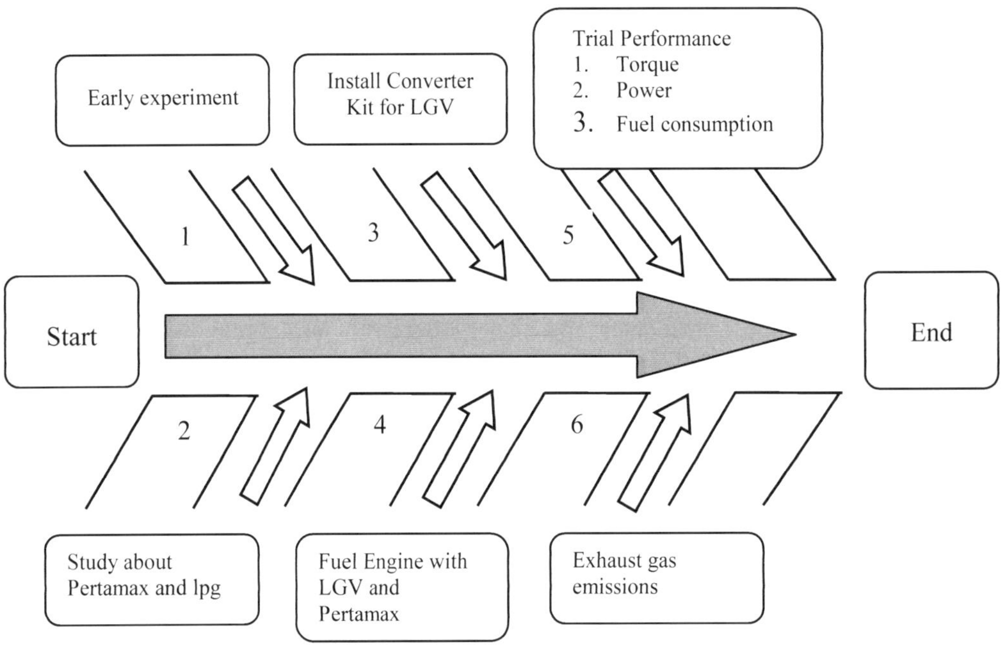

Figure 2. Flow of the experiment

Engine specification can be found in table 1, below:

Table 1. Engine Specification

Type	4 cylinder (in line),SOHC.
Cylinder Volume	2000 cc
Power	102 HP/136 PS/5000 rpm
Torque	200 Nm/18.6 kgm/4000 rpm
Cooling System	Radiator
Weight	1,480kg

3.1. Installment Converter Kit

When we installing the Converter Kit for Dual Fuel in Vehicle. We can see as as shown at Figure 3. When we use Common Fuel, the flow will direct to solenoid and goes to Fuel pump, through in

Carburator. But when we are going to use Vigas then the solenoid will off and the flow will go to Reducer. And through in to carburator by mixer first.

The Reducer can reduce the pressure so the vehicle use Vigas is not in high compression.

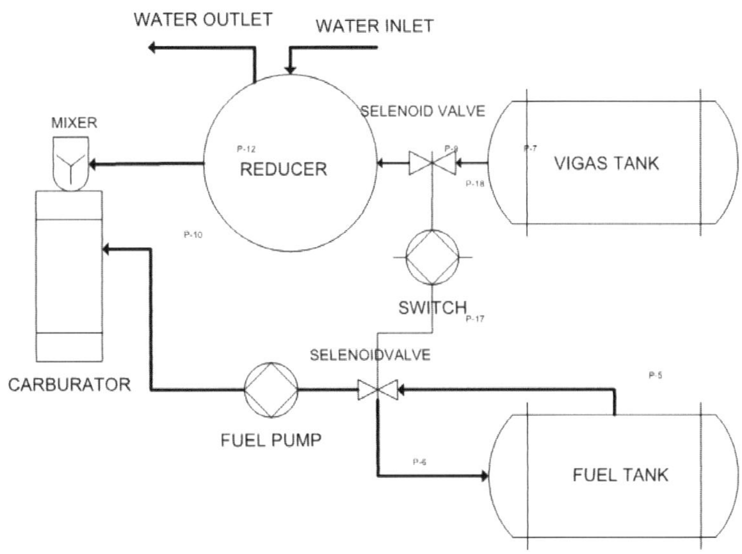

Figure 3. Installment Dual Fuel system in Vehicle

3.2. Measuring instrument

Measuring Instrument already calibrate, each used type as follow :

- Dynometer Static
- Emissions measurement tools.
- Digital Stopwatch
- Buret / Measuring Instrument solar fuel consumption with the capacity 30ml

4. Results and Discussion

4.1 Power

Follow the result above Dyno Static Test, trend of Maximum Power in motorcycle use Vigas is 42.3 HP with radius 4782 rpm, even the motorcycle with use Pertamax give the Maximum Power is 35 HP with radius 4211 rpm. So using Vigas can increase Maximum Power 20.86%, and average using Vigas can increase Power 14.1%.

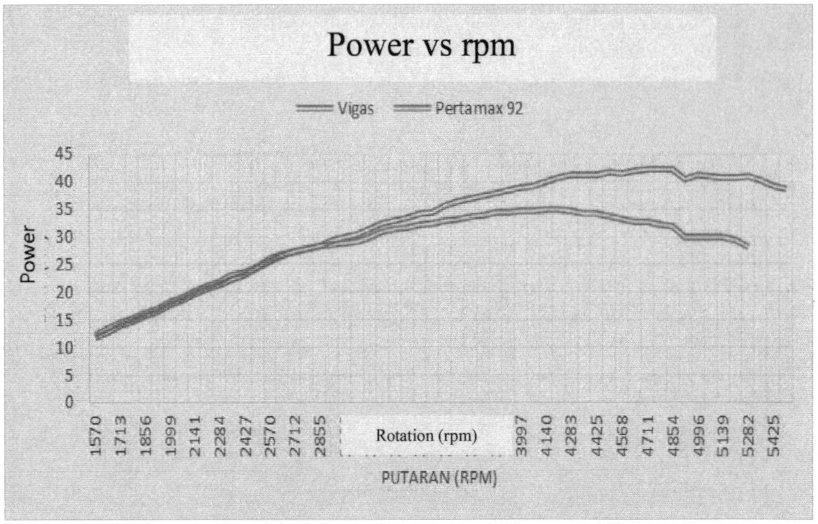

Figure 4. Graphic Motor Power use Vigas and Pertamax

The relationship rotation of engine (rpm) and power can be found at figure 4. It was found that Vigas has better perform on 3000—4800 rpm then another engine with Pertamax 92. However start from 1570 to 3000 rpm, the performance was similiar.

4.2 Torque of engine

Average happened Torque in motorcycle use Vigas more big than motorcycle use Pertamax because the combustion more stable dan radius more high so didn't knocking. The average increase torque 7.57%, but the maximum torque in Vigas Motorcycle smaller than Pertamax Motorcycle, the result show torque decreased 0.94%, and torque trend as shown at Figure 5.

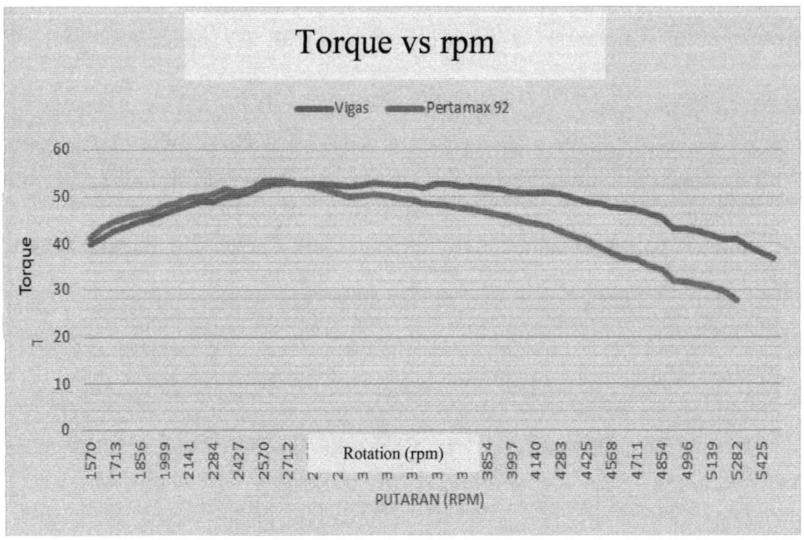

Figure 5. Torque Graphic for Vigas and Pertamax Motor

4.3 Fuel Consumption

Measuring of Fuel Consumption as shown at below Table 2

Table 2. Fuel Consumption

Step of Measurement	Vigas (km/l)	Pertamax 92 (km/l)
1	7.5	7.3
2	7.8	7.2
3	7.7	7.4
4	7.9	7.2
5	7.8	7.3
6	7.8	7.2
7	7.6	7
8	7.8	7.2
9	7.9	7.2
10	7.5	7.3
Total	77.3	72.3
Average	7.73	7.23
Incremental	6.9	

As shown in table 2, Vigas has more distance than Pertamax. Distance average can reach by Vigas motorcycle 7.73 km/l, even Pertamax motorcycle 7.23 km/l, so if compare by increase distance Vigas Motorcycle is 6.9%. It can received because in Vigas Motorcycle when combustion can direct inject to Carburator without smoke.

The result shows Air Fuel Ratio (AFR) parametes for both of them, still below from the standard, that its means fat Combustion (too rich mixture), or still waste fuel especially when low rpm of engine.

4.4. Emission

The experiment result for Exhaust gas emissions as shown at Table 3 [7].

Table 3. Emission

Parameter	Vigas	Pertamax92	Verge according to LH 04/2009
CO (g/km)	0.95	0.12	2.2
HC (g/km)	0.01	0.006	-
NOx (g/km)	0	0.027	-
CO_2 (g/km)	182.5	206.8	-

As shown on table 3, using Vigas at Motorcycle can reduce effect of Glass House CO_2 13.3% compare than the engine using Pertamax 92.

5. Conclusion

Follow the experimental result we get some points like this :

- Using Liquid Gas For Vehicle (LGV) with Commercial brand Vigas can increase Maximum Engine Power 20.86% and Average Power 14.1%
- Maximum Torque from Vigas Motorcycle smaller than Pertamax Motorcycle, reduce around 0.94%
- Using Vigas at motorcycle can increase distance around 6.9% compare than motor with Pertamax.

- Air Fuel Ratio (AFR) for both of this fuel still below the standard, so it happens waste of fuel, especially in low radius.
- Using Vigas at Vehicle can reduce Exhaust gas emissions especially CO_2.

References

[1] Qadri Munzir, et.all (2013), Analyze Comparison between Emission LGV Fuel with Premium on Daihatsu Grand Max Standar, National Symposium RAPI XII, UMS (in bahasa, original title: Analisa Perbandingan Emisi Gas Buang Bahan Bakar LGV Dengan Premium Pada Daihatsu Grand Max Standar)

[2] Sumartono, H, Time Studi for solar injection to Diesel engine perform using Double fuel CNG – Diesel, (2003), Tesis Program Studi Teknik Mesin Universitas Indonesia. (in Bahasa, original title: Studi Pengaruh Waktu Injeksi Solar Terhadap Performa Mesin Diesel Ber-bahan Bakar Ganda CNG-Solar)

[3] Ishiyama. T, H Kawanabe, K Ohashi, M Shioji, S Nakai, A study on premixed charge compression ignition combustion of natural gas with direct injection, (2005), International Journal of Engine Research ISSN: 1468-0874 DOI: 10.1243/146808705X30459 Volume 6, Number 5, Pages: 443 – 451

[4] Chalilullah Rangkuti, Comparasion Air Fuel from Emission analyze from engine, (1995), ISSN :0853.5175, Edisi No. 002 (in bahasa, original title: Perbandingan Udara Bahan-bakar dari Analisa Gas-buang pada Suatu Motor-bakar)

[5] John B Heywood, Internal Combustion Engine Fundamental, (1988), Mc Graw Hill, Singapore.

[6] LIM Pei Li, The Effect of Compression Ratio on the CNG-Diesel Engine, (2004), University of Southern Queensland Faculty of Engineering and Surveying. http://www.usq.edu.au/, diakses September 2006.

[7] Newnan. D G, Engineering Economic Analysis, (1991), Binarupa Aksara, Jakarta,

[8] Roger A. Strehlow, Combustion Fundamentals, (1985), Mc Graw Hill Book Company.

[9] Susanti.V, Hartanto.A, Subekti.R.A, Saputra.H.M, (2011), Governor rules about convertion Oil Fuel to Gas Fuel for Vehicle, LIPI Press . (in Bahasa, original title: Kebijakan Nasional Konversi BBM ke BBG Untuk Kendaraan)

[10] Widodo B Santoso, Achmad Praptiyanto, Arifin Nur, (2006), Affect CNG – Ratio to combustion in diesel engine with double fuel Diesel – CNG, Prosiding Seminar Nasional Tenaga Listrik dan Mekatronik, hal. 283-287, ISBN :979-26-2441-4. (in Bahasa, original title: *Pengaruh CNG-Ratio Terhadap Proses Pembakaran Pada Motor Diesel Berbahan bakar Ganda Solar-CNG)*

[11] Yusaf T.F., Abd. Halim S., Yusoff A. & Ahmad Faris I. (1999), A study of dual fuel system using compressed natural gas for commercial diesel engine, International Journal of Power and Energy Systems, Vol. 19, No. 2. (ms 163-167)

[12] Zhiqiang Lin and Wanhua Su, (2011), A study on the amount of pilot injection and its effects on rich and lean boundaries of the premixed CNG/air mixture for a CNG/diesel dual-fuel engine, International Journal of Global Energy Issues, Vol. 20, Number 3/2003, Pages 290-301

IOP Conf. Series: Journal of Physics: Conf. Series **962** (2018) 012025 doi:10.1088/1742-6596/962/1/012025

Computational Simulation on Facial Expressions and Experimental Tensile Strength for Silicone Rubber as Artificial Skin

Andi Amijoyo Mochtar
Hasanuddin University, Departement of Mechanical Engineering, Jl Perintis Kemerdekaan KM10, Makassar, Indonesia

Corrresponding author: andijoyo@unhas.ac.id

Abstract. Applications of robotics have become important for human life in recent years. There are many specification of robots that have been improved and encriched with the technology advances. One of them are humanoid robot with facial expression which closer with the human facial expression naturally. The purpose of this research is to make computation on facial expressions and conduct the tensile strength for silicone rubber as artificial skin. Facial expressions were calculated by determining dimension, material properties, number of node elements, boundary condition, force condition, and analysis type. A Facial expression robot is determined by the direction and the magnitude external force on the driven point. The expression face of robot is identical with the human facial expression where the muscle structure in face according to the human face anatomy. For developing facial expression robots, facial action coding system (FACS) in approached due to follow expression human. The tensile strength is conducting due to check the proportional force of artificial skin that can be applied on the future of robot facial expression. Combining of calculated and experimental results can generate reliable and sustainable robot facial expression that using silicone rubber as artificial skin.

***Keywords*:** facial expresions, tensile strength, silicone rubber, artificial skin, proportional force, FACS

Introduction

Now a days, the robotics application becomes important and world-wide in human life activity. Robot mimics have been developed by following human mimics in facial expression. In order to follow human facial expression, robot needs apply the facial muscle area, muscle movement directions, and the magnitude of force which depends on the facial expression type [1]. This feature of the robot is required for a specific skill in developing communication either verbal or nonverbal [2,3]. In order to imitate a human gesture, several steps should be followed such as face registration and facial expression recognition [4]. The facial object is not necessary to be life which some parts of face i.e, mouth, nose, and eyes will be used in initial face registration. Then the image classification is applied by matching the facial expression with the facial action coding system (FACS) pattern [5]. This method must be integrated in order to solve the problem in detecting facial expression motion. Generally, there are six basic expressions that most people used for facial expression i.e., happiness, sadness, surprise, anger, disgust, and fear[1, 6, 7]. There are 44 actions of motion of the facial expression that found in FACS including several combinations of the different of facial expression which also known as action units (AUs)[8, 9]. The FACS is not only representing a general facial expression, but also it covers some specific of emotional expression. The AUs is affected the facial expression by the contraction or relaxation of one or more facial muscle in a specific area of the face. Therefore, human can recognize the robot mimics by using that facial expression.

As represented artificial muscle in the robot, some researchers using shape memory alloy (SMA) or DC motor as an actuator when the force needed to be larger[10, 11]. The direction magnitude of force area are determined by type of AUs in facial expression. The boundary condition around the edge of face including at the end-point of eyes is set-up as fixed 6 degree of freedom (DOF)[10, 12].

This study will describe computational simulation of facial expression and experimental tensile strength for slicone rubber as artificial skin and analysis. Firstly, the tensile test was carried out on silicone rubber (RTV2 VP 7550) used as artificial skin of robot face. Secondly, the mechanical properties of artificial skin were identified using the experimental results. Thirdly, the facial shape was measured by 3D scanning. In addition, the facial model was made by meshing the 3D scanning data.

1. Methodology of tensile strength experiment and calculation for artificial skin

1.1 Experimental set-up

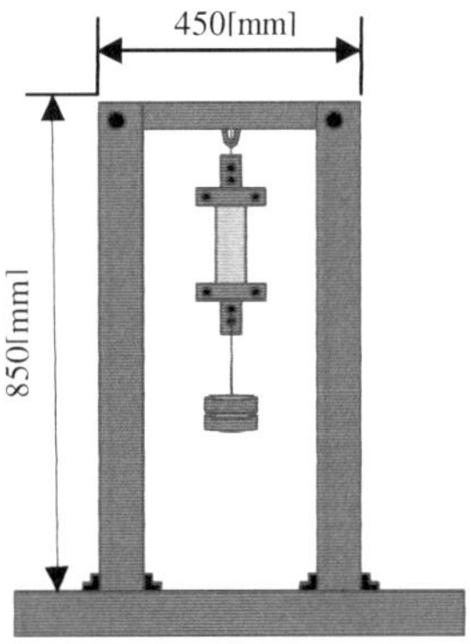

(a) Design of experimental tensile strength

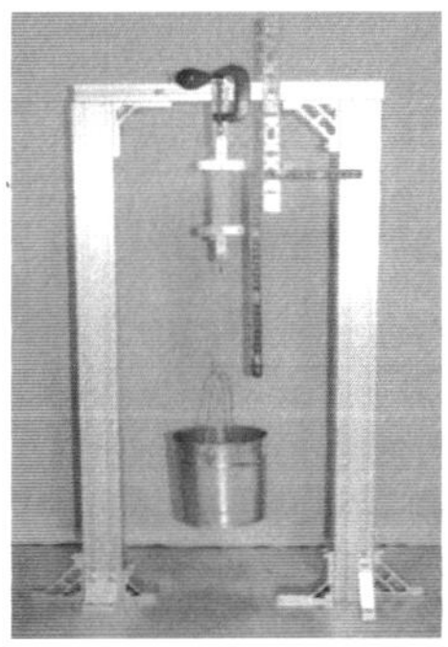

(b) Photograph of experimental tensile strength

Figure 1. Experimental set-up of tensile strength

Tensile strength experimental set-up consists of clamps in two end sides, load, bucket as weight container, ruler scale. All of them are mounted in the aluminium frame with dimension 450mm width and 850mm high. The mass weight is 53gr for each load and the length of test specimen is measured by using aluminium plate ruler, respectively, as shown below in Figure 1.

1.2 Experimental methods

There are seven data (A-G) of tensile strength experiments with different test pieces as shown in Table 1. The length of test specimen was recorded after 10 [sec] from start loading, respectively. The test specimen was clamped vertically in both sides. The top side was stationary and the bottom side is free moving to direction of gravity center. Load the test specimen was continued until the test specimen is broken.

Silicone rubber as the facial artificial skin that will be used as facial robot skin needs test due to define the maximum force that can be applied in facial expression. The dimension of test piece in calculation is $60 \times 30 \times 3$ [mm^3]. There are two parameter in material properties in used, which are E

= 0.072 for Modulus Elasticity and υ = 0.48 for Poisson's ratio, respectively. The number of element in sample test is 200 and the number of nodes is 281. We fixed boundary condition (6 DOF) in one of the end-side for the test piece due to stationary. In force condition, we applied variety magnitude of forces which starts from 0.52 [N] until 15.60 [N], respectively. Since the material is silicone rubbe (RTV2 VP 7550) which has high elasticity therefore nonlinear analysis is required in determine th strain of test pieces.

Table 1. Experimental stresses-strains data for silicon rubber under w = 0.053 kgf, respectively

No	Data A		Data B		Data C		Data D		Data E		Data F		Data G	
	(ε)	(σ) ×10^{-2} [MPa]	(ε)	(σ) ×10^{-2} [Mpa]	(ε)	(σ) ×10^{-2} [MPa]	(ε)	(σ) ×10^{-2} [MPa]	(ε)	(σ)x10^{-2} [MPa]	(ε)	(σ) ×10^{-2} [MPa]	(ε)	(σ)x10^{-2} [MPa]
0	0	0	0	0	0	0	0	0	0	0	0	0	0	0
1	0.07	0.52	0.07	0.49	0.07	0.57	0.07	0.59	0.05	0.60	0.08	0.55	0.07	0.60
2	0.12	1.03	0.12	0.99	0.18	1.15	0.17	1.17	0.12	1.19	0.17	1.10	0.18	1.19
3	0.20	1.55	0.20	1.48	0.27	1.72	0.28	1.76	0.27	1.79	0.27	1.65	0.32	1.79
4	0.30	2.06	0.28	1.98	0.40	2.30	0.40	2.34	0.38	2.38	0.37	2.20	0.38	2.39
5	0.42	2.58	0.38	2.47	0.53	2.87	0.55	2.93	0.53	2.98	0.48	2.75	0.48	2.98
6	0.52	3.09	0.52	2.97	0.67	3.44	0.68	3.52	0.67	3.58	0.62	3.30	0.68	3.58
7	0.62	3.61	0.63	3.46	0.80	4.02	0.82	4.10	0.82	4.17	0.75	3.85	0.82	4.18
8	0.78	4.12	0.78	3.96	0.93	4.59	0.95	4.69	0.97	4.77	0.88	4.4	1.00	4.77
9	0.93	4.64	0.93	4.45	1.10	5.17	1.08	5.27	1.10	5.36	1.02	4.95	1.17	5.37
10	1.07	5.15	1.05	4.95	1.22	5.74	1.22	5.86	1.23	5.96	1.15	5.50	1.32	5.97
11	1.20	5.67	1.18	5.44	1.35	6.32	1.33	6.45	1.35	6.56	1.27	6.05	1.47	6.56
12	1.35	6.18	1.32	5.94	1.43	6.89	1.43	7.03	1.45	7.15	1.38	6.60	1.57	7.16
13	1.45	6.70	1.45	6.43	1.52	7.46	1.52	7.62	1.55	7.75	1.50	7.15	1.68	7.76
14	1.57	7.21	1.55	6.92	1.60	8.04	1.60	8.21	1.62	8.35	1.57	7.70	1.78	8.35
15	1.68	7.73	1.65	7.42			1.68	8.79	1.72	8.94	1.65	8.25	1.87	8.95
16	1.80	8.24	1.77	7.91			1.77	9.38	1.80	9.54	1.75	8.80	1.97	9.55
17	1.90	8.76	1.87	8.41			1.85	9.96	1.87	10.13	1.83	9.35		
18	2.00	9.27	1.95	8.90			1.92	10.55	1.95	10.73	1.90	9.90		
19							1.98	11.14			1.98	10.45		

1.3 Results and discussion of tensile strength experiment

The calculated result shows that the length of test pieces are different from each other. It depends o the material properties including Modulus elasticity and Poisson's ratio. The test pieces were give force start from 0.52 [N], respectively. The test pieces were treated with different Poisson's ratio which start from 0.11 until 0.49. From the perspective of view, it shows large deformation effects and the elasticity effects larger for the low Poisson's ratio. For test piece which has Poisson's ratio 0.20 and given force 0.52 [N] generates Normal stress 7.08x10^{-3} [MPa]. Therefore, the Nominal strain from the calculated is 0.08. In another test case, when the test piece has Poisson's ratio 0.49 and given force 0.52 [N] generates Normal stress 4.38x10^{-3} [MPa]. Hence, the nominal strain from the calculation is 0.04. Both cases aforementioned can be explained that the Poisson's ratio as mechanical properties has determining the magnitude of strain that generated.

The shaped deformation of test piece can be seen clearly when the large force is given. For the test pieces which have Poisson's ratio 0.20 and 0.11, the maximum force that given in the calculation of test piece was 11.43 [N]. The average method in calculation was used to approximate Poisson's ratio for the test piece in the experiment. Table 1 shows several stresses in vary axis (σ_x, σ_y,

σ_z) including σs which determine Mises stress. The purpose of Mises stress is important to consider in analyzing the stress-strain for 3D in order to apply the proportional forces in test pieces. The maximum value of Mises stress was the standard value to determine whether the stress was still in the proportional area for strain. Designation of strain will determine the value of Poisson's ratio that acquired.

In tensile strength experiment, we provides seven test pieces where the condition of the experiments were similar including the room temperature, loading time method, and also material properties of test pieces.

Figure 2 shows that data A was the highest nominal strain among all the experiment data, $\varepsilon = 2$, and nominal stress , $\sigma = 9.77 \times 10^{-2}$ [MPa], as the first tensile test. However, data C was the shortest nominal strain among all the experiment data, $\varepsilon = 1.6$, and nominal stress, $\sigma = 8.04 \times 10^{-2}$ [MPa]. Data A has reached the maximum strain for the 18th loads, but the maximum strain for data C occurs on the 16th loads. These phenomenon have proved that the strain for all specimen tests were not the same because of coincide broken. The test specimens also have creep during loading time, where the test specimen keep stretching toward the gravity center even without adding weight. Figure 2 shows that the characteristics of tensile strength for all the data are almost similar where these represented in polynomial graph.

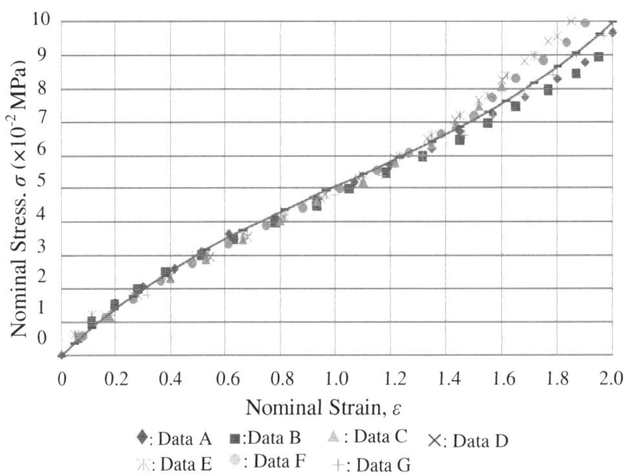

Figure 2. Graph of tensile strength experiment for silicone rubber which shown all data experiments

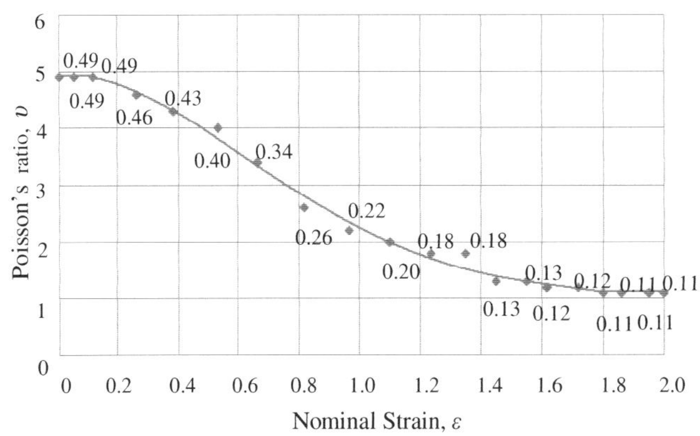

Figure 3. Determined value of poisson's ratio based on nominal strain for silicone rubber

Figure 3 shows the relationship between Nominal strain and Poisson's ratio where the strain value is determined by the stress value. The Poisson's ratio of silicone rubber was 0.48 that found after pass several tests in calculation to approach the approximate Poisson's ratio for the test pieces in experiment.

Figure 4 shows the comparison deformation of test pieces in calculation and experimental. The shapes of test pieces in calculation and experimental are tend to be similar until the 8th loads. However, the shapes deformations between calculation and experiment were changed after the 9th loads. Since we cannot define the poisson's ratio on the experiment, therefore we approached poisson's ratio on the calculation by try-and-error methods.

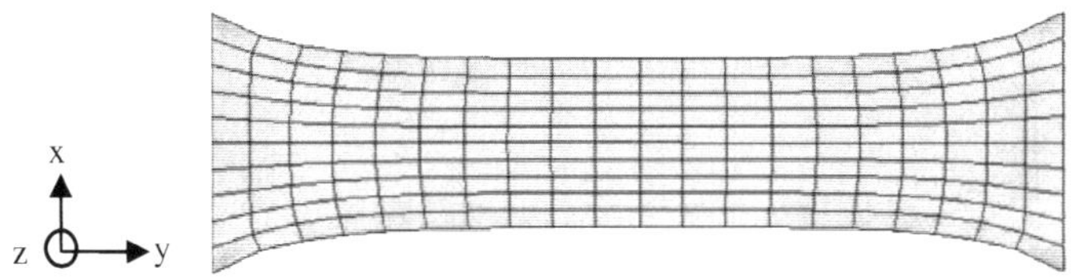

(a) Calculated results for normal stress in y-axis $\sigma_y = 6.06\times10^{-2}$ [Mpa] and $\upsilon = 0.48$

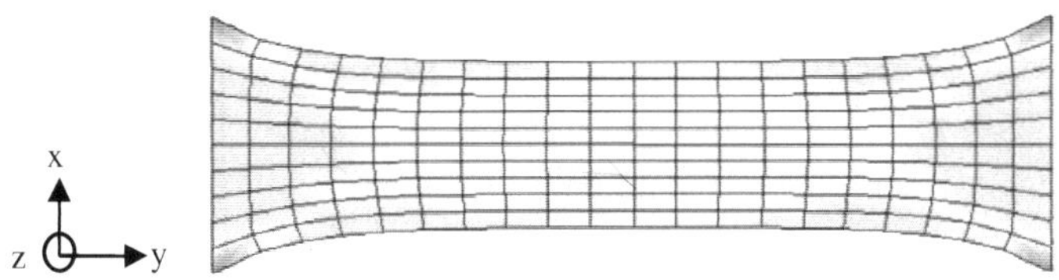

(b). Calculated results for mises stress in all axis $\sigma_s = 5.86\times10^{-2}$ [Mpa] and $\upsilon = 0.48$

(c). Experimental results at the 8th loads under f = 4.16 [N]

Figure 4. Condition of test pieces in calculated and experimental results when it loaded

2. Methodology of facial shape measurement

2.1 Experimental Set-up

In order to obtain an image from a facial data, we capture images using a 3D line laser. The geometry and color information has been setup in 3D by employing 3D laser scanning. The experimental devices consist of a 3D line laser, a charge coupled device (CCD) video camera, and a calibration panel background with a scale point to adjust the size of an object. The 3D laser scanning software allows to manipulate different angle points of facial surface. The position of the scanned object is in front of the background board. The positions of the video camera and 3D line laser are parallel to each other and in front of the scanned object as shown in Figure 5.

The dimension of the background board is 240 [mm], which is based on the size of facial object. Calibration process performed in the software determines the proportional image results and the vividness of image surface. The laser scanner is attached to a tripod and can be moved up and down during scanning process. The CCD video camera is attached to another tripod which is positioned next to the tripod for the laser scanner. The recording of facial image is started as soon as the laser scanner start working.

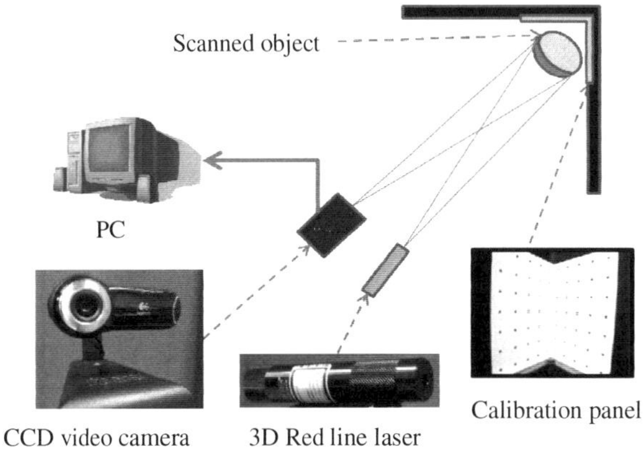

(a) Arrangement of scanning system

(b) Scanned object and calibration panel

Figure 5. Experimental setup for facial scanning

In order to obtain vivid images from the scanning process, the face of the object need to be cleaned before scanning so as to reduce reflection. One way of doing this is by using cosmetic foundation on the face of the object which can make the color of the skin uniform and flawless. In addition, the eyes of the object need to be shielded using eye protection made of paper. The reason for using eye protection is not only to avoid eye injuries but also to define the shape of the eyes. The below figure shows the experimental setup including the arrangement and position of scanning device as well as the object of scanned.

2.2 3D scanning method

Scanning of the object face is done in front of a white background. The distance between the object face and the CCD video camera is approximately 1000 [mm]. Before scanning is done, a picture of the object is taken under a bright light condition. This picture will be used as a reference feature of the object. After taking the picture, laser scanning is then done under a dark condition. Calibration of object face is needed to synchronize the color of the target face with the color obtained from laser scan. The CCD camera connected to a computer capture images through scanning.

The initial process is start scanning process under a dark condition. During the scanning process, the laser scanner is manually moved up-and-down and is positioned at certain angle. The trajectory of the scanning process can be seen on the monitor process as shown in Figure 6.

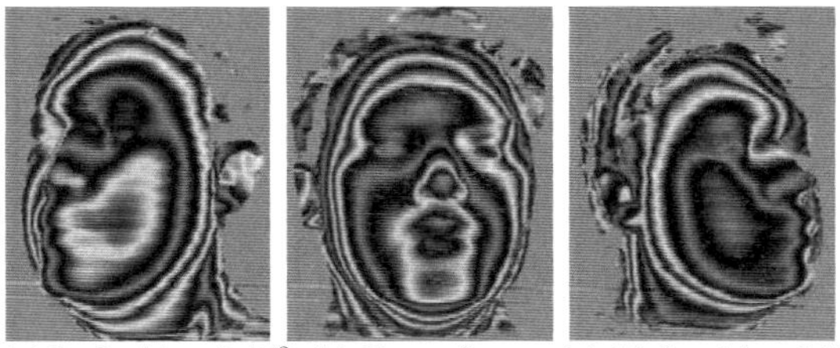
(a) Left side of 45° (b) Front face (c) Right side of 45°

Figure 6. 3D scanning views with depth images

It is important to note that in this simulation we need to obtain data of the image for the whole face. Therefore, scanning of the object face is done from three directions, i.e. from normal direction, 45° left side facial direction, and 45° right side facial direction.

The depth of the image color that appears on the monitor during scanning represents image in a real time. The time frequency of scanning depends on the area of the face that has been covered. A face shape is a relative model and consist of multiple points that need to be scanned in a consistent speed.

2.3 Results and discussion of measurement

Results of the scanning process are shown below. The image consist of three parts: front face, 45° right side, and 45° left side. Figure 7 shows the result in 3D scanning for the three scanning angles used in the experiment. Scanning results from the three viewing angles are collected and combined to create a 3D image with a proportional dimension. Shape fusion step allows scan results to align conveniently and to fill the gap if there is a minor imprecision during merge of the scans. During the registration process, the alignment and merging for among several distinct shape images are indicated by many surface penetration points. Sometimes the first time registration process is not perfect (i.e. overlap surfaces or a coordinate axis not match), but it can be repeated using randomized algorithm in software until we get the best result.

After the registration process is completed, we then define the resolution of the image. In this experiment, we use 120 image resolutions (with 20,865 nodes and 41,026 elements) from the image resolution range of 50-400. If we use the lowest image resolution, the fusion process is much more robust against noise and imprecise alignment. However, applying the highest resolution has a consequence in a longer processing time.

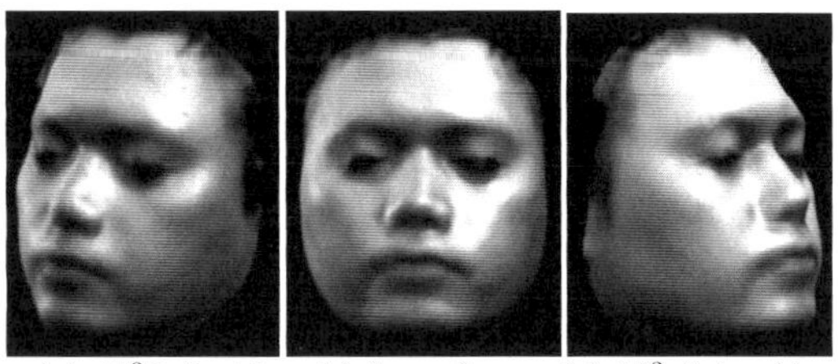
(a) 45° left side (b) front face (c) 45° right side

Figure 7. Merged results of photo image and scanning data

3. Making meshes of facial model

3.1 Meshing by scanned 3D facial data

The image data is retrieved from the shape fusion result where it needs meshing in the surface of the object face. The meshing procedure consists of a several steps that start from analyzing the geometry model until optimizing the meshing in all the surface of elements. Meshing the facial model is determined by the shape object surface. The quadrilateral also known as quad type is applied in meshing face because of the best quality surface. The quad type consists of four nodes (grid points) and four lines as shell element.

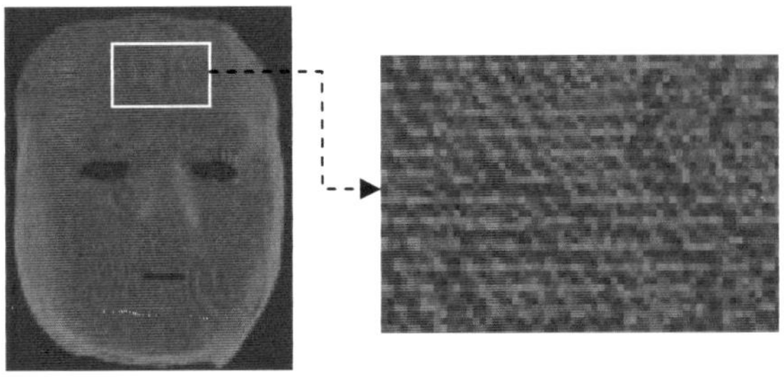

(a) Before meshing

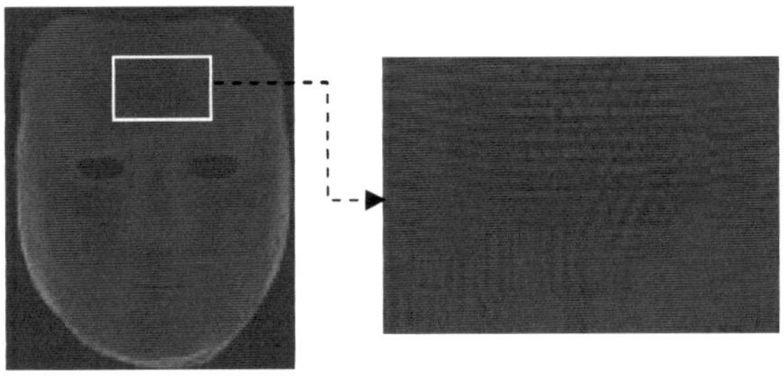

(b) After meshing

Figure 8. Comparison of facial model between initial condition and final model

The image resolution of meshing is start from 80, 100, 120, and 250. We selects the resolution 120, smoothness grade 7 and removal value 5, which is reliable in model shape element and the ability of computer to run the mesh program. The angle degree of quad type is not always 90^0, but the range is between 45^0 and 135^0 due to pass the parameter editor including maximum-minimum interior angle quad and skew criteria. The edge of facial object can be reshaped and smoothed including around the mouth and eyes circle as shown in figure 8.

Making the quad type of element needs several remove and merge line methods. It starts from one element with triangle type and then two type of it will be merged. The empty element sometimes found and need to be recovered by creating a new element.

3.2 Result sand discussion of mesh optimization

Recondition needs to apply after meshing, since there are still bad element sector left behind on the surface as shown in Figure 9. This is caused by overlapping element in one area, so that it needs to remove the bottom or the top element. Another cause of element dent is disconnection of line between one element and the other one. Therefore, it needs weld step for the line connection.

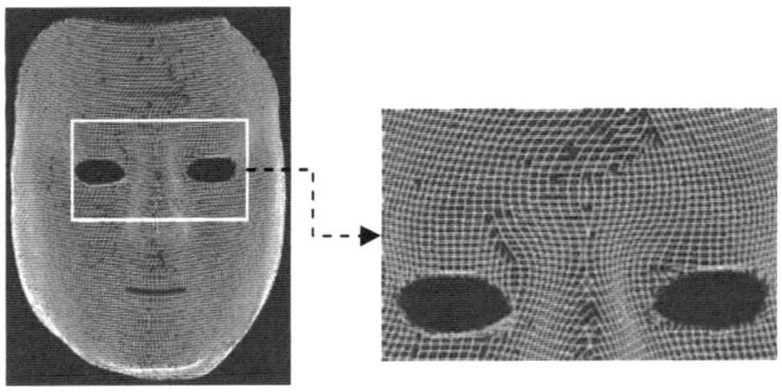

Figure 9. Red spots indicate defect elements after meshing

The optimized facial object can be done by selecting the proportional image resolution where the best image resolution that we apply is 120. We can see the difference smoothness of element surface where the image resolution affect the sharpness image with a shape proportionally.

Conclusion

The facial expressions for robot in calculation is not exactly similar with the human facial expression. However, this research is applying FACS due to imitate the muscle face of human based on the facial expression in require for robot. A Facial expression robot is determined by the direction and the magnitude external force on the driven point which depends on the facial expression type of robot. Silicone rubber as artificial skin for robot face has been tested by conducting tensile strength. The reason for conducting the tensile strength for is due to examine the capability in hold the large deformation of artificial skin for facial robot.

References

[1] M. Shayganfar, C. Rich, and C. L. Sidner, "A Design Methodology for Expressing Emotion on Robot Faces," 2012.
[2] H. Search, C. Journals, A. Contact, M. Iopscience, I. O. P. Conf, and I. P. Address, "Facial expressions recognition with an emotion expressive robotic head," vol. 012086.
[3] J. Broekens, "Emotion and Reinforcement : Affective Facial," pp. 113–132, 2007.
[4] S. Boucenna, P. Gaussier, and P. Andry, "A robot learns the facial expressions recognition and face / non-face discrimination through an imitation game," pp. 1–19.
[5] Paul Ekman, "Facial Expression." Lawrence Erlbaum Association, New Jersey, pp. 1–20, 1977.
[6] S. S. Ge, C. Wang, and C. C. Hang, "A Facial Expression Imitation System in Human Robot Interaction."
[7] M. I. Khan, "Facial Expression Recognition for Human-Robot Interface," vol. 9, no. 4, pp. 300–306, 2009.
[8] S. V Ioannou, A. T. Raouzaiou, V. A. Tzouvaras, T. P. Mailis, K. C. Karpouzis, and S. D. Kollias, "Emotion recognition through facial expression analysis based on a neurofuzzy network," vol. 18, pp. 423–435, 2005.
[9] R. Mittal, P. Srivastava, A. George, and A. Mukherjee, "Autonomous Robot Control Using Facial Expressions," vol. 4, no. 4, 2012.
[10] V. Raghavaiah, C. Kode, M. C. Cavusoglu, and M. T. Azar, "Design and Characterization of a Novel Hybrid Actuator using Shape Memory Alloy and D . C Motor for Minimally Invasive Surgery Applications," no. July, pp. 416–420, 2005.
[11] F. C. Janzen, A. M. Tusset, V. Piccirillo, and J. M. Balthazar, "Motion and vibration control of a slewing flexible structure by SMA actuators and parameter sensitivity analysis," vol. 3054, pp. 3041–3054, 2015.
[12] M. Moosaei, C. J. Hayes, and L. D. Riek, "Performing Facial Expression Synthesis on Robot Faces : A Real-time Software System."

An Analysis of Radiation Penetration through the U-Shaped Cast Concrete Joints of Concrete Shielding in the Multipurpose Gamma Irradiator of BATAN

Tanti Ardiyati, Bang Rozali, Kasmudin

Center for Nuclear Facilities Engineering, National Nuclear Energy Agency of Indonesia (BATAN), Tangerang Selatan, Indonesia

e-mail: tanti.ardiyati@batan.go.id

Abstract. An analysis of radiation penetration through the U-shaped joints of cast concrete shielding in BATAN's multipurpose gamma irradiator has been carried out. The analysis has been performed by calculating the radiation penetration through the U-shaped joints of the concrete shielding using MCNP computer code. The U-shaped joints were a new design in massive concrete construction in Indonesia and, in its actual application, it is joined by a bonding agent. In the MCNP simulation model, eight detectors were located close to the observed irradiation room walls of the concrete shielding. The simulation results indicated that the radiation levels outside the concrete shielding was less than the permissible limit of 2.5 μSv/h so that the workers could safely access electrical room, control room, water treatment facility and outside irradiation room. The radiation penetration decreased as the density of material increased.

Keywords: radiation penetration, cast concrete joints, multipurpose gamma irradiator.

1. Introduction

The National Nuclear Energy Agency of Indonesia (BATAN) has a multipurpose gamma irradiator. It is a panoramic type, IAEA category IV irradiator whose radiation source, Cobalt-60 (Co-60), is stored in a pool filled by demineralized water when not in use [1][2]. In the irradiation room, the Co-60 source is lifted by a pneumatic system to the irradiation position. It is then lowered by gravity to the storage pool after the completion of irradiation process. The gamma irradiator is loaded with a 300-kCi Co-60 source and is designed to accommodate a radiation source with a maximum activity of up to 2 MCi.

BATAN's multipurpose gamma irradiator, or henceforth referred to simply as "the irradiator" or "the multipurpose gamma irradiator" for brevity, has been designed to be very safe following the requirements of the IAEA Basic Safety Standards. The facility should keep the individual dose, the number of people exposed, and the probability of incurring exposures as low as reasonably achievable and below permissible dose limits [3][4]. Therefore, the use of appropriate radiation shielding that encircles the irradiation room should be optimized to diminish direct radiation exposures toward workers and members of the public to the most effective levels [5].

Based on the Chairman of BAPETEN Regulation No. 11/Ka-BAPETEN/VI-99, the radiation shielding for irradiation facilities must be designed sufficiently thick to shield against radiation exposures of up to one and a half times the maximum source activity [6]. The implementation of the Regulation to the irradiator means that its radiation shielding should be able to protect against radiation exposures from source activities of up to 3 MCi, so the radiation shielding can reach 2 m in thickness [2].

The appropriate shielding material should be chosen to protect and ensure the safety of the workers and members of the public from radiation exposures. The radiation shielding construction must consider the seismic load [6], economic value of the material used, and building construction capacity. Concrete, earth fill, steel, lead, and iron may be used as radiation shielding in irradiation facility [5][6]. Among the materials mentioned above, concrete is frequently used as material for constructing radiation shields [5].

The radiation shielding of the multipurpose gamma irradiator should have a thickness of 2 m, which means that the construction requires concrete casting procedure and must be undertaken in several steps due to the large volume of the concrete needed. Since the construction of the radiation shielding must be undertaken in several steps, the shielding wall has joints. This cast concrete joints could be a problem in the future since it may lead to structural failure [7] and radiation penetration if the joints are filled with lower-density materials. The existence of the joints in the cast concrete enables the possibility that radiation could penetrate the horizontal planes of the joints up to a certain depth. Therefore, concrete shielding without joints is safer for the workers.

The work on the radiation penetration of the concrete shielding has been performed by Ardiyati, Rozali, and Damayanti in 2016 [8]. They analyzed the possible radiation penetration through the joints of the cast concrete shielding of the multipurpose gamma irradiator at BATAN. In the work, several detectors were simulated to be placed close to the observed area of the concrete shielding. Models of the cast concrete joints of the concrete shielding were formed from U-shaped cast concrete joints. The simulated cases were the one where there were no gaps between the joints and where a 0.2-mm thick air-filled gap was present. They reported that the concrete shielding construction was safe for both cases since the radiation exposures were below the permissible limit of 2.5 μSv/h.

In this current work, we analyzed the radiation penetration in the U-shaped cast concrete joints of concrete shielding in the multipurpose gamma irradiator. The objective of this analysis was to ascertain that the radiation exposure passing through the concrete shielding to the environment would not exceed the maximum permissible limit. The work was performed by measuring the radiation penetration on the cast concrete joints that were filled with bonding agent when the Co-60 source was on irradiation position and comparing it with the radiation penetration on the cast concrete shielding without joints and with joints filled with air.

2. Theory

The concrete shielding of the multipurpose gamma irradiator had to be designed very thick compared to ordinary walls so that be able to stop the radiation from a source with activity of up to 2 MCi. For use as a radiation shield, concrete must have a minimum density of 2400 kg/m^3 [6] and capable of receiving pressures of up to 210.9 kg/cm^2 [6]. Moreover, the maximum permissible limit for radiation exposures on the outer walls of the building that are not supervised is 2.5 μSv/h [6].

In the concrete casting with large volumes, the casting process cannot be undertaken continuously from bottom to top, and should be undertaken in two steps instead. Joining old and new concrete requires joints because the construction of concrete shielding wall as a monolithic unit cannot be done in one placement of concrete [7]. The construction joints that were used in this concrete shielding wall can serve as stress relief [9]. The construction joints usually use bonding agent to compound the old and new concrete. Cement grout or polyvinyl acetate are usually used as bonding agents. The construction joints should be made and positioned in the right place so that the structural strength is not reduced [10].

An analysis of radiation penetration can be performed using deterministic methods and Monte Carlo methods (probabilistic methods). Deterministic methods compute the transport equation for the average particle behavior [11]. The calculation using deterministic methods is much simpler; however, it needs nuclear material data that is difficult to obtain. Monte Carlo methods calculate individual particles and record the information of some aspects (tally) of their average behavior [11]. These methods need large calculation resources to simulate every movement and interaction of particles with matter.

Monte Carlo N-Particle Transport Code (MCNP) is a stochastic computer simulation program that randomly traces an appropriate number of particles to obtain particle transport equation with seven-dimensional phase space of time, position, direction, and energy [12]. To execute MCNP, system geometry, material and cross sections, particle source, and particle tally must be defined as inputs. The geometry was represented in the X, Y, and Z coordinates where, in this work, the X and Y coordinates were defined relative to the midpoint of the source storage pool and the Z coordinate was relative to the surface of the irradiation room floor. Material definitions include material number, material composition, and cross section compilations. Tallying is used for scoring the quantity of interest [12]. There are various tally type, i.e., F1 (surface current), F2 (average surface flux), F4 (average flux in a cell), FMESH4 (track-length tally over 3D mesh), F5a (flux at a point or ring), FIP5 (pin-hole flux image), FIR5 (planar radiograph flux image), FIC5 (cylindrical radiograph flux image), F6 (energy deposition), F7 (fission energy deposition in a cell), and F8 (pulse height distribution in a cell). The most commonly used tallies are F1, F2, F4, and F5 [12].

3. Methodology

The analysis in this work was performed by placing several detectors close to the observed irradiation room walls of the concrete shielding. This method was to examine the radiation penetration through the U-shaped cast concrete joints of the concrete shielding of the multipurpose gamma irradiator. It started with the modeling of the measurable factors needed for MCNP simulation. The simulation itself was performed by measuring the radiation penetration on the cast concrete joints filled with bonding agent when the Co-60 sources was on irradiation position and compared it with the radiation penetration on the cast concrete shielding without joints and with joints filled with air.

In the current work, the joints were filled with a bonding agent. System geometry, material, and particle source were defined as inputs to the program. The information related with concrete shielding, construction joints, detectors, and source were modeled for the geometry inputs and material definitions.

The Co-60 sources in the irradiator were cylindrical and arranged inside source pencils. The source pencils were to be placed in source modules that contained 40 source pencils each. The facility has four source modules and three source racks that are arranged side by side. In total, the multipurpose gamma irradiator has 480 source pencils that are distributed into the three source racks.

The calculation modeling in the MCNP was made conservative by ignoring source racks and irradiated product racks construction. The only material in the irradiation room that was modeled for the MCNP calculation was the 480 source pencils with equally-distributed activity in each pencil. The source pencils were modeled as having a length of 43.7 cm and diameter of 0.7 cm. They were encapsulated in AISI 304 stainless steel containers with 45.1 cm length, diameter of 1.1 cm, and density of 7897.68 kg/m^3. The density of the Co-60 source was 8900 kg/m^3.

Co-60 was set to have a maximum activity of 2 MCi since the radiation shielding must be designed to safely block the radiation from up to 3 MCi. Therefore, the maximum activity for the calculation modeling was 3 MCi. The maximum permissible limit for radiation exposures on the outer walls of the building was 2.5 μSv/h.

The radiation shielding of the multipurpose gamma irradiator was made from concrete. In the simulation, its geometry was defined as the actual size of the concrete shielding of the multipurpose gamma irradiator. Its wall thickness was varied between 150 cm to 200 cm and represented in the X, Y and Z coordinates. The concrete and air densities were 2320 kg/m^3 and 1.294 kg/m^3, respectively.

The structure of the concrete shielding had be made strong enough to withstand with all conditions including earthquakes. Therefore, the construction joints of the concrete shielding were made U-shaped since it was a new design in the concrete construction in Indonesia. The U-shaped cast concrete joints were also used for limiting direct gamma ray penetration through the gap of the cast concrete joints. The gamma rays that penetrate to the horizontal planes of the joints will be stopped by the concrete in the middle and outer part of the cast concrete joints.

To connect the old and new concrete structure, the facility used a bonding agent. The bonding agent is mainly composed of polyvinyl acetate and has density of 1100 kg/m^3. In the MCNP modeling, the height of the joints was set to 2 mm that made the simulation easier for both bonding agent-filled and air-filled joints. The detail of the joints was based on the research presented in [8] and can be seen in Figure 1.

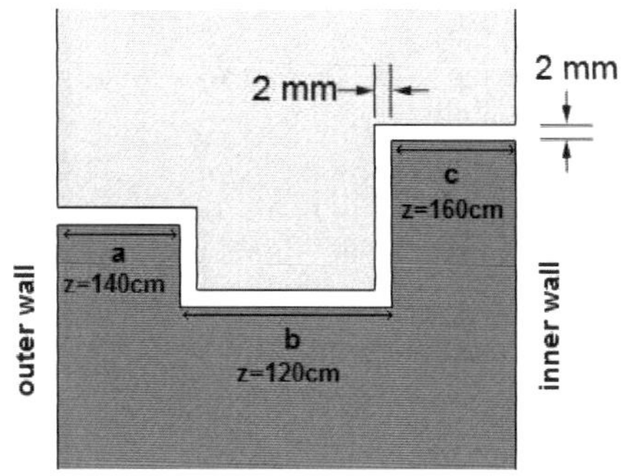

Figure 1. Illustration of the U-shaped cast concrete joints of the concrete shielding in the multipurpose gamma irradiator [8].

A number of detectors were placed close to the irradiation room walls of the concrete shielding. The locations of the detectors were based on the research presented in [8] and can be seen in Figure 2 below. The observed walls were chosen as they are in areas frequently accessed by the workers. The detectors' locations were represented by their X, Y, and Z coordinates and they were modeled as spheres in the MCNP simulation.

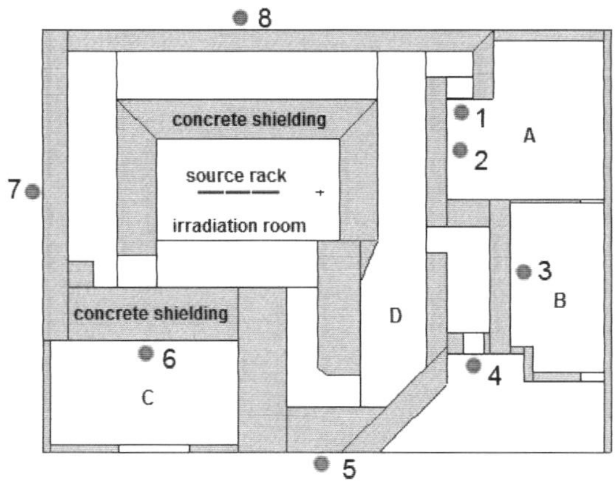

Legends:
1. Workers access doors detector
2. Electricity room detector
3. Control room detector
4. Goods access door detector
5. In front of maze detector
6. Water treatment facility detector
7. Outside west wall detector
8. Outside north wall detector

A. Electricity room
B. Control room
C. Water treatment facility
D. Goods maze

Figure 2. Location of the detectors around the irradiation room [8].

4. Results and Discussions

Based on the simulation results using MCNP, the radiation penetration on the concrete shielding can be obtained. The simulations were performed for three conditions of the concrete shielding, namely concrete shielding without joints, cast concrete joints filled with bonding agent, and cast concrete joints filled with air. It was found that the radiation levels outside the concrete shielding were less than the permissible limit of 2.5 μSv/h, as can be seen in Table 1.

Table 1. Radiation penetration results from MCNP

No	Detector locations	Detector Positions in MCNP			Radiation Penetration on the concrete shielding (μSv/h)		
		X	Y	Z	Without joints	Joints filled with bonding agent	Joints filled with air
1	Workers access door	830	449	140	3.17216E-06	8.51086E-05	3.64966E-04
2	Electricity room	781	250	140	3.06209E-05	4.38672E-05	2.09450E-04
3	Control room	1021	-220	140	1.13869E-06	1.15122E-06	3.04382E-05
4	Goods access door	880	-511	140	4.49958E-06	2.75530E-05	5.67009E-03
5	In front of maze	230	-881	140	1.02415E-02	1.35463E-02	1.40748E-01
6	Water treatment facility	-380	-461	140	2.41924E-04	2.60304E-04	1.11635E-03
7	Outside west wall	-751	100	140	1.31771E-06	4.04498E-06	1.09653E-05
8	Outside north wall	0	711	140	1.35980E-04	3.08053E-04	1.30822E-03

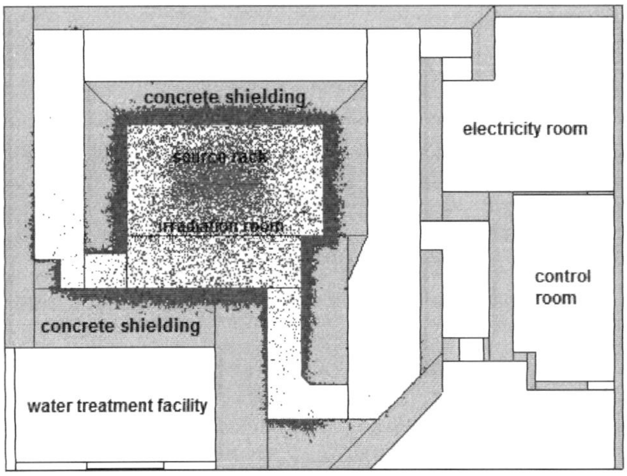

Figure 3. Particle tracking result on the concrete shielding with joints filled with air

The particle tracking figure above was generated from the MCNP modeling of the air-filled cast concrete joints. The other conditions resulted in the same particle tracking as Figure 3. From the particle tracking figure, it can be seen that the radiation particle from Co-60 could not pass through the U-shaped air-filled and bonding agent-filled cast concrete joints of the concrete shielding. It only penetrated a small portion of the inner concrete shielding wall. Therefore, from the MCNP calculation and particle tracking result in Figure 3, it could be assured that the workers could safely access control room, electricity room, water treatment facility, and the outside of the irradiator building.

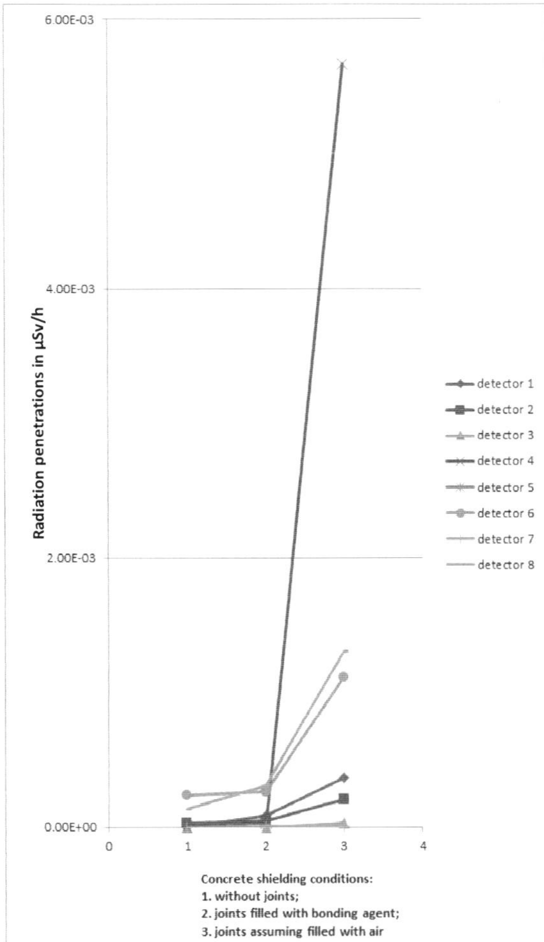

Figure 4. Radiation penetration on the concrete shielding wall for each detector

Based on Table 4.1, Figure 4 graphs the radiation penetration for the three conditions of concrete shielding. It can be concluded that concrete shielding without joints resulted in least penetration, while the one with air-filled joints resulted in most, largely because of the differences in material density: the density of shielding concrete was 2320 kg/m^3, bonding agent 1100 kg/m^3 and air 1.294 kg/m^3. Radiation cannot penetrate as far for denser materials. However, for all the three scenarios, the concrete shielding were safe for workers to have activity outside irradiation room even when in the irradiation mode.

5. Conclusions

U-shaped cast concrete joints of concrete shielding were a new design in massive concrete construction in Indonesia and it would increase the safety level of the facility. The study simulated three conditions of the concrete shielding, namely concrete shielding without joints, cast concrete joints filled with bonding agent, and cast concrete with air-filled joints. The results of the simulations were the radiation levels outside concrete shielding were less than the permissible limit of 2.5 µSv/h.

The radiation penetration decreased as the density of material increased. The three conditions of concrete shielding is safe for workers outside the irradiation room, namely in the control room, electricity room, water treatment facility, and outside of the irradiator, even when the irradiation is taking place.

Acknowledgment
The authors would like to thank Dr. Muhammad Dhandhang Purwadi for his help, assistance, and fruitful discussion throughout this work.

References

[1]. IAEA 1996 Manual on Panoramic Gamma Irradiators (Categories II and IV) (Vienna: IAEA) p 13
[2]. IAEA Gamma Irradiators for Radiation Processing (Vienna: IAEA) pp 10-20
[3]. IAEA 1996 International Basic Safety Standards for Protection against Ionizing Radiation and for the Safety of Radiation Sources Safety Series No. 115 (Vienna: IAEA) p 22
[4]. IAEA 1996 Manual on Shielded Enclosures (Vienna: IAEA) p 7
[5]. IAEA 2010 Radiation Safety of Gamma, Electron and X Ray Irradiation Facilities Specific Safety Guide no. 8 (Vienna: IAEA) p 40
[6]. BAPETEN 1999 Chairman of BAPETEN Regulation No. 11/Ka-BAPETEN/VI-99 about Irradiator's Construction and Operation License p 16 (in Indonesian)
[7]. CCAA 2004 Joints in Concrete Buildings Technical Note (NSW:CCAA) pp 1-2
[8]. Ardiyati T, Rozali B, and Damayanti R 2016 Proc. Natl. Conf. on Nuclear Technology Utilization (SENPATEN) vol. 1 ISSN: 2540-8062 (Tangerang Selatan: PRFN-BATAN) pp 180-188 (in Indonesian)
[9]. ACI Committee 2001 Joints in Concrete Construction ACI 224.3R-95 (Detroit: ACI) p 32
[10]. BSN 2013 Requirements for Structural Concrete for Buildings SNI 2847:2013 (Jakarta:BSN) p 48 (in Indonesian)
[11]. X-5 Monte Carlo Team A 2008 General Monte Carlo N-Particle Transport Code, Version 5 vol. 1 (Oak Ridge: Los Alamos National Laboratory) ch. 1 pp 1-2
[12]. Hussein E Monte Carlo Particle Transport with the MCNP Code pp 2-67
[13]. Shultis J K and Faw R E 2011 An MCNP Primer Department of Mechanical and Nuclear Engineering (Manhattan: Kansas State University)

Static, Dynamic, and Fatigue Analysis of the Mechanical System of Ultrasonic Scanner for Inservice Inspection of Research Reactors

Muhammad Awwaluddin, Kristedjo K., Khairul Handono, Ahmad H.

Center for Nuclear Facilities Engineering, National Nuclear Energy Agency, PUSPIPTEK Area, South Tangerang, Indonesia 15310

Abstract. This analysis is conducted to determine the effects of static and dynamic loads of the structure of mechanical system of Ultrasonic Scanner i.e., arm, column, and connection systems for inservice inspection of research reactors. The analysis is performed using the finite element method with 520 N static load. The correction factor of dynamic loads used is the Gerber mean stress correction (stress life). The results of the analysis show that the value of maximum equivalent von Mises stress is 1.3698E8 Pa for static loading and value of the maximum equivalent alternating stress is 1.4758E7 Pa for dynamic loading. These values are below the upper limit allowed according to ASTM A240 standards i.e. 2.05E8 Pa. The result analysis of fatigue life cycle are at least 1E6 cycle, so it can be concluded that the structure is in the high life cycle category.

Keywords: Ultrasonic Scanner, Research Reactor, Mechanical System, Fatigue, Static.

1. Introduction

The ultrasonic scanner is one of the tools that comprise the sensors for ultrasonic testing (UT) used to inspect the reactor pool walls. The UT inspection system can be used to detect aging and the various types of degradation of a research reactor system, structure and components (SSCs). Presently, all existing research reactors in Indonesia are experienced an aging of their structures; therefore, it is necessary to examine their SSCs. Our research team has designed the mechanical system an ultrasonic scanner for inservice inspection of research reactor and needs to analyze the mechanical safety of the design before constructing and testing it in reactor pool.

A mechanical systems design for the ultrasonic scanner for research reactors has been analyzed. The analysis includes static, dynamic, and fatigue life analyses on the structure of the mechanical systems. The mechanical system analyzed consists of arm, column, and connection systems. The analysis used the finite element method (FEM) because this analysis method is commonly used for performing structural and static load structure evaluation as well as determination of the fatigue life of the structure. Here, FEM was used to test the effects of static and dynamic loads on the stress distribution of arm structures and column connections.

The purpose of this analysis to find the fatigue life of the structure used the Gerber correction factor because this correction factor is suitable for ductile materials [1]. The ductile materials used in the mechanical system for research reactor ultrasonic scanner are 6061 aluminium alloy (Al 6061) and 304 stainless steel (SS304) [2].

2. Theory

2.1. Finite Element Modeling (FEM)

Finite element modeling is required for analysis with FEM. To create the model, the CAD geometric model (as shown in Figure 7.) is discretized into smaller and simpler elements, with the result shown in Figure 1. [3]. The meshing of the elements is performed for various sizes, from relatively large to the smallest, to ensure the convergence of the analysis results.

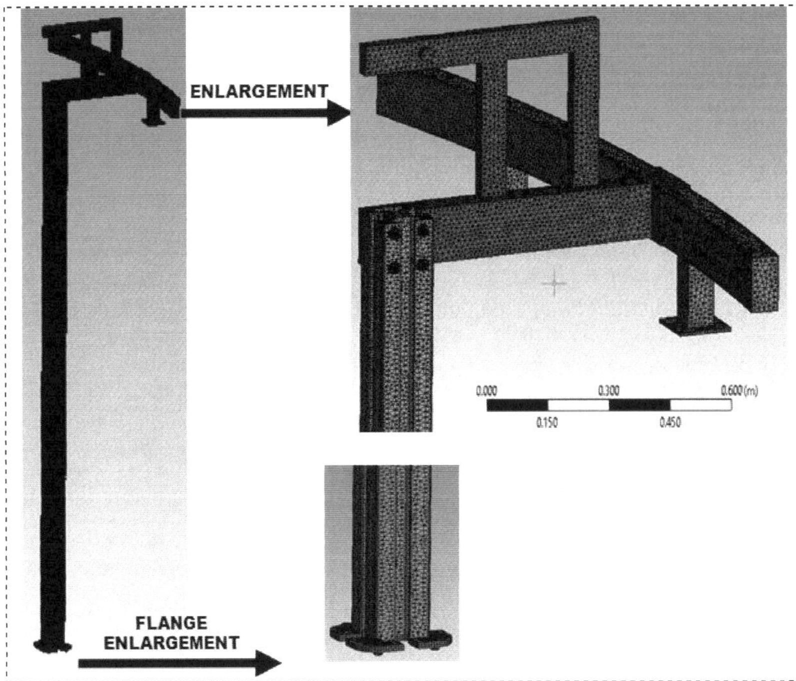

Figure 1. Finite Element Model of an ultrasonic scanner mechanical system

In this modeling, the mechanical properties of the materials used are also inputs of the analysis. They can be seen in Table 1.

Table 1. Mechanical Properties of material [4][5][6].

Material	Young's modulus (Pa)	Poisson's Ratio (ν)	Yield Strength (Pa)
6061 Aluminium	69E9	0.33	1.2408E8
304 Stainless Steel	190E9	0.29	2.05E8

2.2. Loading Conditions

Static and dynamic analysis of the ultrasonic scanner mechanical system for inservice inspection of a research reactor should be carried out to ascertain the safety of the design. In this case, the load and constraints provided shall be in accordance with the loading conditions at the time of operation. The load is in the form of static and dynamic load within a certain period of time. From the load, the fatigue life can also be known. The dynamic load effect is calculated because it will add about 8-77% of the load that can cause fracture or fatigue which can lead to structural failure [7]. To do so, the effects of static and dynamic load on the ultrasonic scanner mechanical system need to be determined. Thus, the system's safety can be ascertained, as the age at which the structure experiences fatigue is known.

2.3. Fatigue Analysis

A design for the mechanical system of an ultrasonic scanner for inservice inspection of a research reactor shall either satisfy the minimum limit of fatigue life or design for infinite life. This can only be confirmed by physical tests or fatigue analysis. In this analysis, the fatigue life of the ultrasonic scanner mechanical system on the finite element stress analysis is predicted using computer code from ANSYS Workbench. Fatigue calculations are performed for materials SS 304 and Al 6061 based on the S-N curves of the materials. The S-N curve shown in Figure 2. and Figure 3. used as the reference for SS 304 and Al 6061 materials.

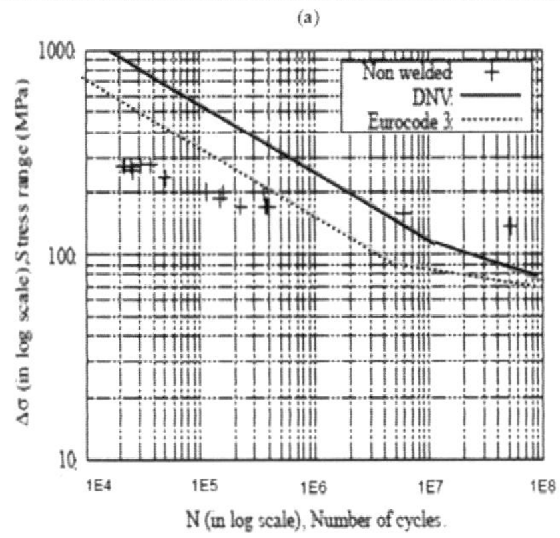

Figure 2. S-N curve of SS304, based on EUROCODE 3 EN-1993-1-9 and DNV-RP-C203 [8],

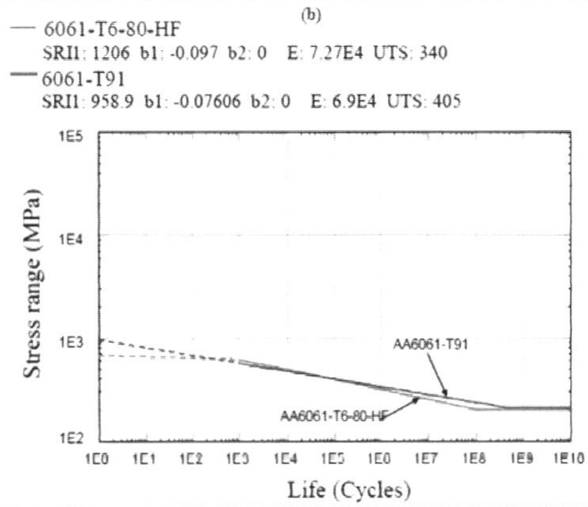

Figure 3. S-N curve of Al 6061 [9].

Fatigue analysis can be perform after static analysis showed that the safety factor criteria were fulfilled i.e., value is 1 – 10. In the fatigue analysis, the load factor given is 1 so that the von Mises (σ_{ym}) stress found from the static analysis represents the maximum stress (σ_{max}) occurring in the structure. The maximum stress is formulated as [10]:

$$\sigma_{max} = \sigma_{ym} = \sigma_m + \sigma_a \quad (1)$$

Here, σ_m is the average stress in units Pa. and σ_a is the alternating stress in the unit of Pa. The assumption given is that the amplitude of the alternating stress σ_a is 10% of the average stress σ_m, so the maximum stress σ_{max} = σ_m, + $0.1\sigma_m$, =$1.1\sigma_m$, meaning that a given load is 10% greater than the static load . The stress ratio (R) is defined as [10]:

$$R = \sigma_{min}/\sigma_{max} \quad (2)$$

Where σ_{min} is the average stress minus the alternating stress amplitude in Pa. ($\sigma_{min} = \sigma_m - \sigma_a$. The alternating stress is assumed to have constant period constant amplitude; it is illustrated in Figure 4.

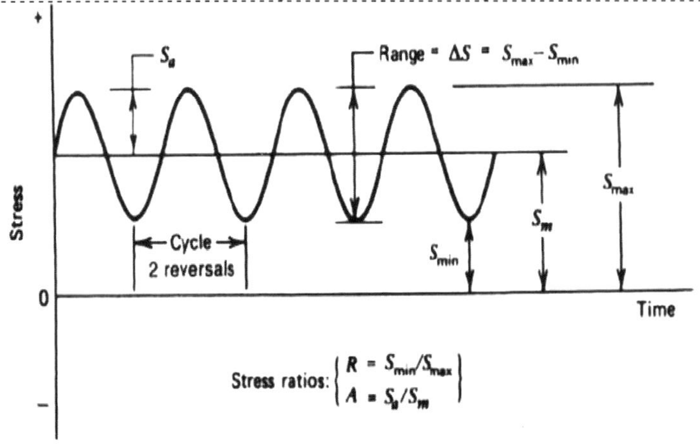

Figure 4. Graph of loading with constant amplitude [11] [12].

The data on the alternating stress σ_a, which is generated at the input stage, can not be fed directly to the S-N curve. Therefore, it takes a theory of correction to determine an effective alternating stress ($\sigma_{a\ effective}$). Some of the correction approaches that can be used are Goodman, Gerber, and ASME elliptic corrections. The limits of stress to avoid metal fatigue with each of the correction factors can be seen in Figure 5. The present analysis used the Gerber mean stress correction theory to determine $\sigma_{a\ effective}$ that occurs at the average stress (σ_m correction theory) because it is suitable for ductile materials. To calculate $\sigma_{a\ effective}$ based on correction theory, the formula 3 is used, with $\sigma_{a\ effective}$ being equal to σ_a after the correction factor [1][10] :

$$\frac{\sigma_a}{S_e} + \left[\frac{\sigma_m}{S_{ut}}\right]^2 = 1 \tag{3}$$

Where S_u is ultimate tensile strength of the SS 304 and the Al 6061. The $\sigma_{a\ effective}$ resulting from interpolation was then used as the reference parameter on the S-N curve to determine the fatigue life. The diagram of stress limit versus metal fatigue is shown in Figure 5.

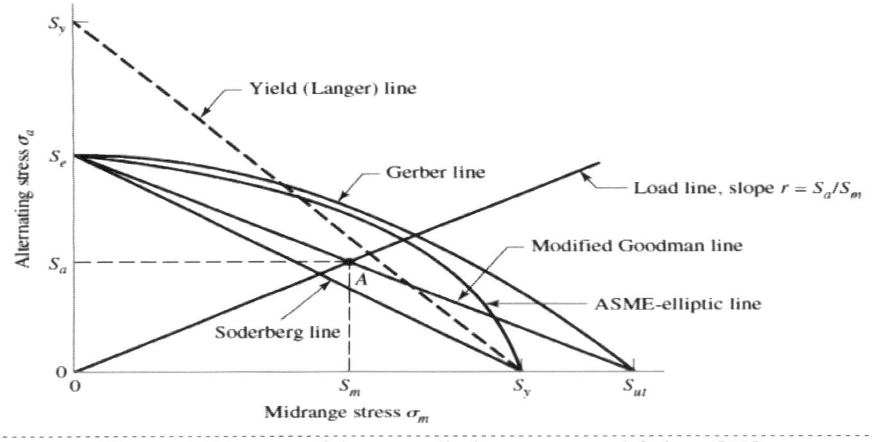

Figure 5. Diagram of the stress limit versus metal fatigue [10].

3. Methodology

The static, dynamic, and fatigue analysis of the mechanical system for the ultrasonic scanner for insevice inspection research reactor using FEM using software package followed the methodology presented on Figure 6:

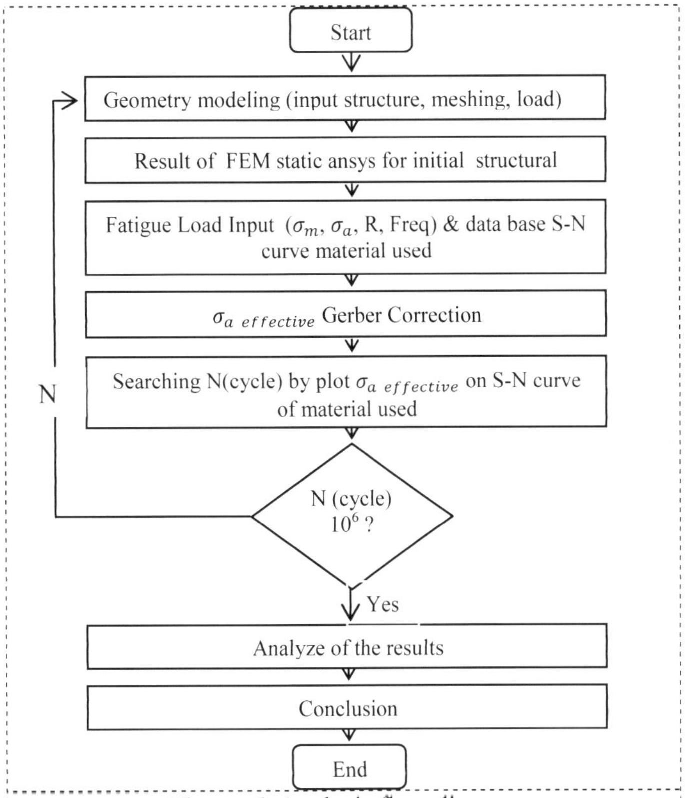

Figure 6. Analysis flow diagram

The 3-D model of the ultrasonic scanner mechanical system consists of the guide for the movement of the column retaining arm, columns, column shafts, and connections system used in this analysis. The curved column is restrained by a column guard, allowing movement with a radius of 2950 mm. The hollow, curved column is made of SS 304 with of 40 mm × 80 mm size, 1.20 mm thick, as shown in Figure 7. The system was designed for the radial data retrieval process on the research reactor pool wall. The column holder is made of SS 304 also, with the same cross-section of 40 mm width, 80 mm height, and a column thickness of 1.20 mm thickness, while the ultrasonic handle's cross-section has a width of 20 mm, height of 40 mm, a thickness of 1.20 mm. The handle is connected with a wire sling. The column is made of Al 6061 profile with a cross-section size of 80 mm × 80 mm. The design was intended to facilitate the vertical movement on the research reactor pool wall [2].The 3-D modeling of the mechanical system was performed with the Solidworks 2015 CAD software package. The model is shown in Figure 7.

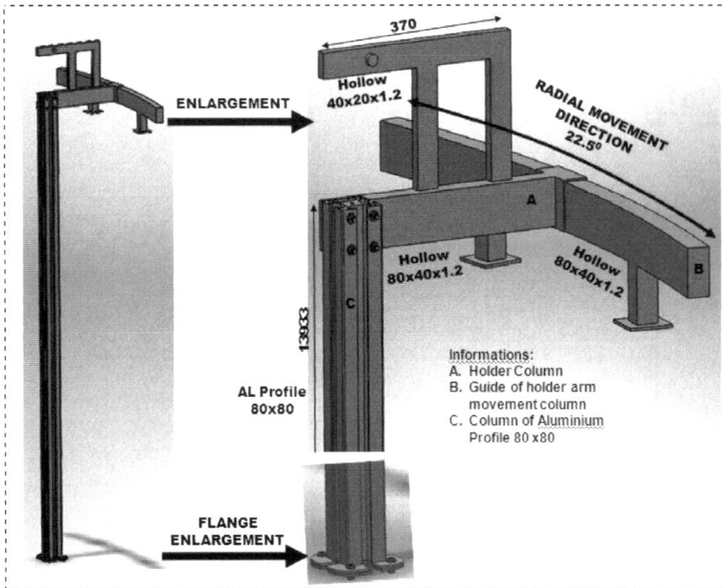

Figure 7. Model of ultrasonic scanner mechanical system

The geometric model for FEM consisted of 419998 four tetrahedron element nodes and 758497 nodes and with quality of mesh is fine.

The loads in this mechanical system were 200 N, representing the load of the ultrasonic sensor holder and the weights located on the top of the column holder, and 320 N, representing the load of the aluminum profile column located at the bottom of the column holder [2]. Aluminum column loads are assumed not to be submerged in reactor pool water. The constraint in this analysis is to use the fixed support that is placed on the stand guide column of the column retaining arm. The boundary conditions of FEM are shown in Figure 8.

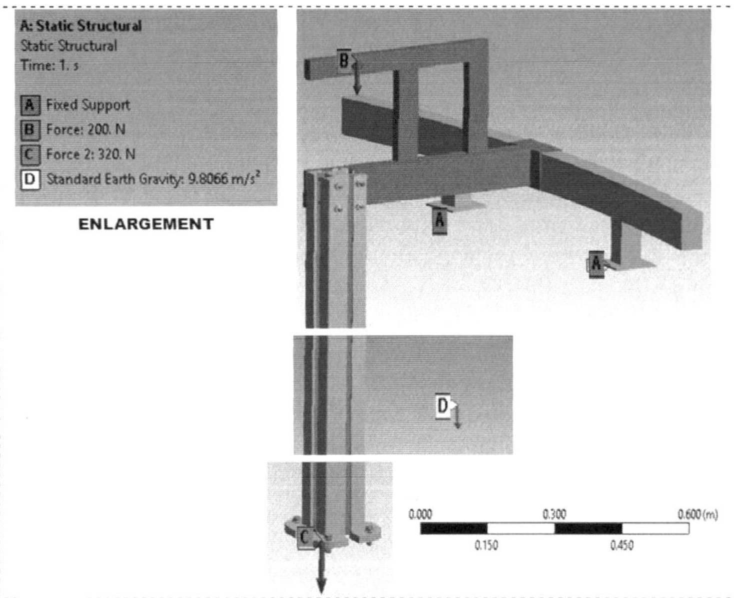

Figure 8. Application of loading and boundary conditions in FEM

For dynamic loads, it was assumed that repeated loading occured for 45 minutes with 60 cycles at 1 Hz lasting for one minute, with loading tolerance of ±10%. that is, the amplitude of the dynamic or alternating loads is 10% of the static load magnitude. The correction theory used was the Gerber mean stress theory because this correction theory is suitable with material used i.e. SS 304 and Al 6061. This is equivalent to 2700 cycles per day or 1E6 cycles per year. The ambient temperature is assumed to be uniform 22 °C. The ratio, frequency, and correction of dynamic loading can be seen in Figure 9.

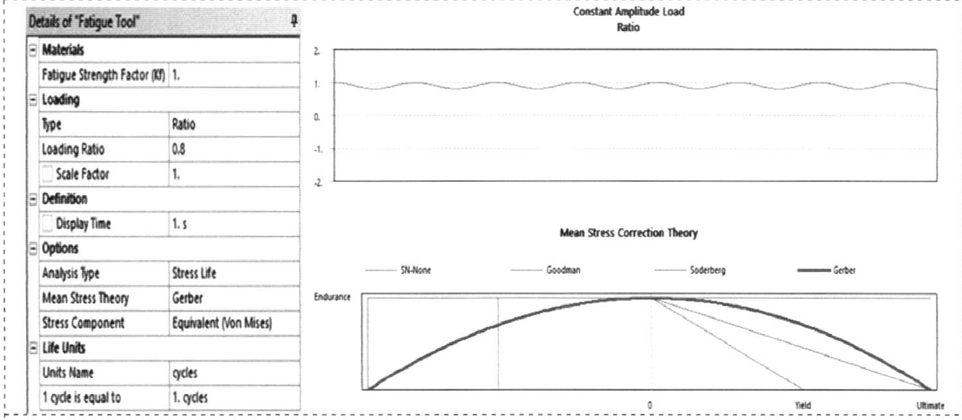

Figure 9. Ratio, frequency, and correction of dynamic loading.

The final stage of the process is to output the fatigue life cycle, safety factor, and equivalent alternating stress on a form in the finite element software display.

4. Result And Discussion

The result of static loading analysis using equivalent (von Mises) stress showed that the minimum stress was 4.9298 Pa and the maximum stress was 1.3698E8 Pa, occurring in the part of the structure made from SS 304. The results show that the stress occurring in the structure is still below the material limit of SS304 maximum yield strength of 2.05E8 Pa. Therefore, the process of analysis continued with fatigue loading analysis by giving the load parameters repeatedly with frequency of 1 Hz. The results of Dynamic loading analysis using equivalent Alternating Stress was the minimum stress 0.49298 Pa and the maximum stress value of 1.4758E7 Pa. The result of static loading analysis can be seen in Figure 10. and the result of dynamic loading analysis can be seen in Figure 11.

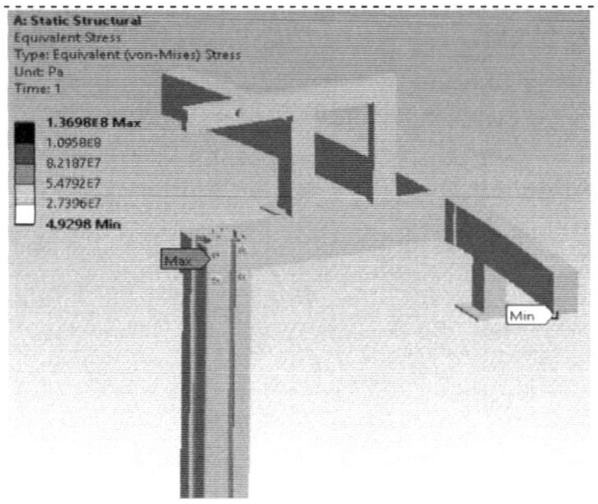

Figure 10. Results of static loading analysis

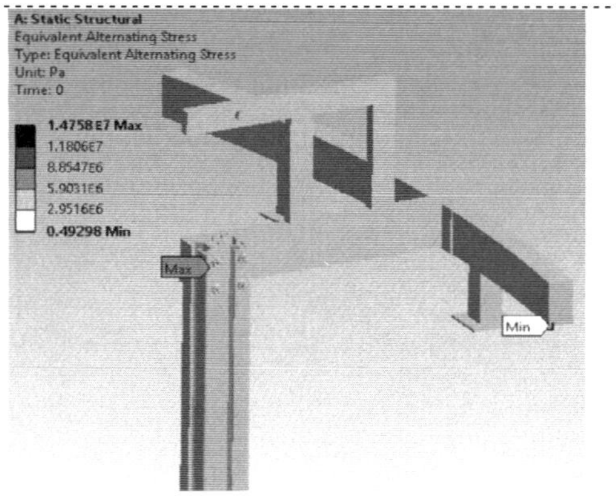

Figure 11. Results of dynamic loading analysis

The result of the security factor analysis shows that the minimum security value of the structure due to static loading was 1.8251 and occurring on the column support bolt. The value is allowed by the design recommendations because the minimum limit of the security factor for static loading is 1. The result of security factor the structural analysis due to dynamic loading was minimum 2.7856 and the value is likewise allowed by the design recommendations because the minimum limit of the security factor for dynamic loading is 2. The maximum value of the security factor of the structure of the ultrasonic scanner mechanical system based on the static and dynamic analysis of the finite element software was 15 which means that the security factor was high. The results of security factor analysis due to static loading can be seen in Figure 12. and the results of security factor analysis due to dynamic loading can be seen in Figure 13.

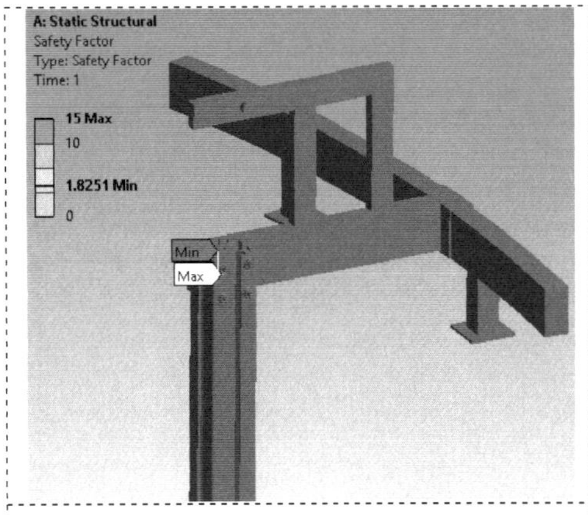

Figure 12. Results of safety factor analysis for static loading.

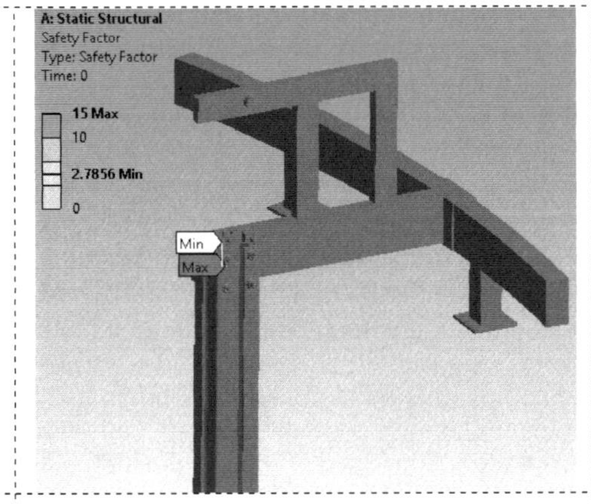

Figure 13. Results of safety factor analysis for dynamic loading.

The result of fatigue life cycle analysis was performed with load described in Figure 9., namely a 1-Hz load of ±10% of 520 N, and using the Gerber mean stress theory as correction. It was found that minimum or shortest fatigue life cycle occuring on any part of the system was 1E6 life cycles and the maximum or longest was 1E10 life cycles. The parts of the system with the shortest fatigue life were the column retaining bolt and the ultrasonic scanner holder connection, so the solution to lengthen the life cycle would be by strengthening the welding on each connection and enlarging the column support bolts. Based on the results obtained, the design of mechanical structure was found suitable for the design requirements of mechanical structure design having to fulfill the high life cycle number or the minimum limit of life cycle being 1E6. Thus, the safety of the operation is ensured. The results of the fatigue life cycle analysis can be seen in Figure 14.

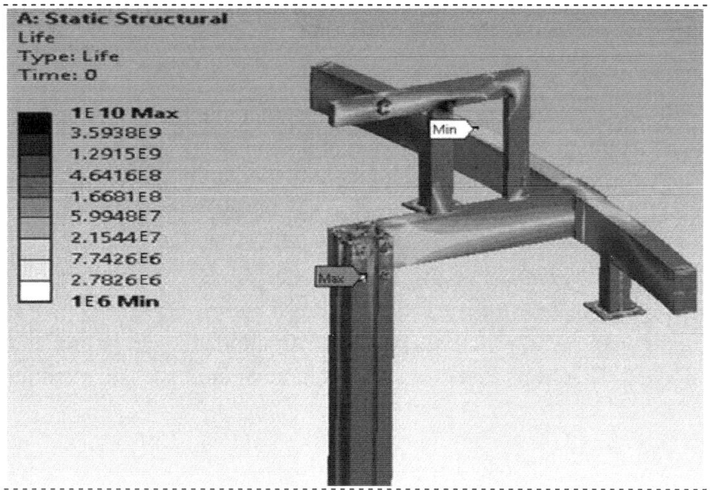

Figure 14. Life cycle analysis results

5. Conclusion

A reseach on the static, dynamic, and fatigue lifes of the mechanical system of the ultrasonic scanner for inservice inspection of research reactor has been conducted. The results of the analysis can be summarized as follows: the maximum equivalent von Mises stress for static loading analysis is 1.3698E8 Pa and its minimum is 4.9298 Pa. The result of the equivalent alternating stress for dynamic loading analysis which its maximum is 1.4758E7 Pa and its minimum is 0.49298 Pa in the SS 304 structure. Analysis result of safety factor for mechanical structure found a maximum value of 15 and a minimum of 1.8251 for static loading and a minimum of 2.7856 for dynamic loading, both of which fulfilled the criteria of structure safety of 1-10 for static loads and 2-10 for dynamic loads. The design of ultrasonic scanner mechanical system fulfills the high life cycle category that is at least 1E6 life cycles. Its maximum fatigue life is 1E10 cycles. As these results are still bellow of stress limit according to ASTM A240 standards is 2.05E8 Pa for SS 304, so the structure of the ultrasonic scanner mechanical system can be used safely.

6. Acknowledgment

The authors would like to thank the engineering team of ultrasonic scanners for inservice inspection of research reactors that have assisted in data collection as well as collection of materials and PTIM-BPPT for ANSYS computer and software tools used so this paper can be completed.

References

1. L. Raymond., E. P. Browell. "Predicting Fatigue Life with ANSYS Workbench", International ANSYS Conference May 2-4, 2006.
2. Muhammad Awwaluddin, Khairul Handono, Mechanical System Design of Ultrasonic Scanner for Research Reactor, National Seminar on Utilization of Nuclear Technology, South Tangerang , 2016.
3. Moaveni, Saeed. Finite Element Analisis: Theory And Application With ANSYS. Pearson Prentice Hall : United States of America, 2008.
4. American Standard Testing And Material (ASTM) International Standards Catalog , 2016.
5. ASHBY F. MICHAEL, Materials Selection in Mechanical Design, Third edition, Pergamon Press, 2005.
6. Oberg, Erik., Jones, Franklin., Horton, Holbrook., Ryffel, Henry., McCauley, Christopher., Machinery's Handbook 29th Edition, 2012.
7. Awwaluddin M., Hardjanto Tri, Sumanto Joko, "Fatigue Life Analysis of Mechanical Structure of Patients on Integrated Renograp Devices Using Finite Element Method, Journal of Nuclear Devices, Vol 09, No. 01, June 2015. page 37-48. (Analisa Fatigue Life Struktur Mekanik Pasien Pada Perangkat Renograf Terpadu Menggunakan Metode Finite Element, Jurnal Perangkat Nuklir, Vol 09, No. 01, Juni 2015. Hal 37-48).
8. Soren Heide Lambertsen, Lars Damkilde, Anders Schmidt Kristensen, Ronnie Refstrup Pedersen, "Estimation of Fatigue Life of Laser Welded AISI304 Stainless Steel T-Joint Based on Experiments and Recommendations in Design Codes", World Journal of Mechanics, 2013, 3, 178-183
9. Rahman M. M., Ariffin A. K., Jamaludin B. N., Haron C. C. H. C., "Fatigue Behaviour Of 6000 Series Aluminum Alloys On Cylinder Block Of a Free Piston Linear Engine Using Total Life Approach, Journal of Mechanical, Faculty Of Engineering University Kebangsaan Malaysia, June 2006, No 21, 1-15.
10. Budynas G. Richard, Nisbett Keith J. "Shigley`s Mechanical Engineering Design, ninth edition, McGraw-Hill Companies, 2011.
11. R. Kieselbach, "Curious cases of failure", Engineering Failure Analysis, Vol. 11, 2004, pp. 501-513.
12. Jung-Kyu Kim, Dong-Suk Shim, "The variation in fatigue crack growth due to the thickness effect", International Journal of Fatigue, Vol. 22, 2000, pp. 611-618.

The Mechatronic System Design Of Ultrasonic Scanner For Inservice Inspection Of Research Reactor

Khairul Handono, Kristedjo K., M. Awwaluddin and Ihsan Shobary.

Center for Nuclear Facilities Engineering, National Nuclear Energy Agency, PUSPIPTEK Area, South Tangerang, Indonesia 15310

Abstract. The mechatronic system design of ultrasonic scanner for inservices inspection of Research Reactor has been conducted. The requirement designed must be reliable operated, safety to personnel and equipments, ease of maintenance and operation, protection of equipment mechanically, interchangeability of equipments and addition of the several model of probe immersion ultrasonic tranducer. In order to achieve the above goals and obtain the desired results, a mechatronic design based on mechanical and electronic practical experiences will be needed. In this paper consist of the mechanical design and the system mechanical movement using stepper motor control. The criteria and the methods of designs of mechanical and electronic equipments of the system have been discussed and investigated. A mechanical and instrumentation control system drawing and requirement of design will be presented as the outcome of the design. The designed of mechanical system is consequently simulated by solidwork software. The intention of the above research is to create solutions in different ways of inservice inspection of integrity of Reactor.

Keywords : mechatronic, ultrasonic scanner, inservice inspection and research reactor

1. Introduction

Indonesia has operated three research reactors. The first is Triga Mark 2000 kWth which has been operated since January 1965 and located in Bandung, West Java. The second is Kartini Reactor 450 kWth which has been managed since 1979 and the reactor is located in Yogyakarta, Centre of Java. The last and largest one is the Multipurpose Reactor G.A. Siwabessy (RSG-GAS) 30 MWth, located in Serpong, Banten and it has been operated since August 1987. In fact, Reactor GA Siwa Bessy is now 30 years old so there has been a reactor component aging. Due to the aging management becomes an important issue in the operation of old reactors. Nuclear Energy Regulatory Agency as a regulatory body utilization of nuclear energy in 2015 issued Regulation No. 2 Bapeten on Aging Management Systems, Structures and Components (SSCs) for Research Reactors [1].

The Regulation contains the obligation of every owner of non-power nuclear reactors required to publish and implement aging management program or the data of related periodic degradation of the material in the reactor components such as fuel, the reactor tank, cooling pipes, and more. In the internal scope of RSG-GAS operating license will soon expire. As an extension of the operating license requirements, Perka BAPETEN No. 7 in 2015 requires the results of the study of aging research reactors and have to report data component aging RSG-GAS reactor area [1]. At higher regulations contained in the IAEA safety standards for standardization in the management of aging research reactors contained in the IAEA Safety Series No. Standard SSG-10.[2].

To realize the SSCs aging management at the research reactor, construction manipulator ultrasonic scanner system to retrieve the data integrity of the research reactor components should be done. This paper presents the design of an ultrasonic scanner manipulator system in mechatronics (mechanical design and control, and electronics). In the mechanical system is a manipulator arm, where the sensors, actuators support in the form of stepper motors and mechanical systems support [3].

Mechanical system design requirements of ultrasonic scanners for the research reactor in adjusted to actual conditions and SAR research reactor consists of: ultrasonic manipulator can be used for data retrieval radially and vertically on the walls of the reactor GA Siwabessy [3]. Then, the equipment can be used for sampling data on the reactor core, containment structure. Mechanical materials used have a composition similar to the material that is installed in the reactor GA Siwabessy. Other requirements should be flexible in terms of easy assembly, the enhanced mobility so easily moved out of the reactor and is easy to operate [2,4].

Electronic design and control requirements also refer to the requirements of the SARs and Safety Standard Series includes : the control system must be able to control all the mechanical system components safely. Electronic components must have the reliability to radiation and waterproof and equipment instrumentation and control systems must be calibrated.

2. System Description

A research reactor GA Siwabessy technical specifications is an open-pool type reactor cooled, among others : 5000 mm diameter reactor pool with a depth of 13,750 m. Components inside the reactor, among others : the rack element fuel, spent fuel racks, piping systems, support fuel rack, reinforcing the walls of the reactor. The in-service inspection of SSCs were carried out by using an ultrasonic scanner system. This ultrasonic system is important, because the complexity of the components in the reactor tank, as of the main component inspection need consistent with the requirements and guaranteed to be safe in operation. The design of ultrasonic scanner system as shown in Figure 1.

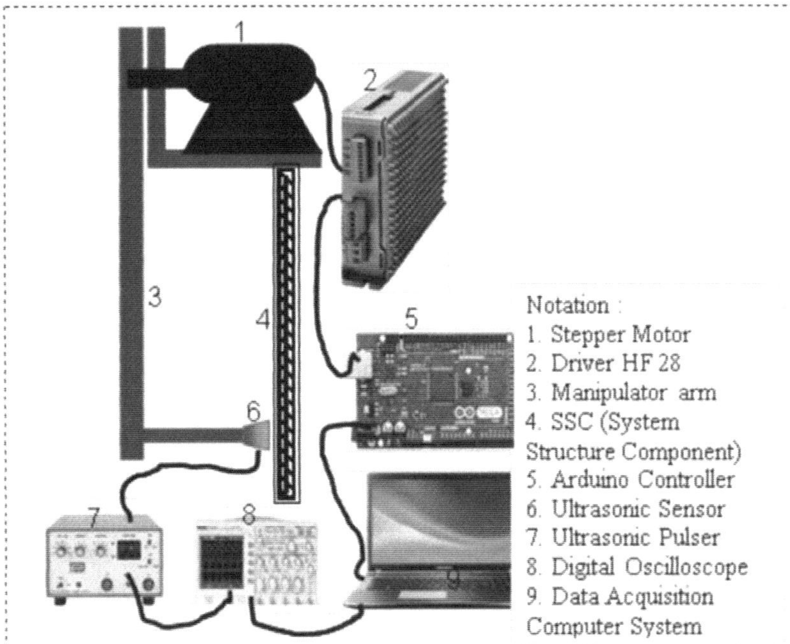

Figure 1. Ultrasonic Scanner System

The main component of mechanical system is manipulator arm as shown in Figure 1 the component (3), support structur of stepper motor (1), support structure of ultrasonic sensor cover and set of gopro camera. The material of manipulator arm is selected. The material requirements of mechanical systems is a first priority because in order to comply with Regulation of Bapeten No. 2 of 2015 on the Establishment List Information Design and also Regulation of the Bapeten No. 1 of 2009 on the Provision System of Physical Protection of Installations and Nuclear Material where in the metal material will be at risk of exposure to radiation [1]. So that the mechanical material will be selected is Aluminum 6061 with high purity. While the use of the SS component is permissible under the requirements of pretty no time limit in accordance with radiation safety regulations [5]. Mechanical ultrasonic scanner system to adjust the physical condition of the reactor pool RSG GAS and its contents, in which components are in such number of pipe is an important consideration. Manufacture of these components to be quite complicated so that the required mechanical simulation using the solidwork software. Ultrasonic sensor is emerson type (no 6) has 30 mm of diameter and ultrasonic scanner is operated upright to the object (SSCs, no 4) in scanning with a maximum distance of 1 inch respectively [6]. Generating ultrasonic waves sent by the ulrasonic pulser (component no.7), then the wave came to be measured using a digital oscilloscope (8), stored by the data acquisition computer system until to obtain the condition of the object being observed. Ultrasonic scanner manipulator arm is driven by a stepper motor controlled by an Arduino control system (component no 5) by using HF driver 28 (component no 2) [7,8].

3. Methodology

The step of mechatronic system design for ultrasonic scanner based on the flow chart is shown in Figure 2.

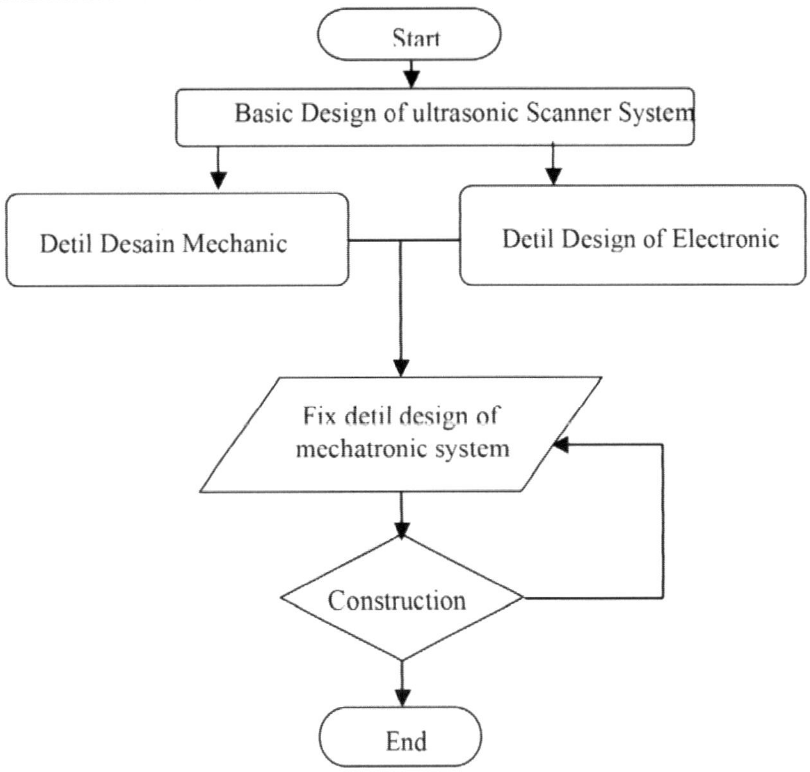

Figure 2. Flow chart of ultrasonic scanner system design

4. Result And Discussion

Mechatronics system design consists of the design of mechanical systems and electronic systems design. The main design priciples are defined to meet the project objective, including technical feasibility, safety and reability, flexibility, and later-upgrade sustainability [3,9].

4.1. Design Characteristic of Mechanical System

The mechanical system is improvment from conseptual design by using solidwork software design [3]. The evaluation detil degn of mechanic system has been conducted by review base on detil design mechanical system . The design mechanical system of ultrasonic scanner as shown in Figure 3, and the design of mechanical support system shown in figure 4.

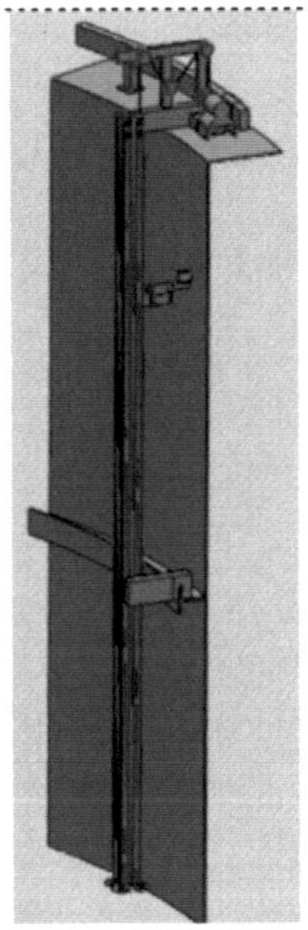

Figure 3. Design mechanical system of ultrasonic scanner

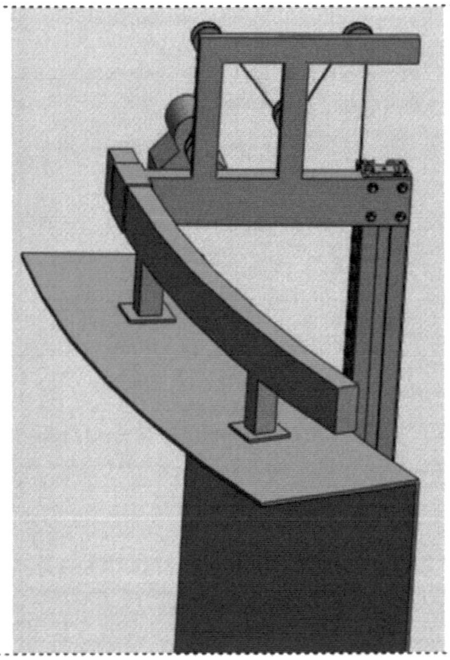

Figure 4. Design mechanical component support system of ultrasonic scanner

Based on the results of solidwork design in Figure 3 that the results of the general system design of the mechanical ultrasonic scanner system as a whole consists of mechanical systems for vertical movement, mechanical systems to move radially and the mechanical system holder ultrasonic sensor. Distance retrieval of data for a process based on the design in Figure 3. Directions Recording such as vertically up and down and radially. A maximum of 15 m with the distance between the point of taking 10 cm and radially is maximum 30^0 respectively.

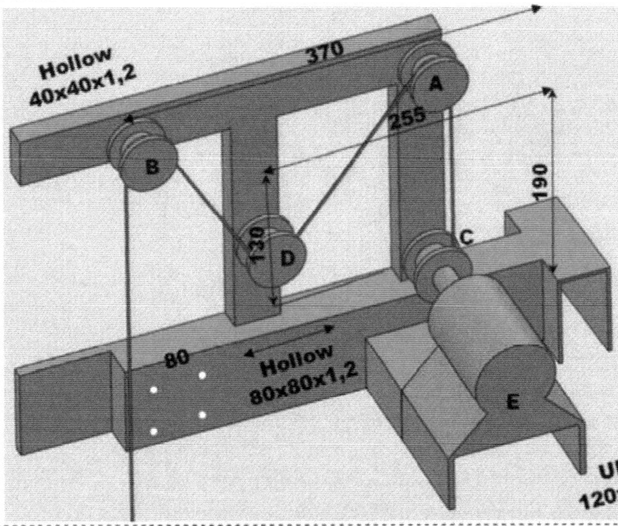

Figure 5. Design mechanical roller system of ultrasonic scanner

Stepper motors used for vertical movement of the mechanical system in Figure 5. The ultrasonic holder is a stepper motor, torque load calculated a value of 3 Nm. With the torque load of the specification of at least 2 A. Ampere motor is used to drive the wire sling load 20 Kg. From Figure 5, the fix roller A and B only used for the sliding guide wire slings, roller C is used for rolling wire sling and connected by a motor, this roller will be as a drum with grooves so that the wire sling arrangement would be neat. Roller fasteners D can be positioned towards vertical by 60 mm and is used to adjust the tension wire sling as expected by the operator. Figure 5, E is a stepper motor in accordance with the specifications mentioned above and adjustable speed. For determination of the movement distance at the time of data collection vertically then used a rotary encoder that will be affixed to the roller C and given a display that can be read distance.

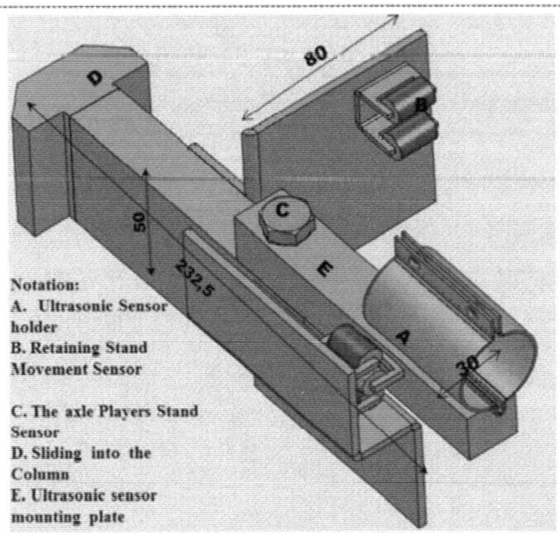

Figure 6. Design mechanical support of ultrasonic sensor

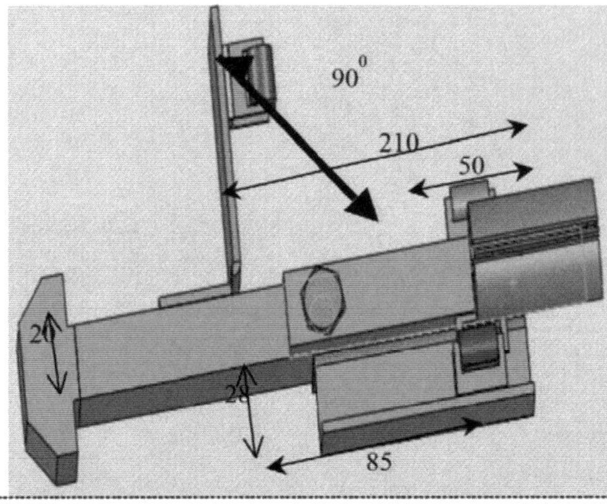

Figure 7. The ultrasonic sensor scanner plate holder .

From Figure 6. It can be explained that : A is the ultrasonic sensor scanner plate holder made from 6061 aluminum plate 1mm thick with a diameter of 30 mm and a length of 50 mm. Sensor ultrasonic scanner is placed inside the front of the sensor surface, then fastened using the screws M4 x 1 mm material SS 316L. B is retaining motion sensor holder made of teflon material and serves as an anchoring plate holder movement sensor holder ultrasonic scanner at the time of data collection and avoid reinforcing walls of the reactor. C is the sensor holder players with a diameter of 10 mm and made of 316L SS which serves as a swivel motion of ultrasonic sensor mounting plate scanner. D is the foundation as well as sliding into a column made of aluminum profile series 6061 and functions as the foundation of vertical motion in the column. From Figure 7, the ultrasonic sensor scanner plate holder made from 6061 series aluminum with a size of 20 x 20 mm and serves as the foundation of the plate holder and the sensor can be moved radially 90^0.

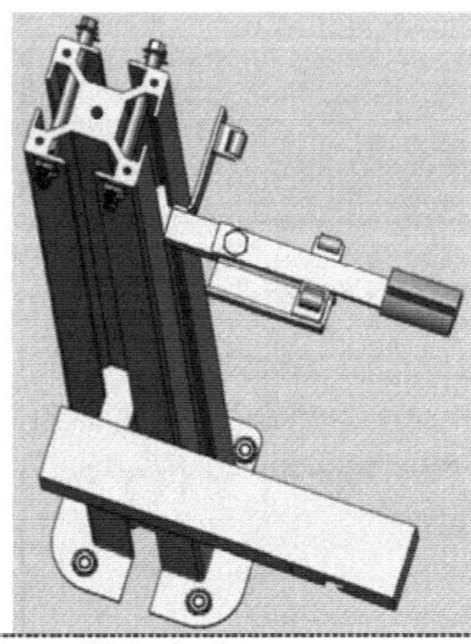

Figure 8 Manipulator arm systems for vertical direction

From Figure 8 can be explained that the mechanical system holder ultrasonic sensor scanner can be moved vertically up and down using a sling wire diameter of 2 mm made of SS 316L and is connected with the stepper motor. The follow of movements in the groove or profile contained in the column.

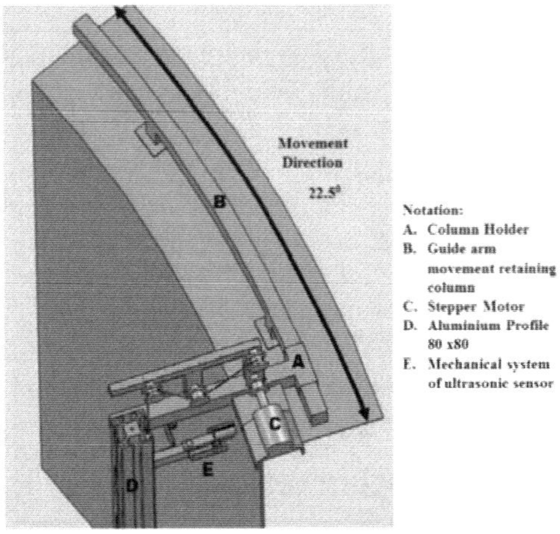

Figure 9. Mechanical systems movement towards radial

From Figure 9 can be explained that the movement of the mechanical system (mechanical systems and base column ultrasonic sensor scanner) using a radial direction with sliding movement system using hollow SS 316L 40 x 80 x 1.2 mm. Radial movement can be done as far as 22.50 considering the room space and the number of components in the reactor pool. The mechanical system movements toward radial will be installed on the top surface of the outer side of the pool reactor utilizing the existing screw holes on the surface. The position can move following circle movement in reactor wall [10].

4.2. Design Characteristic of Electronic and Control System

Design of computerized control system and data acquisition system in the format of block diagram shown in Figure 10.

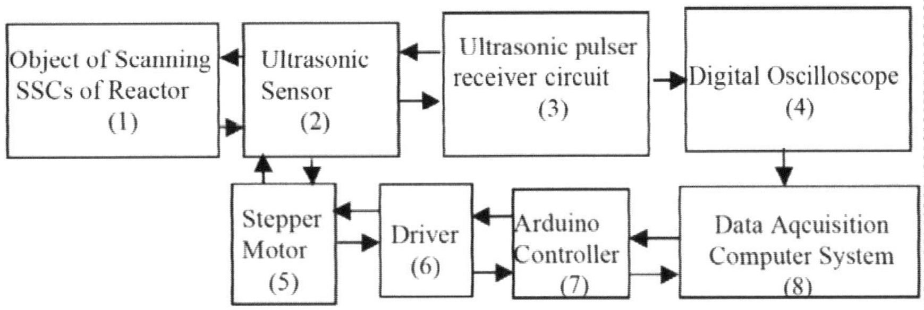

Figure 10. Schematic Diagram of Ultrasonic Scanner Data Acquisition System

From figure 10, the object of scanning SSCs (1) is wall of reactor pool, primary pipe, and reactor core. The ultrasonic sensor (2) is immersion type. Typical frequency up to 25 MHZ for routine applications.[6,7,11,12]. The ultrasonic wave generator is from subsystem no (3) also as function of ultrasonic pulser riciever system direct send to digital oscillocope (4). The data ultrasonic scanner from oscilloscope to save using data acquisition computer system

(8). The ultrasonic sensor and gopro camera (2) movement by stepper motor (5). The characteristic of motor stepper is shown in Table 1.

Table 1. The Stepper motor characteristic

Switch No.	**Resolution**	**Step Angle**
0	1	0.72°
1	2	0.36 °
2	4	0.18 °
3	5	0.144 °
4	8	0.09 °
5	10	0.072 °
6	16	0.045 °
7	20	0.036 °
8	25	0.0288 °
9	40	0.018 °
A	50	0.0144 °
B	80	0.009 °
C	100	0.0072 °
D	125	0.00576 °
E	200	0.0036 °
F	250	0.0288 °

From Table 1 shown that switch number from 0 to F is hexadecimal is the characteristic of digital switch of stepper motor. Select the step angle (motor rotation angle) per 1 pulse. The set step angle is dividing basic step angle of stepper motor by setting value. The calculation formula of step angle is basic step angle divided by resolution. The resolution of motor from 1 to maximum resolutoin 250 produce the step angle from 0.72 ° until smalest step 0.0288 ° respectively. The switch number connecting and conversion and coding using encoding stepper motor to transmit and reicieved using driver and control by Arduino controller (7). The characteristic of motor torque and driver is shown in Figure 11.

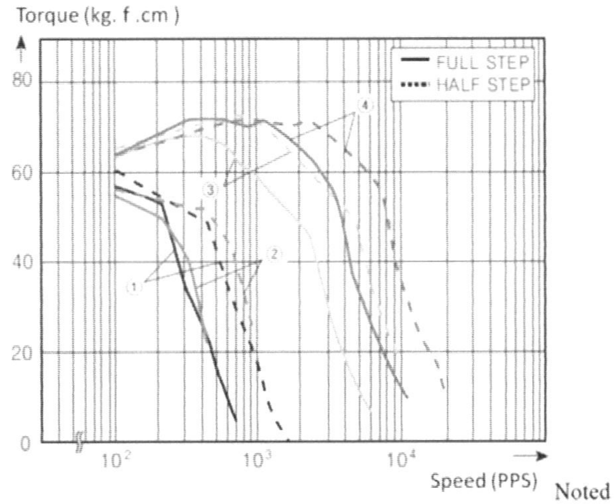

Noted : 1 &2. Driver MD5-HD14, Current 1.4Amp/phase, 3. Driver MD5-HF14 setting current 1.4 Amper/phase and 4.Driver MD5-HF28, setting current 2.8 Amper/phase

Figure 11. Characteristic of Stepper motor torque

From figure 11, the optimum torque of stepper motor using driver MD5-HF 28.
Human Machine Interface (HMI) of ultrasonic scanner design is shown in Figure 12

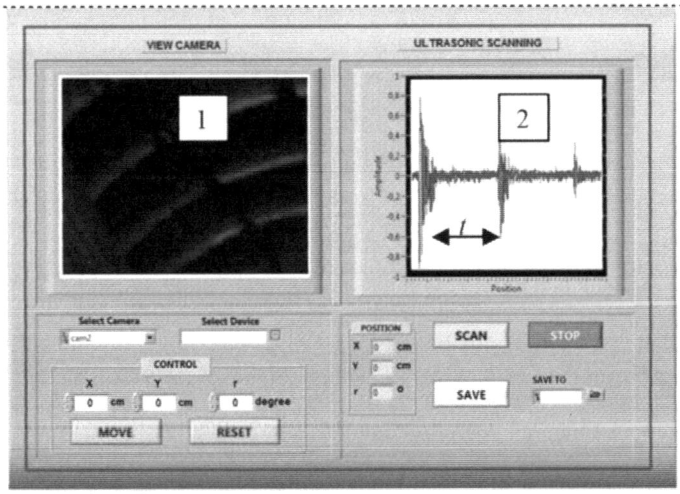

Figure 12. HMI Ultrasonic Scanner Design

From Figure 12 shown that for monitor ultrasonic sensor by using gopro camera, result of this monitor is shown from few camera (1) and to monitor ultrasonic signal from ultrasonic scanning (2),*t* is the thickness of SSCs material testing . All of the data to save using the data acquisition computer system by using Labview.

5. Conclusion

The mechatronic system design of ultrasonic scanner for inservices inspection of Research Reactor has been conducted. The criteria and the methods of designs of mechanical and electronic equipments of the system have been discussed and investigated. A mechanical and instrumentation control system drawing and requirement of design has been presented as the outcome of the design. The designed of mechanical system is drawing by solidwork software. The electronic and control system adjust to the design of mechanical systems. The mechanical system components can be controlled by electronic system. Electronically using computerized data acquisition system, a data of ultrasonic scanner can be loaded.

6. Acknowledgment

The authors would like to thank the team of mechatronic ultrasonic scanners for inservice inspection of research reactors that have assisted in the research especially the team of mechanics and the team of instrumentation control system.

7. Reference

1. Head of Bapeten, Regulation No. 2 Bapeten on Aging Management Systems, structures and components (SSCs) for Research Reactors, Jakarta, 2015.
2. Ageing Management for Research Reactors, IAEA Safety Standard Series No. SSG-10, Viena, 2010.
3. Muhammad Awwaluddin, et al Mechanical Conseptual Design Ultrasonic Scanner for Research Reactor, National Seminar on Utilization of Nuclear Technology, South Tangerang , 2016.
4. ASTM Standard, Vol 03.03, Standard Guide for Evaluating Characteristics of Ultrasonic Search Units: United States of America, 2008
5. Ashby F. Michael, Materials Selection in Mechanical Design, Third edition, Pergamon Press, 2005.
6. Patankar V. H., Chaurasia Reetesh, Nair Pradeep Design and Development of Instrumentation For Air-Coupled Ultrasonics, JNDE, Vol.8, Issue 4, March 2010, pp 52-56.
7. Lalwani SK, Randale GD, Patankar VH, Agashe AA, Jain RK, Chaurasia R, Jyothi P and Pithawa CK, "Design and Development of Modular, Configurable 8-Channel Ultrasonic Pulser-Receiver for NDT of Materials", JNDE, Vol. 11, Issue 4, March 2013, pp 38-42.
8. Chi-Shiang Cho, Wei-Ho Chung, Sy-Yen Kuo, Measurement and analysis of the leak tightness of reactor containment vessels: experiences and results, Elsevier Journal, Nuclear Engineering and Design pp 292 (2015) 112–122, 2015
9. Masanori Ando, Shigeru Takaya, Creep-fatigue evaluation method for weld joint of Mod.9Cr-1Mo steel Part II: Plate bending test and proposal of a simplified evaluation method, Elsevier Journal, Nuclear Engineering and Design,pp 310 (2016) 217–230, 2017.
10. K.V. Rajkumar, Govind K. Sharma, R. Dhayalan, T. Anish Kumar, An ultrasonic methodology for in-service inspection of shell weld ofcore support structure in a sodium cooled fast reactor, Elsevier Journal, Nuclear Engineering and Design, pp282 (2015) 83–92, 2015.
11. Paritosh Nanekara, N. Jothilakshmia, T. Jayakumarba, Ultrasonic phased array examination of circumferential weld joint inreactor pressure vessel of BWR, Elsevier Journal, Nuclear Engineering and Design, pp 265 (2013) 366– 374, 2013.
12. Jung-Ryul Leea, Hyomi Jeonga, Chia Chen Cianga, Dong-Jin Yoonb, Application of ultrasonic wave propagation imaging method to automatic damage visualization of nuclear power plant pipeline, Elsevier Journal, Nuclear Engineering and Design, pp 240 (2010) 3513–3520, 2010

Neutron dose rate analysis on HTGR-10 reactor using Monte Carlo code

Suwoto, H. Adrial, A. Hamzah, Zuhair, S. Bakhri, G. R. Sunaryo

Center for Nuclear Reactor Technology and Safety – National Nuclear Energy Agency of Indonesia (BATAN), PUPSPIPTEK Complex, Office Buliding No. 80, Serpong, Tangerang Selatan 15310, Indonesia, Telp. (021)756-0912, Fax. (021)756-0913,

E-mail: suwoto@batan.go.id

Abstract. The HTGR-10 reactor is cylinder-shaped core fuelled with kernel TRISO coated fuel particles in the spherical pebble with helium cooling system. The outlet helium gas coolant temperature outputted from the reactor core is designed to 700 °C. One advantage HTGR type reactor is capable of co-generation, as an addition to generating electricity, the reactor was designed to produce heat at high temperature can be used for other processes. The spherical fuel pebble contains 8335 TRISO UO_2 kernel coated particles with enrichment of 10% and 17% are dispersed in a graphite matrix. The main purpose of this study was to analysis the distribution of neutron dose rates generated from HTGR-10 reactors. The calculation and analysis result of neutron dose rate in the HTGR-10 reactor core was performed using Monte Carlo MCNP5v1.6 code. The problems of double heterogeneity in kernel fuel coated particles TRISO and spherical fuel pebble in the HTGR-10 core are modelled well with MCNP5v1.6 code. The neutron flux to dose conversion factors taken from the International Commission on Radiological Protection (ICRP-74) was used to determine the dose rate that passes through the active core, reflectors, core barrel, reactor pressure vessel (RPV) and a biological shield. The calculated results of neutron dose rate with MCNP5v1.6 code using a conversion factor of ICRP-74 (2009) for radiation workers in the radial direction on the outside of the RPV (radial position = 220 cm from the center of the patio HTGR-10) provides the respective value of 9.22E-4 μSv/h and 9.58E-4 μSv/h for enrichment 10% and 17%, respectively. The calculated values of neutron dose rates are compliant with BAPETEN Chairman's Regulation Number 4 Year 2013 on Radiation Protection and Safety in Nuclear Energy Utilization which sets the limit value for the average effective dose for radiation workers 20 mSv/year or 10μSv/h. Thus the protection and safety for radiation workers to be safe from the radiation source has been fulfilled. From the result analysis, it can be concluded that the model of calculation result of neutron dose rate for HTGR-10 core has met the required radiation safety standards.

Keywords: HTGR-10, neutron dose rate, MCNP5v1.6, ICRP-74.

1. Introduction

Nuclear energy is seen as the best candidate in the future as a potential energy sources to meet energy demand. Safety is one of the problems facing the nuclear reactor technology that is most crucial at this time, especially after the Fukushima incident. Fear of accidents due to human error or due to natural disasters cannot be avoided, but the high temperature gas cooled reactor (HTGR) as one type of Generation IV reactor is designed to have a mechanism of passive safety in its operations, and able to cover the problems of safety concern.

The Indonesia Experimental Power Reactor (HTGR-10)[1] designed by RENUKO (Rekayasa Engineering-Nukem-Kogas) is one type of HTGR that will soon be built in PUSPIPTEK area, Serpong Indonesia. The HTGR-10 refers and adopt the design of HTGR technology coming from Germany which applied to Chinese HTR-10 reactor[2,3,4]. The fuel of HTGR-10 reactor is designed using UO_2 TRISO coated fuel particles[5] which formed in the spherical fuel ball called pebble. Theoretically, the HTGR-10 reactor core can use the various fuel of kernel TRISO coated fuel particles such as uranium dioxide (UO_2)[6,7], thorium oxide (ThO_2/UO_2)[8,9,10] or plutonium dioxide (PuO_2)[11,12] without core shape and geometry changing. However, in this study only discusses fuel of uranium dioxide with two differences enrichment i.e. 10% and 17% of U^{235}/U^{238}.

The outputted thermal power of HTGR-10 reactor is about 10 MW and outlet temperatures around 700°C and expected to generate electricity up to 3 MWe. In addition, the HTGR-10 reactor expected can be used as co-generation plant such as sea water desalination and another industrial processes using high temperature application like hydrogen production using thermo-chemical water splitting system and coal liquefaction.

The HTGR-10 is 10 MWth reactor should be met with the operation standard safety limits of a nuclear reactor, so that the research on calculation of neutron dose rate is very important related to design of radiation shielding. The technical safety limit parameter of HTGR reactor should be met with the operation safety limit is also intended to safety limit of radiological aspect[13]. Radiological safety limit values are required for the HTGR-10 reactor include the dose rate of neutron radiation produced by the reactor core. The safety limit values should not be exceeded for safety purposes for radiation workers, communities and the surrounding environment.

The kernel of TRISO coated fuel particles, spherical fuel balls (pebble) and modeling of reactor core include reflector, radiation and radiological shielding of HTGR-10 core are fully 3-D modeled using Monte Carlo MCNP5v1.6 code. Tally card utilization on the Monte Carlo MCNP5v1.6 code in modeling of radiation sources, the modification to the desired output by adding normalized tally of neutron source strength and energy tally (DE, Dose Energy card) is performed. Tally is a language MCNP to mention the physical desired quantities. The addition of this tally will provide the output of the energy spectrum of the radiation source. All calculations are done on initial core of HTGR-10 that use fresh fuel. Status of previous studies have been calculated the neutron dose rate on the reactor type of pebble-bed HTGR 200 MWth (RGTT200K)[14] and analysis of neutron source strength and neutron dose rate calculation of RDE initial core[15] with results in accordance with BAPETEN Chairman's Regulation Number 4 Year 2013 on Radiation Protection and Safety in Nuclear Energy Utilization.

Monte Carlo MCNP5v1.6 code is a powerful probabilistic method code to simulate particles trip of neutrons, photons, electrons and others. This code has been developed by Monte Carlo X-5 team from LANL (Los Alamos National Laboratory) - United States. To perform the analysis of neutron dose rate calculation on the HTGR-10 reactor, Monte Carlo MCNP5v1.6 code with continuous neutron cross-section energy library (ACE files) taken from ENDF/B-VII file was used extensively in this research. The neutron energy group structure (EGS) in accordance with HTGR type reactor spectrum that can be made using EGS99304 utility code. The coated fuel particle with TRISO layered, spherical pebble fuel, the reactor core, control rod channel, cooling helium gas channel and a reflector and radiation and biological shielding[16] dimension and geometry are inputted as an input parameter.

The neutron dose rate calculation of using Tally Flux Detector (F5) combining of Tally Dose (DE) and Tally Energy DF(E) which is used to convert from the flux to dose with certain factors in accordance with the amount of conversion factors derived from the reference of ICRP-74 (2009)[17].

The addition of this tally will provide the output of the dose rate (Sv/h or rem/h) generated at a certain position on the geometry of the HTGR-10 core.

The main purpose of this research was to determine neutron dose rates distribution on HTGR-10 reactors. Furthermore the results of this research can be used to design of the radiation and the biological shielding thickness. Therefore the distribution of neutron radiation that spread to the whole area around the reactor can be well predicted. To meet the radiation safety requirements, especially radiation workers, then around the reactor core is installed safe radiation and biological shielding. The reactor operation meet of the radiation safety standards required by applicable law is fulfilled.

2. Description of Fuel And Reactor Core

Brief illustration of the fuel and reactor core of HTGR-10 is follow the geometry of the HTR-10 reactor in China. Schematic and technical data of the TRISO coated fuel particle of uranium dioxide (UO_2) as given in figure 1 and table 1.

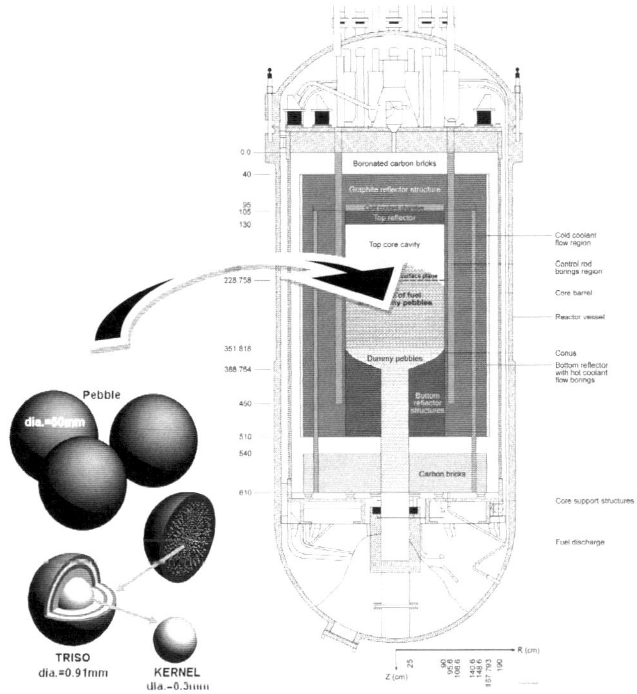

Figure 1. Illustration of fuel and core used in the HTGR-10 reactor[18]

Table 1. Core and fuel data parameter of HTGR-10 reactor[1,18,19]

PARAMETER	Value
REACTOR CORE	
▪ Thermal Reactor Power	10
▪ Equivalent Core Diameter, cm	180
▪ Equivalent Core High, cm	197
▪ Reactor Core Volume, cm^3	5.00
▪ Volumetric filling fraction of balls in the core	0.61
▪ Height of the empty cavity above the pebble bed, cm	41.7
▪ Diameter of fuel discharging tube, cm	50
FUEL ELEMENT	
SPHERICAL FUEL PEBBLE	
▪ Diameter of ball, cm	6.00

▪ Diameter of fuel zone, cm.	5.00
▪ Density of graphite in matrix and outer shell	0.50
▪ Heavy metal (uranium) loading (weight) per ball, g	5.00
▪ Density of graphite in matrix and outer shell, g.	1.73
TRISO Coated Fuel Particle	
▪ Fuel kernel material	UO_2
▪ Kernel diameter, cm	0.05
▪ Enrichment, % (U-235)	17
▪ Kernel density, g/cm^3	10.40
Coating layer	
▪ Coating layer materials (starting from kernel)	C/IPyC/SiC/OPyC
▪ Coating layer thickness, cm	0.095/0.040/0.035/0.04
▪ Coating layer density (g/cm^3)	1.1/1.90/1.38/1.90
▪ Kernel total diameter + TRISO coating layer, cm	0.092
SIDE REFLECTOR	
▪ Equivalent thickness of the side reflector graphite, cm	77.8
▪ Equivalent thickness of the side boronated carbon brick, cm	22.2
▪ Number of control rod channels	10
▪ Diameter of the control rod channel & radial coordinate position of the control rod guide channel, cm	13 & 102
▪ Number of cold helium flow channels	20
▪ Diameter of the helium flow channel & radial coordinate position of the cold helium flow channel centre, cm	8 & 144.6
CORE BARREL	
▪ Inner radius of core barrel, cm	190
▪ Thickness of core barrel, cm	5
▪ Base Stainless Steel Material	SA516-70
RPV (Reactor Pressure Vessel)	
▪ Inner radius of RPV, cm	210
▪ Thickness of RPV, cm	10
▪ Base Stainless Steel Material	SA516-70
BIOLOGICAL SHIELD[1]	
▪ Inner radius of biological shield, cm	320
▪ Thickness of biological shield, cm	250
▪ Base Material	Concrete

Table 2. Concrete wall specifications used in biological shielding[20]

Concrete density: 2.3000 g/cm^3 (NIST Ordinary Concrete)				
Nuclide	**MCNP-ID**	**Weight fraction**	**Atomic fraction**	**Atomic Density (atom/barn.cm)**
H	1001	0.022100	0.305330	0.030369
C	6000	0.002484	0.002880	0.000286
O	8016	0.574930	0.500407	0.049773
Na	11023	0.015208	0.009212	0.000916
Mg	12000	0.001266	0.000725	0.000072
Al	13027	0.019953	0.010298	0.001024
Si	14000	0.304627	0.151042	0.015023
K	19000	0.010045	0.003578	0.000356
Ca	20000	0.042951	0.014924	0.001484
Fe	26000	0.006435	0.001605	0.000160
Amount		0.999999	1.000000	9.9464E-02

Material composition data in the biological shielding of 250 cm concrete thickness used in the research was taken from the National Institute of Standards and Technology (NIST) - the USA, as presented in table 2.

The average active height and diameter of HTGR-10 cylindrical reactor core are 197 cm and 180 cm, respectively. The reactor core divide into 10 zones in axial directions and 10 zones in axial directions, as presented in figure 2. The lower axial part in the core is zone number: 900, 901, 902, 903, 904, 905, 906, 907, 908 and 909. The center axial part in the core is zone number: 950, 951, 952, 953, 954, 955, 956, 957, 958 and 959. The top axial part in the core is zone number: 990, 991, 992, 993, 994, 995, 996, 997, 998 and 999.

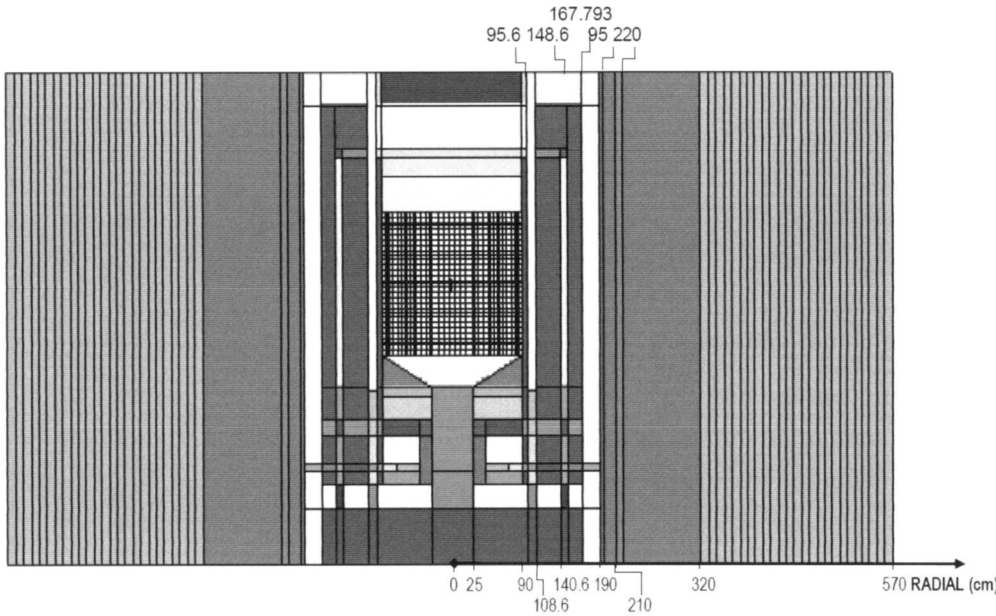

Figure 2. The geometry modelling of HTGR-10 core with MCNP5v1.6 code

3. Calculation Methodology

TRISO kernel coated fuel particles and pebble fuel geometry modelling are involved packing fraction consideration[21]. Determination of the number of the group structure of the HTGR-10 reactor spectrum is done using EGS99304 utility code. MCNP5v1.6 code was used extensively in this research by utilizing of continuous energy cross-sectional data taken from ENDF/B-VII file.

Modelling of kernel TRISO coated fuel particles is shown in figure 3. TRISO coated fuel particles in the Monte Carlo MCNP/MCNPX code do with fully heterogeneity model on kernel TRISO i.e. kernel UO_2, high porosity carbon buffer layer, inner pyrolytic carbon layer, silicon carbide layer, outer pyrolytic carbon layer and graphite matrix. Spherical fuel zone containing of 8335 UO_2 kernel TRISO coated fuel particles with diameter 2.5cm using packing fraction of 5%. The spherical fuel zone covered with graphite outer shell of 0.5 cm thickness, thus finally, pebble fuel formed with a diameter of 6 cm is shown in figure 4.

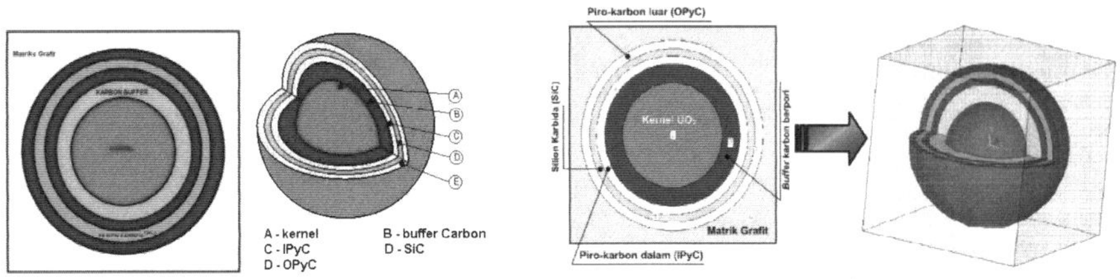

(a) **(b)**

Figure 3. Modelling of kernel TRISO coated fuel particle in Monte Carlo MCNP5v1.6 code
(a). Heterogeneous cells model of kernel TRISO coated fuel particles
(b). Kernel cell model at the center of a simple cubic (SC)

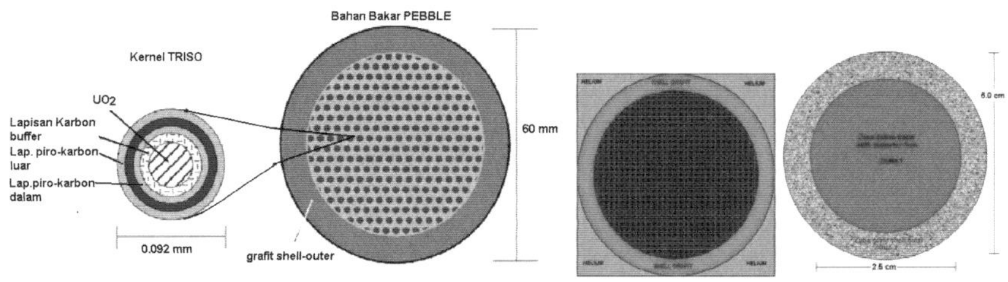

Figure 4. Heterogeneous modelling of kernel TRISO in the formation of spherical fuel pebble using Monte Carlo MCNP5v1.6 code

Spherical fuel (pebble) is then inserted into the reactor core with lattice model BCC (Body Centered Cubic), such that the packing fraction obtained by 61% (optimal value for HTGR with pebble fuel). The calculation of the radiation field by proper tallies of MCNP5v1.6 code, was carried out by calculating the neutron track length estimate over cell flux along the radial of the reactor core, using the F5 tally (particle flux at a ring detector). The detectors position in the reactor system are placed at the radius of 28.4605cm, 40.2492cm, 49.2950cm, 56.9210, 63.6396cm, 69.7137cm, 75.2994cm, 80.4984cm, 85.3815cm, 90cm, 95.60cm, 108.60cm, 148.60cm, 167.79cm, 190cm, 195cm, 210cm, 220cm, 320cm, 330cm, 340cm, 350cm, 360cm, 370cm, 380cm, 390cm, 400cm, 410cm, 420cm, 430cm, 440cm, 450cm, 460cm, 470cm, 480cm, 490cm, 500cm, 510cm, 520cm, 530cm, 540cm, 550cm, 560cm and 570cm from the reactor center.

The calculation of neutron average neutron fluxes generated reactor core using the following equation:

$$\Phi\left[\frac{neutron}{cm^2.\sec}\right] = Tally\ F5\,(\Phi_{F5})(1/cm^2) * NSS\,(\text{neutron/sec}) * \frac{1}{k_{eff}} \tag{1}$$

where:

- Tally F5 (Φ_{F5}) is the result of the calculation output Monte Carlo MCNP code, (1/cm^2),
- k_{eff} is a constant value of neutron effective multiplication factor,
- NSS is Neutron Source Strength, neutron/sec.

Determination of the neutron dose rates produced by reactor core are calculated using equation (1), then converted into neutron dose rates with specific conversion factor based on the ICRP-74 (2009) reference as presented in table 3. Doses are derived from neutron fluxes obtained with detector F5 tallies by means of the Dose Energy card (DEn), and the Dose Function card (DFn).

Table 3. Conversion factor from neutron fluxes into neutron dose rates[23]

ICRP-74 (2009)			
Neutron energy (MeV)	**DF(E) (mrem/h)/(n/cm^2.s)**	**Neutron Energy (MeV)**	**DF(E) (mrem/h)/(n/cm^2.s)**
1.00E-09	2.3760E-03	0.1500	4.7520E-02
1.00E-08	3.2400E-03	0.2000	6.1200E-02
2.53E-08	3.8160E-03	0.3000	8.3880E-02
1.00E-07	4.6440E-03	0.5000	1.1592E-01
2.00E-07	4.8600E-03	0.7000	1.3500E-01
5.00E-07	4.8960E-03	0.9000	1.4400E-01
1.00E-06	4.7880E-03	1.0000	1.4976E-01
2.00E-06	4.6440E-03	1.2000	1.5300E-01
5.00E-06	4.3200E-03	2.0000	1.5120E-01
1.00E-05	4.0680E-03	3.0000	1.4832E-01
2.00E-05	3.8160E-03	4.0000	1.4688E-01
5.00E-05	3.5640E-03	5.0000	1.4580E-01
1.00E-04	3.3840E-03	6.0000	1.4400E-01
2.00E-04	3.2040E-03	7.0000	1.4580E-01
5.00E-04	2.9880E-03	8.0000	1.4724E-01
1.00E-03	2.8440E-03	9.0000	1.5120E-01
0.0020	2.7720E-03	10.000	1.5840E-01
0.0050	2.8800E-03	12.000	1.7280E-01
0.0100	3.7800E-03	14.000	1.8720E-01
0.0200	5.9760E-03	15.000	1.9440E-01
0.0300	8.5320E-03	16.000	1.9440E-01
0.0500	1.4796E-02	18.000	2.0520E-01
0.0700	2.1600E-02	20.000	2.1600E-01
0.1000	3.1680E-02		

4. Results and Discussion

The calculation of the neutron dose rates are done using a combination of a couple of Tally F5 (particle flux at a ring detector) by Tally DE (Dose Energy Card) and Tally DF (Dose Function Card) which is used to convert from the neutron flux to dose. The conversion factors from neutron fluxes to neutron dose rates in accordance with the value of the conversion factor was taken from International Commission on Radiation Protection (International Commission on Radiological Protection, ICRP), the ICRP-74. With the pair combined tally DE and DF will provide the output of the dose rate (Sv/h or rem/h) generated at a certain position on the geometry of HTGR-10 core with the normalization to value of neutron source strength has been derived from it (as tally FM).

All calculations of neutron dose rates generated by HTGR-10 core is performed with UO_2 fuel enriched ^{235}U at 10% and 17%. The conversion factor of neutron flux to dose rates based on ICRP-74 (2009) reference is fully used.

The calculated results of neutron dose rate using Monte Carlo MCNP5v1.6 code equipped with tally detector cards (Tally F5) and conversion factors Flux-to-Dose multiplied by neutron source strength has been obtained. Neutron dose rates distribution on HTGR-10 core as shown in figure 5 for the enrichment of 10% and figure 6 for the enrichment of 17% are clearly observed. All calculations using MCNP5v1.6 with the conversion factor of the neutron flux in the neutron dose of ICRP-74 (2009) are performed.

The latest regulations from BAPETEN Chairman's Regulation Number 4 Year 2013 on Radiation Protection and Safety in Nuclear Energy Utilization, average dose limit value effective for radiation workers are set at 20 mSv (2000 mrem) per year or equivalent to 10 μSv/h.

From the analysis, it can be seen on figure 5 and figure 6 that neutron dose rates on the axial direction of the center part of the core has higher than those in the bottom and top of HTGR-10 core.

Using the conversion factor of ICRP-74 (2009), at a radial distance of 168 cm from the center of HTGR-10 core (the outermost part of reflector zone), neutron dose rate is weakened of 2.08 μSv/h and 2.44 μSv/h for 10% and 17% enrichment, respectively. Those both values are geatly met with requirements of BAPETEN Chairman's Regulation Number 4 Year 2013.

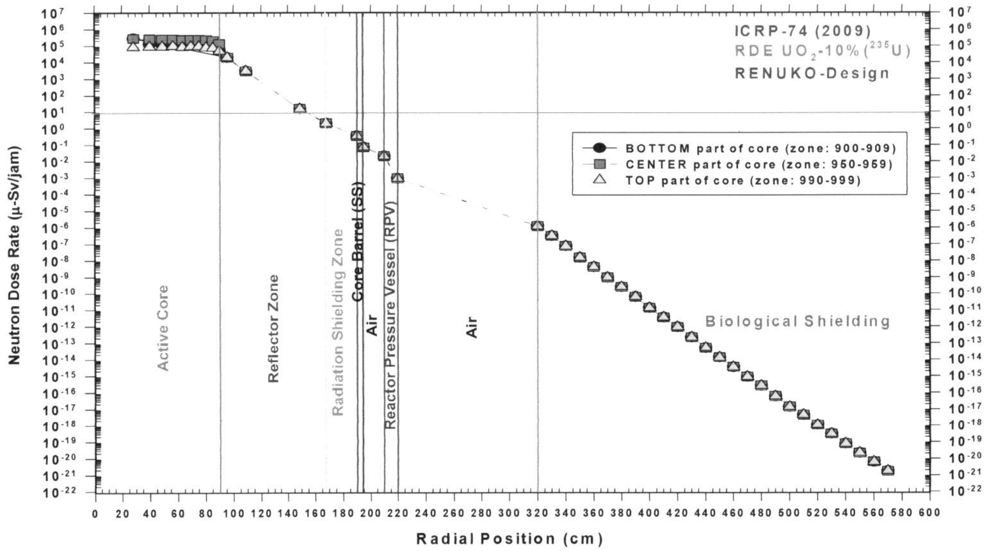

Figure 5. The calculated results of neutron dose rate distribution on HTGR-10 core with UO_2 fuel enrichment 10% (^{235}U) using MCNP5v1.6 code

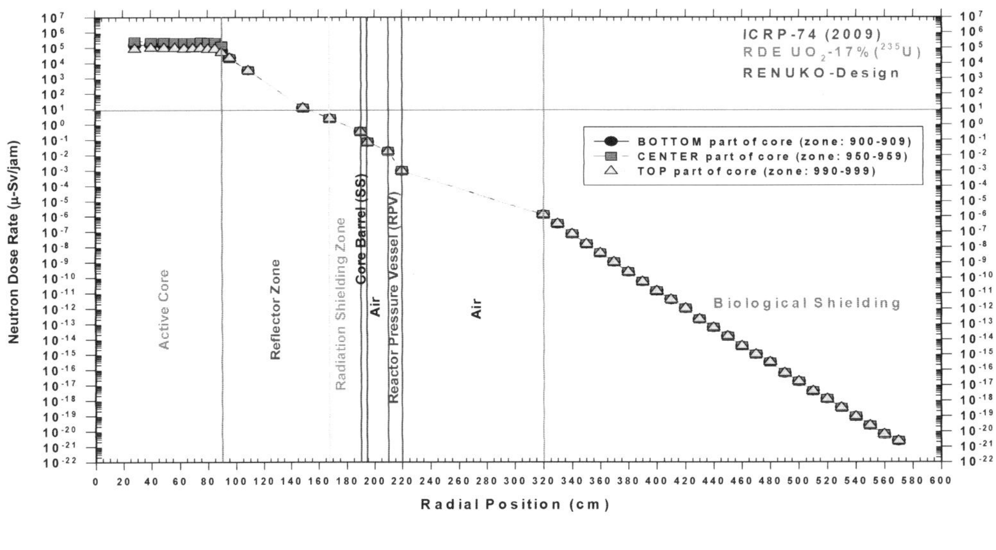

Figure 6. The calculated results of neutron dose rate distribution on HTGR-10 core with UO_2 fuel enrichment 17% (^{235}U) using MCNP5v1.6 code

The MCNP5v1.6 calculation results of the neutron dose rates at a distance of 220 cm from the center of core (outer part of reactor pressure vessel, RPV), the neutron dose rates value of 9.22E-4 μSv/h and 9.58E-4 μSv/h are obtained for enrichment of 10% and 17%, respectively.

The both calculation results of neutron dose rate calculations performed on biological shield of 250 cm thickness (radial distance from the core center of 570 cm) are clearly safe. In accordance with the MCNP5v1.6 calculations results on 0 cm thickness of the biological shielding, the value of neutron dose rates have been dropped up to 1.18E-06 μSv/h (enrichment 10%) and 1.33E-06 μSv/h (enrichment 17%) by using ICRP-74 (2009) flux-to-dose conversion factor.

Using a conversion factor flux-to-dose ICRP-74 (2009) used in compliance with the applicable requirements are BAPETEN Chairman's Regulation Number 4 Year 2013. So it can be concluded that the design of modelled biological shielding thickness is sufficient and safe for radiation workers, so that the worker is protected from nuclear radiation and can work safely in a nuclear reactor environment.

5. Conclusion

Overall calculation results of neutron dose rates on HTGR-10 core are performed with Monte Carlo MCNP5v1.6 code by using a conversion factor of neutron fluxes-to-dose dose rates based on ICRP-74 (2009) are clearly observed. The calculation analysis results for the neutron dose rates on HTGR-10 core outside the radial position of 220 cm from the center of HTGR-10 reactor core (outer part of RPV) are giving each value of 9.22E-4 μSv/h and 9.58E-4 μSv/h for enrichment UO_2 10% and 17%, respectively. Those both values are greatly met with the requirements specified by BAPETEN Chairman's Regulation Number 4 Year 2013, so that the radiation workers have safe and protected from neutron radiation sources produced reactor core.

Acknowledgment

The author would like to thank Syaiful Bakhri Ph.D. as Head of Reactor Physics and Technology Division and Dr. Pande Made Udiyani as research coordinator, on suggestion and improvement of this paper. This study was fully funded by DIPA PTKRN-BATAN 2016.

References

[1] Tim Penyusun Dokumen Spesifikasi Teknis RDE-RENUKO, 2014. Dokumen Spesifikasi Teknis Reaktor Daya Eksperimental, Nomor Idetifikasi Dokumen: **DT.002.KRN.2014**, Rev.0, tanggal: 20-Nov-2014, PTKRN-BATAN, Tangerang Selatan.

[2] Wang, M. J., et al., 2014. Criticality calculations of the HTR-10 pebble-bed reactor with SCALE6/CSAS6 and MCNP5, Annals of Nuclear Energy **64**, 1–7.

[3] Chen, F., et al., 2009. Benchmark Calculation for the Steady-State Temperature Distribution of the HTR-10 under Full-Power Operation, Journal of Nuclear Science and Technology, Vol. **46**, No. 6, p. 572–580.

[4] Rosales, J,. et al., 2014. Computational Model for the Neutronic Simulation of Pebble Bed Reactor's Core Using MCNPX, Hindawi Publishing Corporation, International Journal of Nuclear Energy, Volume **2014**, Article ID 279073, 12 pages, http://dx.doi.org/10.1155/2014/279073.

[5] Kania, M. J., Nabielek, H. And Nickel, H., 2015. Coated Particle Fuels for High-Temperature Reactors, Materials Science and Technology, **DOI: 10.1002/9783527603978**.mst0449, p.1–183.

[6] Kania, M. J., Nabielek, H., Verfondern, K., Allelein, H.J., 2013. Testing of HTR UO_2 TRISO Fuels in AVR and in Material Test Reactors, Journal of Nuclear Materials 441, 545–562.

[7] Tran, H. N., Hoang, V. K., 2012. Neutronic characteristics of an OTTO refueling PBMR, Nuclear Engineering and Design **253**, 269 – 276.

[8] Allelein, H.J., Kania, M. J., Nabielek, H., Verfondern, K., 2014. Thorium Fuel Performance Assessment in HTRs, Nuclear Engineering and Design **271**, 166–170.

IOP Conf. Series: Journal of Physics: Conf. Series **962** (2018) 012029 doi:10.1088/1742-6596/962/1/012029

[9] Trinuruk, P., Obara, T., 2015. Particle-Type Burnable Poisons for Thorium-Based Fuel in HTGR, Energy Procedia **71**, 22 – 32.

[10] Wols, F. J., Kloosterman, J. L., Lathouwers, D., van der Hagen, T.H.J.J, 2014. Core Design and Fuel Management Studies of a Thorium Breeder Pebble Bed High Temperature Reactor, Nucl. Technol., **186**, pp. 1–16.

[11] Jonnet, J. J., Kloosterman, J. L., Boer, B., 2010. Performance Of TRISO Particles Fueled With Plutonium and Minor Actinides in A PBMR-400 Core Design, Nuclear Engineering and Design **240**,1320–1331.

[12] Acir, A., Coskun, H., 2012. Neutronic Analysis of The PBMR-400 Full Core Using Thorium Fuel Mixed With Plutonium or Minor Actinides Annals of Nuclear Energy **48**, 45–50.

[13] Valentin, J. (Ed.), 2012. The 2007 Recommendations of the International Commission on Radiological Protection, p.98 (5.10. Dose Limits), Recommendation No. **244**, Elsevier, submitted: March 29.

[14] S. Suwoto and Z. Zuhair, 2016. Analysis of Neutron Dose Rates on RGTT200K Core Using MCNP5, Jurnal Sains dan Teknologi Nuklir Indonesia **17** (2), 15.

[15] S. Suwoto, H. Adrial, Z. Zuhair. 2017. Analisys of Neutron Source Strength and Neutron Dose Rate Calculation of RDE Intial Core, Urania. Scientific Journal of Nuclear Fuel Cycle, 23(1), 33.

[16] Remetti, R., Andreoli, G., Keshishian, S., 2012. Monte Carlo Calculation of The Neutron Effective Dose Rate at The Outer Surface of The Biological Shield Of HTR-10 Reactor, Nuclear Engineering and Design **243**, 148–152.

[17] Ward, D. C., 2009. Impact of Switching To The ICRP-74 Neutron Flux-To-Dose Equivalent Rate Conversion Factors, at The Sandia National Laboratory Building 818 Neutron Source Range, SANDIA report, SAND2009-**1144**, March (2009).

[18] Seker V., Çolak U., 2003. HTR-10 full core first criticality analysis with MCNP, Nuclear Engineering and Design **222**, 263–270.

[19] Terry, W. K., et al.**,** 2006. Evaluation of the HTR-10 Reactor as a Benchmark for Physics Code QA, ANS Topical Meeting on Reactor Physics, PHYSOR.

[20] Mcconn, R.J., Jr., et al., 2011. Compendium of Material Composition Data for Radiation Transport Modeling, PIET-43741 -TM-963, **PNNL-15870** Rev. 1, March 4.

[21] Türkmen, M., Çolak, U., 2012. Effect of Pebble Packing on Neutron Spectrum and the Isotopic Composition of HTGR Fuel, Annals of Nuclear Energy **46**, 29–36.

Evaluation of the AP1000 delayed neutron parameters using MCNP6

T.M. Sembiring, J. Susilo, S. Pinem

Center for Nuclear Reactor Technology and Safety, National Nuclear Energy Agency of Indonesia (BATAN), Kawasan Puspiptek Gd. No. 80 Serpong , Tangerang Selatan 15310 Indonesia

Email: tagorms@batan.go.id

Abstract. The MCNP6 code contains numerous features, one of those is to determine the delayed neutron parameters. The accuracy of calculated delayed neutron parameters affect the accuracy of transient or dynamic condition. The objective of this paper is to determine the delayed neutron parameters of the advance PWR reactor, AP1000, using MCNP6 code with the recent ENDF/B evaluated nuclear data file ENDF/B-VII.1. The MCNP6 calculation results shows that the maximum difference occurred in the β_i and λ_i parameters are 38.30% and 45.63%, respectively. The superiority of MCNP6 can be seen in the change of prompt neutron life time (ℓ) parameters that cannot be obtained by the deterministic code, so it can be used in the sensitivity analysis of the delayed neutron parameters. Based on this research work, the accident analysis of the AP1000 reactor use the effective delayed neutron fraction (β_{eff}) of 0.0051 and the prompt neutron life time (ℓ) of 19.5 μs for the first cycle.

Key-words: PWR, AP1000, delayed neutron parameters, MCNP6, ENDF/B-VII.1

1. Introduction

Recent advances in PC based high performance computation technologies enable us to apply the Monte Carlo codes to large scale problems, such as detailed three-dimensional whole-core calculations for commercial power reactors within acceptable computation time at low cost. One of Monte Carlo codes which contains numerous new features not found in either code is MCNP6 version 1 [1]. The code has capability to determine the delayed neutron parameters, either effective delayed neutron fraction (β_{eff}), effective neutron generation time (Λ_{eff}), delayed neutron precursor fraction (β_i) and effective delayed neutron precursor decay constant (λ_i).

The delayed neutron parameters are fundamental to describe the kinetic and dynamic response of both critical and subcritical nuclear systems to internal or external perturbation. The accuracy of calculated delayed neutron parameters affect the accuracy of transient or dynamic condition. The MCNP code is widely used for the PWR reactor analysis [2-7]. The MCNP6 code as well as the ENDF/B-VII.1 have been validated for many types of nuclear reactor [8-11] that show some improvements.

The delayed neutron parameters of AP1000 reactor [12] has been determined by using deterministic codes in the previous research works [13-14]. The research work in Ref. [13] evaluated the delayed neutron parameters for BOC (beginning of cycle) at the first core cycle. However, Ref [14] evaluated the delayed neutron parameters as a function of burn-up from BOC to EOC (end of cycle) at the first core cycle. Those research works used the old nuclear data and delayed neutron data, such as ENDF/B-

VII.0 and JENDL-3.3. Therefore, the research work based on the recent ENDF/B evaluated nuclear data and new version of MCNP code is needed to evaluate the delayed neutron data for complex core configuration such as the AP1000 reactor. The objective of this paper is to evaluate the delayed neutron parameters of the AP1000, such as β_{eff}, prompt neutron life time (ℓ), β_i and λ_i, for BOC and EOC at the first core cycle using MCNP6 code. The results of this work can be used in the accident analysis for the first cycle of AP1000 reactor.

In this research work, the atomic densities of fission product nuclides for 16 types of fuel assemblies are determined by using PIJ module in the SRAC2006 code [15]. The detail geometry model of AP1000 reactor for the MCNP6 code input is adopted from the previous research work that has been validated with the design value [16]. The core calculations are carried out for the BOL (fresh core) and EOL for the given boron concentrations. The average burn-up of AP100 fuel assemblies at the EOL condition are obtained from Ref [17]. The evaluation is carried out by comparing with the design and previous research works [13-14,18].

2. Brief description of AP1000 reactor

The Westinghouse Advanced Passive PWR AP1000 is an 1100 MWe class PWR, which is conservatively based on proven PWR technology, but with an emphasis on safety features that rely on natural forces [12]. The thermal power of 3400 MW is generated by 157 fuel assemblies with three radial regions with the enrichments of 2.35 w/o, 3.40 w/o and 4.45 w/o, as shown in Fig. 1. There are 16 types of fuel assemblies based on the enrichment and the number IFBA and Pyrex [18]. The main data and specification of core and fuel configuration of AP1000 reactor are described in the Ref [18], completely.

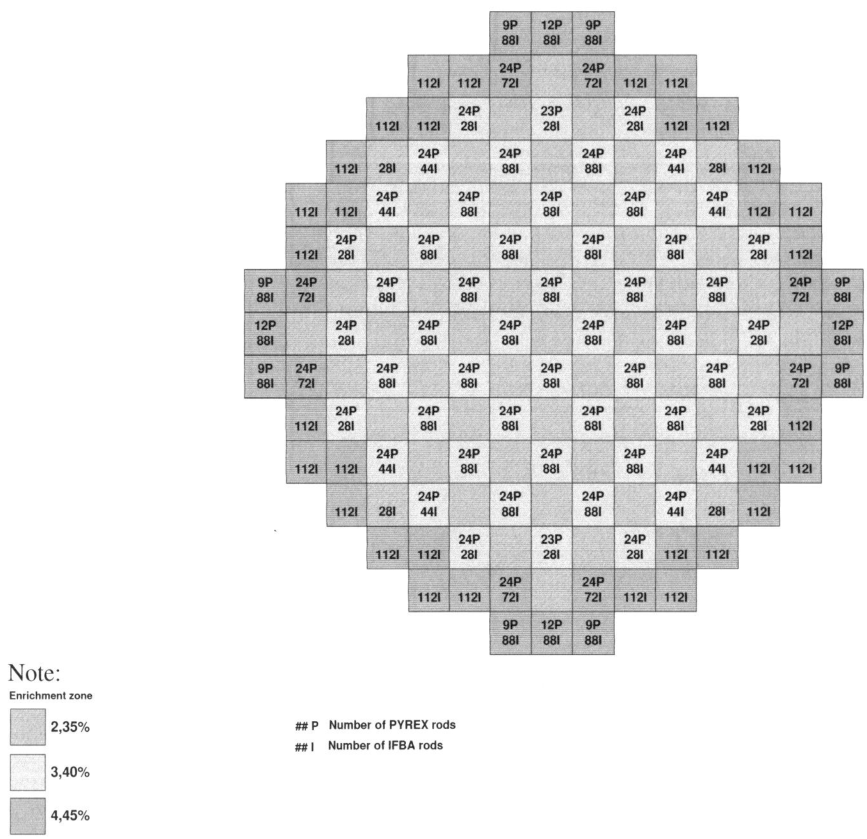

Figure 1. AP1000 core configuration [9][13]

Based on the design document [19], the delayed neutron fraction ($\beta_{\rm eff}$) typically yield values no less than 0.0075 at beginning of cycle (fresh core) and 0.0050 at end of cycle for the first cycle. For the next cycles, the $\beta_{\rm eff}$ of 0.0049 (BOC) and 0.0044 (EOC) are used in the accident analysis. It is noted that the prompt neutron life time (ℓ) of AP1000 reactor is 19.8 μs as described in Ref [18].

3. Methodology

The delayed neutron parameters that can be calculated by MCNP6 are the effective delayed neutron fraction ($\beta_{\rm eff}$), effective neutron generation time ($\Lambda_{\rm eff}$), delayed neutron fraction of precursors in group i (β_i) as well as effective delayed neutron precursor decay constant in group i (λ_i). Those parameters are calculated by using the data of the delayed neutron data ($\nu_d(E)$) and the delayed neutron spectra ($\chi_d(E)$) from the ENDF/B-VII.1. In this present work, the calculation objects are the BOC and EOC cores at the first cycle, means all fuel assemblies at BOC are fresh. The delayed neutron parameters are calculated in the given boron concentration for BOC and EOC[18].

The following calculation steps are used in this research work:

1. The detail geometry model of AP1000 reactor are adopted from Ref [16] as shown in the Fig 2.

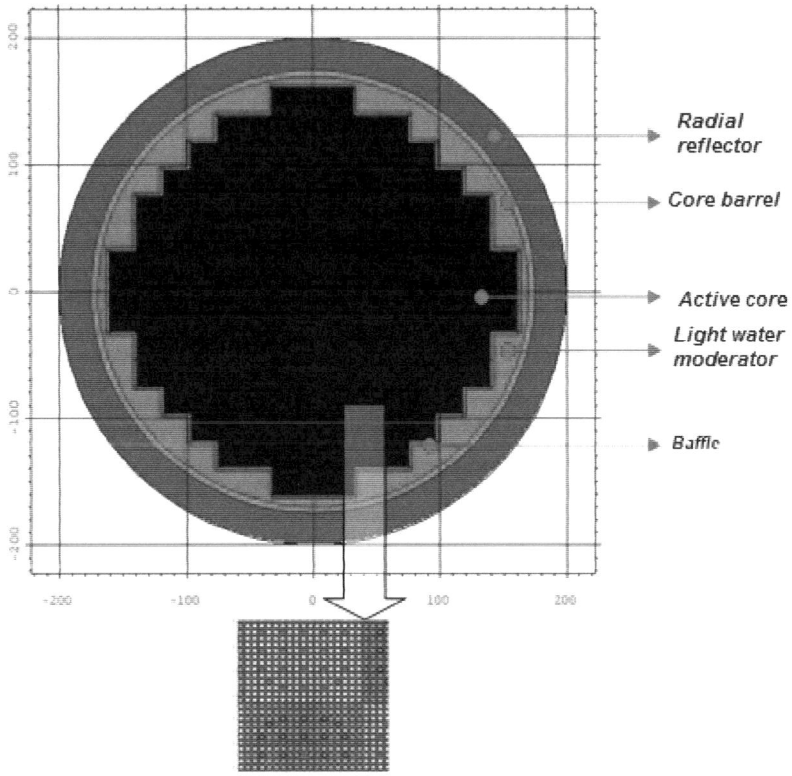

Figure 2. Detail 3-dimension model of the AP1000 core in X-Y plane (radially) [16]

2. The atomic densities of light water moderator are determined based on the temperature, pressure and the boron concentration. For the BOC case [18], the moderator temperatures are 293K (cold) and 600K (hot) as well as the fuel temperatures are 293 K (cold) and 600K (hot). The pressure is 158.19 bar for hot condition. The boron concentrations for cold and hot (equilibrium xenon) at BOC and EOC are 1574 ppm and 827 ppm, respectively.

3. The average burn-up of EOC core at first cycle is assumed by 20 GWD/MTHM[17]. The atomic densities of fission product nuclides at 20 GWD/MTHM is determined by the PIJ module of SRAC2006 code.
4. The atomic densities changes of burnable absorbers, such as IFBA and PYREX, were calculated by the PIJ module in SRAC2006 code.
5. The evaluation was carried out by comparison with the design values [18] and the deterministic codes of BATAN-3DIFF and CITATION-SRAC2006 results [13-14].

The MCNP6 calculations used 600 cycles with 100,000 neutrons per cycle in the KCODE card. The first 100 cycles are skipped in order to avoid source convergence problem. The delayed neutron parameters are calculated by the KOPTS card. The $S(\alpha,\beta)$ thermal neutron scattering for light water is chosen as same as the temperature condition.

4. Results and discussions

Table 1 shows the calculated effective delayed neutron fraction (β_{eff}) and prompt neutron life time (ℓ) parameters for the AP1000 reactor at BOC and EOC with boron concentration of 1574 ppm and 827 ppm, respectively. For the BOC, the MCNP6 gives a βeff value that is relative difference by -7.33% from the β_{eff} value obtained by design value [18]. The deterministic code calculations by using BATAN-3DIFF and CITATION-SRAC have the relative difference by -2.67% and -5.33%, respectively. In addition, for EOC, the calculation results of MCNP6 and CITATION-SRAC have relative difference with design value by 3.80% and 2.00%, respectively.

Using 3σ uncertainty (99% confidence level), at BOC condition, the results of MCNP6 is within margin with those of BATAN-3DIFF and CITATION-SRAC, but out of range from the design reference. On the other hand, at EOC condition, the calculation result of MCNP6 is within margin with those of CITATION-SRAC and design reference. It seems the different delayed neutron data and the number of neutron energy group being used in those codes do not contributes significantly to the calculated β_{eff}. It is noted that the BATAN-3DIFF, CITATION-SRAC and MCNP6 use the evaluated nuclear data based on Brady-England, JENDL-3.3 and ENDF/B-VII.1 [13-14], respectively. Beside that the number of neutron energy group for macroscopic cross-section in the BATAN-3DIFF, CITATION-SRAC and MCNP6 codes are 2,10 and continuous energy, respectively.

The value of β_{eff} at EOC condition are dominated by the fission products, such as ^{239}Pu, ^{240}Pu, ^{241}Pu and ^{242}Pu (as shown in Fig 3), so, compared with the BOC condition, the value of β_{eff} decrease by 30%, 28.2% and 25.3% in the design value, the CITATION-SRAC and the MCN6 results, respectively. The neutron spectrum at EOC condition become harder than at BOC condition. Therefore, the calculated β_{eff} is sensitive in the softer neutron spectrum of BOC condition. The possibility that the boron concentration give effect to the calculated β_{eff} needs to be analyzed in the future work.

Table 1. The calculated β_{eff} and ℓ values for AP1000 reactor at BOC and EOC

Parameters	Design [18]	BATAN-3DIFF [13]	CITATION-SRAC [14]	MCNP6
	BOC (cold, zero power, broron concentration of 1574 ppm)			
β_{eff}	0.0075	0.0073 (-2.67%)[a]	0.0071 (-5.33%)[a]	0.00695±0.00013 (-7.33%)[a]
ℓ, μs	19.8	14.1 (-28,79%)[a]	20.8 (5.05%)[a]	19.53±0.04 (-1.36%)[a]
	EOC (hot, zero power, broron concentration of 827 ppm)			
β_{eff}	0.0050	-	0.0051 (2.00%)[a]	0.00519±0.00012 (3.80%)[a]
ℓ, μs	-	-	20.8	20.44±0.04

Note: a: relative differece with design value

Table 1 shows that the prompt neutron life time (ℓ) parameters obtained by the MCNP code give a very good agreement with the design value compared with the deterministic codes since the relative difference is -1.36%. The ℓ parameter at EOC is higher than BOC by 4.66%. It means the ℓ is not influenced by the burn-up condition or neutron spectrum change. The MCNP6 codes can evaluate the small change of ℓ parameter while the CITATION-SRAC code give same value for BOC and EOC. Based on the calculation results of BATAN-3DIFF code, the ℓ is sensitive to the number of neutron energy group.

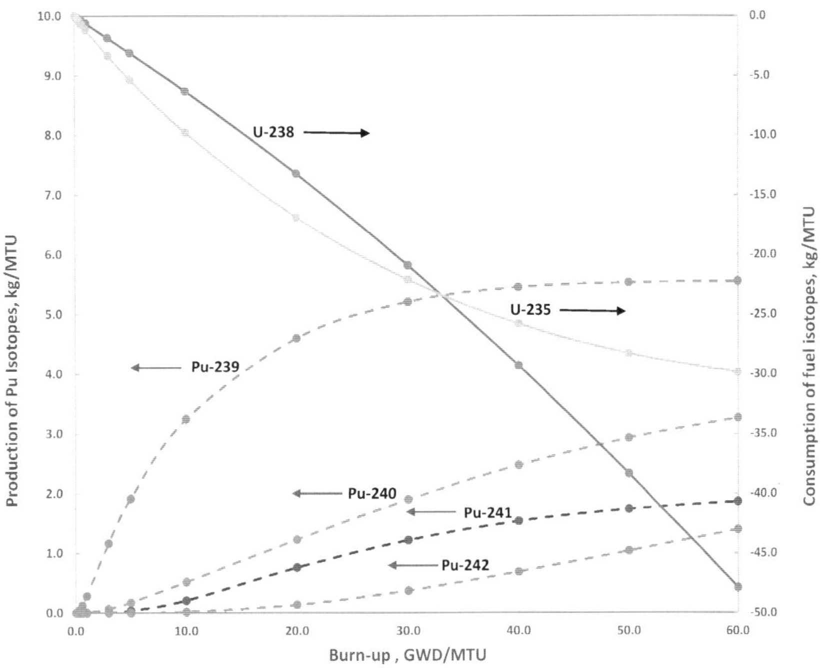

Figure 3. The consumption fuel isotopes and production of Pu isotopes of AP1000

Table 2 shows the calculated β_i and λ_i values obtained by using BATAN-3DIFF, CITATION-SRAC and MCNP codes. If we used the MCNP result as the reference, it is clear that the two deterministic codes give a large different since the difference is in the range of -27.87% - 38.30% and -31.79% - 19.57% at BOC and EOC, respectively. The maximum difference occurred in the 6th group. It seems that the calculation results is strongly depend on the delayed neutron fraction of precursors in group *i* and absolute delayed neutron yield for each fission product nuclides.

The calculated effective delayed neutron precursor decay constant in group *i* (λ_i) in Table 2 shows same situation as the β_i. For the BOC, the difference is in the range of -0.31% - 14.73% and -6.36% - 40.19% by using the BATAN-3DIFF and CITATION-SRAC codes, respectively. The difference between BOC and EOC is relatively same since the range is in -5.53% - 45.63% at EOC by using CITATION-SRAC code. As discussed above, the λ_i is depend on the delayed neutron decay constant of delay group *i* for each fission product nuclide, too.

Table 2. The calculated β_i and λ_i values for AP1000 reactor at BOC and EOC

Group Precursor, i	BATAN-3DIFF [13] β_i	BATAN-3DIFF [13] λ_i	CITATION-SRAC [14] β_i	CITATION-SRAC [14] λ_i	MCNP β_i	MCNP λ_i
			BOC			
1	2.77×10^{-4}	1.330×10^{-2}	2.11×10^{-4}	1.251×10^{-2}	$2.40\times10^{-4} \pm 2\times10^{-5}$	$1.336\times10^{-2} \pm 0.0$
2	1.40×10^{-3}	3.250×10^{-2}	1.45×10^{-3}	3.071×10^{-2}	$1.21\times10^{-3} \pm 6\times10^{-5}$	$3.260\times10^{-2} \pm 0.0$
3	1.20×10^{-3}	1.219×10^{-1}	1.36×10^{-3}	1.150×10^{-1}	$1.22\times10^{-3} \pm 6\times10^{-5}$	$1.211\times10^{-1} \pm 0.0$
4	2.50×10^{-3}	3.169×10^{-1}	2.80×10^{-3}	3.100×10^{-1}	$2.68\times10^{-3} \pm 8\times10^{-4}$	$3.059\times10^{-1} \pm 2\times10^{-5}$
5	1.27×10^{-3}	9.886×10^{-1}	9.64×10^{-4}	1.208	$1.27\times10^{-3} \pm 6\times10^{-5}$	$8.617\times10^{-1} \pm 9\times10^{-5}$
6	6.50×10^{-4}	2.954	3.39×10^{-4}	3.238	$4.70\times10^{-4} \pm 4\times10^{-5}$	$2.894 \pm 5.0\times10^{-4}$
			EOC			
1	-	-	1.42×10^{-4}	1.263×10^{-2}	$1.50\times10^{-4} \pm 2\times10^{-5}$	$1.337\times10^{-2} \pm 0.0$
2	-	-	1.10×10^{-3}	3.060×10^{-2}	$9.20\times10^{-4} \pm 5\times10^{-5}$	$3.185\times10^{-2} \pm 0.0$
3	-	-	9.64×10^{-4}	1.196×10^{-1}	$8.30\times10^{-4} \pm 5\times10^{-5}$	$1.197\times10^{-1} \pm 2\times10^{-5}$
4	-	-	1.92×10^{-3}	3.209×10^{-1}	$1.95\times10^{-3} \pm 7\times10^{-5}$	$3.068\times10^{-1} \pm 3\times10^{-5}$
5	-	-	7.35×10^{-4}	1.268	$9.40\times10^{-4} \pm 5\times10^{-5}$	$8.707\times10^{-1} \pm 1.1\times10^{-4}$
6	-	-	2.66×10^{-4}	3.274	$3.90\times10^{-4} \pm 3\times10^{-5}$	$2.914 \pm 7.9\times10^{-4}$

5. Conclusions

The evaluation of the AP1000 delayed neutron parameters performed by the recent ENDF/B evaluated nuclear data file ENDF/B-VII.1 with the MCNP6 code leads to very satisfactory results. The superiority of the MCNP6 code can be seen in the change of prompt neutron life time (ℓ) parameters that cannot be obtained by the deterministic code, so can be used in the sensitivity analysis of the delayed neutron parameter. The maximum difference occurred in the β_i and λ_i parameters by the range of 38.30% and 45.63%, respectively. The accident analysis of the AP1000 reactor for the first cycle can be carried out using this research work results because the results have a very good agreement with the designn value.

Acknowledgments

Research work of Center for Nuclear Reactor Technology and Safety of BATAN was supported by the fiscal year of 2016 (DIPA 2016).

References

[1] Goorley T, James M, Booth T, Brown F, Bull J, Cox L J, Durkee J, Elson J, Fensin M, Forster R A, Hendricks J, Hughes H G, Johns R, Kiedrowski B, Martz R, Mashnik S, Mckinney G, Pelowitz D, Prael R, Sweezy J, Waters L, Wilcox T and Zukaitis T 2016 Features of MCNP6 *Ann. Nucl. Energy* **87** 772–83

[2] Kahler A C, MacFarlane R E and Chadwick M B 2014 Integral data testing of ENDF/B-VII.1

IOP Conf. Series: Journal of Physics: Conf. Series **962** (2018) 012030 doi:10.1088/1742-6596/962/1/012030

files - success stories and need to improve stories *Nucl. Data Sheets* **118** 410–3
[3] Pecchia M, Vasiliev A, Ferroukhi H and Pautz A 2017 A methodology for evaluating weighting functions using MCNP and its application to PWR ex-core analyses *Ann. Nucl. Energy* **105** 121–32
[4] Pecchia M, Wicaksono D, Grimm P, Vasiliev A, Perret G, Ferroukhi H and Pautz A 2016 Validation of Monte Carlo based burnup codes against LWR-PROTEUS Phase-II experimental data *Ann. Nucl. Energy* **97** 153–64
[5] Sargeni A, Burn K W and Bruna G B 2016 The impact of heavy reflectors on power distribution perturbations in large PWR reactor cores *Ann. Nucl. Energy* **94** 566–75
[6] Vasiliev A, Ferroukhi H, Pecchia M and Pautz A 2016 Localized neutron flux assessment and verification studies using MCNPX PWR full core model *Ann. Nucl. Energy* **92** 317–32
[7] Maiorino J R, Laranjo G, Moreira J M L, Rossi P C R and Santos T A 2017 Feasibility to convert an advanced PWR from UO 2 to a mixed U / ThO 2 core – Part I : Parametric studies *Ann. Nucl. Energy* **102** 47–55
[8] Bécares V, Pérez-Martín S, Vázquez-Antolín M, Villamarín D, Martín-Fuertes F, González-Romero E M and Merino I 2014 Review and comparison of effective delayed neutron fraction calculation methods with Monte Carlo codes *Ann. Nucl. Energy* **65** 402–10
[9] Santamarina A, Blaise P, Erradi L and Fougeras P 2012 Calculation of LWR $\beta_{\rm eff}$ kinetic parameters: Validation on the MISTRAL experimental program *Ann. Nucl. Energy* **48** 51–9
[10] Perret G, Blaise P, Messaoudi N, Geslot B, Wagemans J, Baeten P, Kochetkov A and Pautz A 2016 Validation of Monte-Carlo methods for generation time and delayed neutron fraction predictions *Ann. Nucl. Energy* **97** 165–70
[11] Kahler A C, MacFarlane R E, Mosteller R D, Kiedrowski B C, Frankle S C, Chadwick M B, McKnight R D, Lell R M, Palmiotti G, Hiruta H, Herman M, Arcilla R, Mughabghab S F, Sublet J C, Trkov A, Trumbull T H and Dunn M 2011 ENDF/B-VII.1 Neutron Cross Section Data Testing with Critical Assembly Benchmarks and Reactor Experiments *Nucl. Data Sheets* **112** 2997–3036
[12] Schulz T L 2006 Westinghouse AP1000 advanced passive plant *Nucl. Eng. Des.* **236** 1547–57
[13] Sembiring T M and Pinem S 2012 The evaluation of delay neutron parameters in the AP1000 PWR reactor *Prosiding Seminar Nasional ke-18 Teknologi dan Keselamatan PLTN Serta Fasilitas Nuklir* (BATAN) pp 67–75
[14] Susilo J and Suparlina L 2013 Analysis of the AP1000 core Kinetics Parameter as a Function Burnup Fraction *Prosiding Seminar Nasional ke-19 Teknologi dan Keselamatan PLTN Serta Fasilitas Nuklir* (Yogyakarta: BATAN) pp 13–22
[15] Okumura K, Kugo T, Kaneko K and Tsuchihashi K 2007 *SRAC2006 : A comprehensive neutronics calculation code system*
[16] Sembiring T M 2011 Analysis of the 3-dimensional core model for evaluation of criticality parameters of the Advanced PWR 100 MW class *Tri Dasa Mega* **13** 78–95
[17] Susilo, J, Sitorus Pane J 2016 Fuel burn-up distribution and transuranic nuclide contents produced at the first cycle operation of AP1000 *Tri Dasa Mega* **18** 101–11
[18] Westinghouse 2011 *AP1000 Design Control Document, Chapter 4. Reactor* (Westinghouse)
[19] Westinghouse 2011 AP1000 Design Control Document, Chapter 15 . Accident Analyses (Westinghouse) pp 1–40

The change of radial power factor distribution due to RCCA insertion at the first cycle core of AP1000

J Susilo, L Suparlina, Deswandri, G R Sunaryo
Center for Nuclear Reactor Technology and Safety
PUSPIPTEK Complex O.B. No.80, Setu, Tangerang Selatan, 15310

shushilo@batan.go.id

Abstract. The using of a computer program for the PWR type core neutronic design parameters analysis has been carried out in some previous studies. These studies included a computer code validation on the neutronic parameters data values resulted from measurements and benchmarking calculation. In this study, the AP1000 first cycle core radial power peaking factor validation and analysis were performed using CITATION module of the SRAC2006 computer code. The computer code has been also validated with a good result to the criticality values of VERA benchmark core. The AP1000 core power distribution calculation has been done in two-dimensional X-Y geometry through ¼ section modeling. The purpose of this research is to determine the accuracy of the SRAC2006 code, and also the safety performance of the AP1000 core first cycle operating. The core calculations were carried out with the several conditions, those are without Rod Cluster Control Assembly (RCCA), by insertion of a single RCCA (AO, M1, M2, MA, MB, MC, MD) and multiple insertion RCCA (MA + MB, MA + MB + MC, MA + MB + MC + MD, and MA + MB + MC + MD + M1). The maximum power factor of the fuel rods value in the fuel assembly assumedapproximately 1.406. The calculation results analysis showed that the 2-dimensional CITATION module of SRAC2006 code is accurate in AP1000 power distribution calculation without RCCA and with MA+MB RCCA insertion.The power peaking factor on the first operating cycle of the AP1000 core without RCCA, as well as with single and multiple RCCA are still below in the safety limit values (less then about 1.798). So in terms of thermal power generated by the fuel assembly, then it can be considered that the AP100 core at the first operating cycle is safe.

Keywords: Power Factor,RCCA, AP1000, SRAC2006

1. Introduction

Neutronic parameters of nuclear reactor core design safety analysis can be done through a computer program simulation. Neutronic safety design parameters of the reactor core include criticality, reactivity control rods, power distribution and others. To obtain an accurate calculation, then the computer program used must be validated. The data result of measurement through experimentation and data calculation results in the case of the benchmark can be used to validate computer codes. Validation study for several computer codes ever conducted, among others;

- Validation of the Monte Carlo code RMC with C5G7 benchmark [1],
- Validation of the Serpent-ARES codes sequence using the MIT-BEAVERS benchmark [2],

- Benchmarking of the WIMS9/PARCS/TRACE code system for neutronic calculations of the Westinghouse AP1000TM reactor [3],
- Development of a MCNP-ORIGEN burn-up calculation code system and its accuracy [4],
- Comparison of nuclear data uncertainty propagation methodologies for PWR burn-up simulations [5],
- Validation of full core geometry model of the nodal3 code in the PWR transient benchmark problems [6].

Once validated, in the next step, the computer programs can be used as a tool for analyzing the safety against an existing reactor core design or to make a new core design. Several studies of nuclear reactor core analysis using a computer codes include:

- Analysis of the three-dimensional VENUS-2 MOX core benchmark using the Monte Carlo code TRIPOLI-4 [7]
- PWR experimental benchmark analysis using WIMSD and PRIDE codes [8]
- Qualification of the Taiwan Power Company's pressurized water reactor physics methods using CASMO-4/SIMULATE-3 [9].

In this research, performed two-steps of study. The first step is to perform validation of the computer code SRAC2006 module of CITATION[10] against the value of radial power factor distribution. The second step is to analyze the changes in fuel assembly power peaking factor due to RCCA insertion. The computer code has been validated against the criticality of the core in the Virtual Environment Reactor Analysis (VERA) benchmark with goodresults[11]. As an object, AP1000 reactor core at the first operating cycle was used in this calculation. The AP1000 nuclear reactor is a pressurized water reactor (PWR) generation III+ type[12]. The nominal power that can be generated is 1000 MWe. As moderator and coolant are light water (H_2O). In the AP1000, the H_2O pressurized about 2250 psia or 153 atm to prevent boiling at the high temperatures ($T \approx 325\ ^\circ C$). The reactor core is composed of 157 fuel assemblies. The fuel assembly provided with 24 guide tubes used for Pyrex or Rod Cluster Control Assembly (RCCA) insertion. The inserted RCCA position is among the 256 fuel rods in the fuel assembly. The insertion of RCCA in the core causing material changescontained in the fuel assembly These material changes will affect to the distribution of radial power factor generated.

Material changes before and after RCCA insertion (H_2O into Ag-In-Cd/SS304)will also influence to the power peaking factor. Changes in maximum power factor need to be known so that the reactor core can operate safely ascertained by meeting the safety criteria design that have been set.The of power factor distribution calculation was carried out on 2-dimensional geometry ¼ part model of the core using of CITATION module. As input to the calculation module, in addition to the data size and geometry of core, fuel assembly macroscopic cross-sectional data is also needed. The cross-section data obtained through the calculation using PIJ module. As the output from the core calculation, it will obtain the radial thermal power at the each of fuel assemblies. Power factor calculation was done on several core conditions. The first is the AP1000 core without RCCA. The second is the AP1000 core with single RCCA insertion (AO, M1, M2, MA, MB, MC, and MD). And, the thirdly is the AP1000 with multiple insertion condition of RCCA (MA + MB, MA + MB + MC, MA + MB + MC + MD and MA + MB + MC + MD + M1). As a validation, the calculation results of the core without RCCA conditions and with the bank of multiple RCCA (MA + MB) were compared with an existing data. From the analysis of the distribution of the power factor, it is expected to know the level of accuracy of the SRAC2006 computer program and the safety level of AP1000 core on the first operating cycle power factor value in terms of maximum power generated by fuel assembly.

2. Methodology

In this research, the radial power distribution calculating the fuel assemblies in the first operating cycle AP1000 core. The calculation was done using the computer program SRAC2006 CITATION module. The calculation stepsare the preparation of fuel assembly data model, macroscopic cross-section calculation, reactor core modelling,and core calculation.

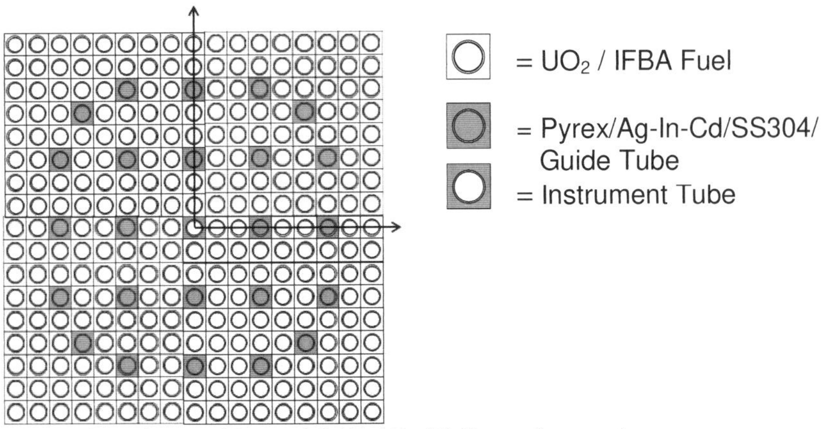

Figure 1.Fuel assembly model 17×17 of 2 dimension section

2.1. Calculation of the macroscopic cross-section of the fuel assembly

Calculation of fuel assembly macroscopic cross section is done using the SRAC2006 code moduleof PIJ. The calculation is done by condensing the energy groups of 107 into 8 groups. The cross-section library data used is SRACLIB-EDF70. The calculations were done by modeling ¼ part of the fuel assembly. As a result of the calculation is the tables of macroscopic cross- section data in the 8 energy groups. As input calculation is the atoms density of the materials, temperature, dimensional and the size fuel assemblies, mesh division and others. A 2-dimensional model of AP1000 fuel assemblies shown in Figure 1. The size of the fuel assemblies is 21.504 cm × 21.504 cm. The fuel assembly shaped 17 × 17 grid cell consists of 264 fuel lattice cells, 24guide tube, one instrument tube and light water (H_2O) as a moderator. In the guide tube will be filled with H_2O, Pyrex or absorber Ag-In-Cd / SS304, depending on operating conditions of core. As shown in Figure 2, fuel lattice cells in the AP1000 reactor core has a square shape with a size of 1.260 cm x 1.260 cm. Standardfuel lattice cell composed from UO_2 pellets (enrichment of 2.350%, 3.400% or 4.450%) with a radius of 0.410 cm, gab 0.426 cm, cladding ZIRLO 0.483 cm and the outer part is the moderator H_2O. While in Integrated Fuel Burnable Absorber (IFBA), then there ZrB_2 thin layer on the outside of UO_2 pellets. ZIRLO cladding composition is Zr of 97.850%, Fe of 0.150%, Sn of 1.000%, and Nb of 1.000%. Pyrex lattice cell composed from the inner tube, the absorbent material that is borosilicate glass (B_2O_3-SiO_2), the outer tube, and the guide tube. Control rod of lattice cells consists the absorbent material Ag-In-Cd / SS304, cladding, and guide tube. Material composition and size of the control rods are shown in Table 1. At the time of insertion RCCA, then guide tube cell lattice would have 24 control rods Ag-In-Cd (black absorber), or 12 AG-In-Cd + 12 SS304 (gray absorber).

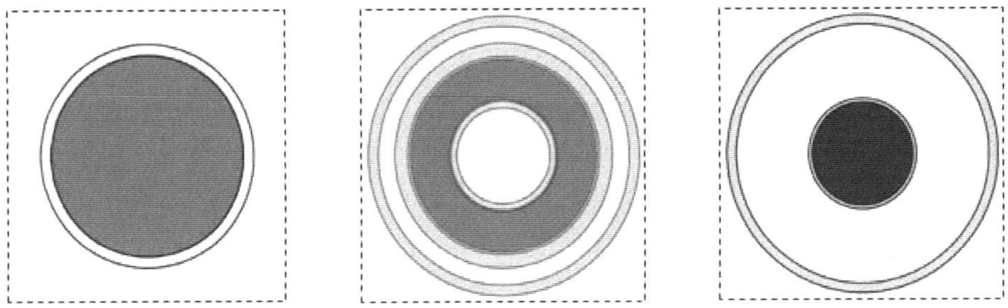

a. Fuel b. Pyrex c. Control Rod

Figure 2. Lattice cell model of fuel, pyrex dan control rod

IOP Conf. Series: Journal of Physics: Conf. Series **962** (2018) 012031 doi:10.1088/1742-6596/962/1/012031

Table 1. Material compositionandsize of control rod

	Pyrex	Black Absorber	Gray Absorber
Material Composition (Number)	B_2O_3-SiO_2	AgInCd 80%15%5% (24)	AgInCd (12) SS304 (12)
Poison Density (g/cm^3)		10.158	10.158 7.889
Inner tube (cm)	0.461 (OD)	-	-
Inner tube material	SS304	-	-
Poison Radius (cm)	0.461 (ID) / 0.968 (OD)	0.433	0.4064
Cladding material (cm)		SS304	
Clad thicknes (cm)		0.047	
Guides Tube inner diameter (cm)		1.008	
Guides Tube Outer Diameter (cm)		1.224	

2.2. The Thermal Power Factor Calculation

Figure 3 shows a model of the ¼ part on the first AP1000 operating cycle with the distribution of fuel assemblies and control rod cluster. The first line shows the type of fuel that the fuel assemblies with enrichment UO_2 2.350%, 3.400% or 4.450%. The second line is the amount of fuel IFBA located on each of the fuel assemblies. The third line shows the RCCA type or number of Pyrex contained in fuel assemblies.

In the AP1000, control rod bank classification can be divided into 11 groups:

1. MA (Mechanical Shim Gray Bank A): 4 Clusters
2. MB (Mechanical Shim Gray Bank B): 4 Clusters
3. MC (Mechanical Shim Gray Bank C): 4 Clusters
4. MD (Mechanical Shim Gray Bank D): 4 Clusters
5. M1 (Mechanical Shim Black Bank 1): 4 Clusters
6. M2 (Mechanical Shim Black Bank 2): 8 Clusters
7. AO (Axial Offset Control Bank): 9 Clusters
8. SD1 (Shutdown Bank 1): 8 Clusters
9. SD2 (Shutdown Bank 2): 8 Clusters
10. SD3 (Shutdown Bank 3): 8 Clusters
11. SD4 (Shutdown Bank 4): 8 Clusters

In accordance with the document control rev.19, the AP 1000 core calculation performed with the insertion RCCA condition as follows; MA + MB, MA + MB + MC, MA + MB + MC + MD, and MA + MB + MC + MD + M1. In addition, the core calculation with the conditions of each insertion RCCA includes AO, M1, M2, MA, MB, MC, and MD. While RCCA SD1 to SD4 were not be calculated. Those RCCA have to serve the shutdown of the reactor. At the time of shut-down reactor, the reactor core power is 0.

	H	G	F	E	D	C	B	A
8	AO	3.40 88I 24P	2.35 MD	3.40 88I 24P	2.35 M1	3.40 28I 24P	2.35 MB	4.45 88I 12P
9	3.40 88I 24P	2.35 SD1	3.40 88I 24P	2.35 SD3	3.40 88I 24P	2.35 SD2	4.45 72I 24P	4.45 88I 9P
10	2.35 MD	3.40 88I 24P	2.35 MA	3.40 88I 24P	2.35 AO	3.40 28I 24P	4.45 112I SD4	
11	3.40 88I 24P	2.35 SD3	3.40 88I 24P	2.35 SD1	3.40 44I 24P	4.45 112I M2	4.45 112I	
12	2.35 M1	3.40 88I 24P	2.35 AO	3.40 44I 24P	2.35 28I MC	4.45 112I		
13	3.40 28I 23P	2.35 SD2	3.40 28I 24P	4.45 112I M2	4.45 112I			
14	2.35 MB	4.45 72I 24P	4.45 112I SD4	4.45 112I				
15	4.45 88I 12P	4.45 88I 9P						w/o IFBA Pyrex/RCCA

Figure 3. RCCA distribution at the ¼part model of the AP1000core

Calculation of fuel power distribution of fuel assembly in the first operating cycle of AP1000 was carried out on the two-dimensional geometry of ¼ part core model. The core model, in addition to the fuel assembly, also consist of another material. The material from the inside to the outside of core are the baffle reactor, cooling made form light water H_2O, barrel reactor, and the reactor vessel. Baffle reactor is outside the blanket that surrounds the entire fuel assembly with the SS304 material. Core barrel reactor is a tube bulkhead separation between the coolant flow and the inner reactor with size d_{in} / d_{out} 339.725 cm / 349.885 cm, made from SS304 material. And the reactor vessel is a container from all parts of the reactor core material with a size d_{in}/d_{out} 398.800 cm / 420.100 cm SS304 material.

3. Results and Discussion

Figure 4 shows a comparison of radial power factor distribution at the first cycle of AP1000 core without RCCA condition. The first line shows the type of fuel. Number 1, 2 and 3 are the UO_2 fuel enriched about 2.350%, 3.400% and 4.450%, respectively. While the letter P and I are the numbers of Pyrex and IFBA. The second line is the value of the power factor reference data are the result of calculations with computer program ANC (A Westinghouse Advanced Nodal Computer Code.) The third line is the value of the distribution of power factor calculation results with computer program SRAC2006 CITATION module. The fourth line is the value of the difference between the reference data with the value of the calculation result. While row five is percent difference value.

	H	G	F	E	D	C	B	A
8	1.232 1.279 -0.047 -3.815	24P88I 1.150 1.154 -0.004 -0.348	1.225 1.268 -0.043 -3.510	24P88I 1.138 1.137 0.001 0.088	1.215 1.254 -0.039 -3.210	24P88I 1.151 1.161 -0.010 -0.869	0.941 0.957 -0.016 -1.700	12P88I 0.572 0.541 0.031 5.420
9	24P88I 1.151 1.154 -0.003 -0.261	1.229 1.273 -0.044 -3.580	24P88I 1.142 1.142 0.000 0.000	1.211 1.250 -0.039 -3.220	24P88I 1.119 1.113 0.006 0.536	1.138 1.168 -0.030 -2.636	24P72I 0.927 0.913 0.014 1.510	9P88I 0.454 0.436 0.018 3.965
10	1.225 1.268 -0.043 -3.510	24P88I 1.143 1.142 0.001 0.087	1.212 1.250 -0.038 -3.135	24P88I 1.119 1.111 0.008 0.715	1.178 1.203 -0.025 -2.122	24P88I 1.028 1.026 0.002 0.195	112I 0.829 0.815 0.014 1.689	
11	24P88I 1.138 1.137 0.001 0.088	1.212 1.250 -0.038 -3.135	24P88I 1.119 1.111 0.008 0.715	1.169 1.193 -0.024 -2.053	24P88I 1.047 1.033 0.014 1.337	112I 1.081 1.041 0.040 3.700	112I 0.584 0.561 0.023 3.938	
12	1.216 1.254 -0.038 -3.125	24P88I 1.120 1.113 0.007 0.625	1.179 1.203 -0.024 -2.036	24P88I 1.047 1.033 0.014 1.337	0.861 0.859 0.002 0.232	112I 0.671 0.630 0.041 6.110		
13	24P88I 1.153 1.161 -0.008 -0.694	1.140 1.168 -0.028 -2.456	24P88I 1.030 1.026 0.004 0.388	112I 1.082 1.041 0.041 3.789	112I 0.672 0.630 0.042 6.250			
14	0.944 0.957 -0.013 -1.377	24P72I 0.930 0.913 0.017 1.828	112I 0.831 0.815 0.016 1.925	112I 0.585 0.561 0.024 4.103				
15	12P88I 0.576 0.541 0.035 6.076	9P88I 0.457 0.436 0.021 4.595						

Number of Pyrex&IFBA
Calc. Result (1)
Reff. Data (2)
Difference(1-2)
Diff. (%)

Figure 4. Radial power factor distribution at the first cycle of AP1000 core without RCCA

The fuel assembly radial power peaking factor value from the reference data is 1.232, namely fuel with enrichment 2.350% at H-8 position. In the same position, the calculation results show the value of 1.279 or the difference in -0.047 (-38.150%). It is due to the H-8 position in the center of the core. In addition,although the fuel assemblies having a low enrichment (2.350%), but there are no Pyrex and IFBA absorbent material inside. Fuel assembly at the center of core, in addition to the abssence of of strong absorbentmaterials, the largest thermal flux neutrons are also produced. Most of the thermal neutrons absorbed by the fuel will react with U-235 fission. So the fuel assembly in the position of H-8 produces the greatest thermal power.

The highest power factor value is still below the design limit. Design limitations specified value is hot channel factor of 2.62. The value is the highest power of the fuel rods divided by average power value of all fuel rod in the cores. Based on reference data, then at the beginning of the cycle indicates the value of the fuel rod radial power peaking factor in the fuel assembly is approximately about 1.406. So that when multiplied by the highest radial power factor produces by fuel assembly become a value of 1.798. That is still far below the maximum limit value of 2.600. On AP1000 core with a power of 3400 MW, the average linear power is 18.760 kW / m. While the peak value of the fuel rod linear power is 49.260 kW/m. So that the power factor of the fuel rods is about 2.60.At the near beginning of life, unrodded, hotfullpower, and no xenon condition of core, then the F-DELTA-H value is1.406.F-DELTA-H is defined as the ratio between the thermal power integral fuel rods divided by the average power of all fuel rod.

Lowest power factor value is equal to 0.454 at position A-9. This occurs because although the UO_2 fuel assemblies with enrichment 4.450%, but is at a position on the edge of the IFBA number in which there were 112 among 264 pellets. The highest difference between the value of the reference data and the calculation results amounted to 6.250% (grid position D-13) is a device UO_2 fuel enrichment 4.45% with 112 IFBA. So it can be said that the SRAC2006 code CITATION module valid for the calculation of the radial power factor of AP1000.

	H	G	F	E	D	C	B	A
	1.336	1.187	1.302	1.166	1.292	1.127	0.725	0.469
	1.252	1.158	1.224	1.152	1.237	1.110	0.720	0.491
	0.084	0.029	0.078	0.014	0.055	0.017	0.005	-0.022
8	6.287	2.443	5.991	1.201	4.257	1.508	0.690	-4.691
	1.187	1.294	1.120	1.270	1.142	1.176	0.854	0.399
	1.158	1.213	1.092	1.207	1.145	1.140	0.871	0.415
	0.029	0.081	0.028	0.063	-0.003	0.036	-0.017	-0.016
9	2.443	6.260	2.500	4.961	-0.263	3.061	-1.991	-4.010
	1.302	1.120	0.994	1.094	1.250	1.052	0.821	
	1.225	1.092	0.960	1.089	1.225	1.077	0.856	
	0.077	0.028	0.034	0.005	0.025	-0.025	-0.035	
10	5.914	2.500	3.421	0.457	2.000	-2.376	-4.264	
	1.166	1.270	1.094	1.224	1.078	1.088	0.587	
	1.152	1.208	1.089	1.195	1.114	1.168	0.628	
	0.014	0.062	0.005	0.029	-0.036	-0.080	-0.041	
11	1.201	4.882	0.457	2.369	-3.340	-7.353	-6.985	
	1.292	1.142	1.250	1.078	0.926	0.676		
	1.238	1.146	1.226	1.114	0.936	0.736		
	0.054	-0.004	0.024	-0.036	-0.010	-0.060		
12	4,180	-0.350	1.920	-3.340	-1.080	-8.876		
	1.127	1.176	1.052	1.088	0.676			
	1.111	1.141	1.078	1.168	0.736			
	0.016	0.035	-0.026	-0.080	-0.060			
13	1.420	2.976	-2.472	-7.353	-8.876			
	0.725	0.854	0.821	0.587				
	0.722	0.873	0.857	0.628				
	0.003	-0.019	-0.036	-0.041				
14	0.414	-2.248	-4.385	-6.985				
	0.469	0.399					Reff. Data (1)	
	0.495	0.417					Calc. Result (2)	
	-0.026	-0.018					Difference(1-2)	
15	-5.544	-4.511					Diff. (%)	

Figure 5. Radial power factor distribution at the first cycle of AP1000 core with multiple (MA+MB) RCCA inserted

Radial power factor distribution at the first cycle of AP1000 core with the insertion of multiple RCCA (MA + MB) showed in Figure 5. Radial power factor value of the fuel assembly reference data is shown on the first line, the result of the calculation using CITATION module of SRAC2006 on the second line. Differences are both shown in the 3rd row, percent of differences shown in the 4th row. The lowest differences power factor value is 0.003 (H-13) with a difference ratio of 0.414%. At the position of H-13 and B-8 is a 3.400% enrichment UO_2 fuel assemblies with 24 Pyrex 88 IFBA, and their insertion of RCCA is bank MB. A number of absorbent materials that cause a little difference between reference data and calculation results. At the RCCA bank, MA insertion position F-10 that there is a difference greater than the position RCCA bank MB. In addition to differences in the position of MA more inner of the core, also due to the absence of absorbent material such as Pyrex and/or IFBA on the position. The biggest difference between the reference data and the calculation results shown by the fuel assemblies in the position of H-8 is about 0.084. Position H-8 is located right in the mid of core, and an enrichment UO_2 fuel assembly is 2.350% without Pyrex and IFBA. This causes the thermal neutron flux is higher towards the middle of the core. While the ratio of the largest difference occurs at position C-12 and D-13 that is about -8.876%. So that the power factor calculation AP1000 core with multiple RCCA insertions conditions. Calculation using CITATION module of the SRAC2006 computer code can also be said to be valid.

	H	G	F	E	D	C	B	A
8	1.232 0.659/AO 1.413 1.687	1.150 1.116 1.294 1.572	1.225 1.389 1.285 1.605	1.138 1.282 0.964 1.396	1.215 1.362 0.539/M1 1.389	1.151 1.368 0.899 1.246	0.941 1.216 0.880 0.997	0.572 0.780 0.563 0.605
9	1.151 1.116 1.294 1.572	1.229 1.311 1.368 1.653	1.142 1.272 1.212 1.465	1.211 1.267 1.147 1.446	1.119 1.071 0.924 1.225	1.138 1.244 1.017 1.168	0.927 1.146 0.896 0.937	0.454 0.605 0.452 0.467
10	1.225 1.390 1.286 1.605	1.143 1.273 1.212 1.465	1.212 1.275 1.299 1.466	1.119 0.978 1.177 1.226	1.178 0.571/AO 1.212 1.121	1.028 0.926 1.059 0.848	0.829 0.926 0.865 0.691	
11	1.138 1.284 0.964 1.397	1.212 1.268 1.148 1.446	1.119 0.979 1.177 1.227	1.169 0.975 1.287 1.125	1.047 0.828 1.165 0.809	1.081 0.997 1.206 0.463/M2	0.584 0.610 0.647 0.348	
12	1.216 1.365 0.539/M1 1.391	1.120 1.074 0.925 1.226	1.179 0.572/AO 1.213 1.122	1.047 0.829 1.166 0.810	0.861 0.788 0.990 0.578	0.671 0.648 0.775 0.369		
13	1.153 1.373 0.901 1.249	1.140 1.248 1.019 1.171	1.030 0.928 1.060 0.850	1.082 0.999 1.207 0.464/M2	0.672 0.649 0.776 0.369			
14	0.944 1.223 0.883 1.001	0.930 1.152 0.898 0.940	0.831 0.930 0.867 0.693	0.585 0.611 0.648 0.349				
15	0.576 0.787 0.567 0.610	0.457 0.611 0.455 0.471						

Without RCCA
AO Inserted
M1 Inserted
M2 Inserted

Figure 6. Radial power factor distribution at the first cycle of AP1000 core with single AO, M1, M2 RCCA inserted

Radial power factor distribution at the first cycle of the AP1000 core with the insertion of single RCCA ((AO, M1, M2) and (MA, MB, MC, MD)) showed at the Figure 6 and 7, respectively. In Figure 6, line 1, 2, 3 and 4 row are AP1000 power factor in the current conditions without RCCA, with the insertion of RCCA AO, M1, and M2, respectively. In Figure 7, line 1, 2, 3 and 4 are AP1000 power factor in the current conditions without RCCA, with insertion RCCA of MA, MB, MC, and MD, respectively.

As it is known that the AO, M1, and M2 is a black absorber with 24 control rods are Ag-In-Cd. While the MA, MB, MC and MD are gray absorber with 12 control rods SS304 and 12 Ag-In-Cd. Of the two pictures can be seen changes in the value of the power factor of the fuel assemblies in each of the control rod positions before and after insertion. At the time of insertion AO at position H-8, D-10, and F-12 changes the value of the power factor respectively amounted to 46.510%, 51.484%, 51.528% and. At the time of insertion M1 (D-8 and H-12) changes the value of the power factor in a row and amounted to 55.638%, 55.674%. At the time of insertion M2 (C-11 and E-13) changes the value of the power factor in a row and amounted to 57.169%, 57.116%. Although the material is inserted at (Ag-In-Cd), but at each position will produce different power factor. The analysis showed that more toward the outside of the core, then change the value of the power factor will be slightly larger. It is caused by the influence of the magnitude of the thermal neutron flux at each of these positions.

From Figure 6b, it is known that at the time of insertion MA of F-10 changes the value of the power factor of 25.743%. At the time of insertion MB at position B-8 and H-14 changes the value of the power factor in a row and amounted to 29.224%, 29.237%. At the time of insertion, MC position D-12 changes the value of the power factor of 25.668%. At the time of insertion of the MD position F-8 and -10 changes the value of the power factor in a row and amounted to 30.939%, 30.939%. Judging from the position RCCA inside the core, then from the inside to the outside is MD, MB, MA, and MC. When viewed from the power factor changes, then the MD and MB greater than the MA and MC. As it is known that RCCA MA, MB, MC, and MD are gray absorbers which are 12 Ag-In-Cd and 12 SS304. So apart from a neutron absorbing material, effect SS304 to reducing the amount of moderator in the fuel assemblies. So that the neutron moderated from the rapid energy to thermal energy will also be reduced. As a result of fuel assemblies in the position of insertion RCCA MA and MC which only amounted to one RCCA will be slightly smaller than the position RCCA MD and MB.

From Figure 6.a and 6.b it can be seen that the power factor changes on the fuel assemblies in the position of insertion RCCA AO, M1 and M2 showed a larger value then MA, MB, MC, and MD. As mentioned before that RCCA AO, M1, and M2 named with black absorbers that the material consists of 24 rods of Ag-In-Cd. While RCCA MA, MB, MC, and MD are named with gray absorbers, that are consist of 12 rods of Ag-In-Cd and 12 rods of SS304.

	H	G	F	E	D	C	B	A
8	1.232 1.141 1.342 1.314 0.987	1.150 1.060 1.247 1.222 0.904	1.225 1.137 1.312 1.290 *0.846 / MD*	1.138 1.100 1.188 1.185 1.023	1.215 1.239 1.213 1.256 1.223	1.151 1.218 1.050 1.185 1.220	0.941 1.018 *0.666 / MB* 0.969 1.023	0.572 0.625 0.449 0.590 0.628
9	1.151 1.060 1.247 1.222 0.904	1.229 1.114 1.328 1.300 0.998	1.142 1.016 1.220 1.196 0.968	1.211 1.151 1.267 1.253 1.136	1.119 1.135 1.127 1.145 1.137	1.138 1.203 1.078 1.162 1.210	0.927 1.000 0.807 0.948 1.008	0.454 0.495 0.379 0.466 0.498
10	1.225 1.137 1.312 1.291 *0.846 / MD*	1.143 1.017 1.220 1.196 0.969	1.212 *0.900/MA* 1.287 1.250 1.124	1.119 1.036 1.171 1.129 1.104	1.178 1.193 1.205 1.170 1.223	1.028 1.083 1.019 1.020 1.101	0.829 0.890 0.794 0.828 0.902	
11	1.138 1.101 1.188 1.186 1.024	1.212 1.152 1.267 1.253 1.137	1.119 1.037 1.171 1.129 1.105	1.169 1.138 1.220 1.130 1.202	1.047 1.071 1.082 0.967 1.108	1.081 1.140 1.101 1.015 1.168	0.584 0.624 0.584 0.563 0.637	
12	1.216 1.240 1.214 1.257 1.225	1.120 1.137 1.127 1.146 1.138	1.179 1.194 1.205 1.171 1.224	1.047 1.071 1.082 0.967 1.109	0.861 0.897 0.892 *0.640/ MC* 0.926	0.671 0.709 0.692 0.577 0.728		
13	1.153 1.221 1.051 1.188 1.223	1.140 1.205 1.079 1.164 1.213	1.030 1.085 1.020 1.022 1.103	1.082 1.141 1.101 1.016 1.169	0.672 0.709 0.692 0.577 0.729			
14	0.944 1.022 *0.668 /MB* 0.973 1.027	0.93 1.003 0.808 0.951 1.011	0.831 0.892 0.795 0.830 0.904	0.585 0.625 0.594 0.564 0.638				
15	0.576 0.630 0.452 0.594 0.633	0.457 0.499 0.382 0.469 0.502						

Without RCCA
MA Inserted
MB Inserted
MC Inserted
MD Inserted

Figure 7. Radial power factor distribution at the first cycle of AP1000 core with single RCCA (MA, MB, MC, MD) inserted

Radial power factor distribution at the first cycle of AP1000 core with the insertion of the multiple RCCA showed in Figure 8. Row 1, 2, 3, 4, 5 and 6 show power factor value in the first operating cycle of AP1000 without RCCA insertion RCCA banks MA, MA + MB, MA + MB + MC, MA + MB + MC + MD and MA + MB + MC + MD + M1, respectively. Power peaking factor value on each condition without RCCA is 1.232 (position H-8), with insertion of MA RCCA is 1.240 (position H-12 symmetrical with the D-8), with insertion of MA + MB is 1.252 (position H-8), with insertion of MA + MB + MC is 1.347 (the position of the D-8), with insertion of MA + MB + MC + MD is 1.274 (the position of the G-8, symmetrical with C-9) and with insertion of MA + MB + MC + MD + M1 is 1.323 (position E-11). The biggest difference is at the insertion of MA + MB + MC

RCCA from 1.232 (condition without RCCA) to 1.347 or an increase of 9.334%. It is caused by the insertion position RCCA MC is on the D-12 which is located alongside a relative inactive of the core. So that the thermal neutron flux will be enlarged in the center of the center of core position. At each addition RCCA an increase in maximum power factor value than before insertion RCCA. The maximum value of the power factor is still meet the safety, well below the thermal design power factor of the fuel assembly (approximately 1.8).

	H	G	F	E	D	C	B	A
8	1.232 1.141 1.252 1.347 1.063 1.195	1.150 1.060 1.158 1.241 0.961 1.058	1.225 1.137 1.224 1.299 *0.887/MD* *0.898*	1.138 1.100 1.152 1.207 1.082 0.869	1.215 1.239 1.237 1.284 1.293 *0.564/M1*	1.151 1.218 1.110 1.146 1.216 0.956	0.941 1.018 *0.720/MB* *0.743* *0.810* *0.772*	0.572 0.625 0.491 0.508 0.561 0.568
9	1.151 1.060 1.158 1.241 0.961 1.058	1.229 1.114 1.213 1.294 1.035 1.127	1.142 1.016 1.092 1.152 0.964 0.995	1.211 1.151 1.207 1.254 1.175 1.081	1.119 1.135 1.145 1.176 1.198 0.987	1.138 1.203 1.140 1.166 1.244 1.131	0.927 1.000 0.871 0.892 0.974 0.964	0.454 0.495 0.415 0.427 0.471 0.484
10	1.225 1.137 1.225 1.300 *0.887/MD* *0.898*	1.143 1.017 1.092 1.152 0.965 0.995	1.212 *0.900/MA* *0.960* *0.995* *0.918* *0.968*	1.119 1.036 1.089 1.100 1.091 1.142	1.178 1.193 1.225 1.217 1.272 1.320	1.028 1.083 1.077 1.069 1.153 1.210	0.829 0.890 0.856 0.855 0.937 1.003	
11	1.138 1.101 1.152 1.207 1.083 0.869	1.212 1.152 1.208 1.255 1.176 1.081	1.119 1.037 1.089 1.100 1.092 1.142	1.169 1.138 1.195 1.152 1.193 1.323	1.047 1.071 1.114 1.026 1.095 1.233	1.081 1.140 1.168 1.095 1.193 1.356	0.584 0.624 0.628 0.605 0.666 0.755	
12	1.216 1.240 1.238 1.285 1.295 *0.565/M1*	1.120 1.137 1.146 1.177 1.199 0.988	1.179 1.194 1.226 1.218 1.273 1.321	1.047 1.071 1.114 1.027 1.095 1.234	0.861 0.897 0.936 *0.693/MC* *0.751* *0.876*	0.671 0.709 0.736 0.631 0.692 0.814		
13	1.153 1.221 1.111 1.148 1.218 0.958	1.140 1.205 1.141 1.167 1.274 1.133	1.030 1.085 1.078 1.070 1.154 1.212	1.082 1.141 1.168 1.096 1.194 1.357	0.672 0.709 0.736 0.632 0.692 0.814			
14	0.944 1.022 *0.722/MB* *0.745* *0.813* *0.775*	0.930 1.003 0.873 0.894 0.977 0.967	0.831 0.892 0.857 0.857 0.939 1.005	0.585 0.625 0.628 0.606 0.667 0.756				
15	0.576 0.630 0.495 0.512 0.565 0.573	0.457 0.499 0.417 0.429 0.475 0.487						

Without RCCA
MA Inserted
MA+MB Inserted
MA+MB+MC Inserted
MA+MB+MC+MD Inserted
MA+MB+MC+MD+M1 Inserted

Figure 8. Radial power factor distribution at the first cycle of AP1000 core with multiple RCCA inserted

4. Conclusion

From the calculation result analysis, it can be concluded that the SRAC2006 code CITATION module valid for the calculation of the radial power factor of AP1000. Insertion of RCCA caused increasing power peaking factor of the fuel assembly. The value of the fuel rod radial power peaking factor in the fuel assembly is assumed approximately about 1.406.Then the value of fuel assembly power peaking factor in the core still has below the design criteria, 2.600. In the case of power thermal produced by fuel assembly, therefore the AP1000 core can be operated safely.

Acknowledgement
This research has been completed under financially support through government budget of DIPA2016-PTKRN-BATAN, and authors would like to thank Head of PTKRN for the facility and senior researchers for support and advice.

References
[1] GaoB, MaXB, ChenYX, YuH2013 Validation the Monte Carlo code RMC with C5G7 benchmark *Nuclear Engineering and Design* **265** 64
[2] Leppänen J, Mattila R, Pusa M 2014 Validation of the Serpent-ARES codes sequence using the MIT-BEAVERS benchmark – initial core at HZP conditions*Annals of Nuclear Energy* **69** 212
[3] ElsawiMA, HraizAS B 2015 Benchmarking of the WIMS9/PARCS/TRACE Code System for Neutronic Calculations of The Westinghouse AP1000TM Reactor *Nuclear Engineering and Design* **293** 249
[4] Zheng M, Tian W, Wei H, Zhang D, Wu Y, Qiu S, Su G2014 Development of a MCNP–ORIGEN burn-up calculation code system and its accuracy assessment*Annals of Nuclear Energy* **63** 491
[5] DiezCJ, Buss O, Hoefer A, Porsch D, Cabellos O2015 Comparison of nuclear data uncertainty propagation methodologies for PWR burn-up simulations*Annals of Nuclear Energy* **77** 101
[6] Sembiring T M, Pinem S, Liem P H2015 Validation of full core geometry model of the nodal3 code in the PWR transient benchmark problems *Journalof Nuclear Reactor Technology TRI DASA MEGA* **17** 141
[7] Savva P, Varvayanni M, Catsaros N2014 Analysis of the three-dimensional VENUS-2 MOX core benchmark using the Monte Carlo code TRIPOLI-4 and the ENDF/B-VI.4, ENDF/B-VII.0 and JEFF-3.1 nuclear data sets *Nuclear Engineering and Design* **273** 215
[8] Arshad F, Ahmad S, Haq I2014PWR experimental benchmark analysis using WIMSD and PRIDE Codes *Annals of Nuclear Energy* **72** 11
[9] Lin H C, Yaur S J, Lin T Y, Kuo W S, Shine J Y, Huang Y L2012 Qualification of the Taiwan Power Company's Pressurized Water Reactor Physics Methods Using CASMO-4/SIMULATE-3 *Nuclear Engineering And Design* **253** 71
[10] Keisuke O, Teruhiko K, Kunio K and Keichiro T 2007 SRAC2006: A Comprehensive Neutronics Calculation Code System JAEA-Data/Code-2007-004 Japan Atomic Energy Agency
[11] Susilo J 2014 Verification of MVP-II and SRAC06 Code for the Case VERA Benchmark Reactor Core *Journal of Nuclear Reactor Technology TRI DASA MEGA* **16**75
[12] Anonim AP1000 Design Control Document Revision 19 Chapter 4.3 Nuclear Design Westinghouse Electric, USA Available from website: https://www.nrc.gov/docs/ML1117/ML11171A445.pdf Accessed: February 20, 2017.

Preliminary study for alternative conceptual core design of the MTR research reactor

T. Surbakti, S. Pinem., T.M. Sembiring, M. Subekti, G.R. Sunaryo.

Center for Nuclear Reactor Technology and Safety, BATAN,
Kawasan PUSPIPTEK Gd. No. 80 Serpong, Tangerang Selatan, 15310 Indonesia

Email: tukiran@batan.go.id

Abstract. The utilization of the research reactor is increasingly widespread, especially for radioisotope production and testing of advanced materials and preference to use a compact core. The reactor core design has been determined on the maximum thermal flux in the middle of the core per MW. BATAN has designed several alternative research reactor cores. The purpose of this research reactor is to obtain the optimum reactor core configurations with the criteria to have a thermal neutron flux in the centre of the core with minimum of 1.0×10^{15} n/cm^2 s. Power level of the research reactor is 60 MWthwith U9Mo/Al fuel 85 cm of height. Design of plate-type fuels with a higher core results in the heat transfer to the coolant optimal. All 16 fuel assemblies and the 4 control rodsare inserted into the core for this reactor. The core design calculations were carried out with the WIMSD-5B and BATAN-FUEL codes. Conceptual design calculation results show that the core configuration with 5 × 5 grids, all the fresh fuel, fuel loading of 470 g, a D_2O reflector, a maximum thermal neutron flux in the central core is 1.09×10^{15} n/ cm^{2}s and the cycle length is 33 days. The reactor core design is the most optimal for MTR type. For the equilibrium core, a fuel loading of 600 g results in the maximum thermal flux of 1.07×10^{15} n/cm^{2}s and the two safety rods should be used in the core.

Keywords: conceptual design, U9Mo/Al fuel, research reactor, WIMSD-5B code , BATAN-FUEL code.

1. Introduction

Batan has 3 old research reactors, two reactors are TRIGA type and one is RSG-GAS, MTR (Material Testing Reactor) type. TRIGA type reactoris also no longer produced its fuel meanwhileRSG-GAS research reactor uses silicide fuel. In the future, the silicide fuel will not be used anymore due to very limited density that can be loaded in an assembly[1]. The world's research reactor will switch to new fuels that are manufactured using high-density molybdenum uranium fuels higher than it's silicides.
RSG-GAS also uses a wide core and future reactors tend to use compact core for the purpose of obtaining high flux neutrons to irridiate radio isoptopes taking a shorter time. To anticipate the above problem, it is necessary to do a new research reactor core design, where the fuel, core configuration and facilities adjustthe needs of the future. To build a new research reactor, there are several steps that

IOP Conf. Series: Journal of Physics: Conf. Series **962** (2018) 012032 doi:10.1088/1742-6596/962/1/012032

need to be done, such as, study the functions, objectives and benefits of the reactor, as well as the conceptual design, basic design and detailed design.

An early conceptual design of Material Testing Reactor (MTR) type from neutronics aspect has been derived by T. Surbakti., et al.[2] in which the main characteristics of core configuration as specified as follows: The RRI research reactor has a nominal power of 20 MW, uses a uranium-molybdenum alloy, U-9Mo/Al fuel with geometry adopting from RSG-GAS' fuel. The core configuration of 5 x 5 latticewhich is made up of 20 fuel elements and 5 irradiation positions is able to produce thermal neutron flux in the order of 2.87x 10^{14} n/cm^{2}s. This neutron flux value is however lower than one stated as acceptance criteria of MTR type reactor in which the reactor should have a maximum thermal neutron flux in irradiation positions and that in the reflector region at least in the order of 1.0x10^{15} and 0.5x10^{15} n/cm^{2}s, respectively.

The second conceptual design was proposed by I. Kuntoro. et al.[3] where the grid core is the same but the height of the fuel and reactor power are different. To fulfil the acceptance criteria, the core was designed with the fuel element with 70 cm high, the power level of 50 MW and the others size are the same with the lattices of reactors previously designed. The core configuration is specified by the number and positions of fuel elements and irradiation positions. After finding the optimum core configuration met the acceptance criteria that the value of neutron flux in the centre of the core is not less than 1.0x10^{15} n/cm^2 s, it is however needed to analyzethe alternative design of RRI core.

The alternative design of RRI core for this research is using MTR fuel with mass loading variations and the purpose of this research is to obtain the optimum reactor core configurations having a thermal neutron flux in the centre of the core 1,0x10^{15} n/cm^{2}s at 60 MWt power with fuel 85 cm high. Design of plate-type fuels with a higher core will facilitate the heat transfer to the coolant optimal. Based on previous research on 5x5 core configuration with the number of fuel 16 and the control rod 4 is the most efficient fuel. The core design calculations were carried out with the WIMSD-5B and BATAN-FUEL. Reactivity and neutron flux calculations have been done using the computer code BATAN-FUEL [4]. Some core models have been performed to calculate those on previous research [5]. For this year, the research will be focused on estimating the characteristic core with a control rod core configurations 5 x 5 and the core 85 cm high. Generation of macroscopic cross section of the fuel and material is used WIMSD-5B code with nuclear data file from ENDF/B-VII.0 [6]. Calculations was done by searching the equilibrium core with a new type of U9Mo/Al fuel. While alternative conceptual design core was obtained using neutron diffusion method program code 2-D, Batan-2DIFF, the generation of cross section of the fuel was carried by the neutron transport code.
The initial stage of the study has been conducted to design research reactors in several countries in the world using high density uranium fuel, such as, JMTR in Japan, CARR in China, MPRR-30 in India, HFR in the Netherlands and OPAL in Australia [7-8]. Those are also used new fuels of U9Mo/Al which is being irradiated. The density of uranium in the U9Mo/Al fuel can reach density up to 16gU/cm^3[9]. From these studies, it has also done a study of the design of the reactor core of MTR research reactor in Indonesia using U9Mo-Al fuel of 360 - 700 g.

This paper presents the results of research on the alternative conceptual design of a new research reactor from neutronic aspects. The conceptual design is of the first step to gain a thorough conceptual design. The purpose of this research is to earn the optimum research reactor core configurations design with the criteria to have a thermal neutron flux of minimum 1.0x10^{15} n/cm^{2}s on a 60 MW power with85 cm of core height, a length of one cycle of operation more than 20 days and use the most efficient fuel. In this conceptual design,the applied safety criteria to control reactivity is if one stuck rod, the reactor remains subcritical even the control rods having the biggest reactivity failed shut down.

2. Fuel Description

Fuel plays an important role in the design of a reactor and based on the UCD (User Criteria Document), RRI reactor uses the same fuel type with RSG-GAS reactor, straight plate type with dimensions of 8.05 cm × 7.6 cm × 60 cm (active high). The reactor core lattice of RRI is 8.1 cm x 7.71

cm. To control the reactivity, control rods in the reactor RRI using fork-type (fork type) is equal to the RSG-GAS reactor with material Ag-In-Cd placed in 2 (two) sides of the outer fuel element control. Number plates of fuel in the standard fuel elements (FE) and control element (CE) each 21 plates and 15 plates, respectively. Core configuration with 5x5 grids can be seen in Figure 1. Table 1 presents data and FE and CE geometry used in the MTR reactor. Figures 2 and 3 respectively show the FE and CE used in the MTR reactor adopted from RSG-GAS reactor.

The fuel used in the MTR reactor is U9Mo-Al with 19.75% enrichment. The reactor can be operated with a long cycle length, the level of loading of uranium (^{235}U) in the design of selected higher than that used in the reactor RSG-GAS for fresh core,namely : 1. 360 g ^{235}U or the meat uranium density of 4.27 g / cm^3; 2. 410 g ^{235}U or the meat uranium density of 4.86 g / cm^3; 3. 470 g ^{235}U or the meat uranium density of 5.57 g / cm^3.For equilibrium core : 1. 500 g ^{235}U or the meat uranium density of 5.93 g / cm^3; 2. 600 g ^{235}U or the meat uranium density of 7.12 g / cm^3; 3. 700 g ^{235}U or the meat uranium density of 8.30 g / cm^3

In connection with the reflector, alternative conceptual reactor core is designed for heavy water (D_2O). For this research, the design of MTR reactor core only uses heavy water reflector. The possible irradiation facilities are provided in the core and to control aspects of reactivity, it is possible using safety rod (control rod safety) the same as the control elements without fuel plates but using dummy alluminum plates. Finally, safety control rods are placed on the core.

3. Methodology

Analysis of neutronics safety aspect for MTR reactor core with high loading is aim to determine the optimum core configuration to meet safety margin and acceptance criteria on the neutronic parameter. In this study, it is also determined the maximum burn up discharged in every end-cycle of operation. From this calculation, it can be determined a maximum irradiation time or cycle length in the core. Safety analysis from neutronic aspect of view is done by several calculationsusing WIMSD-5B, and Batan-FUEL computer codes. The WIMSD-5B code is used to generate cell calculation to achieve cross section constant, such as, neutron diffusion, sigma absorption and sigma fission for U9Mo-Al fuel plates. The U9Mo-Al fuel with mass loading of uranium between 360 – 700 gram was used inthis calculation.The cross section constantsof U9Mo-Al fuel is of a burn up function in 17 steps of burn up. The method used in this calculation is chosenmulti-slabs which is available in WIMSD-5B code. The cross section is formed with geometry of the core as an input to core calculation by BATAN-FUEL code to determined core reactivity change, radial and axial peaking power factor distribution and flux neutron as well. One of the modules in the BATAN-FUEL code is Batan-2DIFF to calculate the core reactivity. Core calculation from neutronics aspects for alternative conceptual design of MTR requires several steps, namely:

3.1. Calculation of the cell.

In this study, the program packages WIMSD-5B was used to generate all the material diffusion constant across the core in 4 (four) groups of neutron energy. The boundaries of the neutron energy is 10 MeV, 0.821 MeV, 5.531 keV, 0.625 eV and 1×10^{-5}eV. Especially for the fuel generation group constants are generated as a function of the loading mass of ^{235}U in the core (360-700 g), temperature (cold and hot) and conditions of Xe (free and equilibrium). The generation of neutron diffusion constants group is under ambient conditions (20°C) and as a function of burn-up. The multi slab input model for fuel that available in WIMSD-5B code can be shown in Figure 4.

3.2. Preparation of core configurations.

The selected core configurations is 5 × 5 (25 lattice cores), prepared to keep classes burn-up symmetrical to ¼ core for equilibrium core. Thus, the facility for irradiation position and a safety control rod arranged so that symmetrical core is maintained.

3.3. Core calculations

Core calculation were performed to achieve the fresh and equilibrium core of the MTR reactor. Core calculations were performed using BATAN-FUEL code. Flowchart of core calculation of BATAN-FUEL code is shown in Figure 5.The core was modelled in two-dimensional X-Y geometry model. In this calculation, the reactor power is assumed fixed at 60 MWthand to determine the effects of the configuration of the neutron flux and a long operating cycle. Because the cycle length are parameters that affect the reactivity over the core, the core calculation was performed for a long range so that the cycle of rod stuck criteria are met. Referred to stuck rod, if a control rod that has the greatest reactivity values entered unsuccessfully into the core, the core must be in the sub-critical condition. If the rod stuck criteria are not met, it is possible to wear a safety control rod on the core. For power peak factor (PPF), if the value of the maximum PPF exceeds the set limit, the reconfiguration of the position and location of the position of fuel irradiated in the core needs to be done so that the maximum value of PPF is reduced. If not, the configuration is then however the core can not be designed safely. The core calculations were done for each level core and mass loading of uranium using D_2O reflector.The assumptions used in this study are as follows:Core shroud is not used, so that reflectors directly influence the neutron fluxes in the core area.Core cooling using H_2O.In the hot conditions (full power), all non-fuel materials are in room temperature conditions.
Further core calculation with Batan-2DIFF code. Core burn up distribution at the beginning of cycle (BOC) was obtained from the results from the previous studies. Constant group of neutron diffusion as a function of temperature and density in accordance with core burn-up obtained by interpolation. By using the constant cross-section, it has been interpolated calculation reactivity coefficient of fuel and moderator. The expected result of this calculation, all have negative reactivity coefficients. If the value is not negative, the design of the core configuration will be rejected because it is very riskyfor the safe operation of the reactor.
The fresh and equilibrium core configuration can be accepted if they meet safety criteria, namely:The minimum margin reactivity when stuck rod is -0.5%$\Delta k/k$;FPD maximum value is 1.4 and thermal neutron flux is more than 1.0 x10^{15} n/cm^{2}s.

Table 1. Fuel material geometries data for MTR reactor core

Parameter	Values
Grid dimension of the core (cm)	7.71 × 81 × 85
Thickness of fuel plate, (cm)	0.13
Coolant channel width,(cm)	0.255
Number of plates for fuel element	21
Number of plates for control element	15
Cladding material	AlMg2
Edge plate material	AlMg1
Thickness of fuel cladding(cm)	0.038
Active zone dimension (meat), cm	0.054 × 6.275 × 85
Fuel material	U9Mo-Al
Fuel loading of Uranium, (gram)	360, 410, 480, 500, 600, 700
Absorber material	Ag-In-Cd
Thickneess of absorber (cm)	0.338
Absorber cladding material	SS-321
Thickness of absorber cladding (cm)	0.085

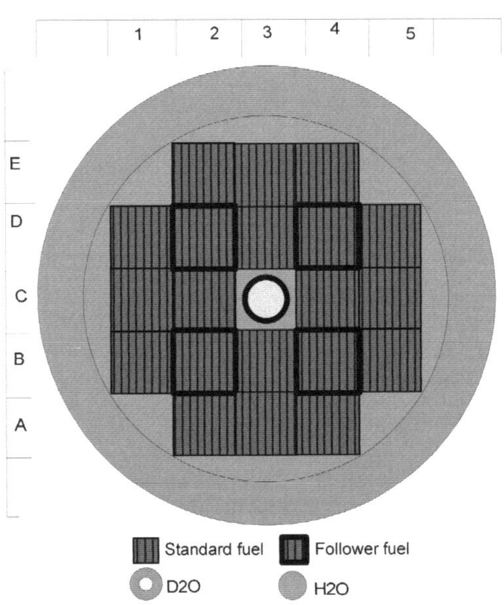

Figure 1. Core configuration of MTR reactor

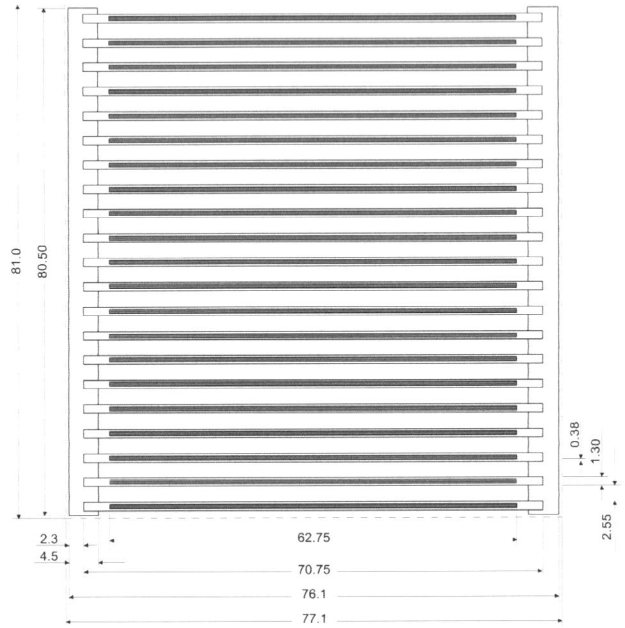

Figure 2. Fuel element of MTR reactor

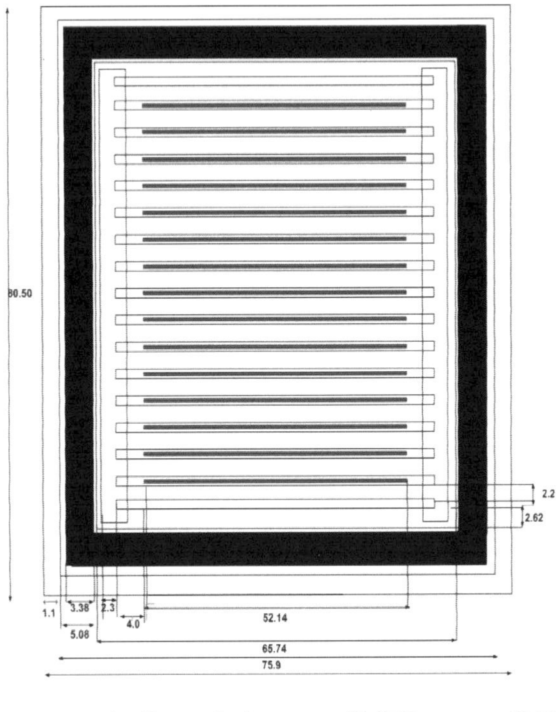

Figure 3. Control element of MTR reactor[10]

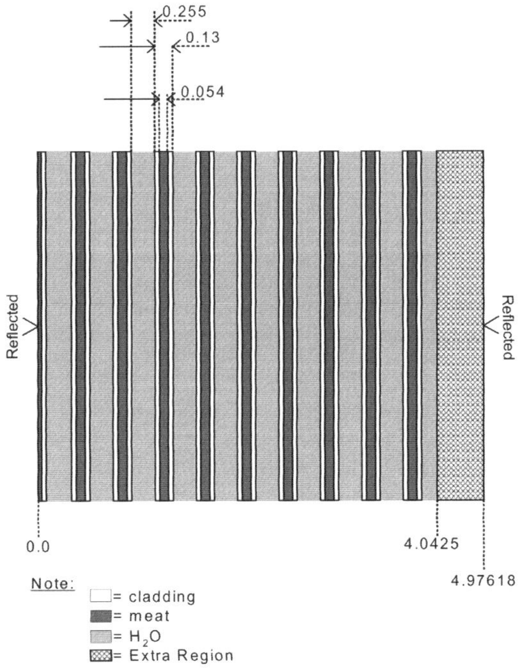

Figure 4. Cell model of U9Mo-Al fuel

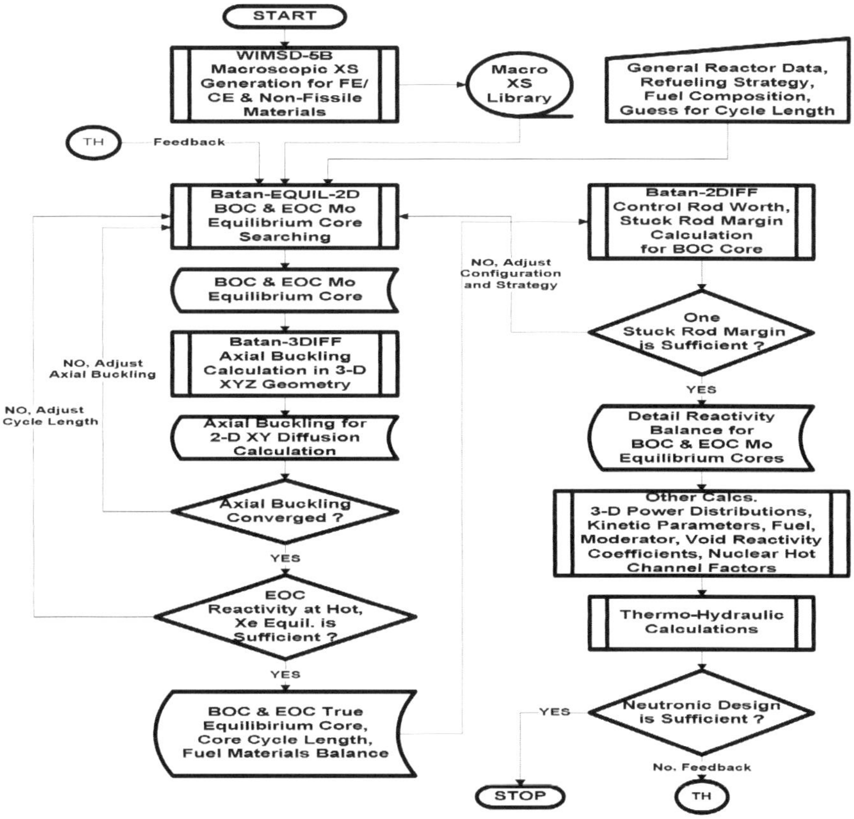

Figure 5. Flowchart of core calculation for Batan-FUEL code [11]

Table 2. Nuclear basic data of MTR reactor compare to CARR core

Parameter	Unit	CARR core	MTR core
Reactor power	MWt	60	60
Core height (active)	Cm	85	85
Number of fuel assemblies	-	21	20
Number of control element	-	4	4
Coolant	-	Light water	Light water
Fuel meat material	-	U_3Si_2-Al	U9Mo/Al
235U Enrichment	%	20	19.75
Uranium density	gram/cm^3	4.3	4.27; 4.86;5.69
$_{235}$U Loading	Kg	11.1	-
Absorber material	-	Hf	Ag-In-Cd
Fuel assembly size	mm^3	76.2×76.2x85	77.1x81.0x85
Grid	-	5 x 5	5 x 5

IOP Conf. Series: Journal of Physics: Conf. Series **962** (2018) 012032 doi:10.1088/1742-6596/962/1/012032

4. Results and discussion

4.1. Neutronic parameter calculation results for fresh core

Based on the calculations, the configuration of the fresh core is expressed in Figure 1. Core configurations 5 × 5 the same with the CARR core (China). The data of CARR reactor is showed in Table 2. The result of the calculation for the fresh core of the MTR reactor, the mass of ^{235}U and each loading fuel are stated in Table 3. It is noted that if the minimum shutdown margin or shutdown reactivity is positive (+), the parameter neutronic from calculation results are not shown in Table 3. Core configurations of 5 × 5 consists of 16 standard fuel elements (FE) and 4 control fuel elements (CE) and the four irradiation positions in the core (IP1, IP2, IP3and IP4) and the centre irradiation position in the core is IP0. The D_2O reflector is used, and the reactor requires an additional two (2) safety control rods (safety rods), SR1 and SR2 outside of the core.

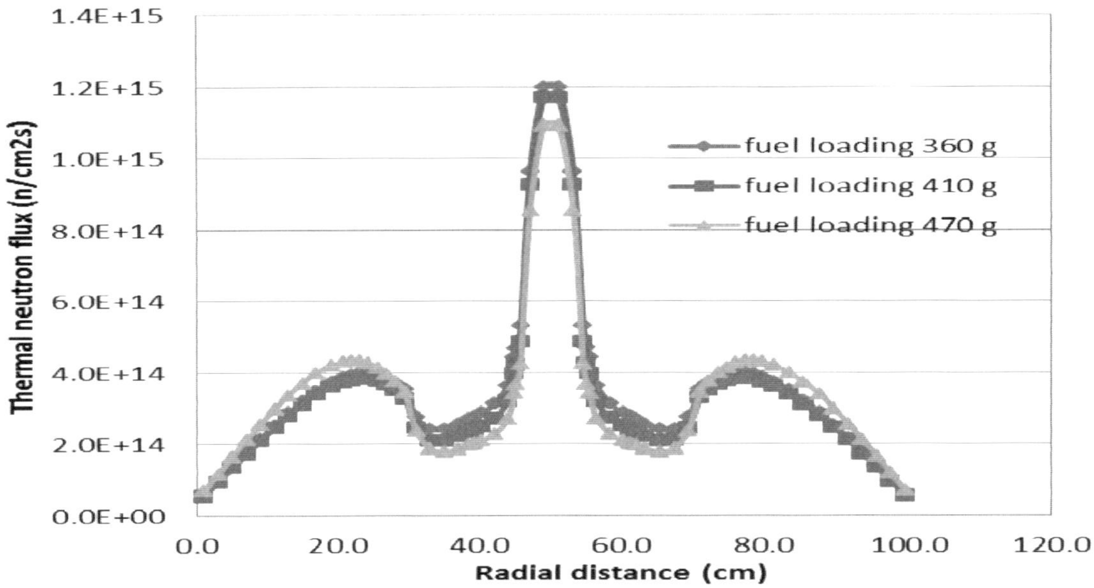

Figure 6. Thermal neutron flux for fresh core

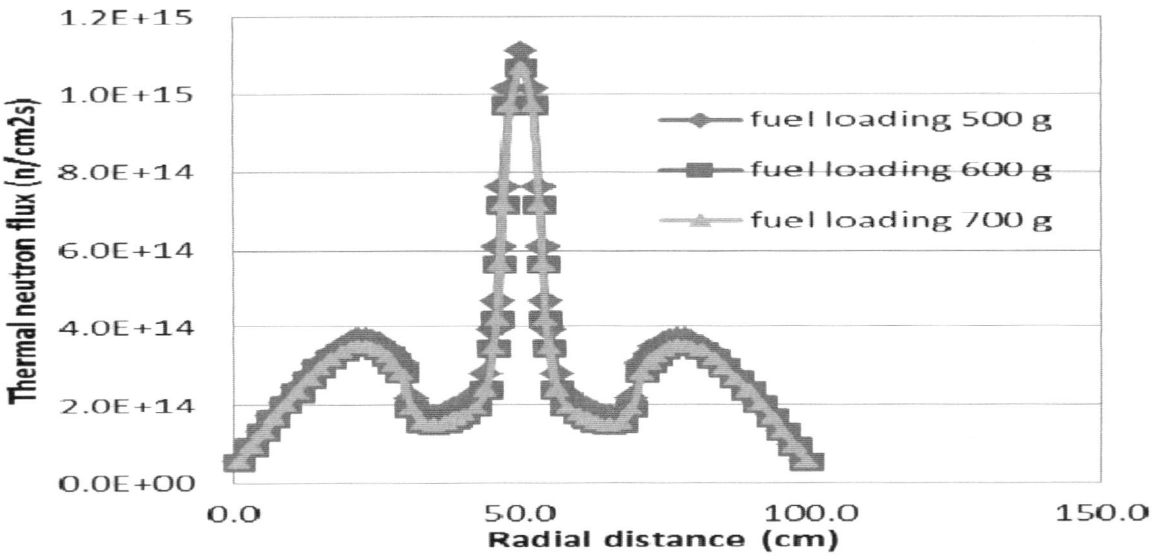

Figure 7. Thermal neutron flux for equlibriumcore

Table 3 also shows that from the aspect of control of reactivity, a core with a reflector D_2O and 360 g of fuel loading has a minimum value shutdown margin (stuck rod condition) compared with other cores. This is due to the reactivity core with this configuration without safety rods. Table 3 also shows that the safety criteria for minimum shutdown margin for all cores are met the safety criteria namely-0.5% Δk/k. The cycle lenght for this core is not meet with the acceptance criteria less than 20 days. This is due to a shift in the neutron spectrum increasingly hard (harder) due to rising levels of ^{235}U fit in reflector D_2O. As a result, the absorptive ability of the control rods for thermal neutrons decreases.

From the aspect of the operation of the reactor, Table 3 shows that the cycle length is the greatest for the core with fuel loading 470 g, but the shutdown margin is smaller than the core with loading fuel 410 g with the cycle length of 25 days. FPD value changes due to the fuel loading 235U and the type of reflector. Therefore the primary factor to determine the maximum value of FPD depends on burnup distribution around the FE and CE irradiation position (IP). The maximum radial FPD value for all cores are less than the value of safety criteria 1.4.

The maximum burn up discharged in every ending cycle of operation for all cores is less than the safety margin namely 70%. The maximun burn up for fresh core in this analysis is 36.22%.
Figure 5 showed that the thermal neutron flux for fresh core. All core neutron fluxes are meet with the neutron flux more than 1.0×10^{15} n/cm^{2}s at the center of the core.

Table 3. Neutronic parameter of MTR core for fresh core

No	**Reactivity balance**	**Fuel loading per assembly for fresh core**		
		360 gram	**410 gram**	**470 gram**
1	Power (MW)	60	60	60
2	One cycle reactivity (%Δk/k)	4.14	4.17	4.22
3	Xenon reactivity xenon (%Δk/k)	3.76	3.83	3.85
4	Samarium reactivity (%Δk/k)	0.30	0.31	0.32
5	Hot to cold reactivity (%Δk/k)	0.57	0.58	0.59
6	Excess reactivity (%Δk/k)	8.17	8.92	11.21
7	Shutdown reactivity (%Δk/k)	-4.52	-4.22	-3.23
8	Total control rods worth reactivity (%Δk/k)	-12.69	-13.14	-14.44
9	Shutdown margin reactivity (%Δk/k)	-1.06	-2.74	-1.07
10	Cycle length (days)	17	25	34
11.	Maximum burn-up (%)	24.21	29.32	36.22
12	Average radial PPF	1.30	1.31	1.32
13	Average power density (W/cm^3)	630	630	630

4.2. Neutronic parameter calculation results for equilibrium core

For equilibrium core, the fuel loading is varied from 500 g, 600 g and 700 g. The first equilibrium core, fuel managements used is 1/4 model which means that for end of cycle 1 CE and 4 FEs are out and inserted for the fresh fuels. The core model I can be shown in Figure 8. The second is 2/8 models which mean that for end of cycle 2 CEs and 8 FEs are out and inserted for the fresh fuels. The core model II can be showed in Figure 9. Table 4 shows that for all cores the safety criteria are met but there is one core is not meet acceptance criteria because the cycle length less than 20 days.

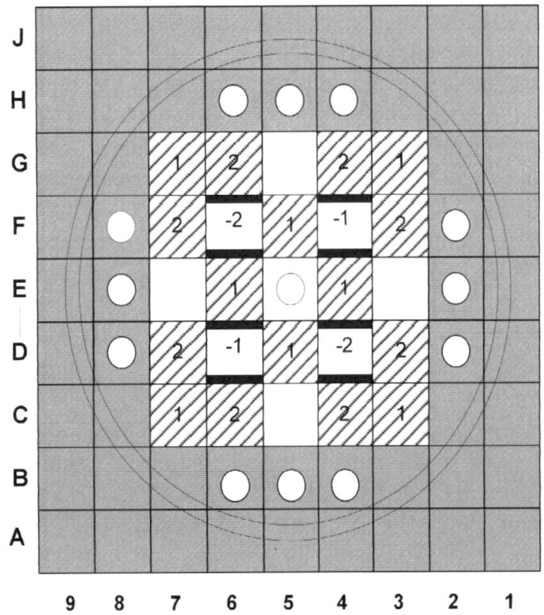

Figure 8. Core model I

where :
number 1 and 2 in the core are burnup classes,
-1 and -2 are control rods

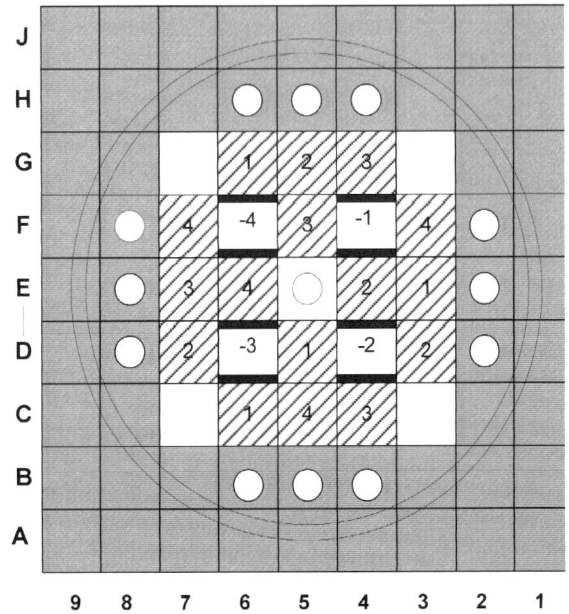

Figure 9. Core model II

Where :
number 1, 2, 3 and 4 in the core are burnup classes,
-1, -2, -3 and -4 are control rods

Tabel 4 shows that the whole neutronic parameters, related to reactivity control, excess reactivity, shutdown margin for 5 × 5 core configurations have the same tendency to three (3) other fresh core configurations. All neutronic parameters are met the safety criteria and the PPF value is the main parameter for neutronic aspect consideration, because it is very sensitive to the configuration of the core especially for the position of the irradiation facility on the core (IP). The core configurations have higher PPF values compared to other core configurations. The maximum value could reach 1.36 for ¼ fuel management model with 700 g fuel loading and still smaller than that of of 1.4.
Figure 7 shows the maximum radial neutron flux at the centre position (IP0) in core for thermal energy ($\leq$ 0.625 eV). It is clear due to consistent trend for the neutron flux in a central position (IP). The maximum thermal neutron flux is available in the core when the fuel loding 500 g. The trend is the same for all levels of fuel loading ^{235}U. The decrease of the neutron flux is due to rising levels of ^{235}U and the same power flux neutron inversely proportional to the increase of cross-sectional $\nu\Sigma f$ (which is equivalent to the increase in the level of fit ^{235}U).The maximum burn up discharged in every end-cycle of operation for all cores is less than the safety margin of 70% and the maximun burn up for equilibrium in this analysis is 56.78%.

Table 4. Neutronic parameter for MTR core for equilibrium core

No.	Reactivity balance	Fuel loading per assembly					
		500 gram		**600 gram**		**700 gram**	
		¼	**2/8**	**1/4**	**2/8**	**1/4**	**2/8**
1.	Power (MW)	60	60	60	60	60	60
2.	One cycle reactivity (%Δk/k)	2.01	3.02	2.31	3.23	2.43	3.32
3.	Xenon reactivity (%Δk/k)	2.18	2.24	2.45	2.43	2.71	2.55
5.	Samarium reactivity (%Δk/k)	0.56	0.55	0.56	0.53	0.53	0.58
6.	Hot to cold reactivity (%Δk/k)	0.43	0.41	0.44	0.42	0.45	0.44
7.	Excess reactivity (%Δk/k)	5.82	7.68	6.54	8.90	7.07	9.88
8.	Shutdown reactivity (%Δk/k)	-2.01	-2.05	-2.09	-2.04	2.02	-2.01
9.	Total control rods worth reactivity (%Δk/k)	-25.79	-24.41	-23.80	-22.47	-21.62	-20.29
10.	Shutdown margin reactivity (%Δk/k)	-19.97	-16.73	-17.26	-13.57	-14.55	-10.41
11.	Cycle length (days)	14	24	20	33	27	44
12.	Maximum burn-up (%)	49.45	44.52	57.64	50.29	65.43	56.78
13.	Average radial PPF	1.32	1.24	1.34	1.26	1.36	1.28
14.	Average power density (W/cm^3)	523	523	523	523	523	523

Neutron fluxes in the irradiation positions can be shown in Table 5. The thermal neutron fluxes for core model I are met the acceptance criteria, more than $1.0x10^{15}$ n/cm^2s for all fuel loading. The higher of thermal neutron flux implies the lower fuel loading and hence decreasingthe cycle length. The optimal one is chosen for second core for 600 g of fuel loading.

Table 5. Neutron flux in the irradiation positions of the core model I

Neutron energy group	**Maximum neutron flux 10^{15} (neutron cm^{-2} s^{-1})**		
	500 g	**600 g**	**700 g**
in the centerof the core			
Fast> 0.821 MeV	0.24	0.23	0.22
Epithermal, 0.625 eV<*E*< 0.821 MeV	0.51	0.50	0.49
Thermal, < 0.625 eV	1.11	1.07	1.06
at irradiation position			
Fast> 0.821 MeV	0.21	0.19	0.18
Epithermal, 0.625 eV<*E*< 0.821 MeV	0.48	0.46	0.36
Thermal, < 0.625 eV	0.53	0.50	0.48

5. Conclusion

The results of this study have shown three (3) fuel loadings of MTR reactor for fresh core configurations with powered thermal capacity of 60 MW. The optimum cores that met with safety use U9Mo-Al fuel with a loading of uranium ^{235}U of 470 g and a cyclelength of 34 days (2040 MWD). Based on the results for the maximum capacity utilization, the equilibrium core configuration with 5 × 5 with a load rate of 600 g ^{235}U using 2/8 model of fuel management is the best equilibrium core. Indeed, the maximum thermal neutron flux in the centre of the core is of 1.07×10^{15} n / cm^2s, minimum shutdown margin -2.04% Δk/k (stuck rod) and a cycle length of 33 days, and the core has an irradiation facility with the maximum thermal neutron flux of 5.0×10^{14} n / cm^2s.

Acknowledgement
The author acknowledges the head of BTKR, Dr. Syaiful Bakhri, as well as his staffs, who has assigned the author to do this research analysis. The author also thanks to the committee of guidance for researcher to review and finally develop this paper qualified. Indeed, the author is also very appreciative to Head of Center for Nuclear Reactor Technology and Safety hence enabling the author to develop this paper being possibly worthwhile for nuclear community in BATAN.

References

[1]. Villarino. E. A2013 Core Performance Improvements Using High Density Fuel in Research Reactors RRFM St Petersburg Rusia68-75
[2]. Tukiran, PinemS, Sembiring TM, Suparlina L. and Susilo J.2012 Neutronic Conceptual Design of the Innovative Research Reactor Core Using Uranium Molibdenum Fuel Tri Dasa Mega**14** 3178-191
[3]. Kunturo I. and Sembiring TM2014 Alternative Core Design for the Innovative Research Reactor (RRI) from Neutronics Aspects Tri Dasa Mega, **16** 11-10
[4]. Tukiran and Sembiring TM 2016Neutronics Analysis on Mini Test Fuel in the RSG-GAS Core Tri Dasa Mega**18** 129-40
[5]. Smith DL 2011 Evaluated Nuclear Data Covariances: The Journey From ENDF/B-VII.0 to ENDF/B-VII.1Nucl. Data Sheets**112** 12 3037-3057
[6]. Chusova I A, Shelegova A S, and Kochnovb O Y 2016 Design Features of Water-cooled Research Reactors Nuclear Energy and Technology 2 4 287–293
[7]. Guohai Wei, Songbai Han, Hongli Wang, Lijie Hao, Meimei Wu, Linfeng He, Yu Wang, Yuntao Liu, Kai Sun, Dongfeng Chen 2013 Design of the testing set-up for a nuclear fuel rod by neutron radiography at CARR Physics Procedia **43** 307 – 313
[8]. Raheleh K 2014 Sensitivity Analyses of the Use of Different Neutron Absorbers on the Main Safety Core Parameters in MTR Type Research Reactor Nucl. Eng. Technology **46**4 513–520
[9]. Rokhmadi and Tukiran 2013Fuel Density Effect on Parameter of Reactivity Coefficient of the RRICore Tri Dasa Mega**15** 2 77-89
[10]. Setiyanto and Tukiran2016Analysis of Gamma Heating at TRIGA Mark Reactor Core BandungUsing Plate Type Fuel Tri Dasa Mega**16**3 160
[11]. Suparlina S and Tukiran 2014 Analysis of Fuel Management Pattern of Research Reactor Core of the MTRType Design Tri Dasa Mega**16**289-99

Cooling Performance Analysis of ThePrimary Cooling System ReactorTRIGA-2000Bandung

I.D. Irianto, S. Dibyo, S. Bakhri, G.R. Sunaryo
Center for Nuclear Reactor Technology and Safety
National Nuclear Energy Agency of Indonesia
Puspiptek Area, Building 80, Serpong,Tangerang 15310, Indonesia

Email: igndjoko@batan.go.id

Abstract. The conversion of reactor fuel type will affect the heat transfer process resulting from the reactor core to the cooling system. This conversion resulted in changes to the cooling system performance and parameters of operation and design of key components of the reactor coolant system, especially the primary cooling system. The calculation of the operating parameters of the primary cooling system of the reactor TRIGA 2000 Bandung is done using ChemCad Package 6.1.4. The calculation of the operating parameters of the cooling system is based on mass and energy balance in each coolant flow path and unit components. Output calculation is the temperature, pressure and flow rate of the coolant used in the cooling process. The results of a simulation of the performance of the primary cooling system indicate that if the primary cooling system operates with a single pump or coolant mass flow rate of 60 kg/s, it will obtain the reactor inlet and outlet temperature respectively 32.2 °C and 40.2 °C. But if it operates with two pumps with a capacity of 75% or coolant mass flow rate of 90 kg/s, the obtained reactor inlet, and outlet temperature respectively 32.9 °C and 38.2 °C. Both models are qualified as a primary coolant for the primary coolant temperature is still below the permitted limit is 49.0 °C.

Keywords: Cooling System, the primary pump, mass flow rate, temperature.

1. Introduction

The Bandung TRIGA Mark II reactor is one of the research reactors in Indonesia that is used for research activities related to nuclear technology as well as for the production of isotopes. As a research reactor, this reactor is used as a neutron source to irradiate material, while the heat energy from the reactor is discharged through the cooling system. This reactor is a swimming pool type reactor in which the reactor core is placed at the bottom of the pool with anatural convection cooling system. As a common TRIGA reactor, the reactor fuel type is rod type fuel. This reactor has been upgraded to be able to achieve the thermal power 2000 kW [1], then is called reactor TRIGA-2000 Bandung. To support this upgrade process,some studies to accomplish this upgrading have been done. The reactor coolant system including the modification of the heat exchanger, the core or the neutronic design [2] and the reactor core support structures need to be redesigned to be able to cool the reactor core which the thermal power is 2000 kW. The new design of support system also consists of new reactor core assembly i.e. top and bottom grid plates, which fit within the existing reflector assembly.

The next upgrading is to convert the reactor fuel type from the rod-type into the plate-type. Replacement the rod-type fuel assembly into the plate-type fuel assembly is an innovation to find the solutions for the stop production of the rod-type fuel assembly of TRIGA reactor in the world. The use of plate-type fuel assembly in the reactor TRIGA-2000 Bandung demanding changes the cooling system from natural circulation mode into forced circulation mode. Sudjatmi[3] reported that to operate the reactor TRIGA-2000 Bandung using plate-type fuel assembly with the thermal power more than 600 kW, the temperature at the fuel cladding temperature is 108.88 °Cand at the reactor outlet coolant temperature is 82.39 °C. Therefore, to operate the reactor TRIGA-2000 Bandung using plate-type fuel assembly, the primary cooling system has to be set into forced circulation mode. To support this fuel type conversion, the entire support system design of the reactor is needed to be redesigned. The fixing bar at the top of the standing fuel assembly has to be constructed to avoid the impact cases to the fuel assembly such as an impact from a dropping fuel assembly[4].

In reactor TRIGA-2000 Bandung, the reactor coolant system consists of two loops of the cooling system, i.e. primary cooling system and secondary cooling system. According to the modification of the cooling system, the flexibility of cooling pipe has to be considered [5]. The primary cooling system serves to take heat from the reactor to be moved to the secondary coolant and then discharged into the environment. The main components of the primary cooling system which be forced circulation mode are a delay tank which is mounted directly on the outlet of the reactor, primary pump and a heat exchanger which connected the primary cooling system and the secondary cooling system[6,7]. The function of delay tank is to lower the content of N-16 in the cooling water by extending the cooling trajectory before going to the other components. This is important because the N-16 are radioactive. N16 is radioactive which be produced from the interaction of the fission neutron with the oxygen atom which presents in the water around the reactor core [8].

This work is conducted in the design of the primary cooling system in the reactor TRIGA-2000 Bandung that has been upgraded to the thermal power 2000 MW and has to convert the reactor fuel type from the rod-type into the plate-type. The modification to improving the core cooling system of reactor TRIGA-2000 Bandung has to be studied by numerical simulation. Analysis of the cooling performance of the primary cooling systemreactor TRIGA-2000 Bandung has been done by using computer code ChemCad. The ChemCad software has been widely used in the modeling and simulation of the cooling system [9,10] and in energy conversion system simulation [11]. The purpose of this study was to determine the minimum mass flow rate of the primary cooling system that is able to cool the reactor core of reactor TRIGA-2000 Bandung which uses plate-type fuel.

2. Current Reactor Description

The reactor TRIGA-2000 Bandung has had a license from the regulatory body for steady state operation at the nominal power 2000 kW[12]. The main function of this reactor is for isotope production such as Mo-99, I-131, and P-32. This reactor is a swimming pool type surrounded by a concrete shield. The configuration of the reactor is a hexagonal grid plate surrounded by a graphite reflector. The vertical cross section of the reactor TRIGA-2000 Bandung is shown in Figure 1[13].

The reactor coolant system of reactor TRIGA-2000 Bandung consists of two coolant system loops, i.e. primary coolant system, and secondary coolant system. The primary coolant system was designed to take a bulk water out from the reactor tank, circulate it through a heat exchanger and return it to the reactor tank. The function of the heat exchanger is to transfer heat from the reactor tank to the secondary coolant system. The schematic diagram of the primary coolant system is shown in Figure 2 [14].

The maximum temperature allowed for the primary cooling fluid is 49 °C, this temperature is the maximum temperature limit for resins used in water purification systems unit. The temperature difference between the water outlet of reactor tank and the water inlet reactor tank at normal operation in the power 2000 kW is 10 °C. The outlet coolant water temperature from the heat exchanger is 32.2 °C, and the inlet coolant temperature to the heat exchanger is 42.2 °C [14].

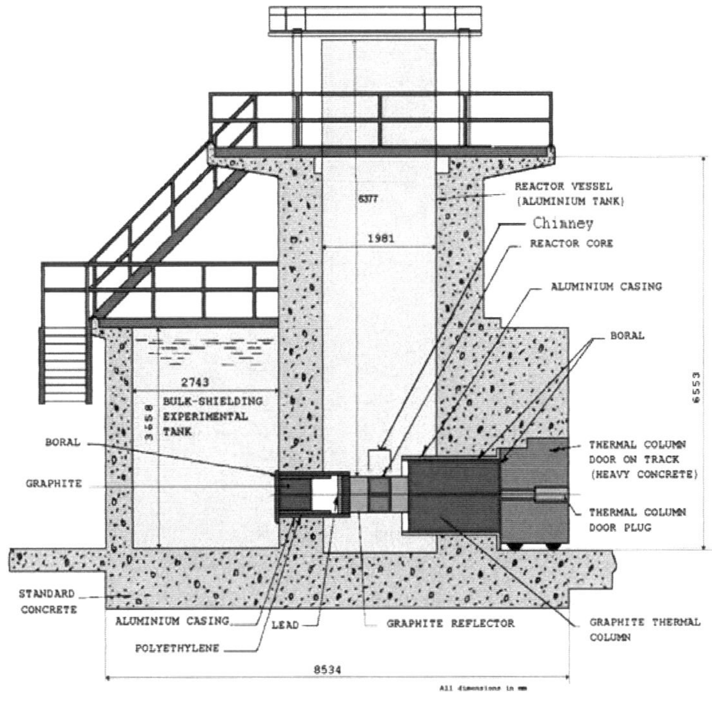

Figure 1. Vertical cross section of the reactor TRIGA-2000 Bandung[13]

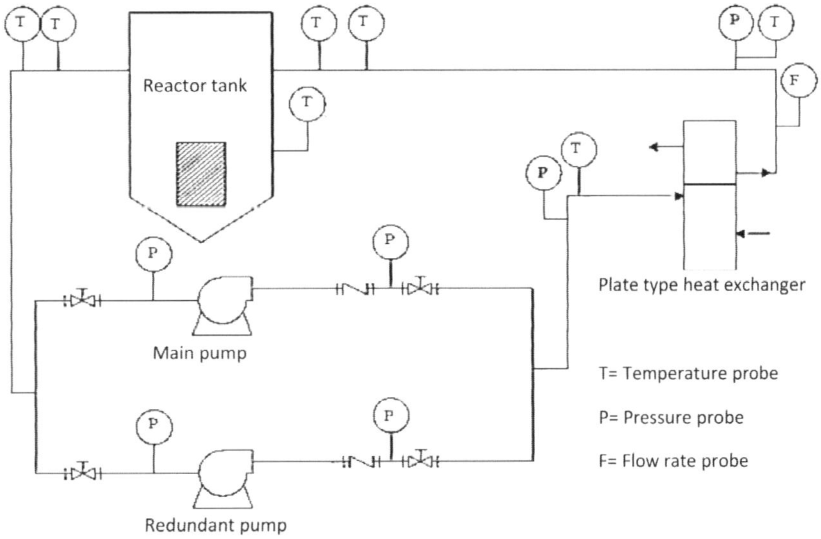

Figure 2. Schematic diagram of the primary coolant system [14]

3. Theory

3.1. Pressure drop calculation.

The pressure loss due to friction with wall and the turbulence in the channel of uniform cross section is normally calculated as follows [15]:

$$\Delta P = f\ \frac{L}{D}\left(\frac{\rho\ u_{mean}^2}{2}\right) \tag{1}$$

where ΔP is the pressure drop (Pa) between two points in a pipe of diameter D (m) that are separated by an axial distance L (m). u_{mean} stands for the mean flow speed (m/s), and ρ is the fluid density (kg/m^3), and f is the friction factor. The frictional pressure drop coefficient could be taken to be proportional to the length of the flow channel and inversely proportional to the channel diameter. So in practice, the pressure drop is [15]:

$$\Delta P = \sum_i f_i \frac{L_i}{D_e}\left(\frac{\rho_i u_i^2}{2}\right) \tag{2}$$

where f_i is dimensionless friction factor, L_i is channel length, and D_e is the equivalent diameter of the channel.

3.2. N-16 Production.

During reactor in operation, the exposure of neutron radiation and gamma ray radiation around the reactor are in the very high level. Gamma-ray radiation emanating from the N-16 gave a significant radiation exposure around the reactor. N-16 is radioactive derived from the interaction between oxygen in water with a neutron, according to the following reaction:

$$^{1}_{0}n + {}^{16}_{8}O \rightarrow {}^{1}_{1}p + {}^{16}_{7}N \tag{3}$$

$$^{16}_{7}N \xrightarrow{7.2\,s} {}^{16}_{8}O + {}^{0}_{-1}\beta + \gamma\ (6.13\ MeV) \tag{4}$$

$^{16}_{7}N$ produce and radiate gamma rays (6.13 MeV) and β particles during its decay chain. Nitrogen-16 ($^{16}_{7}N$) is an isotope in the primary coolant system of TRIGA reactor which has a very short half-life time ($T_{1/2}$ = 7.2 s)[8].

4. Methodology

Thepurpose of this study was to determine the minimum mass flow rate of the primary cooling system that is able to cool the reactor core of reactor TRIGA-2000 Bandung as a result of the modification of the type of fuel rod into a plate type and assesses the opportunities for modification.The redesign of the reactor coolant system of reactor TRIGA-2000 Bandung shall refer to the condition of the cooling system that is currently installed. The change of the convection mode in the cooling process such as natural convection mode into forced convection mode is necessary to add a component such as a delay tank.

The cooling performance analysis of the primary cooling system for reactor TRIGA-2000 Bandung is done by simulation calculations using the program package ChemCad 6.1.4. The step calculation can be shown below:

- First, prepare the model of theprimary cooling system for reactor TRIGA-2000 Bandung.
- Then provide primary cooling system data input for reactor TRIGA-2000 Bandung model which includes the geometry of the primary cooling system.
- Then give the thermal-hydraulic input parameters to the primary cooling system including mass flow rate, temperature, and pressure.
- After all, inputs are completed, set the boundary conditions in the reactor cooling system reactor TRIGA-2000 Bandung such as ambient temperature.
- Once ready, program execution to calculate the thermal-hydraulic parameters of the primary cooling system.
- Finally, the analysis of the output calculations.

5. Results and Discussion

The cooling performance analysis of the primary cooling system is intended to measure the performance of the reactor coolant system for reactor TRIGA-2000 Bandung, especially the performance due to the fuel element conversion. There are some modifications or additions to the main component in the reactor coolant system of reactor TRIGA-2000 Bandung. One of the main components of the primary cooling system in reactor TRIGA-2000 Bandung is a delay tank. As shown in Equation (3), in the primary coolant will be formed N-16 as a result of the reaction of oxygen with neutrons. N-16 is a radioactive compound that emits gamma rays. Delay tank is used to remove or lower the radiation exposure of radiation N-16 prior to entry into the primary pump. Working principle of the delay tank is to extend the travel time so that the activity of N-16 radioactive decays. In her research report, Wardani[16]managed to model a delay tank using a curved pipe with length 118.0760 m.Theresults showed that the optimum time required to decrease the activity of N-16 to 3% was 5 times the half-life of N-16 or 36.7047 seconds.

The thermal energy generated from the reactor core were taken through the primary cooling system and transferred through a heat exchanger to the secondary cooling system for further discharged into the environment. As a consequence of the conversion of fuel elements, the cooling system needs to be changed from natural convection mode into a forced convection mode. In addition, the performance of the cooling system must be improved. The cooling system performance can be improved, among others, modification of the heat exchanger as an intermediary heat transfer between the primary cooling system and a secondary cooling system [17,18]. Nazar [19] reported the results of his research that the addition of the chimney and the plate in the heat exchanger capable of lowering the temperature of the reactor primary coolant reactor TRIGA-2000 Bandung. In addition to modifications with the addition of plate heat exchangers, the cooling performance of the primary cooling systems can also be improved by adjusting the mass flow rate of the primary coolant [17].

Analysis of the performance of the cooling system of reactor TRIGA-2000 Bandung is carried out by means of simulations using a computer code ChemCad. Two models of the primary cooling system are configured while considering the cooling system configuration conditions attached. The first model is the model of the primary cooling system with one main pump, while the spare pump is not operating as shown in Figure 3. In the second model, the two primary pumps are operated in parallel as shown in Figure 5. To measure the performance of the primary cooling system in the reactor TRIGA-2000 Bandung, thermal power of the reactor was set at 2000 kW. With the thermal power of the reactor constant, the primary coolant mass flow rate varied from 40 kg/s up to 100 kg/s. The capacity of the existing primary pump is 950 gpm[19].

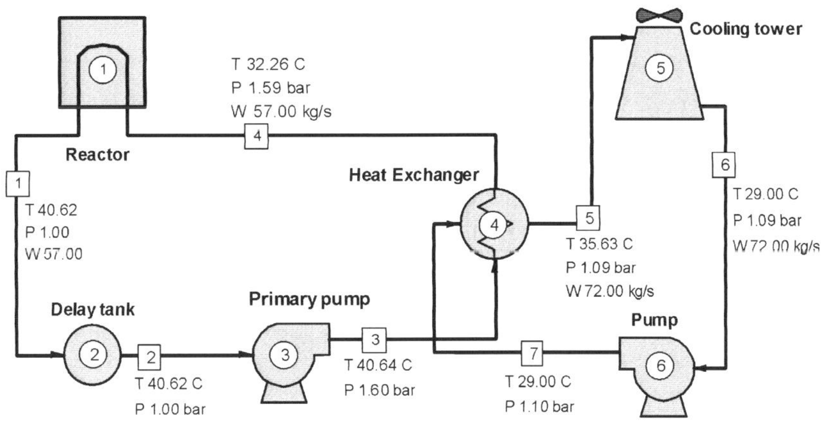

Figure 3. The simulation model of the reactor cooling system of reactor TRIGA 2000 Bandung with one primary pump using ChemCad.

The simulation results for first primary cooling system model which using one primary pump with a capacity of 950 gpm or 59.9 kg/s as shown by the graph in Figure 4. The first model of the cooling system is the model as on the condition of the installed systems now is the one that operates the primary pump. Data input to the maximum pump capacity is 950 gpm or 59.9 kg/s, set at 10 MW power reactor and coolant mass flow rate varied from 40.0 kg/s up to 60.0 kg/s with a range of 5.0 kg/s.

IOP Conf. Series: Journal of Physics: Conf. Series **962** (2018) 012033 doi:10.1088/1742-6596/962/1/012033

Figure 4 shows the simulation results of the primary cooling system on the first model that shows the graph of inlet and outlet temperature of the reactor as a function of the mass flow rate of the primary coolant. In Figure 4 it appears that at maximum pump capacity with the primary coolant mass flow rate of 59.9 kg/s, an inlet and outlet temperature of the reactor was 32.2 °C and 40.2 °C. For comparison in a study reported by Daeseong Jo[20] on a Canadian McMaster Nuclear Reactor (MNR) reactor, at 2 MW of power with a cooling mass flow rate of 75 kg/s, the inlet, and reactor temperature are 30.0 °C and 36.4 °C.This temperature is lower than the maximum allowable temperature in the primary cooling system. In the safety analysis report of reactor TRIGA 2000 Bandung, the maximum allowable temperature in the primary cooling system is 49.0 °C. Determination of the maximum temperature is based on the maximum temperature limit for resins used in water purification systems unit. If the cooling fluid temperature is greater than 49.0 °C will result in the resin is not effective for use, meaning the reactor has to be stopped. With this result, conceptually it can be said that the primary cooling system with one primary pump installed is still possible to cool the TRIGA reactors with thermal power 2000 kW.

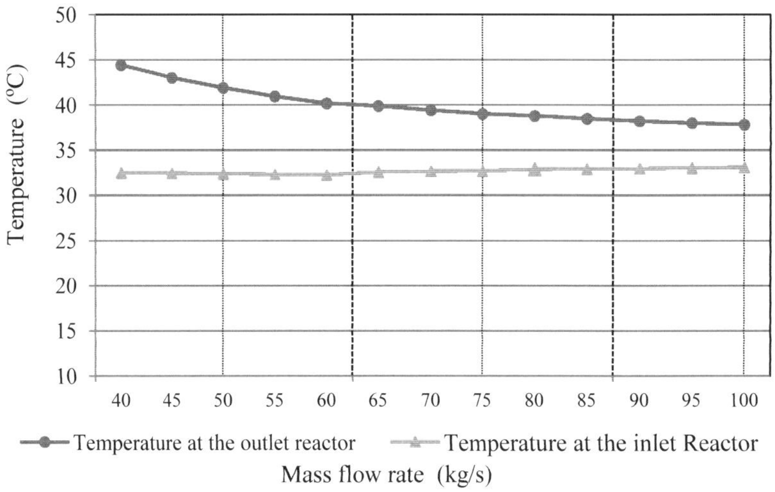

Figure 4. Inlet and outlet temperature of the reactor as a function of the coolant mass flow rate

Model of the second primary cooling system is a model with two primary pumps are installed in parallel and operated jointly. Thus the coolant mass flow rate of the primary coolant can reach 100 kg/s. For the purposes of the simulation, the model primary cooling system with two primary pumps mounted parallel is shown in Figure 5. Characteristics of all the major components on the second models are same as the characteristics of all the major components in the first model. In the second model, both the primary pump operates with the same capacity is 45 kg/s or 75% of the maximum capacity. Simulations are carried out by setting the reactor operates at a maximum power of 10 MW thermal. The primary coolant mass flow rate is varied from 40 kg/s up to 100 kg/s with a range of 5 kg/s. Keep in mind that to obtain the mass flow rate of the primary coolant of more than 60 kg/s, in the primary system has to be done by operating two primary pumps of 59.9 kg/s. From the simulation results as shown in Figure 6 the inlet and anoutlet temperature of the reactor are respectively 32.9 °C and 38.2 °C.

IOP Conf. Series: Journal of Physics: Conf. Series **962** (2018) 012033 doi:10.1088/1742-6596/962/1/012033

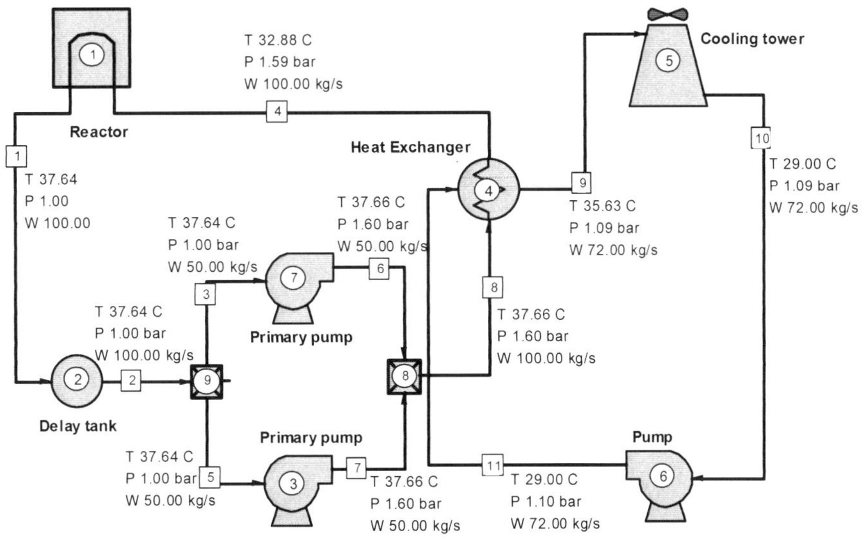

Figure 5. The simulation model of the reactor cooling system of reactor TRIGA 2000 Bandung with two primary pumps using ChemCad.

On the condition of the system that is mounted as shown in Figure 4, with one pump and one spare pump operates, conceptually it is still possible to cool the TRIGA reactor with a thermal power of 2 MW. The maximum pumping capacity of the pump is 950 gpm or 59.9 kg/s of water. The simulation results, as shown in Figure 5 and Figure 6, the maximum pump capacity (59.9 kg/s) inlet and an outlet temperature of each reactor are 32.2 °C and 40.2 °C.

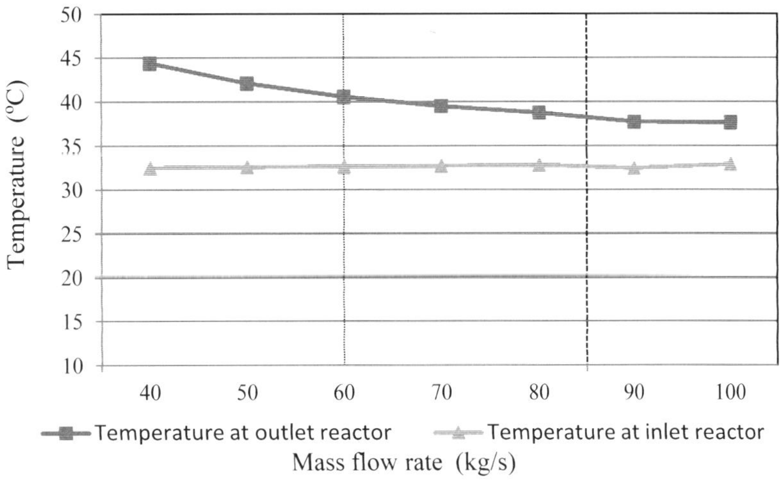

Figure 6. Inlet and outlet temperature of the reactor as a function of the coolant mass flow rate

From the simulation results obtained reactor inlet and outlet temperature respectively 32.9 °C and 38.2 °C. Keep in mind that to obtain a coolant mass flow rate of 90 kg/s it is necessary to pump replacement due to the capacity of existing pumps a maximum of 59.9 kg/s. As an alternative to simulation is done both by operating both pumps as shown in Figure 6. The simulation results obtained reactor inlet and outlet temperature respectively of 32.4 °C and 37.7 °C for a refrigerant mass flow rate of 90 kg/s.

6. Conclusions

The simulation of the primary cooling system of TRIGA 2000 reactor Bandung has been done. There are two primary pumps that exist in the primary cooling system TRIGA 2000 Bandung, each with a capacity of 60 kg/s. In case two primary pumps are operated simultaneously at 75% capacity or a mass flow rate of 90 kg/s, the inlet and outlet temperatures of 32.9 °C and 38.2 °C are obtained. Conversely, if it operates only one primary pump with the primary coolant mass flow rate of 60 kg/s, it obtains the inlet and outlet temperatures of 32.2 °C and 40.2 °C respectively. The reactor temperature which is still lower than the maximum permitted is 49.0 °C. Basically by operating one primary pump which the primary coolant mass flow rate of 60 kg/s is enough to cool the reactor TRIGA 2000 Bandung with 2000 kW thermal power.

Acknowledgments

The authors would like to acknowledge the Center for Nuclear Reactor Safety and Technology and all member of BATAN Bandung TRIGA plate reactor team for supporting this research using the funding of DIPA PTKRN BATAN FY 2016.

References

[1] Umar E and Fiantini R 2010 Modification of the Core Cooling System of TRIGA 2000 Reactor *The 2nd International Conference on Advances in Nuclear Science and Engineering* pp 224–31
[2] Basuki P, Yazid P I and Suud Z 2014 Desain Neutronika Elemen Bakar Tipe Pelat Pada Teras TRIGA 2000 Bandung *Indones. J. Nucl. Sci. Technol.* **15** 169–80
[3] Sudjatmi K A, Hastuti E P, Widodo S and Naza R 2015 Analysis of Natural Convection in TRIGA Reactor Core Plate Types Fueled Using Coolood-N2 *Tri Dasa Mega* **17** 67–78
[4] Kim H, Yim J, Lee B, Oh J and Tahk Y 2014 Drop Impact Analysis of Plate-Type Fuel Assembly in Research Reactor *Nucl. Eng. Technol.* **46** 529–40
[5] Rahardjo H P 2011 Piping Flexibility Analysis of the Primary Cooling System of TRIGA 2000 Bandung Reactor due to Earthquake *Atom Indones.* **37** 83–9
[6] Dibyo S, Hastuti E P and Irianto I D 2015 Analisis Desain Proses Sistem Pendingin Pada Reaktor Riset Inovatif 50 MW *Tri Dasa Mega* **17** 19–30
[7] Dibyo S, Juarsa M and Irianto I D 2016 Analysis On Thermal Characterictic Of Heater In Loop FASSIP-01 Experimental Facility *Seminar Nasional Teknologi Energi Nuklir 2016* pp 545–51
[8] Sadeghi N 2010 Estimation of reactor power using N 16 production rate and its radiation risk assessment in Tehran Research Reactor (TRR) *Nucl. Eng. Des.* **240** 3607–10
[9] Solanki K and Patel N 2014 Process Optimization Using CHEMCAD *Int. J. Futur. Trends Eng. Technol.* **1** 47–51
[10] Cormos C-C 2015 Biomass direct chemical looping for hydrogen and power co-production: Process configuration, simulation, thermal integration and techno-economic assessment *Fuel Process. Technol.* **137** 16–23
[11] Irianto I D 2016 Design And Analysis Of Helium Brayton Cycle For Energy Conversion System Of RGTT200K *Tri Dasa Mega* **18** 75–86
[12] Umar E, Kamajaya K, Suwono A, Tandian N P and Hardianto T 2005 An Experimental Study Of Natural Convection In The Hottest Channel Of TRIGA 2000 kW Reactor *Indones. J. Nucl. Sci. Technol.* **6** 1–12
[13] Daryoko M 2014 Perhitungan biaya dismantling reaktor TRIGA Mark II-Bandung dalam daerah terkontrol *J. Teknol. Bahan Nukl.* **10** 74–84
[14] Umar E and Kamajaya K 2006 Redesign, Construction and Characteristic Operation of the Primary Cooling System of TRIGA 2000 *Pacific Basin Nuclear Conference* pp 472–9
[15] Martínez-cuenca R, Mondragón R, Hernández L, Segarra C, Jarque J C and Hibiki T 2016

Forced-convective heat-transfer coefficient and pressure drop of water-based nanofluids in a horizontal pipe *Appl. Therm. Eng.* **98** 841–9

[16] Wardhani V I S and Rahardjo H P 2016 Pengaruh Bentuk Routing Perpipaan Sistem Pendingin Primer Reaktor TRIGA Konversi Terhadap Penurunan Aktivitas N-16 Di Permukaan Tangki Reaktor *Tri Dasa Mega* **18** 145–54

[17] Irianto I D, Sudadiyo S and Dibyo S 2015 Performance Analysis On RGTT200K Cogeneration System For Changes In The Reactor Coolant Mass Flow Rate *Seminar Nasional Teknologi Energi Nuklir 2015* (Bali) pp 15–6

[18] Dibyo S, Irianto I D and Pujiarta S 2016 Evaluation of Operating Performance of the Reactor Coolant System of RSG-G AS using Chemcad 6.1.4 *Knowl. E* **1** 45–54

[19] Nazar R 2011 Perolehan Suhu Air Pendingin Primer Reaktor TRIGA 2000 Ketika Penambahan Cerobong Dan Pelat Penukar Panas *Tri Dasa Mega* **13** 161–72

[20] Jo D, Park J and Chae H 2014 Development of thermal hydraulic and margin analysis code for steady state forced and natural convective cooling of plate type fuel research reactors *Prog. Nucl. Energy* **71** 39–51

IOP Conf. Series: Journal of Physics: Conf. Series **962** (2018) 012034 doi:10.1088/1742-6596/962/1/012034

Analysis of helium purification system capability during water ingress accident in RDE

Sriyono, Rahayu Kusmastuti, Syaiful Bakhri, Geni Rina Sunaryo

Puspiptek Area Building 80, Serpong, South Tangerang City, 15310, INDONESIA

sriyono@batan.go.id

Abstract. The water ingress accident caused by steam generator tube rupture (SGTR) in RDE (Experimental Power Reactor) must be anticipated. During the accident, steam from secondary system diffused and mixed with helium gas in the primary coolant. To avoid graphite corrosion in the core, steam will be removed by Helium purification system (HPS). There are two trains in HPS, first train for normal operation and the second for the regeneration and accident. The second train is responsible to clean the coolant during accident condition. The second train is equipped with additional component, i.e. water cooler, post accident blower, and water separator to remove this mixture gas. During water ingress, the water release from rupture tube is mixed with helium gas. The water cooler acts as a steam condenser, where the steam will be separated by water separator from the helium gas. This paper analyses capability of HPS during water ingress accident. The goal of the research is to determine the time consumed by HPS to remove the total amount of water ingress. The method used is modelling and simulation of the HPS by using ChemCAD software. The BDBA and DBA scenarios will be simulated. In BDBA scenario, up to 110 kg of water is assumed to infiltrate to primary coolant while DBA is up to 35 kg. By using ChemCAD simulation, the second train will purify steam ingress maximum in 0.5 hours. The HPS of RDE has a capability to anticipate the water ingress accident.

1. Introduction

Nuclear energy is one of the most promising energy sources to satisfy energy security, environmental protection and efficient supply [1]. BATAN has a plan to build an Experimental Power Reactor which is called RDE. RDE10 is designed based on HTGR technology [2,3]. It uses graphite as moderator and structural material. The reactor coolant is helium gas that flows in the fuel gap from the bottom to the upper part of reactor core. Helium is an inert gas; it will not chemically react with any compounds [3]. Additionally, exposing helium to neutron radiation does not make it radioactive, unlike most other possible coolants [4]. The RDE has an indirect energy conversion system the heat from the core is transferred to the secondary coolant by steam generator (SG). This reactor is designed to generate electricity; otherwise the heat will be used to support hydrogen production process, the coal gasification and liquefaction, etc. [3].

Ruptures in part or all of SG tubes are a hypothetical water ingress accident [5]. The ingress of larger amounts of water or steam into the primary circuit of RDE is seen to be an accident of considerable safety relevance. The two impacts of water to the graphite structures of core and fuel elements are: moderation and graphite corrosion effect [6,7]. The changing of the concentration moderated nuclei the neutron balance is affected and by chemical oxidation of the graphite the structural integrity of the core is attacked and explosive gases may be produced. The overall destructive potential of these effects depends strongly on the quantities of the invading material as well as on the physical and operational boundary conditions of the reactor system at the time of the ingress [8,9].

Water ingress may be caused by ruptures inside the steam generator as a consequence of the much higher pressure in the secondary loop. The core can only be affected if larger amounts of steam or water will be transported from the steam generator to the reactor. The status of the primary blower plays an important role in this field and the condensation of steam on cold surfaces and within the gas volume is of major significance [10]. The system pressure will increase and passive safety valves or burst-disks actions come into play. The neutronic reaction depends strongly on the fuel element design, i.e. moderation ratio of the core. In extreme scenarios positive reactivity feedback effects by temperature induced steam concentration reductions in the core after massive water ingress can be thought of. Finally a heavy corrosion of the fuel elements will change the geometry of the core and thereby influence the reactivity behaviour additionally. These effects will depend on the effectiveness of the shutdown systems [11]. At normal operation, the presence of water in the primary coolant is limited and it is controlled by helium purification system (HPS) [12,13]. HPS is designed to purify the impurities in helium coolant. HPS has two trains, first train is for normal operation, and the second one is for the accident. In previous paper the purification process of HPS was discussed during normal operation so the acceptable limits of impurities were achieved [8,9,10]. The first HPS train components are cartridge filter, CuO bed, water separator, molecular sieve bed, and cryogenic active charcoal bed [12,13]. During water ingress accidents, the second train must be capable of mitigating the steam/water that ingresses to the primary system [14]. Therefore, the capability of HPS during water ingress accidents must be analyzed.

The goal of this study is to determine the total amount of water that can be mitigated by HPS during the accident. The time required to purify will be calculated during the accident. Helium mass amounts vary depending on the pipes break. The HPS process and modeling by ChemCAD computer code is used in this study. This research focused on the second train of HPS which is consisted of blower, water cooler, and water separator unit operations. The steam and water mixture mass flow rate into the primary coolant will determine the time required to clean the total amount of water ingress.

2. Theory

2.1 Experimental Power Reactor (RDE)

RDE is designed with an indirect cycle, and uses pebble fuel type. The reactor has 10 MW of thermal power which equals to ~ 3 MW of electrical power [2,3,14]. RDE is built for experimental research, generates electricity and supplies the heat for supporting hydrogen production and other research facilities. The heat is transferred from the core to secondary system via steam generator (SG). The secondary coolant is demineralized water. Secondary cooling water turns to dry steam moving towards the turbine and generator to produce electricity. The output steam from turbine is condensed using a condenser. Heat extracted from the secondary cooling by tertiary cooling will be discharged into the environment through cooling towers. The tertiary coolant uses river water after mechanical and chemical treatment. The energy conversion system diagram of RDE is shown in Figure 1 [14]. The HPS inlet flow is bypass flow from SG inlet pipe with high pressure of helium. The mass flow rate is 5% of the main primary coolant mass flow rate. After flowing to HPS, the pure helium is stored in the storage tank and will be injected to the coolant when it is needed. The primary system of the RDE is connected to a two-train helium purification system and is assigned to one purification train. The second purification train is available in the event of failure, maintenance or regeneration of the assigned train.

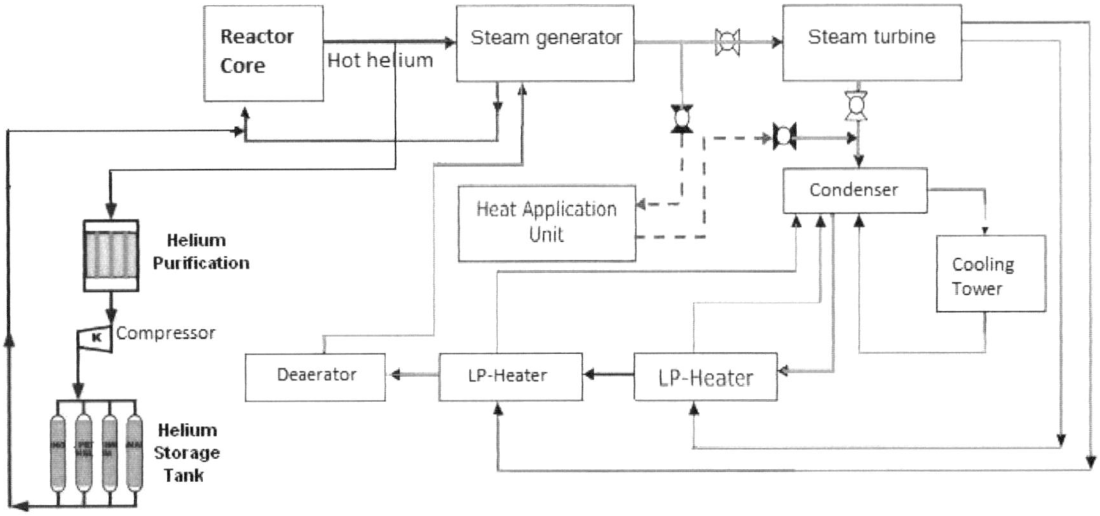

Figure 1. The energy conversion system diagram of RDE [14]

Table 1. Technical specification of RDE [14]

Parameter		Value	units
Core design			
Fuel type	:	*Pebble* (bola)	
Core height (active)	:	~ 2	m
Core diameter	:	~ 1.8	m
Core heat transfer area	:	305.4	m^2
Electrical power (*gross*)	:	~ 3	MWe
Electrical power (*net*)	:	~ 2.9	MWe
Core thermal power output	:	~10	MW
Primary cooling system			
Primary coolant	:	Helium	
Primary coolant mass flow rate	:	4.4	kg/s
Primary coolant pressure	:	30	bar
Core inlet temperature	:	250	°C
Core outlet temperature	:	700	°C
Average core temperature	:	450	°C
Primary gas blower	:		
Total static head	:	approx. 1.5	bar
Nominal delivery	:	approx. 4.3	kg/s
Outlet pressure	:	approx. 30.3	bar
Outlet temperature	:	250°	C
Secondary cooling system			
Nominal steam mass flow rate	:	~4.0	kg/s
Nominal helium mass flow rate	:	~4.4	kg/s
Nominal feed mass flow rate	:	~4.0	kg/s
Steam temperature	:	~ 530	°C
Steam pressure	:	~ 60	bar
Water feed temperature	:	~ 160	°C
Secondary pump power	:	223	liter per minute

2.2 Helium Purification System

The helium purification system of RDE consists of two trains. The first train is used for normal operation, and the second one is used for regeneration and accident conditions. The additional components used for the accident conditions are: commissioning filter, post-accident cooler, post-accident water separator, and post-accident blower as seen in Figure 2. The technical specification of each train of HPS is shown in Table 2 [14,15]. HPS has main components i.e. filter, CuO bed, molecular sieve bed and cryogenic active charcoal bed.

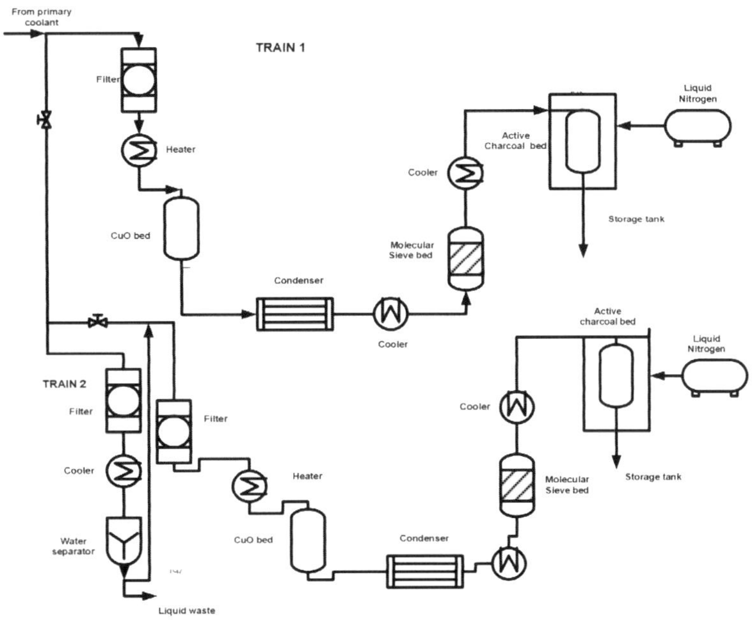

Figure 2. Helium purification system trains [14]

The helium purification system and the helium supporting systems of RDE fulfil the following functions i.e. removal of chemical and particulate contaminants from the primary coolant to maintain at a specified values, supplying purified helium to systems filled with helium as helium make up system, removal of helium from the primary system and the helium filled supporting systems and storage in the purified gas store, relieving pressure and dumping helium from helium filled auxiliary and supporting systems and, possibly storing of radioactively contaminated helium [14].

Table 2. The technical specification of two trains of HPS [14]

***Train* 1**		
Parameter		**Value**
Inlet pressure	:	1 − 39 bar
Inlet temperature	:	−180 - 300°C
Purification fraction rate (purification constant)	:	0.05/hour
Helium purification mass flow rate in the normal operation (250°C, 30 bar)	:	~ 10.5 kg/hour
One purification cycle (before regeneration)	:	1000 hour
***Train* 2**		
Inlet pressure	:	1−39 bar
Inlet temperature	:	30-300 °C
Purification fraction rate (purification constant)	:	1/hour
Mass flow rate purification	:	~ 245 kg/hour

IOP Conf. Series: Journal of Physics: Conf. Series **962** (2018) 012034 doi:10.1088/1742-6596/962/1/012034

3. Methodology

A water ingress accident can be caused by a rupture of one or two heat exchange pipes of the steam generator (SG) with 2F (twice pipe area) [16]. The locations of the rupture are assumed at the superheating and the water preheating part. After the one or two pipes rupture occurred, water and steam from the pipes will release into the primary system, and then the helium blower will force the steam-helium gas mixture into the reactor [17]. The water ingress process is analysed using the ChemCAD computer code. The process of simulating model of the SG is shown in Fig 3, which includes the volumes and pipes representing different parts of the SG. The broken pipe and the normal pipes are simulated separately. After 15 seconds of the accident, the moisture at a measuring point exceeds 800 vpm, the reactor protective system detects the water ingress accident delaying 10 seconds, then the protective managements are put into start with different delay time. It is assumed that the mass flows of helium and water remain constant within the first 25 seconds, then reduce exponentially and open the discharging valve of the SG. In the following, the part of the broken pipe between the rupture position and the entrance of feed water is defined as cold leg, the other part as hot leg.

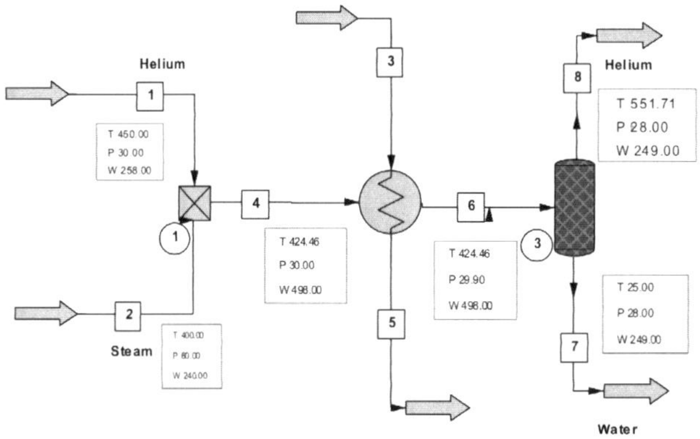

Figure 3. The condenser simulating model of HPS of RDE

In the event of steam generator tube rupture (SGTR) accident, regarding to the safety analysis report, the purification function is changed manually from the first train to the second train. There are two possible impact scenarios of the water ingress. The first impact is assumed that the totally entire mass of water from the ingress can be extracted by HPS. There is no water interacting with graphite in the core. And the second scenario is assumed that some of entire mass of water from the ingress has already entered to the core and interacted with the graphite. The mass is assumed between 10% and 30% of the total water ingress. The total water ingress assumed caused by STGR is seen in Table 3. Ideally the second condition is what will happen.

Table 3. The total water ingress assumed caused by steam generator tube rupture

Parameter	units	Superheating part of rupture		Preheating part of rupture	
Number of tube rupture		1	2	1	2
Iniate water ingress mass flow rate	kg/sec	0.59	1.18	0.43	0.86
Total water ingress	kg	17	34	12	24
Total water ingress discharging values		87	104	82	94

Figure 3 shows that stream 1 is primary coolant flow at 500 °C, 30 bar and 0.215 kg / sec of mass flow rate (5% of the main stream), while the stream 2 is water ingress at a pressure of 60 bar, with the assumption that the mass flow rate is 1.1 kg / sec. Both streams are mixed in the coolant then separated using a condenser, with stream 4 is output in the form of water and a little helium gas still remaining. Both mixtures are then separated with water separator to determine the amount of water that can be condensed. Helium gas water separator output is then saved to a storage tank for reuse when the reactor is operating normally.

4. Results and Discussion

There are two scenarios which have been simulated in the ChemCAD model. First scenario is water ingress accidents in Beyond Design Basis Accident (BDBA) and the second is Design Basis Accident (DBA). In BDBA is assumed that all tubes of SG are rupture, but in DBA only maximum two of steam generator tubes are broken. During the tube rupture water flows from the superheating and preheating section. The total mass flow from both accidents is shown in Table 3.

This released water must be cleaned by HPS. The accident/second train of HPS is responsible to this condition. The accident train technical specification is shown in Table 2. After the water steam enters the primary cooling system, it will spread to the whole primary coolant flow system. Water mass flow rate is highly dependent on break size. In this condition, water steam is mixed with helium gas in the temperature of 500°C and 30 bar of pressure. Commonly, the mixture is dry vapour. Dry vapour phase condition is dependent on the temperature and pressure of the system. The influence of water ingress into the core does not have significant impact on the area outside the core, but it affects the material inside the core like fuel pebble and graphite reflector.

During SG tube break, water and steam from a ruptured tube will infiltrate into the primary coolant and is mixed together with helium. Helium gas-steam mixture is circulated by compressor flows through the side reflectors canals. The reactor will operate in over moderated, the power and temperature increase. The chemical effect of water/steam to the graphite material is oxidation corrosion. After a few seconds, the second train of purification operates and cleans the steam. Usually the steam is stagnant in the inlet of header vessel because of large volume. The steam volume entering the reactor core is very little because the train automatically anticipates the steam ingress after the accident.

At normal operation, the presence of water (H_2O) in primary coolant is relatively low so that they can be handled by first helium purification system train. This first train capability is lower than the second train, just only 10.5 kg/hour. But in a severe accident of water ingress the presence of water in the reactor coolant becomes so huge that it is not able to be handled by first train and so automatically the operation is changed in to the second one. The second train has capability of 245 kg per hour of water ingress.

The main component of the second train is a filter, condenser, and water separator. Condenser serves to condense the steam coming then separated with a water separator. Water is separated then removed and managed as liquid waste, while the helium gas is passed into the first train of purification system to be further purified. Pure helium gas which resulted from the purification system then stored in the storage tanks to be reused when needed. The amount of water mass release to the primary coolant is assumed to be from 20 kg to 200 kg. This value is used to simulate the ChemCAD. Figure 4 illustrates simulation results of second train capability during DBDA and DBA. Figure 4 illustrates duration of purification needed when a BDBA and DBA accident occur. The simulation indicates that the BDBA consumed longer duration compare to the DBA. The maximum total amount of water in BDBA is 104 kg and during the DBA is 35 kg.

In the BDBA simulation, it takes up to 0.5 hours to purify the total mass of water ingress but lesss than 0.5 hours in the DBA. In both simulations, it is assumed that the steam does not react with graphite and 100% steam can be removed by second train of HPS. It is widely known that the graphite components in the RDE are subjected to impurity oxidants in the helium cooling during normal reactor operation or they might be exposed to air or water vapour in accident that leads depressurization. Graphite is porous material with most of the accessible surface in the interior of the pores. Graphite oxidation is controlled by the following three major steps i.e. mass transfer through the boundary layer between the exterior surface of the solid and the main gas stream, diffusion through the internal pores, and chemical reaction in the pores. There will be a chemical reaction between water and graphite. The stoichiometric reactions are as follows:

$$C + H_2 \rightarrow CO + H_2 \tag{1}$$

$$C + CO_2 \rightarrow 2CO \tag{2}$$

Beside the reaction between water and graphite, graphite oxidation can also be caused by O_2. If the graphite structures are oxidized continuously, the oxidized graphite structures may loose the structural integrity and the produced carbon monoxide may detonate. The effect of oxidation on mechanical properties specifically tensile strength of graphite is significant. The reaction is $2C + O_2 \rightarrow 2CO$. The CO is more stable below 700°C comparing to CO_2 [18].

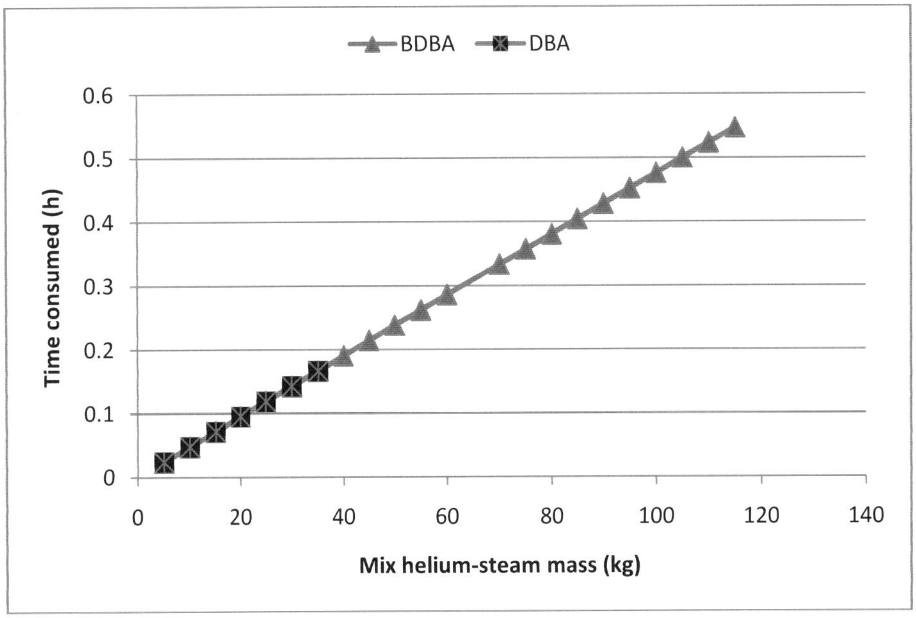

Figure 4. The BDBA and DBA scenario at water ingress in RDE.

According to the chemical reactions (1) and (2), the products are hydrogen and carbon monoxide gas. The reaction between graphite and water is a spontaneous reaction. It means without any necessary energy addition from outside the reaction has occurred. This reaction directly takes place after water infiltrate to the reactor core at 700 °C. The high temperature condition accelerates the reaction so that in a very short time the water will run out and produce hydrogen gas and carbon monoxide [17]. The quantity of hydrogen and carbon monoxide gas of the reaction product is dependent on the much amount of water infiltration to the core. Figure 5 illustrates the correlation between hydrogen and carbon monoxide which is resulted from the interaction between water and graphite in the core. The total amount of water is from 3.3 kg to the 35 kg of mass. From the Arrhenius equation the corrosion reaction rate is calculated using: $R_e = 4.5 \times 10^4 \exp\left(-\frac{113}{RT}\right)$ for the temperature below 1200°K (923°C). But for the temperature above 1200 °K the equation is different i.e. $R_e = 1.8 \times 10^{10} \exp\left(-\frac{251}{RT}\right)$. R_e represents the reaction rate (weight loss per unit exterior surface area per unit time, mg/cm^2.h). The oxidation rates due to O_2 and H_2O are much faster than the one by the CO_2. The oxidation rate by H_2O was 3 to 5 times faster than by CO_2 in the temperature range of

800–1000°C [18]. For this reason and because CO_2 should be at very low concentration in RDE systems, as it is controlled as low as possible by the purification system, the influence of CO_2 was neglected.

The hydrogen gas generation during water ingress is shown in Figure 5. The graph shows the hydrogen gas generation in the reactor coolant progressively with increasing H_2O in the core, and in the worst conditions (assuming the total amount of incoming water is 110 kg or equal to with 33.33 kmol) the amount of hydrogen gas generated is as much as 12.2 kg, equivalent to 33.34 kmol H_2 gas and carbon monoxide gas as much as 171.1 kg which is equivalent to 33.33 kmol of CO. Regarding to the calculation the hydrogen generation is larger and must be considered to the reactor safety impact. The hydrogen is an explosive gas.

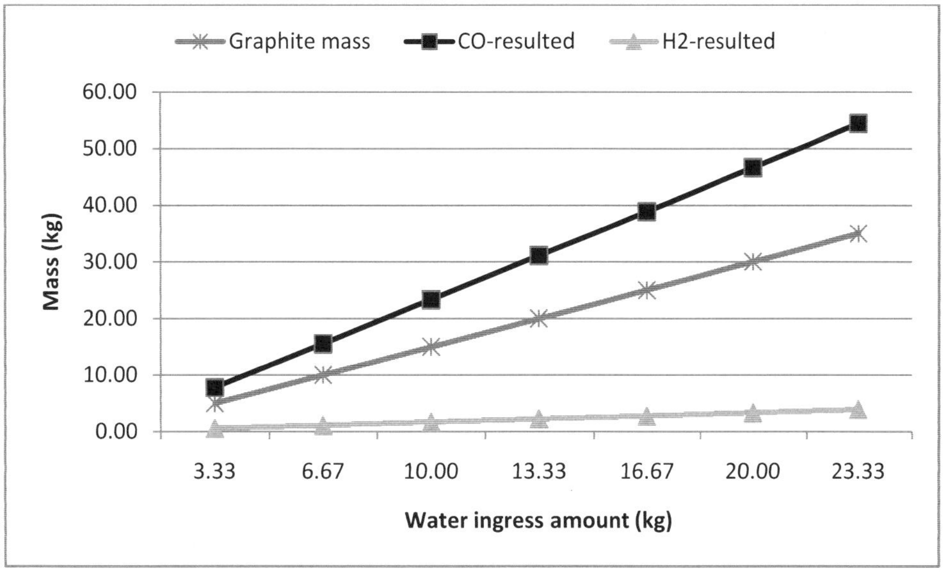

Figure 5. The relation between water ingress and the amount of CO, H_2 resulted in the reactor core.

5. Conclusion

The second train of HPS has been modeled to simulate the water ingress event in RDE. In the SGTR accident, the purification function was changed to the second train. It assumed the BDBA and DBA scenario. In BDBA scenario the total amount of water ingress is assumed to be up to 110 kg, but in the DBA the total mass amount of water coming into the primary coolant is 3-35 kg. It takes up to 0.5 hours to clean up the water ingress in BDA scenario accident as a cathastropic event. If any portion of the water that interacts with graphite in the core, it produce carbon monoxide gas and hydrogen gas. The latter is highly dependent gas to the mass amounts of water interact. In the most severe water infiltration conditions (assuming the number of incoming water is 110 kg), the amount of hydrogen gas generated is as much as 12.2 kg, equivalent to 33.3 kmol H_2 gas and carbon monoxide gas is as much as 171.1 kg equivalent to 33.33 kmol CO gas.

Acknowledgment

This research is supported by PTKRN annual budget of the year 2017 (DIPA PTKRN 2017). The authors thank to the Head of Centre for Nuclear Reactor Technology and Safety – BATAN

References

[1] Zuhair and Suwoto 2015 The Effect of Applying Silicon Carbide Coating on Core Reactivity of HTR Pebble Bed in Water Ingress Accident *National Seminar Proceeding on Nuclear Energy Technology* ISSN : 2355-7524 pp 16-22

[2] Ptkrn 2014 Dokumen Justifikasi Teknis Pembangunan Reaktor Daya Eksperimental, *No. Dokumen : DT 001.KRN.2014*, BATAN, pp 1-10 Indonesian

[3] Ptkrn 2014 Spesifikasi Teknis Reaktor Daya Eksperimental *No. Dokumen : DT 002.KRN.2014*, pp 14-15 Indonesian

[4] Wu Zongwin et.al., 2012 The Design Feature of HTR-10 *Nuclear Engineering and Design* vol 218, 2002, pp 25–32

[5] Zuoyi Zhang and Yujie Dong 2007 Assessment of Water Ingress Accident in A Modular High Temperature Gas-Cooled Reactor *Tsinghua University Institute Of Nuclear Technology*, Beijing 100084, China pp 23-29

[6] Katscher W and Moormann R 1990 Fission Product And Graphite Corrosion Under Accident Conditions in The HTR *Nuclear Engineering and Design*, North Holland, pp 219-225

[7] Natesan K. A. Purohit and S.W. Tan 2003 Material Behavior in HTGR Environments, Argonne National Laboratory, *NUREG/CR-6824 ANL-0237*, pp 43-55

[8] Sumijanto and Sriyono 2016 Optimasi Laju Alir Massa dalam Purifikasi Pendingin RGTT200K Untuk Proses Konversi Karbonmonoksida *Jurnal Teknologi Reaktor Nuklir TRI DASA MEGA*, Volume 18, Nomor 1, ISSN 1411-240X, pp 11-20 Indonesian

[9] Sumijanto 2010 Kajian Dampak Gas Pengotor Pendingin Primer Terhadap Integritas Material Struktur RGTT *Majalah Ilmiah Teknologi Keselamatan Nuklir Sigma Epsilon*, Volume 14 Nomor 2, pp 5-12 Indonesian

[10] Sriyono 2010 Analisis Termodinamika dan Penentuan Rasio Pengotor Gas Helium Terhadap Degradasi Material Pendingin RGTT 200 MW *Prosiding PPI – PDIPTN*, Pustek Akselerator dan Proses Bahan – BATAN, pp 205-212 Indonesian

[11] Gao Zuying, Jiang Zhiqiang, Li Baoyan, and Wang Chunyun 1996 Safety Analysis of Accident Scenarios for the HTR-10 *Tsinghua Science and Technology*, ISSN 1007-0214 07/19 Vol 1, No. 1, March 1996, pp 27-31

[12] Yao MS 2012 The Helium Purification System of HTR-10 *Nuclear Engineering and Design* 218, Institute of Nuclear energy Technology, Tsinghua University, Beijing, pp 163-167

[13] Hamamoto Shimpei and Shimazaki Y et.al 2014. Investigation of chemical Characteristics of Primary Helium Gas Coolant of HTTR (High Temperature Engineering Test Reactor)", *Nuclear Engineering and Design* vol 271, pp 487-491

[14] Batan 2015 Final Report : Safety Analysis Report (SAR), Reaktor Daya Eksperimental, *RDE-DS-WBS 06, Documents Design Rev 02*, December, 2015 pp 10-50.

[15] Yan Wang , Yanhua Zheng, Fu Li, and Lei Shi 2014 Analysis on blow down transient in water ingress accident of high temperature gas-cooled reactor *Nuclear Engineering and Design* vol. 271, pp 404-410

[16] Iaea Tecdoc-784 1993 Response of fuel, fuel elements and gas cooled reactor cores under accidental air or water ingress conditions *Proceeding of a Technical Committee meeting*, Beijing, pp 10-128

[17] Z. Yanhua , S. Lei , and W. Yan 2010 Water-Ingress Analysis for The 200MWe Pebble-Bed Modular High Temperature Gas-cooled Reactor, Nuclear Engineering and Design vol 240, 3095-3107, pp 3095-3107.

[18] Rik Van Bremen 2013 Water ingress scenario analyses of a thorium fuelled HTR *Bachelor Thesis*, Dept. Radiation Science & Technology, Faculty of Applied Sciences, TU Delft, Netherlands, pp -10-57

Analysis of radiation safety for Small Modular Reactor (SMR) on PWR-100 MWe type

P M Udiyani, I Husnayani, Deswandri, and G R Sunaryo

Center for Nuclear Reactor Technology and Safety, National Nuclear Energy Agency of Indonesia-BATAN Serpong-Tangerang Selatan, Indonesia

Email: pmade-u@batan.go.id

Abstract. Indonesia as an archipelago country, including big, medium and small islands is suitable to construction of Small Medium/Modular reactors. Preliminary technology assessment on various SMR has been started, indeed the SMR is grouped into Light Water Reactor, Gas Cooled Reactor, and Solid Cooled Reactor and from its site it is group into Land Based reactor and Water Based Reactor. Fukushima accident made people doubt about the safety of Nuclear Power Plant (NPP), which impact on the public perception of the safety of nuclear power plants. The paper will describe the assessment of safety and radiation consequences on site for normal operation and Design Basis Accident postulation of SMR based on PWR-100 MWe in Bangka Island. Consequences of radiation for normal operation simulated for 3 units SMR. The source term was generated from an inventory by using ORIGEN-2 software and the consequence of routine calculated by PC-Cream and accident by PC Cosyma. The adopted methodology used was based on site-specific meteorological and spatial data. According to calculation by PC-CREAM 08 computer code, the highest individual dose in site area for adults is 5.34E-02 mSv/y in ESE direction within 1 km distance from stack. The result of calculation is that doses on public for normal operation below 1mSv/y. The calculation result from PC Cosyma, the highest individual dose is 1.92.E+00 mSv in ESE direction within 1km distance from stack. The total collective dose (all pathway) is 3.39E-01 manSv, with dominant supporting from cloud pathway. Results show that there are no evacuation countermeasure will be taken based on the regulation of emergency.

1. Introduction

Indonesia consists of around 17,500 big, medium and small islands, the big islands are: Java, Sumatra, Kalimantan, Sulawesi and Papua. Total area is about 1.9 million square miles including the ocean. Population is around 250 millions people with rate of 1.21% (2015). Based on that condition, Indonesia needs high demand and supply of energy in the future. National economic and energy assessment has shown that nuclear energy would be part of energy sources in national energy mix policy simultaneously with fossil energy source, gas, water and other renewable energy. Due to high demand and supply of energy in the future, thus it has been chosen that SMR technology is much more appropriate for small islands to support their development. Preliminary technology assessment on various SMR have been started, indeed the SMR is grouped into Light Water Reactor, Gas Cooled Reactor, and Solid Cooled Reactor and from its location it is group into Land Based reactor and Water Based Reactor. Based on Technology Assessment, the type of SMR that has a high probability to be built in Indonesia is PWR (Pressurized Water Reactor) type. The SMRs have been designed as PWR type in the world such as CAREM-25 (100MWt), ACP-100 (310MWt), FLEXBLUE (530MWt), KLT-40 (150MWt) and SMART-300 (100MWe) [1]. Research on various types of PWR have been done, which are about the reactor design, reactor core, and the reactor [2,3,4].

Fukushima accident made people doubt about the safety of Nuclear Power Plant (NPP), which impact the public perception on the safety of Nuclear Power Plants. Lesson leant from Fukushima accident is that the safety of nuclear power plants need to be highlighted back to improve the public acceptance of nuclear safety and security in Indonesia. To prove that the operation of nuclear power plant in Indonesia required analysis of radioactive dispersion into the environment under normal conditions and postulated accident. The objective of this paper is to assess of radiation consequences on site for routine operation and DBA (Design Basis Accident) postulation of SMR PWR-100 MWe in Bangka Island. Consequences of radiation for normal operation were simulated for 3 (three) units SMR. The source term was generated from an inventory activities released by using ORIGEN-2 software and the consequence calculated by PC-Cream (routine release) and PC Cosyma (accident release). The adopted methodology used in this research work was based on the predominant site-specific meteorological and spatial data.

2. Methodhology

The assessment approach in this work was based on atmospheric dispersion calculation [5-14]. The reactor inventory used ORIGEN2.0 [15], the dose calculation used PC-CREAM 08 for normal operation [5,12] and PC Cosyma for DBA calculation [7-9]. Based on inventory, release assumption each compartment and postulation, the source term would be calculated. Through various pathways, source terms release will get into the human body depending on the type and behaviour of nuclides as well as meteorological and environmental conditions.

2.1. Inventory and Source term Calculation

The series of studis performed on the first core and transition core configurations 2 until the equilibrium core is reached. The initial step is to make an input of the first core configuration with composition consists of 57. Each Fuel Assembly (FA) consists of a cylindrical UO_2-Gd_2O3 fuel in a 17x17 matrix. Fuel Assembly in the reactor core has 2 levels of uranium enrichment of 2.82% and 4.88%. The FA has average burn-up of 36 MWD/KGHM with the one reactor cycle is 1000 days operation [1].

The source terms were calculated for routine and accident releases. The calculation of source terms was begun with calculation of reactor inventory. After the inventory was obtained, the calculation of source term that release from inventory to environment were conducted. The routine release is assumed that the release of fission products from pinholes reached 0.1%.. In addition to the fission products of the porosity of the cladding, the fission products in the primary coolant contamination also comes from impurities natural uranium and enriched uranium on the outer surface of the cladding. Uranium contaminants in surface cladding can reach 10 microns uranium weight [12].

For the accident cases, scenarios were postulated by Large Break Loss of Coolant Accident (LB-LOCA). The source term was calculated based on reactor inventory with the assumption core damaged reaches approximately 33% which is the optimal damaged based on results of the study and an agreement for LB-LOCA accident [9,13-14]. Estimation of source term were calculated by using the following assumptions: 33% cores damaged, the gap release of noble gases; Kr = 7.5% and Xe = 2.15%, for Iodine = 0.65%, other nuclides = 0.0051. For release core inventory: Iodine is 0.22%, Cs-137 is 0.5%, and other nuclides 0.06%. Reduction in containment for nuclide Iodine is 0.46. The efficiency of the filter in the chimney of the reactor was taken to 0% noble gases, iodine (organic) 90%, and other nuclides (Br, Te, Cs, Rb) 99% [7,9].

2.2. Doses calculations

The mathematical models used to get the generic data sets are atmospheric dispersion using a Gaussian plume model, dry and wet deposition. A Gaussian Modification (Segmented Gaussian) atmospheric dispersion code PC Cream and PC Cosyma were used to estimate the radiological doses of the released radionuclides [5-9]. The segmented Gaussian plume model is accurate on the 0-50km scale. The model requires the following input data and parameters: time and location, atmospheric data (wind speed and wind direction, air temperature, relative humidity, cloud cover, mixing layer height), stability class of the atmosphere (proposed by the model based on meteorological data), roughness length and information about the release (air pollutant, duration of release, amount of released material, source height). The distribution of Pasquille Guifford atmospheric stability classes as an annual base of the West Bangka site were processed hourly. Radiouclides are released through the ventilation stack of 60 meters height to the environment. The dose were calculated by area within 20 km around the site for nine distances (0.5, 1,2, 3,,4, 5, 10, 15, 20, 30, 40 and 50 km) and 16 directions/ sectors at different wind speed.

3. Result and Discussion

3.1. Inventory and source terms

Table 1. Core inventory, routine source term and accident of SMR-100 MWe

Nuclide	Core Inventory (Bq)	Routine source term (Bq/y)	Accident source term (Bq)	Nuclide	Core Inventory (Bq)	Routine source term (Bq/y)	Accident source term (Bq)
Kr -85m	8.76E+16	2.89E+13	1.97E+15	Cs-137	3.38E+16	2.79E+07	1.02E+09
Kr- 88	2.45E+17	8.09E+13	5.52E+15	Te-132	4.85E+17	1.60E+08	3.71E+10
Xe-133	3.76E+17	1.24E+14	2.43E+15	Ba-139	6.19E+17	5.11E+08	4.74E+10
I-131	3.37E+17	1.67E+09	6.66E+10	Ba-140	6.19E+17	5.11E+08	4.78E+10
I-132	4.93E+17	2.44E+09	9.74E+10	Sr -90	2.55E+16	2.11E+07	1.95E+09
I-133	6.99E+17	3.46E+09	1.38E+11	Mo -99	6.36E+17	5.25E+08	4.87E+10
I-134	7.77E+17	3.85E+09	1.53E+11	Ru-106	1.53E+17	5.04E+07	1.17E+10
I-135	6.67E+17	3.30E+09	1.32E+11	Rh 105	3.23E+17	1.07E+08	2.47E+10
Rb- 88	4.80E+14	3.96E+05	3.68E+07	Y- 90	2.66E+16	2.20E+07	2.04E+09
Cs-134	4.48E+16	3.69E+07	3.43E+09	Y- 91	4.51E+17	3.72E+08	3.45E+10

Table 1 gives the activity of core inventory, source term routine and accident for SMR-100 MWe with PWR type. Based on inventory data, the source term for routine released and for LB-LOCA accident have been calculated. The result estimation is shown on the same table. The source term calculation in the normal/routine operations are activity for a year. Based on activity, and a half-life nuclide, and toxicity then are selected 20 nuclides as the source term. The routine source term were used as input data for PC CREAM, and accident source term were used for PC Cosyma.

3.2. Routine doses

The results of the mean effective individual dose for adult (all pathway and nuclide) from sums stack of three SMR calculations are given in Fig 1 and Fig.2. The term dose for normal dose used in the paper is therefore the sum of the annual external and internal effective doses to individuals received over 1 y. The individuals doses at each sector were assessed assuming that the individuals occupancy indoors is 80-90%, depend on habit and occupancy of population. Trend of radiation dose in Fig. 1 showed that the individual dose decreases with increasing distance from the stack. The data also show that the highest doses for all distances is 3.78E-02 mSv/y within 1km distance from all stacks. The dominant nuclide is Kr-88 for Gamma from Plume pathway and Rb-88 for Beta from Ground pathway

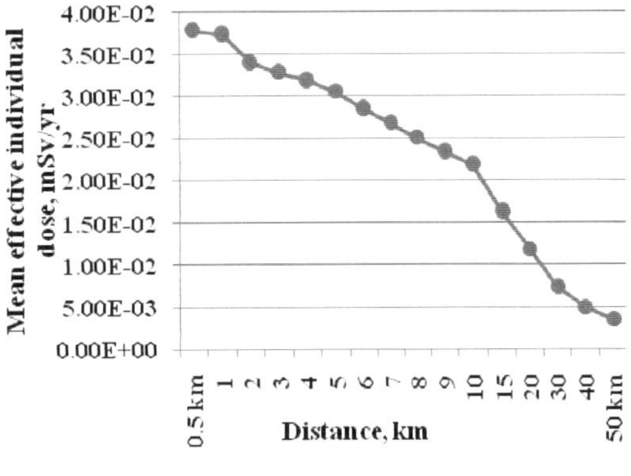

Fig.1. Mean effective individual dose by distance

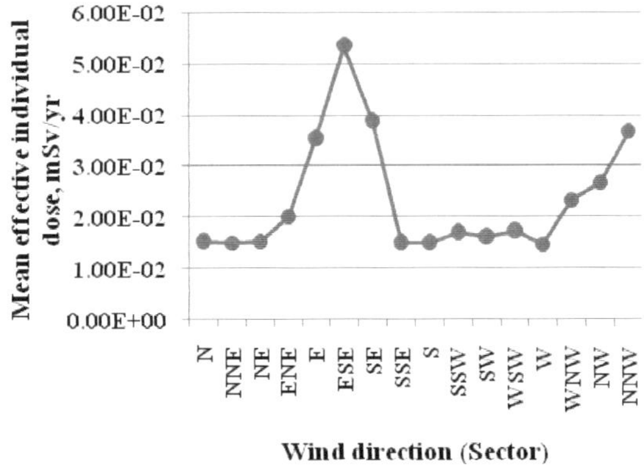

Fig.2. Mean effective individual dose by sector

Figure 2 show the mean effective dose for adult (all pathway and nuclide) from sums stack of three SMR calculations by wind direction or sectors. The wind directions divided into 16 sectors. According to calculation by PC-CREAM 08 computer code, the highest mean effective individual dose in terrestrial for adults is 5.34E-02 mSv/y in ESE direction within 1 km distance from stack. (Fig.1). It can also be concluded that the estimated effective doses are lower than the dose constraint of 0.3 mSv/y (BAPETEN, 2014) associated with this plant.

3.3. Accident Doses

The dose from accident was calculated by LB LOCA accident postulation. Based on accident source term on Table 1, dose was calculated by PC. Cosyma. The calculation result is shown on Fig. 3 and Fig.4.

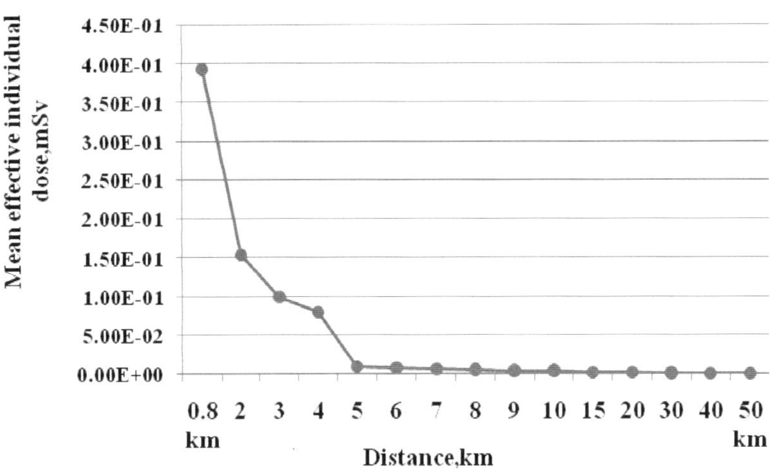

Fig.3. Mean effective individual dose for accident by distance

Mean effective individual dose for accident by distance is shown on Fig 3, and Mean effective individual dose for accident by wind direction is shown on Fig 4. The dose was calculated by 20 source term nuclide which were dispersed in atmosphere and deposed on ground surface. The dominant dose from air is supported by noble gases Xe-133 and Kr-88. Nuclides of I-131, Sr-90, and Te-132 support dose from ground surface.

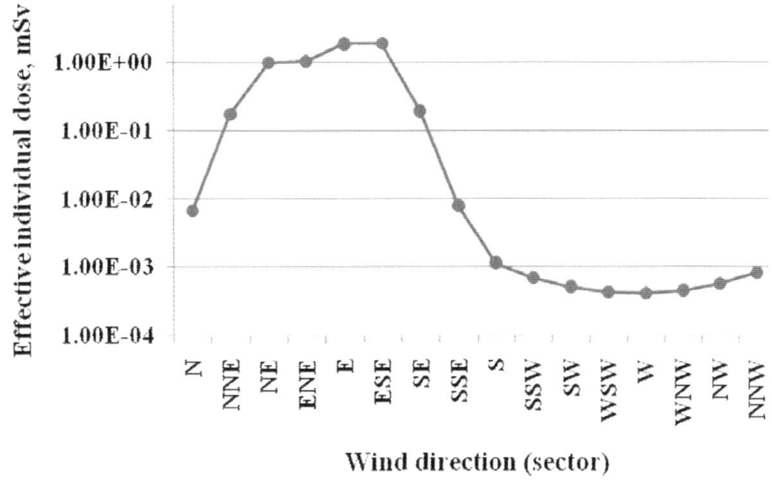

Fig.4. Mean effective individual dose for accident by wind direction (sector)

Fig. 3 and Fig. 4 show the highest individuals effective dose for LBLOCA is 1.92.E+00 mSv in ESE direction within 1km distance from stack. Results show that there are no evacuation countermeasure will be taken based on the regulation of emergency. Evacuation shall be taken if the dose have accepted by public is ≥ 50 mSv (BAPETEN). From Fig. 3 that the individual dose decreases with increasing distance from the stack

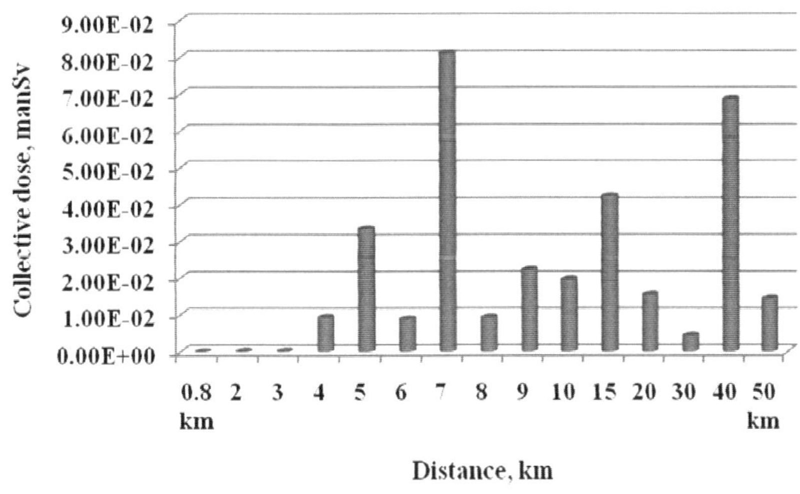

Fig.5. Collective dose for accident by distance

Collective dose is based on individual dose and population. Fig.5 shows on the collective dose for LBLOCA accident by distance. The total collective dose (all pathway) is 3.39E-01 manSv, with dominant supporting from cloud pathway. The results show that for LBLOCA postulated accident at 100 PWR-MWe, it does not lead to evacuation countermeasures on public in site. The assessment have been done to prove that the construction of SMR is safe and meets the safety standard for normal operation and the impact in DBA conditions.

4. Conclusion

The result of analysis radiation safety for SMR with PWR-100 MWe: the highest individuals dose for routine discharge of normal operation for adults is 5.34E-02 mSv/y in ESE direction within 1 km distance from stack. It can also be concluded that the estimated effective doses are lower than the dose constraint of 0.3 mSv/y associated with this plant (BAPETEN, 2014). The highest individual effective dose for LBLOCA is 1.92.E+00 mSv in ESE direction within 1km distance from stack. The total collective dose (all pathway) is 3.39E-01 manSv, with dominant supporting from cloud pathway. Results show that there are no evacuation countermeasure will be taken based on the regulation of emergency. The assessment have been done to prove that the construction of SMR -100MWe in site is safe and meets the safety standard for normal operation and the impact in DBA conditions.

Acknowledgments

The study is supported by PTKRN-BATAN through the facilities of Government Budgets in fiscal years 2015 as part of the research activity on safety analysis of SMR. Authors thank to colleagues from Reactor Safety Technology Division who have helped and encouraged author to complete this study.

References

[1] Kuntjoro S 2016 Radionuclide inventory analysis of the SMR Smart 330 MWt reactor *Proceding of SENTEN* (Batam-Indonesia)

[2] Ingersolla DT and Houghtona ZJ, 2014 NuScale Small Modular Reactor for co-generation of electricity and water *Journal of Desalination* **340** 84–93

[3] Liu Z and Fan J 2014 Technology readiness assessment of Small Modular Reactor (SMR) designs *Progress in Nuclear Energy* **70** 20-28

[4] Ashiqa M, Ilyasa M and Ahmad SI 2016 Optimization of PWR design parameters for Implementation in SMRs *Annals of Nuclear Energy* **94** 123–28

[5] Sohrabi M, Parsouzi R, Amrollahi R et al. 2013 Public exposure from environmental release of radioactive material under normal operation of unit-1 Bushehr nuclear power plant, *Annals of Nuclear Energy* **55** 351–54

[6] Katata G, Ota M, Terada H, et al, 2012 Atmospheric discharge and dispersion of radionuclides during the Fukushima Dai-ichi nuclear power plant accident. Part I: Source term estimation and local-scale atmospheric dispersion in early phase of the accident, *J. Env. Rad.* **109** 103-13

[7] Udiyani PM, Kuntjoro S, and Widodo S 2016 A Backward method to estimate the Dai-ichi reactor core damage using radiation exposure in the environment, *Aij. J.* **42** 2 63-70, DOI: http://dx.doi.org/10.17146/aij.2016.497

[8] Birikorang S A, Abrefah R G, Sogbadji RBM 2015 Ground deposition assessment of radionuclides following a hypothetical release from Ghana Research Reactor-1 (GHARR-1) using atmospheric dispersion model, *Prog. in Nuc. E. J.* **79** 99-101

[9] Udiyani PM, and Widodo S 2012 Estimation of atmospheric dispersion coefficients for the analysis of PWR accidents in Indonesia *Tri Dasa Mega J.* **14**:2 (2012), 122-27. (in Indonesia).

[10] Leelossy Á, Mészáros R, and Lagzi I, 2011Short and long term dispersion patterns of radionuclides in the atmosphere around the Fukushima Nuclear Power Plant *J. Env. Rad.* **102**, 1117- 21

[11] Petr P, and Pechova E 2014 An unconventional adaptation of a classical Gaussian plume dispersion scheme for the fast assessment of external irradiation from a radioactive cloud, atmospheric environment **89** 298-308

[12] Udiyani P.M, 2012 The evaluation of radionuclide release of AP-1000 to environment on routine operation *Proceeding National Conference of Nuclear Energy Development V*, (Jakarta) (in Indonesian)

[13] Kang D G, Ahn SH and Chang S H 2013 A combined deterministic and probabilistic procedure for safety assessment of beyond design basis accidents in Nuclear Power Plant: application to ECCS performance assessment for design basis LOCA redefinition, *Nuclear Engineering and Design* **260** 165–74

[14] Rahim FC, Rahgoshay M, and Mousavian SK 2012 A Study of Large Break LOCA in the AP1000 Reactor Containment. *Prog. Nucl. Energy J.* **54: 1** 132-37

[15] Croff AG.,et al,' 1983 ORIGEN2, A Versatile Computer Code For Calculating The Nuclide Compositions And Characteristics of Nuclear Materials', Oak Ridge National Laboratory Chemical Technology Division, P.O. Box, Oak Ridge, Tennessee, 37830, USA

Master Logic Diagram: An Approach to Identify Initiating Events of HTGRs

J H Purba

Center for Nuclear Reactor Technology and Safety, National Nuclear Energy Agency of Indonesia (BATAN), KawasanPuspiptek, Serpong, Tangerang Selatan – Banten, Indonesia

Email: purba-jh@batan.go.id

Abstract. Initiating events of a nuclear power plant being evaluated need to be firstly identified prior to applying probabilistic safety assessment on that plant. Various types of master logic diagrams (MLDs) have been proposedforsearching initiating events of the next generation of nuclear power plants, which have limited data and operating experiences. Those MLDs are different in the number of steps or levels and different in the basis for developing them. This study proposed another type of MLD approach to find high temperature gas cooled reactor (HTGR) initiating events. It consists of five functional steps starting from the top event representing the final objective of the safety functions to the basic event representing the goal of the MLD development, which is an initiating event. The application of the proposed approach to search for two HTGR initiating events, i.e. *power turbine generator trip* and *loss of offsite power*, is provided. The results confirmed that the proposed MLD is feasiblefor finding HTGR initiating events.

Keywords: Master logic diagram, initiating events, high temperature gas cooled reactor, probabilistic safety assessment.

Introduction

Probabilistic safety assessment (PSA) has been considered as an important tool to evaluate safety systems of nuclear power plants (NPPs). It can ensure the safety of NPPs in relation to potential initiating events [1, 2]. To deal with the limitation of the conventional PSA due to lack of historical failure data for evaluating the reliability of a safety system, fuzzy numbers have also been introduced intoNPP PSA [3, 4]. When PSA results suggest that changes should be made to the design to improve safety, the changes should be applied where reasonably applicable.

To perform PSA, all initiating events need to be firstly identified and evaluated.An initiating event is an event, which can affect the normal operation of a plant and potentially lead to an accident when plant-mitigating systems do not properly work as expected. Initiating events include component failures, human errors, internal hazards and external hazards. Because core damage is unlikely to happen, the goals of the PSA for high temperature gas cooled reactors (HTGRs) should not be the estimation of a core damage frequency (CDF) and a large early release frequency (LERF) as in light water reactors (LWRs) [5]. The objective of HTGR PSA is to calculate frequencies of HTGR-specific plant state corresponding to all licensing basis events (LBEs).

Due to the unique concept of HTGR technologies whosedesigns are very different from LWR technologies, lists of existing initiating events of LWRs are not relevant. Therefore, initiating events, which are specific to HTGR technology design, needs to be identified. International Atomic Energy Agency (IAEA) recommendsapplying a systematic approach to search for initiating events of a nuclear power plant[6].

We encounter from literatures that master logic diagram (MLD) approacheshave been applied to identify nuclear power plant initiating events in the past [7] and have been proposed forsearching initiating events of the next generation of nuclear plants (NGNPs) [8].MLD is a logic diagram, which is similar to a fault tree but without formal mathematical properties[9].It can represent the effects of a failure in a complex system [10] and provides a great degree of certainty for identifying initiating events [7]. Furthermore, Ahn, et al. [11]found that this approach can logically identify additional initiating events and group initiating events. Each MLD approach proposed in those previous studies has different stepsand basis in finding initiating events. The purpose of this study is to propose and discuss another type of MLD approach to search for HTGR initiating events. The proposed MLD approach consists of five steps to finally find an initiating event. To show the process of the proposed MLD,searching processes of two HTGR initiating events, i.e. *power turbine generator trip* and *loss of offsite power*, are provided and discussed.

HTGR Safety Philosophy

Currently, there are two designs of HTGRs, namely, pebble bed reactors (PBRs) and prismatic block reactors (PMRs) [12, 13]. PBRs consist of fuel in the form of pebbles, which are stacked together in a cylindrical pressure vessel. Meanwhile, PMRs refer to a prismatic block core configuration in which hexagonal graphite blocks with fuel elements are stacked to fit in a cylindrical pressure vessel. Both designs are shown in Figure 1[14].

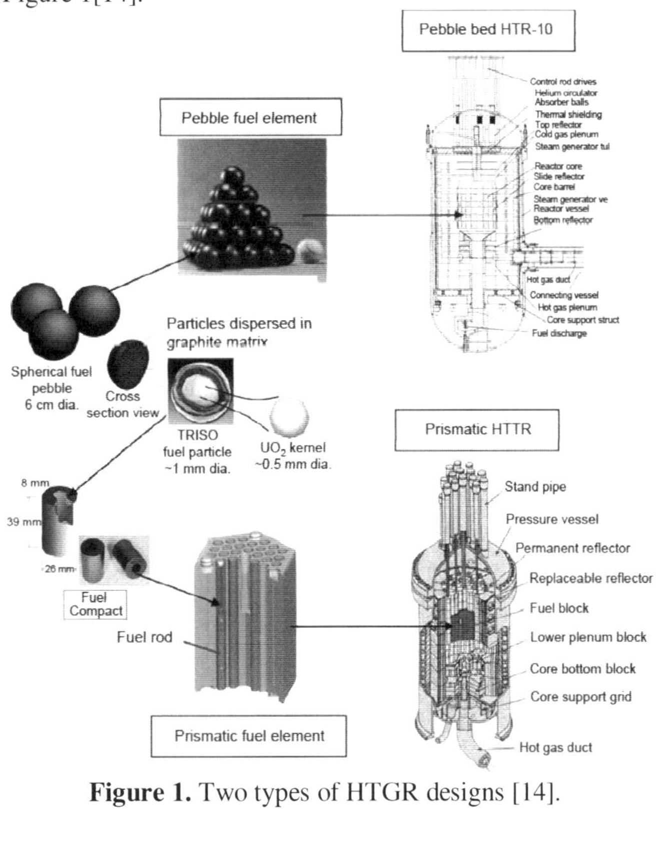

Figure 1. Two types of HTGR designs [14].

IOP Conf. Series: Journal of Physics: Conf. Series **962** (2018) 012036 doi:10.1088/1742-6596/962/1/012036

HTGRs have been designed to be inherently safe reactors in which active safety systems and operator intervention are not needed when unexpected events happen [15]. This safety enhancement is achieved through the implementation of inherent safety features and engineered passive safety features [16, 17]. Inherent safety features are realized through Tri Iso-Structural (TRISO) – coated fuel particles, graphite moderator, and helium coolant. The TRISO-coated fuel particles in which three ceramic coating layers, i.e. inner pyrocarbon (IPyC), silicon carbide (SiC), and outer pyrocarbon (OPyC), surround fissionable fuel kernel as shown in Figure 2 [18]. TRISO-coated fuel particles can withstand extremely hightemperature up to 2000 ^{0}C [19].

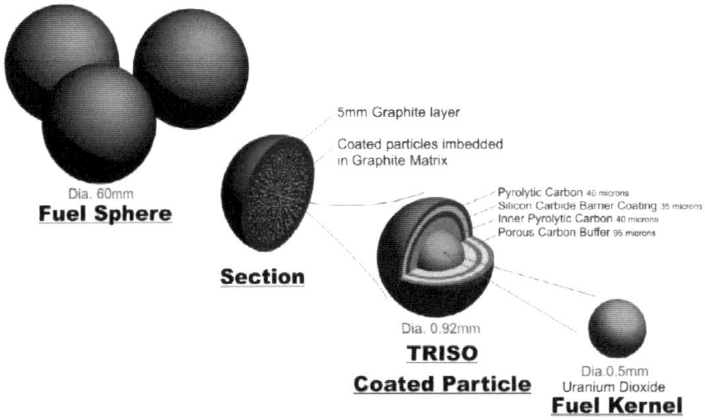

Figure 2. TRISO-coated fuel particle [18].

Meanwhile, the engineered passive safety features are characterized by passive heat removal capability and inherent shutdown capability. The passive heat removal feature is achieved through the low-power-density core with a relatively large height-to-diameter ratio within an un-insulated steel reactor vessel. The inherent shutdown capability is achieved by large negative temperature coefficient and large thermal margin to enable reactor to deal with scram failure.

Methodology

MLDapproach proposed in this study consists offivefunctional steps as graphically shown in Fig. 3. In the sequel, the detail of each functional step is elaborated.

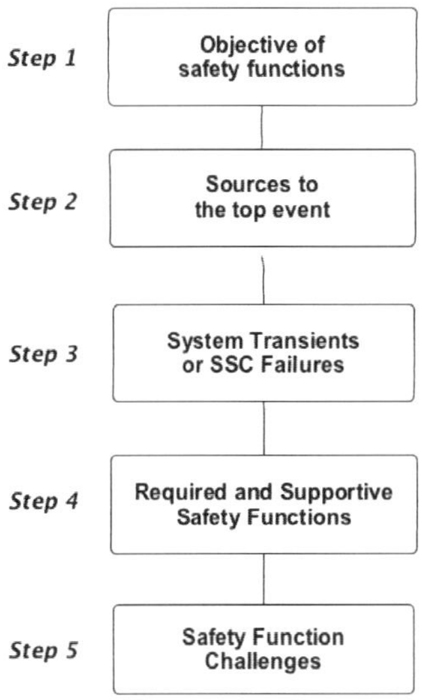

Figure 3. Proposed master logic diagram approach.

Step 1:
This functional step generates the final objective of the safety functions of a plant, which is being investigated. The output of this step is, then, set as the top event of the proposed master logic diagram.

Step 2:
This functional step identifiessources, which can cause the top event identified in *Step 1*. The output of this step isa list of sources to the top event when transients occur to the system or failures happen to systems, structures and components (SSCs).

Step 3:
This functional step identifies transients, which could occur to the system or failures, which could happen to SSCs of the plant being investigated. The output of this step is a list of system transients or SSC failures.

Step 4:
This functional step identifies safety functions, which are required to perform, to avoid system transients or SSC failures. The output of this step is a list of required and supportive safety functions.

Step 5:
This functional step identifies challenges to safety functions defined in *Step 4*, which could cause system transients or SSC failures defined in *Step 3*. The output of this step is a list of initiating events.

Results and Discussion
To obtain an operating license, possible initiating events and all sequential events following those initiating events of the nuclear power plant being proposed to be constructedneed to be provided. HTGRs as the next generation of nuclear power plants, which have limited operating data and experiences, also need to provide a list of initiating events, which could cause the release of radioactive materials into environment. This section discusses the process of finding two HTGR initiating events, i.e. *power turbine generator trip* and *loss of offsite power*, to illustrate the implementation of the proposed MLD approach described in Section 3.

Step 1:
The final objective of HTGR safety functions is to prevent or mitigate radioactive material releases into environment. This study, then, justifies that *material radioactive releases into environment* is the top event of the proposed MLD as graphically shown in Fig. 4.

Step 2:
Sources of radioactive materials in HTGRs can be classified into two categories, i.e. *from inside of the helium pressure boundary (HPB)* and *from outside of the HPB*. Both sources are shown in Table 1[8].

Table 1. Sources of radioactive materials in HTGRs[8].

Categories	Material radioactive sources
From Inside of HPB	• Fuel elements in the core • Intact coated particles • Failed or defected coated particles • Uranium contamination outside coated particles • Materials attached to graphite components • Dust and plate-out on the surface of the HPB • Circulating primary coolant activity
From Outside of HPB	• Fuel elements in the storage systems • Gas-borne activities in the helium purification system (HPS) • Solid and liquid rad-waste systems

Those two source categories in Table 1 are, then, presented as the sources of the top event in the proposed MLD in Fig. 4.

Step 3:
Radioactive materials could be released from the inside of the helium pressure boundary (HPB) if the possibleconditions defined below happen to HTGRs[8]:

a. Plant transients with intact primary system HPB;
b. Energy conversion system transients with intact HPB and reactivity addition;
c. Primary system HPB leaks and breaks;
d. HPB heat exchanger failures.

In condition *a*, fission products could be vented from HPBthrough the vessel relief valve in response to the possibility of the primary circuit to be over pressured if helium is not released. This condition could be triggered by a water ingress event [20]. Helium in the reactor building is, then, released to environment through reactor building vent. For illustration purposes only, the *plant transients with intact primary system HPB* is the only system transient shown in the proposed MLD in Fig. 4.

Step 4:
The conditions of the required and supportive safety functions, which could cause *plant transients with intact primary system HPB*, are[8]:

a. Main loop and shutdown cooling system (SCS) are still capable of forced cooling operation
b. Main loop system fails but SCS is still capable of operation
c. SCS fails but main loop system is still capable of operation
d. Main loop and SCS are not capable of operation.

To simplify the proposed MLD, only the first two conditions are presented in Fig. 4, which relate to initiating events illustrated in this study.

Step 5:
The cause of the plant transients with intact HPB in *Step 3* in which the main loop and shutdown cooling system are still operated by forced coolingin *Step 4*is power turbine generator trip[8]. Therefore, *power turbine generator trip* is an initiating event for HTGRs. Meanwhile, the cause of the plant transients with intact HPB in *Step 3*in which main loop system fails but shutdown cooling system is still in operationin *Step 4*is loss of offsite power[8]. Therefore, *loss of offsite power* is another initiating event for HTGR.

All these processes and their corresponding results are graphically shown in the proposed MLD in Fig. 4. Following the same procedures, other HTGR initiating events should be able to be identified.

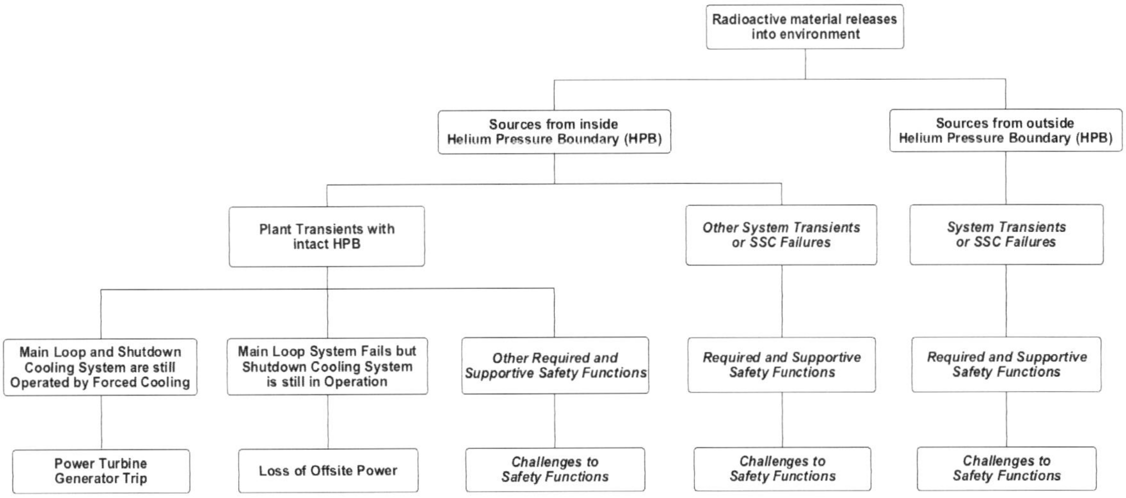

Figure 4. Master logic diagram for power turbine generator trip and loss of offsite power.

It can be seen from Fig. 4 that *power turbine generator trip* and *loss of offsite power* are two of many initiating events, which should be developed for HTGRs. Generally, these results have demonstrated that the proposed MLD can be implemented to search for HTGR initiating events.

Conclusion

The proposed master logic diagram approach has been applied to search for HTGR initiating events. It consists of five main steps. To illustrate the applicability of the proposed MLD approach, searching for two initiating events that could initiate the release of the radioactive materials into environment, i.e. *power turbine generator trip* and *loss of offsite power*, have been discussed. The results of the case study confirmed that the proposed MLD is feasible for finding HTGR initiating events. Nevertheless, the proposed MLD still needs to be further investigated to see how effective it will search for HTGR initiating events.

Acknowledgments

Author acknowledges assistance and encouragement from colleagues at Reactor Safety Technology Division and financial support from the Center for Nuclear Reactor Technology and Safety through DIPA of the financial year of 2017. Special thanks also go to the Budget Authority (Director of the Center for Nuclear Reactor Technology and Safety) and the Chancellor of the Term of Reference for Design and Safety Assessment of Experimental Power Reactor for the release of the funding.

References

1. Jamali, K., *Use of risk measures in design and licensing of future reactors.* Reliab. Eng. Syst. Saf., 2010. **95**(9): p. 935-943.
2. Guimaraes, A.C.F., et al., *Fuzzy uncertainty modeling applied to AP1000 nuclear power plant LOCA.* Ann. Nucl. Energy, 2011. **38**(8): p. 1775-1786.
3. Purba, J.H., et al. *Probabilistic safety assessment in nuclear power plants by fuzzy numbers.* in *The 9th International FLINS Conference*. 2010. Chengdu: World Scientific.
4. Purba, J.H., J. Lu, and G. Zhang, *Fuzzy failure rate for nuclear power plant probabilistic safety assessment by fault tree analysis*, in *Computational Intelligence Systems in Industrial Engineering*, C. Kahraman, Editor. 2012, Atlantis Press: Paris, France. p. 131-154.

5. Tong, J., et al., *Development of Probabilistic Safety Assessment with respect to the first demonstration nuclear power plant of high temperature gas cooled reactor in China.* Nucl. Eng. Des., 2012. **251**: p. 385–390.
6. IAEA, *Safety of Nuclear Power Plants: Design*, in *Specific Safety Requirement SSR-2/1 (Rev. 1).* 2016, International Atomic Energy Agency: Vienna - Austria.
7. Papazoglou, I.A. and O.N. Aneziris, *Master Logic Diagram: method for hazard and initiating event identification in process plants.* J. Hazard. Mater., 2003. **97**(1-3): p. 11-30.
8. INL, *Next Generation Nuclear Plant Probabilistic Risk Assessment White Paper.* 2011, U.S. Department of Energy: Idaho.
9. Ahn, M.H., K.J. Lee, and S.Y. Jeong, *Development of integrated dose assessment program for accident analysis of LILW managed in temporary storage facility.* Prog. Nucl. Energy, 2010. **52**(8): p. 837-849.
10. Jenab, K., et al., *Dynamic MLD analysis with flow graphs.* Reliab. Eng. Syst. Saf., 2012. **106**: p. 80-85.
11. Ahn, M.H., K.J. Lee, and K.W. Choi, *Establishment of risk-based accident scenarios using the master logic diagram related to LILW management in the temporary storage facility.* Ann. Nucl. Energy, 2008. **35**: p. 2420-2425.
12. Ferng, Y.M. and K.Y. Lin, *CFD investigation of thermal-hydraulic characteristics in a PBR core using different contact treatments between pebbles.* Ann. Nucl. Energy, 2014. **72**: p. 156-165.
13. Purba, J.H. *Preliminary studies on developing PSA framework for HTGRs: Relevant events to be considered.* in *Seminar Nasional Teknologi Energi Nuklir.* 2015. Denpasar, Bali.
14. Pioro, I.L., ed. *Handbook of Generation IV Nuclear Reactors.* Woodhead Publishing Series in Energy. 2016, Woodhead Publishing: Cambridge - USA.
15. Reitsma, F., K. Kunitomi, and H. Ohashi. *The IAEA coordinated research project on modular HTGR safety design: Status and outlook.* in *International Topical Meeting on High Temperature Reactor Technology.* 2016. Las Vegas, NV: Omnipress.
16. INL, *Modular HTGR Safety Basis and Approach.* 2011, U.S. Department of Energy: Idaho.
17. Vierow, K., et al., *Application of dynamic probabilistic risk assessment techniques for uncertainty quantification in generation IV reactors.* Prog. Nucl. Energy, 2014. **77**: p. 320-328.
18. Delley, A.O., et al., *Feasibility and Alternatives for Receipt, Storage, and Processing of HTGR Pebble Fuel at SRS.* 2014, U.S. Department of Energy National Laboratory: Savannah River Site.
19. García-Berrocal, A., C. Montalvo, and J. Blázquez, *Temperature transients in TRISO type fuel.* Ann. Nucl. Energy, 2015. **76**: p. 172-176.
20. INL, *Scoping analysis of source term and functional containment attenuation factors.* 2012, U.S. Department of Energy: Idaho Falls, Idaho.

Main steam line break accident simulation of APR1400 using the model of ATLAS facility

A S Ekariansyah, Deswandri, Geni R. Sunaryo

Center for reactor nuclear safety and technology (BATAN), Puspiptek area, Setu, Tangerang Selatan

andi_se@batan.go.id

Abstract. A main steam line break simulation for APR1400 as an advanced design of PWR has been performed using the RELAP5 code. The simulation was conducted in a model of thermal-hydraulic test facility called as ATLAS, which represents a scaled down facility of the APR1400 design. The main steam line break event is described in a open-access safety report document, in which initial conditions and assumptionsfor the analysis were utilized in performing the simulation and analysis of the selected parameter. The objective of this work was to conduct a benchmark activities by comparing the simulation results of the CESEC-III code as a conservative approach code with the results of RELAP5 as a best-estimate code. Based on the simulation results, a general similarity in the behavior of selected parameters was observed between the two codes. However the degree of accuracy still needs further research an analysis by comparing with the other best-estimate code. Uncertainties arising from the ATLAS model should be minimized by taking into account much more specific data in developing the APR1400 model.

Keywords: main steam line break accident, ATLAS facility, APR1400, best-estimate, RELAP5

1. Introduction

The Advanced Power Reactor 1400 (APR1400) is a pressurized water reactor (PWR) design, which was developed mainly by Korea Engineering Power Corporation (KEPCO) in 2001. Its design is basically an enhancement and improvement of previous Korea's major nuclear power plant model of OPR1000 [1]. 4 units of APR1400 are under construction in 2 sites of Korea for planned operation from 2016 to 2018, plus additional 4 more at United Arab Emirates for scheduled operation by 2020 [2]. The APR1400 is a two-loop pressurized water reactor supplying its nuclear steam supply system (NSSS) to a rated thermal output of 4000 MW or 1455 MW electrical output. The safety of APR1400 against various design bases accidents (DBA) has been described in the Design Control Document, which can be assessed via U.S.NRC website as part of design certification application review [3]. Related to the APR1400 accident analysis contained in that document, not many academic research papers have written the results of analyses nor a direct comparison using different tools or codes with event sequences of all DBAs in the document. Most academic papers associated the APR1400 accident analysis with the experimental results performed using a thermal-hydraulic test facility, which is called as ATLAS (Advanced Thermal-Hydraulic Test Loop for Accident Simulation) facility. Such accident analyses using the ATLAS involved several initiating events, which was conducted by W. Baek et al [4] for LOCA events, by Y. Kim et al [5] for station blackout event, and by K. Kang et al [6] for steam generator tube rupture event. Those researches did not relate the experimental results with the APR1400 accident analysis described in the Design Control Document since the ATLAS facility is regarded as a scaled down representation of the APR1400 design [7], except by Ki-Yong Choi et al [8], who compared the APR1400 accident analysis, ATLAS experimental results, and MARS code prediction. Nevertheless, there are many initiating events described in the APR1400 Design Control Document, which are not yet studied by benchmarking the APR1400 accident analysis with the safety analysis codes such as MARS, RELAP5, etc.

The purpose of this paper is to discuss a benchmark activity of one selected accident event using the RELAP5 code from a model of the ATLAS facility. The ATLAS facility has been modelled using RELAP5 code by other researcher [9], but this paper will associate the ATLAS model using the similar code developed from available ATLAS facility data, that has been used for the best estimate and uncertainty method analysis of direct vessel injection (DVI) line break [10]. The initiating event to be analyzed is the main steam line break (MSLB) in the APR1400 as described in the document. The reason of selecting is the availability of several MSLB simulations using different codes such MARS [11] and SPACE [12]. Both researches described the simulation results with the ATLAS experiment but not with the results contained in the document. The MSLB event will be simulated by the RELAP5/SCDAP/Mod3.2 using the previous ATLAS nodalization [10], so that the accuracy of the ATLAS model can be assessed. At the end, this research activity will explore more the main sequences related to the MSLB event and the level of safety of the APR1400 design to cope the accident.

2. Main steam line break (MSLB) in APR1400

Main steam line break (MSLB) event in the APR1400 is classified as postulated accident in the design basis accident definition. The event is defined as steam system piping failures or shortly steam line break (SLB) occuring inside or outside the containment. The NSSS response of the APR400 has been simulated using the CESEC-III [3] as a licensing conservative code with simple thermal-hydraulic nodalization. Two important safety concerns of MSLB accidents considered in the APR1400 Design Control Document is among others the return to power (RTP) due to the positive moderator reactivity feedback after reactor trip from a overcooling of the core [13] and fluid flashing in the reactor upper head and hot legs [14]. MSLB cases leading to the RTP after reactor trip are large (double-guillotine) MSLB occurred inside the containment during full or zero power operation. On the other side, MSLB cases occurred outside the containment are chosen to maximize the potential of fuel degradation before reactor trip, in which one of those during the full-power operation is selected in this analysis. The case to be validated using RELAP5 is an MSLB before the main steam isolation valve (MSIV) on one of two main steam piping lines outside the containment, which is occurred with a loss of offsite power (LOOP) following a reactor trip. Assumptions and initial conditions before the event are that the core is in full power level of 4,062 MWt, pressurizer pressure of 163.45 bar, core inlet temperature of 296 °C, two safety injection pumps are inoperation, and offsite power is lost concurrent with reactor trip. The sequences simulated using the CESEC-III as described in the Design Control Document are showed in Table 1 along with the setpoint value of APR1400 protection and mitigation system.

Table 1. Sequences of MSLB outside containment in the APR1400 [3]

Event	Setpoints of protection and mitigation systems or condition
Steam line break starts	-
Overpower trip signal reached	121.0 %
Auxiliary feedwater flow (AFW) started from overpower trip signal	-
Delayed reactor trip after overpower trip signal	-
Loss of offsite power (LOOP) initates Turbine trip and reactor cooling pump (RCP) trip	after reactor trip
MSIV close signal from steam generator pressure setpoint	52.73 bar
Main feedwater isolation valves (MFIV) close from steam generator pressure setpoint	-
Safety injection actuation signal from pressurizer pressure setpoint	109.32 bar
Point shutdown cooling initiated	Core inlet temperature 176.7 °C Pressurizer pressure 31.64 bar

Overpower is caused by the reactor cooling system (RCS) cooldown, which initiates an increase in core reactivity based on the negative moderator and doppler reactivity coefficients. That RCS cooldown is caused by a decrease in temperature and pressure in the reactor cooling system (RCS) from increased heat transfer from the primary system to steam generators secondary side [15]. Those events presented in Table 1 has been judged distorted by other researcher [16], therefore a best-estimate simulation is necessary to evaluate those event sequences using the ATLAS test loop facility as described in this research paper.

The ATLAS itself is a test loop facility representing a scaled-down primary and secondary cooling system of the APR1400 consisting similar loop arrangement of 2 hot legs, 4 cold legs, 2 steam generators, and 4 reactor cooling pump (RCP). It also incorporates safety injection features similar to the APR1400 such as 4 safety injection tanks (SITs) as passive emergency core cooling system (ECCS) and 2 high and low-pressure safety injection pumps for performing direct vessel injections [17]. The ATLAS dimensions are a half in height and length and 1/288 in volume of the APR1400 and capable for a full pressure simulation up to 20 MPa with maximum 10 % of the scaled nominal core power [7]. In this paper, the MSLB event will be conducted using the ATLAS model, which has been developed completely by the RELAP5/SCDAP/ Mod3.4 code and been used for the best estimate and uncertainty method analysis of direct vessel injection (DVI) line break [10]. Additional nodalizations are needed in order to simulate the MSLB event according to the event sequences as determined in the APR1400 design control document.

3. Methodology

The MSLB simulation requires a model of steam lines, MSIV, and valve for turbine isolation, which have to be incorporated in the basic ATLAS model. The steam line piping data of the ATLAS and its related components are not available, therefore a scaled-down calculation has been conducted in order to obtain proper dimensions based on the APR1400 piping data. Based on APR1400 Design Control Document [3], the largest effective steam blowdown area on each steam line is around 30 % of the steam line cross-section area or 0.119 m2. It means that the 100 % steam line cross-section area is 0.3967 m2. From that value, the main steam line cross-section area of the ATLAS model is calculated from the scaling parameter formula [17] to be 0.00275486 m2. In the steam generator upper-head, a flow restrictor is installed to limit the steam discharge at the SLB with an area of 0.001170211 m2 [12]. In order to simulate a double-ended guillotine break, the SLB nodalization consists of 3 quick opening valves modelled in steam line B. One valve is an normally open valve connecting the steam line B during normal operation, which will be closed when the SLB starts. Two other normally close valves will connect the other ends of the steam line to simulate steam flows from both direction discharged in to a volume with atmospheric pressure. On each steam line A and B, a normally open valve is modelled to simulate the MSIV before it is be merged in a joined steam line (steam header) with assumed double steam line area, passed through a turbine trip valve in to a discharged volume with the steam generator operating pressure of 7.7 MPa. The length of particular steam line is determined by assumption. Figure 1 shows the ATLAS nodalization using the RELAP5 for the SLB simulation referring the steam generator upper dome.

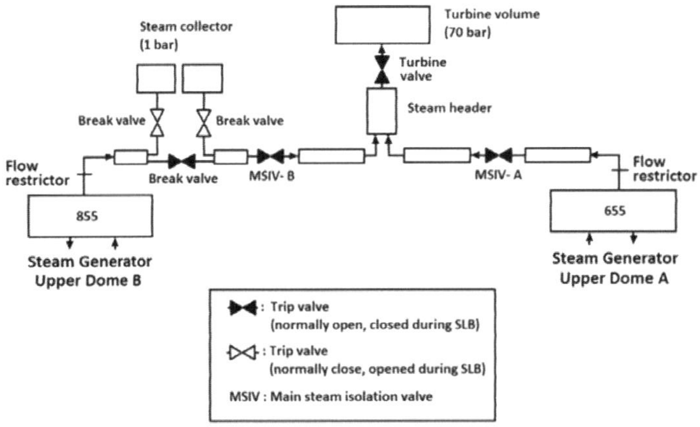

Figure 1. ATLAS nodalization in the steam generator for double-ended guillotine MSLB simulation using RELAP5

Before the MSLB simulation, the ATLAS is operated with its nominal heater power of 1,638 kW. The sequences after the MSLB are that the main feedwater pumps(MFWs) supplying feedwater in to the steam generator secondary side is stopped after the heater power reaches 121 % nominal power, which are replaced by the operation of auxiliary feedwater pumps (AFWs). The AFW volumetric flow rate of APR1400 is 3,596 liter per minute or 59 kg/sec, which is converted by the ATLAS scaling formula to 0.2953 kg/sec. The overpower trip signal will generate reactor trip, in which the heater power is dropped to simulate a decay reactor power. The heater power of the ATLAS will follow the APR1400 core power during the transient as calculated by the CESEC-III code [3]. In the same time, loss of offsite power is assumed, which initiates turbine valve to close and RCP begins to coast down. All safety injections either from passive injection system (4 SITs) and active injection system (2 SIPs) are considered to be operable during the SLB sequences. Parameters to be evaluated are steam generator secondary side pressure in the affected steam line B and affected steam line A, pressurizer pressure, inlet and outlet core coolant temperature, and reactor pressure vessel (RPV) upper head flashing phenomena.

4. Results and discussion

The SLB simulation is performed based on the steady-state condition as described in the methodology. Table 2 shows resulted steady state parameters for the primary and secondary system of ATLAS facility after 1000 sec RELAP5 simulation before the the SLB simulation. There are 4 initial conditions assumed in the APR1400, which are core power, pressurizer pressure, core mass flow rate, core inlet temperature and pressurizer water volume. The APR1400 core power of 4,062 MW is scaled down in the ATLAS facility to a heater power of 1,638 MW. The pressurizer pressure is unchanged of around 163 bar as also the core inlet temperature of 568 K. The pressurizer water volume of APR1400 is 39.9 m3, which is scaled down into 0.1385 m2 for the ATLAS pressurizer. The cold leg mass flow is calculated from the APR1400 core mass flow of 23,619.44 kg/sec, which is scaled down by the ATLAS scaling formula to around 29 kg/sec after divided by 4 cold legs. Those values for the ATLAS facility should be maintained during the ATLAS simulation using the RELAP5 as shown in the Table 2.

Table 2. Steady-state parameter of the ATLAS calculated by RELAP5 before the MSLB simulation

Parameter	ATLAS
Primary system	
Heater power (MW)	1,638.0
Pressurizer pressure (bar)	162.95
Core inlet temperature (K)	567.77
Core outlet temperature (K)	571.08
Cold leg flow rate (kg/sec)	23.44
Seconday system	
SG dome pressure (bar)	77.96
Main feedwater flow rate (kg/sec)	0.433
Main steam flow rate (kg/sec)	0.431
Primary to secondary heat removal (MW) for 1 SG	0.82

From the Table 2, the steam flow rate in the one steam line is around 0.43 kg/sec, which is consistent with the main feedwater flow rate supplying water into the steam generator (SG). The core inlet temperature of 567.7 K is close to the 568 K of the APR1400 initial condition. The cold leg flow rate is selected to be 23 kg/sec, which is not much deviated from the 29 kg/sec. The heat losses from the primary and secondary system into the environment are not considered in the model meaning that adiabatic boundary conditions are applied to the ATLAS model.

The MSLB is initiated by opening the 2 normally close valves and closing 1 normally open valve in the steam line B as shown in Figure 1. Figure 2 shows the break flow rates from the 2 directions of the affected steam line simulating the double-ended break. Those actuations of the break valves will direct the steam line from the SG-B and SG-A into the discharge volume or steam collector with 1 bar pressure.

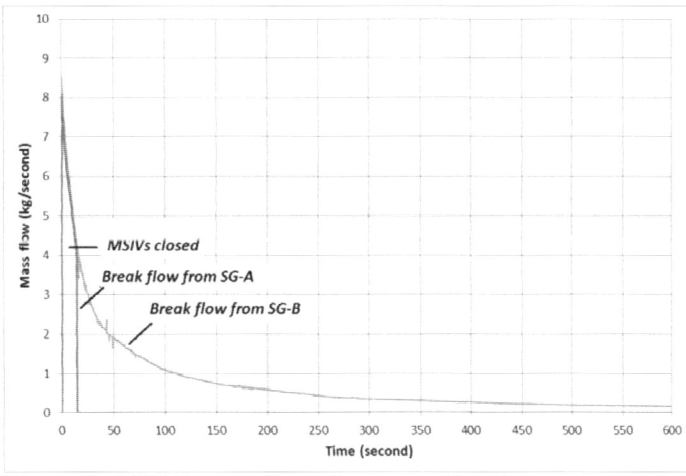

Figure 2. MSLB mass flows from two directions of steam line loop B

The initial break mass flow is calculated to be 13 kg/sec from both directions of steam line. The break flow from the SG-A is interrupted with the actuation of MSIV in the both steam lines, while the break mass flow from SG-B will continue to follow the available steam inventory. There is no comparison of APR1400 break mass flow to evaluate the calculated mass flow in the ATLAS, however a similar SLB experiment conducted in the ATLAS has shown an experimented break mass flow of around 6 kg/sec [11]. Therefore a modified model in the steam line is performed by given different loss coefficient with the result of simulated break flow of 8 kg/sec as shown in Figure 2.

The discharged steam flow will decrease the pressure measured on top of both SGs as shown in Figure 3. There are differences in the SG pressure between the SG-B and SG-A from the initial steam pressure of 77.96 bar.

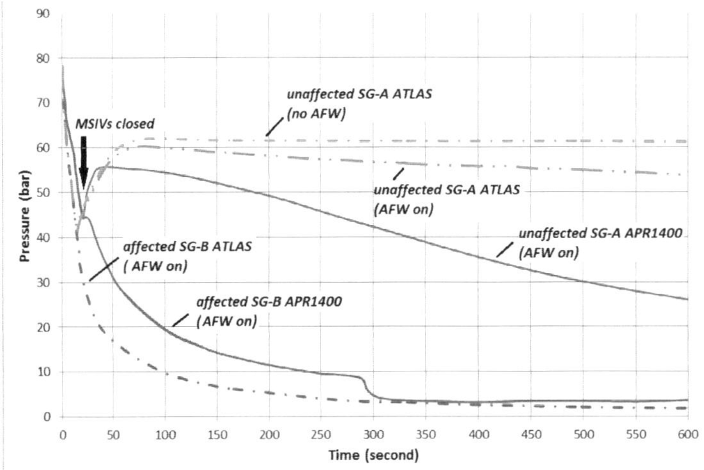

Figure 3. Steam generator pressure due to MSLB with AFW actuations

On Figure 3, the pressure in the SG-A and B are dropped rapidly due to the continuous discharged steam flow through the breaks. As the steam pressure decreases to the value of 52.73 bar, both MSIVs are closed with 6 second time delay. The MSIV-A blocks the discharged steam from the SG-A, while the closening of MSIV-B has no effect on the discharged steam from SG-B because the MSLB occurs upstream of the closed MSIV-B. That is the reason for the increase of SG-A pressure after closing of MSIVs on Figure 3. The SG-B pressure characteristics from the ATLAS model is almost similar with the APR1400 model, however the SG-A pressure is much different. In the APR1400 calculation, a continuous decrease of the SG-A pressure is obvious, while in the ATLAS calculation the SG-A pressure is almost stagnant without the actuation of AFW pump, even the values are slowly decreased. The SG-A pressure of the ATLAS decreases more clearly when dedicated AFW pump is operated to supply water into the secondary system thus cooling down the steam temperature. The SG-A behavior of the APR1400 calculation can only occur, when the operator performs a steam venting via automatic depressurization system (ADS) or operates turbine bypass to depressurize the pressure or additional heat loss is considered in the APR1400 model. That discrepancy of the unaffected SG pressure has been also validated, when a environmental heat loss is taken in to consideration as performed by other ATLAS model simulated by MARS[11] and SPACE code [12].

The primary cooldown due to increased inventory leakage flow rate in the SG secondary system initiates a pressurizer pressure decrease [18]as shown in Figure 4. The primary depressurization calculated in the ATLAS occurs bigger than CESEC-III calculation of APR1400 even though there is no heat loss from the primary to environment.A comparison with the RELAP5 calculation is included in the figure by showing a similar analysis with the MARS code [11].Both best-estimate calculation show an uncertainty in the result involving the ATLAS model. The primary pressure achieves its level back, when the safety injection pumps (SIPs) inject water directly to the vessel after primary pressure setpoint of 109.7 bar has been achieved with some time delay.

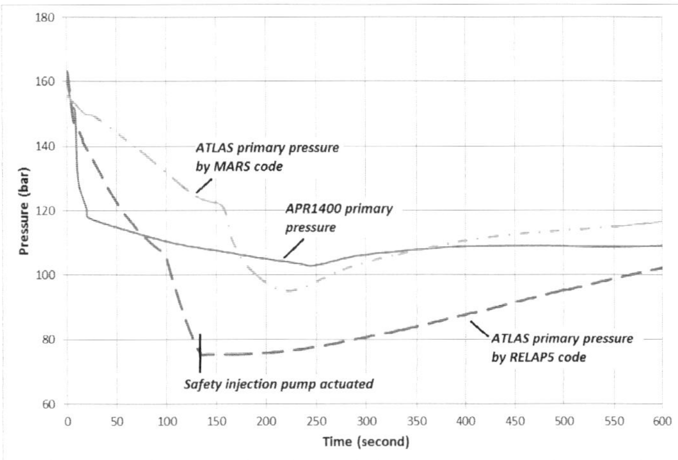

Figure 4. Pressurizer pressure after MSLB

The primary cooldown as shown by the primary pressure decrease will also cause the decrease of saturation pressure. In that event, vapor void is appeared in the upper head of the vessel, where the static pressure is lower compared to other part of the ATLAS vessel. The value of vapor void fraction is shown in Figure 5 between the APR1400 and ATLAS model calculation. The maximum value of vapor void in the ATLAS model will be higher in correlation with the higher MSLB mass flow. The vapor void fraction will increase until additional water injection from the SIPs increase the water density as the vapor void fraction decreases.

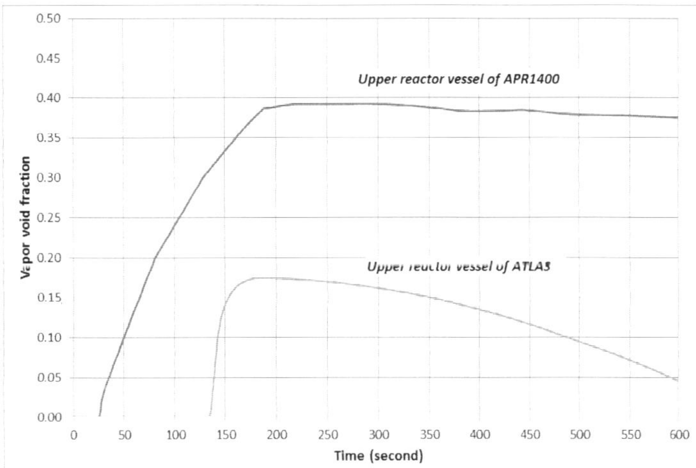

Figure 5. Vapor void fraction in the vessel upper head

The differences in the steam pressures between SG-A and SG-B have effect on the temperature of cold legs taking heat from the secondary system. During the MSLB transient, the coolant temperature in the cold legs downstream of the SG-B is lower than the one of the SG-A. Therefore a non-uniform temperature distribution occurres a the core inlet [19]. Figure 6 presents the characteristics of the core inlet and outlet temperature during the MSLB event.

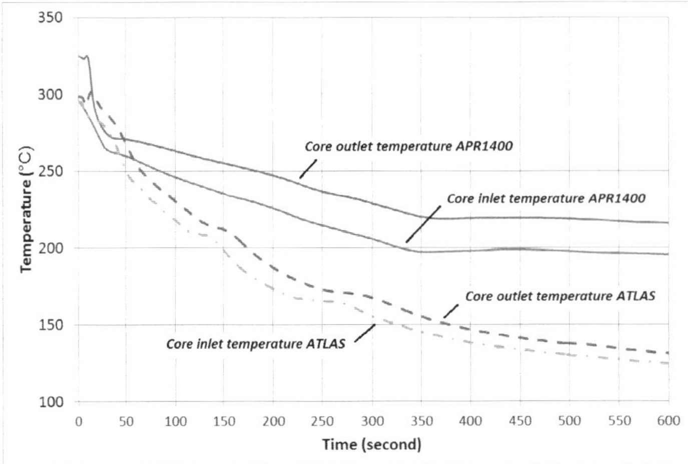

Figure 6. Inlet and outlet core temperature

The behavior of the core inlet temperature in the ATLAS model is close related with the rate of heat removal from the SG-B due to the discharged steam during the MSLB. The primary system takes more heat from the SG-B in the ATLAS model as indicated by the bigger decrease of pressurizer pressure in Figure 4. Therefore the resulted core inlet temperature in the ATLAS is lower than the APR1400 model. The decrease also follows the heater power characteristic after reactor trip.

The overall results of ATLAS simulation using RELAP5 code show a general similarity in the behavior of selected parameters with the APR1400 calculated by CESEC-III. However, the magnitude of the parameter fluctuations is different between the calculation results of the two codes. The uncertainties arising from the RELAP5 calculation as a best-estimate code become higher since the comparison is conducted based on the ATLAS model not on experimental results. A validation with other best-estimate code might be necessary by considering much more data of the APR1400 model.

5. Conclusion

Based on the MSLB simulation using the ATLAS model calculated by the RELAP5/SCDAP/Mod3.2, the overallresults of the secondary pressure, primary pressure, inlet and outlet core temperature, and vessel vapor void show a general similarity with the results of APR1400 model calculated by CESEC-III. Since the CESEC-III code is a code based on a conservative approach, the magnitude of the parameter fluctuations is different with the calculation results of RELAP5 as a best-estimate code. The degree of the calculation results requires afurther validation with other best-estimate code by considering much more data of the APR1400 model. Nevertheless, the research has given a valuable experience regarding the benchmarking process of a safety analysis described in a safety analysis report.

Acknowledgments

This work was fully funded by BATAN research project (DIPA) of the year 2016, which was managed by Center for Nuclear Reactor Safety and Technology (PTKRN). The author is very thanksful to Mr. Surip Widodo for his valuable assistance in developing the model of ATLAS facility.

References

[1] Lee S S, Kim S A, Suh K Y 2009 *Nucl. Eng. Tech.* **41** 995

[2] Li Y, Chen C, Xin P, Chen Y, Hou H 2017 *Proc. 20th. Pacific Basin Nucl. Conf. (Beijing)* vol 3 (Singapore Springer) p. 321

[3] Korea electric power corporation &Korea hydro & nuclear power co. ltd 2014 *APR1400 Design Control Document Tier 2: Chapter 15-Transient and Accident Analyses*, downloaded from

https://www.nrc.gov/reactors/new-reactors/design-cert/apr1400/dcd.html on 3/3/2017
[4] Baek W P, Kim Y S, Choi K Y 2009 *Nucl. Eng. Tech.* **41** 775
[5] Kim Y S, Yu X G, Kang K H, Park H S, Cho S, Choi K Y 2013 *Nucl. Eng. Tech.* **45** 179
[6] Kang K H, Park H S, Cho S, Choi K Y 2012 *Nucl. Tech.* **177** 382
[7] Kim YS, Choi K Y, Park H S, Cho S, Kim B D, Choi N H, Baek W P 2008 *Ann. Nucl. Energy* **35** 1791
[8] Choi K Y, Park H S, Cho S, Baek W P 2009 *Nucl. Eng. Tech.* **41** 655
[9] Sharabi M and Freixa J 2012 *Nucl. Eng. Tech.* **44** 709
[10] Ekariansyah A S and Widodo S 2017 *Tri Dasa Mega* **19** 19
[11] Ha T W, Yun B J, Jeong J J 2014 *Journal Energy Eng .* **23** 112
[12] Youn B S 2014 *Journal of Energy and Power Engineering* **8** 1647
[13] Ferroukhia H, Zerkaka O, Chawlaa R 2009 *Ann. Nucl. Energy* **36** 1184
[14] Choi K Y, Kim Y S, Song C H, Baek W P 2012*Sci. Technol. Nucl. Ins.* **2012** ID 375070
[15] Coscarelli E, Del Nevo A, D'auria F 2013 *Sci. Technol. Nucl. Ins.* **2013** ID 128305
[16] Song J H, Chung B D, Jeong J J, Baek W P 2004 *Journal Korean Nucl. Soc.* **36** 338
[17] Kim Y S, Choi K Y, Song C H, Baek W P 2014 *Ann. Nucl. Energy* **63** 509
[18] Reyes J N 2008 *Sci. Technol. Nucl. Ins.* **2008** ID 453621
[19] Zerkak O, Ferroukhi H 2011 *Ann. Nucl. Energy* **38** 60

Analysis on the Role of RSG-GAS Pool Cooling System during Partial Loss of Heat Sink Accident

Susyadi, Endiah P. H, Sukmanto D, Andi S. E, Hendro T. Syaiful B, Geni R.S.

Center for Nuclear Reactor Safety and Technology -BATAN
Kawasan PUSPIPTEK Gedung 80, Setu, Tangerang Selatan -15310

Email: susyadi@batan.go.id

Abstract. RSG-GAS is a 30 MW reactor that is mostly used for radioisotope production and experimental activities. Recently, it is regularly operated at half of its capacity for efficiency reason. During an accident, especially loss of heat sink, the role of its pool cooling system is very important to dump decay heat. An analysis using single failure approach and partial modeling of RELAP5 performed by S. Dibyo, 2010 shows that there is no significant increase in the coolant temperature if this system is properly functioned. However lessons learned from the Fukushima accident revealed that an accident can happen due to multiple failures. Considering ageing of the reactor, in this research the role of pool cooling system is to be investigated for a partial loss of heat sink accident which is at the same time the protection system fails to scram the reactor when being operated at 15 MW. The purpose is to clarify the transient characteristics and the final state of the coolant temperature. The method used is by simulating the system in RELAP5 code. Calculation results shows the pool cooling systems reduce coolant temperature for about 1 K as compared without activating them. The result alsoreveals that when the reactor is being operated at half of its rated power, it is still in safe condition for a partial loss of heat sink accident without scram.

Keywords: RSG-GAS, RELAP5, pool cooling system, loss of heat sink, transient

1. Introduction

The G.A. SiwabessyMulti Purpose Nuclear Reactor (RSG-GAS) is a pool-typed material testing reactor which is at the moment mostly operated for radioisotope production taking advantage of its relatively high neutron fluxes[1-3]. In addition, some other activities were also performed such as experiments for code validations [4-6], reactor fuel studies [7-10], neutronic-thermalhydraulic related investigations[11, 12] and safety analysis [13]. By design, the reactor has rated thermal power of 30 MW, but at present predominantly average operation is at around 15 MW for a cost efficiency reason. During normal operation, reactor core which is made up of plate-type fuel assemblies, provides heat to primary side water coolant that be circulated into two heat exchangers [14] for transferring the heat into secondary side. The inlet and outlet temperatures of the core is about 313 K (40.2 ^{o}C) and 322 K (48.9 ^{o}C). The coolant circulation is driven by two constant speed pumps; each has supply capacity of 430 kg/s. At the secondary side the heat removedfrom the heat exchangers is then dump into the environment through cooling towers using two secondary pumps. During shutdown, where the main pumps are stopped, thecore's decay heat is removedby pool cooling system consisting of three modulescalled JNA 10/20/30 (see Figure 1). These modulesaredemineralizedwater circulating closed circuit that independent of one another. Each module composes of finned pipe submerged in reactor pool for a certain depth, circulation pump and two air cooled heat exchangers located outdoor. Every module is designed to be able to deliver heat of 63 kW.

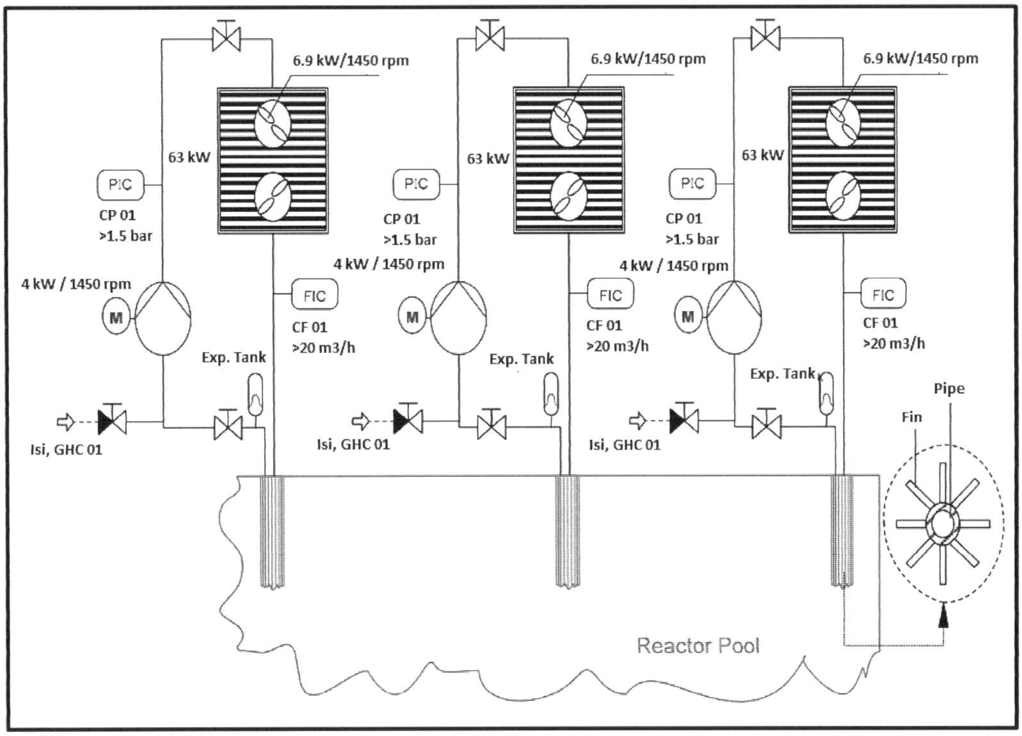

Figure 1. Schematic diagram of reactor pool cooling system JNAs [15]

During an accident, particularly loss of normal heat sink (either due to failure of secondary side pump or cooling tower system), the cooling for reactor coolant is very important as the core still produce energy from decay heat. Investigation conducted by S. Dibyo[16], that performed numerical analysis with RELAP5 using simplified/partial model shows that there is no significant coolant temperature increase if the reactor successfully trips and JNA system works as designed. However, lessons learned from the Fukushima accident revealed that analysis for worse condition needs to be performed as accident can happen not only due to single failure but also from multi failures [17] which could bring condition beyond its design limit. Recall that the reactor and it supporting systems are ageing, then there is an upturn probability that multiple failures accident could happen in RSG-GAS. In this paper, the role of pool cooling system when a partially loss of heat sink accident occurs (i.e. one of secondary-side pumps broken) at steady state operation of 15 MW followed by failure to trip is investigated. As the reactor remains on power, the accident causes a rapidly increase pool temperature because only half of the secondary-side system works. One way to resist the increase is by activating the pool cooling system (JNAs). For that reason, this research studies the dynamic of the coolant temperature. The purpose is to clarify the characteristic of reactor transient and final state of coolant temperature. The method used is through numerical simulation, i.e. by modeling reactor system including its pool cooling system into RELAP5[18]. The simulation performed starts at steady state which then undergo a secondary pump failure but fail to have shutdown. The result of this study is for complementing the document of safety review required by Nuclear Energy Regulatory Agency (BAPETEN) for extension of the operation license.

2. Reactor System Modeling

To be able to analyze the role played by the JNAs during loss of heat sink, the RSG-GAS hydro dynamic and the JNA systems must be modeled in RELAP5. Figure 2 shows diagram of RSG-GAS which includes primary and secondary sides of the reactor cooling system. The JNAsthemselvesarelocated in the reactor pool as shown in Figure 1. Main characteristic of the reactor for input deck development is presented in Table 1.The system shown in Figure 1 and 2 is then represented using generic models available in RELAP5 such as PIPE, VOLUME, BRANCH, JUNCTION, TIME DEPENDENT VOLUME, TIME DEPENDENT JUNCTION and HEAT STRUCTURE. The complete representation is displayed in Figure 3. In addition, connection of JNAs with reactor system is shown in Figure 4.

Table 1. Main Characteristics of GA Siwabessy Reactor.[14]

Parameter	Value
Nominal power	30 MW
Fuel (enrichment)	U_3Si_2-Al, LEU (19.75%)
Total number of plates	960
Plates per standard assembly	21
Plates per control assembly	15
Standard assemblies in core	40
Control assemblies in core	8
Average power per plate	31.25 kW
Average heat flux	0.415 MW/m^2
Core flow rate	3270 m^3/h
Static pressure (top of the core)	0.1997 MPa
Flow direction	Downward

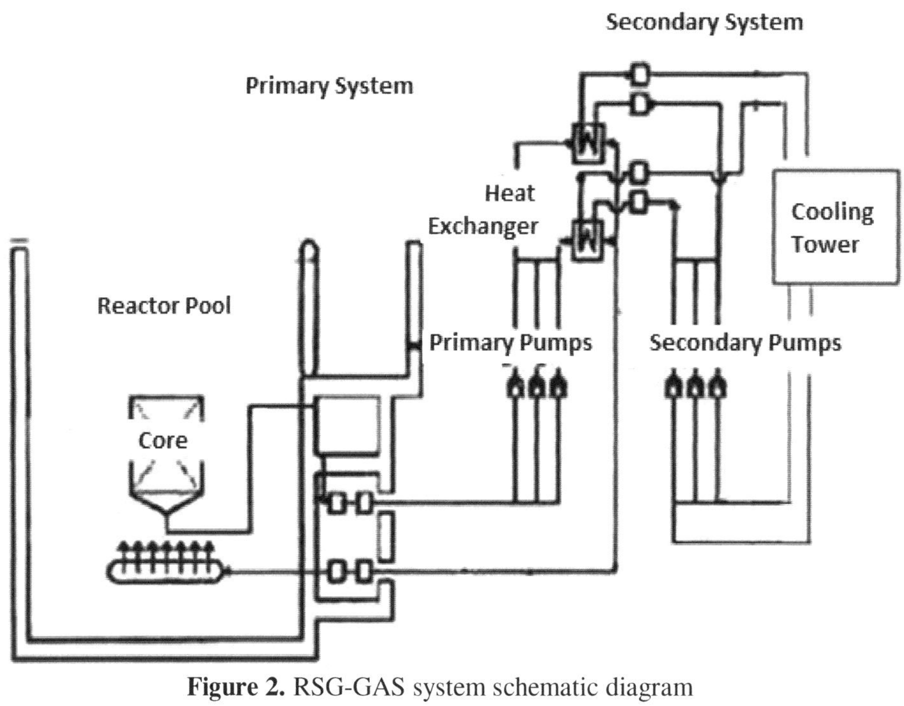

Figure 2. RSG-GAS system schematic diagram

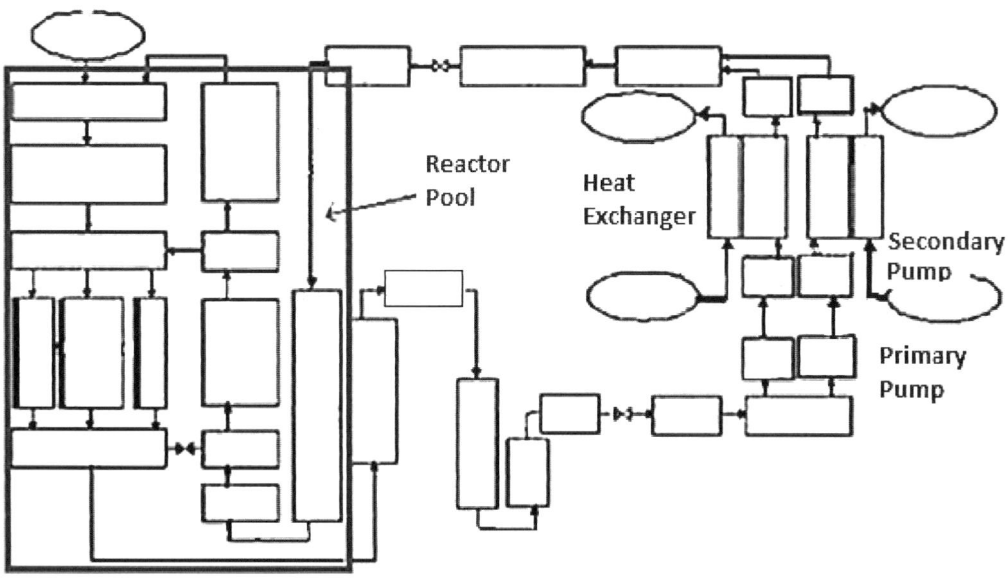

Figure 3. RSG-GAS system representation in RELAP5

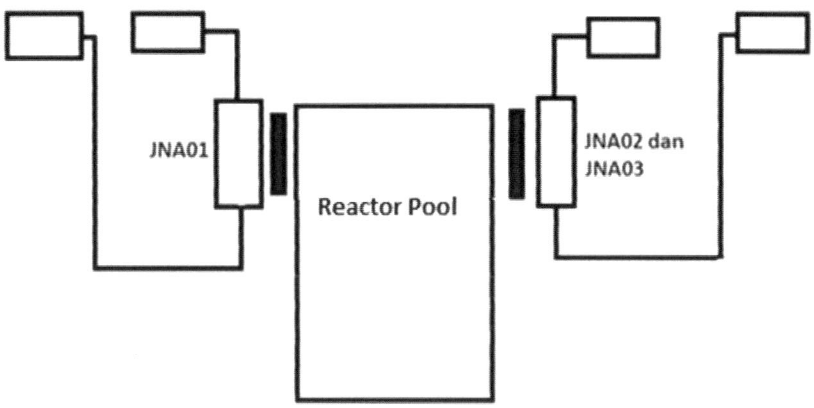

Figure 4. RELAP5 model for Pool Cooling System

Figure 4 shows that the upper part of the reactor pool model is connected with three JNAs. Each JNA consists of 2 boundary volumes (TMDPVOL), 1 coolant driving device (TMDPJUN), 1 connecting pipe (SNGLJUN), onepipe(PIPE) and a model for heat transfer calculation (HEAT STRUCTURE) which has two faces. Oneface is connected to a volume in the upper side of the pool and the other face is with PIPE of pool cooling system. From that model, the input deck was developed where the parameters used are shown in Table 2.Recalling that the focus of this study is fuel and coolant temperatures, and the overall response of reactor then the modeling of JNA is simplified by looking at its capability to remove heat from the pool. Noting that there are three modules, then in the modeling there are also three JNA models connected with reactor system.

Table 2. Main Characteristics of Pool Cooling System

Parameter	Value
Type	Finned tube immersion coolers with 6 straight fins along the tube
Material	Material Al Mg3
Design Operation Temperature	338 K (65 °C)
Mass flow rate	7.6 kg/second
Thermal Power	63 kW
Air dry bulb Temperature	309 K (36 °C)
Cooling tower inlet-outlet temperatures difference (during comissioning)	3.5 K – 4.5 K
Number of finned pipe per modul	16 pipes
Pipe lenght	10 m
Pipe Diameter (ID/OD)	30/35 mm
Effective surface area per module	20 m^2
Operating pressure	5 Bar

2.1. Operating Limit[19]

Safety Analysis Report (SAR) of RGS-GAS describes safety limit and operation limit of the reactor. The safety limit is used for safety analysis in the safest condition, while operation limit is implemented for reactor protection system. The limit described below is limit from the thermal hydraulic point of view. In RELAP5, operation limit is used for the LOGIC modeling for reactor scram.

a. The Safety limit, among others are:
   1. Maximum core inlet temperature is 317.5 K (44.5°C)
   2. Minimum coolant flow to the core is 85% of primary cooling flow rate
   3. Minimumsafety limit of flow instability (S) in overpower operation (114% form nominal power) is 2.67
b. Operation limit, among others are:
   1. Maximum core inlet temperature is 315 K (42°C).
   2. Maximum reactor power is 114% of nominal power
   3. Minimum flow rate id 85% of total primary system flow rate
   4. Minimumsafety limit of flow instability (S) in nominal power operation is 3.38

In this research, it is assumed that trip fails to occur, so then some of the above operation limit is not considered.

3. Calculation Results and Discussion

3.1. Steady state.

To be able to perform a transient analysis on the role of JNA 10/20/20 during accident using RELAP5, a steady state execution is performed first. Noting that at present and in the future reactor is going to be operated at half of nominal power, the execution values are shown in Table 3.

Table 3. Reactor data for steady state calculation at 15 MW.

Parameter	**Value**
Reactor power	15 MW
Primary system flow rate	960 kg/s
Pool coolant level	12 m
Core outlet temperature	314 K (41 °C)
Core inlet temperature	309.5 K (36.5 °C)
Average secondary coolant inlet temp.	305 K (32 °C)
Average secondary coolant outlet temp.	308 K (35 °C)

Data from steady state calculation is for initial condition of the accident scenario.

3.2. Transient

Transient calculation performed is simulation of failure at one of secondary pumps without reactor trip. The calculation was performed two times, i.e. condition with the all JNAs is disabledand activated. The results are then compared in order to look at the role of the JNAs in contributing the heat removal. The first result is shown is Figure 5.

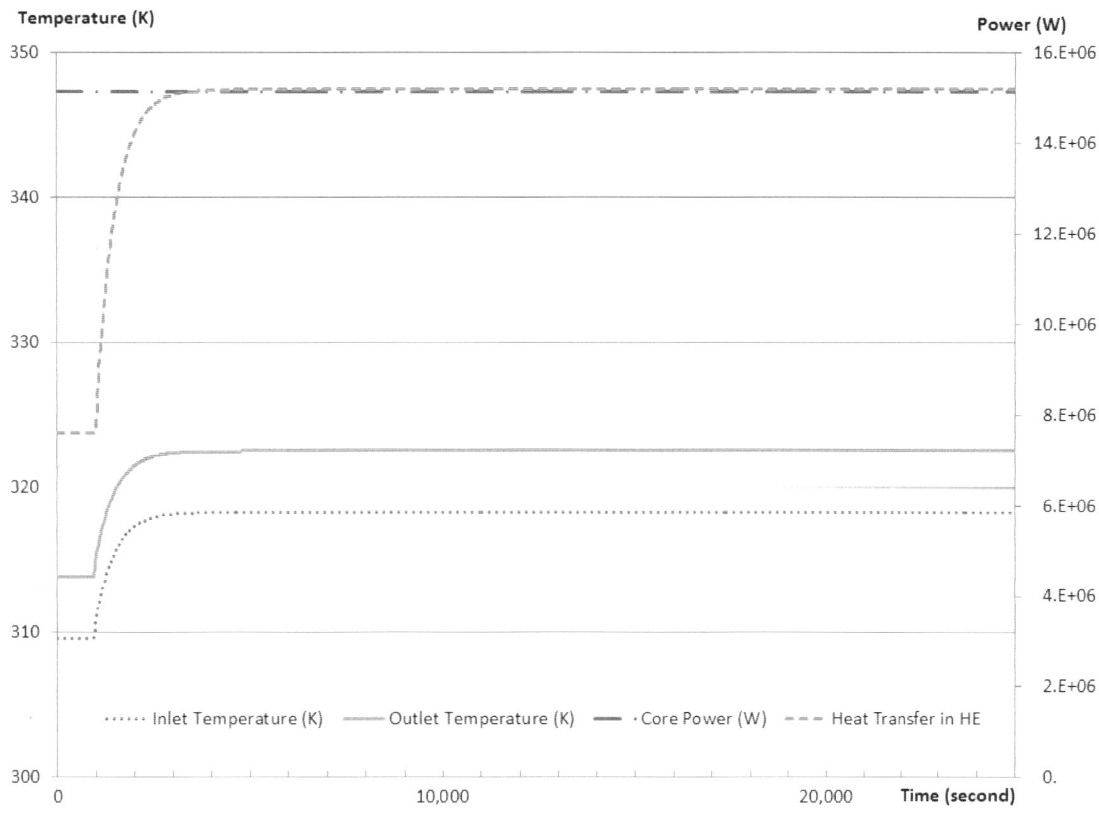

Figure 5. Typical heat transfer and coolant temperature in RSG-GAS without JNA activation

In Figure 5, the reactor is initially at steady state of 15 MW which is then at t= ~1000 seconds having failure in heat sink system. It can be seen that since the loss of heat sink partially, the coolant temperature gradually increased, both in core's inlet and outlet. This increase happens in consort with the losing of heat removal capability due to in loss of heat transfer in one of heat exchanger. The increase is finally reaches equilibrium when the heat transferred by the remaining active HE equals to the core power. The equilibrium temperature at the inlet and outlet of the core is about 318 K and 323 K.

In the second simulation, all JNAsis activated. It was assumed that the operator takes such action manually within 15 minutes after the pump failure. Result of the calculation is shown in Figure 6.

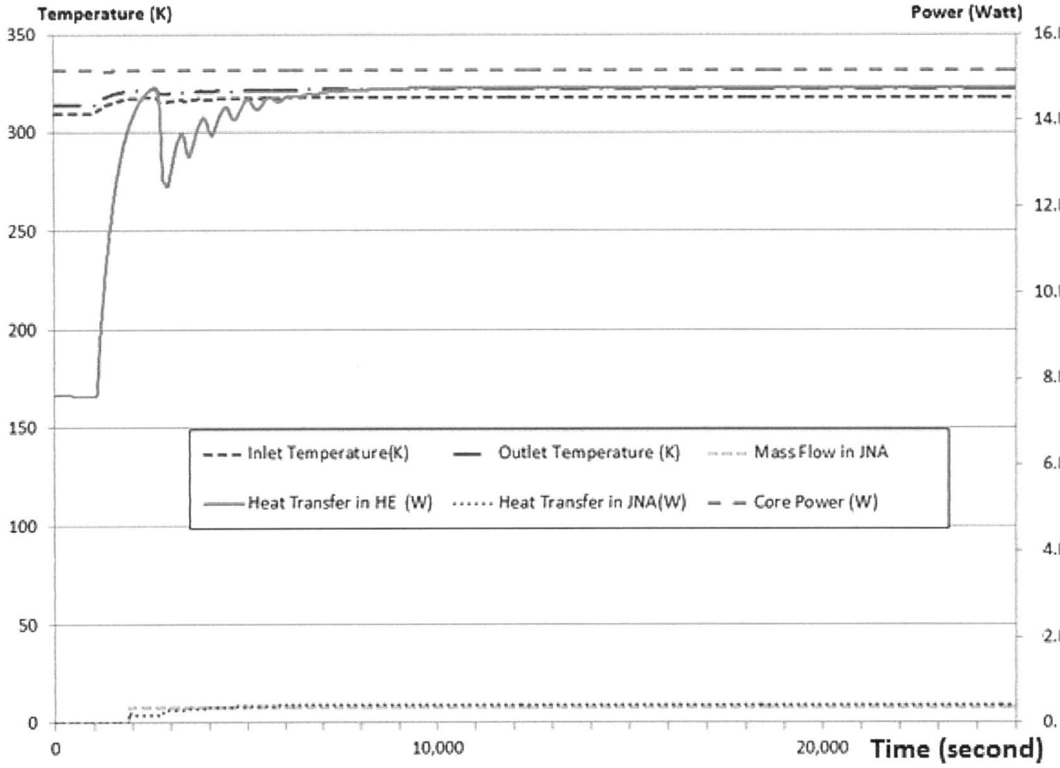

Figure 6. Typical heat transfer and coolant temperature in RSG-GAS with JNA activation

Figure 6 shows the dynamic of heat transfer and coolant temperature in the system with activated pool cooling system. It can be observed that some of the heat is removed through JNAs. At the beginning of the transient, heat tranfer in the active HE oscilates as a result of temperature fluctuation from JNA activation. However, the trend of temperature coolant increase follow the shape of condition without JNA activiation. From this figure, the role of JNA for heat removal is not significant, that is about 420 kW. Furthermore, equilibrium temperature of core inlet and outlet is 1 K lower than if the JNA is activated, or around 318 K and 322 K. In both scenario discussed above, the fuel temperature is still below operational limit and there is no possibility of core damage. From those conditions, it can be concludedthat even there is no reactor trip during partially loss of heat sink at power of 15 MW, the safety of reactor remains intact.

4. Concluding Remark

Investigation on the role of pool cooling system ofRSGGAS for a partial loss of heat sink accident followed by trip failure has been conducted. The investigation was performed through simulation using numerical model in RELAP5 which includes reactor system and pool cooling system (JNA). Calculation runs for two transient scenarios at power level of 15 MW,i.e. with and without JNAs activation. The result shows that under prescribed condition the reactor still under safe condition where the final core inlet and outlet temperatures are 318 K and 323 K. Furthermore, the calculation also shows that the use of JNA would contribute to reduce temperature for about 1 K.

Acknowledgement
This work was supported by the national project fund (DIPA) 2016 and the author would like to acknowledge the National Atomic Energy Agency of Indonesia (BATAN) for providing facilities to perform the calculation.

References

1. Pinem, S., J. Susilo, and T. Sembiring, Optimization of Radioisotope Production at Rsg-Gas Reactor Using Deterministic Method. Jurnal Teknologi Indonesia (JTI), 2015. **35**(2).
2. Kuntjoro, S., Analisis Aktivitas Isotop MO-99 di Reaktor RSG-GAS. SIGMA EPSILON-Buletin Ilmiah Teknologi Keselamatan Reaktor Nuklir, 2017. **20**(1): p. (in Indonesian).
3. Luhur, N., Implementation on controlling post irradiated material at the GA Siwabessy multi purpose Reactor RSG-GAS. Buletin Pengelolaan Reaktor Nuklir, 2014. **11**(1): p. 41-51.
4. Hong, L.P., et al., Study on the control rod interaction effect in RSG GAS multipurpose reactor (MPR-30). Annals of Nuclear Energy, 2002. **29**(6): p. 701-716.
5. Ekariansyah, A.S., Susyadi, and S. Dibyo, Aplikasi Program RELAP5/MOD3. 2 Untuk Simulasi Beam Tube Rupture RSG-GAS. TRI DASA MEGA-Jurnal Teknologi Reaktor Nuklir, 2006. **8**(3): p. (in Indonesian).
6. Chatzidakis, S., et al., A comparative assessment of independent thermal-hydraulic models for research reactors: The RSG-GAS case. Nuclear Engineering and Design, 2014. **268**: p. 77-86.
7. Liem, P.H., et al., Nondestructive burnup verification by gamma-ray spectroscopy of LEU silicide fuel plates irradiated in the RSG GAS multipurpose reactor. Annals of Nuclear Energy, 2013. **56**: p. 57-65.
8. Ginting, A.B. and P.H. Liem, Absolute burnup measurement of LEU silicide fuel plate irradiated in the RSG GAS multipurpose reactor by destructive radiochemical technique. Annals of Nuclear Energy, 2015. **85**: p. 613-620.
9. Pinem, S., et al., Fuel element burnup measurements for the equilibrium LEU silicide RSG GAS (MPR-30) core under a new fuel management strategy. Annals of Nuclear Energy, 2016. **98**: p. 211-217.
10. Liem, P.H. and T.M. Sembiring, Design of transition cores of RSG GAS (MPR-30) with higher loading silicide fuel. Nuclear Engineering and Design, 2010. **240**(6): p. 1433-1442.
11. Pinem, S., T. Sembiring, and P. Liem, Neutronic and Thermal-Hydraulic Safety Analysis for the Optimization of the Uranium Foil Target in the RSG-GAS Reactor. Atom Indonesia, 2016. **42**(3).
12. Subekti, M., M.D. Isnaini, and E.P. Hastuti, Analisis Distribusi Kecepatan Pendingin Dalam Elemen Bakar Tipe Pelat Menggunakan Metode CFD Untuk Reaktor Riset RSG-GAS. TRI DASA MEGA-Jurnal Teknologi Reaktor Nuklir, 2015. **15**(2): p. (in Indonesian).
13. Khakim, A., Safety analysis of MTR type research reactor during postulated beam tube break inducing positive reactivity. Annals of Nuclear Energy, 2016. **87, Part 2**: p. 793-798.
14. BATAN, Laporan Analisis Keselamatan RSG-GAS rev 10., 2011, PRSG-BATAN: Serpong.
15. Catur, A.S. and Djunaidi. Degradasi Kemampuan Sistem Pendingin Darurat Kolam Reaktor JNA 10/20/30. in Seminar Nasional III SDM Teknologi Nuklir. 2007. Yogyakarta (in Indonesian).

16. Dibyo, S., Simulasi Pemodelan Termohidrolika Sistem Pendingin Kolam RSG-GAS, in Lokakarya Komputasi dalam Sains dan Teknologi Nuklir2010, BATAN: BSD City, Indonesia (in Indonesian).
17. Yang, J.-E., Fukushima Dai-Ichi accident: lessons learned and future actions from the risk perspectives. Nuclear Engineering and Technology, 2014. **46**(1): p. 27-38.
18. Carlson, E., RELAP5/MOD3 code manual, Volumes I, II, III (draft), US Nuclear Regulatory Commission Report, 1990, NUREG/CR-5535.
19. Hastuti, E.P. Analisis Keselamatan Moda Satu Jalur Pendingin Teras Silisida RSG-GAS. in Seminar Hasil Penelitian P2TRR tahun 2002. 2003. Serpong, Indonesia: BATAN (in Indonesian).

IOP Conf. Series: Journal of Physics: Conf. Series **962** (2018) 012039 doi:10.1088/1742-6596/962/1/012039

Study RELAP5 Helium Properties for HTGR Thermal Hydraulic Analysis

Surip Widodo, Anis Rohanda, Muhammad Subekti, Topan Setiadipura, Syaiful Bakhri, Geni Rina Sunaryo

Center for Nuclear Reactor Technology and Safety, BATAN, Puspiptek Area Building No 80, Serpong, Tangerang Selatan 15310, Indonesia

e-mail:suripw@batan.go.id

Abstract. The system codes non-specific for HTGR such as RELAP5 has been utilized for HTGR thermal hydraulic analysis even helium gas property is not based on KTA 3102.1. However, those RELAP5 applications for HTGR above are merely based on the assumption that RELAP5 helium properties are comparable to the helium properties in the KTA 3102.1. Therefore, the study for comparing the helium properties used in RELAP5 and the helium properties in KTA 3102.1 is required. The objective of this paper is to study the appropriateness' helium properties in RELAP5 code for high temperature gas reactor (HTGR) thermal hydraulic analysis. There has been an inclined interest in the scientific community in the study of the application RELAP5 for HTGR thermal hydraulic analysis. The KTA 3102.1 provides the helium properties that are the most commonly use for the HTGR thermal hydraulic analysis. For this study, the RELAP5 helium properties are compared with the helium properties in KTA 3102.1. The comparison results showed that the RELAP5 helium properties are satisfactory for the HTGR thermal hydraulic analysis.

Keywords: Helium, KTA 3102.1, RELAP5, HTGR Thermal Hydraulic.

1. Introduction

Thermal-hydraulics plays a very important role in the design, operation, performance and safety of nuclear power plants. The thermal hydraulic analysis of nuclear reactors is commonly accomplished by what are known as "System Codes"[1]. These codes calculate the flows in the complex network of piping systems, pumps and vessels that together form the thermal hydraulic systems of a nuclear reactor. Thermo-physical properties of reactor coolant are required by the system codes to calculate the thermal-hydraulic behaviour of the nuclear reactor.

The Helium gas is used as coolant in the high temperature gas reactor (HTGR) technology for both prismatic [2] and pebble [3] core arrangements. For the HTGR thermal hydraulic analyses purpose, helium properties in the KTA 3102.1 [4] should be use in the calculation as suggested by IAEA [5]. Specific computer codes for HTGR thermal hydraulic analyses have been developed and used[6–9]. Those codes use helium properties that provided in KTA 3102.1.

In the mean time, the system codes non-specific for HTGR such as and RELAP5[10–14]has been utilized for HTGR thermal hydraulic analysis even the helium gas property is not based on KTA 3102.1. The RELAP5 has been proposed as part for HTGR pebble bed design analysis [15], A simulation of helical steam generator for HTGR has been performed by using RELAP5-3D [16],

RELAP5 has been used to perform Intermediate Heat eXchanger [IHX] as well[17].Moreover, steady state and transient thermal simulations of HTR-10 has been carried out using RELAP5-3D[18].

However, those RELAP5 applications for HTGR above are merely based on the assumption that RELAP5 helium properties are comparable to the helium properties in the KTA 3102.1. There have been no controlled studies that compare the helium properties used in the RELAP5 and the helium properties in the KTA 3102.1. The objectives of this study are to determine whether or not the RELAP5 is appropriate for HTGR thermal hydraulic analyses regarding on the helium properties. In this study, the helium properties that are concerned include the mass density, specific heat, dynamic viscosity and thermal conductivity. For this study, the helium properties data were collected by executing the RELAP5 for helium pressure ranges between 1 and 100 bar and temperature ranges between 25 and 1500 ^{O}C. Those helium properties data were compared with the associate values in the KTA 3102.1 based on the pressure and temperature. The findings should make an important contribution to justify the application RELAP5 for HTGR thermal hydraulic analyses. It is beyond the scope of this study to examine the HTGR modeling system.

2. Theory

The helium properties in the KTA 3102.1 are shown in Table 1. The values are depending on the pressure and temperature or only in temperature.

Table 1. Material properties for helium [4]

Properties	**Value**	**Standard deviation (%)**
Mass Density$\left(\frac{kg}{m^2}\right)$	$48.14 \times \frac{p}{T}\left(1 + 0.4446 \times \frac{p}{T^{1.2}}\right)$	$0.03 \times \sqrt{p}$
Specific Heat$\left(\frac{J}{kg \times K}\right)$	5195 for constant pressure 3117 for constant volume	$0.05 \times p^{\left(0.6-0.1 \times \frac{T}{T_0}\right)}$
Dynamic Viscosity$(Pa \times s)$	$3.674 \times 10^{-7} \times T^{0.7}$	$0.0015 \times T$
Thermal Conductivity$\left(\frac{W}{m \times K}\right)$	$2.682 \times 10^{-3}(1 + 1.123 \times 10^{-3} \times p) \times T^{0.71 \times (1-2 \times 10^{-4} \times p)}$	$0.0035 \times T$

where :
p is a gas pressure in bar, T is a gas temperature in K and T0 = 273.16 K.

The above equations applicable for the following ranges:
1 bar < p < 100 bar and,
293 K < T < 1773 K.

In RELAP5, the water (h_2o) is a default coolant, while helium is provided as a non-condensable gas. The non-condensable gas will mix with default coolant. The nearly pure helium can be realized by put 0 (zero) in the initial gas fraction of non-condensable gas [19]. To get the helium properties in RELAP5, the following variables minor edit are used: rhog for vapor density, csubpg for vapor specific heat, visg for vapor viscosity and thcong for vapor thermal conductivity.

3. Methodology

An imaginary Tank was use to simulate the numerical experiment using RELAP5. The model boundary was defined to provide the controlled parameters namely pressure and temperature. Figure 1 shows the RELAP5 nodalization of the imaginary Tank model used in the present study. The nodalization follows the nodalization guideline [19]. The Tank comprised of three parts: inlet, testing part, and outlet. The node in the testing part was modeled using a single volume component of RELAP5. The boundary conditions of temperature and pressure were set at the inlet and outlet of flow path TMDPVOL 110 and TMDPVOL 130 respectively. Table 2 shows the variation of pressure and temperature in this study, therefore there were 81 pairs of pressure and temperature condition. Those pressure and temperature were chosen to variety typical operation of HTGR[20]. The time dependent junction TMDPJUN 115 controls the flow rate. The helium properties were taken from SNGLVOL 120. Those properties then compared with the helium properties on the KTA 3201.1 based on pressure and temperature.

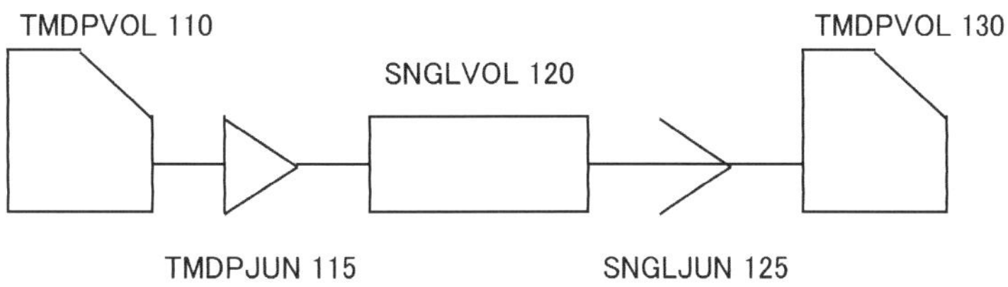

Figure 1. The nodalization of imaginary Tank to get helium properties

Table 2. Variation of pressure and temperature for boundary conditions

Controlled parameters	Values								
Pressure (bar):	1.1	2	5	10	30	40	60	70	99
Temperature (^{O}C):	25	100	250	500	700	900	1000	1200	1500

4. Results and Discussion

4.1. Helium mass densities

Fig. 2 shows the comparison of helium mass densities in RELAP5 and in KTA 3102.1 in the range of pressure and temperature of interest 1.1 to 99 bars and 25 to 1500 ^{O}C respectively. The low values of helium mass density found in the low pressure and high temperature conditions, whereas the high values of helium mass density found in the high pressure and low temperature conditions. The highest different mass density between RELAP5 and KTA 3102.1 is about 5% and it is occur at high pressure and low temperature (i.e. 99 bar and 25 ^{O}C). For typical current and future HTGR [20], the helium working pressure and temperature are 30 ~ 70 bar and 250 ~ 900 ^{O}C respectively. In such a conditions, the helium mass densities are around 1.23 to 6.33 kg/m^3. From Fig. 2, it can be seen that at mass density 1.23 to 6.33 tkg/m^{3}he RELAP5 has helium mass density that is very close to the helium mass density in KTA 3102.1. Therefore, the helium mass density in RELAP5 can be use to analyses in HTGR.

4.2. Helium specific heat at constant pressure

Fig. 3 shows the helium specific heat at constant pressure in RELAP5 and in KTA 3102.1 as function of helium temperature. For helium temperature under 600 ^{O}C, RELAP5 has lower values of specific heat than KTA 3102.1 has. For helium temperature over 600 ^{O}C, RELAP5 has higher values of specific heat than KTA 3102.1 has. For typical HTGR operational coolant temperature ranges (i.e. 250 ~ 900 ^{O}C), the RELAP5 provides helium specific heat that not only in side the allowable plus-minus a sigma error but also very close to the 5195 J/kg-K. Therefore, the helium specific heat at constant pressure in RELAP5 can be use to analyses in HTGR.

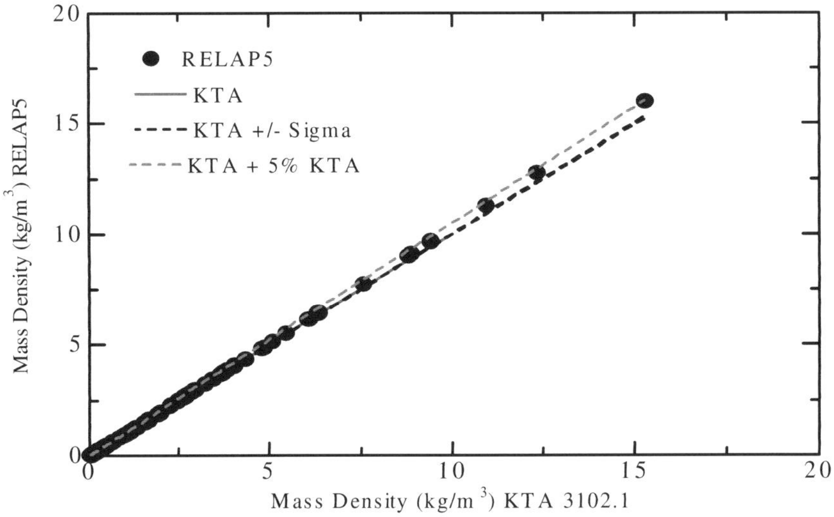

Figure 2. Helium mass density in RELAP5 Vs. KTA 3102.1

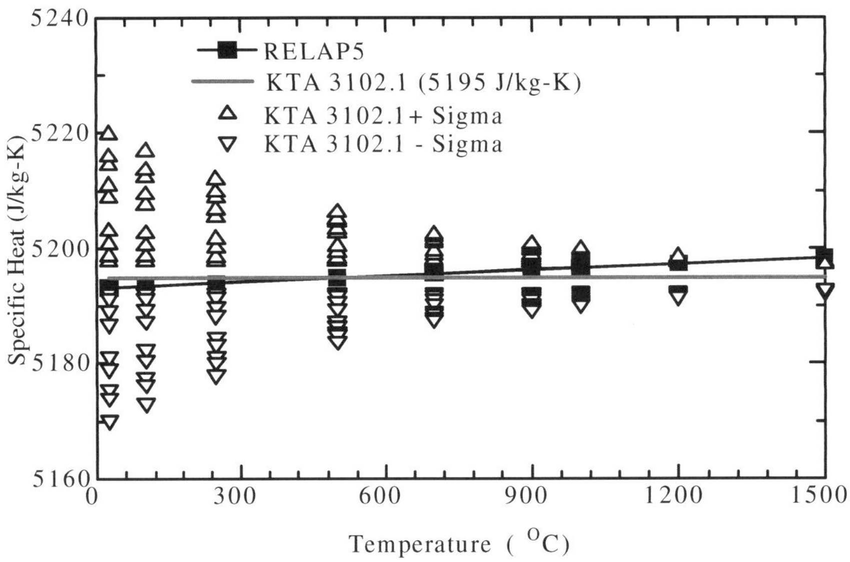

Figure 3. Helium specific heat at constant pressure in RELAP5 and KTA 3102.1

4.3. Helium dynamic viscosity

A Fig. 4 shows the comparison of helium dynamic viscosity in RELAP5 and in KTA 3102.1. The data are based on the helium pressure and temperature in the range 1.1 to 99 bars and 25 to 1500 ^{O}C respectively. The low values of helium dynamic viscosity are corresponding with the low temperature and vice verse. The highest different dynamic viscosity between RELAP5 and KTA 3102.1 is about 14% and it is occur at high temperature (i.e. 1500°C). In the range of HTGR coolant operational temperatures, the RELAP5 has the highest under value about 9% of dynamic viscosity than KTA 3102.1 has and it is occur at 900 ^{O}C. In the low temperatures, the helium dynamic viscosity in RELAP5 is closer to the helium dynamic viscosity in KTA 3102.1. Therefore, the helium mass density in RELAP5 can be use to analyses in HTGR with caution at higher temperature.

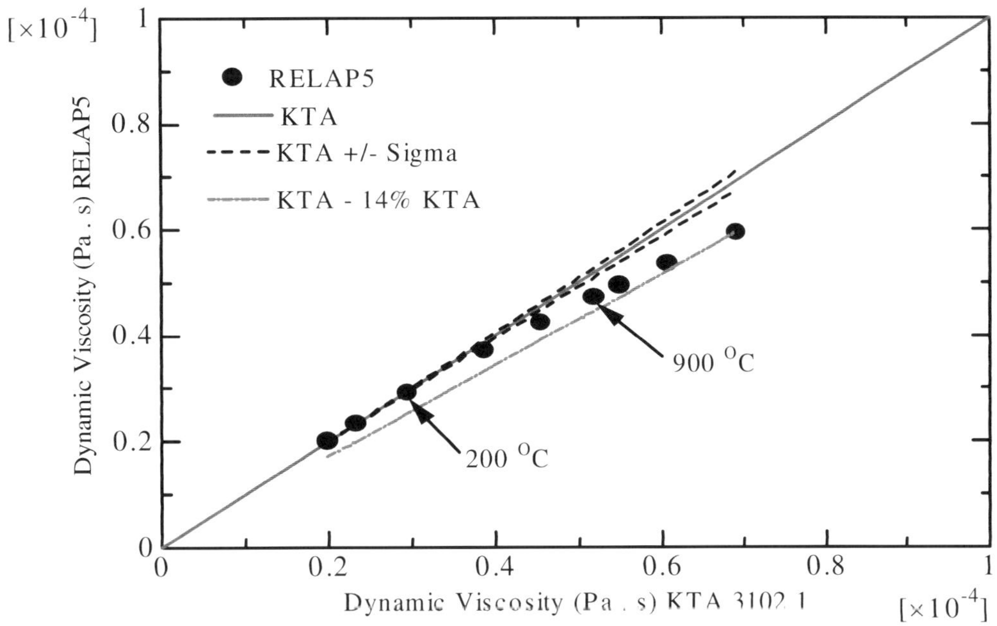

Figure 4. Helium dynamic viscosity in RELAP5 Vs. KTA 3102.1

4.4. Helium thermal conductivity

The comparison of helium thermal conductivity in RELAP5 and in KTA 3102.1 can be seen in Fig. 5. The data in this figure are based on the helium pressure and temperature in the range 1.1 to 99 bars and 25 to 1500 ^{O}C respectively. The high values of thermal conductivity are corresponding with the high temperature and vice verse. For all temperatures, RELAP5 predicts helium thermal conductivity slightly lower, about 3%, than thermal conductivity in KTA 3102.1. From Fig. 5 it also can be seen that the helium thermal conductivity in RELAP5 are well in side the allowable plus-minus a sigma of KTA 3102.1. Therefore, the helium thermal conductivity in RELAP5 can be use to analyses in HTGR.

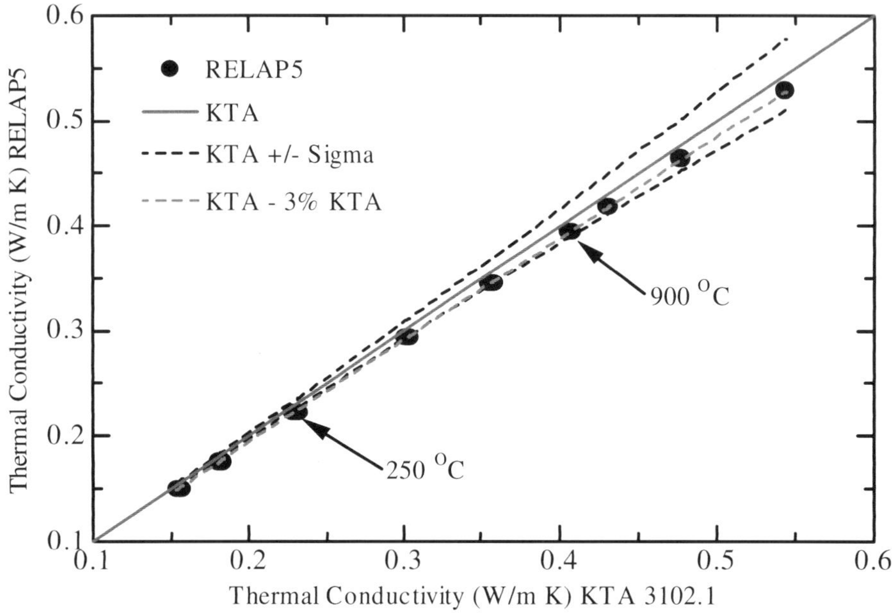

Figure 5. Helium thermal conductivity in RELAP5 Vs. KTA 3102.1

In general, the difference of helium properties in RELAP5 and KTA 3102.1 is increasing with the increase of the helium temperature. However, RELAP5 has helium properties that very close to the helium properties in KTA 3102.1. Some values of helium properties in RELAP5 are slightly higher and some values are slightly lower than corresponding values in KTA 3102.1. Therefore, the helium properties in RELAP5 can be use to analyses in HTGR

5. Conclusion

The study of helium properties for HTGR analysis was performed for RELAP5 code. In the range of HTGR coolant operational pressures and temperatures, it is found that the RELAP5 helium properties such as mass density, specific heat, dynamic viscosity and thermal conductivity are very close to the helium properties in KTA 3102.1. The difference values between helium properties in RELAP5 and in KTA 3102.1 are within standard deviation of the helium properties in KTA 3102.1, except for the helium dynamic viscosity in which the largest error is near 9%. As conclusion of the present study, it was found that the RELAP5 helium properties are appropriate for HTGR thermal hydraulic analyses.

Acknowledgments

The work presented in this paper was funded by DIPA of the Center for Nuclear Reactor Technology and Safety, BATAN (2016 - 2017)

References

[1] Zohuri B and Fathi N 2015 *Thermal-hydraulic analysis of nuclear reactors* (Springer)
[2] Kadak A C 2016 The Status of the US High-Temperature Gas Reactors *Engineering***2** 119–23
[3] Zhang Z, Wu Z, Wang D, Xu Y, Sun Y, Li F and Dong Y 2009 Current status and technical description of Chinese 2×250MWth HTR-PM demonstration plant *Nucl. Eng. Des.***239** 1212–9
[4] Commission N S S 1986 Reactor Core Design of High-Temperature Gas-Cooled Reactors, KTA Standards: 3102.1 (1978), 3102.2 (1983), 3102.3 (1981), 3102.4 (1984), 3102.5 (1986).
[5] IAEA 2003 Evaluation of High Temperature Gas Cooled Reactor Performance: Benchmark

Analysis Related to Initial Testing of the HTTR and HTR-10 *Int. At. Energy Agency, Rep. No. IAEA-TEDOC-1382*

[6] Hossain K, Buck M, Said N Ben, Bernnat W and Lohnert G 2008 Development of a fast 3D thermal-hydraulic tool for design and safety studies for HTRS *Nucl. Eng. Des.***238** 2976–84

[7] Nouri-Borujerdi A and Tabatabai Ghomsheh S I 2015 An improved porous media approach to thermal-hydraulics analysis of high-temperature gas-cooled reactors *Ann. Nucl. Energy***76** 485–92

[8] Seubert A, Sureda A, Bader J, Lapins J, Buck M and Laurien E 2012 The 3-D time-dependent transport code TORT-TD and its coupling with the 3D thermal-hydraulic code ATTICA3D for HTGR applications *Nucl. Eng. Des.***251** 173–80

[9] Setiadipura T, Pane J S and Zuhair Z 2016 Studi Awal Desain PEBBLE BED Reactor Berbasis HTR-PM Dengan Skema Resirkulasi Bahah Bakar Once-Through-Then-Out *J. Pengemb. Energi Nukl.***18** 59–65

[10] Ekariansyah A S, Widodo S and Susyadi D T 2011 Sony Tjahjani, Hendro Tjahjono. Pengembangan Model untuk Simulasi Keselamatan Reaktor PWR 1000 MWe Generasi III+ menggunakan Program Komputer RELAP5 *Tri Dasa Mega***13** 50–62

[11] Isnaini M D, Dibyo S, Suroso S, Hastuti E P and Subekti M 2012 Evaluasi Parameter Desain Termohidrolika Teras dan Sub Kanal PLTN AP1000 Pada Kondisi Tunak *Tri Dasa Mega***14**

[12] Ekariansyah A S and Widodo S 2016 Performance analysis of AP1000 passive systems during Direct Vessel Injection (DVI) line break *Atom Indones.***42** 79–88

[13] Susyadi S, Tjahjono H, Dibyo S and Pane J S 2016 Investigasi Karakteristik Termohidrolika Teras Reaktor Daya Kecil Dengan Pendinginan Sirkulasi Alam Menggunakan RELAP5 *Tri Dasa Mega***18**

[14] Widodo S and Ekariansyah A S 2016 AP1000 partial and complete loss of flow accident analysis using RELAP5 *Seminar Nasional Teknologi Energi Nuklir* (Batam)

[15] Huda M Q and Obara T 2008 Development and testing of analytical models for the pebble bed type HTRs *Ann. Nucl. Energy***35** 1994–2005

[16] Hoffer N V, Sabharwall P and Anderson N A 2011 *Modeling a helical-coil steam generator in RELAP5-3D for the next generation nuclear plant* (Idaho National Laboratory (INL))

[17] Sato H, Ohashi H, Nakagawa S, Tachibana Y and Kunitomi K 2014 Validation and application of thermal hydraulic system code for analysis of helically coiled heat exchanger in high-temperature environment *J. Nucl. Sci. Technol.***51** 1324–35

[18] Scari M E, Costa A L, Pereira C, Velasquez C E and Fortini Veloso M A 2016 HTR steady state and transient thermal analyses *Int. J. Hydrogen Energy***41** 7192–6

[19] Fletcher C D and Schultz R R 1995 RELAP5/MOD3 code manual volume V: User's Guidelines *Idaho Natl. Eng. Lab. Lockheed Idaho Technol. Company, Idaho Falls, Idaho***83415**

[20] Yan X L 2016 3 - Very high-temperature reactor A2 - Pioro, Igor L. BT - Handbook of Generation IV Nuclear Reactors *Woodhead Publishing Series in Energy* (Woodhead Publishing) pp 55–90

Steady State Temperature Distribution Investigation of HTR Core

Sudarmono, Suwoto, Syaiful Bakhri, Geni Rina Sunaryo

Center for Nuclear Reactor Technology and Safety, Bld 80 PUSPIPTEK Area, Serpong, South Tangerang, 15310, Indonesia

aksel@batan.go.id

Abstract. Reactor operation safety is highly related to fuel temperature parameter in the reactor core. To maintain the fuel in good integrity (no crack or melting), fuel temperature should be continuously within licensing criteria. Fuel temperature parameter is one of safety parameters in nuclear reactor operation. Fuel temperature is determined by local heat flux. High heat flux generation will cause a change in heat equilibrium and consequently result a change in fuel temperature. In the present study, steady state temperature distribution investigations of pebble bed for HTR-10 have been performed. The calculation is performed by using VSOP'94 code to do the calculation by which the core is divided into 50 components to represent the positions of various material compositions and to model fuel into 5 layers. The analysis is carried out based on the input data of reactor parameters, core specification, and fuel specifications as well as layer data of HTR-10. The results of this simulation in HTR-10 core at steady state indicated that the maximum temperature is 969.9°C at the solid material and 1031.3 °C in the fuel at the steady state condition. These temperatures are much lower than that of the maximum safety margin of fuel pebble, i.e. 1600°C, therefore, it can be concluded that TRISO fuel is able to contain all radioactive fission products.

Keywords: HTR-10 core, TRISO fuel temperature, solid material temperature, steady state

1. Introduction

Following the failure of heat removal system in the boiling water reactor (BWR) of Fukushima, Daiichi, Japan, professionals in nuclear reactor technology give more attention to high temperature gas-cooled reactor. It is believed that this reactor has high inherent safety feature due to its fuel characteristics, which withstands at relatively high temperature, so that it is categorized as an advanced nuclear reactor design. In order to implement BATAN's vision in 2015 – 2019, BATAN has to develop and operate an experimental power reactor with capacity of 10 MW_{th} using high temperature gas-cooled reactor (HTGR) technology, which is called Experimental Power Reactor-10 [1].

Currently, the analyses on this reactor type are being carried out and the supporting documents for construction license are being prepared. Many studies on core physics design and heat transfer in HTGR have been performed. F. Chen et al simulated anticipated transient without scram (ATWS) on HTR-10 at power level of 3 MW using VSOP'94 computer code. This analysis was aimed to verify inherent safety characteristics in modular HTGR. The simulation results show that the maximum temperature at the fuel center is always less than 1600°C [2].

A transient three-dimensional compressible CFD model has been developed to investigate the thermal-hydraulic characteristics within the core under steady-state and accident conditions. Thermal-hydraulic characteristics and hydrogen production following steam generator rupture were investigated by Y.M. Ferng et al using CFD. In a postulated accident, water that is released into the core reacts with graphite on fuel surface, causing H_2 and CO generation. His calculation results show that H_2 concentration, generated by the chemical reaction between carbon and water, is not sufficient to induce hydrogen explosion in a postulated accident. In addition, the cooling peak temperature during transient is estimated lower than the specified limit margin. This study confirms the safety of HTR-10 core [3]. Experiments on UO_2 pellet treatment to produce HTR-10 fuel were performed by Wuhua Duan et al [4]. Heat removal analysis on HTR-10 spent fuel using natural convection mode was carried out by WANG Jin-hua et al. His study indicates that fuel spent and tank temperature is still within the permitted margin. Moreover, his study provides options in residual heat removal method during fuel loading and interim storage [5]. Meng-Jen Wang et al investigated material homogenization effect on criticality of HTR-10 using MCNP5 v11.51 and SCALE6 computer code. His study indicates that MCNP5 v11.51 and SCALE6 give good agreement results for geometry homogenization [6]. Meanwhile, Rainer Reimerta et al investigated heat supply process of modular HTR in aluminum calcination industry. His investigation show that industries requiring power as high as 100 MW and temperature between 950 – 1000°C are better supplied by HTR with modified process. Technically, this study was considered feasible [7]. Many studies on heat removal in high temperature gas-cooled reactor have also been conducted by BATAN's researchers. Sudarmono carried out heat removal analysis on RGTT200K solid material [8]. In addition, investigation on thermal flow characteristics in HTGR core using VSOP'94-KONVEK and VSOP'94 has been performed [9]. This paper is aimed to identify the effect of coolant flow rate on TRISO fuel temperature distribution in HTR-10 core. There are several factors influencing the fuel temperature distribution, i.e. generated power, helium flow rate, and operational pressure. This study is aimed to investigate significant factors affecting HTR-10 fuel temperature performance using VSOP'94 computer code.

2. Theory

HTR-10 is a high temperature reactor with capacity of 10 MW_{th}. The reactor achieved its first criticality in December 2000. Its main system consists of pebble bed, steam generator, hot gas duct and helium gas circulator. The major structure of HTR-10 comprises fuel pebbles. The 98.6-cm thick graphite structure covers the reactor core and the cavity above the reactor core is 41.698 cm high. Fuel pebble enters the core through the inlet at under the vessel and fuel pebble that has good integrity is transferred back to the core through the upper part of reactor vessel. Therefore, it is apparent that fuel pebble is not static, but dynamically and slowly moves from the top and bottom of reactor vessel. The graphite reflector is placed on top, side, and bottom part of reactor vessel. In the side reflector, there are 10 control rods, 7 channels for absorber spheres, 3 experimental pipes, and 20 channels for cold helium coolant. This reactor type has two independent shutdown systems: control rod system and small absorber sphere system. These two shutdown systems are placed in channel located on the side of pebble bed. These systems run based on natural circulation cooling mechanism [10]. The helium temperature at the inlet and outlet of the core can reach from 250°C up to 700°C. The testing plant adopted a sub-C steam turbine cycle for energy conversion. The primary helium circulation and the second steam turbine circulation were coupled by the steam generator (SG) that serves as the heat exchanger. The reactor core was designed to be installed in the reactor pressure vessel while the steam generator, the helium blower with its driving motor were installed in another separate pressure vessel, called SG pressure vessel. These two pressure vessels were arranged side by side and were connected by the horizontal coaxial hot gas duct. The hot helium from the reactor core passes through the gas duct and is fed into the SG and flows around the heating tube bundles downward. During this process, heat the water/steam on the other side, while the hot helium is cooled. Steam generated in the SG enters the inlet of a steam turbine and drives it to do work. The testing plant reached its full power in January 2003, and the outlet temperature is 700°C. The SG provides the steam turbine with water steam temperature of 435°C and inlet pressure of 3 MPa at a mass flow rate of 4.3 kg/s. Figure 1 describes the schematic diagram of the side - by- side arrangement of pressure vessels and the flow path of the primary loop coolant. A series of arranged tests and experiments has been performed on HTR-10 to verify its inherent safety features and to obtain operational characteristics. It also survived a power supply shutdown accident.

VSOP'94 is Design and Safety Analysis Code For Pebble Bed Reactor [11]. VSOP'94 was originally written with a simple 1-D thermal fluid solver to estimate local fuel and coolant temperatures in the core. That solver is still available but the 2-D (R-Z) THERMIX-KONVEK code has since been coupled to VSOP'94 to provide a more accurate and flexible thermal fluid analysis capability. THERMIX-KONVEK treats the pebble bed as a porous medium but also has an internal pebble fuel model to estimate peak fuel temperatures.

THERMIX-KONVEK code is heat transfer calculation code developed by KFA-Julich. The main correlation and parameters of different materials available in the Thermix code have been evaluated in many studies in Germany and improved for wide uses in other high temperature reactors, such as HTRs, HTR-PM and HTR-10 [12,13,14]. In addition, Thermix code was also used in the development of HTR-PM design for thermohydraulic transient studies [15]. The structure of Thermix code includes calculation of heat transfer based on conduction and convection mechanism at steady state and transient condition using basic and technical solution. Thermix code consists of 4 modules i.e. GASTEM, KONVEK, STROEM, and TFELD. These codes require reactor data and fuel modelling, core specification, fuel pebble, and TRISO fuel [16].

3. Methodology

To identify of solid material and fuel temperature on thermal flow of helium gas as coolant of HTR-10, temperature distribution investigation of Pebble Bed For HTR-10 in Steady State is performed. The variables used are helium thermal flow rate, core pressure, and core power. During the normal operation of the testing plant, the loop pressure of the primary coolant circulation is 3.0 MPa with a mass flow-rate of 4.32 kg/s, while the inlet temperature of the reactor core is 250°C. The lower structure of pebble bed is cone-shaped and has 1.8 m in diameter and 1.97 m in height. The fuel sphere has diameter of 50 mm and 5-mm graphite cladding. The pebble bed is filled with about 27,000 spherical fuel elements of 6 cm diameter, each of which consists of 12,000 coated fuel kernels. These fuel pebbles are loaded continuously into the core whose volume is 5.015m^3 and power density is 2W/cm^3. Steady State solid material and fuel temperature distribution investigation of Pebble Bed For HTR-10 as a function of position at core height is carried out.

HTR-10 fuel modelling is based on the fuel design consisting of low enriched uranium-triple coated isotropic (LEU-TRISO) contained in a graphite-layered pebble. These layered particles contain uranium dioxide kernel coated with four layers to form a fuel pebble of 6 cm in diameter. HTR-10 fuel pebble comprise a number of fuel (in form of UO_2 kernel coated with porous carbon buffer (C) as the fourth layer, inner pyrolitic carbon (IPyC) as the third layer, silicon carbide (SiC) as the second layer, and outer pyrolitic carbon (OPyC) as the first layer, as shown in Figure 1.

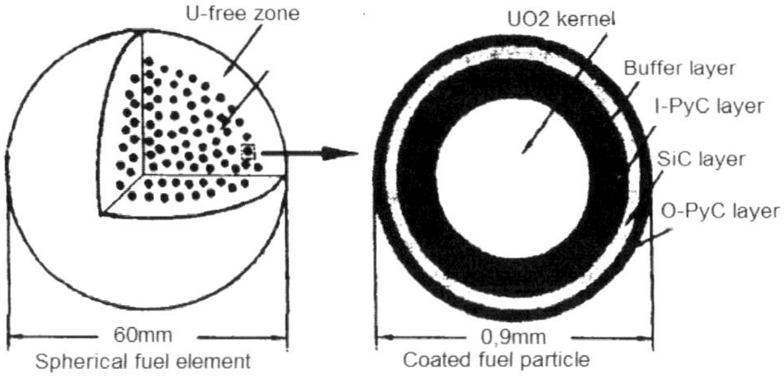

Figure 1. HTR-10 spherical fuel element[10]

For the temperature distribution investigation of Pebble Bed For HTR-10 at steady state, fuel specification and geometry data of TRISO fuel and fuel pebble as well as core geometry data, as tabulated in Table 1, are required.

Table 1. Input Data [10]

PARAMETER		VALUE
Core Geometry		
Core height(cm)	:	197
Number of control rods in side reflector	:	10
Number of absorber ball units in side reflector	:	7
Core diameter(cm)	:	180
Enrichment of fresh fuel element, (%)	:	17
Number of fuel elements in equilibrium core	:	27,000
Pebble Geometry		
Diameter of ball (mm)	:	60
Diameter of fuel zone (mm)	:	50
Density of graphite in matrix and outer shell (g/cm^3)	:	1.73
Enrichment of ^{235}U (weight),%	:	17
Equivalent natural boron content of impurities in graphite (ppm)	:	1.3
Volumetric filling fraction of balls in the core	:	0.61
TRISO Fuel Geometry		
Kernel material	:	UO_2
UO_2 density (g/cm^3)	:	10.40
Kernel diameter(μm)	:	500
Coating layer thickness,(μm)	:	90 /40 /35/40
Coating layer density(g/cc)	:	1.1/1.9/3.18/1.9
Coating material	:	Buffer/I-PyC/SiC/O-PyC

Heat transfer in fuel pebble in the HTR-10 core consists of conduction heat transfer at radial and axial direction (pebble-pebble), convection heat transfer between pebble surface and flowing gas (helium coolant) and radiation heat transfer among fuel pebbles.

To identify the temperature distribution investigation of Pebble Bed For HTR-10, several observation points are selected as shown in Table 2. The position of TRISO fuel in the core is considered as a layer with particular radius.

The assumption used is that fuel sphere is located in the reactor core at position of point $z = 0$ cm, which is in the helium coolant inlet to reactor core. Moreover, point $z = 197$ cm is located at the helium coolant outlet or under the reactor vessel. This is similar to the modelling of conceptual design developed for HTR-10[10].

Table 2. Output Observation Points[8,9]

Position		Observation Points
Radial	:	Solid material and Fuel pebble at 0 to R=90 cm (center)
Axial	:	0 to 197 cm
Zone	:	3 zone Zone 1; at middle of the reactor core as a function of position (R,Z) at radial direction is performed at the middle of core R= 0 cm (centerline); R=7 cm; R=14 cm; R=21 cm; R=28 cm; and R=35 cm. Zone 2; at between middle and side of the reactor core as a function of position (R,Z) at radial direction is performed at the between middle core and side core R= 38.5 cm; R=45.5 cm; R=52.5 cm; R=59.5 cm; R=65.5 cm; and R=70.5 cm Zone 3; at side of the reactor core as a function of position (R,Z) at radial direction is performed at the side core R= 73 cm; R=78 cm; R=82 cm; R=85 cm; R=87 cm; and R=89 cm

The analysis is conducted using VSOP'94 calculation code, in which material properties such as helium density, dynamic viscosity, specific heat capacity, TRISO fuel material heat conductivity, are integrated into the code. Analysis on calculation results is carried out for the hottest position in the core. Observation on solid material and fuel temperature as a function of position (R,Z) at radial direction is performed during the normal operation steady state condition for as a function of position (R,Z) at radial direction in the middle core (zone 1), between middle core and side core (zone 2) and in side core (zone 3). Selection of the data coincides done in order to obtain a good image.

4. Results and Discussion

Solid material and Fuel temperature distribution as a function of reactor core height is observed for during the normal operation, the loop pressure of the primary coolant circulation is 3.0 MPa with a mass flow-rate of 4.32 kg/s, while the inlet temperature of the reactor core is 250°C. Based on the calculation results, fuel temperature distribution at each fuel sphere layer. Kernel temperature and the outer layer of TRISO fuel is discussed. Kernel is the nucleus of fuel pebble material and made of UO_2. Kernel has 500 μm in diameter. Each pebble consists of 12,000 kernels. Kernel is not directly in contact with thermal flow. Heat generated by fission product of kernel is cooled through graphite layer of 0.5 cm thick that become fuel sphere shielding. In addition to solid material and temperature of fuel in the core, temperature for as a function of position (R,Z) at radial direction is also obtained. To identify fuel reactor core and solid material temperature of HTR-10, analysis on temperature characteristics at steady state condition, at which a reactor runs normally, was carried out.

The results could not be obtained immediately, but data selection should be carried out first. The data are then analyzed to obtain the simulation results in form of chart, as shown in Figure 2 to Figure 4. The axial direction is core height, while x direction is temperature in fuel.

The geometrical design of HTR-10 has been carried out with logical approach. It is assumed that absorber sphere is inserted to the reactor core and at position z=0 inlet helium coolant is located. Next, z = 125 cm is located in the helium coolant region, between the inlet and outlet, while z = 197 cm is located at the coolant outlet where it leaves the core. It is different from the one shown in Figure 2 since in point 193 cm is located near the outlet. The difference is due to different visual design in interpreting VSOP'94-THERMIX, but the real meaning is that point 0 cm is inlet point, as simulated. The results of calculation of fuel temperature and solid material temperature distribution during the normal operation at steady state condition for as a function of position (R,Z) at radial direction in the zone 1 are shown in Figure 2.

This graph shows that the fuel temperature distribution at zone 1 as a function of axial direction and temperature difference is well observed in the core. At the outlet, the helium temperatures at the center (z = 193 cm; r = 0 cm), and edge (z = 193 cm; r = 35 cm) of the core are 1031.3°C, and 969.6°C, respectively. Moreover, at the region be the helium inlet and outlet, the temperatures at the center (z = 125 cm; r = 0 cm), and edge (z = 125 cm; r = 35 cm) of the core are 759.5°C, and 715.6°C, respectively. The graph of fuel temperature distribution indicates that the temperature obtained rises as a function of axial distance and a temperature difference in the core is well observed.

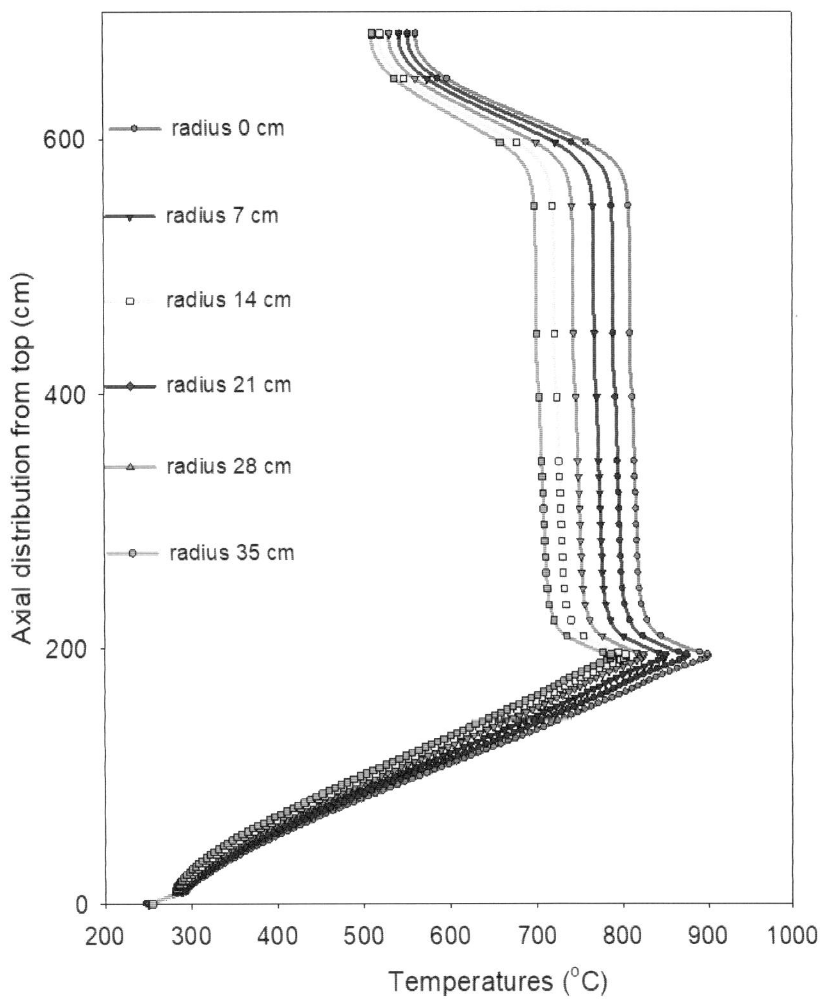

Figure 2. Temperature distribution of fuel and solid material in the middle of core (zone 1).

This graph also shows that the solid material temperature distribution at zone 1. At the outlet, the helium temperatures at the center (z = 193 cm; r = 0 cm), and edge (z = 193 cm; r = 35 cm) of the core are 969.9°C, and 911.9°C, respectively. Temperatures at point (z = 193 cm, r = 28 cm) is 931.9°C. Moreover, at the region be the helium outlet, the temperatures at the center (z = 125 cm; r = 0 cm), and edge (z = 125 cm; r = 35 cm) of the core are 706.2°C, and 666.2°C, respectively.

The results of calculation of fuel and solid material temperature distribution during the normal operation steady state condition for as a function of position (R,Z) at radial direction is performed in between the middle and side core (zone 2) are shown in Figure 3. The results of calculation of fuel temperature distribution during the normal operation steady state condition for as a function of position (R,Z) at radial direction is performed at R= 38.5 cm; R=45.5 cm; R=52.5 cm; R=59.5 cm; R=65.5 cm; and R=70.5 cm.

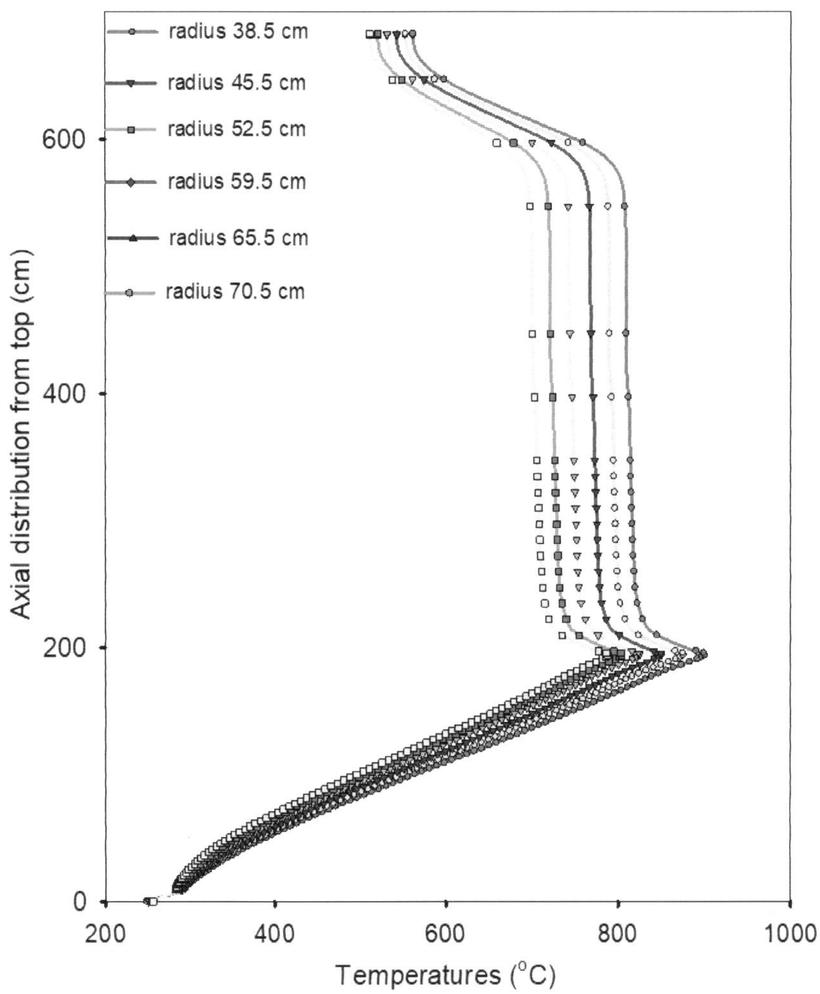

Figure 3. Temperature distribution of fuel and solid material in the between middle core and side core (zone 2).

This graph shows that the fuel temperature distribution at between middle and side of the reactor core as a function of axial direction and temperature difference is well observed in the core, as shown in Figure 6 At the outlet, the helium temperatures at the center (z = 193 cm; r = 38.5 cm), and edge (z = 193 cm; r = 70.5 cm) of the core are 957.6°C and 836.7°C, respectively. Temperatures at point (z = 193 cm, r = 65.5 cm) is 854.1°C. Moreover, at the region be the helium inlet and outlet, the temperatures at the center (z = 125 cm; r = 38.5 cm) and edge (z = 125 cm; r = 70.5 cm) of the core are 707°C, and 620.4°C, respectively.

This graph also shows that the solid material temperature distribution at zone 2 of the reactor core as a function of axial direction and temperature difference is well observed in the core. At the outlet, the helium temperatures at the center (z = 193 cm; r = 38.5 cm), and edge (z = 193 cm; r = 70.5 cm) of the core are 900.7°C, and 788°C, respectively. Temperature at point (z = 193 cm, r = 65.5 cm) is 804.2°C. Moreover, at the region be the helium inlet and outlet, the temperatures at the center (z = 125 cm; r = 38.5 cm), and edge (z = 125 cm; r = 70.5 cm) of the core are 658.4°C, and 577.9°C, respectively.

The results of calculation of fuel temperature and solid temperature distribution during the normal operation steady state condition for as a function of position (R,Z) at radial direction is performed at R= 73 cm; R=78 cm; R=82 cm; R=85 cm; R=87 cm and R=89 cm in side core (zone 3) are shown in Figure 4.

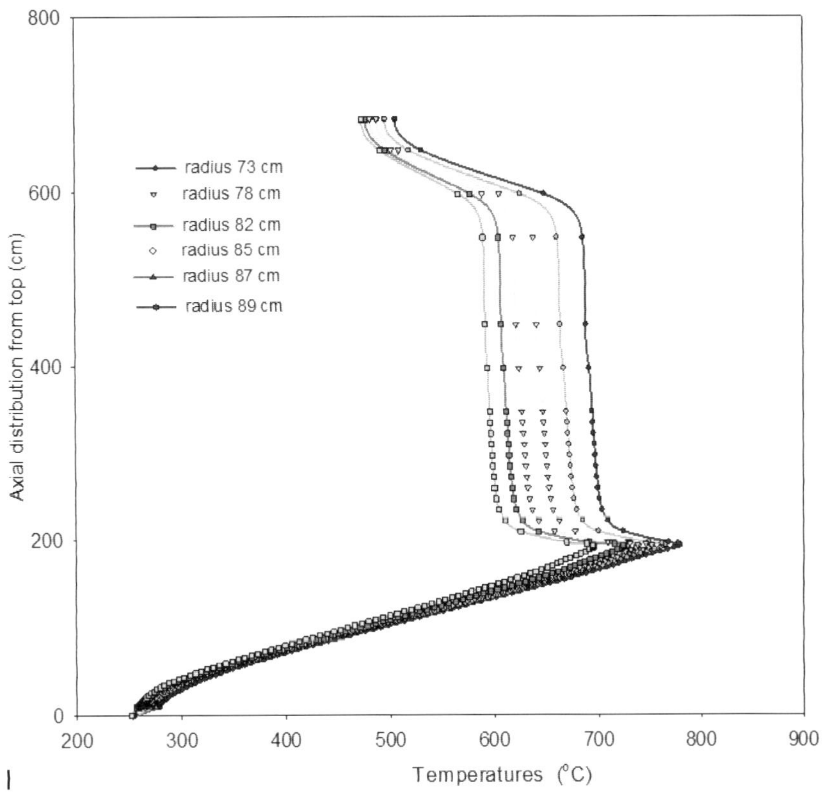

Figure 4. Temperature distribution of solid material in the side core (zone 3).

This graph shows that the fuel temperature distribution at zone 3 as a function of axial direction and temperature difference is well observed in the core. At the outlet, the helium temperatures at the center (z = 193 cm; r = 73 cm) and edge (z = 193 cm; r = 89 cm) of the core are 828.5°C and 693.2°C, respectively. Temperatures at point (z = 193 cm; r = 87 cm) and (z = 193 cm, r = 89 cm) are 769.9°C and 758.7°C. Moreover, at the region be the helium inlet and outlet, the temperatures at the center (z = 125 cm; r = 73 cm) and edge (z = 125 cm; r = 89 cm) of the core are 615.4°C and 543.4°C, respectively.

This graph also shows that the solid material temperature distribution at side of the reactor core (zone 3) as a function of axial direction and temperature difference is well observed in the core, as shown in Figure 4. At the outlet, the helium temperatures at the center (z = 193 cm; r = 73 cm) and edge (z = 193 cm; r = 89 cm) of the core are 780.3°C and 664.4°C, respectively. Temperatures at point (z = 193 cm; r = 87 cm) and (z = 193 cm, r = 89 cm) are 723.4°C and 712.3°C. Moreover, at the region be the helium inlet and outlet, the temperatures at the center (z = 125 cm; r = 73 cm) and edge (z = 125 cm; r = 89 cm) of the core are 573.1°C and 514°C, respectively.

The fuel temperature distribution in the reactor core is affected very much by processes either inside the core. In the core, conduction heat transfer among pebbles and between pebble and the wall occurs directly. The temperature near the reactor wall is lower because there are some moderators surrounding the core so that the number of neutron decreases and heat flow from pebble to the wall occurs due to lower temperature of the wall. The highest temperature in the core is near the helium outlet due to continuous helium flow causing hot flow to go down to the outlet and it unlikely for helium to return to the inlet. In addition, radiation and convection heat transfer occurs. Radiation heat transfer between pebble and other pebble in distance is not influenced by coolant flow, while convection heat transfer between pebble and other pebble or between pebble and the wall in distance is affected by heat flow brought by the coolant.

The temperature distribution of fuel provides a temperature decrease in relation to its height. The farther temperature from the inlet, the lower the temperature. Based on the results, temperature difference in fuel in the reactor core is affected very much by effective thermal conductivity. Thermal conductivity is influenced very much by several properties, such as heat transfer coefficient between solid and gas, pebble zone surface area, and helium fluid temperature.

Table 3. Comparison Results

Parameter	HTR 10 China [17]	HTR Core Using VSOP'94
Thermal power, MW	10	10
Primary helium pressure, MPa	3.0	3.0
Average helium temp. at reactor inlet, ^{o}C	250	250
Primary coolant flow rate, kg/s	4.32	4.32
Core volume, m^3	5.0	5.0
Effective core diameter, m	1.80	1.80
Core height, m	1.97	1.97
Diameter of fuel element, cm	6	6
Number of fuel elements	27,000	27,000
Fuel kernel diameter, μm	500	500
Maximum fuel temp, ^{o}C	918.7	1031.3
Outlet helium temp, ^{o}C	700	759.5
Maximum coolant temp, ^{o}C	818	970
Maximum temp. of side reflector, ^{o}C	666.9	666.2
Maximum temp. at bottom reflector, ^{o}C	789.7	706.2
Limitation of fuel temperature, ^{o}C	1600	1600

Comparison of fuel and solid temperature calculation on HTR core using VSOP'94 and similar calculation of HTR-10 reactor core is shown in table 3. Compared with parameters of HTR reactor at 10 MW, the calculation results of fuel and solid material temperature in zone 1 position of HTR core reactor, seems in agreement, there are under limitation of fuel temperature. Therefore, it can be concluded that steady state temperature distribution of high temperature reactor determination has been verified.

5. Conclusion

The results of this simulation indicate several important points, i.e. the highest temperature in the solid material at the outlet, the helium temperatures at the center (z = 193 cm; r = 0 cm), and edge (z = 193 cm; r = 35 cm) of the core are 969.9°C and 911.9°C, respectively. The highest temperature in the fuel temperature distribution are 1031.3°C and 969.6°C, respectively. Moreover the temperatures resulted in this analysis are compared with that of the previous research and with the maximum safety margin of fuel pebble temperature. i.e. 1600 °C. This value is chosen as a limit so that fuel pebble integrity is maintained and no radionuclide such as Cs-137 causing hazard to human body is released. Therefore. the HTR-10 core design meets the safety criteria. Moreover analysis on the solid material and fuel temperature distribution in HTR-10 core using VSOP'94 code is very beneficial for considering the best and most suitable reactor type for BATAN.

Acknowledgments

The authors would like to appreciate to the Management of PTRKN-BATAN. This research was funded from the PTRKN budget in fiscal year 2015 – 2019.

References

[1] PTKRN-BATAN Strategic Program 2015-2019. Ident No. : Lap.016.KRN.2015. Rev:0 (in Indonesian).
[2] Chen F. et.al. 2009 Post-test analysis of helium circulator trip without scram at 3 MW power level on the HTR-10 *Nuclear Engineering and Design* **239** p1010–1018
[3] Ferng YM, Chen CT 2011 CFD investigating thermal-hydraulic characteristics and hydrogen generation from graphite-water reaction after SG tube rupture in HTR-10 reactor *Applied Thermal Engineering* **31** p2430-2438.
[4] Wuhua D, Liyang Z. Yongjun Z 2011Treatment of UO_2 pellets used for preparing fuel elements of HTR-10 followed by supercritical fluid extraction *Progress in Nuclear Energy* **53** p664-667.
[5] Wang JH 2013 et al. Natural safety analysis of the spent fuel residual heat removal in loading and storage process of HTR-10 *Energy Procedia* **39** p227–239.
[6] Meng-Jen W et.al 2014 Effects of geometry homogenization on the HTR-10 criticality calculations *Nuclear Engineering and Design* **27** p356–360.
[7] Rainer R, Manfred S 2012 Process heat from modularized HTR *Nuclear Engineering and Design* **251** p244–251.
[8] Sudarmono 2014 Heat transfer analysis of RGTT200K solid Material (in Indonesian) *Jurnal Teknologi Bahan Nuklir* **10** p42-52.
[9] Sudarmono 2015 Investigation on thermal-flow characteristics of HTGR core using VSOP'94 code *J. Tek. Reaktor. Nukl.* **17** p41-54.
[10] IAEA-TECDOC-1382 2003 Evaluation of high temperature gas cooled reactor performance: benchmark analysis related to initial testing of the HTTR and HTR-10 p 227-235.
[11] Gougar, Terry WK, OUGOUNG AM et.al Manual PEBBED6 Code *Idaho National Laboratory Nuclear Science and Engineering* April-2015
[12] Dongliyu, et.al 2012 Simulation for the secondary loop of the Chinese 200MWe HTR-PM base on power *Energy Procedia* **16** p1831-1838
[13] Shutang Z, Shaojie L, lei S 2008 A network-based system of simulation control and online assistance for HTR-10 *Annals of Nuclear Energy* **35** p1360–1369
[14] Javier O, et.al 2011 THETRIS: A micro-scale temperature and gas release model for TRISO fuel *Nuclear Engineering and Design* **241** p5018-5032.

[15] Yanhua Z, et.al 2009 Thermohydraulic transient studies of the Chinese 200 MWe HTR-PM for loss of forced cooling accidents *Annals of Nuclear Energy* **36** p742-751

[16] Sudarmono, Suwoto 2014. The effect of coolant flowrate, pressure and power on grafit temperature in RGTT200K core (in Indonesian) ISSN 2355-7524.p651-662

[17] Zuying Gao, Lei Shi, 2002 Thermal hydraulic calculation of the HTR-10 for the initial and equilibrium core, *Nuclear Engineering and Design* **218** p51-64.

Development a computer codes to couple PWR-GALE output and PC-CREAM input

S Kuntjoro, M Budi Setiawan, Nursinta Adi W, Deswandri, G R Sunaryo

Center for Nuclear Reactor Technology and Safety, National Nuclear Energy Agency of Indonesia-BATAN, Kawasan PUSPIPTEK Gd. 80, Serpong-Tangerang Selatan, Indonesia
Email: srikuncoro@batan.go.id

Abstract. Radionuclide dispersion analysis is part of an important reactor safety analysis. From the analysis it can be obtained the amount of doses received by radiation workers and communities around nuclear reactor. The radionuclide dispersion analysis under normal operating conditions is carried out using the PC-CREAM code, and it requires input data such as source term and population distribution. Input data is derived from the output of another program that is PWR-GALE and written Population Distribution data in certain format. Compiling inputs for PC-CREAM programs manually requires high accuracy, as it involves large amounts of data in certain formats and often errors in compiling inputs manually. To minimize errors in input generation, than it is make coupling program for PWR-GALE and PC-CREAM programs and a program for writing population distribution according to the PC-CREAM input format. This work was conducted to create the coupling programming between PWR-GALE output and PC-CREAM input and programming to written population data in the required formats. Programming is done by using Python programming language which has advantages of multiplatform, object-oriented and interactive. The result of this work is software for coupling data of source term and written population distribution data. So that input to PC-CREAM program can be done easily and avoid formatting errors. Programming sourceterm coupling program PWR-GALE and PC-CREAM is completed, so that the creation of PC-CREAM inputs in souceterm and distribution data can be done easily and according to the desired format.

1. Introduction

Radionuclide analysis under normal operating conditions was performed using PC-CREAM program. Required inputs include source term derived from output of PWR-GALE program, population distribution, production and consumption of agricultural products as well as atmospheric stability. The calculation of sourceterm for PWR reactors under normal conditions has been done using the PWR-GALE program [1]. The output of the program will be used as input to the radionuclide distribution program using the PC-CREAM program. PC-CREAM input data is arranged in a certain format and entered manually, so it is very susceptible to errors due to incompatibility with the desired format. To overcome these errors, easy to written and to facilitate the required input data, a coupling program was written using Python language.

Radionuclide analysis under normal operating conditions was performed using PC-CREAM program. Required inputs include source term derived from output of PWR-GALE program, population distribution, production and consumption of agricultural products as well as atmospheric stability. The calculation of sourceterm for PWR reactors under normal conditions has been done using the PWR-GALE program [1]. The output of the program will be used as input to the radionuclide distribution program using the PC-CREAM program. PC-CREAM input data is arranged in a certain format and entered manually, so it is very susceptible to errors due to incompatibility with the desired format. To overcome these errors, easy to written and to facilitate the required input data, a coupling program was written using Python language.

Python programming has been used by several researchers for various problems, among others related to the calculation of physical parameters and reactors, performed by Martensen at al, Ragan-Kelley B at al, Huff K and Shim HJ [2,3,4,5]. Python programming is also used for research on site analysis conducted by researchers Stahlscmidt Z R et al, Matott S L et al and Pogacsas T et al [6,7,8]. It is also used for data control system and uncertainty analysis by White J T et al, Binder J M et al and Enkovaara J et al [9,10,11].

In this work, Python programming language is utilized for coupling of radionuclide dispersion calculation which is source term as the output of program PWR-GALE is used as input data for PC-CREAM. Python is used due to its advantages such as multiplatform (can be used and work in MS Windows, Macintosh, Linux and others). In addition, Python also shaped object-oriented programming, interactive and functional. Commands in Python allow the users to have a direct interaction on the monitor screen, such as a table form that can be filled in directly and executed into a file after the completion of the table.

The objective of this work is to develop computer programs to couple PWR-GALE output and PC-CREAM input. The first step of this work is to examine the data requirements used as input for the radionuclide dispersion in PC-CREAM code. Next step is to design the program to meet those needs and then write a routine using Python language. The programming results are checked whether the results obtained are in accordance with the needs.

2. Methodology

PWR-GALE and PC-CREAM coupling program is a routine to rectify input-output of two data of source term and population distribution. The input of PC-CREAM in the form of source term is shown in Table 1.

Table 1. Source term input data sample case for PC-CREAM

Radionuclide	Activity (Bq)
Ar-41	4.00E-09
Kr-89	3.31E-09
I-131	2.25E-06
Cs-137	3.71E-04

Furthermore, a work was conducted for the PWR-GALE program output, and the output from PWR GALE can be seen in Figure 1.

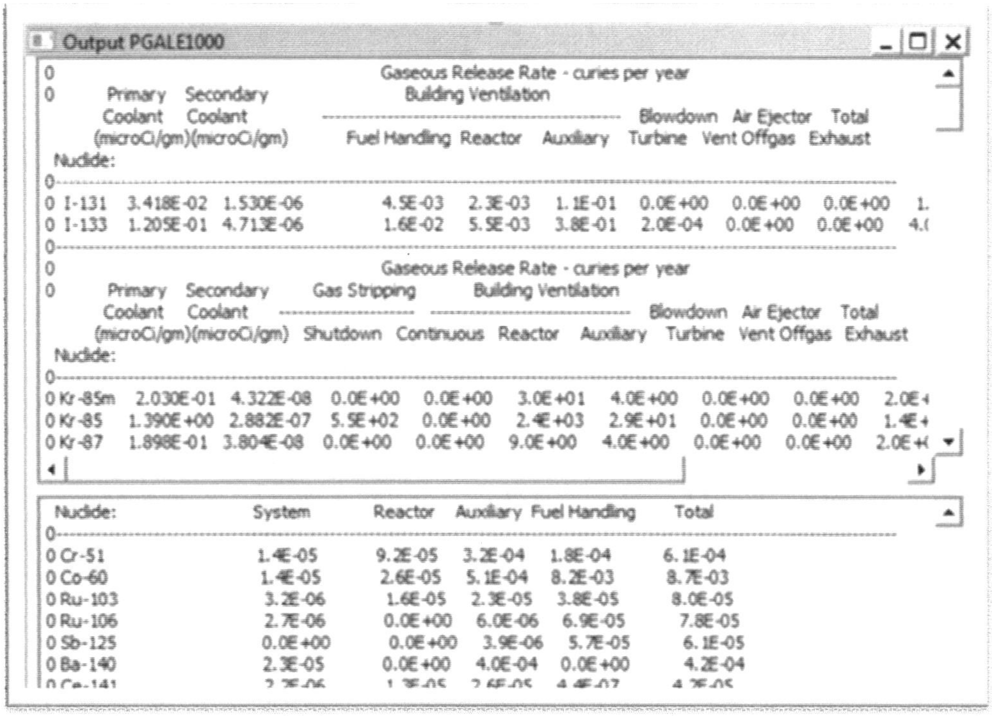

Figure 1. Output data of PWR-GALE

Further work was conducted to programming for population distribution input data is shown in Figure 2.

9 8

300 800 1000 2000 3000 4000 5000 10000 15000

22.5 45.0 67.5 90.0 112.5 135.0 157.5 180.0

	22.5	45.0	67.5	90.0	112.5	135.0	157.0	180.0
300								
800								
1000								
2000				Data	Penduduk			
3000								
4000								
5000								
10000								
15000								

Figure 2. Population distribution sample case

Based on Figure 1, a routine is written to obtain input according to Table 1. The steps taken are: to create a flow chart to read the PWR-GALE program output; and to write the output out of the program whose form and results are similar to the PWR-GALE program output. After that the program is written and executed, if the result is correct, it can be sure that the program reading is correct. Two output files (output PWR-GALE and output Coupling Programming) are written which are the same content and furthermore making programming output file containing only the name of nuclide and the activity of nuclide in accordance with Table 1.

For programming to written the population distribution, the initial step was to create a flow chart to create a display on the screen, which shows how to create a population distribution table. Furthermore programming is written so that it obtain a Table that can be filled as needed. When the table is executed will get a file containing the population distribution function of distance and tangential direction according to PC-CREAM format.

3. Results and Discussions

The results obtained are flow diagrams for writing coupling program from PWR-GALE program output with PC-CREAM program.

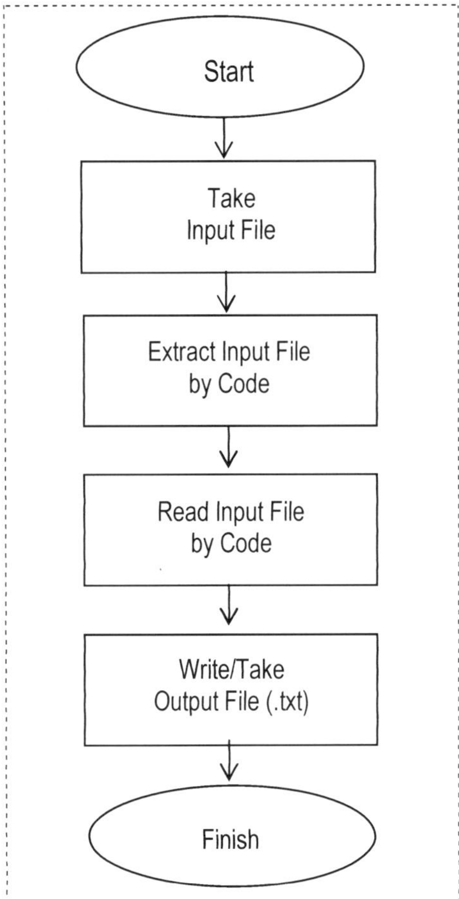

Figure 3. Flowchart for programing coupling of Source term PWR-GALE and PC-CREAM

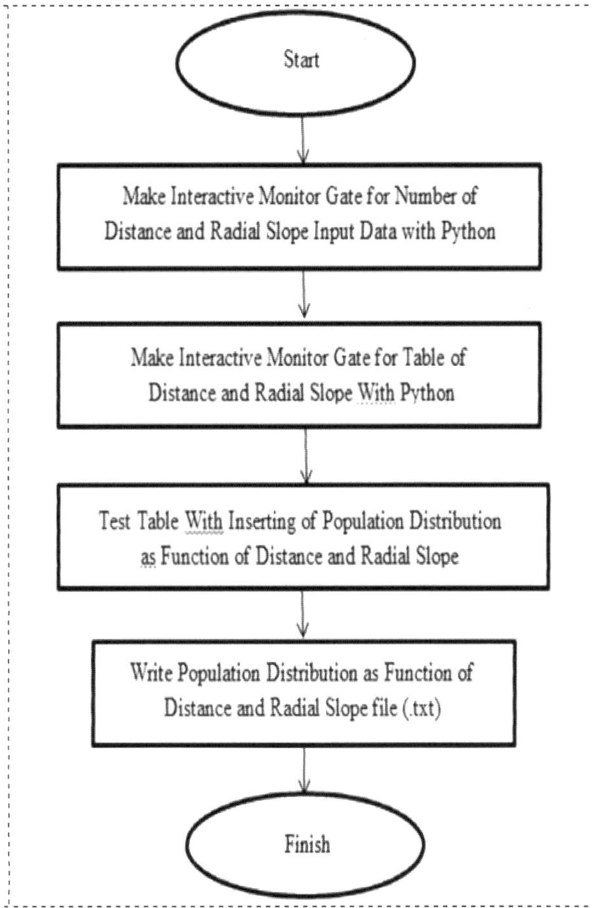

Figure 4. Flowchart for programing population distribution function of distance and radial slope

The flowcharts above show that the coupling programming is done gradually, beginning with opening the output file PWR-GALE program, and then reading the output data line by line. Then a routine is written in accordance with the output program PWR-GALE. After programming is done the program is executed, and the result is compared. When the result of the created program is the same as the file being read, then the programming is done. If the result is not the same, improvement is made to the program, and done the same process with the previous one. Furthermore, a routine was written to couple the source term from PWR-GALE and PC-CREAM. The results obtained are as shown in Figure 5.

```
C:\USPEN2015\USPEN-PAK SRI KUNCORO\output.txt                                                   Buka

0                                          Gaseous Release Rate - curies per year
0       Primary   Secondary                  Building Ventilation
        Coolant   Coolant         ------------------------------------------  Blowdown  Air Ejector  Total
     (microCi/gm)(microCi/gm)     Fuel Handling  Reactor   Auxiliary   Turbine  Vent Offgas  Exhaust
 Nuclide:
0--------------------------------------------------------------------------------------------------------
0 I-131   3.418E-02  1.530E-06        4.5E-03   2.3E-03   1.1E-01   0.0E+00   0.0E+00   0.0E+00   1.
0 I-133   1.205E-01  4.713E-06        1.6E-02   5.5E-03   3.8E-01   2.0E-04   0.0E+00   0.0E+00   4.(
0--------------------------------------------------------------------------------------------------------
0                                          Gaseous Release Rate - curies per year
0       Primary   Secondary     Gas Stripping        Building Ventilation
        Coolant   Coolant    ------------------   ----------------------------  Blowdown  Air Ejector  Total
     (microCi/gm)(microCi/gm) Shutdown  Continuous  Reactor   Auxiliary   Turbine  Vent Offgas  Exhaust
 Nuclide:
0--------------------------------------------------------------------------------------------------------
0 Kr-85m   2.030E-01  4.322E-08  0.0E+00   0.0E+00   3.0E+01   4.0E+00   0.0E+00   0.0E+00   2.0E+
0 Kr-85    1.390E+00  2.882E-07  5.5E+02   0.0E+00   2.4E+03   2.9E+01   0.0E+00   0.0E+00   1.4E+
0 Kr-87    1.898E-01  3.804E-08  0.0E+00   0.0E+00   9.0E+00   4.0E+00   0.0E+00   0.0E+00   2.0E+(

 Nuclide:                   System     Reactor   Auxiliary  Fuel Handling    Total
0--------------------------------------------------------------------------------------------------------
0 Cr-51                     1.4E-05    9.2E-05   3.2E-04    1.8E-04       6.1E-04
0 Co-60                     1.4E-05    2.6E-05   5.1E-04    8.2E-03       8.7E-03
0 Ru-103                    3.2E-06    1.6E-05   2.3E-05    3.8E-05       8.0E-05
0 Ru-106                    2.7E-06    0.0E+00   6.0E-06    6.9E-05       7.8E-05
0 Sb-125                    0.0E+00    0.0E+00   3.9E-06    5.7E-05       6.1E-05
0 Ba-140                    2.3E-05    0.0E+00   4.0E-04    0.0E+00       4.2E-04
0 Ce-141                    2.2E-06    1.3E-05   2.6E-05    4.4E-07       4.2E-05
```

Figure 5. Programming output file result of souceterm from PWR-GALE

This programming to read the output of PWR-GALE has been in accordance with the planning. Furthermore, a routine was added on top of the coupling program to write the output of the data file nuclide and total activity, seen in the last 7 lines of the program output. The output file is used as input for the PC-CREAM program, and the results obtained are as shown in Figure 6.

Aktivitas Radionuklida.txt

Nuclide	Activity (Bq)
Cr-51	6.1E-04
Co-60	8.7E-03
Ru-103	8.0E-05
Ru-106	7.8E-05
Sb-125	6.1E-05
Ba-140	4.2E-04
Ce-141	4.2E-05

Figure 6. Radionuclide activity for the PC-CREAM input

IOP Conf. Series: Journal of Physics: Conf. Series **962** (2018) 012041 doi:10.1088/1742-6596/962/1/012041

From Figure 6. It shows that to make PC-CreAM input easy. It can only running a coupling program that has been created with input from the PWR-GALE program, and generate input in accordance with the PC-CREAM format.

Futher programming is conducted tomake programming to written the population distribution as a fucntion of distance and radial direction. The first step isto create an interactive table on the monitor screen and the results are shown in Figure 7 and Figure 8.

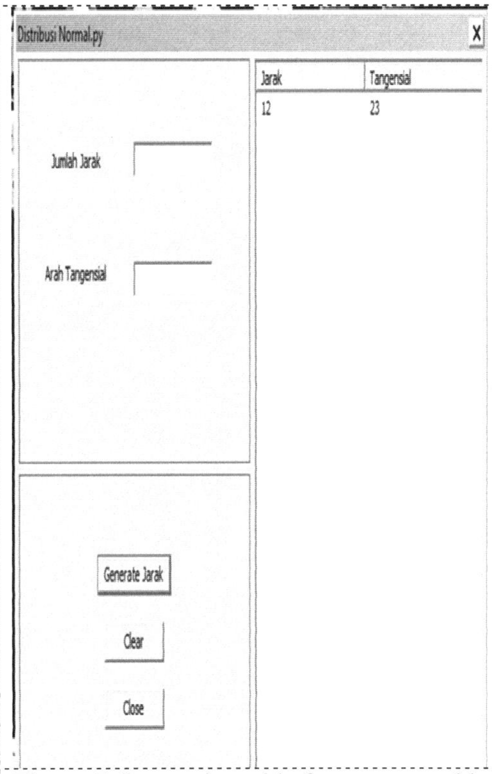

Figure 7. Interactive table for generate table as show in Figure 8.

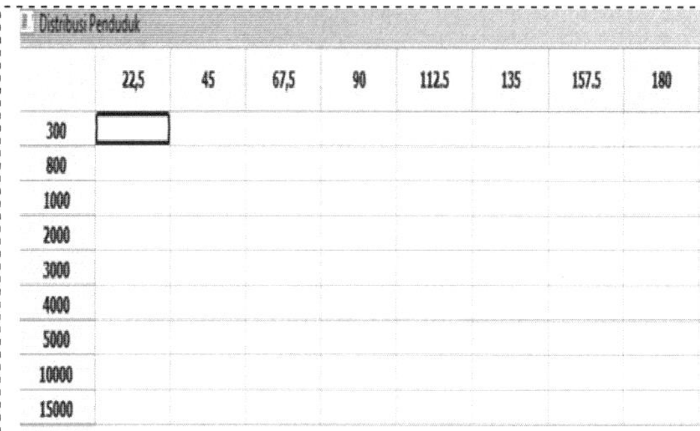

Figure 8. Interactive table for make population data as function of distance and tangential angel.

Figure 7 shows an interactive table to create a table as shown in Figure 8. By filling the tangential distance and direction in Figure 7 it will appear on the screen as shown in Figure 8. Figure 8 is an interactive table on the monitor screen, where by filling in the columns in Figure 8, and then by executing the program, it will get the population distribution file with the format as in Figure 2. By running the program of population distribution data, data input will be obtained for the PC-CREAM program according to the desired format, and avoid the error if done manually. Based on the results obtained, the programming for the coupling source term and written input population distribution is completed as planned.

4. Conclusion

A coupling programming was completed for source term coupling and written of data population distribution inputs. The routine was made to facilitate the inputs for PC-CREAM program, so it will avoid errors in the input format and simplify the data entry work compare to manual entry data. The use of Python language in programming provides advantages because it facilitates interaction with the monitor, so that the results made can be seen directly so as to minimize errors in programming.

Acknowledgments
This research funded by PTKRN-BATAN DIPA budget for fiscal year 2015. Our gratitude goes to the Head of PTKRN for use the DIPA budget.

References

[1] S Kuntjoro 2009 Analisis Suku Sumber Reaktor Jenerik AEC 3568 MWth Pada Kondisi Operasi Normal, Prosiding Seminar Nasional Ke-15 Teknologi dan Keselamatan PLTN Serta Fasilitas Nuklir
[2] ortensen M at al 2016 High performance Python for direct numerical simulations of turbulent flows *Journal Computer Physics Communications* **203** 53-65
[3] Ragan-Kelley B at al 2014 Optimizing physical parameters in 1-D particle-in-cell simulations *Journal Computer Physics Communications* 185 2487–94
[4] Huff K 2015 PyRK: A Python package for nuclear reactor kinetics *Proc. of the 14th Python in science Conf. (SCIPY)* 87-93
[5] Shim H J at al 2015 McCARD for neutronics design and analysis of research reactor cores *Journal Annals of Nuclear Energy* **82** 48–53
[6] Stahlschmidt Z R at al 2011 The role of temperature and humidity in python nest site selection *Journal Animal Behaviour* **81** 1077-81
[7] Matott L S at al 2011 Application of MATLAB and Python optimizers to two case studies involving groundwater flow and contaminant transport modelling *Journal Computers & Geosciences* **37** 1894–99
[8] Pogácsás R at al 2016 ArcGIS scripting - Generating unique hydrogeological maps *Journal of GeoPython* **1** 8-10
[9] White J T at al 2016 A python framework for environmental model uncertainty analysis *Journal Environmental Modelling & Software* **85** 217-28
[10] Binder J M at al 2017 Qudi: A modular python suite for experiment control and data processing *Journal SoftwareX* **6** 85–90
[11] Enkovaara J at al 2011 GPAW - massively parallel electronic structure calculations with Python-based software *Journal Procedia Computer Science* **4** 17–25

Evaluation on Creep Properties of Type 316SS Series

Sri Nitiswati, Sudarno, Andryansyah, Deswandri, Geni Rina Sunaryo

Center for Nuclear Reactor Technology and Safety - National Nuclear Energy Agency of Indonesia, Puspiptek Area Building 80, Serpong, South Tangerang City, 15313, INDONESIA

nitis@batan.go.id

Abstract. Evaluation on creep properties of type 316 stainless steel (SS) series has been done. Type of 316SS series is a candidate material for major structural components of high temperature reactor such as liquid metal reactor (LMR) because it has a good mechanical property in high temperature, compatibility with the coolant sodium, and adequate welding ability. The objective of this research is to obtain and compare creep properties between type 316SS and type 316LN SS materials. The method use by conducted creep rupture tests in ranges temperature of 600°C, 700°C and 800°C and under a constant applied stress level of 200 MPa (393 N). Research result obtained that time to rupture of type 316SS less than that of type 316LN SS. The values of extension (mm), creep strain (mm/mm), creep strain rate (mm/mm/hrs), and rupture elongation (%) of type 316SS are higher than that of type 316LN SS for the same temperatures and applied constant stress level. It is concluded that type 316LN SS has a good creep resistance than that of type 316SS.

Keywords : Creep properties, type 316SS series, high temperature reactor

1. Introduction

Type of 316stainless steel (SS)seriesis a candidate material for majorstructural components of high temperature reactor such as liquid metal reactor (LMR) because it has good mechanical properties in high temperature, compatibility with the coolant sodium, and adequate welding ability [1,2]. There are some types of 316SS series such as: 316SS, 316LN and 316FR.Since the LMR components are used for more than 30 years at a high temperature up to 600°C, one of the most important properties of the components is the creep behaviour [1,2]. It is essential to achieve reliable long-term test data beyond its design life to know a creep rupture of these components. However, it is difficult to actually obtain long-term creep data because of time consuming work and it is an expensive process. Therefore some designs of creep test have been carry out, one of them is called accelerated creep test means creep rupture test by increasing the test temperature or by increasing the test stress beyond its temperature design or stress design [3].

Research on creep properties of 316SS series for special purposes for example study on creep crack growth, analysis of stress effect on creep damage, predicting the creep life, creep fatigue life evaluation of nuclear component in ranging temperature of 550°C up to 950°C have been performed worldwide [1-5]. Research for a certain purpose to obtain and compare creep properties between type 316SS and type 316LN SS has never been done.

This paper presents creep properties obtained from the results of creep rupture tests of the type 316SS series such as type 316SS and type 316LN. Objective of this research is to obtain and compare creep properties such as: time to rupture, creep strain, creep strain rate, and rupture elongation between both of material. The method use by conducted creep rupture tests in ranges temperature of 600°C, 700°C and 800°C and under a constant applied stress level of 200 MPa (393 N).

2. Methodology

The methodology used by conducted creep-rupture test, such as a test in which progressive specimen deformation and the time for rupture are both measured. In general, deformation is much larger than that developed during a creep test [6]. The test temperatures in different condition such as 600°C, 700°C and 800°C and under a constant applied stress level of 200 MPa (393 N)for both of the materials.

Basic equation relating creep strain such as [7]:

$$d\varepsilon / dt = f(\sigma, \varepsilon, t, T) \tag{1}$$

where σ is applied stress, ε is creep strain such as the time-dependent total strain (extension plus initial gage length) produced by applied stress during a creep test [8], t is time of creep test, and T is creep test temperature.

Creep rupture test were carried out using constant load machines with a 10/1 level ratio. The temperature is controlled by a thermocouple that is applied directly on the specimen and was maintained constant within 2°C maximum during the period of the test. The specimen is fitted with an LVDT (Linear Vertical Differential Transformer) for continuous of its elongation, if required. All specimens were held at the test temperature for about 1 hour minimum before starting the test. All creep data was automatically acquired through a data logger. Material used for this research are commercial materials such as type 316SS and type of 316LN SS. Creep specimens were a cylindrical form of a 25 mm gage length, and 5 mm diameter as shown in Figure 1, and chemical composition of the type 316SS and type 316LN SS is given in Table 1.

Table 1. Chemical Composition of Type 316 SS and type 316 LN SS (% weight)

Material	C	Si	Mn	P	S	Ni	Cr	N	Fe
316SS	0.480	0.568	1.243	0.032	0.015	10.750	16.770	1.025	Balance
316LN SS	0.022	0.680	0.989	0.030	0.025	12.270	17.150	0.090	Balance

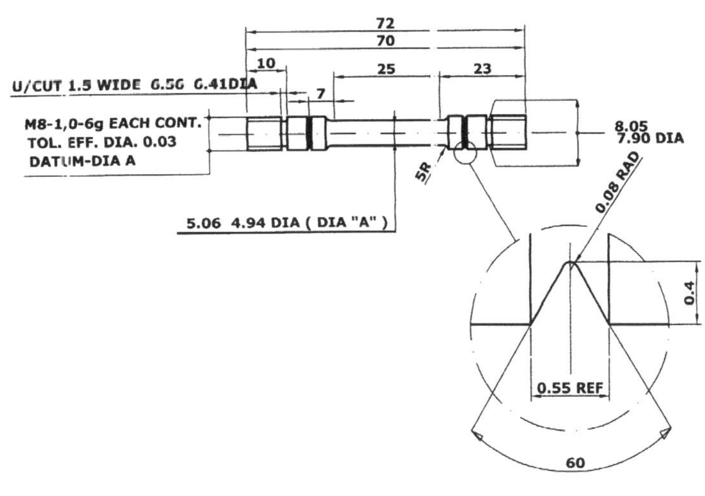

Figure 1. Creep specimen [6,9]

3. Results and Discussion

3.1. Creep rupture properties

To understand the creep properties such as: time to rupture, creep strain, strain rate, and rupture elongation of type 316SS and type 316LN SS, have been investigated through a series of creep rupture test at three difference test temperatures of 600°C, 700°C, and 800°C and under a constant applied stress level of 200 MPa (393 N). Long-term creep rupture data reached approximately1,268hours, 760hours, and 384hours for type 316SS at 600°C, 700°C, and 800°C respectively. For type 316LN SS at 600°C, 700°C, and 800°C reached approximately 3,880hours, 2,350hours, and 950hours respectively.

The two creep curves such as extension (mm) vs. time to rupture (hours) and creep strain (mm/mm) vs. time to rupture (hours) obtained from three creep rupture tests data for each of both materials type 316SS and type 316LN SS are shown in Figure 2 up to Figure 5.

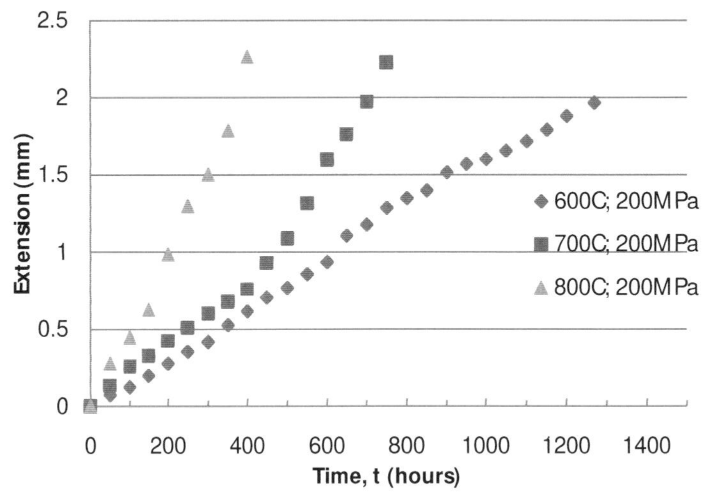

Figure 2. The profile of extension base on time to rupture of type 316SS

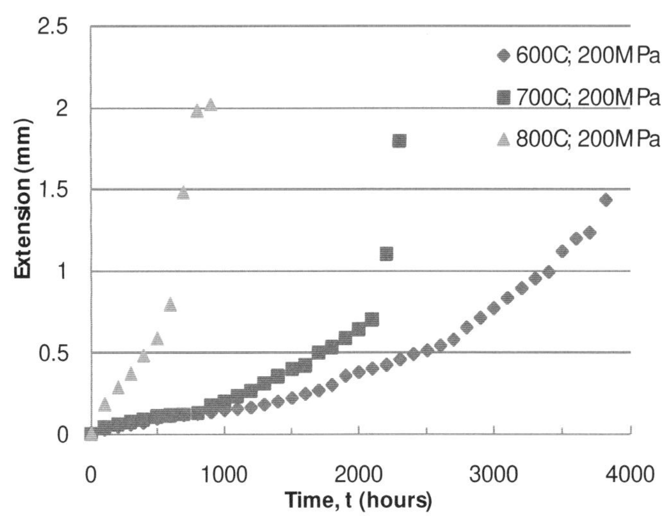

Figure 3. The profile of extension base on time to rupture of type 316LN SS

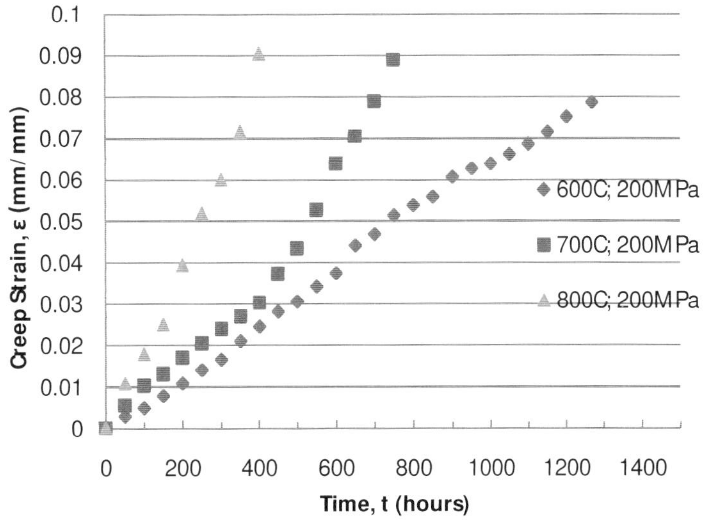

Figure 4. The profile of creep strain on time to rupture of type 316SS

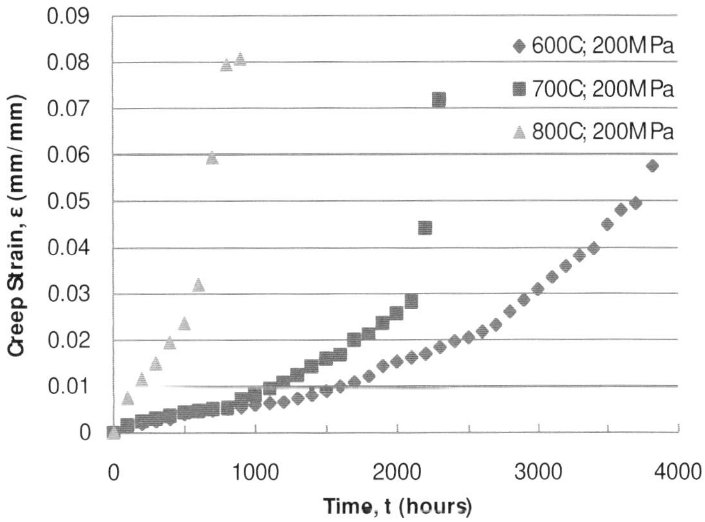

Figure 5. The profile of creep strain on time to rupture of type 316LN SS

Figure 2 and Figure 3 shown curves of extension (mm) vs. time (hours) for type of 316SS and type of 316LN SS respectively. From these figures we obtained that time to rupture (hours) of type 316SS in three difference temperature of 600°C, 700°C, and 800°Cand under applied stress constant of 200 MPa (393N) are less than that of time to rupture (hours) of type 316LN SS at the same temperatures and applied stress. These is indicated that type 316SS has creep-life less than that of type 316LN SS.

Figure 4 shows the curve of creep strain (mm/mm) vs. time (hours). From the figure 2 and 4, we can compared that the three values of extension (mm) and the three values of creep strain (mm/mm) obtained from the three values of extension (mm) divided by a 25 mm initial gage length of creep specimen of type 316SS are higher than that of the three values of extension (mm) and the three values of creep strain (mm/mm) of type 316LN SS for the same condition as shown on Figure 3 and Figure 5. Similar calculation of creep strain rate (mm/mm/hrs) obtained from creep strain (mm/mm) divided by time to rupture (hrs) for both of type materials obtained that creep strain rate (mm/mm/hrs) of type

316SS is higher than that of type 316LN SS for the same condition. In other hand, time to rupture of type 316SS for the three condition are less than that of type 316LN SS. It is means that type 316LN SS has a good creep resistance. This is has a good agreement with a theory as states in the reference [1].

3.2. Rupture elongation
Rupture elongation can be measured by fit the ends of the fractured specimen together carefully and measure the distance between gage marks or overall length to the nearest 0.2 mm at room temperature. When the gage length is marked on the reduced section of a specimen having a nominally uniform cross-sectional area, the elongation is equal to the gage length after fracture minus the original gage length, the difference expressed as a percentage of the original gage length. The method for calculating the rupture elongation shows in Figure 6.

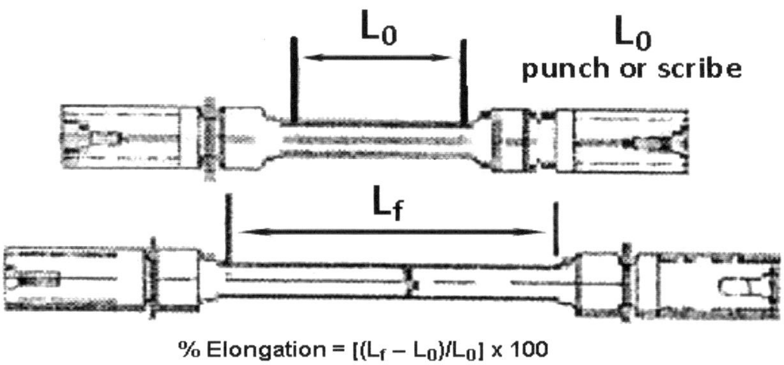

Figure 6. Method of calculating elongation when the gage length is marked on the reduce section of the specimen [6]

The value of time to rupture, creep strain, creep strain rate and rupture elongation have been calculated by formula (1). The summary of creep rupture test results is shown in Table 2 for both of type 316SS and type 316LN SS.

Table 2. Creep rupture properties data of the type 316SS and type 316LN SS

Material	Test Temperature (°C)	Applied Stress (MPa)	Time to Rupture (hours)	Extension (mm)	Creep Strain (mm/mm)	Creep Strain Rate (mm/mm/hrs)	Rupture Elongation (%)
Type 316SS	600	200	1,268	1.965	0.0786	6.198E-05	7.86
	700		760	2.225	0.0890	1.171E-04	8.90
	800		384	2.260	0.0904	2.354E-04	9.04
Type 316LN SS	600	200	3,880	1.435	0.0574	1.479E-05	5.74
	700		2,350	1.797	0.0719	3.059E-05	7.19
	800		950	2.022	0.0809	8.515E-05	8.09

4. Conclusion

It is concluded that type 316LN SS used in this research has a good creep resistance than that of type 316SS. This is because type 316LN SS has a very low carbon (0.022% weight) and nitrogen (0.090% weight) contains. It is very well known that a material especially with very low carbon has a good creep resistance than that of a material with higher carbon contains. The requirement of type 316SS as a "LN" grade is shall be have carbon and nitrogen contains maximum are 0.03% and 0.1% weight respectively. Material with this specification is superior in creep resistance due to the "L" grade of low carbon. If it is compared in between both of these material, type 316SShas ten times higher in carbon contain than that of type 316LN SS.

Acknowledgment

This work was performed under long-term research and development program funded by Center for Nuclear Reactor Technology and Safety, National Nuclear Energy Agency. The author express very much appreciate and thank you to all people whom support this research.

References

[1] Woo Gon Kim, et.al. 2005 Application of Creep Ductility Model for Evaluating Creep Crack Growth of Type 316SS Series *Materials Science Forum*vol.474-479, pp.1433-6

[2] Woo Gon Kim2007Creep Rupture Properties of Type 316LN Stainless Steel Welded by the SAW Process *Solid State Phenomena* vol. 119 pp.91-4

[3] S. Nitiswati, et.al. 2016 Analysis on The Stress Effect of SUS 316 Creep Damage Parameter *Proc. of Nat. Seminar on Nuclear Energy Tech.* pp.305-10

[4] Song-Nan Yin 2008 Creep Curve Modeling to Generate the Isochronous Stress-Strain Curve of Type 316LN Stainless Steel *Key Engineering Materials* vol.385-387, pp 705-708

[5] Kim Woo-Gonet. et.al. 2005Application and Standard Error Analysis of the Parametric Methods for Predicting the Creep Life of Type 316LN SS *Key Engineering Materials* vol. 297-300, pp.2272-7

[6] American Standard for Testing and Materials 2006 Metals-Mechanical Testing; Elevated and Low-Temperature Tests *Metallography* Vol. 03.01, E.139-06-Standard Test Method for Conducting Creep, Creep-Rupture, and Stress-Rupture Tests of Metallic Materials, pp.162-3

[7] Kim Woo-Gon 2000 Creep Design Using Damage Mechanics Concepts *Proc. of The Symposium on Nuclear Materials and Fuel* Korea pp. 162 - 5, August

[8] Jason L Rhoads Basic Explanation of Creep Processes *Dep. of Nuclear Engineering University of California* pp. 235-7, NE-161 Report

[9] Anonymous 1989 MAYES Model Mark II TC Creep Testing Machine-Instruction Manual *MAYES Company*, pp.36

Analysis of JKT01 Neutron Flux Detector Measurements In RSG-GAS Reactor Using LabVIEW

Rokhmadi[1], Agus Nur Rachman[1], Sujarwono[2], Taswanda Taryo[1], Geni Rina Sunaryo[1]

[1]Center for Nuclear Reactor Technology and Safety - BATAN, Kawasan PUSPIPTEK Gd. 80 Serpong Tangerang Selatan 15310
[2]Center for Multipurpose Reactor – BATAN, Kawasan PUSPIPTEK Gd. 80 Serpong Tangerang Selatan 15310

e-mail: rokh_rsg@batan.go.id

Abstract. The RSG-GAS Reactor, one of the Indonesia research reactors and located in Serpong, is owned by the National Nuclear Energy Agency (BATAN). The RSG-GAS reactor has operated since 1987 and some instrumentation and control systems are considered to be degraded and ageing. It is therefore, necessary to evaluate the safety of all instrumentation and controls and one of the component systems to be evaluated is the performance of JKT01 neutron flux detector. Neutron Flux Detector JKT01 basically detects neutron fluxes in the reactor core and converts it into electrical signals. The electrical signal is then forwarded to the amplifier (Amplifier) to become the input of the reactor protection system. One output of it is transferred to the Main Control Room (RKU) showing on the analog meter as an indicator used by the reactor operator. To simulate all of this matter, a program to simulate the output of the JKT01 Neutron Flux Detector using LabVIEW was developed. The simulated data is estimated using a lot of equations also formulated in LabVIEW. The calculation results are also displayed on the interface using LabVIEW available in the PC. By using this simulation program, it is successful to perform anomaly detection experiments on the JKT01 detector of RSG-GAS Reactor. The simulation results showed that the anomaly JKT01 neutron flux using electrical-current-base are respectively, 1.5×,1.7× and 2.0×.

1. Introduction

The regulation of the head of BAPETEN article 6 paragraph 1 point a, b, c, d states that the safety of the non-power reactor deals with the safety assessment during the operation phase including the design of the reactor, the current condition of the structure, system and component (SSC), equipment qualification and ageing [1]. The GA Siwabessy Multipurpose Reactor (RSG-GAS reactor) is a research reactor which had the first criticality in 1987 and achieved full power of 30 MW in 1992[2]. It is therefore needed to evaluate each SSC to ensure the safety of reactor operation for both operator and environmental safety.

The previous evaluation by the researchers showed that the reactor safety system appeared good performance and reliable so that when there is interruption or even failure of operation or disruption of reactor usage, reactor can be extinguished automatically by reactor protection system[3]. The occurrence of reactor operation occurs in the scale of zero and scale one where no radiological impact. In a study of secondary water coolant, a RSG-GAS reactor to operate until 2017 had the quality of secondary water coolant at very good conditions in suppressing the growth of microbes [4]. Due to the existence of some disturbances in the operating cycle of 61-75, namely the power loss of 57 times due to electric trips from the state electricity company (PLN) and 7 times error occurred by the operator, RSG-GAS reactor remains reliable [5].

One of the important parameter in the nuclear reactor is the measurement of neutron flux. In RSG-GAS reactor the neutron flux measurements are divided into three measuring ranges, start up range (JKT01), intermediate range (JKT02) and power range (JKT03) [6]. The development of neutron flux monitoring system has been done, the result shown that the system is able to display the measurement result of neutron flux detector in RSG-GAS [7]. In previous research focused only on the way to collect digital data from the measurements of neutron flux detector and has not yet perform an analysis of neutron flux measurements.

The measurement of neutron flux is important, especially in start up condition. If the start up go too fast the reactor will automatic shutdown. To anticipate the failure measurement of neutron flux detector, need to be analitic. In this research, simulated the anomaly on the measurement neutron flux detector on start up range. The aim of the simulation is to predict and to show the deference between the normal condition and anomaly condition that occur on the neutron flux detector JKT01. To simulate the anomaly condition on neutron flux detector, did by used the LabVIEW software [8]. Using LabVIEW software we can enter the measurement data of neutron flux from RSG-GAS reactor to evaluate and also simulate it to the anomaly condition.

2. Theory

Flux neutrons in a reactor core are important information to monitor the measurement system including: shutdown, subcritical, reactor coolant, and start-up check and reactor operation in all spectrum loads. The neutron flux in the core has the following equations [9]:

$$\emptyset = nv \tag{1}$$

$\emptyset$ = neutron flux ($n\ /scm^2$).
n = amount of neutrons (n/cm^3)
v= neutron speed in the core ($cm/\ s$)

To measure the amount of neutron flux density in the core, required reliable instrumentation. The instrumentation that used to monitor the number of neutrons in the core is a neutron flux detector. The neutron flux detectors inside the RSG-GAS reactor core consist of JKT01, JKT02, JKT03 and JKT04, each detector have different roles in monitoring the number of neutron fluxes when the reactor operates. The number of neutron flux is monitored through electrical signals in the form of electric current in the order of milliampere (mA). The electrical signals are proportional to the reaction count per second (CPS) or the magnitude of the neutron flux. The accuracy of the number of neutron flux inside the reactor core depends on the quality of the electrical signals. If there is detector disturbance, it will result in electrical signals from the detector will experience anomaly.

Detector

The RSG-GAS Reactor detector consists of three types: JKT01, JKT02 and JKT03. JKT01 neutron flux detector acts as RSG-GAS reactor start-up and Fission Chamber (FC) type, schematics as shown in Figures 1a and 1b.

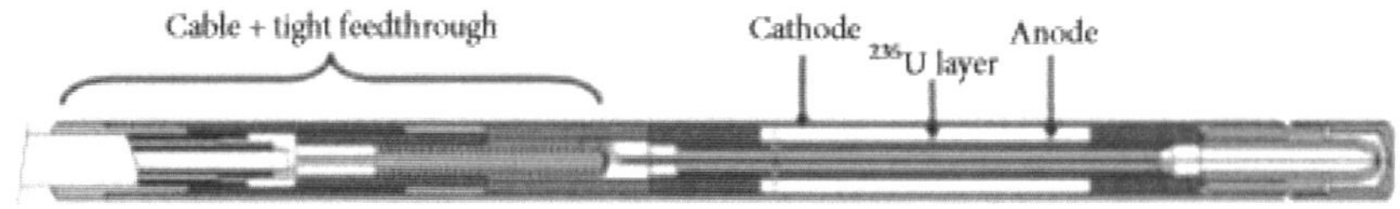

Figure 1a. Overall plan of Fission Chamber with fissile element [8]

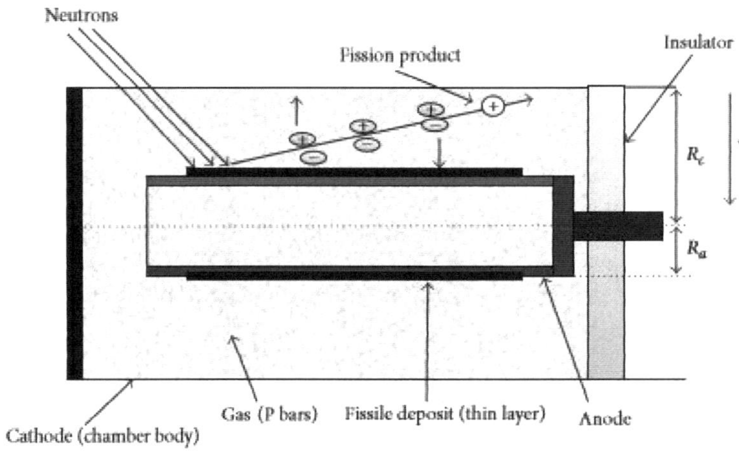

Figure 1b. Simplified schema of cylindrical Fission Chamber [10]

Linear Regression

Linear regression attempts to model the relationship between two variables by fitting a linear equation to observed data. One variable is considered to be an explanatory variable, and the other is considered to be a dependent variable [11].

The equation has the form[12]:

$$y = a + bx \tag{2}$$

where is the dependent variable (that's the variable that goes on the ***y*** axis), ***x*** is the independent variable (i.e. it is plotted on the ***x*** axis), ***b*** is the slope of the line and ***a*** is the y-intercept.

$$b = \frac{n\sum xy - \sum x \sum y}{n\sum x^2 - (x)^2} \tag{3a}$$

$$a = \frac{\sum y \sum x^2 - \sum x \sum xy}{n\sum x^2 - (x)^2} \tag{3b}$$

Correlation coefficient of linear regression[13] :

$$r = \frac{n\sum xy - \sum x \sum y}{\sqrt{[n\sum x^2 - (\sum x)^2][n\sum y^2 - (\sum y)^2]}} \tag{4}$$

The value of r is such that $-1 \leq r \leq +1$. The + and – signs are used for positive linear correlations and negative linear correlations, respectively [14]. A correlation greater than 0.8 is generally described as *strong*, whereas a correlation less than 0.5 is generally described as weakand similar on negative linear correlations.

3. Methodology

To research begin with take data form the instrumentation system in the reactor. The diagram block of instrumentation system on neutron flux detector is shown in Figure 2 below.

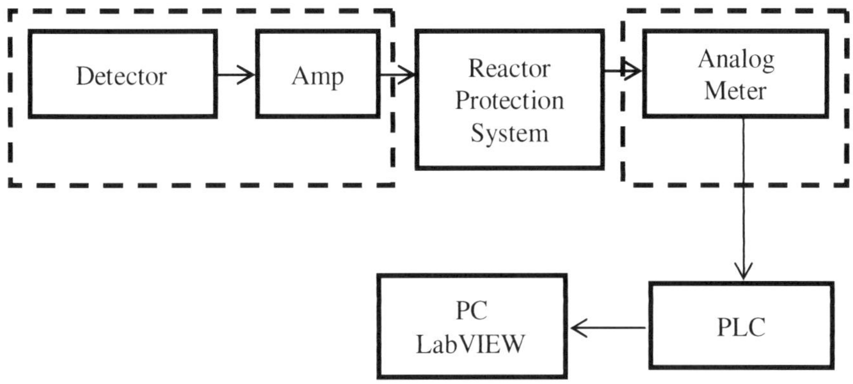

Figure 2. Diagram block of instrumentaton system on neutron flux detector

Neutron flux detector JKT01 detects neutron flux in the reactor core and converts it into electrical signals. The electrical signal is forwarded to the amplifier to become the input of the Reactor Protection System (RPS). Output from the RPS forwarded to the Main Control Room (RKU) to display in the analog meter as an indicator that used by the operator to monitor and control the reactor condition. Analog signal from the analog meter become input for the Programable Logic Controller (PLC). PLC convert analog data to digital data so that can be read by computer. LabVIEW is used to display digital data of neutron flux which has been processed by PLC. Neutron flux measurement data is saved on the computer so that the data can be analys, even the reactor is not on operation.

Data that has been save on computer can be called using LabVIEW programming. Labview can be used to process data and perform mathematical calculations. So that LabVIEW can simulate for the anomalı measurement data. In this research anomali data set to 1.5×, 1.7× and 2.0× from the original data. The value is set on that condition because it will see how the system response when the anomaly occur on a fairly minor error. In this research the anomali is set on the JKT01 CX811. Using the multiple program in LabVIEW, it can manipulate the original measurement data from JKT01 to the 1.5×, 1.7×, and 2,0×

4. The Results and Discussions

Table 1 shown the results of neutron flux value (CPS) versus current (mA) on JKT01CX811. From the Table 1 can be know that the patern of increasing the neutron flux on the reactor is logaritmic. Base on this data, it can be found the equation.

Tabel 1. Neutron Flux Value (CPS) Versus Current (mA) on JKT01CX811

No.	Current (mA)	CPS
1.	0	0
2.	0.52	10E-1
3.	2.86	10E+0
4.	5.72	10E+1
5.	8.56	10E+2
6.	11.42	10E+3
7.	14.28	10E+4
8.	17.14	10E+5
9.	20	10E+6

The data calculation are formulated using equation (2) and (3) and those are then obtained by linear regression equation and the correlation and its graph are shown in Figure 3.

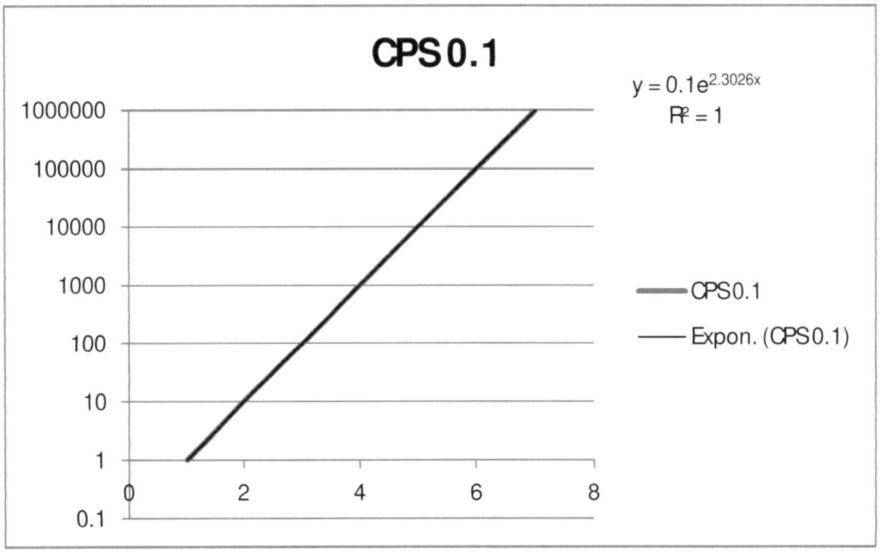

Figure 3. Graph of logarithmic of current (I) vs of CPS

Based on the data seen in Table 1 and developed using regression equations, simulations of anomalous electrical current are 1.5×, 1.7× and 2.0× respectively. The results of the normal electric current can be seen in Figures 4, 5, and 6 as follows:

.

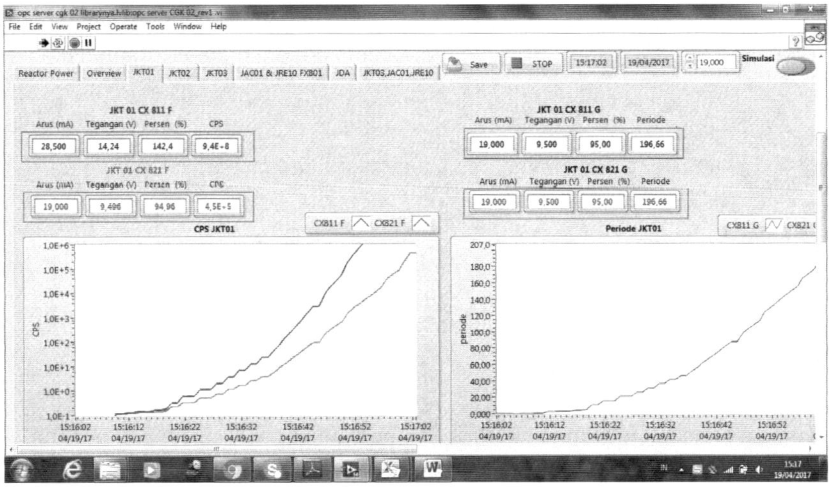

Figure 4. Simulation on 1.5 by normal current

Figure 4 show the simulation of anomali that ocure on JKT01 CX811 neutron flux masurement. The output current on JKT01 CX811 get simulate multiply 1.5 time than normal condition. The measurement results show a drift signal on the appointment of JKT01CX811 compared to JKT01 CX821. In line with the increase in current, the drift signal is greater than the normal signal.

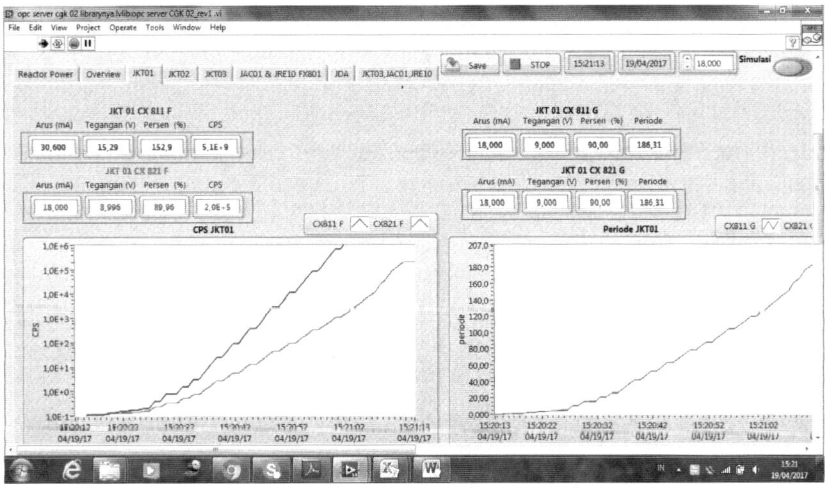

Figure 5. Simulation on 1.7 by normal current

Figure 5 show the simulation of anomali that ocure on JKT01 CX811 neutron flux masurement. The output current on JKT01CX811 get simulate multiply 1.7 time than normal condition. The measurement results show a drift signal on the appointment of JKT01CX811 compared to JKT01 CX821. At first of the measurement, the difference is not visible. But in line with the increment of input current, the drift signal is greater than the normal signal.

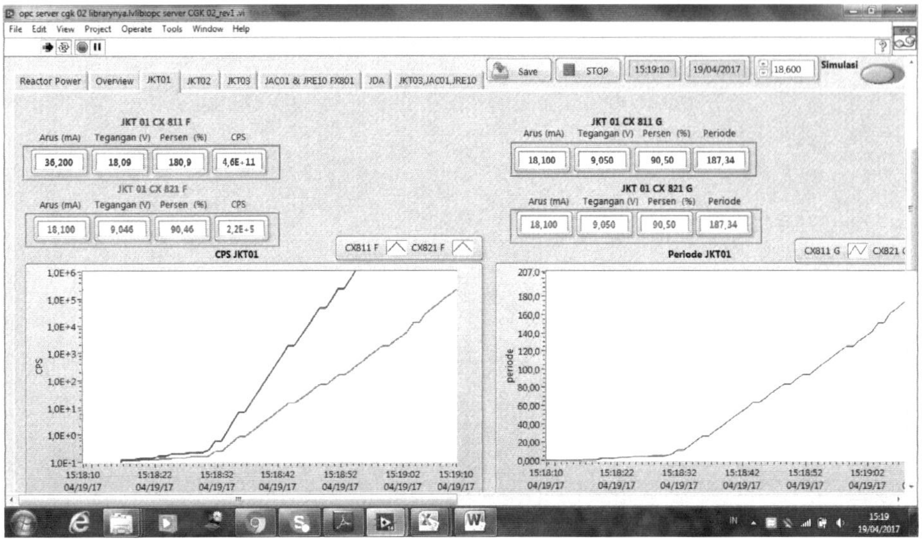

Figure 6. Simulation on 2.0 by normal current

Figure 6 show the simulation of anomali that ocure on JKT01CX811 neutron flux masurement. The output current on JKT01CX811 get simulate multiply 2.0 time than normal condition. The measurement results show a drift signal on the appointment of JKT01CX811 compared to JKT01 CX821. At this simulation the drift signal is visible from the begining of measurement. It cause the multiply of the signal is double than the normal. The drift signal is getting bigger when the input current is increase her paragraphs are indented (Body text Indented style).

5. Conclusions

The measurement of neutron fluxes in RSG-GAS reactor core through the detector JKT01 CX811 and CX821 has been done in this work. A program to simulate the output of the JKT01 neutron flux detector using LabVIEW was developed. The simulated data is estimated using a lot of equations also formulated in LabVIEW. The calculation results are also displayed on the interface using LabVIEW available in the PC.

Using this simulation program, it is successful to perform anomaly detection experiments on the JKT01 detector of RSG-GAS Reactor. The simulation results showed that the anomaly JKT01 neutron fluxes using electrical-current-base are respectively, 1.5×, , 1.7× and 2.0×.

Acknowledgment

The authors expresses their appreciation to Director of Center for Nuclear Reactor Technology and Safety as well as all staff of Division of Reactor Safety Technology and hence making the author to develop this paper to be presented in ICoNETS 2017 in Makassar. The authors are also very appreciative to the committee of guidance for researcher to review and finally establish this paper qualified. The authors really expected this paper being possibly useful for nuclear community in Indonesia

References

[1] Perka BAPETEN No.2 Th. 2011.
[2] SUNARYO, Geni Rina et.al., “Recent Status RSG-GAS Ageing Management”, https://www.google.co.id/?gws_rd=cr,ssl&ei=GfT4WIjwC4HlvgTBj41o#q=rsg-gas+geni+rina+sunaryo diakses 21 April 2017.
[3] KUNTORO, Iman, Evaluasi Kinerja Sistem Keselamatan Reaktor RSG-GAS Selama Operasi 25 Tahun *Buletin Pengelolaan Reaktor Nuklir. Vol. XI, No. 1, April 2014: 1- 10*

[4] SUNARYO, Geni Rina, "Surveilance Management For Secondary Water Cooling Quality Of RSG-GAS", Indonesian Journal of Nuclear Science and Technology Vol. 18, No 1, Februari 2017; 39 – 4, Bandung (2017)

[5] Wiranto, Slamet, Gangguan Operasi RSG-GAS Pada Siklus Operasi 61-75, SEMINAR NASIONAL SDM TEKNOLOGI NUKLIR VII YOGYAKARTA, 16 NOVEMBER 2011

[6] Safety Analysis Report of Multipurpose Reactor, Rev 3, 2011

[7] Agus Nur Rachman et al, "DIGITALISASI SISTEM PENGUKURAN FLUKS NEUTRON REAKTOR RSG-GAS BERBASIS LABVIEW" PROSIDING SEMINAR NASIONAL TEKNOLOGI ENERGI NUKLIR 2016 Batam 4-5 Agustus 2016 **ISSN : 2355-7524**

[8] Halvorsen, Hans-Petter, OPC and Real Time System in labVIEW , Department of Electrical Engineering Information Technology and CybermeticsTelemark University College, Norwegia, 2012

[9] http://www.nuclear-power.net/nuclear-power/reactor-physics/nuclear-engineering-fundamentals/neutron-nuclear-reactions/neutron-flux-neutron-intensity/ download in April 24, 2017.

[10] HananeArahmane, El-MehdiHamzaoui andRajaaCherkaouiElMoursli, Hindawi , Science and Technology of Nuclear Installations Volume 2017, Article ID 5369614

[11] http://www.stat.yale.edu/Courses/1997-98/101/linreg.htm, down load in April 24, 2017

[12] http://www.statisticshowto.com/how-to-find-a-linear-regression-equation/ down load in April 24, 2017

[13] https://www.google.co.id/?gws_rd=cr,ssl&ei=flzWOn6NMjrvgTngrHQAQ#q=regresi+dan+korelasi, *riskayanto.staff.gunadarma.ac.id/Downloads/.../Modul+7+ +Regresi+dan+Korelasi.do...*STATISTIKA 2 – *Regresi&Korelasi. REGRESI&KORELASI* SEDERHANA, download in April 25, 2017

[14] https://mathbits.com/MathBits/TISection/Statistics2/correlation.htm, down load in April 25, 2017

Neutron Fluence And DPA Rate Analysis In Pebble-Bed HTR Reactor Vessel Using MCNP

Amir Hamzah, Suwoto, Anis Rohanda, Hery Adrial, Syaiful Bakhri and Geni Rina Sunaryo

Center for Nuclear Reactor Technology and Safety – National Nuclear Energy Agency of Indonesia (BATAN), PUPSPIPTEK Complex, Office Buliding No. 80, Serpong, Tangerang Selatan 15310, Indonesia, Telp. (021)756-0912, Fax. (021)756-0913,

email: ahamzah@batan.go.id

Abstract. In the Pebble-bed HTR reactor, the distance between the core and the reactor vessel is very close and the media inside are carbon and He gas. Neutron moderation capability of graphite material is theoretically lower than that of water-moderated reactors. Thus, it is estimated much more the fast neutrons will reach the reactor vessel. The fast neutron collisions with the atoms in the reactor vessel will result in radiation damage and could be reducing the vessel life. The purpose of this study was to obtain the magnitude of neutron fluence in the Pebble-bed HTR reactor vessel. Neutron fluence calculations in the pebble-bed HTR reactor vessel were performed using the MCNP computer program. By determining the tally position, it can be calculated flux, spectrum and neutron fluence in the position of Pebble-bed HTR reactor vessel. The calculations results of total neutron flux and fast neutron flux in the reactor vessel of $1.82\text{x}10^8$ n/cm^2/s and $1.79\text{x}10^8$ n/cm^2/s respectively. The fast neutron fluence in the reactor vessel is $3.4\text{x}10^{17}$ n/cm^2 for 60 years reactor operation. Radiation damage in stainless steel material caused by high-energy neutrons (> 1.0 MeV) will occur when it has reached the neutron flux level of $1.0\text{x}10^{24}$ n/cm^2. The neutron fluence results show that there is no radiation damage in the Pebble-bed HTR reactor vessel, so it is predicted that it will be safe to operate at least for 60 years.

Keywords: Pebble-bed HTR reactor, fast neutron fluence, radiation damage, reactor vessel.

1. Introduction

Neutron flux and neutron spectrum and neutron fluence in the core and in the reactor vessel are the important parameters in reactor technology and safety. If the characteristics of these parameters have been determined accurately, the other important parameters of the nuclear reactor could be analyzed to ensure the sustainability and safety of the reactor operation. In this study is to analyze neutron fluence in the pebble-bed HTR reactor vessel using MCNP program [1]. In the previous research, neutron spectrum analysis has been done in the core and in the PWR reactor vessel using MCNP [2-4]. In addition, the neutron flux and the neutron spectrum analysis in the core and in irradiation facilities of the RRI-50 concept reactor have been conducted using MCNP program [5]. Based on this experience, this paper presents the results of research related to neutron fluence analysis in the pebble-bed HTR reactor vessel. This study has to be done because, like the PWR NPP reactor, the distance between the vessel and the reactor core is very close, but the existing media are graphite and He gas, so it is estimated that there are still many fast neutrons reaching the reactor vessel. The fast neutrons with high-energy (E > 1 MeV) have enough energy to move or displace the reactor vessel material atoms from its lattice position and insert into the adjacent lattice. More and more atoms that move from their lattice positions when accumulated will cause vacansions and are inserted in the adjacent lattice when it accumulates into intertitions. This abnormal material condition at some point will degrade the properties of the material known as radiation damage [6-8]. The purpose of this study was to estimate the magnitude of neutron fluence in the pebble-bed HTR reactor vessel during its lifetime.

The Pebble-bed HTR reactor core is modeled into 5 (five) core segments in the axial direction according to burn-up level. The determination of the level of fuel burn-up is performed using the ORIGEN2.1 program [9]. The composition and geometry size of all components of the reactor becomes the main input in the MCNP. Based on the model, the calculation of neutron flux distribution and neutron spectrum from the reactor core to reactor vessel using MCNP has been carried out successfully. The calculation of the neutron flux distribution and the neutron spectrum is done by utilizing the F5 and/or F4 tally card in the kcode mode. The neutron fluence and atomic displacement rate per atom per year (dpa) were performed based on neutron flux calculation results in HTR Pebble-bed reactor vessel.

2. Methodology

2.1. HTR Pebble-bed Core Modeling

The composition and geometry data of each reactor core component is required to model 3-dimensional HTR Pebble-bed reactor core. The core composition and other components of the Pebble-bed HTR reactor are based on the reactor specification data contained in Table 1.

Table 1. Main specifications of HTR Pebble-bed reactor core.

Parameters	Values
Core radius, cm	90
Critical core height, cm	123.06
Height of core cavity, cm	221.818
Operational core height, cm	197
Height of conus, cm	36.946
Outer diameter of graphite reflector, cm	380
Height of graphite reflector, cm	610
Radius of fuel discharge tube, cm	25
Height of fuel discharge tube, cm	610
Diameter of fuel pebble, cm	6.0
Thickness of buffer layer, cm	0.009
Thickness of IPyC layer, cm	0.004
Thickness of SiC layer, cm	0.0035
Thickness of OPyC layer, cm	0.004
Uranium fuel loading, g/bola	5
Density of graphite matrix in fuel pebbed, g/cm^3	1.73
Boron in fuel element, ppm	1.3
Density of graphite matrix in reflector, g/cm^3	1.76
Density of boron in reflector graphite, ppm	4.8366
Ratio of O to U in kernel	2.0
Density of buffer, g/cm^3	1.1
Density of IPyC layer, cm	1.9
Density of SiC layer, cm	3.18
Density of OPyC layer, cm	1.9
Boron in kernel, ppm	4
Boron in dummy pebble, ppm	0.125

Boron in carbon bricks, atom/b-cm	3.46349e-3
Pebble packing fraction	0.61

The reactor core with a nominal power of 10 MWth is divided into 5 (five) segments in the axial direction with the difference of the fuel burn-up level to approximate the actual conditions. Each segment at beginning of cycle (BOC) condition, starting from the uppermost segment is assumed as fresh fuel, then the second segment underneath has been burned 2320 mega watt days (MWD) and so on until the fifth segment has been burned as much as 9280 MWD. At the end of cycle (EOC), the top segment has been burned as much as 2320 MWD and so on until the fifth segment has been burned 11600 MWD. The calculation of the composition of the burned fuel including the activation, actinide and daughters and fission products corresponding to all segments is carried out using the ORIGEN2.1 program [9].

The calculation of the criticality and distribution of neutron flux and neutron spectrum is accomplished by meeting the success criteria of MCNP program results such as relative error (RE) <0.05, variance of the variance (VOV) <0.1 and figure of merit (FOM): fixed. If the criterion has not been reached, then the calculation is repeated by adding the number of particles or the cycle number on the kcode card until the criteria are met. Criticality calculations and the distribution of neutron flux and neutron spectrum on the Pebble-bed HTR reactor core using MCNP program with a cross-section library compiled from ENDF/B-VII.1 [10-15] for neutrons, protons and photo-nuclear interactions for up to 100 MeV.

2.2. Calculation of Neutron Flux Distribution and Spectrum

The neutron flux and neutron spectrum distribution analysis in HTR Pebble-bed reactor were conducted in radial direction starting from the reactor core axis. The neutron flux distribution analysis was carried out in 4 energy groups, i.e. thermal neutrons (E < 0.625 eV), epithermal neutrons (0.625 eV < E < 5.53 keV), advanced neutrons (5.53 keV < E < 0.1 MeV) and fast neutrons (E > 0.1 MeV). The analysis of the neutron spectrum was carried out as much as 640 [10,15-20] energy groups according to the structure of available energy groups in the cross-section library ranging from 10^{-11} MeV to 20 MeV. The calculation of the neutron flux distribution and the neutron spectrum is also carried out by meeting the criteria of accuracy as described above. Analysis of neutron flux distribution and neutron spectrum is done by utilizing F4 and/or F5 tally cards in kcode mode. The results obtained from the MCNP are still in normalized flux values, and a conversion factor is needed to obtain the absolute neutron flux value. The magnitude of the conversion factor is obtained by the equation: [1]

$$FM = 3.15E10 \,.\frac{P.\eta}{k_{eff}} \quad \text{......} \quad (1)$$

with:

P = reactor power (10 MWth),
η = neutrons produce in one fission reaction,
k_{eff} = neutron multipcation factor,

2.3. Neutron Fluence and Displacement per Atom (dpa) Calculations

The neutron fluence in the reactor vessel is the multiplication between the neutron flux and the duration of the reactor operation. Radiation damage in stainless steel materials caused by high-energy neutrons (> 1.0 MeV) could occur when neutron fluence has exceeded the levels of 1.0 x 10^{24} n/cm^2 [10,12]. Displacement per atom (dpa) is an atom that moves from its lattice position because it is exposed to radiation that has considerable energy. However, the calculation of the rate of dpa per second is theoretically carried out using the equation :

$$dpa = \int_0^{\infty} \emptyset(E)\, \sigma_{dpa}(E)\, dE \quad \text{......} \quad (2)$$

where:

$\phi(E)$ = differential neutron flux density, n/cm^2/s/MeV.

$\sigma_{dpa}(E)$ = cross-section of displacement per atom, cm^2,

3. Results and Discussion

3.1. HTR Pebble-bed Core Modeling

The results of tracing the HTR pebble-bed core data in the form of reactor specifications tabulated in Table 1 above. As can be seen in Table 1, the radius and height of the active core during operation and cone height underneath the pebble-bed HTR reactor is 90 cm, 197 cm, and 36.946 cm respectively. The 3-dimensional model of the pebble-bed HTR reactor core including reactor vessel can be seen in Figure 1. At BOC condition, the core segments burn-up have been calculated using the ORIGEN2.1 program, starting from the top position are the 1st, 2nd, 3rd, 4th, and 5th segments are 0%, 11.2%, 20.1%, 27.7%, and 34.5% respectively. At EOC condition, the 1st, 2nd, 3rd, 4th, and 5th segments burn-up are 11.2%, 20.1%, 27.7%, 34.5%, and 40.6%. All of these fuel compositions are taken into account in the analysis including activation, actinide and daughters and fission products are about 137 elements and actinide isotopes correspond to the level of its burn-up.

The 3-dimensional core model including reactor vessel has been made based on geometry data and composition of the core and other components. As shown in Figure 1, each reactor core segment is modeled in homogeneous form. The reactor core is surrounded by a neutron reflector made of graphite. The outside of reflector is covered with stainless steel barrel. All of the core components are inside the stainless steel reactor vessel.

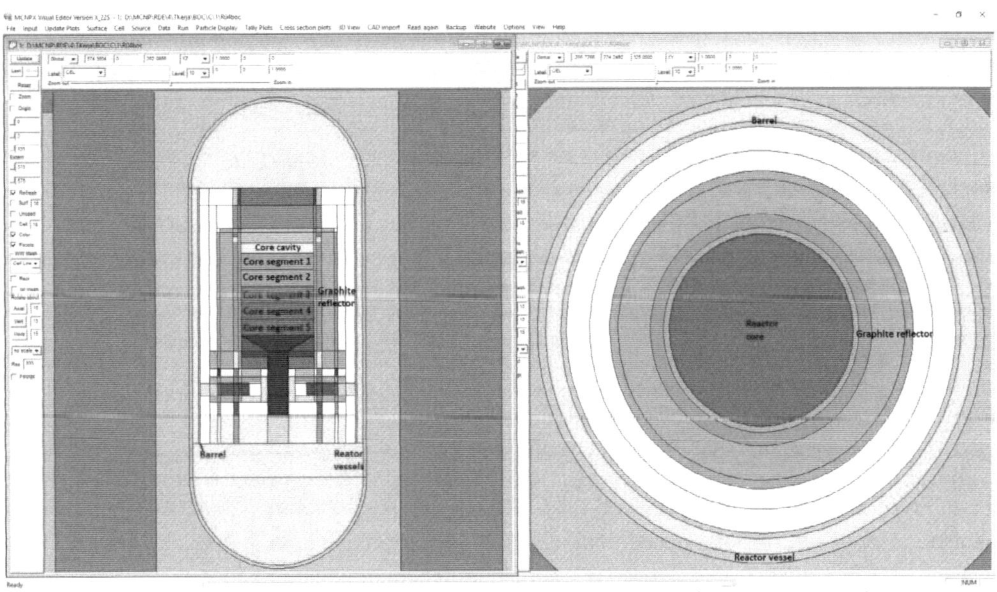

Figure 1. The model of the HTR Pebble-bed core to the reactor vessel.

The result criticality calculations of HTR Pebble-bed in operational condition with 197 cm core height resulted in effective neutron multiplication factor at BOC and EOC are 1.08586±0.00018 and 1.04257±0.00017 respectively. This result can be reasonably believed to be true because the calculation of k-eff in critical condition with a 123.06 cm core height is 0.99673±0.00039 which is very close to 0.9973 of benchmark results for HTR-10 [21]. The above calculation still assumes that the entire control rod is pulled out from the reactor core. Thus it can be seen that the HTR Pebble-bed core model could be considered true.

3.2. Neutron Flux Distribution and Neutron Spectrum

Calculation of neutron flux distribution and neutron spectrum in Pebble-bed HTR reactor core using MCNP program based on the model that has been made. MCNP program simulation is conducted using kcode card option with 5000 particles as much as 5000 cycles with additional 100 cycles skipped to obtain good statistical accuracy and meet the success criteria of analysis result.

The calculation results of neutron flux distribution in the HTR pebble-bed reactor core with the power of 10 MWth is plotted in Fig. 2. The thermal and epithermal neutron flux value in the reactor core is around $3.5\text{x}10^{13}$ n/cm^2/s and almost evenly distributed towards the radial direction. Fig. 2 shows that epithermal neutron flux with energy 0.625 eV < E < 5.53 keV is slightly higher than that of thermal neutron flux. It shows that neutrons in the reactor core are a little bit less moderated. The thermal neutron flux (E < 0.625 eV) is highest compared to the epithermal neutron flux and the fast neutrons in the reflector region. It shows that neutrons have been well moderated in the reflector region until a distance of about 170 cm from the core axis.

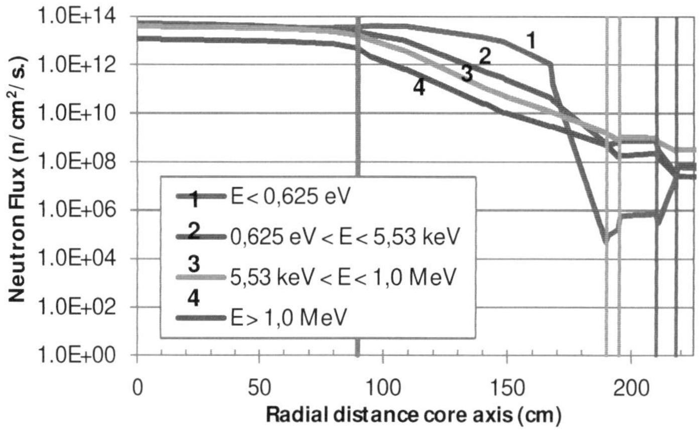

Figure 2. Radial neutron flux distribution from Pebble-bed HTR core to reactor vessel.

In Fig. 2 it can be seen that the thermal neutron flux has a considerable decrease in barrel media and so do in the reactor vessel. It is because of the stainless steel material in the core barrel and the reactor vessel has high neutron absorption cross-section. In the area between the barrel and the reactor vessel containing the He gas cooler, the thermal neutron flux is slightly increased. This is due to the accumulation of thermal neutron flux as a result of moderation of fast neutron, intermediate neutrons and epithermal in the reflecting medium entering the region between the core barrel and the reactor vessel.

In Fig. 2 it is also seen that the fast neutron flux that reaches the reactor vessel is still quite high. That is because the distance between the reactor core and the reactor vessel is relatively close so that many fast neutrons are not yet moderated perfectly become lower energy neutrons. The exposure of fast neutron fluence with energy more than 1.0 MeV during its lifetime when reaching the order of $1\text{x}10^{24}$ n/cm^2 [10,12] can cause radiation damage to the reactor vessel.

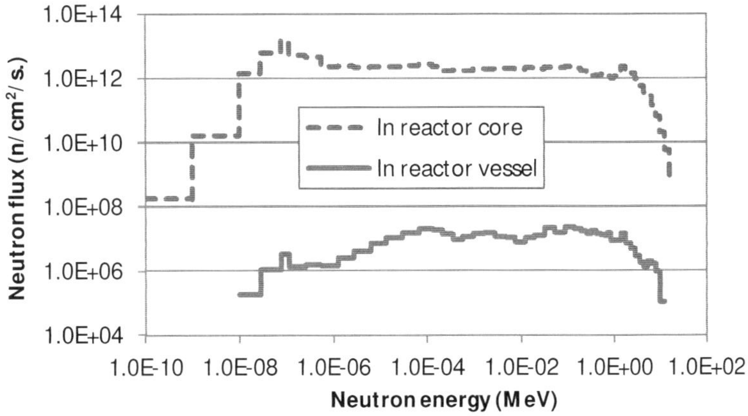

Figure 3. Neutron spectrum in the Pebble-bed HTR core and in reactor vessel.

The results of neutron spectrum calculations in the core and in the Pebble-bed HTR reactor vessel using the MCNP program are plotted in Fig. 3. The calculation of the neutron spectrum uses a neutron energy group structure of 640 groups [10,15-20] taken from the energy distribution in the cross-section library. When compared to the neutron spectrum in the core (Fig. 3), the neutron spectrum in the reactor vessel has been distorted because it has been absorbed by the materials it passes including by the reactor vessel itself. The calculation results of thermal neutron flux and fast neutron flux in the reactor core are 3.5×10^{13} n/cm^2/s and 1.5×10^{13} n/cm^2/s respectively. While the thermal neutron flux and fast neutrons in the reactor vessel are 1.82×10^{8} n/cm^2/s and 1.79×10^{8} n/cm^2/s. The thermal neutron flux in the reactor vessel decreased more than the fast neutrons significantly. The greater amount of thermal neutron flux decreasing indicates that there are many thermal neutrons absorbed in the medium through which it travels. The fast neutron decreasing in the reactor vessel is not as much as the decreasing of thermal neutron flux which suggests that the process of moderation and fast neutron absorption is not as much as the absorption of thermal neutrons in the medium through which it travels. This is a characteristic of Pebble-bed HTR reactor core that uses graphite as a moderator. Compared with a water-based reactor (H2O) as a moderator, the moderation process of the fast neutron on the reactor core with graphite moderator is not as much as on the water-moderated reactor core.

3.3. Neutron Fluence and dpa rate Calculation Results in the Reactor Vessel

The calculation results of the thermal and fast neutron flux in the Pebble-bed HTR reactor vessel described above are 1.82x 0^{8} n/cm^2/s and 1.79x 0^{8} n/cm^2/s respectively. For 1-year reactor operation, the thermal and fast neutron fluence in the reactor vessel is 5.75x 0^{15} n/cm^2 and 5.65×10^{15} n/cm^2 respectively. When the reactor is operated for 60 years, the fast neutron fluence in the reactor vessel is 3.4×10^{17} n/cm^2. That neutron fluence is still very small compared to the fluence limit of 1.0×10^{24} n/cm^2. That fact is making it safe to operate the reactor for 60 years or even longer.

The dpa cross-section for SS-316 stainless steel material is plotted in Fig. 4. In Fig. 4, the dpa cross-section is quite high at low energy and decreases sharply and rise again at higher energy with the lowest energy of about 5×10^{-4} MeV. While the neutron flux in thermal energy in the reactor vessel (Fig. 3) is very low that the calculation results dpa rate becomes small. The result of calculation of annual dpa rate caused by fast neutron flux in reactor vessel using equation (2) is 2.6×10^{-6}. The annual dpa rate contribution caused by the epithermal and thermal flux neutrons in the reactor vessel is 2×10^{-7}. The magnitude of total dpa in the Pebble-bed HTR reactor vessel during even 60 years of operation is a very small (1.7×10^{-4}) for the occurrence of radiation damage.

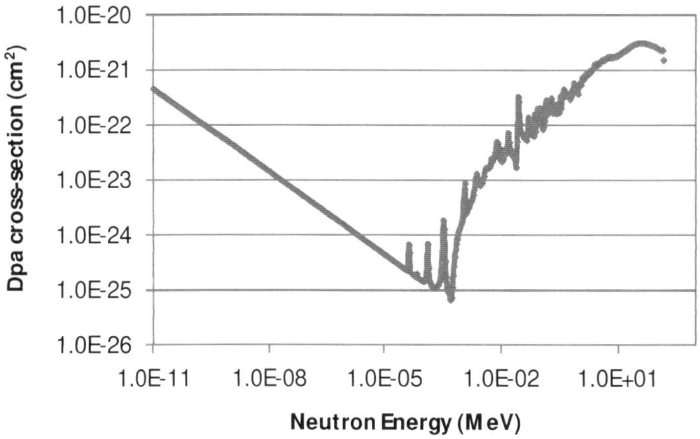

Figure 4. Dpa cross-section for SS316 stainless steel. [10]

4. Conclusion

Fast neutron fluence with energy above 1.0 MeV in a Pebble-bed HTR reactor vessel during 60 years of operation is $3.4x10^{17}$ n/cm^2 which is a small fraction of the threshold of radiation damage ($1.0x10^{24}$ n/cm^2). The annual dpa rate in the Pebble-bed HTR reactor vessel is very small ($2.6x10^{-6}$). The total dpa in the reactor vessel during 60 years of operation is $1.7x10^{-4}$ for the occurrence of radiation damage. It could be concluded that the Pebble-bed HTR reactor vessel is safe from radiation damage for the operation of 60 years or even more.

Acknowledgment

The author is very grateful to the head of PTKRN – BATAN, and head of BFTR which have facilitated the author in doing this work. The author also would like to thank all of the colleagues who have supported the finishing of this paper.

References

[1]. Luka S and Matjaž R 2006 Calculation of Power Density with MCNP in TRIGA reactor, International Conference Nuc. Energy for New Europe (Portorož: Slovenia Elsevier) p 18

[2]. Amir H 2008 Analysis of neutron flux distribution and axial offset of PWR 1300 MWe reactor core using MCNP program "in Indonesian" Prosiding Seminar Nasional ke-14 Teknologi dan Keselamatan PLTN serta Fasilitas Nuklir (Bandung: PRKRN–BATAN) p 188–96

[3]. Amir H 2010 Fast neutron spectrum analysus at the pressure vessel of PWR using MCNP "in Indonesian" Prosiding Seminar Nasional ke-16 Teknologi dan Keselamatan PLTN Serta Fasilitas Nuklir (Surabaya: PRKRN–BATAN) p 29–35

[4]. Amir H 2012 Analysis of distribution of neutron flux and spectrum of 1000 MWe PWR reactor using MCNP "in Indonesian" Prosiding Seminar Nasional ke-18 Teknologi dan Keselamatan PLTN Serta Fasilitas Nuklir (Bandung: PRKRN–BATAN) p 76–86

[5]. Amir H 2015 Neutron energy-spectrum analysis in the irradiation facilities of the conceptual RRI-50 reactor National Seminar Proceeding on Nuclear Energy Technology (Denpasar: PRKRN–BATAN) p 336–41

[6]. Kristina L, Magnus B, Krystyna S, Pål E and Mattias T 2017 Evolution of precipitation in reactor pressure vessel steel welds under neutron irradiation J. Nuc. Mat. **488** 222

[7]. Mustafa U and Teyfik D 2008 Investigation on the radiation damage behavior of various alloys in a fusion reactor using thorium molten salt Mat. and Design **29** 852

[8]. Dong Y, Sencer B H, Garner F A and Marquis E A 2015 Microchemical and microstructural evolution of AISI 304 stainless steel irradiated in EBR-II at PWR-relevant dpa rates J. Nuc. Mat. **467** 692

[9]. Zheng M, Tian W, Wei H, Zhang D, Wu Y, Qiu S and Su G 2014 Development of a MCNP–ORIGEN burn-up calculation code system and its accuracy assessment Ann. Nucl. Energy **63** 491

[10]. Hogenbirk A 2008 An easy way to perform a radiation damage calculation in a complicated geometry Fus. Eng. and Design **83** 1828

[11]. Mohamed E S 2012 Damage parameters of structural materials in fusion environment compared to fission reactor irradiation Fus. Eng. and Design **87** 551

[12]. Simakov S P and Fischer U 2011 Displacement damage induced in iron by gammas and neutrons under irradiation in the IFMIF test cell J. Nuc. Mat. **417** 1321

[13]. Vladimir K 2016 First wall material damage induced by fusion-fission neutron environment Fus Eng. and Design **109–111** 7

[14]. Mehtap G 2013 Investigation of radiation damage in structural material of APEX reactor by using Monte Carlo method Ann. Nuc. Energy **53** 59

[15]. Kryukov A, Debarberis L, von Estorff U, Gillemot F and Oszvald F 2012 Irradiation embrittlement of reactor pressure vessel steel at very high neutron fluence J. Nuc. Mat. **422** 173

[16]. Igor L, Alja˘z C, Sean C, Paola B, Sergey P, Luka S, Brian S, Rafael V, Gediminas S and JET Contributors, 2016 Radiation damage and nuclear heating studies in selected functional functionalmaterials during the JET DT campaign Fus. Eng. and Design **109–111** 1011

[17]. Mehtap G, Basar S and Hızır K 2014 The effect on radiation damage of structural material in a hybrid system by using a Monte Carlo radiation transport code Ann. Nuc. Energy **63** 157

[18]. Gilbert M R, Marian J, Sublet J –Ch 2015 Energy spectra of primary knock-on atoms under neutron irradiation J. Nuc. Mat. **467** 121

[19]. Juan G-A, Juan L F-L, Fortunato A-He 2016 Estimation of the fast neutron fluence in Laguna Verde RPV-Steel specimens simulating their irradiation in a TRIGA reactor Progress in Nuc. Energy **88** 264

[20]. Marco A L, Javier C P-H, Javier O-V, Javier O-V, Vicente J X-M and Armando M G-T 2010 Fast neutron fluence calculations as support for a BWR pressure vessel and internals surveillance program Nuc. Eng. and Design **240** 1271

[21]. IAEA 2003 Evaluation of high temperature gas cooled reactor performance: Benchmark analysis related to initial testing of the HTTR and HTR-10 *IAEA-TECDOC-1382* (Austria: IAEA)

Determining Coolant Flow Rate Distribution In The Fuel-Modified TRIGA Plate Reactor

Endiah Puji Hastuti, Surip Widodo, M. Darwis Isnaini, Geni Rina S., Syaiful B.

Center for Nuclear Reactor Technology and Safety, Bld 80 PUSPIPTEK Area, Serpong, South Tangerang, 15310, Indonesia

endiah@batan.go.id

Abstract. TRIGA 2000 reactor in Bandung is planned to have the fuel element replaced, from cylindrical uranium and zirconium-hydride (U-ZrH) alloy to U_3Si_2-Al plate type of low enriched uranium of 19.75% with uranium density of 2.96 gU/cm^3, while the reactor power is maintained at 2 MW. This change is planned to anticipate the discontinuity of TRIGA fuel element production. The selection of this plate-type fuel element is supported by the fact that such fuel type has been produced in Indonesia and used in MPR-30 safely since 2000. The core configuration of plate-type-fuelled TRIGA reactor requires coolant flow rate through each fuel element channel in order to meet its safety function. This paper is aimed to describe the results of coolant flow rate distribution in the TRIGA core that meets the safety function at normal operation condition, physical test, shutdown, and at initial event of loss of coolant flow due power supply interruption. The design analysis to determine coolant flow rate in this paper employs CAUDVAP and COOLODN computation code. The designed coolant flow rate that meets the safety criteria of departure from nucleate boiling ratio (DNBR), onset of flow instability ratio (OFIR), and ΔT onset of nucleate boiling (ONB), indicates that the minimum flow rate required to cool the plate-type fuelled TRIGA core at 2 MW is 80 kg/s. Therefore, it can be concluded that the operating limitation condition (OLC) for the minimum flow rate is 80 kg/s; the 72 kg/s is to cool the active core; while the minimum flow rate for coolant flow rate drop is limited to 68 kg/s with the coolant inlet temperature 35°C. This thermohydraulic design also provides cooling for 4 positions irradiation position (IP) utilization and 1 central irradiation position (CIP) with end fitting inner diameter (ID) of 10 mm and 20 mm, respectively.

Keywords: CAUDVAP code, COOLODN code, flow distribution, modified reactor, TRIGA plate

1. Introduction

TRIGA 2000 reactor in Bandung is planned to have its fuel element replaced, from cylindrical uranium and zirconium-hydride (U-ZrH) alloy to U_3Si_2-Al plate type of low enriched uranium of 19.75% with uranium density of 2.96 gU/cm^3. Its reactor power is maintained at 2 MW. This change is planned to anticipate the discontinuity of TRIGA fuel element production by CERCA, which is a joint venture company with CERCA of France, as the supplier of international TRIGA fuel element. The selection of this plate-type fuel element is supported by the fact that such fuel type has been produced in Indonesia and used in MPR-30 safely since 2000 [1]. It is not the first time that TRIGA fuel element experienced modification. Previously, TRIGA reactor in Dalat, Vietnam converted its fuel element into cylindrical plate type [2].

The fuel element conversion of TRIGA reactor from cylindrical into plate type has consequences in cooling mode in its reactor core, which is initially based on natural circulation cooling mode and changed into forced circulation at normal condition, and natural circulation during physical test; and changed a transition from forced circulation to natural circulation mode when its cooling pump stops. Moreover, the plate-type-fuelled core configuration that is completed with parallel coolant channel requires cooling flow rate through each fuel element plate channel in order to meet safety requirement. Determination of flow distribution in the plate-type-fuelled TRIGA reactor core has to fulfill safety feature design, which supports the basic safety functions of the reactor facility. The safety functions include that the reactor core should be able to be shut down and maintained at safe conditions, either at operational condition or during accident. Therefore, the reactor core should be equipped with sufficient heat removal system, either after shutdown or during accident to minimize radioactive material releases into environment.

To meet this safety requirement, flow distribution for heat removal of the primary cooling system pump that supplies coolant flow into each position of 5×5 grid in accordance with the specified design configuration [3]. In addition, the flow distribution is designed so that operational safety and cooling availability for reactor utilization is fulfilled. The utilization meant here is cooling availability at irradiation position INPUT located at the most outer angles (4 positions) and central irradiation position CIP at the center of the core (1 position).

Some researches that address coolant flow distribution in plate-fuelled TRIGA reactor include Anwar Ilmar Ramadhan et al. (2017) who carried about a preliminary design on the use of MTR-type fuel element using Computational fluid dynamic (CFD) performed calculation on flow distribution in the plate fuel element, in which the flow in the core configuration was modeled without considering both power peak factor and technical factor of nuclear fuel fabrication [4]. Research on the performance of the flow rate in the plate-type fuel by using CFD has also been conducted by several researchers [5,6,7]. Meanwhile research on reactor design and safety, is also being conducted [8,9,10]. This paper aimed to complete the design before, is describe the result of the design of coolant flow distribution in the reactor core that meets the safety function at normal operation, physical test, shut down and during an accident of loss of coolant flow due to power supply interruption. The primary cooling system that cools the fuel element should fulfill the safety margin in the forced convection cooling mode in spite of free convection. The computation codes used to determine this coolant flow distribution are CAUDVAP and COOLODN.

2. Theory

2.1. Primary Cooling System of Plate-Type-Fuelled TRIGA Reactor

The reactor cooling system has a function to remove fission product heat generated at full load using downward forced cooling circulation mode and transfer the heat to secondary cooling system. The primary cooling system in the reactor core is also used to remove decay heat by natural circulation, when the primary cooling system pump experiences failure. The reactor primary cooling system is connected to the reactor pool water, primary piping circuit, decay tank, primary cooling pumps, heat exchanger, valves, and pipe fitting components. This primary cooling system functions to remove heat from the core at operational condition and during accident while maintaining the reactor in safe condition. Heat is removed by coolant water flow through the core by forced circulation at normal power or by natural circulation when the core is used for physical test, shutdown, or refueling. The heat removed from the core is then transferred to the secondary cooling system and released to environment. The primary cooling system section located in the reactor core has a function to cool the core at normal operation and at free convection cooling condition, which is defined as Safety Category 1 of engineered safety feature. Meanwhile, the primary cooling system section located outside of the reactor pool that is not directly needed to cool the core using natural circulation mode is classified as

Safety Category 2 system. The cooling flow in the reactor core is directly responsible for heat removal in the fuel element as defined as Safety Category 1 of engineered safety feature.

This plate-type-fuelled TRIGA reactor utilizes demineralized water as coolant and moderator. Its reactor core consists of 5×5 lattices. The reactor pool is 8.5 cm deep, while the core is 5.5 m under water. The coolant water flows downward through fuel element gaps to remove 2 MW heat generated by the fuel element and makes N-16 decayed into a delay chamber, and then transfers the heat to a heat exchanger. The reactor core lattice comprise 25 identical holes to hold 16 fuel element, 4 control element, 1 irradiation position at the center and 4 irradiation positions at the core edge, as shown in figure 1.The element ends are inserted to the lattice holes. Additional smaller holes are provided between the main lattice holes in order to fuel element surface.

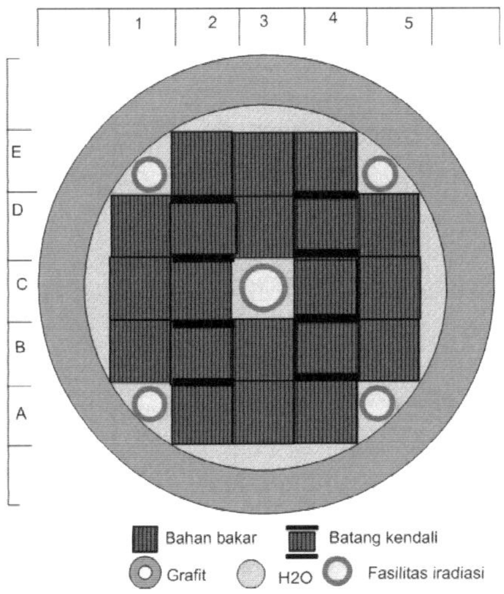

Figure 1. Core configuration of modified TRIGA plate reactor [2]

The fuel elements used are made based on MTR technology, as shown in figure 2. Each standard fuel element consists of a handle at the upper part, two side plates and 21 fuel element plates, and end fitting at the bottom part. Each fuel plate comprises AlMg2 frame and cover plates, which enclose U_3Si_2-Al dispersion fuel element. The control elements are designed to be inserted by fork-type absorber. The sections containing fuel element in the control element is similar to those of the fuel element. As many as 15 fuel element plates are supported by two side plates. Three fuel element plates are taken out at each end of the section containing fuel element in order to provide space for inserting absorber blade. Aluminum plates replace two out of three fuel element plate taken out. The absorber component consists of two Ag-In-Cd blades coated by stainless steel (1.4541, similar to SS321) as shown in figure 3.

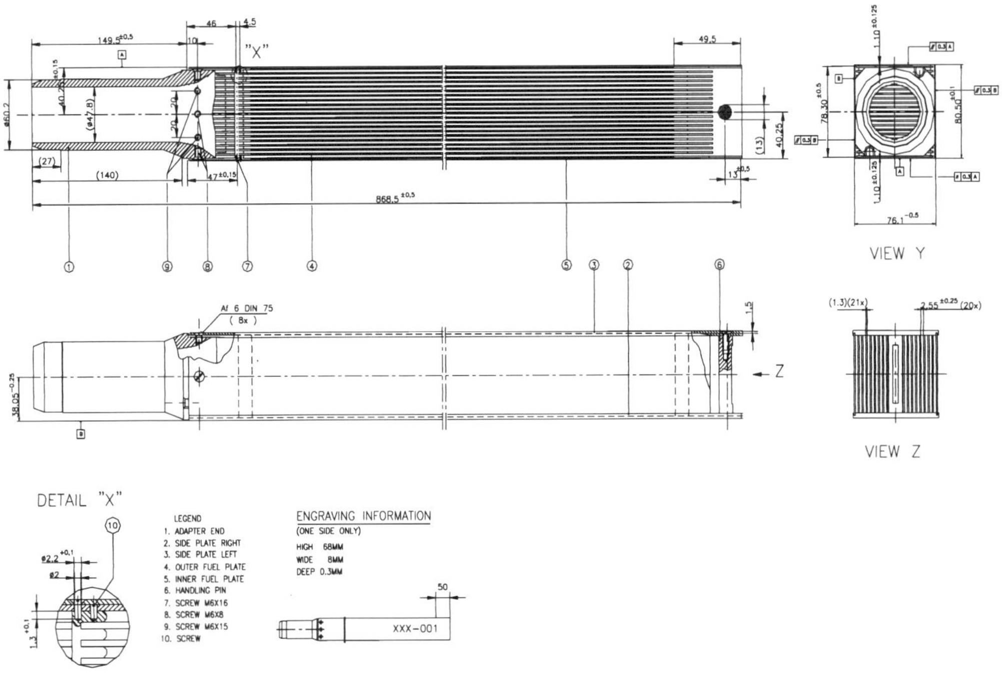

Figure 2. Geometry of MTR Fuel Type [11]

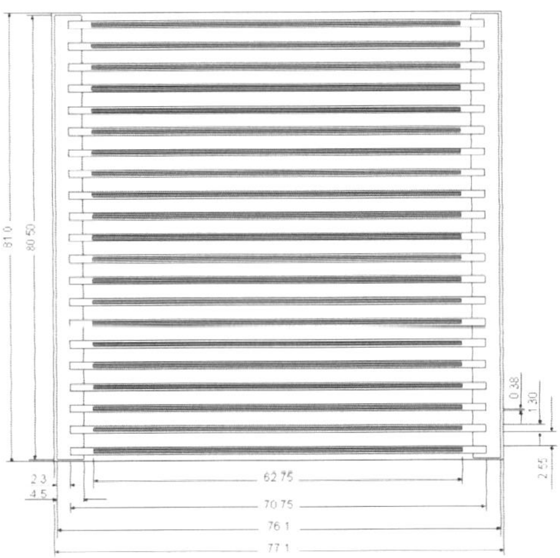

Figure 3.a. Fuel element geometry [10]

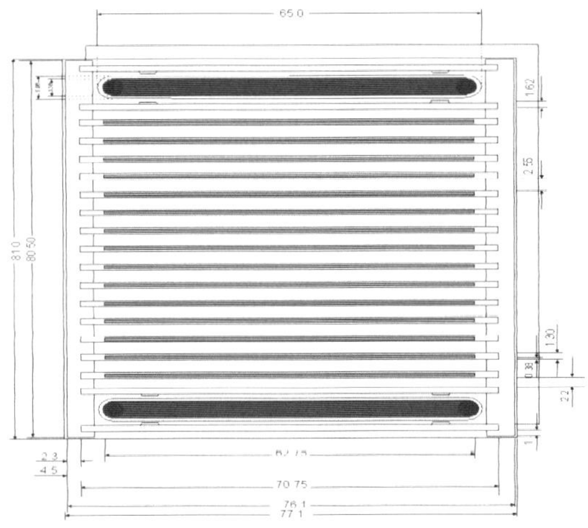

Figure 3.b. Control element geometry [10]

2.2. Safety criteria

The primary cooling system is designed to meet design basis accident, which includes:

a) To provide cooling flow to remove heat from the core at power condition. Distribution of the coolant flow is determined by thermal hydraulic design.
b) To maintain inlet coolant temperature not exceeding normal operation condition.
c) To remove decay heat when primary cooling system pump does not operate at natural circulation.
d) To ensure thermal margin availability of fuel cladding during postulated transient.
e) To maintain the reactor core integrity during accident occurring in abnormal condition.

In order to fulfill this safety design basis, safety margin for the plate-type fuelled TRIGA reactor should be determined. The safety criteria implemented to the plate-type fuelled core are as follows [2]:

1. There is no occurrence of onset of nucleate boiling (ONB), as indicated by the safety margin for the initial nucleate boiling, where ΔT ONB ≥ 0 °C;
2. There is no occurrence of onset of flow instability (OFIR), OFIR ≥ 1;
3. There is no occurrence of departure from nucleate boiling (DNBR), as indicated by DNBR value at all points along the fuel element, DNBR ≥ 1.5.

2.3. CAUDVAP computation code

CAUDVAP v2.60 is a computation code developed by Nuclear Engineering Division, INVAP SE, Argentine. This code is used to calculate flow distribution and pressure drop in reactor core. CAUDVAP v2.60 is developed for research reactor core hydraulics. This code computes distribution of flow rate for steady state condition through various cooling channels, inter-connected in parallel, between plenum inlet and outlet. The calculation parameters include total pressure drop of reactor core with input of total flow, flow distribution at various cooling channel, and friction, as well as flow loss at each flow section [5].

2.4. COOLODN Computation Code

COOLODN is a thermal hydraulic computation code developed by JAERI (Japan Atomic Energy Research Institute). This code has been well verified and used to analyze thermal hydraulic characteristics of JRR-4 reactor core of Japan and RSG-GAS in Serpong. Meanwhile, COOLODN-N has capability to calculate thermal hydraulic characteristics of research reactor with plate-type or rod-type fuel element with either forced or natural convection cooling mode [12]

3. Methodology

To fulfill the safety criteria of plate-type fuelled TRIGA reactor, distribution of coolant flow rate is calculated by knowing the effect of flow rate to safety factor using computation code COOLODN. Meanwhile, primary coolant flow fraction through active core fuel element is computed using CAUDVAP. The calculation results are then input to COOLODN code. The minimum coolant flow through fuel element is calculated by considering minimum flow rate condition as operation condition limit that will make the reactor scram if this value is exceeded. In addition, flow rate that cools the reactor core should also take into account a condition in which the reactor operates at full power.

3.1. Core geometry and fuel element

The geometry of TRIGA reactor pool does not change even though there is fuel element conversion from cylindrical to plate-type. In accordance with the modified core configuration as shown in figure 1. Table 1 gives core geometry and channel geometry that will be passed by primary coolant flow.

Table 1. Coolant channel geometry for core flow distribution calculation

No.	Parameter	Value
1.	Reactor Pool	
	Pool height,m	8.5
	Reactor core position under the pool water level, m	5.5
	Reactor pool diameter, m	1.9812
2.	Reactor core	
	Matrix grid	5 x 5
	Geometry grid, mm	81x77.1
3.	Standard fuel element	
	Number of standard fuel element	16

No.	Parameter	Value
	Number of plate	21
	Number of coolant channel	20
	Fuel element cross section geometry, mm	75.7x67.1x51.5
	Coolant channel gap, mm	2.55
	Channel geometry between 2 fuel elements, mm	2.70x57.1
	Channel geometry between plate and core shroud, mm	2.85x67.1
	Plate geometry, length x width x thick, mm	625x70.75x1.3
	Meat geometry, length x width x thick, mm	600x62.75x0.54
3.	Control Element	
	Number of control elements	4
	Number of plate	15
	Number of coolant channel	14
4.	Irradiation Position (IP),	
	Number of IP	4
	ID end fitting, mm	10
5.	Central Irradiation Position (CIP)	
	Number of CIP	1
	ID end fitting, mm	20

3.2. Coolant flow division in the reactor core for CAUDVAP modeling

Distribution of coolant flow in the reactor core is shown in figure 4. It shows many kinds of coolant flow related to sub-channel geometry. Primary coolant system flows to cool fuel element and control element, including the most outer fuel element plate through bypass flow and irradiation target by providing holes at the end fitting target, either in CIP or IP.

Distribution of primary coolant flow in the plate-type-fuelled TRIGA reactor is divided into active core, bypass, and reflector. Flow division through coolant channel is modeled in more detailed, as shown in figure 4.

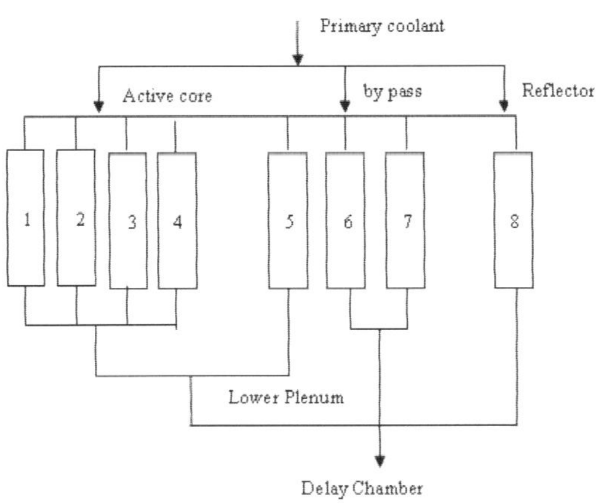

Figure 4. Coolant flow distribution model in modified TRIGA reactor core

Remarks:

Flow in the active core
1. Coolant channel in the fuel elements
2. Coolant channel in the fuel control
3. Outer channel of fuel element
4. Channel in side blade of control element

By pass flow
5. Coolant channel between 2 fuel element
6. Coolant channel in the end fitting of IP
7. Coolant channel in the end fitting of CIP

Reflector
8. Coolant channel in reflector

4. Results and Discussion

To identify the effect of the flow rate of primary cooling system to changes of reactor safety margin, initial estimation is first performed. This initial estimation is selected below and above the debit of the existing pump available in plate-type-fuelled TRIGA reactor in Bandung. The assumption used is 90% of the flow rate of the primary cooling system will cool the active core, consisting of fuel element and control element. It is assumed that the coolant inlet temperature is 35°C, while the power peak factor distribution is obtained from neutronic calculation, where criticality is achieved when the control element is withdrawn as high as 30 cm. The calculation results based on coolant flow rate of 40 – 400 kg/s are shown in table 2 and figure 5.

Table 2. Precalculation results of primary coolant system flow

Flowrate (kg/s)	T coolant outlet (°c)	T clad (°c)	T meat (°c)	ΔT ONB (°c)	T sat (°c)	OFIR	DNBR
40	78.64	122.74	123.66	-5.07	113.45	0.82	1.01
50	69.93	118.57	119.48	-1.44	113.37	1.03	1.26
60	64.12	111.53	112.45	3.06	113.26	1.23	1.51
70	59.96	103.05	103.97	10.16	113.14	1.44	1.76
80	56.84	96.46	97.39	17.1	113	1.64	2.02
90	54.42	91.1	92.03	22.73	112.85	1.85	2.27
100	52.48	86.65	87.59	27.39	112.68	2.05	2.52
200	43.74	64.84	65.79	49.57	110.1	4.1	5.04
300	40.83	56.45	57.41	55.57	105.71	6.16	7.56
400	39.37	51.94	52.9	54.84	98.85	8.21	9.09

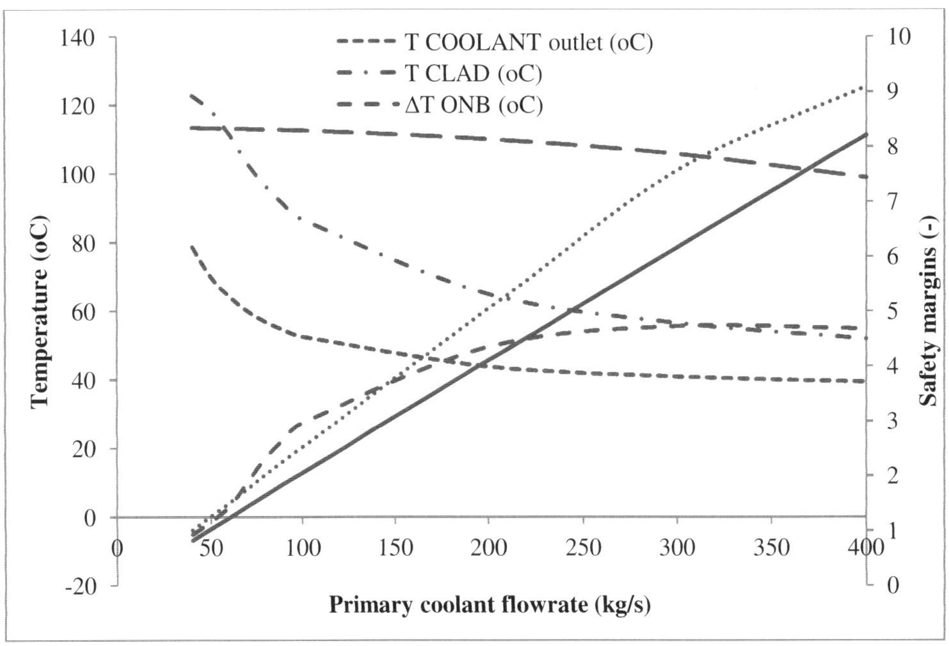

Figure 5. Thermal hydraulic characteristic base on primary coolant flowrate

It is shown in table 2 that at reactor power of 2 MW the primary coolant flow rate meets the safety criteria if it is ≥ 60 kg/s. The safety margin for nucleate boiling and DNBR is exceeded if coolant flow rate is less than that figure. The safety margin for OFIR and DNBR increases significantly for coolant flow rate in the core ≥ 200 kg/s. Moreover, plate and coolant channel temperatures drop significantly. However, this coolant flow is not necessary since it will give an over design figure. Estimation of primary coolant flow rate needed is in the range of 60 - 100 kg/s at 2 MW power when the reactor operates at steady state condition.

To fulfill the minimum flow rate criteria that will trigger reactor protection system if loss of coolant flow accident occurs, a percentage of minimum flow rate used as a limit of operation condition needs to be determined. Taking into account RSG GAS experience, safety margin due to a decrease in flow rate is limited to 85%, under which reactor protection system will make the reactor scram so that the reactor is put in safe condition.

To anticipate primary cooling system disturbance causing loss of coolant flow accident, flow rate that cools the active core of only 85% is calculated. When the flow rate achieves this limit, reactor protection system starts to function and the reactor is protected by reactor scram. This figure is expected to be able to remove residual heat due to loss of coolant flow accident. The postulated worst case condition is that loss of primary coolant flow accident when the reactor operates at nominal power of 2 MW.

Table 3 and figure 6 indicate the calculation results of reactor safety margin of plate-type-fuelled TRIGA reactor based on 85% and the flow rate is in a range of 60 – 100 kg/s at operating power of 2 MW at steady state condition. The calculation results show that there is a shift of minimum flow rate needed so that the minimum safety criteria, which is 70 kg/s, achieved.

Table 3. Thermalhydraulics characteristic base on minimal flowrate

Flowrate (kg/s)	Nom. Flow (kg/s)	T Coolant outlet (^{O}C)	T Clad (^{O}C)	T Meat (^{O}C)	ΔT ONB (^{O}C)	T Sat (^{O}C)	OFIR	DNBR
51	60	69.25	118.07	118.99	-1.04	113.36	1.05	1.29
59.5	70	64.36	112.01	112.93	2.8	113.27	1.22	1.5
68	80	60.69	104.58	105.51	8.58	113.17	1.4	1.71
76.5	90	57.84	98.61	99.54	14.84	113.05	1.57	1.93
85	100	55.56	93.65	94.58	20.05	112.93	1.74	2.14

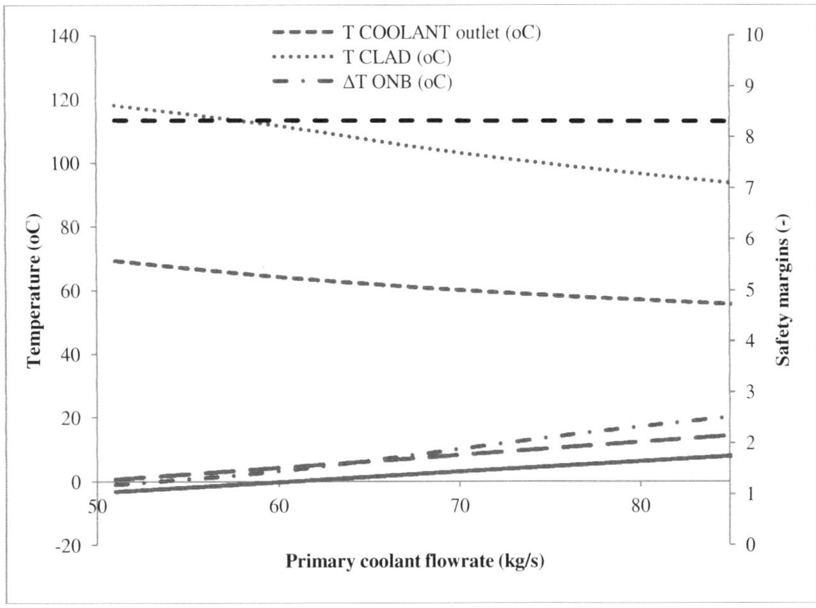

Figure 6. Thermal hydraulic characteristic of 85% primary coolant flowrate

The reactor power will possibly rise during an accident. To anticipate this event, the reactor will be initiated by protection system to scram. Therefore, to specify the setting point for the protection system at accident condition, over-power based calculation should be conducted, where safety margin is still met.

Table 4 shows parameters of over-power-based thermal hydraulics that might occur early during loss of flow accident (LOFA). This calculation results in safety margin in severe accident. The assumptions used include that the flow rate achieves 85% of the minimum flow rate specified as high as 68 kg/s, the inlet temperature is 35°C, the accident occurs at initial cycle, where peak factor of axial and radial power has the highest value. Figure 7 shows characteristics of over-power-based safety margin. The analysis results indicate that over-power-based operation condition, where DNBR margin not exceeded is 110% of the nominal power or 2.2 MW.

Table 4. Thermalhydraulics characteristic based on reactor over power

Power (MW)	Flowrate (kg/s)	T Coolant outlet (^{O}C)	T Cladding (^{O}C)	T Meat (^{O}C)	ΔT ONB (^{O}C)	T Sat (^{O}C)	OFIR	DNBR
2	68	60.69	104.58	105.51	8.58	113.17	1.4	1.71
2.1	68	61.98	107.67	108.64	5.62	113.17	1.33	1.63
2.2	68	63.26	110.72	111.73	3.57	113.17	1.27	1.56
2.4	68	65.83	115.86	116.96	0.6	113.17	1.16	1.43

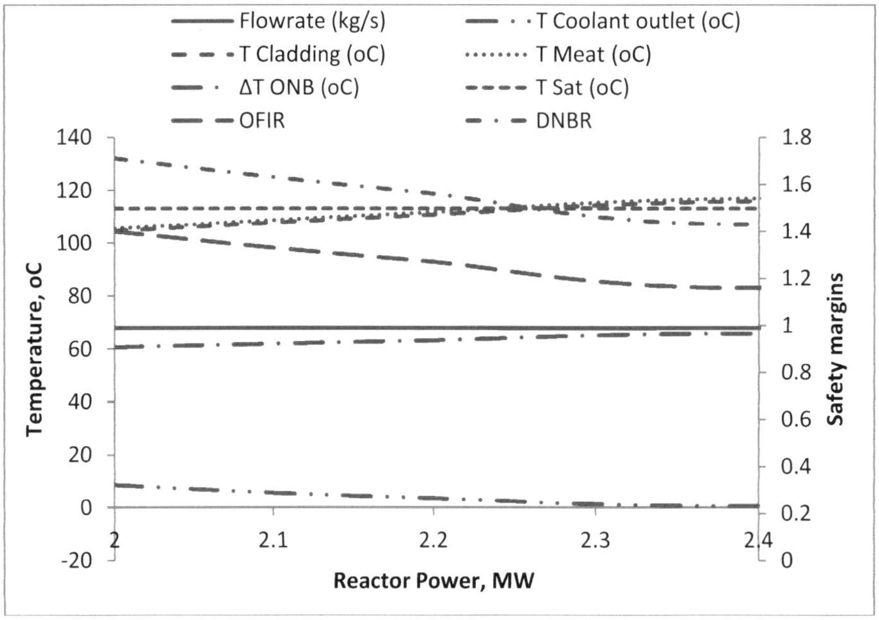

Figure 7. Characteristic of thermalhydraulic and safety margins at over power

The calculation results of plate-type-fuelled TRIGA core using COOLODN has been compared with those of flow rate distribution calculation using CAUDVAP code, as shown in table 5. The calculation results of benchmarking between COOLODN and CAUDVAP indicate that there is an agreement of the two codes in a range of 2 – 8.7%. CAUDVAP is used to compute flow distribution in each sub-channel of the plate-type-fuelled TRIGA, as shown in figure 6. The calculation results of flow distribution show that coolant flow through the reactor core is divided to 88.2%, 4.8%, and 7%, respectively for cooling the active core, bypass, and reflector. In addition, in order to manage the largest coolant percentage flowing through the active core, this design also provides cooling for reactor utilization. For the purpose of target irradiation in the core, cooling for 4 IPs with inner diameter (ID) of 10 mm each, which is located at end fitting irradiation target, is provided. Meanwhile, for CIP position, ID 20 mm is used.

Table 5. Comparison result between Coolodn and Caudvap

Parameter	CoolodN	Caudvap	Deviation (%)
Primary coolant flow, kg/s	80	80	-
Flow through active core, kg/s	90% = 72	88.2% =70.56	2
Coolant velocity in fuel element, m/s	1.123	1.085	3.38
Total loss through fuel element, kg/cm^2	0.06134	0.0561	8.57

Table 6. Flow distribution in TRIGA plate core elements by Caudvap

No.	Channel Type	Flow Distribution	
		Kg/s	%
1.	Active core	73.57	88.20
	Fuel element	51.58	64.47
	Fuel Plate in control element	12.86	16.07
	Outer channel of fuel plate	6,13	7.66
	Blade in control element	3.84	4.80
2.	By Pass		
	Between fuel element and others	3.19	3.99
	Iradiation positions, IP	0.55	0.68
	Central iradiation position, CIP	0.11	0.13
3.	Reflector	5.60	7.00

Table 7. Comparison between TRIGA Plate Innovation and similar reactor

Parameter	JRTR [13]	JRR-4 [14]	MNR [14]	TRIGA Plate Modified
Power (MW)	5	3.5	2	2
Site	Jordanian	Japan	Canada	Indonesia
Reactor Type	Open pool	Open pool	Open pool	Open pool
Fuel Type	Plate	Plate	Curve plate	Plate
Fuel Material	U_3Si_2	U_3Si_2	U_3Si_2	U_3Si_2
Fuel Enrichment , %-U235	19.75	19.75	19.75	19.75
Fuel Density, g/cm^3	4.8	3.8	0.74	2.96
No of FE/CE	18/-	20/-	28/6	16/4
Average heat flux, kW/m^2	173	150	45	72.5
Coolant	H_2O	H_2O	H_2O	H_2O
Total flow rate, kg/s	170	135	75	80
Flow Direction	Downward	Downward	Downward	Downward
Coolant velocity in core, m/s	2.5	1.45	0.69	1.123
Inlet coolant temperature, ^oC	37	35	30	35
Average core ΔT, ^oC	9.2	-	6.4	7.0

Comparison of flow rate needed to cool down plate-type-fuelled TRIGA and similar reactor is shown in table 7. Compared with thermal hydraulic parameters of plate-type fuelled research reactor at 2 – 5 MW, total flow rate through plate-type-fuelled TRIGA reactor seems in agreement. It is shown by coolant flow rate and inlet-outlet temperature difference. Therefore, it can be concluded that determination of coolant flow rate in plate-type-fuelled TRIGA reactor has been verified.

Conclusion

Design and analysis have been performed to determine coolant flow rate in innovated plate-type fuelled TRIGA reactor core that meets the safety criteria of reactor operation at steady state condition and initial transient, as well as over power. The design of flow rate that fulfils safety criteria of DNBR, OFIR, and ΔT ONB indicates that minimum flow rate required to cool plate-type-fuelled TRIGA reactor at 2 MW power is 80 kg/s. Therefore, it can be concluded that operating limitation condition (OLC) for minimum flow rate is 80 kg/s; 72 kg/s is to cool the active core; while the minimum flow rate for coolant flow rate drop is limited to 68 kg/s with the coolant inlet temperature 35 °C. This thermohydraulic design provides cooling for IP utilization as many as 4 positions and 1 CIP with end fitting ID of 10 mm and 20 mm, respectively.

References

[1] Endiah PH, Subekti M, Sukmanto D, Darwis I *2015* Thermal hydraulic design and cooling system optimization of the high Power Innovative Research Reactor "in Indonesia" *J. Tek. Reaktor. Nukl.* **17** p127-140

[2] Sudjatmi KA, Endiah PH, Surip W, Reinaldy N 2015 Analysis of natural convection in TRIGA reactor core plate types fueled using Coolod-N2 "in Indonesia" *J. Tek. Reaktor. Nukl.* **17** p67-78

[3] Prasetyo B, Putranto IY, Zaki S 2014 The neutronic design of fuel plate type for TRIGA 2000 Bandung reactor core "in Indonesia" *J. Sci. Tek.Nukl. Indonesia* **15** p69-80

[4] Anwar IR 2017 Comparative analysis of preliminary design core of TRIGA Bandung using fuel element plate MTR in Indonesia, *AIP Conference Proceedings* p1799

[5] Subekti M, Darwis I, Endiah PH 2013 The analysis of coolant-velocity distribution in plate-typed fuel element using CFD method for RSG-GAS research reactor "in Indonesia" *J. Tek. Reaktor Nuklir,* **15** p67-76

[6] Amin D, Seyed M, Ebrahim A 2015 Three-dimensional analysis of flow blockage accident in Tehran MTR research reactor core using CFD *Progress in Nuclear Energy* **85** p605-612

[7] Xia GL, Su GH, Peng MJ 2016 Analysis of flow distribution in plate-type core affected by uneven inlet temperature distribution *Annals of Nuclear Energy* **92** p333–344

[8] Jian Z, Xiuzhong S, Yasuyuki F, Tadafumi S, Toshihiro Y, Nakajima K 2015 Experimental study on the safety of Kyoto University Research Reactor at natural circulation cooling mode *Annals of Nuclear Energy* **76** p410–420

[9] Safaei AS, Khalafi H, Mirvakili SM 2015 Preliminary thermal-hydraulic safety analysis of Tehran Research Reactor during fuel irradiation experiment *Progress in Nuclear Energy* **79** p32-39

[10] Albarhoum M 2010 Thermal–hydraulics features of the use of LEU U_3Si_2 fuel in low-power reactors *Annals of Nuclear Energy* **37** p1351–1355

[11] BATAN Indonesia 2011 *Safety Analisis Report of RSG GAS Multipurpose Reactor rev.10*

[12] Boulaicha Y,Nacira B, El Bardouni T, Zoubair M, El Bakkari B, Merroun O, El Younoussi C, A. Hteta A, Boukhal H., Chakir E 2011 Steady-state thermal-hydraulic analysis of the Moroccan TRIGA MARK II reactor by using PARET/ANL and COOLOD-N2 codes *Nuclear Engineering and Design* **241** p270–273

[13] Al-Yahia OS, Ho L, Daeseong J 2016 Transient analyses of the Jordanian 5 MW research reactor under LOEP accident *Annals of Nuclear Energy* **87** p575-583

[14] Daesong J, Jonghark P, Heetaek C 2014 Development of thermal hydraulic and margin analysis code for steadystate forced and natural convective cooling of plate type fuel research reactors *Progress in Nucl. Energy* **71** p39-51

Acknowledgment

The author would like to appreciate to the Management of PTRKN-BATAN. This research was funded from the PTRKN budget in fiscal year 2010 – 2014.

Probabilistic analysis on the failure of reactivity control for the PWR

D T Sony Tjahyani, Deswandri, G R Sunaryo

Center for Nuclear Reactor Technology and Safety, BATAN
Puspiptek Complex, Building no. 80, Serpong, Tangerang Selatan 15310, Indonesia

dtsony@batan.go.id

Abstract. The fundamental safety function of the power reactor is to control reactivity, to remove heat from the reactor, and to confine radioactive material. The safety analysis is used to ensure that each parameter is fulfilled during the design and is done by deterministic and probabilistic method. The analysis of reactivity control is important to be done because it will affect the other of fundamental safety functions. The purpose of this research is to determine the failure probability of the reactivity control and its failure contribution on a PWR design. The analysis is carried out by determining intermediate events, which cause the failure of reactivity control. Furthermore, the basic event is determined by deductive method using the fault tree analysis. The AP1000 is used as the object of research. The probability data of component failure or human error, which is used in the analysis, is collected from IAEA, Westinghouse, NRC and other published documents. The results show that there are six intermediate events, which can cause the failure of the reactivity control. These intermediate events are uncontrolled rod bank withdrawal at low power or full power, malfunction of boron dilution, misalignment of control rod withdrawal, malfunction of improper position of fuel assembly and ejection of control rod. The failure probability of reactivity control is 1.49E-03 per year. The causes of failures which are affected by human factor are boron dilution, misalignment of control rod withdrawal and malfunction of improper position for fuel assembly. Based on the assessment, it is concluded that the failure probability of reactivity control on the PWR is still within the IAEA criteria.

Keywords: Fundamental safety function, Probabilistic, Safety analysis, Reactivity control, PWR

1. Introduction

Safety assessment on the power reactor in accordance with international and national regulation is carried out by the probabilistic and deterministic analysis to obtain the site permit, as well as other permits especially the construction and operation permits [1]. The safety analysis is initiated by constructing event sequences with the purpose to determine initiating events and mitigation system consisting of safety system and operator actions. The mitigation system is required to overcome the initiating event, so that the core damage is not occurred [2]. Therefore, the safety analysis especially the probabilistic analysis, is important to analyze the frequency or probability of the initiating event and the failure of mitigation system. Hereinafter, the possibility of the occurrence of core damage can be determined. Based on this analysis, the safety design level of the power reactor can be evaluated.

The initiating event is determined by five group initiating events, which are used to evaluate the fundamental safety functions of reactivity control, heat removal from the reactor or the fuel storage, and confinement of the radioactive materials as also radiation shielding. To control operation, the important parameter considered is to control reactivity for all operating conditions and it needs to be analyzed. On the light water reactor, the reactivity control is the first stage for preventing the occurrence of core damage or the more severe accident propagation.

At present, the pressurized water reactor (PWR) is a type of nuclear power plants, which are operated mostly in the word. The Advanced Passive Pressurized Water Reactor 1000 (AP1000) is one of PWR type, that applies absolutely passive system on the passive core cooling system (PXS) with high reliability [3, 4]. Therefore, it is important to analyze the characteristic of AP1000 from the aspect of reactivity control.

The analysis indicated that neutronic characteristics more important than the thermal hydraulic characteristic at the beginning cycle of operation [5]. Also in general, the level 1of probabilistic safety assessment (PSA) results showed that human action and dependent failures had significant effect on accident and its consequences [6]. The control rod as part of the reactivity control system might experience deterioration, which results in a stuck control rod or in advertent control rod drop [7]. Reactivity initiated accidents (RIAs), which cover a sudden and rapid control rod insertion, comprise the positive and negative reactivity [8]. The rod drop accident is an important event of RIA in LWR safety analysis, because this case can cause its adjacent regional power to increase rapidly and also the increase of fuel enthalpy [9]. Based on these researches, it showed that the reactivity control was generally carried out by deterministic analysis for the rod drop event. Therefore, it is very important to analyze the reactivity control from the several view points such as the mechanical failure, the human error, the procedure error and the others.

The purpose of this paper is to analyze the failure probability of the reactivity control and to determine the failure contributions, which cause the occurrence of the reactivity control failure in AP1000. The analysis is done by using deductive analysis, which is fault tree analysis of the reactivity control based on the system analysis. The system analysis is carried out to determine the intermediate events, which cause the top event. Furthermore, to calculate the probability or to quantify the failure of reactivity control and the intermediate events, the generic failure data of component, system, human error and operator error are used based on the published document [10, 11].

2. The reactivity change on the reactor operation

Based on the safety fundamental, there are ten principles to achieve the safety objective. One of those principles is to prevent and mitigate the accident and its implementation is to apply defence-in-depth concept. That concept is applied on the reactor design with the goal to fulfill the three fundamental safety functions, which are reactivity control, heat removal from reactor and fuel storage, and confinement of radioactive material.

To evaluate the fulfillment of the fundamental safety function, initiating events are postulated so that the evaluation can be done on how to mitigate all initiating events [12, 13]. Generally for the PWR design, there are five the initiating event groups to be evaluated, which are increase in heat removal from the primary system, decrease in heat removal by the secondary system, decrease in reactor coolant system flow rate, reactivity and power distribution anomalies, increase/decrease in reactor coolant inventory and radioactive release from subsystem or component. Therefore, to evaluate the first fundamental safety function on design and to obtain contribution factor to initiating events, it is important to analyze the reactivity control.

In general, the reactivity change on the PWR is caused by some events such as control rod ejection, change of boron concentration, and addition of cold water to the reactor cooling system. Those events will effect the moderation. Accordingly, the reactivity control failure is caused by the system failure that can make those three cases difficult to control [14].

The initiating event having effect significantly on the group of the reactivity control failure is caused by six cases, which are uncontrolled condition during the control rod withdrawal at low power, uncontrolled condition during the control rod withdrawal at full power, boron dilution malfunction, misalignment during control rod withdrawal, improper positioning of fuel assembly and control rod ejection.

The uncontrolled reactivity during the control rod withdrawal is caused by malfunction of reactor control system or the failure of control rod system. The drive mechanisms of rod cluster control assembly (RCCA) are grouped into bank configurations. The purpose of grouping is to prevent automated withdrawal of other respective banks. Each bank grouping has different power supply, so that no more than two banks are withdrawn at the same time. The drive mechanisms of RCCA are the magnetic latch type and coil actuation having function to regulate speed travel. The reactor control system is initiated by the protection and safety monitoring system consisting of the source range high neutron flux reactor trip, intermediate range high neutron flux reactor trip, power range high neutron flux reactor trip, power range high neutron flux reactor trip and high nuclear flux rate reactor trip. During the full power, the protection and monitoring system consist of five events, which cause trip to the reactor. Those are 2 out of 4 channels for exceeding the overpower setpoint, 2 out of 4 channels for exceeding the over temperature ΔT setpoint, 2 out of 4 channels for exceeding the overpower ΔT setpoint, 2 out of 4 channels for exceeding the pressure at pressurizer, and 2 out of 4 channels for exceeding the water level at pressurizer.

On the normal operation, the boron dilution is manually under strict administrative controls requiring close operator surveillance. The administrative procedure is done to limit the rate and duration of the dilution. A boric acid blend system allows the operator to meet the makeup water boron concentration to the reactor cooling system during the normal charging. An inadvertent boron dilution is caused by failure of the chemical and volume control system (CVCS) due to the failure of controller, operator and mechanical.

Each RCCA has the position indicator that is checked at monitor. The failure of control rod withdrawal can be caused by malfunction of system and operator error to interprete the position or installation error after maintenance. The error of fuel assembly leading to an improper position is generated by several causes such as error in placing the fuel assembly position, error in determining the fuel pellets enrichment, and error of enrichment degree of fuel assembly. The control rod ejection accident is induced by the mechanical failure of control rod mechanism pressure housing, so that RCCA is ejected and the drive shaft is failed. The consequences of those failures contribute to the rise of rapid positive reactivity insertion, together with adversed core power distribution and finally it causes localized fuel rod damage.

3. Methodology

The principle of safety analysis using probabilistic method as the safety evaluation is to determine all initiating events considered in design, which are already identified. Also, the event probability is quantified so that the total probability can be calculated. On the evaluation, it is assumed that the smaller the probability of events at design, the higher the design safety level. The reactivity control is the first stage of the fundamental safety function. Therefore, the other fundamental safety function can function effectively if the probability of the first stage is small.

The failure probability of the reactivity control can be calculated based on determination of intermediate events as much as possible having contribution to the reactivity control. In this case, the failure of reactivity control is assumed as the top event. The cause of failure is determined by the system analysis. Furthermore, each failure reason is analyzed in detail by deductive analysis using the fault tree analysis to determine the failure cause based on basic events classified as the component failure, the operator or procedure execution error. The data of the failure or error probability used are based on generic data of the published document [10, 11]. After that, the failure probability of reactivity control is calculated as also the determination of contribution of the failure cause. The analysis diagram to be performed is shown in figure 1.

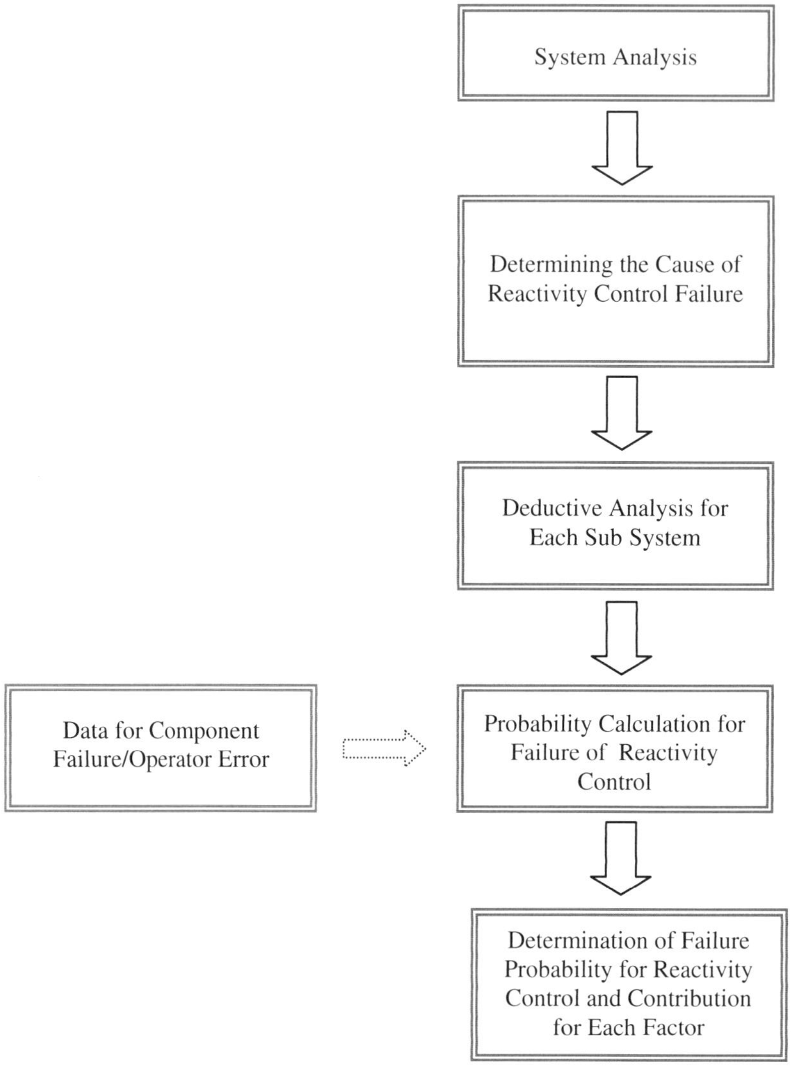

Figure 1. Flowchart of analysis.

4. Results and Discussion

The Initiating event identification based on the system analysis and the failure of reactivity control using deductive analysis are shown in figure 2. The uncontrolled condition during the control rod withdrawal at low power is generally generated by mechanical and electric factor (signal to control). The mechanical factors are events which include the malfunction of control rod system, while the signal factor to move the control rod is classified as malfunction of reactor control system. The malfunction of reactor control system is caused by failure of one of the trip systems, so that the control rod can not shutdown the reactor. The trip system observation is based on neutron flux condition. For the uncontrolled condition during the control rod withdrawal at power, the logic that occurrs is similar with the uncontrolled during the control rod withdrawal at low power. However, the trip to move the control rod or to shutdown the reactor is different as shown in figure 3.

In the same way, it is known that the boron dilution malfunction is caused by the system failure or the human error. The failure of the demineralized water transfer and storage system as also the CVCS are included as the system failure, while the improper dilution from the procedure and the carelessness of dilution supervision are caused by human error or operator error. The control rod withdrawal misalignment is generated by malfunction of system and human error as operator. The error is due to the position interpretation or installation error after maintenance. Significantly, improper positioning of fuel assembly is dominated by human error especially during placement of fuel assembly position according to the core management, enrichment determination in the pellet fabrication, or determination of enrichment degree of fuel assembly. The event of the control rod ejection is dominated by the mechanical failure in the control rod mechanism pressure housing or drive shaft.

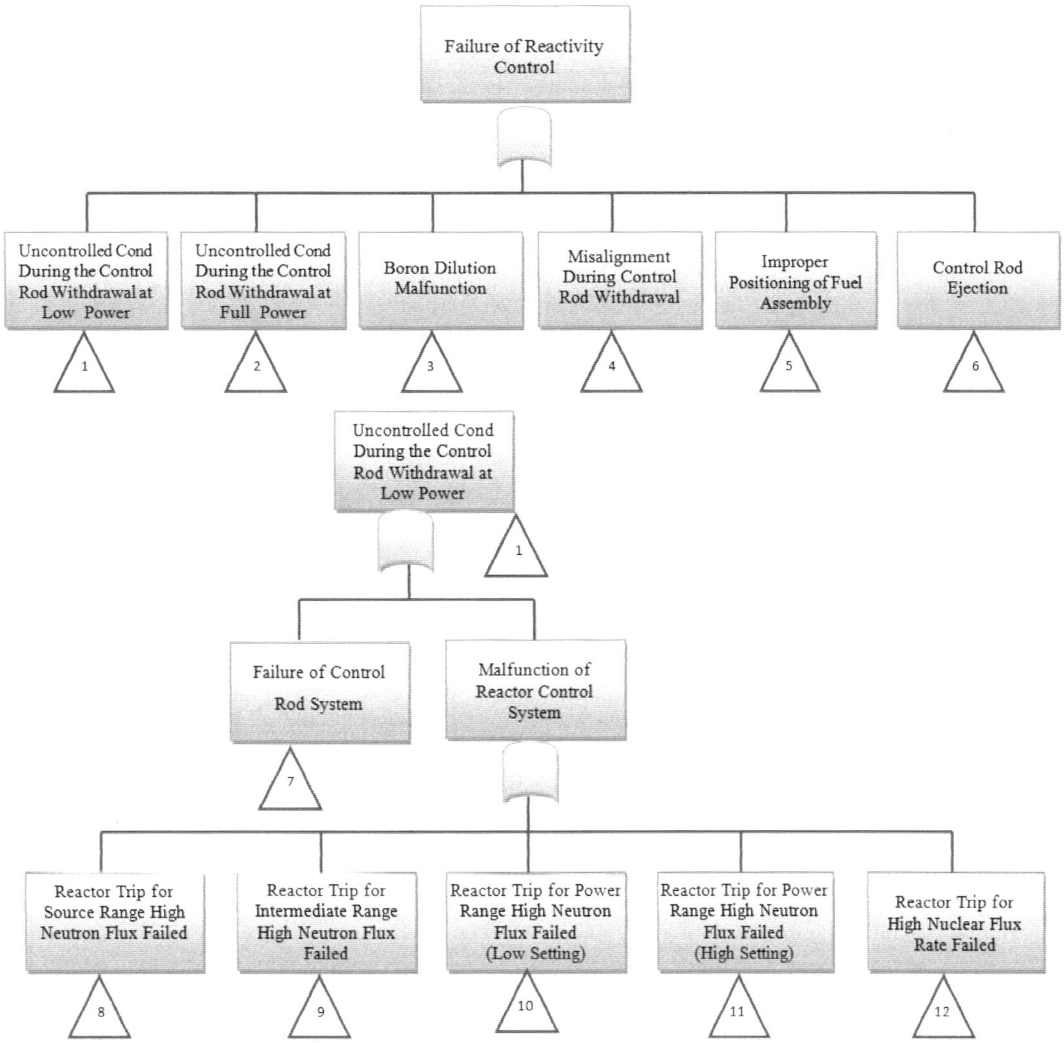

Figure 2. Fault tree for the failure of reactivity and the control rod withdrawal at low power.

By using deductive analysis (figure 2 and 3), the calculation of reactivity control failure is carried out. The probability of intermediate events which contribute to the failure of reactivity control is showed on table 1. Based on those six intermediate events, the failure probability of reactivity control is 1.493E-03 per year as the summed probability of intermediate events. This result is small enough if it is compared with core damage frequency (CDF). The CDF is determined by multiplying the reactivity control probability and the mitigation failure probability [15]. Therefore, the PWR design has capability to mitigate the failure of reactivity control. Also, table 1 shows that the boron dilution malfunction has largest contribution to the reactivity control failure event of 34.82%. This event is considered as Condition II or faults of moderate frequency, so that it is not classified as an accident. The other events which are classified as Condition II are uncontrolled condition during the control rod withdrawal at low power and full power. In this analysis, the event of Condition II is 41.33%.

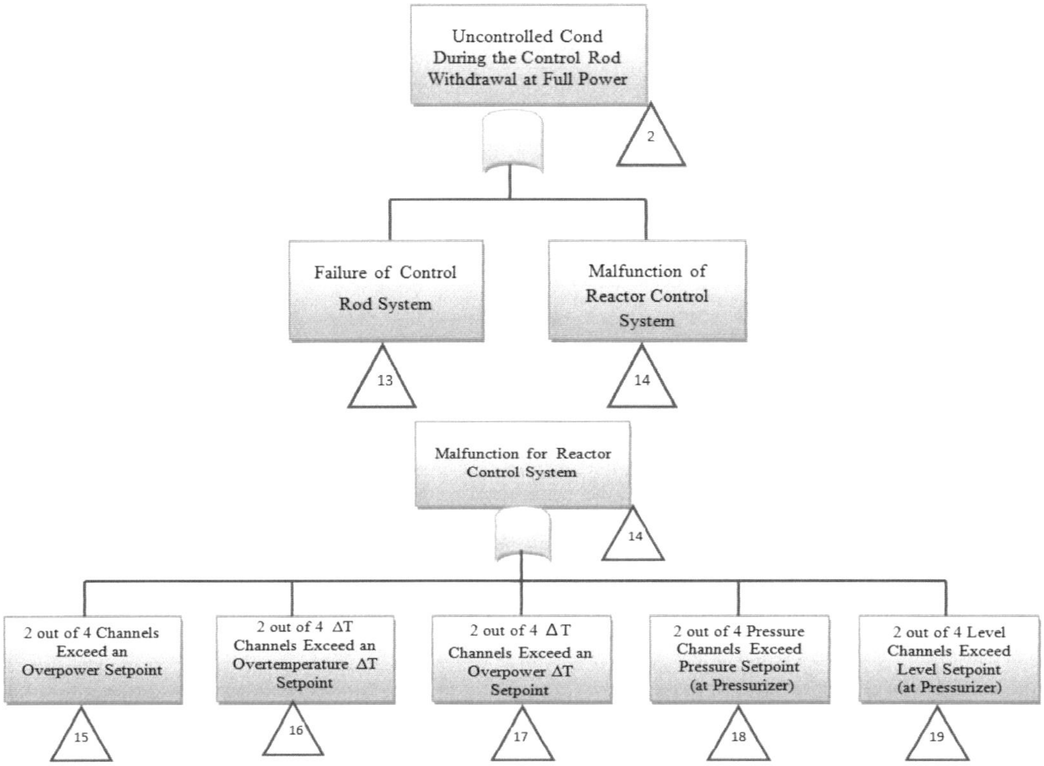

Figure 3. Fault tree for the control rod withdrawal at full power.

Table 1. Calculation of the failure cause probability for the reactivity control.

No.	Intermediate event	Probability, per year	Contribution	Condition Classification
1.	Uncontrolled condition during the control rod withdrawal at low power	2.80E-06	0.19%	II
2.	Uncontrolled condition during the control rod withdrawal at power	9.40E-05	6.32%	II
3.	Boron dilution malfunction	5.18E-04	34.82%	II
4.	Misalignment during the control rod withdrawal	4.43E-04	29.78%	III
5.	Fuel assembly in an improper position	4.29E-04	28.86%	III
6.	Control rod ejection	4.36E-07	0.03%	IV

The second largest contribution is the misalignment during the control rod withdrawal, which is 29.78% and is classified as condition III or infrequent faults. The fuel assembly in improper position is also condition III, so that its contribution of Condition III to the failure of reactivity control is 58.46%. In this case, it needs consideration because these events are dominated by human factor.

The control rod ejection is included in Condition IV or limiting fault, which is defined as faults which are not expected but are postulated because it has the potential of the release of significant amounts of radioactive material. The result showed that the contribution of this event is very small of approximately 0.03%. The probability of this event is very small but it is included as an accident, so that it is eliminated in the design and one of the factors that is used for development of the PWR design.

Table 1 also showed that the three biggest contributions to the failure of reactivity control are affected by human error or administrative observation error. This case indicates that the human factor has influence to the failure of reactivity control, although its probability of error is still acceptable.

5. Conclusion

From the analysis, it is shown that the failure probability of reactivity control on the PWR is relatively small, which is 1.49E-03 per year and it is still within IAEA criteria. The largest contribution to the failure cause of reactivity control is given by events due to affected human factor consisting of boron dilution event, error of the control rod withdrawal, and error of fuel assembly placement. The failure probability of the reactivity control can be reduced by increasing system diversity, system redundancy and strict procedures of operation on the intermediate events.

Acknowledgment

The authors are very grateful for the financial support for this research, which is provided by DIPA 2016 of Center for Nuclear Reactor Technology and Safety.

References

[1] Kang D G, Ahn S H and Chang S H 2013 *Nucl. Eng. Des.* **260** 165
[2] Aldemir T 2013 *Ann. Nucl. Energy* **52** 113
[3] Lioce D, Asztalos M, Alemberti A, Barucca L, Frogheri M and Saiu 2012 *Nucl. Eng. Des.* **250** 538
[4] Kamyab S and Nematollahi 2012 *Nucl. Eng. Des.* **253** 32
[5] Hadad K and Kowsar Z 2016 *Ann. Nucl. Energy* **97** 179
[6] Barati R and Setayeshi S 2014 *Prog. Nucl. Energy* **70** 199
[7] Eisa M, Naguib M and Badawi A 2015 *Ann. Nucl. Energy* **81** 106
[8] Parhizkari H, Aghaie M, Zolfaghari A and Minuchehr A 2016 *Ann. Nucl. Energy* **88** 194
[9] Rabiee A and Atf A 2016 *Prog. Nucl. Energy* **88** 191
[10] Schulz T L 2007 *UK AP1000 Probabilistic Risk Assessment* (Westinghouse Electric Company LLC)
[11] Swain A D and Guttmann H E 1983 *Handbook of Human Reliability Analysis with Emphasis on Nuclear Power Plant Applications* (Nuclear Regulatory Commission)
[12] Marinova B, Groudev P and Papazov V 2010 *Nucl. Eng. Des.* **240** 880
[13] Mehboob K, Park K and Khan R 2015 *Ann. Nucl. Energy* **75** 365
[14] Whiteman J L 2011 *AP1000 European Design Control Document Rev.19* (Westinghouse Electric Company LLC) chapter 15 pp 15.4.1-37
[15] Queral J and Montero-Mayorga 2016 *Ann. Nucl. Energy* **91** 65

Ultrasonic Non-destructive Prediction of Spot Welding Shear Strength

R. Himawan, M. Haryanto, R.M. Subekti, and G.R. Sunaryo

Center for Nuclear Reactor Technology and Safety, BATAN, Puspiptek Area, Building No. 80, Tangerang, 15310 Indonesia

roziqh@batan.go.id

Abstract. To enhance a corrosion resistant of ferritic steel in reactor pressure vessel, stainless steel was used as a cladding. Bonding process between these two steels may result a inhomogenity either sub-clad crack or un-joined part. To ensure the integrity, effective inspection method is needed for this purpose. Therefore, in this study, an experiment of ultrasonic test for inspection of two bonding plate was performed. The objective of this study is to develop an effective method in predicting the shear fracture load of the join. For simplicity, these joined was modelled with two plate of stainless steel with spot welding. Ultrasonic tests were performed using contact method with 5 MHz in frequency and 10 mm in diameter of transducer. Amplitude of reflected wave from intermediate layer was used as a quantitative parameter. A set of experiment results show that shear fracture load has a linear correlation with amplitude of reflected wave. Besides, amplitude of reflected wave also has relation with nugget diameter. It could be concluded that ultrasonic contact method could be applied in predicting a shear fracture load.

Keywords : Non-destructive prediction, Ultrasonic test, Spot welding, Shear strength

Introduction

Ferritic steel has been used as reactor pressure vessel materials due to its superiority in ductility and neutron embrittlement resistance. But, in other hand, this material has a low corrosion resistance against primary cooling water in light water reactor. Thus, as a countermeasure, a high corrosion resistance material such as stainless steel 304 was used as cladding. In fabrication, overlay weld method was applied for joining ferritic steel and stainless steel. Since reactor pressure vessel is a non-replaceable component, pre-service and in-service inspection play an important role to ensure the reliability. The objectives of inspection are to find sub-clad crack, unjoined parts and other inhomogenity. Therefore, an effective inspection method is needed for this purpose.

To simplify the spesimen, two pieces plate of stainless steel are joined by spot welding. As in reactor pressure vessel cladding, spot-welds must be carefully inspected, since the quality of the welds directly influences the strength and durability of the welded body. A various studies regarding to spot welding have been performed. Numerical simulations have been performed to improve the welding process [1-2]. Studies on mechanical and metallurgical of welding spot have been performed in order to qualify the results [3-5]. General practice in determining the quality of spot weld, destructive testing is applied. Due to the destructive character of these testing methods, many non-destructive testing methods were developed to increase the effectiveness and accuracy of testing, such as ultrasonic method, infrared thermography method, and magnetic testing method [6-8]. In ultrasonic methods, various techniques have been studied for this purpose, such as a water immersion method, C-scan method, and the use of lamb wave [8-11]. Intensive studies were also been performed regarding to the signal analysis to quantify the ultrasonic wave parameters [12-14]. In previous studies, an ultrasonic technicques are succesfully applied to monitor fatigue damage process [15]. Unfortunately, most of these studies used a spesific ultrasonic equipment or a generic ultrasonic equipment, by means, it is difficult to be applied in manufacture lines. Thus, the objective of this study is to develop a new ultrasonic testing technique to inspect a spot welding result using a conventional ultrasonic flaw detector. This technique should be suitable for the quality assurance of spot welds in the series production line and it significantly reduces testing time and material cost.

1. Materials and Experiments

1.1. Specimen and material

The specimens were made from austenitic stainless steel 304 by resistance spot welding method. Materials chemical composition and specimen's geometry are shown in table 1 and figure 1, respectively. Spot welding was performed in the centre of overlapping area between two sheets of stainless steel with 3 mm thick.

Table 1. Chemical composition of SS 304 material.

	C	Si	Mn	P	S	Ni	Cr	Mo	Al	Cu
weight (%)	0.21	0.27	0.69	0.005	0.004	0.78	0.38	0.63	0.015	0.16

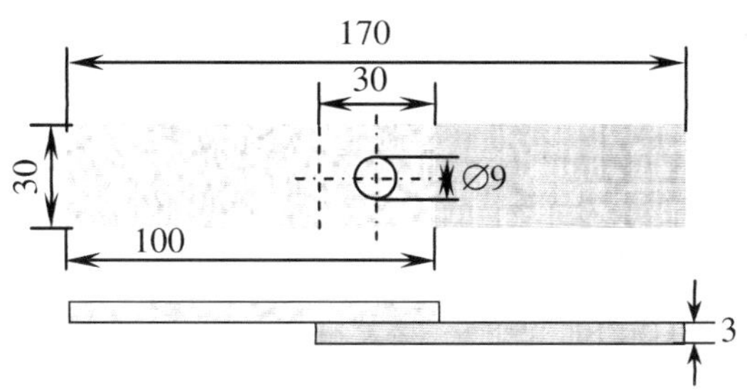

Figure 1. Specimen Geometry

Spot welding process were performed using a single-phase alternating current (AC) 50 Hz equipment. This machine has two electrodes with 25 mm body diameter and 9 mm face diameter. The controlled welding parameters were welding current, electrode force, and force holding time. In order to obtain a variation in quality of spot welding, specimens were welded by various parameters, whilst the electrode force was kept in a constant value, 1.000 N. Five specimens were obtained from each welding parameter. Figure 2 shows an example of a set specimen obtained.

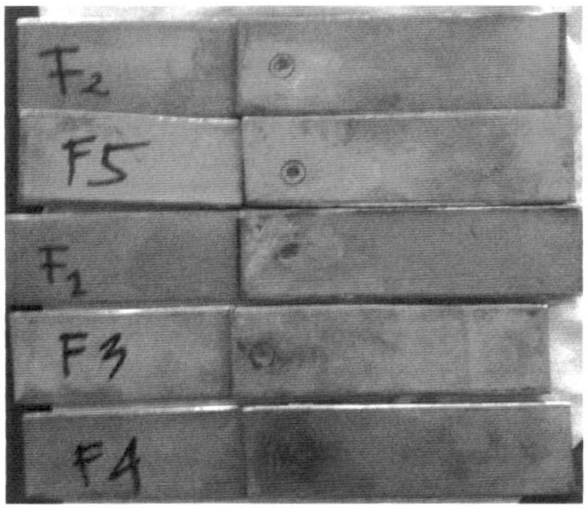

Figure 2. Example of spot welding results

1.2. Ultrasonic testing of welded joint

Figure 3 shows an ultrasonic testing equipment set-up. The equipment consists of ultrasonic transducer, ultrasonic pulser-receiver, digital osciloscopes and computer for data acquisition. The pulse-echo method was applied in this testing and A-scan data type were obtained. The ultrasonic wave was generated by ultrasonic pulser-receiver and transmitted to specimen by ultrasonic probe. The longitudinal type ultrasonic probe with 5 MHz frequency was used in this testing. The ultrasonic transducer has 10 mm in diameter. Ultrasonic back-wall echo signal was received by the same probe and displayed by digital osciloscope. Signal data were digitized in 2 GHz sampling rate and recorded by computer. For analyze purpose, an acquired data could be used to reconstruct the ultrasonic signal using software.

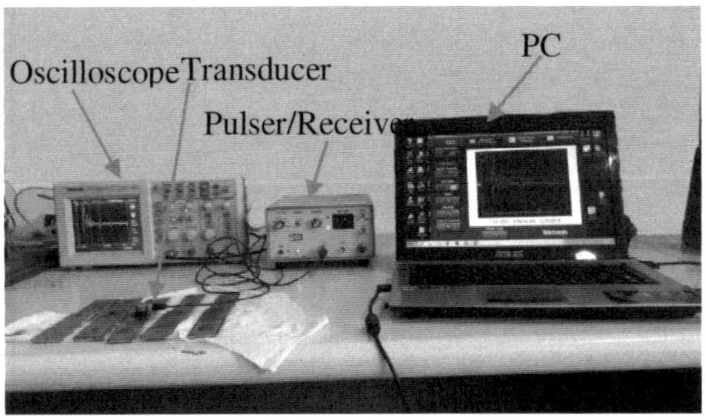

Figure 3. Testing equipment setup.

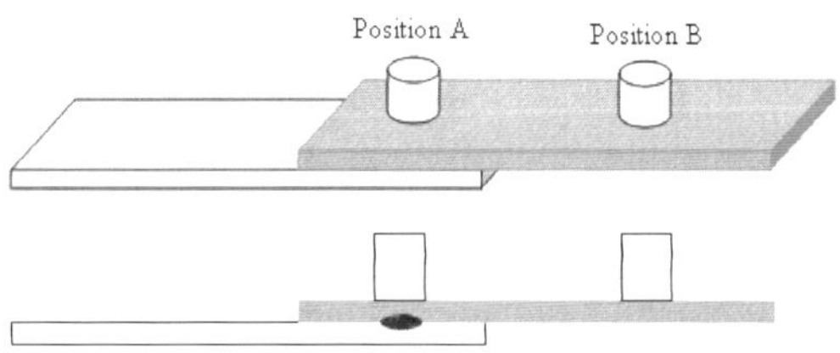

Figure 4. Transducer position in ultrasonic testing.

The ultrasonic testing was performed by placing the transducer in two positions as shown in figure 4. Position A is a position which the transducer is above the welding nugget, whilst position B is a position which the transducer was put on surface outside of overlapping area. Ultrasonic testing results were characterized using reflected wave from interface between two samples, and amplitude was used as parameter.

1.3. Tensile testing

After performing ultrasonic testing, tensile testing was subjected to the spot welded specimens. This test was performed to obtain a real shear fracture load of the welded joint. The test was performed using universal tensile machine with 1 mm/min elongation rate. The tensile machine was equipped with digital control which allows acquiring digital data of load and elongation. After performing tensile testing, observations of fracture surfaces were performed to measure the actual nugget size.

2. Results and discussion

Figure 5 shows a typical ultrasonic reflected signal from position A dan B, respectively. After emanating from transducer and entering sample, ultrasonic wave propagates through the thickness, and when encounters interface, the wave will be reflected. In case of position A, ultrasonic wave will encounter bonding area (nugget) between two sample, therefore, partially will be reflected and another part will be transmitted into the second sample. The intensity of reflected wave depends on how wide the nugget area is. The wider nugget area, the lower ultrasonic intensity is.

In case of position B, there is no bonding area along wave propagation path. Therefore, in this position, almost all of ultrasonic wave is reflected. In this condition, the reflected wave should be the maximum intensity compare to that one from position A. In order to obviate the uncertainty of intensity of reflected wave due to surface condition of each sample and electronic circuit of ultrasonic equipment, to quantify the intensity, then the amplitude of reflected wave from nugget area (A_w) was normalized by amplitude of reflected wave from back wall of sample I (A_n).

Figure 6 and 7 shows the relationship between ultrasonic reflected wave's amplitude and shear strength. In figure 6, a vertical axis shows an original amplitude, whilst in figure 7, a vertical axis shows a normalized amplitude indicates by Aw/An. In figure 6, the data are widely scattered. It indicates, the amplitude of reflected wave (Aw) has no obvious correlation to shear strength. Whereas, in figure 7, there is an obvious correlation between normalized amplitude of reflected wave (Aw/An) and shear strength. The difference trend between an original reflected wave Aw and normalized reflected wave Aw/An indicates an effective treatment (normalization treatment) of ultrasonic signal. The deviation result in figure 6 may be caused by surface condition. Thus, the normalization process resulted in omission of surface condition effect.

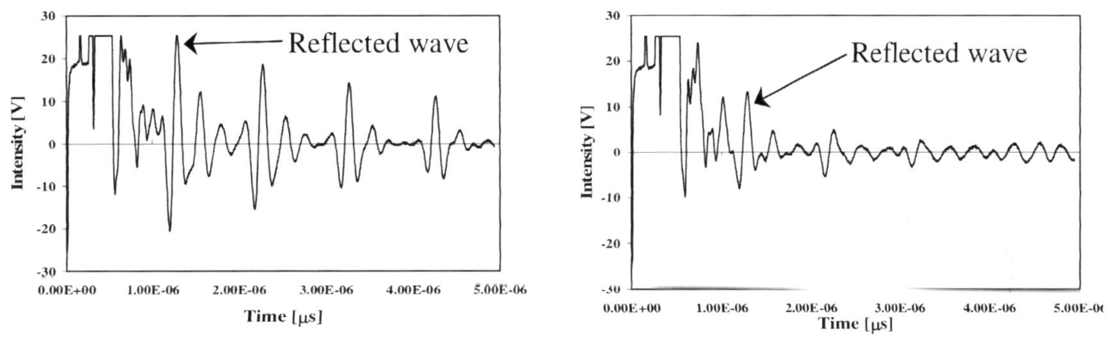

(a) (b)

Figure 5. Typical ultrasonic reflected signal (a) From Position B; (b) From Position A

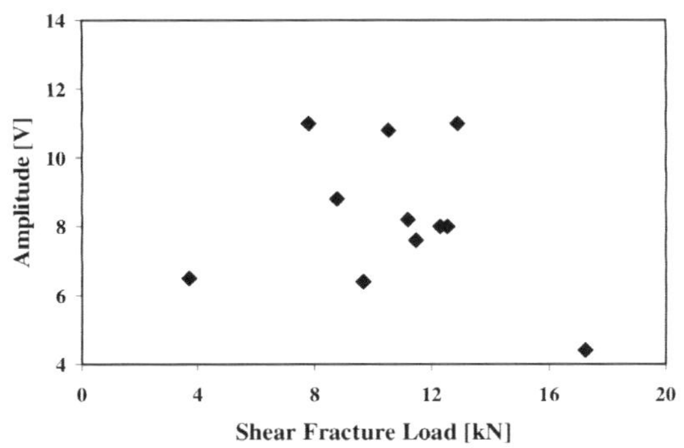

Figure 6. Reflected wave's amplitude from nugget area versus shear strength.

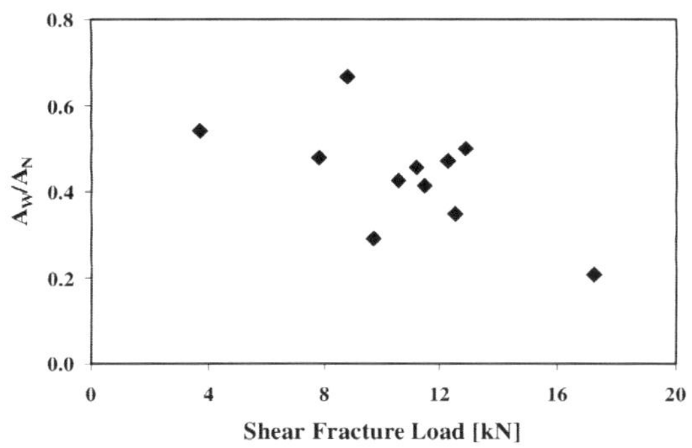

Figure 7. Normalized reflected wave's amplitude from nugget area versus shear strength.

Figure 7 clearly shows that reflected wave decreases as increasing of shear strength. This relationship could be explained as follows. Shear strength of spot welding depends on bonding condition, which is nugget's geometry and metallurgical fusion between two metal sheets. Larger geometry and more perfect fusion provide a high strength. Increasing in nugget geometry and quality of bonding increases portion of transmission wave. Contrarily, the portion of reflected wave decreases. According to figure 7, reflected wave amplitude and shear strength has a linear relationship.

Figure 8 shows the relation between shear strength and nugget size. Nugget size in horizontal axis represents a nugget diameter, since the nugget was assumed as a circle. The diameter was measured as shown in figure 9. It can be seen from the figure that shear strength of the sample increases linearly with the increasing diameter. The data scatter slightly wide. The cause in scattering data could be considered as follows. There is uncertainty in the measurement. As shown in figure 9, measurements were performed using a ruler which has a minimum scale of 1 mm. In this study, the measurements were rounded to millimetre; therefore, the measurement results include a deviation about 0.5 mm. Even the nugget shape is assumed as a circle, an actual shape of nugget is not a circle. Observation of the entire fracture surface, it could be clarified that the shape of nugget are irregular. The assumption affected to the nominal bonding area. The next factor on scattering data is internal discontinuity in the nugget. As shown in figure 9, fracture surface of nugget are not homogeneous. In this figure, in the centre of nugget, there is such a metallurgical discontinuity which could affect the welding strength. The last factor to be considered as reason on scattering data is the thickness of nugget. Nugget is a three dimensional geometry. A thickness increases linearly as diameter increases. Since a planar surface is irregular, the thickness also irregular. Thus, nugget diameter doesn't represent actual nugget geometry.

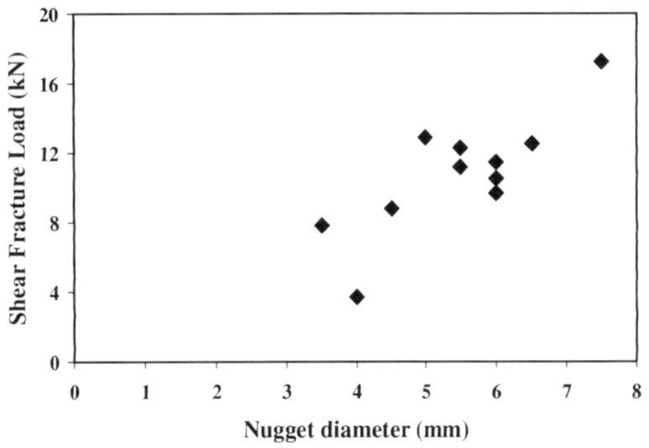

Figure 8. Shear strength versus nugget size.

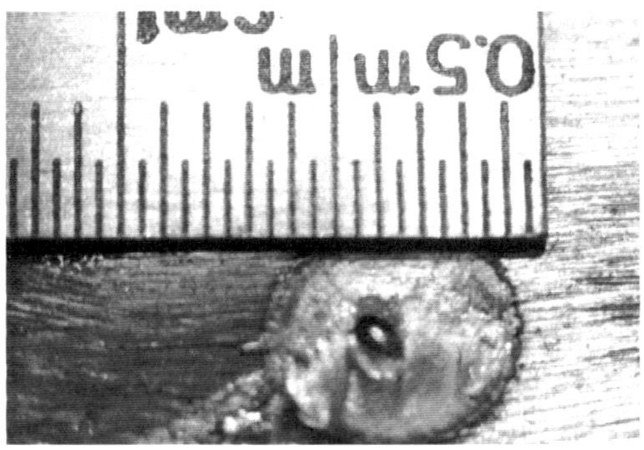

Figure 9. Fracture surface.

Figure 10 shows relation between ultrasonic reflected wave's intensity and nugget size. An open diamond mark represents amplitude of reflected wave from nugget area (left side axis), whilst a closed triangle mark represents amplitude of reflected wave from nugget area normalized by reflected wave from sample's back wall.

It can be seen from the figure that amplitude of reflected wave decreases linearly with nugget size. Both data (with and no normalization) are scatter, but normalized amplitude data gave a smaller scatter compare to without normalization. The reason was already explained in previous discussion for figure 7. If the scatter data in this figure were compared to figure 8, figure 10 has less scatter. In figure 8, a parameter related to nugget has a strong relationship to the shear strength. But, in figure 10, geometrically aspects in nugget have a few relations to reflected wave intensity. The effect of nugget area corresponds to the intensity of reflected and transmission of ultrasonic wave. Therefore, the uncertainty in diameter measurement affected in scattering data.

The main effect to be considered to the scattered data is position of transducer. In this study a circle shape transducer with 10 mm in diameter was used. To achieve a minimum scattered data, the centre of transducer should coincide with the centre of welding nugget. But, it was hard to maintain this condition without special equipment. A shifting between centre of transducer and centre of welding nugget will affect in uncertainty to the result. It can be explained as follow.

Transducer with circle shape, the intensity of incidence wave within the circle is not homogeneous [16]. The intensity decreases in radial direction. From the centre to the half of radii, the intensity almost constant, after that decreases rapidly to almost 10 % compare to the intensity in the centre. Therefore, by considering the diameter of transducer and diameter of nugget, the shift between centre of nugget and centre of transducer will give uncertainty in measurement.

According to these results, the equation for predicting the shear fracture load for spot welding using ultrasonic testing, contact method was proposed as follow.

$$\tau_F = -17\, A_w/A_n + 18.18 \qquad (1)$$

where, τ_F is shear fracture load for spot welding and A_w/A_n is ratio of amplitude of reflected wave from nugget area (A_w) and amplitude of reflected wave from back wall of sample. This equation is valid for spot welding to stainless steel and stainless steel.

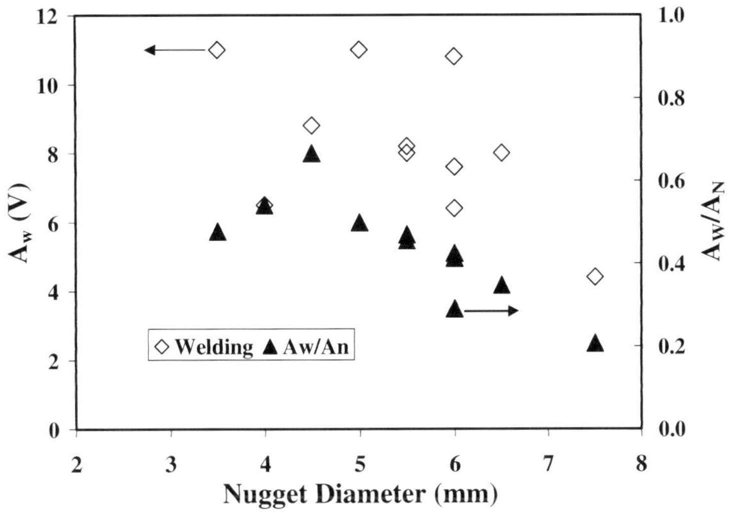

Figure 10. Reflected wave's intensity versus nugget size

3. Conclusion

Ultrasonic tests were performed to spot welded stainless steel plate for predicting shear fracture load. A contact ultrasonic method was applied for this purpose. Experimental results show that there is linear correlation between shear fracture load and ultrasonic reflected wave. The higher shear fracture load, the lower the ultrasonic reflected wave. A low reflected wave due to a wider nugget size constructed in the interface between two plates. A wider nugget size will transmits more ultrasonic energy. An empirical equation was derived from the experiments results to predict shear fracture load by using reflected wave amplitude. This equation can express a good relation between shear fracture load and ultrasonic reflected wave. This equation only valid for stainless steel and stainless steel under welding and testing method described in this paper.

Acknowledgments

The authors are grateful for the financial support from DIPA-BATAN. The authors also thank to Division for Reactor Facility Development's personnel for preparation of sample and conducting shear-tensile test.

References

[1] Mikno Z, Bartnik Z, 2016 *Archives of Civil and Mechanical Engineering* **16 86-100.**

[2] Wana Z, Wang H.P, Wang M, Carlson B.E, Sigler D.R, 2016 *International Journal of Heat and Mass Transfer* **101 749–763**.

[3] Wang S.Q., Patel V.K., Bhole S.D., Wenc G.D., Chen D.L, 2015 *Materials and Design* **78 33–41.**

[4] Floreaa R.S., Hubbardc C.R., Solankid K.N., Bammanna D.J., Whittingtona W.R., Marina E.B. 2012 *Journal of Materials Processing Technology* **212 2358– 2370.**

[5] Patel V.K., Bhole S.D., Chen D.L. 2014 *Materials and Design* **62 124–132**.

[6] Schlichting J, Brauser S, Pepke LA, Maierhofer Ch, Rethmeier M, Kreutzbruck M 2012 *NDT&E International* **48 23**

[7] Tsukada K, Yoshioka M, Kiwa T, Hirano Y, 2011 *NDT&E International* **44 101**

[8] Thornton M, Han L, Shergold M, 2012 *NDT&E International* **48 30**

[9] Safia AM, Salam Akanda MA, Sadique J, Alam MS, 2014 *Procedia Engineering* **90 110**

[10] Rajesh Jesudoss Hynes N, Nagaraj P, Angela Jennifa Sujana J, 2014 *Materials and Design* **62 118**

[11] Takada H, Ozeki T, Ikeda R, Hirose T, 2011 *Materials Transactions* **52 539**

[12] Chen Z, Shi Y, Jiao B, Zhao H, 2009 *Journal of Materials Processing Technology* **209 2329**

[13] Martín Ó, Pereda M, Ignacio Santos , Galán JM, 2014 *Journal of Materials Processing Technology* **214 2478**

[14] Wei L, Huang Z, Que P, 2009 *NDT&E International* **42 430**

[15] Himawan R, Farokhi M, 2008 *Jurnal Teknologi Bahan Nuklir* **4 91**

[16] Roye W, Ultrasonic Testing of Spot Welds in the Automotive Industry, Krautkrämer GmbH & Co. oHG, Hürth, *Technical Note*

Steady state and LOCA analysis of Kartini reactor using RELAP5/SCDAP code: The role of passive system

Anhar R. Antariksawan[1], Puradwi I. Wahyono[2] and Taxwim[2]

[1]Center for Nuclear Reactor Safety and Technology - BATAN, Puspiptek Area Building 80, Serpong, Tangerang Selatan 15310
[2]Center for Accelerator Science and Technology - BATAN, Jl. Babarsari, Yogyakarta

email: anhar@batan.go.id

Abstract. Safety is the priority for nuclear installations, including research reactors. On the other hand, many studies have been done to validate the applicability of nuclear power plant based best estimate computer codes to the research reactor. This study aims to assess the applicability of the RELAP5/SCDAP code to Kartini research reactor. The model development, steady state and transient due to LOCA calculations have been conducted by using RELAP5/SCDAP. The calculation results are compared with available measurements data from Kartini research reactor. The results show that the RELAP5/SCDAP model steady state calculation agrees quite well with the available measurement data. While, in the case of LOCA transient simulations, the model could result in reasonable physical phenomena during the transient showing the characteristics and performances of the reactor against the LOCA transient. The role of siphon breaker hole and natural circulation in the reactor tank as passive system was important to keep reactor in safe condition. It concludes that the RELAP/SCDAP could be use as one of the tool to analyse the thermal-hydraulic safety of Kartini reactor. However, further assessment to improve the model is still needed.

Keywords: research reactor, safety, thermal-hydraulic, LOCA, RELAP5

1. Introduction

A research reactor is an important facility to support the development of nuclear science and technology in a country. That research reactor is in general used for education and training in terms of basic nuclear reactor. A research reactor is also very useful to help the research and development activities. As a neutron source, the research and development in developing new materials could be conducted in a research reactor. Other benefit is that research reactor could be used to produce radio isotopes, which are widely used in industries and medicine. There are 228 type of research reactors in operation in the world today [1]. In Indonesia, there are three research reactors, i.e. two of TRIGA type and one of MTR type. These reactors are all operated by National Nuclear Energy Agency of Indonesia (BATAN).

As other nuclear and radiation facilities, the safety must be assured by the operator. In order to assure the excellent level of safety, the International Atomic Energy Agency (IAEA) has established a fundamental safety principles and specific set of requirements for research reactor [2]. One of the requirements is to conduct safety analysis using mainly deterministic method. In order to comply with this requirement, several thermal-hydraulic analytical codes have been developed. In the past, conservative approach codes were used, but as the need to have more realistic results emerged, best estimate approach codes have been developed. Moreover, most of these codes are developed initially for nuclear power plant, but then they are also validated to be used in research reactor, such as RELAP5 code.

Study to validate RELAP5 code for different type of research reactor has been done since couple decades. Hari et al. used the RELAP5/Mod3.2 code to simulate two transient events without scram on a MTR type reactor named HIFAR [3]. RELAP5/MOD3 has also been qualified against the steady state and transient data from MARIA reactor, a pool type reactor with pressurized fuel channels containing concentric multi-tube assemblies of highly enriched uranium [4]. Other studies on the applicability of RELAP5 computer code for MTR type reactor (RSG-GAS and IEA-R1) have been done by Abdelrazek et al [5], Chatzidakis et al [6], Chatzidakis et al [7] and Hainoun et al [8].

Moreover, some works have been performed to study the applicability of the RELAP5 code to TRIGA type research reactors. The study against IPR-R1, a TRIGA Mark-I type has been done by Costa et al [9] and Reis et al [10,11]. While Marcum et al [12] assessed the validity of RELAP5 for OREGON State TRIGA reactor, and Antariksawan et al studied the applicability of RELAP5 for TRIGA-2000, a TRIGA Mark-II type reactor [13,14].

Since each research reactor has unique features and none of the previous research have been performed for all possible events happen in a research reactor, the present study is aimed to study the applicability of the RELAP5 for Kartini research reactor. The reactor is a pool, water cooled reactor of 100 kW thermal power. The present study differs from previous study especially in terms of the model and the experimental data used, which are specific to the Kartini reactor. On the other hand, the current model is the improvement of the model used in previous study [15], especially in core inlet model, the transient model and the comparison with the latest experimental data. The study presents the results of steady state calculation and also the analysis result of a hypothetical accident known as loss-of-coolant accident (LOCA). This kind of accident is considered as the one of design basis accident. It is important to predict the response of the reactor against such accident in order to know the performance of its safety provisions. In addition, this study is also important in the context of the further use of RELAP5 code as an analytical tool to support the operational safety of Kartini reactor.

2. Methodology

2.1. Reactor description

Kartini reactor is similar to TRIGA Mark-II type reactor, a pool type with U-ZrH cylindrical fuel of about 752 mm length. The core is constituted of 68 fuels and 3 control rods placed in annular configuration. Figure 1 shows the core configuration. The vertical cut view of reactor and core configuration is delineated in figure 2. The core is seated in bottom of reactor tank with about 2 m of diameter and about 5 m beneath water surface. The light water serves as moderator and coolant. The core cooling is done by natural circulation.

The hot water flowing upward from the core is pumped through primary piping system to a shell and tube type heat exchanger in about 6.5 kg/s to transfer the heat to the secondary cooling system. The colder water is circulating back to the reactor tank.

2.2. Code used: RELAP5/SCDAP/MOD3.2 code

The RELAP/SCDAP/MOD3.2 used in this study is developed by Innovative System Software (ISS) based on publically available RELAP5/MOD3.3 code. The code is designed to describe the overall reactor coolant system thermal-hydraulic and core behavior. The code could be used to simulate the normal operating conditions or under design basis or severe accident events. The RELAP5 models calculate the overall thermal-hydraulic response and SCDAP models calculate the core and vessel behavior during normal and accident conditions. It has been validated for a wide range of accident conditions using a variety of experiments and plant data [16].

2.3. Calculation

In this study, the model of Kartini reactor was developed for simulating steady state condition and LOCA transient due to pipe break and beam tube rupture. Nodalization scheme of the core and primary cooling system is described in figure 3.

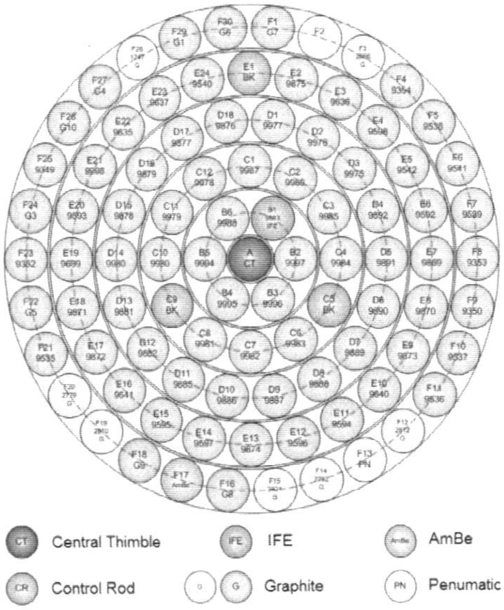

Figure 1. Core configuration of Kartini reactor [15].

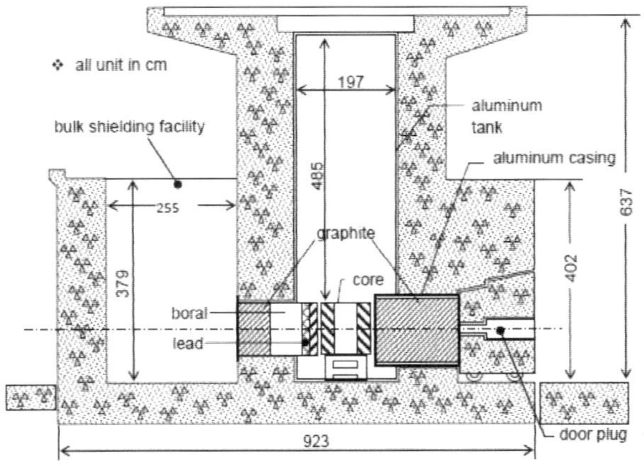

Figure 2. Vertical cut view of Kartini reactor.

Reactor tank is divided into core and coolant region. The last is further differentiated in several regions: lower plenum (B100), core by-pas (P110), mixing region (P145, P150, P160 and P165) and upper plenum (P170). The core is divided in three region: hot channel, average channel and cold channel. The hot channel represents 6 fuels in the most inner ring (P120 and HS120) while the remaining fuel elements is represented by the average channel. The primary piping system and the heat exchanger are included. The secondary cooling side model is simplified by using time dependent volume as a heat sink. In this model, the reactivity feedback effects are not yet accounted for.

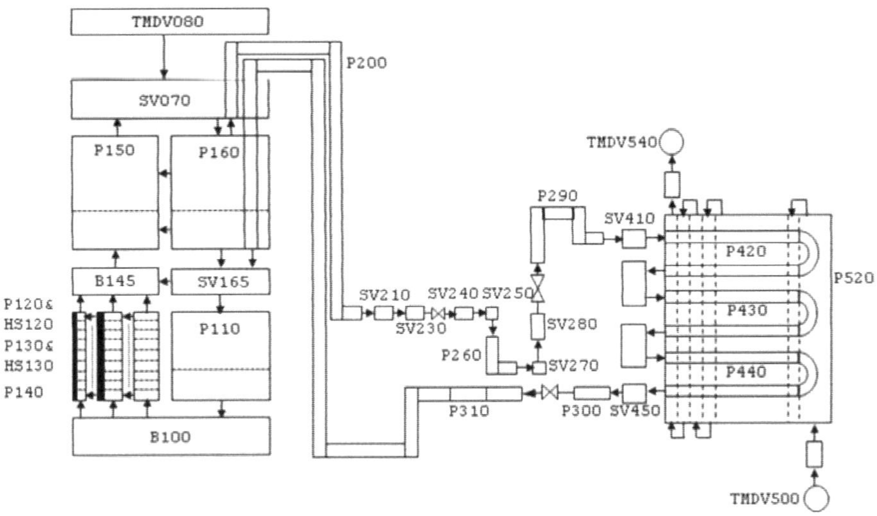

Figure 3. Nodalization scheme of Kartini reactor.

As first step, the steady state calculation is done to validate the model. The calculation results are compared with the available measurement data, especially at the nominal power of 100 kW. Once the model is considered providing good steady state results, the transient calculation is conducted. A short transient due to pump trip without reactor scram is simulated. The calculation result is compared with the measurement data obtained during reactor power calibration. To simulate the LOCA, models for pipe break and beam tube rupture are developed. The pipe break is assumed occur in the pipe circulating cold water to the reactor in the lowest part. A guillotine pipe break is considered to simulate the worst case. In the case of a beam tube rupture, the leak from the beam tube with the size equal to the beam tube diameter is assumed.

3. Results and discussion

3.1. Steady state condition

Figure 4 and 5 show the coolant mass flow rate and fuel and coolant temperature at steady state at 100 kW, respectively. The steady state condition could be achieved after about 3000 s. With regard to the fuel temperature and coolant temperature at the upper tank, calculation obtains 126.67 °C and 35.56 °C, respectively. While, the measurement data indicates the fuel temperature 128.3 °C and the upper poll coolant temperature of 35.4 °C [17]. So, the difference for both is only less than 2%.

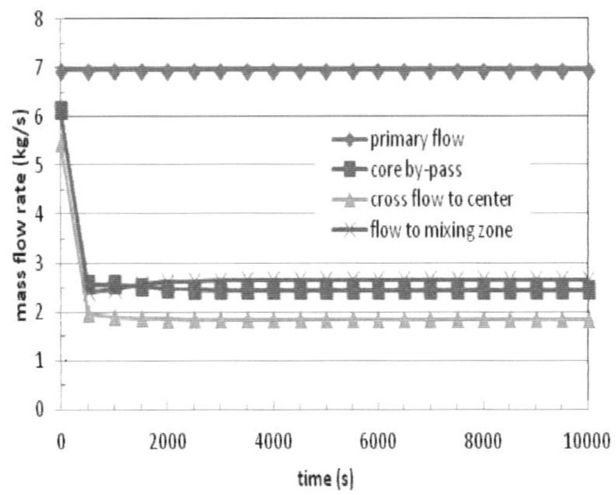

Figure 4. Coolant mass flow rate at 100 kW.

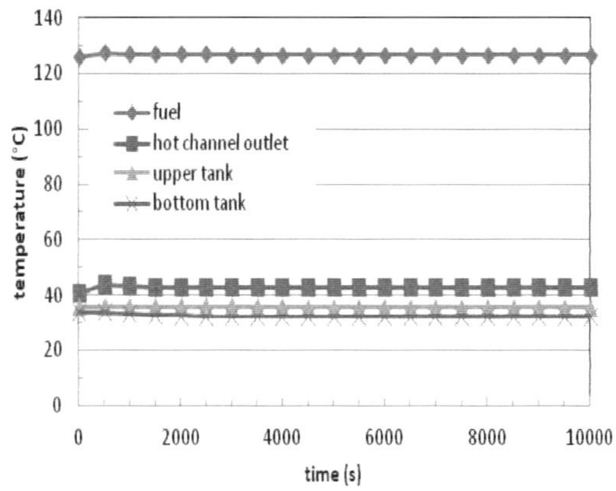

Figure 5. Fuel and coolant temperature at 100kW.

On the other hand, no data of mass flow measurement in the pool is available. However, we could see from the calculation results that when the coolant is flowing out from the inlet pipe at the rate of about 6.5 kg/s, the major part of the flow is going directly to the pool (mixing region), only less than one-third of mass flow coolant rate goes to the core. This is because the reactor core is cooled by natural circulation such that the flow to the core is determined by the reactor power.

Figure 6 shows the fuel temperature increases with the power. Figure 7 provides the calculation results on the increase of the coolant mass flow rate to the core with the reactor power from 10 to 100 kW. The instability at the beginning of the mass flow curve in figure 7 is due to numerical instability (steady state is not yet achieved).

The comparison of calculated and measured fuel temperature is delineated in table 1 and figure 8. The differences between the measured and calculated fuel temperatures at various powers are between -1.67% and +10.5%. The difference is higher in the lower temperature. The difference between measurement and calculation, could be caused by error in power measurement which could attained about 9.7%. Another source of error is from temperature measurement which is approximately 1%. Considering that, the calculation results are acceptable.

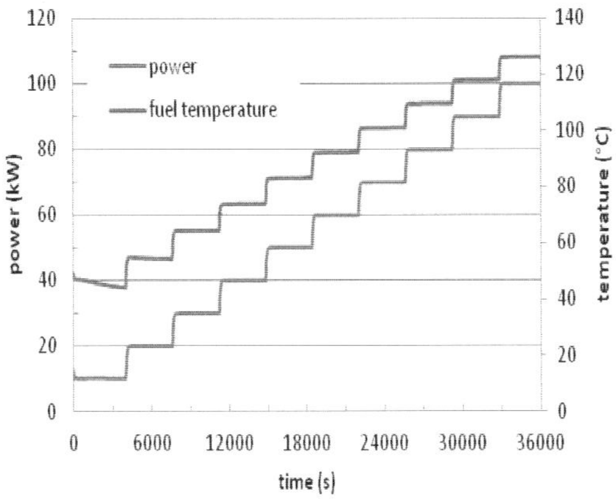

Figure 6. Fuel temperature variation with power

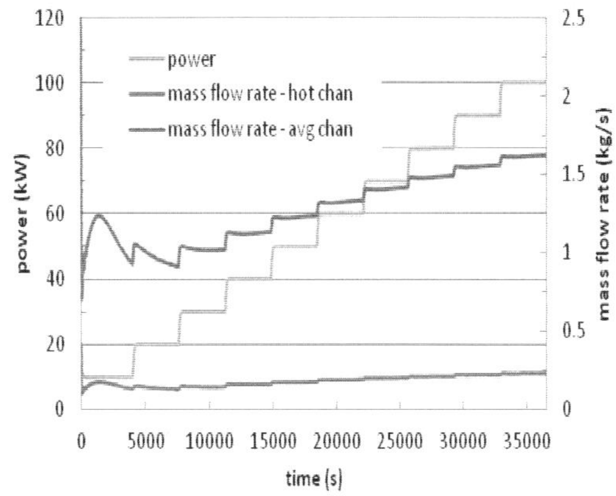

Figure 7. Coolant flow variation with power

Table 1. Fuel temperature variation with power: measured vs calculation,

Power (kW)	Temperature (°C) Calculated	Temperature (°C) Measurement	Difference (%)
10	43.22	39.5	9.39
20	54.28	49.7	9.10
30	64.41	58.3	10.50
40	74.09	68.5	8.10
50	83.38	78.5	6.17
60	92.36	89.3	3.47
70	101.11	97.3	3.90
80	109.64	108.3	1.25
90	117.97	118.8	-0.68
100	126.15	128.3	-1.67

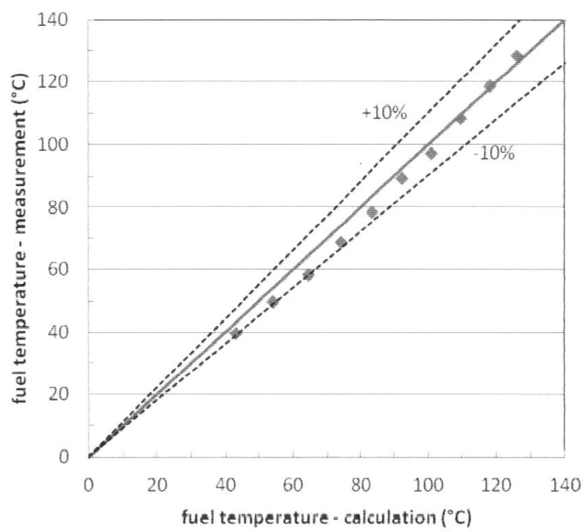

Figure 8. Measured vs calculated fuel temperature.

3.2. Transient during pump trip steady state

In addition, the comparison is also done during transient following primary pump trip. The scenario is as follows: at the steady state operation at 100 kW, primary pump is stopped, but the reactor is not scrammed. During about 60 minutes the coolant temperature at the upper pool is measured once a minute. Figure 9 describes the primary inlet flow to the tank and coolant flows to the core. As in figure 9 it could be shown that after pump trip, the flow to the core is practically unchanged. It is because the flow is dominated by the natural circulation, which in this case the heat generated in the core plays the role pumping the coolant upward. Figure 10 shows the comparison of measured and calculated of the upper pool coolant temperature evolution. The temperature difference of both is quite obvious, about 5 °C. This is probably due to the early measurement of the coolant temperature, i.e. just at a short time after the power reaches 100 kW when the coolant temperature is not yet homogenized. The data from the reactor operation shows that it takes about 7 hours for the coolant temperature, which initially at about 30 °C, to reach steady state at 35 °C. Despite of that difference, the rate of increase of the coolant temperature is identical in both cases, which is about 0.0014 °C/s. It proves that the model of natural convection in the reactor pool is satisfactory.

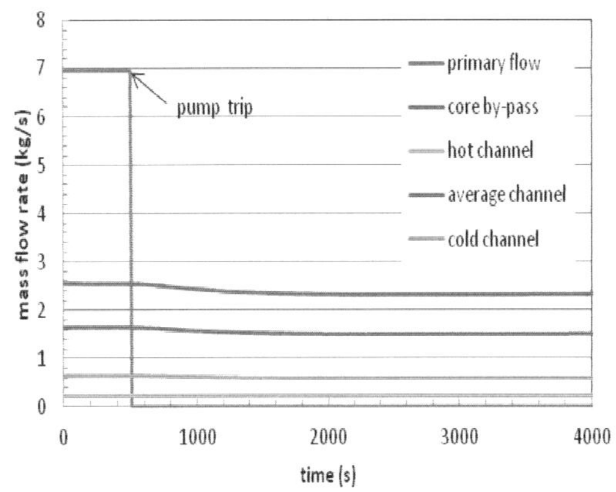

Figure 9. Evolution of flows during pump trip.

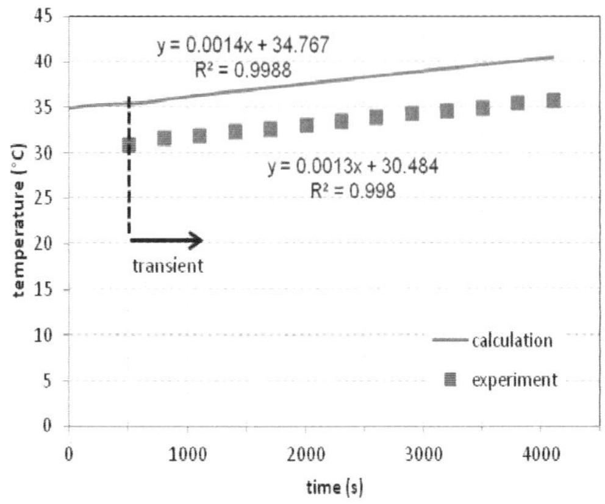

Figure 10. Coolant temperature evolution after pump trip

3.3. LOCA due to a pipe break

It is assumed the lowest part of cold leg experiences double ended break (guillotine break). Following that, the coolant leaks with the rate of about 12 kg/s at the beginning and consequently, the coolant level in tank decreases. As the level reduces up to 20 cm below the normal level, the reactor scram and pump trip are assumed. In the reality, there is a siphon breaker hole in the inlet pipe, but it s not considered in this simulation to obtain the worst case. Figure 11 depicts the flow of primary coolant, leaks and coolant level in tank. The break starts at 500 s of calculation time. At 30 s later, the coolant level is down by 20 cm and it triggers reactor scram and pump trip. The leak continues until 1450 s before stopping when the level reaches about 1.1 m from the bottom of the core or at about the level of the tip of inlet pipe. The leak flow stops because the siphon affect is interrupted. Following to that, the decay heat is removed by natural convection in the remaining coolant. The fuel temperature in hot channel and coolant temperature in various regions in the tank are shown in Figure 12. It could be seen that up to the end of calculation time, the temperatures are stabilize at about 34 °C.

In the field, a siphon breaker hole is added at the inlet pipe. Its location is about 40 cm below the tank's tip. This is a safety facility that is used to prevent water efflux in case of pipe break [18,19]. In such case the coolant loss is reduced and also avoid the core from being uncovered.

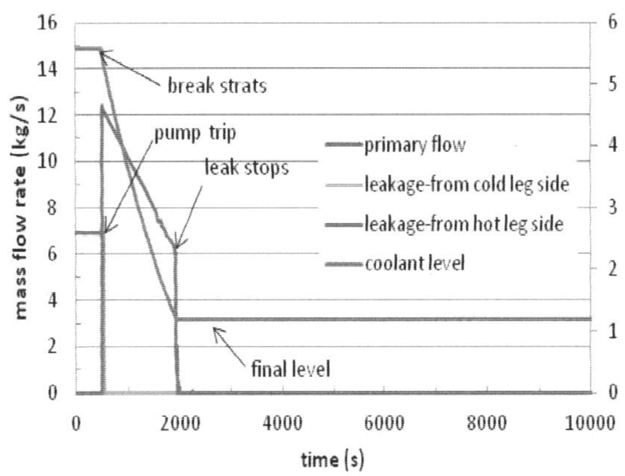

Figure 11. Coolant flow and level during LOCA transient.

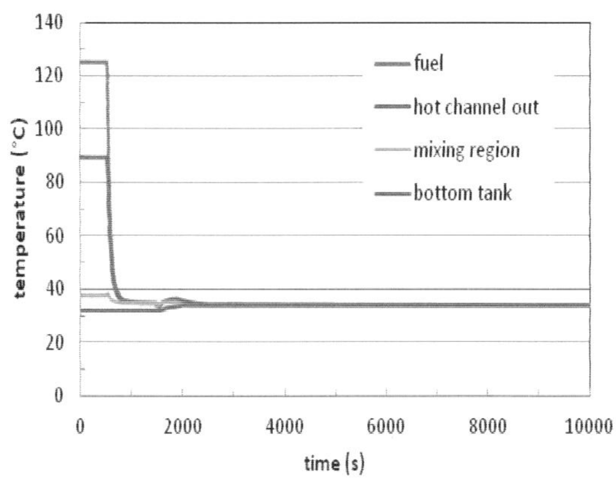

Figure 12. Fuel and coolant temperature during LOCA transient.

3.4. LOCA due to a beam tube

A hypothetical one beam tube rupture accident is simulated to see the characteristic of Kartini reactor following that accident. Figure 13 shows that following the beam tube rupture, coolant leaks from the ruptured beam tube. It causes the coolant level decreases fast and 750 s after the initiation of the leak, the coolant reaches the final level at 0.35 m or at the level of the beam tube. At that time, the leak flowing through the beam tube practically stops. In this scenario of the accident, the reactor scram and pump trip are determined to occur when the coolant level decreases by 20 cm from the normal level. The decay heat remains after reactor scram.

As in figure 14, even the coolant level reaches at mid of the core and upper half of fuel element is uncovered, the remaining coolant, vapor and air could cool the fuel element. In this figure, the fuel temperature, which represents the mid-height of fuel, decreases following the decay heat, but then increases before decreasing again and attains about 50 °C at about 20,000 s after the initiation of the accident. The fuel temperature increases because the coolant cooling capacity decreases, but when the decay heat continues to decrease and the vapor and air could remove the decay heat, the fuel temperature decreases again. The remaining coolant at the bottom part of the core increases slightly and stabilizes at about 40 °C. Other study investigated extreme loss of coolant due to bottom of tank rupture causing complete core uncovered. It showed the fuel temperature increased quickly to more than 1000 °C in about 50 s [20].

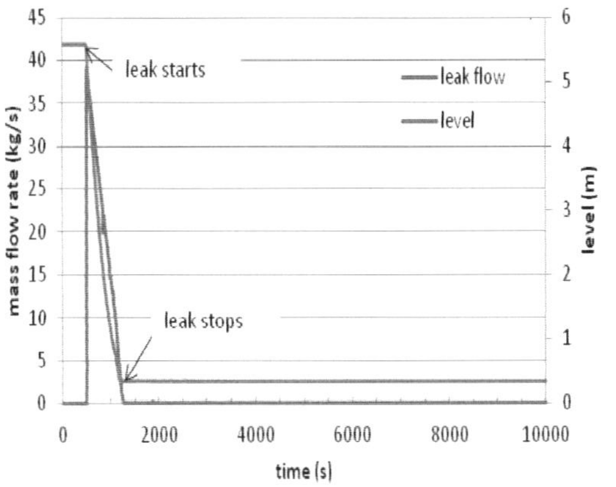

Figure 13. Break flow and coolant level during LOCA due to beam tube rupture.

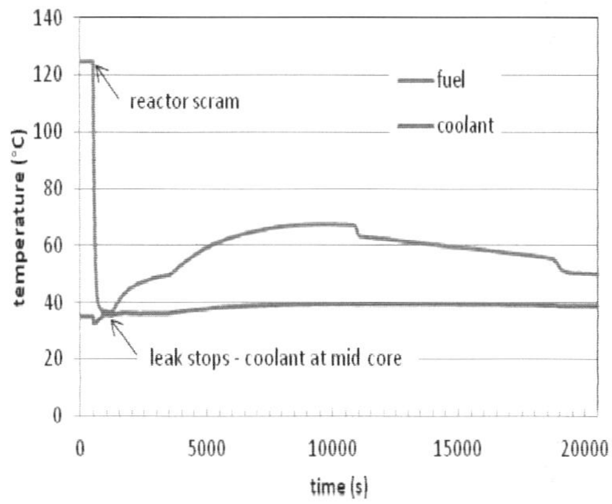

Figure 14. Fuel and coolant temperature during LOCA due to beam tube rupture.

4. Conclusion

A RELAP5/SCDAP model for Kartini research reactor has been developed and used to simulate the steady state operational and transient events. The steady state simulation showed that the RELAP5/SCDAP provided reasonable results when compared with available measurement data. The model could also predict the reactor characteristics following a hypothetical LOCA events. The reactor is predicted that it could keep its good safety performance against LOCA due to worse pipe break and beam tube rupture. The simulation is also shown that the passive system and phenomenon, i.e. siphon breaker hole and natural circulation inside the tank, plays an important role to keep the safety of the reactor during loss of coolant events. The improvement is still needed, especially by extending the comparison with the measurement data. The integration of the reactivity feedback effect would also improve the model. It is more needed to simulate the reactivity insertion accident type which are commonly hypothesized in a research reactor.

References

[1]. IAEA. Research Reactor Data Base [Internet]. 2017 [cited 2017 Mar 7]. Available from:

https://nucleus.iaea.org/RRDB/RR/ReactorSearch.aspx

[2]. International Atomic Energy Agency. Fundamental Safety Principles. Vol. 2, Safety Standards Series. Vienna, Austria; 2006.

[3]. Hari S, Hassan YA, Tu J. Analysis of transient events without scram in a research reactor using the RELAP5/MOD3.2 computer code. Nucl Technol. 2000;130:296–309.

[4]. Szczurek J, Czerski P. RELAP5 / MOD3 model and transient analyses for the MARIA research reactor in Poland. Nukleonika. 2004;49 (4)(December 2003):149–57.

[5]. Abdelrazek ID, Aly MN, Badawi AA, Elnour AGA. Annals of Nuclear Energy Benchmarking RSG-GAS reactor thermal hydraulic data using RELAP5 code. Ann Nucl Energy. 2014;70:36–43.

[6]. Chatzidakis S, Hainoun A, Doval A, Alhabet F, Francioni F. A comparative assessment of independent thermal-hydraulic models for research reactors : The RSG-GAS case. Nucl Eng Des. 2014;268(35):77–86.

[7]. Chatzidakis S, Ikonomopoulos A, Ridikas D. Evaluation of RELAP5 / MOD3 behavior against loss of flow experimental results from two research reactor facilities. Nucl Eng Des. 2013;255(1537):321–9.

[8]. Hainoun A, Doval A, Umbehaun P, Chatzidakis S, Ghazi N, Park S, et al. International benchmark study of advanced thermal hydraulic safety analysis codes against measurements on IEA-R1 research reactor. Nucl Eng Des. 2014;280(January):233–50.

[9]. Costa AL, Amélia P, Reis L, Pereira C, Auxiliadora M, Veloso F, et al. Thermal hydraulic analysis of the IPR-R1 TRIGA research reactor using a RELAP5 model. Nucl Eng Des. 2010;240:1487–94.

[10]. Reis PAL, Costa AL, Pereira C, Veloso MAF, Mesquita AZ, Soares H V, et al. Assessment of a RELAP5 model for the IPR-R1 TRIGA research reactor. Ann Nucl Energy. 2010;37:1341–50.

[11]. Reis PAL, Costa AL, Pereira C, Silva CAM, Auxiliadora M, Veloso F, et al. Sensitivity analysis to a RELAP5 nodalization developed for a typical TRIGA research reactor. Nucl Eng Des. 2012;242:300–6.

[12]. Marcum WR, Woods BG, Reese SR. Experimental and theoretical comparison of fuel temperature and bulk coolant characteristics in the Oregon State TRIGA ® reactor during steady state operation. Nucl Eng Des. 2010;240:151–9.

[13]. Antariksawan A.R., Umar E., Widodo S., Juarsa M. KMH. TRIGA-2000 research reactor thermal-hydraulic analysis using RELAP/SCDAPSIM/MOD3.4. Int J Technol. 2017;(under review).

[14]. Antariksawan A.R. and Widodo S. Parametric Study of LOCA using RELAP5/SCDAP. Tri Dasa Mega. 2017;(under review).

[15]. Antariksawan A.R., Tri Wulan Tjiptono, Juarsa Mulya S. Application of RELAP/SCDAPSIM/MOD3.4 code for thermal-hydraulic calculations of Kartini reactor. In: Proceeding of PPI - PDIPTN. 2006.

[16]. Allison CM, Hohorst JK. Role of RELAP / SCDAPSIM in Nuclear Safety. Sci Technol Nucl Install. 2010;

[17]. PSTA-BATAN. Report on reactor operation for 20 and 50 hours non-stop, LKEG OPR 2016.003.4/RN.00 02/STA 4. Yogyakarta; 2016.

[18]. Lee K, Kim W. Theoretical study on loss of coolant accident of a research reactor. Nucl Eng Des. 2016;309:151–60.

[19]. Kang SH, Ahn HS, Kim JM, Joo HM, Lee K, Seo K, et al. Experimental study of siphon breaking phenomenon in the real-scaled research reactor pool. Nucl Eng Des. 2013;255:28–37.

[20]. Reis PAL, Costa AL, Pereira C. Analysis of an extreme loss of coolant in the IPR-R1 TRIGA reactor using a RELAP5 model. Eng Térmica (Thermal Eng. 2013;12(2):46–50.

Acknowledgement
This study is supported by PTKRN-BATAN and PSTA-BATAN, also INSINAS Program from the Kemenristekdikti under the contract No.: 02/INS-2/PPK/E/E4/2017. The authors thank to Kartini reactor staff for very good collaboration, discussion and exchanging data and views. Thank also goes to Budget Authority (KPA) of PTKRN-BATAN.

The effect of zinc injection on the increasing of Inconel 600 TT corrosion resistances

Febrianto, Sriyono, Surip Widodo, Geni Rina Sunaryo

Puspiptek Area Building 80, Serpong, South Tangerang City, 15310, INDONESIA

febri@batan.go.id

Abstract. Many failures were found in reactor pressure vessel head penetration (RPV) head material. Those failures caused by boric acid corrosion, and from visual examination were found a big hole and white deposit crystal of boric acid during shutdown maintenance at David Besse reactor. Zinc Oxide addition in BWR reactor known as Zinc Injection that has purposed to reduce radiation exposure cause of Hydrogen addition. Beside reducing the radiation exposure, Zinc injection also has an effect in reducing material corrosion. The purpose of study is to determine the effect of zinc addition, boric acid, temperature also the effects of Cobalt Nitrate and Zinc Oxide addition to Inconel 600 TT as RPV head penetration material. The result in the BWR reactor experience will be implementated at PWR reactor, weather zinc oxide addition also has an effect in reducing the corrosion of Inconel 600. The method that used in this research is to observe the corrosion rates for Inconel 600 material using Potentiostat. Examination were conducted in 30 , 40, 60, 70, 80 and 80 $^{\circ}$C using 1000, 1500, 2000, 2500 and 3000 ppm boric acid concentration. The results showed that the corrosion rate for the material were very small, but the highest corrosion rate occurred in 3000 ppm boric acid concentration at 90 $^{\circ}$C with Cobalt Nitrate addition, around 5.210 x 10^{-1} mpy. In the same condition at 3000 ppm boric acid concentration for temperature at 90 $^{\circ}$C, Inconel 600 TT corrosion rate is smaller with Zinc oxide addition, around 4.631 x 10^{-1} mpy.

Keywords : corrosion, zinc injection, radiation, Inconel 600, resistance

1. Introduction

Since the beginning of operational the PWR (Pressurized Water Reactors) reactor, corrosion problems have caused substantial losses of power generation, resulted in large repair and maintenance costs throughout the world. Water chemistry is one of the few methods available to plant operators to control the materials degradation after the reactor being operated, including; steam generators, reactor components, and balance of plant piping. Since corrosion is a major contributor to the O&M costs, utilities must pursue the most cost-effective solutions to their corrosion problems [1,2].

Boric acid could be caused in the degradation of carbon steel and low alloy. Currently, many research programs conduct a study concerning corrosion effect of boric acid to reactor components. Many material failures causing on boric acid corrosion, one of famous failure caused by boric acid corrosion is reactor pressure vessel head leak at David Besse reactor in US. Material degradation that occurs at Davis Besse reactor RPV head is an example of consequence boric acid corrosion [3]. Reactor Beznau unit 1 in Swiss also has an experience of boric acid corrosion at RPV head. It was found boric acid crystal deposit at above part of RPV head. After cleaning of boric acid crystal it was seen hole with 50 mm width and 40 mm depth at Control Rod Drive Mechanism (CRDM) nozzle penetration part [4]. Currently, PWR reactor operates with higher burn up, lithium concentration addition, high coolant temperature to reduce fuel cladding corrosion risk. Many PWR reactors operated at pH between pH 6,9 and 7,4 during fuel cycle. Reactor coolant pH value is controlled by boric acid (H_3BO_3) and lithium hidroxide (LiOH) combination [5].

Lithium maximum concentration at beginning fuel cycle is 2.2 ppm. Higher lithium concentration will inisiate fuel cladding corrosion and nickel alloy primary water stress corrosion cracking (PWSCC). Fuel cladding controll is more complicated, at low pH many corrosion products will deposit at fuel surface and increase of cladding failure risk [6]. At higher pH, corrosion product deposit rate at fuel surface will decrease but rate of corrosion fuel will increase by increasing lithium hidroxide concentration [7,8].

Boric acid is added into reactor coolant to control reactor reactivity and reactor coolant pH. During reactor operation, material and component will be degraded and corroded. From the reactor operation experiences, degradation and corrosion can cause coolant leakage in several degree. One of material degradation example can be initiated by boric acid [1, 3, 5]. This degradation has an potential to reduce system integrity and can initiate loss-of-coolant accident. Material degradation also occurs at Bugey reactor 3 and 4 Reactor Pressure Vessel Head Penetration Material which made of Inconel 600.

The purpose of this study is to determine the effect of boric acid, temperature and Cobalt Nitrate and Zinc Oxide addition to alloy 600 TT corrosion as RPV head penetration material. The method that used in this research to observe the corrosion is Potentiostate technique. Various researches report that the beneficial effects of Zn on the radiation reduction can be attained in PWR also [4].

In this study will be determine the effect of zinc addition and temperature increasing to Inconel 600 corrosion rate. The purpose of this study is to determine the effect of boric acid, cobalt nitrate, zinc oxide addition and temperature to alloy 600 corrosion resistance.

2. Theory

In BWR reactor zinc addition is known as Zinc Injection that has purposed to reduce radiation exposure cause of hydrogen addition. Hydrogen addition is well known as HWC (Hydrogen Water Chemistry). HWC is effective in reducing the corrosion of material, component and reactor structure. Because the hydrogen addition can reduce the coolant oxidizing power. Even if Hydrogen addition can cause of reactor coolant became more reductive but it has an negative impacts such as [5];

1). Increasing the radiation doses at main steam way cause of N-16 change from stabile form (NO_3^- and NO_2^-) to volatile form (NH_3) and goes to turbine area.

2). Radiation dose increase until 3 – 4 times during shutdown and increase radiation exposure risk will be accepted by worker especially during yearly inspection and maintenance activity. This condition occurs cause of change of oxide form Fe_2O_3 (hematite) in oxidation environment to Fe_3O_4 (magnetite) in reduction environment. Change of this oxide form can cause activated corrosion product such as Co-60.

Zinc addition can reduce radiation exposure through [5, 9]

1. Inhibit the corrosion at surface material and resulting the thinner film layer
2. Zinc will compete with cobalt at the same place and reducing cobalt concentration in oxide structure film
3. Repress cobalt release from deposition place into primary coolant.

Boric acid can cause low alloy and carbon steel degradation. Currently, many research programs conduct study concerning boric acid corrosion effect to reactor components. Boric acid is weak acid

but boric acid leak to hot surface material and water will disappear and leave concentric boric acid. Boric acid saturated solution will occur at temperature above 95°C with solution pH around 3 and this solution is very corrosive and causes the local corrosion.
RPVHP reactor structure can be seen at figure 1 and boric acid crystals deposit at figure 2 [10].

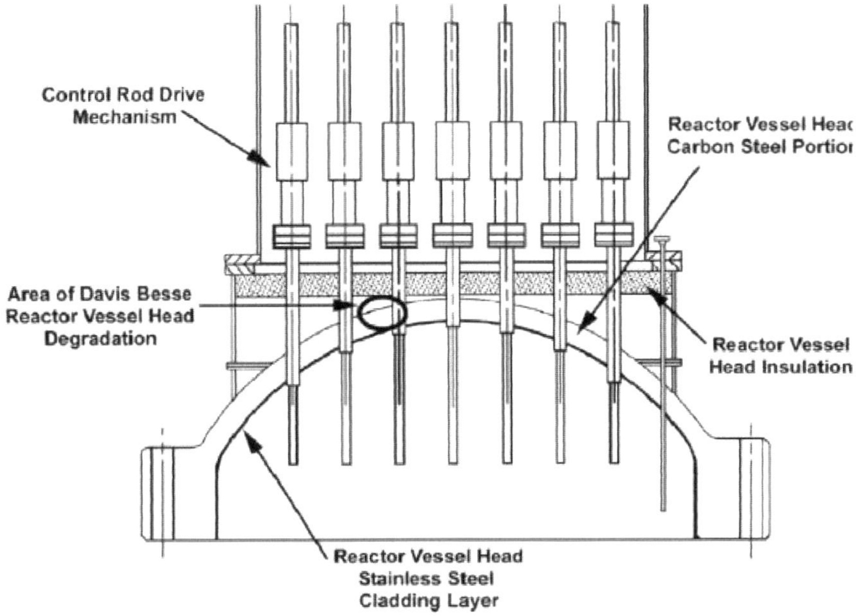

Figure 1. Reactor PWR RPVHP' Structure [11]

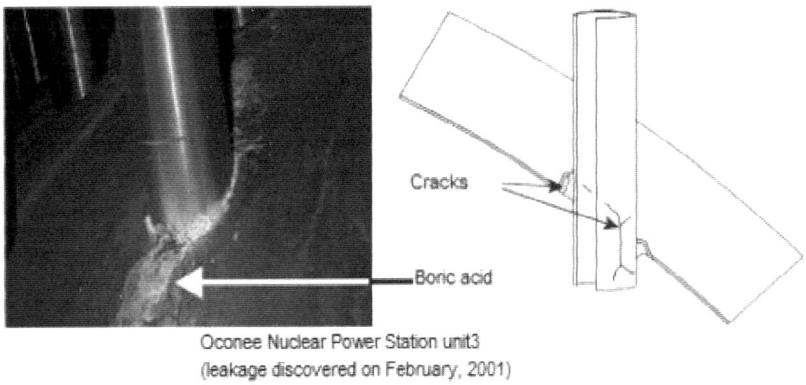

Figure 2. Deposits of boric acid crystals on reactor pressure vessel head from leaking CRDM nozzles [10, 12]

3. Methodology

The experiment was conducted in several steps:
- Specimen preparation
- Inconel 600 TT material corosion test using Potentiostat.
- Data analysis and report.

Inconel 600 TT was used as specimen in this experiment (producted by Inco Alloys Services PTE LTD).The specimen dimension is 19.05 mm in diameter, 1.09 mm in thickness and 10 mm in length. The specimen dimension is the same as steam generator tube dimension. The specimen surface was polished with polished paper from grade 400 ,600, 800 and 1000 and then polished with diamond paste Metadi II. Specimen was conditioned in Autoclave in 100 ppb cobalt nitrate at temperatur 250 °C for 250 hours and to this solution was added 25 ppb zinc oxide at the same temperature for 250 hours.

Specimens that have been conditioned in cobalt nitrate, with zinc oxide and without zinc oxide will be measured its corrosion rate in various boric acid concentrations (1500, 2000. 2500 and 3000 ppm) at various temperatures (30, 40, 60, 70 80 and 90 °C). The potential was automatically varied at a constant rate between two preset levels. Data that found by Potentiostat will be analysed to get corrosion rate.

4. Results and Discussion

4.1 Temperature Effect on Inconel 600 Corrosion Rate

Change of temperature can modify the interaction between the metal and the electrolyte. Increasing temperature will increase corrosion rate by increasing the diffusion of soluble oxygen from solution to the metal surface.

From Figure 3, it is shown that temperature increasing tend to increase corrosion rate of Inconel 600 material. Inconel 600 corrosion rate at 30 °C in 1500 ppm containing boric acid solution is 4.02 x 10^{-6} mpy and corrosion rate at 40 °C in the same boric acid solution concentration is 3.11 x 10^{-5} mpy. At the same boric acid concentration, at 60 °C corrosion rate is 2.69 x 10^{-4} mpy an increase become 4.43 x 10^{-12} mpy at 70 °C. Inconel 600 corrosion rate continue to increase with temperature increasing, is 4.57 x 10^{-1} at 80 °C and 4.74 x 10^{-1} mpy at 90 °C. In the other boric acid concentration also has the same tendency where temperature increasing also increase the corrosion rate of Inconel 600 TT

From the experiment results can be seen that temperature increasing and corrosion rate has linear relation. Increasing temperature tend to increase the corrosion rate. Change of temperature can modify the interaction between the metal and the electrolyte. Increasing temperature will increase corrosion rate by increasing the diffusion of soluble oxygen from solution to the metal surface.

From the experiment results can be seen that the corrosion rate of Inconel 600 TT is small. Inconel 600 TT has a good resistance to corrosion at high temperature and for this reason this material can be used and operated as component and structure at high temperature and pressure. Inconel 600 TT has nickel composition around 58 % so this material can be operated at high temperature and pressure.

Table 1. Inconel 600 corrosion rate in conditioned Cobalt Nitrate at temperature 250 °C for 250 hours at several concentrations of Boric acid and temperature

Material	Temperature (°C)	Boric Acid Concentration (ppm)/Corrosion Rate (mpy)			
		1500	2000	2500	3000
Inconnel 600	30	4.02 x 10^{-6}	5.12 x 10^{-6}	2.16 x 10-^{5}	2.6820 x 10^{-5}
	40	3.11 x10^{-5}	3.53 x 10^{-5}	5.55 x 10^{-5}	1.1961 x 10^{-4}
	60	2.69 x 10^{-4}	3.48 x 10^{-4}	2.18 x 10^{-3}	4.19 x 10^{-2}
	70	4.43 x 10^{-1}	4.47 x 10^{-1}	4.53 x 10^{-1}	4.56 x 10^{-1}
	80	4.57 x 10^{-1}	4.60 x 10^{-1}	4.61 x 10^{-1}	4.63 x 10^{-1}
	90	4.74 x 10^{-1}	4.91 x 10^{-1}	5.12 x 10^{-1}	5.21 x 10^{-1}

Table 2. Inconel 600 corrosion rate in Cobalt Nitrate and Zinc oxide conditioned at temperature 250 °C for 250 hours at several concentration of Boric acid and temperature

Material	Temperature (°C)	Boric Acid Concentration (ppm)/Corrosion rate (mpy)			
		1500	2000	2500	3000
Inconnel 600	30	3.81×10^{-7}	7.21×10^{-7}	7.46×10^{-7}	1.412×10^{-5}
	40	2.69×10^{-5}	3.110×10^{-5}	3.34×10^{-5}	3.3405×10^{-5}
	60	2.12×10^{-4}	3.48×10^{-4}	4.17×10^{-4}	4.19×10^{-2}
	70	3.91×10^{-1}	4.19×10^{-1}	4.42×10^{-1}	4.47×10^{-1}
	80	4.53×10^{-1}	4.54×10^{-1}	4.54×10^{-1}	4.56×10^{-1}
	90	4.59×10^{-1}	4.61×10^{-1}	4.63×10^{-1}	4.631×10^{-1}

4.2 Effect of Boric Acid Addition to Inconel 600 TT Corrosion Rate

From the experiment results can be seen that the Inconel 600 TT corrosion rate increases with increasing boric acid concentration. Inconel 600 corrosion rate at 30 °C in 1500 ppm containing boric acid solution is 4.02×10^{-6} mpy, in the 2000 ppm boric acid solution concentration at the same temperature is 5.12×10^{-6} mpy, in 2500 ppm boric acid concentration, corrosion rate is 2.16×10^{-5} mpy and increase become 2.6820×10^{-5} mpy in 3000 ppm boric acid at 30 °C. Increasing the boric acid concentration also increases the Inconel 600 TT corrosion rate. Increasing the corrosion rate of Inconel 600 with boric acid addition can be explained that increasing the boric acid in will make the solution more acidic. Increasing the solution acidic will make the more aggresive its solution toward corrosion. This would explain the temperature dependence of the corrosion rate shown in Figure 3 as well as its dependence on boric acid concentration. For a thermally activated process, corrosion rates are expected to increase with increasing temperature.

4.3 Cobalt Nitrate and Zinc oxide Addition Effect on Inconel 600 Corrosion rate

Figure 3 and Figure 4 show the measured corrosion rates for Inconel 600 TT in several boric acid solutions after conditioned in Cobalt Nitrate and Zinc oxide.

In this step can be compared the Inconel 600 corrosion rate in two conditions;

1. In several boric acid concentrations after the specimen was conditioned in Autoclave in cobalt nitrate10 ppb at temperatur 250 °C for 250 hours (*A condition*).
2. In various boric acid concentration after the specimen was conditioned in Autoclave in cobalt nitrate 10 ppb at temperatur 250 °C for 250 hours and addition of 25 ppb zinc oxide at the same temperature for 250 hours (*B condition*).

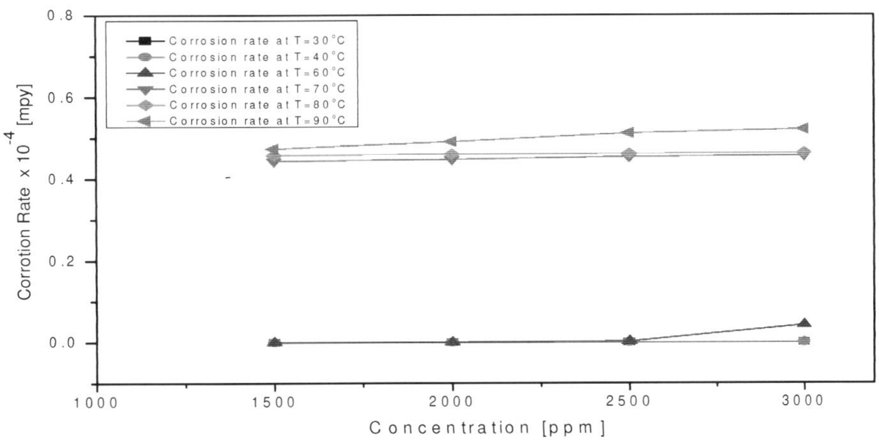

Figure 3. Inconel 600 corrosion rate in conditioned Cobalt Nitrate at temperature 250 °C for 250 hours at various concentrations of Boric acid

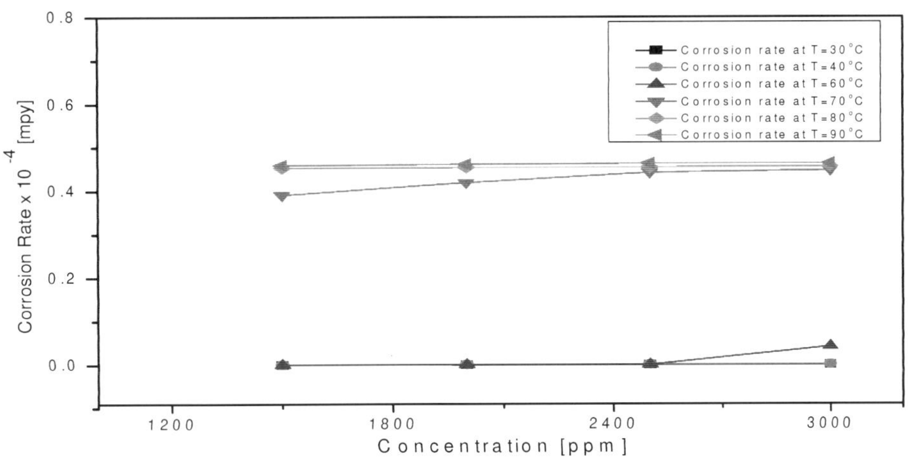

Figure 4. Inconel 600 corrosion rate in Cobalt Nitrate and Zinc oxide conditioned at temperatur 250 °C for 250 hours at various boric acid concentrations

4.4 Zinc oxide Addition Effect on Inconel 600 material Corrosion rate

At temperature 30 °C, Inconel corrosion rate after conditioned with cobalt nitrate 10 ppb addition at temperature 250 °C for 250 hours in 1500 ppm boric acid is 4.02 x 10^{-6} mpy. The Inconel corrosion rate after addition 25 ppb zinc oxide at the same temperature for 250 hours to the above solution is 3.81 x 10^{-7} mpy. Also at temperature 30 °C in 2000, 2500 and 3000 ppm boric acid, the Inconel corrosion rate which conditioned with cobalt nitrate 10 ppb addition(*A condition*) are bigger comparing with the Inconel corrosion rate with zinc oxide addition (*B condition*).

At temperature 40, 60, 70, 80 and 90 °C in the same boric acid concentrations, the Inconel corrosion rate which conditioned with cobalt nitrate 10 ppb addition (*A condition*) are bigger comparing with the Inconel 600 corrosion rate with zinc oxide addition (*B condition*).

A smaller of Inconel 600 TT corrosion rate with zinc oxide addition comparing cobalt nitrate addition can be explained by D. Lister using *Site Preference Energy* view. Ion Zinc (Zn^{2+}) has an ability to fill the same place in oxide layer faster than other oxide, such as Co^{2+}. This behaviour can reduce the other oxide metal with two valence concentration in corrosion layer. Zinc oxide layer is more protective that has tend to to form passive layer and can reduce the next corrosion proccess [13].

From the view of site preference energy concept be explanned that zinc has high bond enegy at tetrahedral interstitial position in crystal lattice comparing with Co^{2+} and other two valence metal ions. Cause of high the value of site preference energy (Table 1), Zin ion can subtitute for other two valence metal ions positions in corrosion film [13]. From the result of experiment found that Inconel 600 TT corrosion rate is smaller with zinc oxide addition comparing with cobalt nitrate addition in the same boric concentration. It can be explained by using *site preference energy.* Zinc has an ability to fill the same place in oxide layer faster than other oxide because zinc has high bond enegy at tetrahedral interstitial position in lattice oxide

Table 3. Value of "Site Preference Energy" from several cations [14]

Ion	Site Preference Energy (kJ/mol)
Fe^{2+}	- 3.35
Ni^{2+}	- 51.04
Co^{2+}	14.64
Zn^{2+}	33.89
Fe 3+	20.51

5. Conclusion

From the result of experiment can be seen that Inconel 600 TT corrosion rate is a relative small. Increasing the temperature and boric acid concentration also increase the Inconel 600 TT corrosion rate. From the experiment data found that Inconel 600 TT corrosion rate is smaller with zinc oxide addition comparing with cobalt nitrate addition in the same boric concentration. It can be explained by using *site preference energy.* Zinc has an ability to fill the same place in oxide layer faster than other oxide because zinc has high bond enegy at tetrahedral interstitial position in oxide lattices. The alloy 600 TT corrosion rate decreases with the addition of zinc oxide in high temperature water. There are some suggested explanations for the improvement of SCC resistance by several researchers. Most arguments are based on the modifications of oxide structure with the impregnation of Zn in the oxide.

The results from the present study, the highest corrosion rate occur in boric acid 3000 ppm solution at temperatur 90 ^{o}C, around 6.31 x 10^{-1} mpy with cobalt nitrate addition.

It can be explained that the decrease of Inconel 600 corrosion rate with the addition of Zn is related to the change of properties of the oxide structure in passive film. By using the Site Preference Energy view, it is possible explanation that the improvement of Inconel 600 corrosion properties with the addition of Zn is due to the substitute the other two valence ion positions in passive film. Kawamura indicates that Zn addition stabilizes the ZnCr2O4 oxide spinel and increases the corrosion resistance of Alloy 600.

Acknowledgment
This work has been carried out under Center for Nuclear Reactor Technology and Safety program, National Nuclear Energy Agency of Indonesia (BATAN).

References

[1] Millett, P.J. and C.J. Wood 1998 Recent advances in water chemistry control at US PWRs in *1998 JAIF international conference on water chemistry in nuclear power plants, proceedings.*

[2] Ford, F. 2006 Environmentally-Assisted Degradation of Structural Materials in *Water Cooled Nuclear Reactors,* vol 11, pp 213-218

[3] Fruzetti, K. and C. Wood 2007 Developments in nuclear power plant water chemistry *Power Plant Chemistry*, vol 9(4): p. 209-215.

[4] Wilson, P.L.A. 2006 Use of primary Water Chemistry in PWRsTo Mitigate PWSCC of Ni-Base Alloys. in *International Conference on water Chemistry of Nuclear Reactor system.* Jeju, Korea, pp 134-140.

[5] Riess, R., et al., 2006 LCC-2 annual report. *Advanced Nuclear Technology International*, pp 25-29

[6] J.-H. Park, O.K.C., K. Natesan, and W. J. Shack 2006. Boric Acid Corrosion of Light Water Reactor Pressure Vessel Materials. in *International Conference on Water Chemistry of Nuclear Reactor system.* Jeju, Korea.

[7] Grimmel, B. and W. Cullen 2005 US Plant Experience with Alloy 600 Cracking and Boric Acid Corrosion of Light Water Reactor Pressure Vessel Materials, *US Nuclear Regulatory Commission*, Office of Nuclear Regulatory Research, Division of Engineering Technology.

[8] Bennetch, J., et al. 2002 Root Cause Evaluation and Repair of Alloy 82/182 J-Groove Weld Cracking of Reactor Vessel Head Penetrations at North Anna Unit 2. in *ASME 2002 Pressure Vessels and Piping Conference* American Society of Mechanical Engineers.

[9] Wolter, D. 2006 Ten Years Zinc Injection for Radiation reduction at The Biblis Power station in *International Conference On Water chemistry of Nuclear Reactor system.* Jeju, Korea.

[10] Series, I.N.E., 2011 Stress Corrosion Cracking in Light Water Reactors: Good Practices and Lessons Learned. *No. NPT-3.13.* IAEA, Vienna, Austria

[11] Park, J.-H., et al. 2005 Boric acid corrosion of light water reactor pressure vessel head materials, Argonne National Laboratory (ANL), USA.

[12] Commission, N.R. 2004 Final Safety Evaluation Report Related to Certification of the AP1000 Standard Design, *NUREG-1793, Initial Report*, USA

[13] Lister, D. 1992 Activity transport and corrosion processes in PWRs, *Water chemistry of nuclear reactor systems,*. Vol 2: p. 49-49.

[14] Navrotsky, A. and O. Kleppa 1967 The thermodynamics of cation distributions in simple spinels, *Journal of Inorganic and Nuclear Chemistry,*. Vol 29(11): p. 2701-2714.

Neutron Radiation Damage Estimation in the Core Structure Base Metal of RSG GAS

S A Santa and Suwoto

Center for Nuclear Reactor Safety and Technology , Puspiptek Complex, Building No. 80, Serpong, Tangerang Selatan, 15310, Indonesia

Email: santas@batan.go.id

Abstract. Radiation damage in core structure of the Indonesian RGS GAS multi purpose reactor resulting from the reaction of fast and thermal neutrons with core material structure was investigated for the first time after almost 30 years in operation. The aim is to analyze the degradation level of the critical components of the RSG GAS reactor so that the remaining life of its component can be estimated. Evaluation results of critical components remaining life will be used as data ccompleteness for submission of reactor operating permit extension. Material damage analysis due to neutron radiation is performed for the core structure components made of $AlMg_3$ material and bolts reinforcement of core structure made of SUS304. Material damage evaluation was done on Al and Fe as base metal of $AlMg_3$ and SUS304, respectively. Neutron fluences are evaluated based on the assumption that neutron flux calculations of U_3Si_8-Al equilibrium core which is operated on power rated of 15 MW. Calculation result using SRAC2006 code of CITATION module shows the maximum total neutron flux and flux >0.1 MeV are 2.537E+14 n/cm^2/s and 3.376E+13 n/cm^2/s, respectively. It was located at CIP core center close to the fuel element. After operating up to the end of #89 core formation, the total neutron fluence and fluence >0.1 MeV were achieved 9.063E+22 and 1.269E+22 n/cm^2, respectively. Those are related to material damage of Al and Fe as much as 17.91 and 10.06 dpa, respectively. Referring to the life time of Al-1100 material irradiated in the neutron field with thermal flux/total flux=1.7 which capable of accepting material damage up to 250 dpa, it was concluded that RSG GAS reactor core structure underwent 7.16% of its operating life span. It means that core structure of RSG GAS reactor is still capable to receive the total neutron fluence of 9.637E+22 n/cm^2 or fluence >0.1 MeV of 5.672E+22 n/cm^2.

Keywords: RSG GAS, critical component, irradiation embrittlement, material damage, ageing component.

1. Introduction

G.A. Siwabessy multipurpose reactor (RSG GAS reactor; previous name MPR-30) is the 3rd owned and operated research reactor by National Nuclear Energy Agency of Indonesia (BATAN) after two research reactor namely TRIGA Mark II type of 2 MW power in Bandung and Kartini Research Reactor of 250 kW in Yogyakarta. RSG GAS reactor was built from 1983 and reached its first criticality on July 29th 1987, then formally inaugurated by the President of the Republic of Indonesia on August 20th, 1987. The power ascension test to achieve nominal power of 30 MWth was reached after completion of 5th core transitions commissioning operation. This full core power configuration was achieved in March 1992. After that achievement from 7th to 30th core configuration the reactor is operated at varying rated power of 10~25 MW according to the needs. Furthermore for efficiency reasons, until now the reactor is operated at haft nominal power of 15 MWth [1-4].

In the RSG GAS reactor, various irradiation tests have been carried out to develop the fuels, materials and also to produce radioisotopes for medical and industrial purposes and other experiments. Therefore a continued and scheduled reactor operation, neutron fluence and spectrum are key parameters in availability aspect and the post irradiation test analysis [5-7]. The readiness to operate RSG GAS reactor with specified accuracy's for neutron fluence in and around core are set of each type of irradiation test as specified according its requirement. In order to meet the needs of continuity and availability of reactor operation, in RSG GAS reactor has implemented the obligations of ageing management program [8-10]. In this ageing management program document, it was mentioned that RSG GAS reactor has selected several critical components to support safety systems among others are core structures and reactor tanks. Research on material damage of reactor critical component due to neutron irradiation stressor has never been done in RSG GAS reactor. The core structure and reactor tank are made of $AlMg_3$ material [11,12].

Aluminum alloys are a common material to be used for core structures and reactors tanks in research reactors due to its high tolerance to radiation damage, short activated total half-life, and relatively low cost. Aluminum is also a good conductor of thermal energy, making it an attractive candidate for reactor core use [13]. Researches of such material damage by highly neutron irradiated under low temperature are of paramount importance due to nuclear research reactor safety concerns [14]. Research on the $AlMg_3$ materials damage due to neutron irradiation is still very rare and is still very little found in scientific publication reports [15]. However, research on the effects of neutron irradiation on core structural materials and pressure vessels on nuclear power reactors has been widely conducted. It has been many experimental data collected related to changes in the mechanical properties of the material at various neutron fluence and temperatures [16,17].

The need for research on the effect of neutron irradiation on various advanced materials continue to grow and become a major concern for the material research society. An accurate experimental techniques for determining displacement per atom (dpa) materials irradiated in the ex-vessel of belt line area have been widely developed in either experimental reactor [18] or power reactors [19]. A simple method to calculate the dpa of SiC an advanced material used in HTGR has also been developed by collaped material cross section into 47 energy group structures and published in a table form as a proposed methodology that can be developed easily for other materials [20]. The correlation of microstructure change and hardening process of Fe and Fe-Cr materials for Gen IV fissions reactor material was investigated by neutrons irradiated as much as 0.01, 01 and 1 dpa at temperatures of 300° C [21]. The effect of neutron irradiation on single cristal pure tungsten was investigated by irradiating it in HIFAR and JMTR reactor. The results show that the hardness and microstructure change a clear dependence on the neutron energy spectrum. The hardness appeared to increase with increasing thermal neutron when fast fluence exceeds 1 E+21 n/cm^2 (E>0.1 MeV) [22]. Other studies have reported the results of Ti_2AlC, combined attractive properties of both ceramics and metals for Gen. IV reactor material due to its resistance to irradiation reason [23].

IAEA through CRP activities seek to establish a database material property for irradiated core structure components for the purpose of continued safe operation and lifetime extension of ageing research reactor [24]. These 4-year term activities have been approved since 2012, and it is participated by 13 member states and is still ongoing until now. There has been quite a lot of important data from each member states has been collected. An intensive literature review has been conducted in [15] to collect research results of various irradiation damage mechanisms of 5xxx and 6xxx series aluminum alloy by neutron irradiation. The displacement damage, transmutation damage, evolution of microstructure and mechanical properties of materials are discussed in detail. An irradiated and unirradiated $AlMg_3$ specimen of BER II research reactor operated in Berlin Germany, in order to assess the allowable operation time of a conic neutron beam tube [25]. Using $AlMg_3$ of archived

IOP Conf. Series: Journal of Physics: Conf. Series **962** (2018) 012050 doi:10.1088/1742-6596/962/1/012050

material and from a highly irradiated decommissioned component of the same material, the tensile specimens have been manufactured and then tested. The experiment result shows that below the designed end of life fluence of 1.0 E+22 n/cm^2 for E>0.1 MeV a uniform elongation of at least 5% was confirmed which is the minimum to ensure sufficient ductility during operation.

The purpose of this paper is to estimate radiation damage of core structure of RSG GAS reactor after it has been irradiated with neutron from the beginning first core to the end of 89th core configuration for almost 30 years RSG GAS reactor operation. Core structure is arranged by core frame support made by $AlMg_3$ materials and some part of reinforcement bolts made by SS304 material. The Al and Fe as metal base of core structure components were considered as main and critical component to support fuels element. The core structure components will experience the highest influence due to neutron irradiation along reactor operation. The displacement per atoms of these base core structure materials are analyzed and estimated as a measure of the remaining operating life of core component by comparing dpa value from equivalent materials. Result estimation will be used as one of the complementary supporting data to get the extension of RSG GAS reactor operating licence.

2. Methodology

2.1. Damage Calculation

Material damage calculation models have been widely developed by some researchers and research institutes [14, 26-27], even some have been standardized by ASME or others standard institution [28]. The radiation damage estimation in core structure under neutron irradiation required the knowledge of: (i) the neutron flux spectrum as a function of neutron energy E and irradiation time t, $\phi(E,t)$, (ii) the atomic displacement cross section for particular material σ_d which is also a funtion of the neutron energy E, $\sigma_d(E)$, and (iii) the neutron irradiation time t_i [14]. The total amount of displacement per atom is obtained by integrating over the exposure time, t_i as follows:

$$dpa = \int_0^{ti} dt \int_0^{\infty} \phi(E,t)\sigma(E)dE \quad (1)$$

Sometime for practical purposes, it is convinient to decompose the neutron flux into a factor that carries the time dependence, $\phi_{tot}(t)$ – time dependent flux intensity, and a normalized flux spectrum that, any time t, equals the unit when integrated over energy, $\Psi(E,t)$:

$$dpa = \int_0^{t_i} \emptyset_{tot}(t)dt \int_0^{\infty} \sigma_d(E)\,\Psi(E,t)dt \quad (2)$$

Equation (2), then can be simplified when consider the flux spectrum to be constant over time exposure as follows:

$$dpa = t_i \emptyset_{tot} \int_0^{\infty} \sigma_d\,(E)\,\Psi(E)dE \quad (3)$$

From equation (3), for given the main features of the operating nuclear reactor and the composition of core structure materials, t_i is known and the decomposed neutron spectrum can be calculated using appropriate neutron physics codes, or can be measure using reactor dosimetry experiment.

In general, the atomic displacement cross section is determines by weighting the result of different processes: (1) nuclear reaction the incident neutron and atomic nucleus of given material and energy transfer from neutron E to target nucleus (T); (2) energy loss acquired by nucleus by interaction of the electrons of the material system (T to T_d); (3) actual displacements per atom production based on equation (1), assuming by given E_d then T_d is known. Material damage cross section for the nuclear reaction obtain from nuclear data libraries which are conveniently available and directly useable from codes calculation such as NJOY, SAND II, etc.[29-33].

2.2. Evaluation Method

The research methodology used to estimate the radiation damage of the base metal of RSG GAS reactor core structure is shown in Figure 1.

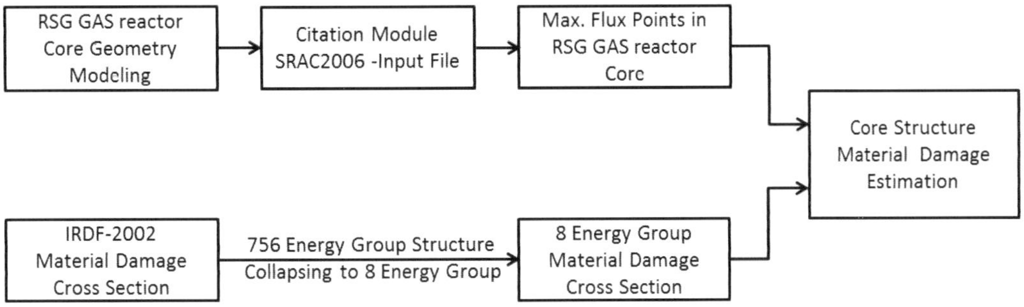

Figure 1. Flowchart research procedure of material damage estimation.

2.3. Assumption

2.3.1. Critical Component Materials

In research reactor, the main concern in ageing management programs due to neutron irradiation stressor typically are pressure tank and connecting pipes, reflector systems, core support structure components, guide tubes, beam related devices higly irradiated portion and shielding structures. Those components are generally fixed and dificult to replace and is classified as an critical component [10,24].

For reasons of RSG GAS reactor remaining life analysis, its convinient by selecting components that was most suffering neutron irradiation such as core structure and reactor tank. Core support structure components are arranged by core support grid and other structural and internal support structure. Core frame, core support grid as well as reactor tank of RSG GAS reactor are made by $AlMg_3$. Core frame and core support grid are constructed to form a core basket with reinforcement bolts made of SS 304 material [11-12]. The effect of neutron irradiation to reactor tank is not be considered since 2950 mm away length distand from core center, therefore the neutron flux intensity is significantly reduced.

In this study it is assumed that the material damage under consideration is limited only due to the neutron irradiation in the base material, therefore displacement per atom of $AlMg_3$ and SS 304 material are reviewed by material damage of Al and Fe element, respectivelly. Life time of Al1100 material of HIFAR reactor tank will be used as reference of life time $AlMg_3$ of RSG GAS reactor core structure [34-35]. Table 1 shows the elemental composition of Al1100, $AlMg_3$ and SUS304, respectively. The content of Al base metal on Al1100 and $AlMg_3$ is so similar that it expected its physical parameters did not deviate too much.

Table 1. Chemical Composition of Core Structure Material

Material	Element	% weight											
		Cu	Mn	Zn	Mg	Fe	Si	Ti	Cr	Ni	Mn + Cr	Other	Base Metal
Al1100[1]	min	0.05	-	-	-	-	-	-	-	-	-	-	99.00
	max	0.20	0.05	0.01	-	0.95	0.95	-	-	-	-	0.05	
$AlMg_3$[2]	min	-	-	-	2.60	-	-	-	-	-	0.10	-	93.80
	max	0.10	0.50	0.20	3.60	0.4	0.40	0.1	0.30	-	0.60	-	
SS304[3]	min	-	-	-	-	-	-	-	17.5	8.0	-	-	66.00
	max	-	2.00	-	-	-	1.00	-	20.0	11.0	-	-	

Note: 1. Al1100 APA Product Data Sheet UNS A91100
2. $AlMg_3$ DIN EN 573-3 Germany Standard
3. Stainless Steel - Grade 304 (UNS S30400)

2.3.2. Neutron Flux Spectrum Calculation
Neutron flux of equilibrium of RSG GAS reactor with U_3Si_2-Al of 2.96 g U/cc density is calculated using Citation module of SRAC2006 code. This program is used to calculate neutron flux at any given points inside and around the reactor core, based on neutron diffusion theory [36]. RSG GAS reactor core is modeled in three dimensional geometry (X,Y,Z) and divided into 69 mesh, 94 mesh and 72 plane of each X, Y and Z axes direction, respectively [37]. The highest neutron flux intensity is located at core center (CIP) near to the fuel element. Calculated neutron flux intensity in this point is used a bases to calculate the material damage as a function of damage cross section and neutron fluence. This assumption means that the fluence parameters such as neutron flux as a function of energy, irradiation time, as well as applied energy groups structure of neutron flux are used as a base case from which materials damage are estimated.

2.3.3. Normalized Irradiation Time
Since RSG GAS reactor reaching the first criticality in 1st core configuration to test power ascention to 30 MWth nominal power till in 6th core configuration, the reactor is operated in variety rated power from a 10.7~30 MWth. After that, from 7th to 30th core configuration, the reactor is operated varying from 10~25 MW rated power. For efficiency reason, then RSG GAS reactor is operating at half of nominal power 30 MW [1,3,6]. Furthermore, the core configuration 31th-89th is operated at a constant rated power of 15 MW. The neutron flux intensity calculations by SRAC2006 code of CITATION module were conducted only at 15 MWth of rated power [37], therefore it is necessary to normalize all core configuration to the rated power of 15 MWth. Total irradiation time is determined as a summation of each number core configuration radiation time after normalizing to 15 MWth rated power.

2.3.4. Material Damage Cross Section
Material damage cross section data was provided from Public Library Document [29-33]. In this research used material damage cross section from IRDF2002. Generally the energy group structure used extends from 0~20 MeV. In orde to calculate the magnitude of displacement per atom (dpa) of material damage it is necessary to use the same energy group structure as used in the neutron flux calculation on the equilibrium core of RSG GAS reactor fueled U_3Si_8-Al. With these assumptions the damage cross section needs to be collapsed in an energy group structure corresponding to the energy group structure to calculate the neutron flux spectrum.

3. Result and Discussion

3.1. Flux Calculation
In core neutron flux calculation were conducted using Citation module of SRAC2006 code applying the energy group structure of the Public Libraries consist of 107 energy groups structure; 74 groups for fast and 48 groups for thermal neutron groups energy, with 12 overlapping groups, covered from 1.00 MeV to 1.0 E-05 MeV. These 107 energy groups structure were then condensed from 74 groups of fast and 48 groups of thermal energy neutron into 5 and 3 energy groups structure, respectively [37].

The RSG GAS reactor core area with a 1.79E+02 cm width; 1.79E+02 cm depth; and 1.79E+02 height is modeled 3D dimensional slab geometry (X,Y,Z). The equilibrium core consist of 40 fuels element (FE), 8 control elements (CE), 4 irradiation position (IP), 4 central irradiation positions (CIP), 4 irradiation positions (IP), 1 pneumatic rabbit system (PNRS), 4 hydraulic rabbit rystem (HYDR), beryllium reflector (BE), beryllium reflector with stopper (BS), and power ramp test facility PRTF-J7 (1) and power ramp test facility PRTF-K7 (2), as seen in Figure 2. The standard fuel element consists of 21 plates of U_3Si_2/Al fuel with U-235 enrichment of 19.75 wt% and density of uranium of 2.96 gU/cc. The nominal uranium weight of each fuel is about 250 g. The control fuel elements CE are basically the same shape with a fuel elements (FE), but 3 outer plates replace by placing absorber blade, so the element has 15 fuel plates. Control elements are made of absorber material that is composed of Ag-In-Cd with composition of 80%, 15%, 5%, respectively [3].

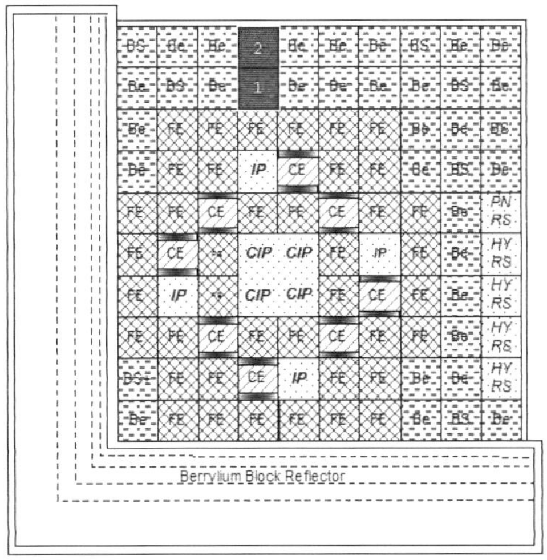

Figure 2. RSG GAS Equilibrium Core with U_3Si_2-Al Fuel [37]

For the purpose of reviewing the critical components ageing, it is preferred to take the maximum value of ageing stressors. The maximum flux neutron spectrum as the most influence material damage effect was taken in the center of core position closest to the fuel, since highest fast neutron flux is located in closer position to the fuel element position. The calculation results show that the maximum thermal neutron flux is located at mid core level at the slab position Z=36 plane, while the maximum fast neutron flux is obtained at the point close to fuel element with the coordinates X=25 mesh and Y=26 mesh, respectively. The calculated neutron flux of each energy group is shown in Table 2. The highest neutron flux is in the thermal neutron energy area at the order of E14 n/cm^2/s and decreases gradually in the epithermal neutron and the fast neutron areas.

Table 2. Calculated Neutron Flux of Each Energy Group at Core Center

Group #	Energy Range (eV) Upper Lower	Neutron Flux (n/cm^2/s)	Energy Group Catagory
1	2.00 E+07 ~ 3.68 E+06	4.535E+12	Fast Neutron
2	3.68 E+06 ~ 8.21 E+05	1.652E+13	
3	8.21 E+05 ~ 6.74 E+04	1.255E+13	
4	6.74 E+04 ~ 1.01 E+02	1.255E+13	Epithermal Neutron
5	1.01 E+02 ~ 6.83 E-01	1.279E+13	
6	6.83 E-01 ~ 1.52 E-01	9.432E+12	Thermal Neutron
7	1.52 E-01 ~ 1.85 E-02	1.578E+14	
8	1.85 E-02 ~ 1.00 E-05	2.753E+13	

3.2. Irradiation Time

At the beginning of operation, RGS GAS reactor uses U_3O_8-Al fuel with a low uranium enrichment ^{235}U of 19.75% and a meat density of uranium of 2.96 gr/cm^3, running from 1st to 34th core configuration of reactor operation. With the advancement of RERTP (Reduced Enriched Uranium for Research and Test Reactor) program, RSG GAS reactor fuel then since 1999, is converted gradually from U_3O_2-Al uranium oxide to U_3Si_2-Al uranium silicide fuel with a meat uranium density of 2.96g/cm^3. Full core conversion can be completed on 45th core configuration. The purpose of this transformation is to extend the reactor cycle operation efficiently with a higher level of burnup. Core conversion to full silicide fuel core was completed in August 2002. Subsequently, on the basis of efficiency reason, the reactor is operated using this silicide fuels.

Since RSG GAS reactor reaching the first criticality in 1st core configuration to test power ascention to 30 MWth nominal power till in 6th core configuration, the reactor is operated in variety rated power from a 10.7~30 MWth. After that, from 7th to 30th core configuration, the reactor is operated varying from 10~25 MW rated power. Furthermore, the core configuration 31th~89th is operated at a constant rated power of 15 MW. The neutron flux intensity calculations by SRAC2006 code of CITATION module were conducted only at 15 MWth of rated power, therefore it is necessary to normalize all core configuration to the rated power of 15 MWth. Total irradiation time is determined as a summation of each number core configuration radiation time after normalizing to 15 MWth rated power. Based on operating data from RSG GAS reactor Operating Report 1st to 89th core configuration [38], the rated power of reactor can be seen in Figure 3. After normalized to the 15 MWth rated power show that cummulative irradiation time was obtained as much as 3.76030 E+08 seconds.

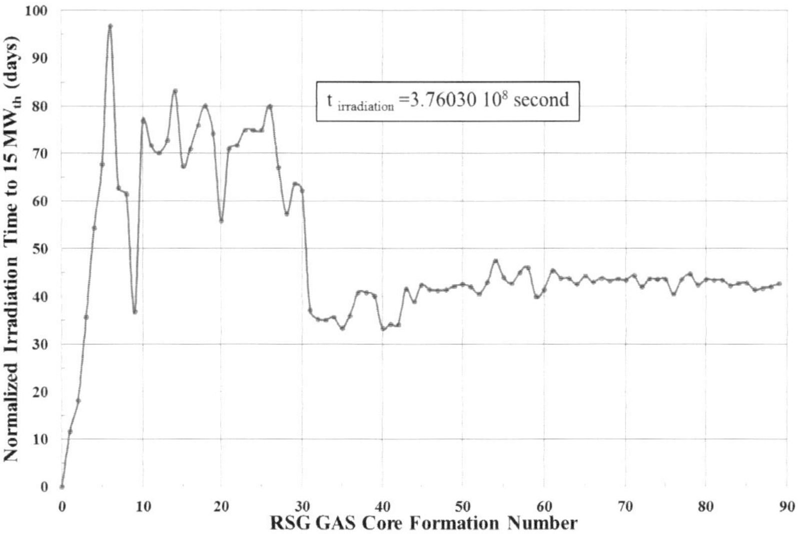

Figure 3. Cummulative Irradiation Time Normalized to the 15 MWth Rated Power

3.3. Material Damage Cross Section

Material damage cross section of base material Al and Fe was taken from IRDF-2002 and were collapsed into 756 energy groups structure as shown in Figure 4. The processed cross section of Al-NRT and Fe-NRT damage cross section data are only available for 0~10 MeV. The damage material cross section pattern between Al and Fe is very similar. Cross section of these two elements decreases to a certain value in the 1.0 E+02 eV energy area and increases continue in the epithermal energy area, then rises sharply and fluctuates in fast neutron energy region reaching to high cross section values.These cross-section data needs to be collapsed in accordance with 8 energy group structures of calculated neutron flux and the results are shown in Table 3 and also illustrated in Figure 4.

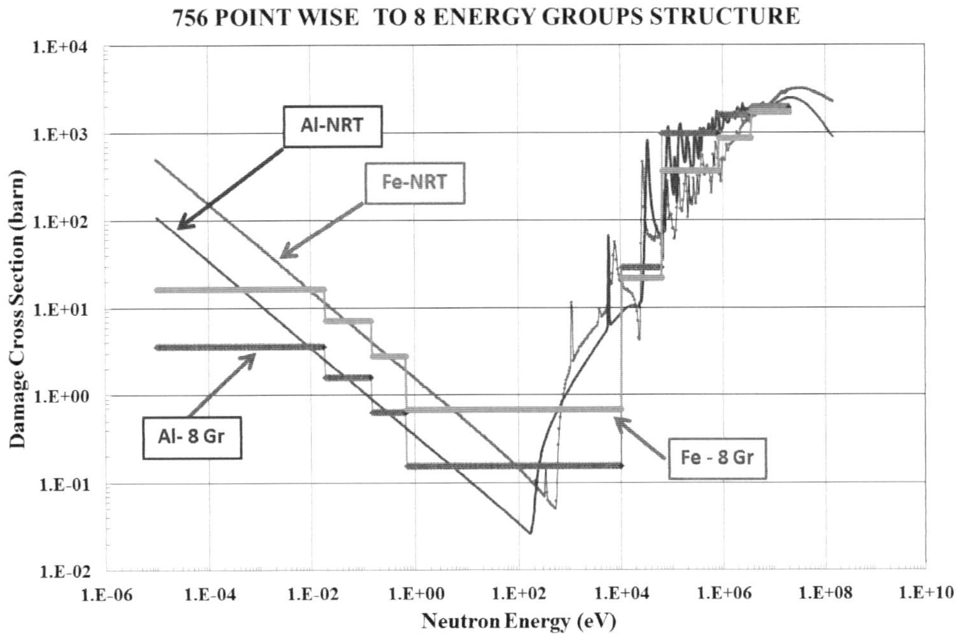

Figure 4. Material Damage Cross Section of Al and Fe in 756 Energy Group Structure and after Collapsing into 8 Neutron Energy Group Structure

Table 3. Damage Cross Section of Al and Fe in 8 Energy Group Structures

Group #	Energy Range (eV) Upper Lower	Damage Cross Section (barn) Al	Fe
1	2.00 E+07 ~ 3.68 E+06	1942.900	1646.300
2	3.68 E+06 ~ 8.21 E+05	1580.800	876.170
3	8.21 E+05 ~ 6.74 E+04	980.030	379.916
4	6.74 E+04 ~ 1.01 E+02	28.624	22.148
5	1.01 E+02 ~ 6.83 E-01	0.155	0.696
6	6.83 E-01 ~ 1.52 E-01	0.630	2.841
7	1.52 E-01 ~ 1.85 E-02	1.590	7.197
8	1.85 E-02 ~ 1.00 E-05	3.610	16.275

3.4. Radiation Damage Estimation

Radiation damage in aluminum core structure is mainly caused by the interaction and capture process. The effect depends of the composition of aluminum, manufacturing circumstances, irradiation temperature and fluence of thermal and fast neutron [13]. In the fast energy range displacement of atom occors and nuclear reaction of the type Al-27(n,α)Na-23 and Al-27(n,p)Mg-26 takes place. In the thermal neutron field, aluminum atoms are transmuted to silicon according to Al-27(n,β)Si-28. The material damage of core structure consists of dislocations, cavities, and Si precipitated that grow with increasing fluence. The reference material for specific study of $AlMg_3$ materials end of life due to neutron irradiation stressor at low temperatures are very rare and not obtainable. Therefore, as a comparison in this study, the reference material will be taken of similar investigation carried out in the reactor HIFAR as shown in Figure 5 [35]. The Al element content in $AlMg_3$ of RSG GAS core structure and Al1100 of HIFAR reactor tank is very similar. The Al1100 specimens used in HIFAR reactor tank irradiated with 1.96 E+14 $n/cm^3/s$ of neutron thermal with a considerable fast component of 1.45 E+13 for E>0.1 MeV and 6.42 E+12 for E>1 MeV has been investigated under consideration of the total power history and operating time of 2.82 E+06 MWh [34]. Si-28 is produced in consentration as much as 2.32 Wt% in the form of precipitated particle and clusters which cause an increase tensile strength and decrease ductility. As shown in Figure 6, the critical value for the elongation representing the end of life value is reach at total fluence of 1.5 E+27 n/m^3 or E>0.1 MeV neutron fluence of 4.8 E+26 n/m^2 coresponding to Si content of 5 Wt% and dpa of 250 cm.

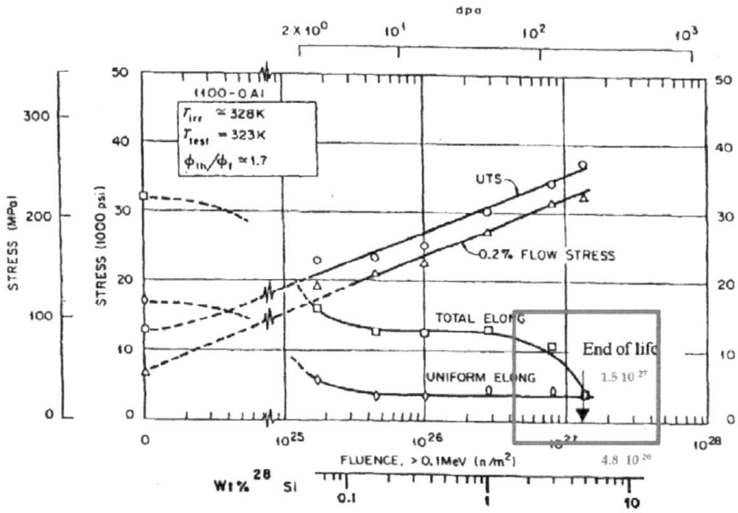

Figure 5. Fluence dependence of tensile poperties of Aluminum [35]

The calculation of radiation damage in $AlMg_3$ core structure was performed by determining which position in reactor core that obtain the highest neutron flux at an equilibrium core configuration of U_3Si_2-Al fuel. The results are shown at Table 4. The calculations of displacement per atom of Al and Fe base material core structure using equation (3) are shown also in Table 4.

Total neutron fluence at core center near the fuel is calculated as much as 1.870 E+23 n/cm^2. In orde to obtain the neutron fluence of E>0.1 MeV, log-log interpolation in (8.21E+05~6.74E+04) MeV energy range was carried out. The result of neutron flux calculation of (8.21E+05~1.00E+05) MeV energy range was 3.760E+08 $n/cm^2/s$, therefore the neutron fluence of E>0.1 MeV as much as 6.941 E+22 n/cm^2. The total dpa damage material for Al and Fe due to total neutron flux were 22.81 and 12.50 cm, respectively. Meanwhile the dpa damage material due to fluence of neutron energy E>0.1 MeV for Al and Fe was calculated at 17.91 and 10.06 cm, respectively.

Based on the research results conducted in HIFAR reactor on Al1100 material, it can be estimated that RSG GAS reactor core has undergone a material operating end of life of 7.16%, or has the equivalent of receiving neutron fluence of 1,269 E+22 n/cm^3. This means that the Al base metal core structure component is still safe to be used and operated in the long run or could be continued for additional E>0.1 MeV neutron fluence of 5.672 E+22 n/cm^2. For more accurate analysis it is necessary to include also the parameter of the amount of material damage in the 10~20 MeV energy range which in this study has not been considered

Table 4. Calculated dpa of Al and Fe in 8 Energy Goups Structure of RSG GAS

Energy Range	Damage Cross Section (Barn)		Neutron Fluence	dpa (cm)	
(eV)	Al	Fe	(n/cm^2)	Al	Fe
2.00E+07 ~ 3.68E+06	1942.900	1646.300	1.705E+21	3.313E+00	2.807E+00
3.68E+06 ~ 8.21E+05	1580.800	876.170	6.212E+21	9.820E+00	5.443E+00
8.21E+05 ~ 1.00E+05	1000.510	379.916	4.776E+21	4.778E+00	1.814E+00
8.21E+05 ~ 6.74E+04	980.030	379.916	4.719E+21	4.625E+00	1.726E+00
6.74E+04 ~ 1.01E+02	28.624	22.148	4.719E+21	4.625E+00	1.726E+00
1.01E+02 ~ 6.83E-01	0.155	0.696	4.809E+21	1.351E-01	1.045E-01
6.83E-01 ~ 1.52E-01	0.630	2.841	3.547E+21	7.455E-04	3.347E-03
1.52E-01 ~ 1.85E-02	1.590	7.197	5.934E+22	2.234E-03	1.008E-02
1.85E-02 ~ 1.00E-05	3.610	16.275	1.035E+22	9.435E-02	4.271E-01

4. Conclusions

Based on the CITATION module of SRAC2006 code modeling of equilibrium core of Al_3Si_2-Al fuels to roughly estimate the neutron spectrum at core center near the the fuels, the material damage of Al and Fe were estimated. The results show that the calculation model can be used to estimate Al and Fe damage material of base metal of core structure. Until the end of 89th core configuration, the core structure has undergone maximum of neutron irradiation as much as 1.870 E+23 n/cm^2 for total neutron fluence and 6.941 E+22 for fast neutron energy E>0.1 MeV, respectively. These means that the $AlMg_3$ of core structure components and the SS304 of bonding strenghthener core structure have undergone 22.81 and 12.50 dpa, respectively. Referring to the assumtion of Al1100 material end of life, it was found that core structure underwent 7.16% of its life span, or still safe to be used for additional E>0.1 MeV neutron fluence as much as 5.672 E+22 n/cm^2. For more accurate analysis it is necessary to collapse neutron energy into more detiled energy group structure and include variable material damage in the 10~20 MeV energy range which in this study has not been considered, although neutron flux in this area is very low.

Acknowledgments

Authors wishing to acknowledge J. Susilo for his continuing assistance on neutron flux calculation using CITATION module of SRAC2006 code, and S. Kuntjoro, R. Himawan and M. Budi Setiawan which have been discussed in depth on this research theme. Acknowledgments are also convey to PRSG colleagues who have supported RSG GAS Operation Report from 1st to 89th core configuration as well as technical drawings of internal core. In particular, we also express our gratitude to M. Subekti as responsible for the term of reference for research work and G. R. Sunaryo as responsible for the utilization of research budget. This work was performed under the auspices of the FY 2017 research budget of Center for Nuclear Reactor Technology and Safety, BATAN.

References

[1] Hong L P, Sembiring T M 2010 Design of transition core of RSG GAS (MPR-30) with higher loading siliside fuel *Nucl Eng Des **240** 1433-42*

[2] Pinem S. Liem L P, Sembiring T M, Surbakti T, 2016 Fuel element burnup measurement for equilibrium LEU siliside RSG GAS (MPR-30) core under a fuel management strategy *An Nucl En* **98** 211-17

[3] Pinem S, Susilo J., Surbakti T, Sembiring TM 2012, Optimization of radioisotope production at RSG GAS reactor using deterministic method, *J. Tek. Ind.* **Spec. Ed**. 37-45
[4] Liem P H, Sembiring T M 2012 Benchmarking the new JENDL-4.0 library o criticality experiment o the a risearch reactor with oxide LEU (20% w/o) fuel, light water moderator and beryllium reflectors , *An Nucl En* **44** 58-60
[5] Liem P H, Amini S, Hutagaol A. G., Sembiring T M 2013 Nondestructive burnup verification by gamma-ray spectrscopy of LEU siliccide fuel plates irradiated in RSG GAS multipurpose reactor, *An Nucl En* **56** 57-65.
[6] Liem P H, Sembiring T M 2013, Development of new ORIGEN2 data library sets of research reactors with ligh water cooled oxide and silicide LEU (20% w/o) fuels based on JENDL-3.3 nuclear data, *Nucl Eng Des* **262** 56-62.
[7] Ginting A B, Liem P H 2015, Absolute burnup measurement of LEU silicide fuel plate irradiated in the RSG GAS multipurpose reactor by destructive radiochemical technique, *An Nucl En* **85** 613-20.
[8] BATAN 2014, Ageing Management Program of RSG GAS Reactor, PRSG DOK No : 002.001/ RN 00 02 / RSG 3. (in Indonesian)
[9] Purba J.W. 2016, Physical Ageing of the Research Reactor Core Structural Materials Due To Neutron Irradiation Exposure: A Review, *J Nucl En Dev* **18** 2 93-100
[10] IAEA, SSG-10, Ageing Management for Research Reactors, 2010.
[11] BATAN-PRSG, Safety Analysis Report of RSG GAS, Rev 10.1, Serpong, BATAN, 2011 (In Indonesian)
[12] BATAN, G.A. Siwabessy – MPR Safety Analysis, Rev 8, Serpong, BATAN 2002.
[13] Reeves J L 2013, Effects of low fluence neutron bombardment on material properties of Aluminum 2024 T-3 and Aluminum wire, *Ph.D Thesis - The University of Utah.*
[14] OECD-NEA 2015, Primary Radiation Damage in Materials, *Nuclear Science* NEA/SNC/DOC **9**, www.oecd-nea.org.
[15] Kolluri M. 2016, Neutron Irradiation Effect in 5xxx and 6xxx Series aluminum Alloys: A Literatue Review, Chapter 15, http://dx.doi.org/10.5772/63294.
[16] Chopra O K, Rao A S 2011, A review of irradiation effects on LWR core internal materials – Neutron embrittlement, *J Nucl Mats* **412**, 195-208.
[17] Meslin E, Lambrecht M, Hernández-Mayoral M, Bergner F, Malerba L, Pareige P, Radiguet B, Barbu A, Gómez-Briceño D, Ulbricht A, Almazouz A 2010, Characterization of neutron-irradiated ferritic model alloys and a RPV steel from combined APT, SANS, TEM and PAS analyses, *J Nucl Mats* **406, 73-83.**
[18] Santa S.A. 2017, Experimental Validation of Ex-Vessel Neutron Spectrum by Means of Dosimeter Materials Activation Method**,** *Atom Indonesia* **43** 1 7-18, http://aij.batan.go.id/
[19] Lee S., Yoo C. 2011, The evaluation of Neutron Exposure Using ex-Vessel Neutron Dosimetry in Comercial Nuclear Reactor Vessel, *J of the Korean Physical Society*, **59** 2, 2067-70.
[20] Chang Y., Cho J.Y., Gii C.S, et al 2014, A Simple Method to Calculate the Displacement Damage Cross Section o Silicon Carbide, *Nucl Eng and Tech*, **46** 4, August 2014, 475-80.
[21] Chen W.Y, Miao Y., Gan J., Okuniewski M A, Maloy S A, Stubbins J F 2016, Neutron Irradiation Effects in Fe and Fe-Cr at 300°C, *Acta Materialia*, 1-10. http://dx.doi.org/10.1016/j.actamat.2016.03.060
[22] Fukuda M., Kiran Kumar N.A.P., Koyanagi T., Garrison L M, Snead L L, Katoh y, Hasegawa A 2016, Neutron Energy Spectrum Influence of Irradiation Hardening and Microstructural Development of Tungsten, *J Nucl Mats*, **479** 249-54.
[23] Liu S., Wang C., Yang T., et al 2017, High Temperature Effect off irradiation Damage of Ti_2AlC, *Nuc Inst Meth* B, http://dx.doi.org/10.1016/j.nimb.2017.01.040
[24] Di Tigliole A B, Rao D V H , Zeman A , Schaaf B V D, Barnea Y, Bradley E , Morris C , Shokr A. 2013, IAEA Coordinated Research Project on the Estabblishment of a Material Properties Database ffor Irradiated Core Structural Components ffor Continues Safe Operation and

Lifetime extention oa Ageing Research Reactor, *Joint IGORR 2013 and IAEA Technology Meeting*; Daejeon (Republic of Korea); 13-18 Oct 2013; https://inis.iaea.org/search/search.aspx?orig_q=RN:4510344

[25] Hein, H., Schnabel H., and Welzal S. 2013, Material Investigations on Highly Irradiated Aluminum Magnesium Alloys for Lifetime Assessment of a Neutron Beam Tube in the BER II Research Reactor, ASTM STP1547-EB, *On STP1547 Effect of Radiation on Nuclear Materials: 25th Volume* – Yamamoto Takuya (ed).

[26] Wootan D 2014, DPA Calculational Methodology used in Fission and Fussion Reactor Materials Application, PNNL, US DoE.

[27] Wirth B D, Monasterio P, Stein W 2008, Calculation of Radiation Damage in SLAC Targets, LLNR-SR-402698.

[28] ASTM, E693 2012, Standard Practice for Characterizing Neutron Exposures in Iron and Low Alloy Steels in Terms of Displacements Per Atom (DPA), E 706(ID), *Active Standard ASTM E693*

[29] MacFarlane, R.E. and A.C. Kahler 2010, Methods for processing ENDF/B-VII with NJOY, *Nuclear Data Sheets* 111, pp. 2739-2890.

[30] Shibata K, Iwamoto O, Nakagawa T, Iwamoto N, Ichihara A, Kunieda S, Chiba S, Furutaka K, Otuka N, Ohsawa T, Murata T, Matsunobu H, Zukeran A, Kamada S, and Katakura J 2011, JENDL-4.0: A New Library for Nuclear Science and Engineering, *J. Nucl. Sci. Technol.*, **48**, 1-30.

[31] Chadwick M B, Oblozinský P, Herman M, Greene N M, McKnight R D 2006, ENDF/B-VII: Next Generation Evaluated Nuclear Data Library for Nuclear Science and Technology, *Nucl. Data Sheets*, **102**, 2931

[32] Santamarina A, Bernard D, Rugama Y 2009, The JEFF-3.1.1 Nuclear Data Library, *JEFF Report* **22**.

[33] Capote R, Zolotarev K I, Pronyaev V G, and Trkov A 2012, Journal of ASTM International (JAI)-Volume **9**, Issue 4, April, JAI104119, in Zsolnay E M, Capote R, Nolthenius H K, and Trkov A. 2012, Technical report INDC(NDS)-0616, IAEA, Vienna.

[34] McDonald N R, Moss C J 1992, The Ageing of HIFAR Aluminum Tank : Chase Study, *IAEA Seminar for Asia and the Pasific on Ageing of Research Reactors*, Bangkok Thailand, ANSTO-Rep. NSU-TN-001.

[35] Harrison R P, McDonals N R, Moss G J 1995, Evaluation of Irradiation Damage of the HIFAR Reactor Aluminum Tank, IAEA-SR-190/35, Hamburg.

[36] Okumura K., Kugo T., Kaneko K., Tsuchihashi K 2007, SRAC2006: A Comprehensive Neutronics Calculation Code System, *JAERI-Data/Code 2007-004*, Japan Atomic Energy Agency, January 2007.

[37] Susilo J, Kuntoro A 2016, Neutron Flux Distribution and Power Generated on UO_2 Irradiated target in PRTF on RSG GAS, *Proc Sem on Nat of Nucl En Tech* 2016, Batam, 4-5 August 2016. (In Indonesian)

[38] PRSG-BATAN 1987~2015, Operation report of RSG GAS 1st-89th Core Configuration, (89 books report all in Indonesian).

Analysis respons to the implementation of nuclear installations safety culture using AHP-TOPSIS

J Situmorang, I Kuntoro, S Santoso, M Subekti, G R Sunaryo

Center for Nuclear Reactor Technology and Safety, BATAN
Puspiptek Complex, Building No.80, Serpong, Tangerang Selatan 15310 Indonesia

E-mail: situmora@batan.go.id

Abstract. An analysis of responses to the implementation of nuclear installations safety culture has been done using AHP (Analitic Hierarchy Process) - TOPSIS (Technique for Order of Preference by Similarity to Ideal Solution). Safety culture is considered as collective commitments of the decision-making level, management level, and individual level. Thus each level will provide a subjective perspective as an alternative approach to implementation. Furthermore safety culture is considered by the statement of five characteristics which in more detail form consist of 37 attributes, and therefore can be expressed as multi-attribute state. Those characteristics and or attributes will be a criterion and its value is difficult to determine. Those criteria of course, will determine and strongly influence the implementation of the corresponding safety culture. To determine the pattern and magnitude of the influence is done by using a TOPSIS that is based on decision matrix approach and is composed of alternatives and criteria. The weight of each criterion is determined by AHP technique. The data used are data collected through questionnaires at the workshop on safety and health in 2015. .Reliability test of data gives Cronbah Alpha value of 95.5% which according to the criteria is stated reliable. Validity test using bivariate correlation analysis technique between each attribute give Pearson correlation for all attribute is significant at level 0,01. Using confirmatory factor analysis gives Kaise-Meyer-Olkin of sampling Adequacy (KMO) is 0.719 and it is greater than the acceptance criterion 0.5 as well as the 0.000 significance level much smaller than 0.05 and stated that further analysis could be performed. As a result of the analysis it is found that responses from the level of decision maker (second echelon) dominate the best order preference rank to be the best solution in strengthening the nuclear installation safety culture, except for the first characteristics, safety is a clearly recognized value. The rank of preference order is obtained sequentially according to the level of policy maker, management and individual or staff.

Keywords: Safety Culture, AHP, TOPSIS

1. Introduction

A variety of ways can be undertaken to evaluate the implementation of safety culture in an organization. Considering that a safety culture is a collection of attributes that characterize the attitudes and behaviors of the worker and organization to safety implementations within the organization. Decision-making on the set of attributes as an alternative can be done by applying multi-attribute or multi-criteria decision-making techniques. The particular emphasis that is expected for the purpose of achieving organizational objectives and / or will be influenced by the organizational culture of the organization, such that there will be a difference between one organization to another, including within the organization itself as between sub organization. In more detail may be there are certain attributes appear more dominant if compared to the other attributes within the organization. The weight of a criterion predicted will also affect how a criterion will contribute to the implementation or implementation.

Taking into account the different perspectives and concerns by each level of the organization, it is necessary to know which level is expected to be an approach that has given the best response to the implementation that has been done. By these ways, the approach to safety culture can be done by using multi attributes or multi criteria consisting of several characteristics and or attributes, and then some indicators at a more detailed level. Each of these characteristics and or attributes, as well as indicators, will indicate their influence in the implementation of the safety culture corresponding individually or collectively. Similarly, therefore, the selection of characteristics and or attributes and indicators that will be used as the most effective factor subject to treatment to obtain a strong safety culture becomes important.

In this paper, the selection of characteristics and or attributes or indicators in the rank of preference order indicates the best alternative solution to obtain a strong safety culture by using TOPSIS (Technique for Order of Preference by Similarity to Ideal Solution). TOPSIS is one of the numerical methods of multi-criteria decision making composed in the form of a decision matrix of alternatives and criteria. For estimation of weight of each criterion will be used technique of AHP (Analitic Hierarchy Process).

As a result, the corresponding responses to the expected implementation of a safety culture will be adopted, based on one of the available alternatives as the best solution. The overall result of the best alternative is based on the overall characteristics. As a result for each characteristic can also be obtained and of course based on all the attributes associated with it.

2. Methodology

TOPSIS was originally developed by Hwang and Yoon in 1981 and developed further by Yoon in 1987 later by Hwang, Lai and Liu in 1993 [1]. The basic concept of TOPSIS is an alternative selection performed on alternatives with the shortest geometric distance from the ideal positive solution and the furthest geometric distance from the ideal solution of negative compensation-based aggregation that compares a set of alternatives by identifying the weight of each criterion. The assumption used in TOPSIS is that the increase or decrease in the meaning of criteria is monotonic[1-3].

The work steps used are[1, 4, 5]:

1. Establish a decision matrix consisting of m alternatives and n criteria, expressed by x_{ij}.

   Composing the decision matrix as an evaluation matrix consisting of the m alternatives and n criteria as the matrix (x_{ij}) of size m x n.

2. Normalize the decision matrix, using:

$$r_{ij} = x_{ij}\left(\sum_{i=1}^{m} x_{ij}^{2}\right)^{-1/2} \tag{1}$$

3. Calculate a weighted normalized decision matrix. Weight calculation is done by AHP calculation with the attribute weights of-j are w_j and $\sum_{j=1}^{n} w_j = 1$. The weighted normalized decision matrix is determined as follows:

$$v_{ij} = (w_j)(r_{ij}) \tag{2}$$

4. Determine the best and worst alternative. Ideal solutions A^+ and Ideal Negative Solutions A^- each corresponding to

$$A^+ = \left\{ \left(\max_i v_{ij} \middle| j \in J \right), \left(\min_i v_{ij} \middle| j \in J' \right), i = 1,2,...,m \right\} = \left\{ v_1^+, v_2^+, ..., v_n^+ \right\}$$

$$A^- = \left\{ \left(\min_i v_{ij} \middle| j \in J \right), \left(\max_i v_{ij} \middle| j \in J' \right), i = 1, 2, ..., m \right\} = \left\{ v_1^-, v_2^-, ..., v_n^- \right\} \tag{3}$$

with:

$J = \{ \text{j} = 1,2,...., \text{n} | \text{j.....criteria indexes} \}$

$J' = \{ \text{j} = 1,2,...., \text{n} | \text{j.....criteria indexes} \}$

5. Calculate the distance to the worst and best state

$$S_i^+ = \sqrt{\sum_{j=1}^{n} \left(v_{ij} - v_j^+ \right)^2}, i = 1,2,...,m$$

$$S_i^- = \sqrt{\sum_{j=1}^{n} \left(v_{ij} - v_j^- \right)^2}, i = 1,2,...,m \tag{4}$$

6. Calculate the similarity against the worst

$$C_i^* = \frac{S_i^-}{\left(S_i^+ + S_i^- \right)}, \quad 0 \langle C_i^+ \langle 1, \quad i = 1, 2, ..., m$$

$$C_i^* = 1 \quad if \quad A_i = A^+$$

$$C_i^* = 0 \quad if \quad A_i = A^- \tag{5}$$

7. Establish an alternative sequence in accordance with the similarity corresponding to C_i^*

The decision matrix in the analysis of responses to the implementation of safety culture using AHP-TOPSIS is based on the consideration that safety culture as a set of attitudes of individuals and organizations that safety is a priority beyond others and that safety should be considered as important. As an explanation it is considered that the characterization of safety culture into characteristic forms will illustrate how the safety culture can be made in the form of application. Furthermore, in more detailed form, these characteristics are described in the form of attributes, and the actual attributes can be broken down into more detailed indicators still. Thus, the safety culture can be stated as multi-attribute state.[6]

In any detailed implementation it was sometimes becomes a subjective consideration, both in terms of its importance and or its understanding. As an explanation of the collective commitment the emphasis on the determination of the significance of an indicator and or attribute is not the same for all levels within the organization. At the policy-making level the commitment to the policy statement can be one of the characteristics, while at the management level clear and clear accountability can be a feature of commitment, while at the individual level one of the characteristics of its commitment is in the form of prudence. Accordingly, the clustering of echelons as policy makers, and management and staff management echoes may generally be distinguished by their response to ongoing implementation.

In keeping with the above explanation, responses to the implementation of safety culture by individuals at all levels of the organization will not always be the same. It was not only because of the very different their competency, but it can also be due to the possibility of differences in perception of the definition of safety culture including its implementation. In accordance with the survey results that suppress the demographics of respondents based only on echelon, the criteria grouping is applied to echelon II, echelon III / IV and Staff.[7, 8]

In accordance with the above explanation, the matrix of decisions in the form of a matrix consisting of m alternatives and n attributes (or criteria) can be arranged on the basis of the characteristics / attributes of safety culture as criteria and groups of echelon or staff as alternatives, Eq. 6

$$A = \begin{array}{c|cccccc|} & x_1 & x_2 & x_3 & . & . & . & x_n \\ A_1 & x_{11} & x_{12} & x_{13} & . & . & . & x_{1n} \\ A_2 & x_{21} & x_{22} & x_{23} & . & . & . & x_{2n} \\ A_3 & x_{31} & x_{32} & x_{33} & . & . & . & x_{3n} \\ . & . & . & . & & & & . \\ . & . & . & . & & & & . \\ . & . & . & . & & & & . \\ A_4 & x_{m1} & x_{m2} & x_{m3} & . & . & . & x_{mn} \end{array} \qquad (6)$$

The decision matrix is composed of 3 alternatives and its criteria are the number of characteristics and or the number of attributes of corresponding characteristics. The criteria are subject to normalization, in order to consider unequal dimensions of the attributes under consideration. The normalization method that is used is the normalization method of the vector according to Eq. 1 to consider the non-linear spacing between the score and the ratio of a single dimension will produce results that more smoothly. In this case, linear normalization is not considered appropriate for the inconsistent dimension of the criterion.

Each criterion on the decision matrix expresses the weakness or strength of the implementation of safety culture and it is considered in monotonically way. The characteristics and or attributes will be used as criteria on the decision matrix. Strengthening or weakening that is considered coming from any characteristic and or any attribute will not necessarily be the same. Likewise the magnitude of the influence of each characteristic and / or corresponding attribute expressed as the weight of each criterion, will also be different for each characteristic and or attribute. The approximate weight of the weights is obtained by using the AHP approach which corresponds to the value of each attribute, and the comparison of each other can be regarded as pairwise. By using the AHP approach a set of weights for the criterion can be obtained.[6, 9]

The calculated weights for each criterion that is based on each characteristic and or each attribute of corresponding characteristic, the weighted normalized decision matrix can be determined, Eq. 7.

$$V = \begin{vmatrix} v_{11} & v_{12} & v_{13} & . & . & . & v_{1n} \\ v_{21} & v_{22} & v_{23} & . & . & . & v_{2n} \\ v_{31} & v_{32} & v_{33} & . & . & . & v_{3n} \\ . & . & . & & & & . \\ . & . & . & & & & . \\ . & . & . & & & & . \\ v_{m1} & v_{m2} & v_{m3} & . & . & . & v_{mn} \end{vmatrix} = \begin{vmatrix} w_1r_{11} & w_2r_{12} & w_3r_{13} & . & . & . & w_nr_{1n} & . \\ w_1r_{21} & w_2r_{22} & w_3r_{23} & . & . & . & w_nr_{2n} & . \\ w_1r_{31} & w_2r_{32} & w_3r_{33} & . & . & . & w_nr_{3n} & . \\ . & . & . & & & & . & \\ . & . & . & & & & . & \\ . & . & . & & & & . & \\ w_1r_{m1} & w_2r_{m2} & w_3r_{m3} & . & . & . & w_nr_{mn} & . \end{vmatrix} \qquad (7)$$

Using a weighted normalized decision matrix Eq. 7 positive and negative ideal solutions can be determined by calculating the relative closeness to ideal solution using Eq. 4 and Eq. 5. While the closest relative distance to the ideal solution is determined by comparison of the distance to the average negative solution with the distance to the positive and negative solutions, Eq. 6. By ranking the relative distance, it can be obtained the rank of preference order which states that the largest as the best solution approach and smallest as the worst solution approach.

3. Results and discussion

The data used are data collected through questionnaires at the annual workshop on safety and health in 2015. Data respondens are 13 Echelon II (noted as ΣEs2), 9 Echelon III / IV (noted as ΣEs34), and 40 staffs (noted as ΣStaf), and resulted in 3,8871 positive Likert scales from 1 to 5. Reliability test of data gives Cronbah Alpha value of 95.5% which according to the criteria is stated reliable. Validity test using bivariate correlation analysis technique between each attribute give Pearson correlation for all attribute is significant at level 0,01. Using confirmatory factor analysis gives Kaise-Meyer-Olkin of sampling Adequacy (KMO) that can be seen in Table 1.

Table 1. Significance test results

KMO and Bartlett's Test[a]		
Kaiser-Meyer-Olkin Measure of Sampling Adequacy.		0.719
Bartlett's Test of Sphericity	Approx. Chi-Square	1726.064
	df	666
	Sig.	0.000

a. Based on correlations

Thus, since the KMO is greater than 0.5 with a degrees of freedom of 666 and a significance level of 0.000, the obtained data can be further processed [10]. Initial information that can be obtained directly through the used Likert scale as the measurement scale states that the highest Likert Scale value between each level is not always at the same level, although overall it is dominated by the echelon 2 level. It can be interpreted that the level associated with the safety policy within collective commitment is decision making level more committed than others, Table 2. Table 2 also can be showed that the first characteristic, i.e. safety is as a clear recognized value, has the highest value that is given by Echelon III/IV or managenment level, while for almost of the characteristics staff gives the lowest value except for characteristics of accountability. The accountability that shows the ownership of safety at the staff level goes beyond the management level.

As a whole it can be stated that on the consideration of decision-level or echelon 2 provides a better response to the strengthening of safety culture followed by echelon 3 & 4 and then staff, i.e. based on the ranks of preferences order 0.8651, 0.5440, and 0.0389 respectively, Table 2. This table shows that Echelon 2 has shown a very positive response to the presence of a strong safety culture, although gave not the best response to the characteristics of safety is as a clear recognized value, instead on this characteristic can be stated ugliest response approach among the three groups of responders. For each characteristic, it can be stated that the best response approach, the majority is given by echelon two or decision makeing level, but not always.

Table 2. Ranking of preference order

Characteristics	Overall	A-Char	B-Char	C-Char	D-Char	E-Char
∑Es2	0.8651	0.4130	0.9920	0.6506	0.6651	0.7055
∑Es34	0.5440	0.6788	0.4544	0.4278	0.6180	0.6380
∑Staf	0.0389	0.4340	0.2665	0.4962	0.2641	0.0061

To further examine the effect of each characteristic can be done by considering the Likert scale. On the first charakteristic. i.e. safety is as a clear recognized value. the smallest result is not give by decision-making level but by staff level. The smallest result is on the staff level and the highest is the management level. When compared to the rank of preference order that placing responses from policy-making levels as a best alternative. it shows that expectations of safety as a higher value in policy making level while management level more focus on determining how to implementation of safety.

Table 3. Weight and Likert Scale of Safety Culture Characteristics

Weight-Char.	0.1986	0.1969	0.1987	0.2041	0.2016
∑Es2	3.8846	4.1231	3.9077	3.8120	4.0220
∑Es34	3.9444	3.9111	3.8444	3.7901	4.0000
∑Staf	3.8750	3.8575	3.8600	3.7000	3.7750
Likert Scale. Av.	3.9014	3.9639	3.8707	3.7674	3.9323

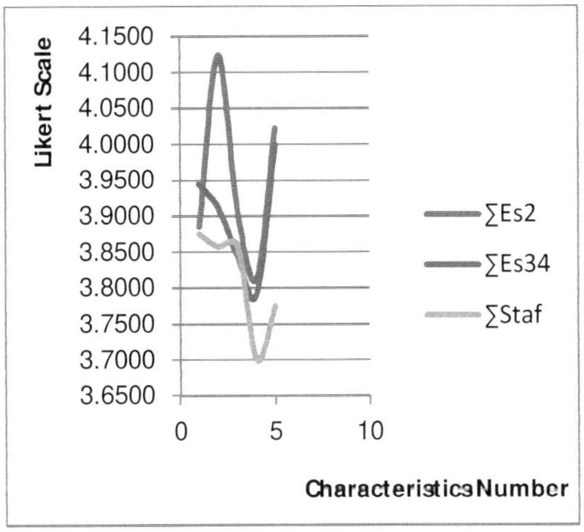

Figure 1. Likert Scale of Characteristics

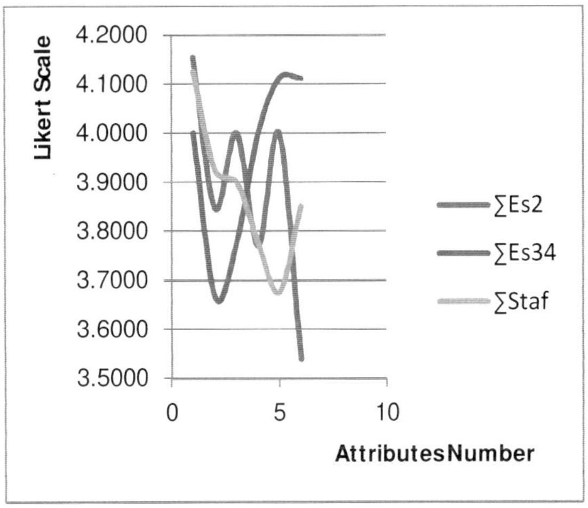

Figure 2. Likert Scale of A Characteristic

Further consideration is imposed on each characteristic. such that can be knew and or understood the shape of the influence of each attribute in each corresponding characteristic. The five characteristics that shall be considered are characteristics of safety is as a clear recognized value represented by characteristic A. the characteristic of leadership for safety is clear and is expressed by B characteristic. the characteristic of accountability for safety is clear and is expressed by the C characteristic. the characteristic of the integrity of safety into all activity is expressed by the D characteristic and the characteristic of safety is the learning driven of is expressed by the characteristics of E.

The results of the response analysis on characteristic of safety is as a clear recognized values are given in Table 4 and Figure 2. The ranks of preference order for this characteristic are for Echelon II. Echelon III/IV. and successive staff are 0.4130, 0.6788. and 0.4340. These results illustrate that the best solution lies on the management level and the worst at the decision-making level. Considering the value of the Likert scale of magnitude of 3.8308, 3.9333. and 3.8250 respectively can be said that the greatest value lies on the management level that may be illustrate to consider more detail the atribut of A proactive and long term approach to safety issues is shown in decision making.

Table 4. Weight and Likert Scale of A Characteristic

Weight-A	0.1580	0.1611	0.1694	0.1702	0.1713	0.1699
∑Es2	4.1538	3.8462	4.0000	3.7692	4.0000	3.5385
∑Es34	4.0000	3.6667	3.7778	4.0000	4.1111	4.1111
∑Staf	4.1250	3.9250	3.9000	3.7750	3.6750	3.8500
Likert Scale. Av	4.0929	3.8126	3.8926	3.8481	3.9287	3.8332

The result of the response analysis on the characteristics of safety leadership (B characteristic) is expressed in Table 5 and Figure 3. The rank of the preference order for these characteristics for Echelon II,. Echelon III/IV, and consecutive the Staff are 0.9920, 0.4544. and 0. 2665 respectively. These results illustrate that the best solution lies on the policy level and the worst at the staff level. The average value of the liker scale of magnitude is 4.1231, 3.9111 and 3.8575 respectively. indicate that the greatest value lies on the policy-making level corresponding to the rank order of preference values so that it can be said that. The attribute of Leadership skills are systematically developed and the attribute of Management has the ability to resolve conflicts as necessary or the 4th and 9th attributes; must be more considered to be developed.

Tabel 5. Weight and Likert Scale of B Characteristic

Weight-B	0.0951	0.1022	0.0994	0.0989	0.1009	0.0965	0.1014	0.0950	0.1075	0.1031
∑Es2	4.3077	3.9231	4.1538	4.0000	4.3077	4.0769	4.3077	4.3077	3.7692	4.0769
∑Es34	3.8889	3.8889	3.8889	3.8889	4.1111	4.0000	3.8889	3.8889	3.7778	3.8889
∑Staf	4.2750	3.8250	3.9000	4.0000	3.7250	4.0500	3.7000	4.0750	3.4500	3.5750
Likert Scale. Av	4.1572	3.8790	3.9809	3.9630	4.0479	4.0423	3.9655	4.0905	3.6657	3.8469

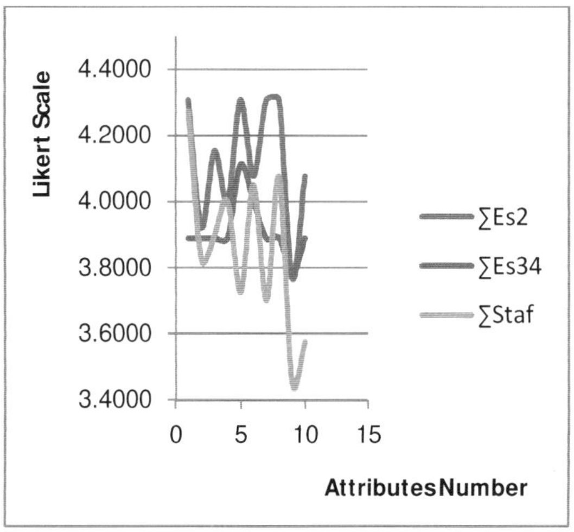

Figure 3. Likert Scale of B Characteristic

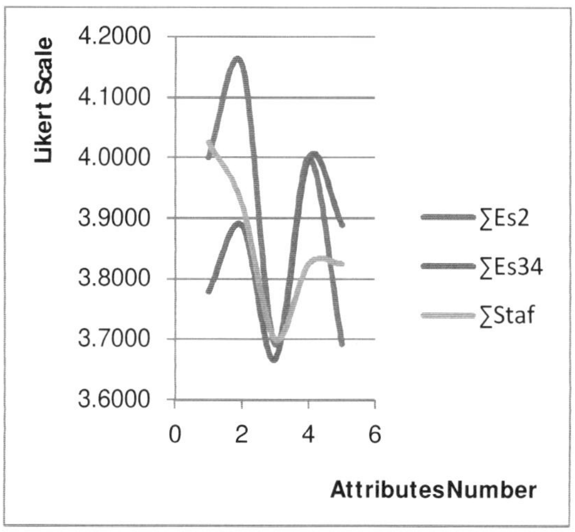

Figure 4. Likert Scale of C Characteristic

The results of the response analysis on the safety accountability characteristics are given in Table 6 and Figure 4. The rank of preference order sequences of this characteristic for echelon II, Echelon III/IV. and successive the Staff are 0.6506; 0.4278. and 0.4962. The best solution pattern lies on response of the decision-making level and the worst on the management level. The average value of the Likert scale on thois characteristic are 3.9077; 3.8444. and 3.8600 respectively. The greatest value is lies on the management level. It can illustrates the consistency of the implementation of the safety to the results that will be obtained. It can be pointed out that the rank of preference order value consistent with the Likert scale value. and then can be said that the fulfillment of requirements on the provision of safety need more attention to attributes of 1. 2 and 3.

Table 6. Weight and Likert Scale of C Characteristic

Weight-C	0.1972	0.1963	0.2100	0.1977	0.1989
∑Es2	4.0000	4.1538	3.6923	4.0000	3.6923
∑Es34	3.7778	3.8889	3.6667	4.0000	3.8889
∑Staf	4.0250	3.9250	3.7000	3.8250	3.8250
Likert Scale. Av	3.9343	3.9892	3.6863	3.9417	3.8021

Table 7. Weight and Likert Scale of D Characteristic

Weight-D	0.1142	0.1118	0.1077	0.1126	0.1104	0.1128	0.1104	0.1107	0.1094
∑Es2	3.7692	3.8462	3.8462	3.8462	3.6923	3.6923	4.0000	3.9231	3.6923
∑Es34	4.0000	3.5556	4.0000	3.6667	3.6667	3.6667	3.8889	4.0000	3.6667
∑Staf	3.5750	3.7500	3.8500	3.6750	3.8000	3.6500	3.6250	3.6000	3.7750
Likert Scale. Av	3.7814	3.7172	3.8987	3.7293	3.7197	3.6697	3.8380	3.8410	3.7113

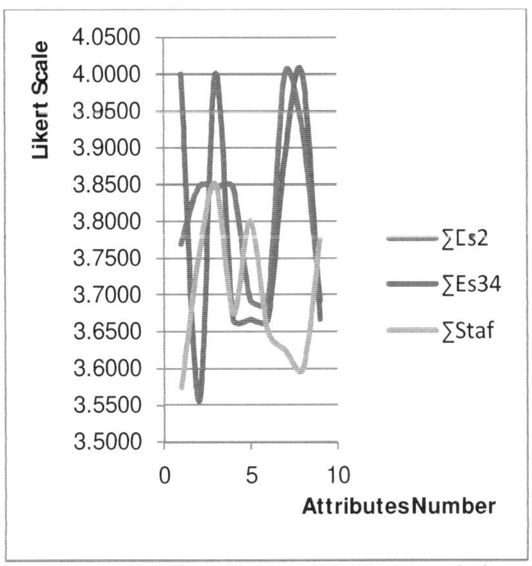

Figure 5. Likert Scale of D Characteristic

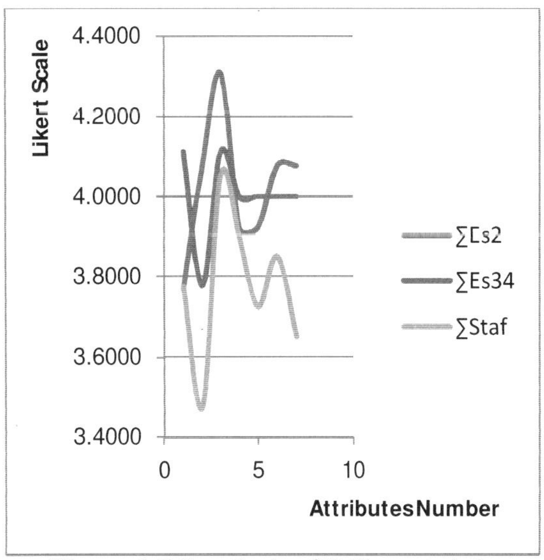

Figure 6. Likert Scale of E Characteristic

The results of the response analysis on the characteristics of safety integrity can be expressed in Table 7 and Figure 5. The rank values of preference sequences for these characteristics for Echelon III,

Echelon III/IV. and successive the staff are 0.6651, 0.6180. and 0.2641. For this characteristics the best solution lies at the policy-making level and worst at the staff level. Likert scale values of magnitude are 3.8120.; 3.7901. and 3.7000 respectively. The greatest value lies on the policy-making level and consistent with the rank of preference order. However, Figure 5 illustrates the inconsistency of safety practices specifically at the management level as the organizer in attributes of 2. 4. 5 and 6 still need to strengthen the implementation

The result of the response analysis on safety characteristics as the driving force for safety learning is given in Table 8 and Figure 6. The rank ratings of the preference sequences for these characteristics for echelon II, Echelon III/IV. and successive the Staff are 0.7055, 0.6380, and 0.0061. The best solution lies on the policy making level of and the worst lies on the staff level. The magnitude of the average value of the Likert scale are 4.0220; 4.0000, and 3.7750 respectively. It is show the consistency with the value of the rank preference order. In this case shall be paid more attention to every attribute that exists in this characteristics. and it can be said that in general the prudence and motivation of staff still need to be improved. Especiaally the emphasis on attributes 1 and 4. see Figure 6 will be usefully.

Table 8. Weight and Likert Scale of E Characteristic

Weight-E	0.1455	0.1533	0.1338	0.1399	0.1438	0.1397	0.1440
∑Es2	3.7692	4.0769	4.3077	3.9231	3.9231	4.0769	4.0769
∑Es34	4.1111	3.7778	4.1111	4.0000	4.0000	4.0000	4.0000
∑Staf	3.7750	3.4750	4.0500	3.9000	3.7250	3.8500	3.6500
Likert Scale. Av	3.8851	3.7766	4.1563	3.9410	3.8827	3.9756	3.9090

4. Conclusion

The AHP -TOPSIS method for ranking sequence preferences can be used practically in such a way that the best and worst solutions of available alternative options can be made. TOPSIS based on decision matrix that composed of alternatives and criteria on multi-criteria decision system. The approach to alternatives is based on levels in the collective commitment to declare safety culture. ie policy making, management, and staff. While the approach is intended as a criterion is the characteristics and or attributes of safety culture. The magnitude of influence in the form of weights contained in each criterion is estimated by AHP technique. Alternative can be expressed as a pattern of how the safety culture is implemented with the particular treatment imposed on every characteristic and or attribute of safety culture. Using of data which derived from questionnaires at a workshop on occupational health and safety 2015. the decision matrix was prepared with Echelon II, Echelon III/IV. and staff as alternatives. and five characteristics and or 37 safety culture attributes were used to be criteria. As a result. there is consistency of Likert scale value in questionnaire with rank of preference order value as available alternative option. Estimated sequence of ranks obtained are 0.8651, 0.5440, and 0.0389 respectively for desicion making level. management level and staff level. The best alternatives option exist at the policy-making level or responses of the Echelon II followed by the management level or responses of the Echelons III/IV and the worst is at the staff level or responses from Staff. Considering the attributes contained in the corresponding characteristics. the best alternative as a response is obtained from the policy making level except on the characteristic of safety is as a clear recognized value. Response patterns as the best successive alternatives of decision-making. management. and staff levels are not always for the overall considered characteristics. and hence to get a more detailed picture it is necessary to consider the elements of the criteria group of the characteristics and or attributes in the corresponding characteristics. Likert scale that provides information about the criteria can be use to establish the influence the corresponding characteristics and or attributes.

Acknowledgment
This research is funded by DIPA PTKRN. The authors thank the BATAN's Safety Culture Team which carried out data gathering during 2015.

References
[1] Pavić Z. and V. Novoselac 2013 *Notes on TOPSIS method.* International Journal of Research in Engineering and Science **1**(2) p. 5-12.
[2] Liu P. 2011 *An Extended TOPSIS Method for Multiple Attribute Group Decision Making Based on Generalized Interval-valued Trapezoidal Fuzzy Numbers.* Informatica (03505596).. **35**(2).
[3] Ozturk D. and F. Batuk 2011 *Technique for order preference by similarity to ideal solution (TOPSIS) for spatial decision problems.* in *Proceedings ISPRS*..
[4] Athawale V.M. and S. Chakraborty 2010*A TOPSIS method-based approach to machine tool selection.* in *International Conference on Industrial Engineering and Operations Management. Dhaka. Bangladesh*..
[5] Chi P. and P. Liu 2013*An extended TOPSIS method for the multiple attribute decision making problems based on interval neutrosophic set.* Neutrosophic Sets and Systems. **1**(1) p. 63-70.
[6] Bhutia P.W. and R. Phipon 2012 *Application of AHP and TOPSIS method for supplier selection problem.* IOSR Journal of Engineering.. **2**(10) p. 43-50.
[7] Situmorang J.2016 *Evaluation of influence factors within implementing of nuclear safety culture in embarking countries International Conference on Human and Organizational Aspects of Assuring Nuclear Safety IAEA p214-215*
[8] Situmorang J 2014 *EVALUASI BUDAYA KESELAMATAN UNTUK PRIORITISASI PENTINGNYA KARAKTERISTIK/ATRIBUT PADA INSTALASI NUKLIR DENGAN TEKNIK AHP (ANALYTIC HIERARCHY PROCESS).* Jurnal Teknologi Pengelolaan Limbah.. **16**(1).
[9] Kallas Z 2011 *Butchers' preferences for rabbit meat; AHP Pairwise comparisons versus a LIKERT scale valuation.* in *Proceedings of the 11st International Symposium on the Analytic Hierarchy Process and Analytic Network Process [ISAHP 2011]*..
[10] Ghozali Imam H. 2005 "Aplikasi Analysis Multivariate dengan program SPSS". edisi 3. Badan Penerbit Universitas Diponegoro. Januari. Semarang

Analysis on Operating Parameter Design to Steam Methane Reforming in Heat Application RDE

Sukmanto Dibyo, Geni Rina Sunaryo, Syaiful Bakhri, Zuhair, Ign.Djoko Irianto

Center for Nuclear Reactor Technology and Safety – National Nuclear Energy Agency of Indonesia (BATAN), PUPSPIPTEK Complex, Office Buliding No. 80, Serpong, Tangerang Selatan 15310, Indonesia, Telp. (021)756-0912, Fax. (021)756-0913,

sukdibyo@batan.go.id

Abstract. The high temperature reactor has been developed with various power capacities and can produce electricity and heat application. One of heat application is used for hydrogen production. Most hydrogen production occurs by steam reforming that operated at high temperature. This study aims to analyze the feasibility of heat application design of RDE reactor in the steam methane reforming for hydrogen production using the ChemCAD software. The outlet temperature of cogeneration heat exchanger is analyzed to be applied as a feed of steam reformer. Furthermore, the additional heater and calculating amount of fuel usage are described. Results show that at a low mass flow rate of feed, its can produce a temperature up to 480°C. To achieve the temperature of steam methane reforming of 850°C the additional fired heater was required. By the fired heater, an amount of fuel usage is required depending on the Reformer feed temperature produced from the heat exchanger of the cogeneration system.

Keywords: RDE, heat application, hydrogen production, steam methane reformer, additional heater.

1. Introduction

The HTGR is one of the next generation reactor types. Currently, the HTGR design is considered one of the leading reactors for the future nuclear power plant which has attractive inherent safety features [1]. Application of nuclear heat for the high temperature of HTGR have been widely studied [2]. The reactor has been developed with various power capacities and can produce electricity and heat applications [3]. The HTR-10 with 10 MWt capacity is the small size of HTGR type reactor that was successfully operated in China [4].

Currently, Indonesia will plan to build a small size HTGR called RDE (Reaktor Daya Eksperimen) that have a thermal power of 10 MWt. High coolant temperature from the RDE is very potential for cogeneration purpose [5]. Therefore, the RDE can be applied for electricity generation, heat generation and for hydrogen production. Utilization of hydrogen is one of the scenarios of Indonesian government for the renewable energy application [6]. Hydrogen is an important raw material for the chemical, the refinery industry, and it may play a future role in the energy sector. Among the existing technologies, most hydrogen production occurs by steam reforming i.e. converting a hydrocarbon-steam mixture into a mixture of mainly hydrogen and carbon dioxide [7]. The steam reforming process is operated at high temperature [8]. Although, an assessment of the steam reforming process utilizing low-temperature nuclear reactor as heat sources have been done [9]. Principally, the products in steam reforming are synthesis gas of carbon monoxide (CO), carbon dioxide (CO_2) and Hydrogen (H_2). This reaction is highly endothermic [10], therefore to achieve such chemical reaction the high temperature with an additional heat is required.

Since the reaction is highly endothermic and is performed in the presence of a catalyst such as nickel or rhodium, in order to achieve near equilibrium conversion, steam reforming in the conventional technology is conducted in a multi-tubular reformer operated at a temperature of 850°C [11, 12]. Meanwhile, the temperature at the reformer inlet is between 450°C and 600°C [13].

The steam methane reformer is one of the facilities for the nuclear process heat application system. In the cogeneration system, the steam can be used for rotate the turbine as well as for the heat application. Furthermore, heat removal on the cooling system can be applied with a cogeneration of heat exchanger (HEx), although the efficiency of direct cycle cogeneration system is higher than in-direct cycle [14].

This study would analyze the feasibility of heat application design of RDE reactor in the steam methane reforming for hydrogen production using the ChemCAD software.

2. Theory

In the RDE during normal operation, the Steam Generator (SG) at the power of 10 MWt transfers the heat of the high temperature of helium gas flowing as primary coolant system to the secondary coolant system. This secondary coolant at the pressure of 4 MPa leaving from the SG [4], the superheated steam produced has no moisture. Therefore, the steam production is higher temperature than saturation temperature. At steam pressure of 4 MPa, the saturation temperature is about of 276 °C [15].

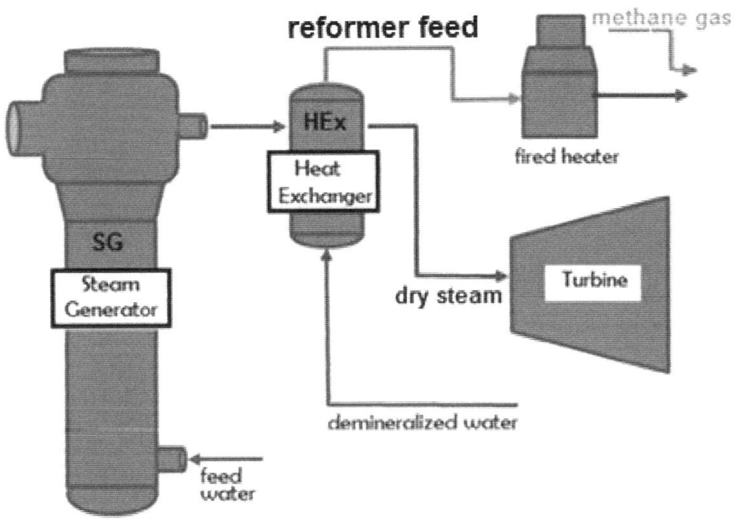

Figure 1. Diagram design of heat application in RDE

The principle in the cogeneration system is the dry steam can be used for rotate the turbine as well as for the heat application. As shown in Figure 1, the analysis on diagram design for heat application in RDE, the HEx is used for heat removal from the steam outlet line of SG. Heat transfer occurs from a hot fluid (steam) to a cooler fluid through a solid wall separating the two. The heat transfers are sensible heat, latent heat, involving a phase change of vaporization.

In accordance with the previous studies [4, 16], the steam outlet temperature produced from the SG was estimated in the range of 440°C-570°C . On the other hand, the steam comes out from the HEx into the fired heater. This heater is used to rise the temperature from the HEx. Furthermore, steam at high temperature reacts with methane gas in the steam methane reformer. Additional heat occurs in the heater so does this heater a fuel gas is required.

The analysis design parameter on heat application uses the CC-Therm heat exchanger module of ChemCAD.6.1.4 code. The ChemCAD.6.1.4 is thermal-hydraulic calculation code that has been very popular and widely used for various applications of heat transfer design, including design operations, the evaluation process equipment manufacturing, systems analysis of unit operation installation process, includes a calculation for the cooling system design. In the code, an equation of state i.e.

mathematical relationship between temperature, pressure and volume or internal energy uses the Redlich-Kwong equation as follows,

$$p = \frac{RT}{Vm-b} - \frac{a}{\sqrt{T}Vm(Vm+b)} \tag{1}$$

$$a = \frac{0.4278\, R^2.Tc^{5/2}}{Pc} \tag{2}$$

$$b = \frac{0.0867\, RTc}{pc} \tag{3}$$

where,

p : pressure (absolute), Pa
V_m : molar volume, the volume of 1 mole of gas or liquid
T: absolute temperature, K
R : ideal gas constant :8.31446 J/(mol·K)
P_c : pressure at the critical point, Pa
T_c : absolute temperature at the critical point, K

3. Methodology

The analysis step of schematic diagram is described in Figure 2. The diagram are consist of design of flow chart, modeling, input data and Output (results).

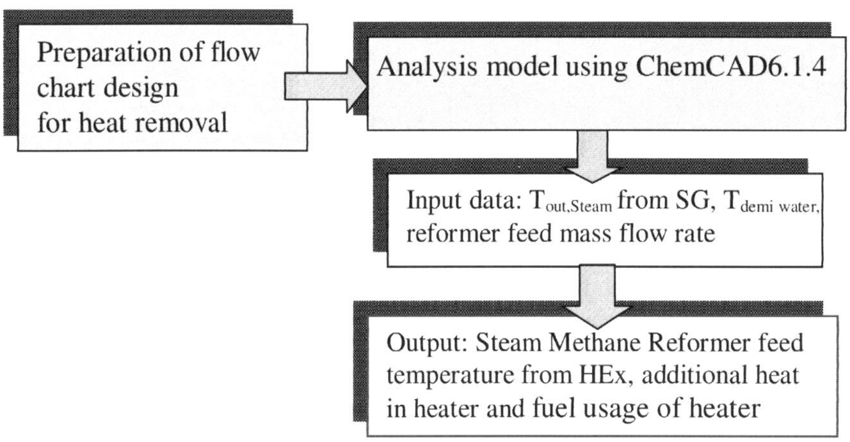

Figure 2. The analysis step of schematic diagram

Following are input data and assumptions used in the analysis,

- steam outlet temperature produced from the SG is determined of 540°C.
- demineralized feed water temperature for HEx (shell-tube type) is 26°C.
- outlet superheated steam from SG at pressure of 4.0 bar.
- input for Steam Methane Reformer feed mass flow rate in the range of 0.8 – 20 kg/s.
- the thermal analysis has been made under steady-state conditions.
- the heat losses in pipelines are negligible.
- natural gas of gross heating value is 1143 (BTU/SCF) [17].

Futhermore the preparation of model for analysis is shown in Figure 3, determination of boundary condition, assumption and input data option were carried out. The output is including the reformer feed temperature (T_{out}) from HEx, heat absorbed from the additional fired heater and fuel usage required.

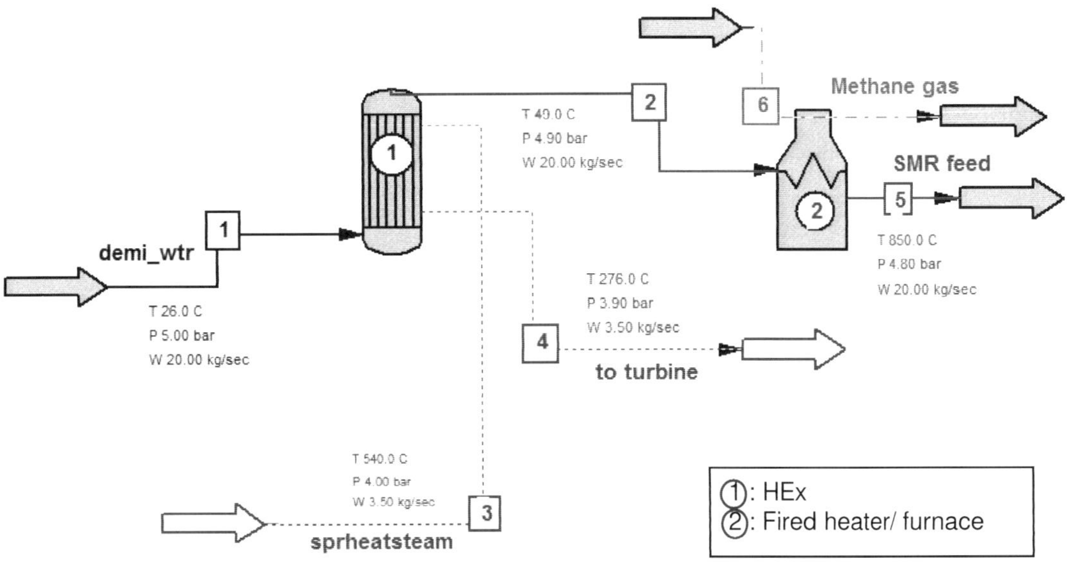

Figure 3. The model of ChemCAD6.1.4

4. Results and Discussion

The analysis of operating parameter design for heat application of Steam Methane Reformer feed has been obtained. The use of a higher temperature of steam superheated is an advantage for application of cogeneration system. Figure 4 shows the relation between mass flow rate against temperature used for feed of Steam Methane Reformer resulted from the HEx. The heat is taken from the hot temperature of outlet pipe line SG. This curve shows that with increasing of feed mass flow rate, so the lower temperature resultes in accordance with the first law of thermodynamics. Therefore, to achieve the perfect reaction between steam and methane gas, its require an additional heat using the fired heater (component no.2 as shown in figure 3). In this case, an additional heat requires an amount of fuel. Figure 4 shows also the curve of the power of the fired heater (in MJ/s) to achieve the temperature of steam methane reforming process i.e. 850°C [11]. For the low outlet temperatures from HEx, the number of fuel usage is required. Figure 5 shows the fuel usage (in SCF) used in a fired heater. In this case, the natural gas with the heating value of 1143 SCF was used.

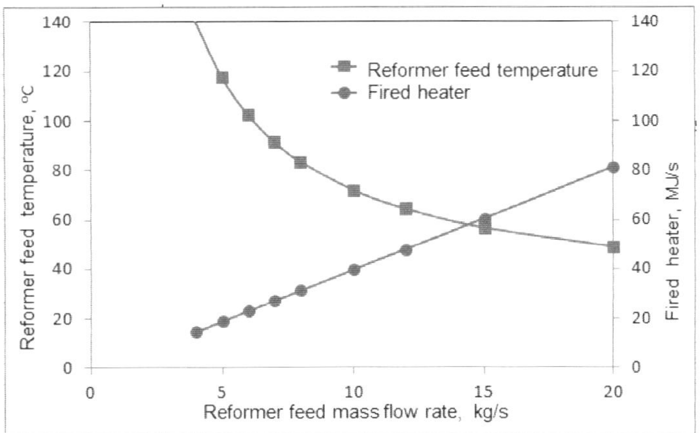

Figure 4. Curve of high reformer feed mass flow rates versus temperature and fired heater power

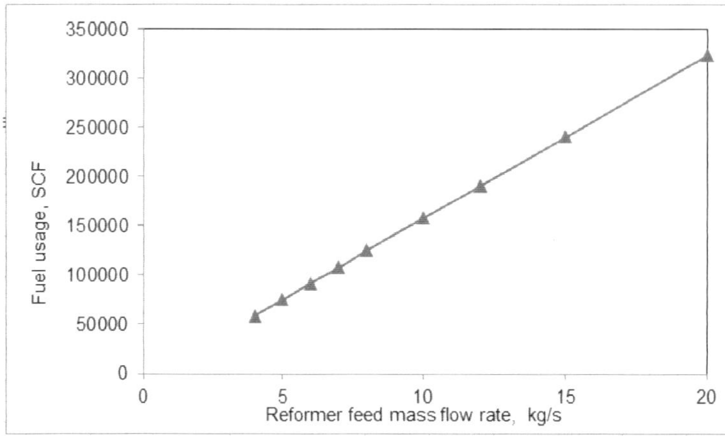

Figure 5. Curve of reformer feed mass flow rates versus fuel usage

Based on the descriptions in Figure 4 and 5, the reformer feed at the low mass flow rate, a small amount of fuel for fired heater was required. Therefore, figure 6 shows the variation of reformer feed at low mass flow rates resulted from the HEx as a function of temperature. It is expected that the reformer feed from HEx into the fired heater is vapor phase (steam). In this analysis, a low mass flow rate can be produced at the temperature up to 480°C. Meanwhile, the outlet temperature of steam from the SG was determined as an input of 540°C. As a notation, that in this heat application using HEx, the pinch temperature causes the HEx can not produce the steam at a temperature above 480°C (pinch zone). The pinch temperature is where the smallest temperature difference occurs between the hot fluids and cold fluids or the temperature between the cold curve and the hot curve is at a minimum.

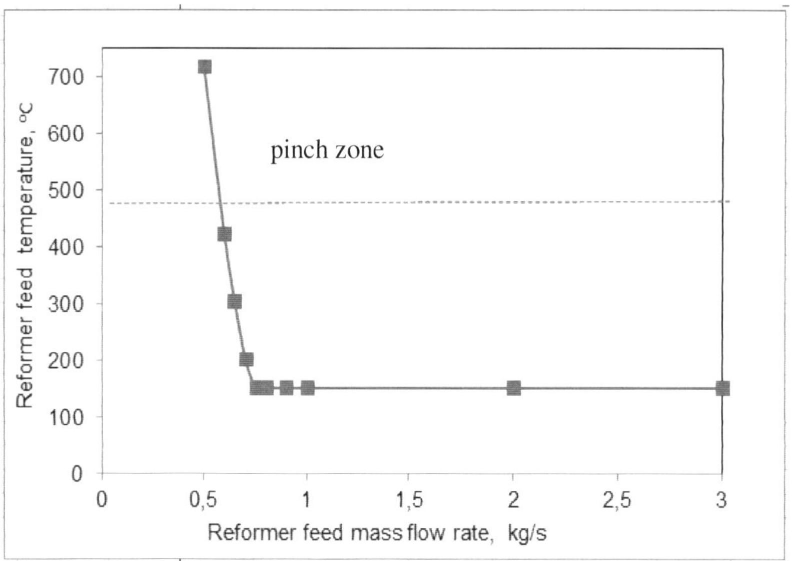

Figure 6. Curve of low reformer feed mass flow rates versus temperature and power of fired heater

5. Conclusion

Analysis on heat application design for feed into the steam methane reformer was carried out. To achieve the temperature of steam methane reforming of 850°C, an additional heat should be required. By the fired heater, an amount of fuel usage is required depending on the steam methane reformer feed temperature produced from the heat exchanger of the cogeneration system. If the steam methane reformer feed in low mass flow rate, therefore the vapor phase (steam) of feed from HEx can be produced. This work is useful to assess the design of heat application of RDE especially for the process of steam methane reforming.

Acknowledgement

The authors would like to acknowledge to the Center for Nuclear Reactor Safety and Technology for supporting this research using the funding of DIPA 2016.

References

[1] Kadak A C 2016 The Status of the US High-Temperature Gas Reactors *Engineering* **2** 119–23
[2] Orhan M F, Dincer I, Rosen M A and Kanoglu M 2012 Integrated hydrogen production options based on renewable and nuclear energy sources *Renew. Sustain. Energy Rev.* **16** 6059–82
[3] Yan X, Noguchi H, Sato H, Tachibana Y, Kunitomi K and Hino R 2014 A hybrid HTGR system producing electricity, hydrogen and such other products as water demanded in the Middle East *Nucl. Eng. Des.* **271** 20–9
[4] Chen F, Dong Y and Zhang Z 2015 Temperature Response of the HTR-10 during the Power Ascension Test *Sci. Technol. Nucl. Install.* **2015** 1–13
[5] Priambodo D, Dewita E and Irianto I D 2015 Analisis energi dan eksergi pada sistem HTR-10 siklus turbin uap *J. Pengemb. Energi Nukl.* **17** 33–43
[6] Kemenristek 2006 *Indonesia 2005-2025: Buku Putih Penelitian, Pengembangan dan Penerapan Ilmu Pengetahuan dan Teknologi Bidang Sumber Energi Baru dan Terbarukan untuk Mendukung Keamanan Ketersediaan Energi Tahun 2025* (Jakarta: Kementerian Negara Riset dan Teknologi Republik Indonesia)
[7] Demir N 2012 Hydrogen production via steam-methane reforming in a SOMBRERO fusion breeder with ceramic fuel particles *Int. J. Hydrogen Energy* **38** 860–853
[8] Khorasanov G L, Kolesov V V and Korobeynikov V V 2015 Concerning hydrogen production based on nuclear technologies *Nucl. Energy Technol.* **1** 126–9
[9] Salimy D H 2010 Produksi Hidrogen Proses Steam Reforming Dimethyl Ether (DME) Dengan Reaktor Nuklir Temperatur Rendah *Jurnal Pengembangan Energi Nuklir* **12** 1–10
[10] Sinaei M, Reza M and Birjandi S 2016 An industrial Steam Methane Reformer optimization using response surface methodology *J. Nat. Gas Sci. Eng.* **36** 540–9
[11] Chibane L and Djellouli B 2011 Methane Steam Reforming Reaction Behaviour in a Packed Bed Membrane Reactor *Int. J. Chem. Appl.* **2** 147–56
[12] Mbodji M, Commenge J M, Falk L, Di Marco D, Rossignol F, Prost L, Valentin S, Joly R and Del-Gallo P 2012 Steam methane reforming reaction process intensification by using a millistructured reactor: Experimental setup and model validation for global kinetic reaction rate estimation *Chem. Eng. J.* **207–208** 871–84
[13] Nieva M A, Villaverde M M, Monzón A, Garetto T F and Marchi A J 2014 Steam-methane reforming at low temperature on nickel-based catalysts *Chem. Eng. J.* **235** 158–66
[14] Irianto I D 2010 Pemodelan sistem konversi energi berbasis kogenerasi reaktor tipe RGTT untuk pembangkit listrik dan produksi hidrogen *Seminar Nasional Pengembangan Energi III, 2010* pp 572–81
[15] Dibyo S and Irianto I D 2017 Design analysis on operating parameter of outlet temperature and void fraction in RDE steam generator *Tri Dasa Mega* **19** 33–40
[16] Kuntjoro S and Udiyani P M 2016 Analisis Inventori Reaktor Daya Eksperimental Jenis Reaktor Gas Temperatur Tinggi *Urania* **22** 53–64
[17] Farag H A A, Ezzat M M, Amer H and Nashed A W 2011 Natural gas dehydration by desiccant materials *Alexandria Eng. J.* **50** 431–9

^{137}Cs Radiological risk estimation of NSD facility at Karawang site by using RESRAD *onsite* application: effect of cover thickness

B Setiawan[1a], **S Prihastuti**[2], **S S Moersidik**[3]

[1]Center for Radioactive Waste Technology-National Nuclear Energy Agency, PUSPIPTEK, Bld 71, 3rd Floor, Serpong-Tangerang 15310, Indonesia

[2]Directorate for Inspection of Nuclear Installation and Material-Nuclear Energy Regulatory Agency, Jl.Gadjahmada No. 8, Jakarta 10120, Indonesia

[3]Department of Civil Engineering-University of Indonesia, Kampus UI, Depok 16424, Indonesia

[a]Corresponding's author: bravo@batan.go.id

Abstract. The operational of near surface disposal facility during waste packages loading activity into the facility, or in a monitoring activity around disposal facility at Karawang area is predicted to give a radiological risk to radiation workers. The thickness of disposal facility cover system affected the number of radiological risk of workers. Due to this reason, a radiological risk estimation needs to be considered. RESRAD onsite code is applied for this purpose by analyse the individual accepted dose and radiological risk data of radiation workers. The obtained results and then are compared with radiation protection reference in accordance with national regulation. In this case, the data from the experimental result of Karawang clay as host of disposal facility such as Kd value of ^{137}Cs was used. Results showed that the thickness of the cover layer of disposal facility affected to the radiological risk which accepted by workers in a near surface disposal facility.

Keywords: near surface disposal, 137Cs, cover system, Kd, radiological risk

1. Introduction

Near surface disposal (NSD) is a disposal facility refers to the emplacement place of solid, or solidified radioactive waste packages containing predominantly short-lived radionuclides which located at or near the land surface [1]. In general, wastes suitable with the types of disposals are those containing a short-lived radionuclides and low concentrations of long lived radionuclides. The NSD type is also encompassing a wide range of disposal options, including disposal in engineered structures on the ground, a few meters of deep disposal in a simple earthen trench, disposal with engineered concrete vaults and also disposal in rock caverns in several tens of meters below the earth's surface such as shown at Fig. 1 [2,3].

The number of factors to the selection of disposal facility depth and the type of facility that is developed will depend on the nature of the waste and the local environmental conditions. Through a series of complementary barriers such as the waste form itself, waste containers, other engineered features associated with the facility design, and the local environment will provide wastes containment and isolation. In some ways, each of barriers serves to prevent the release of radionuclides from the waste form in the facility to the accessible environment as restricted the migration of contaminants area, and also may minimize radiological risk to the worker during operational of facility.

To reduce the associated risks to acceptable levels, radioactive waste is represented as a potential hazard to human health, due to it acts as a source of ionizing radiation must be carefully managed. The offender of disposal activity responsible for the implementation of protection of human health and the environment in accordance with a national system of radiation protection agreed with principles and requirements for radioactive waste management and radiation protection [4]. Through Government Regulation No.61/2013 [5], radioactive waste disposal facility options were classified according to their radioactivity levels as is illustrated at Figure 1, however in this paper the case was focused on the near surface disposal (NSD) facility.

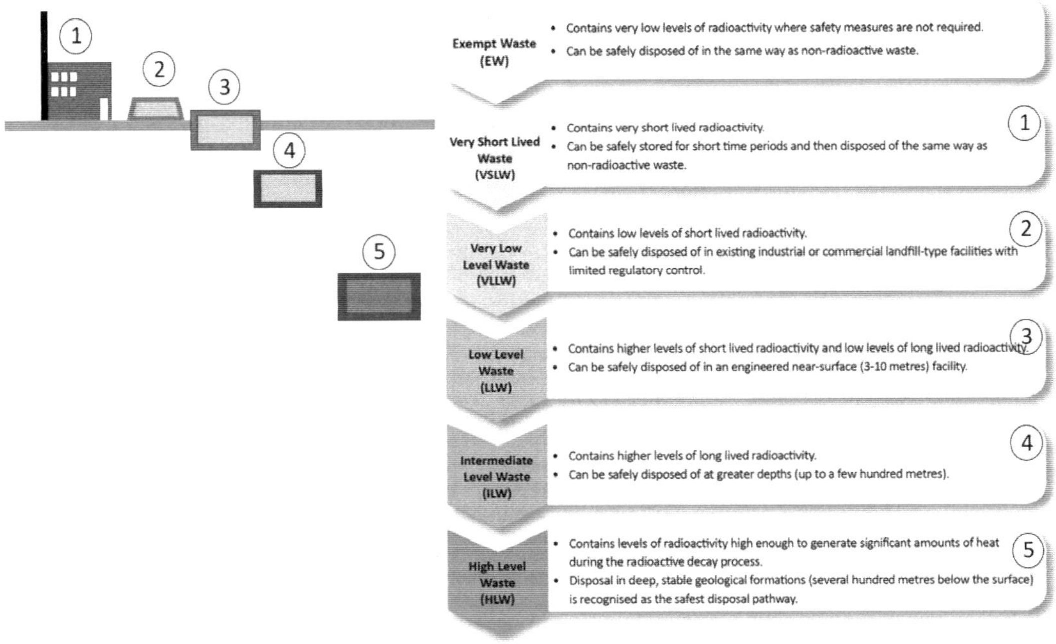

Figure 1. Disposal types option, an illustration [2,3].

Radiation exposure results from all activities related to NSD shall represent the principles and requirements of protection of human health and the environment. The number of doses received by workers shall be kept within dose limits. The various pathways by which humans might be exposed to radiation needed to be given particular attention to the assessment during the operation of a repository and after its closure. As a result of operations at the waste repository facility must be optimized, the radiation protection of persons who are exposed must be optimized and the exposures of individuals must be kept within dose limits.

Karawang area is one of selected area for site candidate of NSD facility at Java Island [6]. In the future on the operational of NSD facility, one things that need to be taken into account properly is the estimation of radiological risk during the operational phase of a repository. In such situation, the operator shall apply the dose limits for occupational exposure for workers and for members of the public. Determination of the cover thickness of NSD facility has become one things that need to be considered to ensure the radiation safety of workers during the operational of facility period.

In this case the ^{137}Cs radionuclide is used as the referenced radionuclide due to it has long half-life, easy to associated with the materials in groundwater and terrestrial, and it make easier to the radionuclide enter into food chain before going into the human body through various pathway. Based on scheme of waste classification in IAEA, radiocesium (^{137}Cs) is included in the categorized as low-level waste with half-live 30 years and placed into a NSD facility [7].

To minimize the level of acceptance dose of worker, the problem of cover thickness of NSD facility need to be estimated, and RESRAD onsite application code can be used. The assessment of radiological risk for the NSD facility which was planned in Java Island almost has been not performed, and the assessment of the effect of cover thickness of NSD facility becomes important to be done. Objective of the activity is to assess the needs of cover thickness on designing of NSD facility to protect the workers from acceptance doses risk during operational or closure stages of NSD. The assessment based on the result of previous activity [8], where Kd values from the previous activity was used as one of input on RESRAD code application. The expected results of the activity are getting an estimated radiological risk when the cover thickness of NSD is implemented.

2. Theory

The RESRAD (RESidual RADioactivity) code is a software tools have been developed by the Argonne National Laboratory, US Department of Energy (DOE) and US Nuclear Regulatory Commission (NRC) since 1989. This code is used to evaluate radiological contaminated sites [9-11] and also to make regulatory decisions to help determine how clean the radioactivity levels at nuclear sites [9]. The RESRAD code calculates both radiological dose and risk. RESRAD allows users to specify the features of their site and to predict the dose received by an individual at time over the next 100,000 years. When estimating the possibility of radionuclide migration and the magnitude of the radiation exposure received by workers and member of public around the NSD facility, scenario shall be created and selected in a few different ways. Selected scenario at safety assessment is a normal radionuclide dispersion scenario through groundwater pathway to radiation workers and member of public as shown schematically at Figure 2. [11,12]

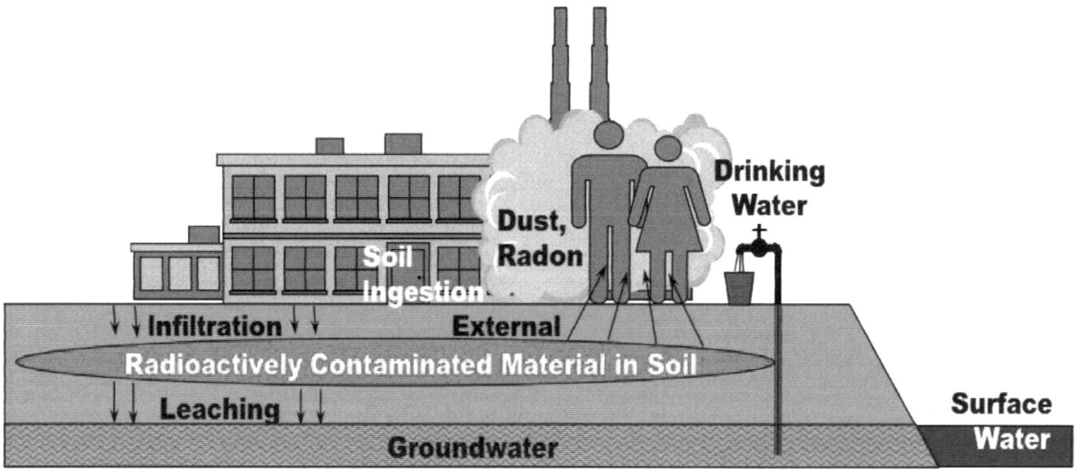

Figure 2. Radiation pathway on RESRAD onsite application [11].

For the purposes of acceptance dose and risk estimation, the individual as objective study is assumed to be standing at above of the disposal facility. The workers are supposed to perform 8 working hours daily at the facility. Drinking water assumed come from water supply, and the worker does not ingestion any foods (plant, meat or milk) grown or raised on the site. At the top of disposal facility were dumped with clay as the disposal cover to minimize radiation exposure on the surface of facility and to reduce the acceptance doses received by workers.

For analysing the results, some inputs were needed on the application, such as radionuclide type, waste activity, specific of Kd value, thickness of contaminated area, density and porosity of cover. This code is used to predict the Total Effective Doses Equivalent, probability excess radiological risk incurred by worker exposed to radioactive emission from waste packages during operational or closure stages of NSD. The acceptable level of radioactive material complies with criteria adopt by regulatory body, are based on the following principles: (1) Total effective dose equivalent should not exceed than 30 mrem/y (0.3 mSv/y), and (2) Doses should be kept as low as reasonably achievable (ALARA) [13].

3. Methodology

The number of received effective doses equivalent depends on the occupancy and shielding factor, the depth of nuclide specific and thickness cover factor, and also specific area factor as shown in Figure 3. A different thickness of cover layer of disposal facility is expected to give a different value of received doses by workers.

$ETF(t) = FO_1 \times FS_1 \times FA_1 \times FCD_1(t)$

where *ETF* : Environmental transport factors
FO : occupancy and shielding factor
FS : nuclide sp.shape factor
FA : nuclide sp.area factor
FCD : nuclide sp.depth and cover factor

Figure 3. Some factors affected to the total effective dose equivalent [11].

Its means that the affected factors would be depend on the fraction of time spent in indoors and outdoors, the γ-shielding, the thickness and density of contaminated area and cover, geometry of area and the place of receptors received the doses factors.

Total effective dose equivalent is the sum of dose equivalent weight of different organ proportioned with risk potential of radiation emission received to each organ. Distribution doses in the human body depends on distribution pattern of radionuclides in the environment, assumed that radionuclide distribution at infinite or semi-infinite areas are homogeneous. To determine the contribution of effective dose equivalent from externally radiation at time t after measurement is,

$$DSR(t) = \sum DCF \times \sum \int_{t}^{t-t_{int}} ETF(t) x SF(t)$$

where
DSR : doses/sources ratio,
DCF : dose conversion factor (mSv/y per-Bq/g), was taken from Federal Guidance Report,
ETF(t) : in time t (g/y),
SF(t) : sources factor at time t, ingrowth, decay, leaching factors of radionuclide.

Radiological risk to the individual that received a radiation exposure while working at a contaminated area is calculated by the equation,

$$Radiological\ risk = \sum_{i=1}^{M} ETF(t) x SF(t) x S(0) x RC x ED$$

where
RC : risk coefficient for externally radiation (risk/y)/(pCi/g),
ED : exposure time (30y),

S(0) : radionuclide concentration in clay at t=0.

A series of input data were used to estimate the doses and risk of the NSD facility to the workers as shown in Table 1.

Table 1. Input data on doses estimation with RESRAD onsite application.

Items	Parameters	Value	
Transport	Kd value of sample-1	21,714 ml/g	**
	Kd value of sample-2	4,035 ml/g	**
Regulation Standard	Federal Guide Regulatory	FGR	
Time	Time of observation	1, 10 and 100 years	
Contaminated zone	Contaminated area	10,000 m^2	
	Thickness of contaminated area	1 m	
	Depth of unsaturated area	10 m	
Uncontaminated zone	Density	1.5 g/cm^3	*
	Total porosity	0.4	*
	Hydraulic conductivity	10 m/y	*
Hydrology on the contaminated zone and cover	Cover thickness	1.2 m	***
	Cover density	1.5 g/m^3	***
	Cover erosion factor	0.001 m/y	*
	Contaminated zone density	1.5 g/cm^3	*
	Erosion rate in contaminated zone	0.001 m/y	*
	Precipitation	1 m/y	*
	Runoff coefficient	0.2	*
External–γ εxposure	Exposure time	30 y	*
	External shielding factor	0.7	*
	Indoor fraction	0.5	*
	Outdoor fraction	0.25	*
	Contaminated zone shape	circular	*

* RESRAD default ** [8] *** [14]

The information of Kd values, we used the specific Kd values of ^{137}Cs from Karawang clay experiments [8]. ^{137}Cs is a long-lived radionuclide, it may accordance in clay samples, accumulated in plants and has high risk potential with time for a certain period. The sample-1 and 2 represented the capability of Karawang clay to absorb ^{137}Cs in a highest and lowest of Kd values, where in higher value implies that the ^{137}Cs will be closely tied to the clay sample and slow release to environment and *vice versa*. Release fraction used solubility factor to estimate maximum concentration of radionuclide leachate. Observation time is done gradually starting from 1st, 10th until the 100th. The results and then are compared with the national radiation protection regulation issued by regulatory body/BAPETEN.

4. Results and Discussion

Kd values were obtained from the previous activity [8], and the result was shown in Figure 4. ^{137}Cs becomes saturated in the sample after a few days contacted, where radiocesium achieved to fulfil the active site of the clay samples with Kd values were 21,714 and 4,035 ml/g for sample-1 and 2, respectively. The values obtained from the experiment is then used as the input of RESRAD application to estimate the doses and radiological risk.

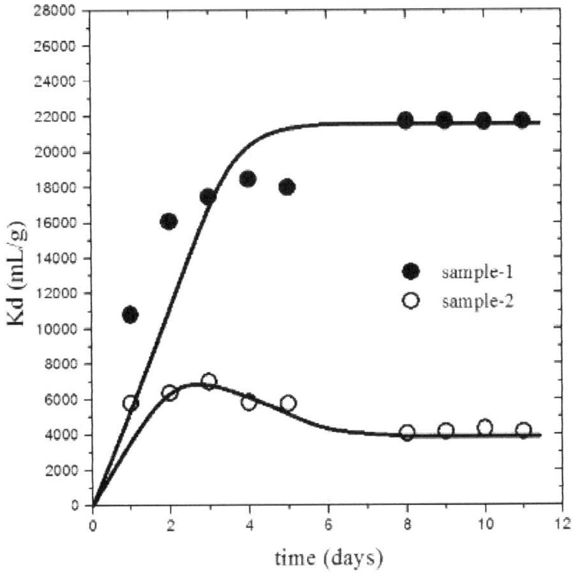

Figure 4. Kd values of ^{137}Cs as a function of time [8].

Total effective dose equivalent for workers at a disposal facility with sample-1 clay host and has cover thickness 0.6 m gave dose value was 2.35 mSv/y at 1st year dispose as shown at Figure 5.

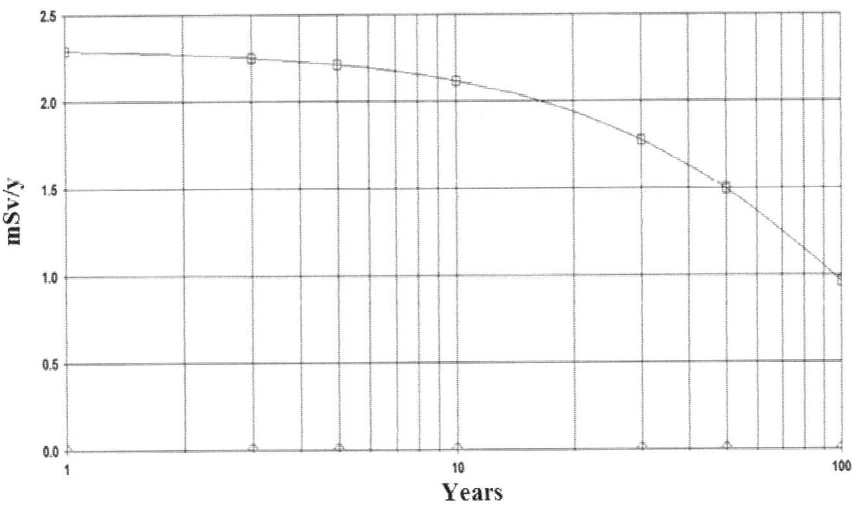

Figure 5. Total effective dose equivalent at a disposal facility at sample-1 clay host, with cover thickness 0.6 m.

The dose came from external radiation exposure way of waste packages decreased with the operation time of waste packages in disposal facility. Decreasing of dose is due to the occurrence of radioactive decay and the influence of nature phenomena such as evaporation, precipitation and release of radionuclide. At the time to the 100th years of disposal operation, the doses received by workers is about 1 mSv/y. If the obtained result is compared with BAPETEN Chairman Regulation No. 4/2013 about Radiation Protection and Safety on Nuclear Energy Application [13] showed that the obtained value is still higher. In the article 23 mentioned that the dose limit to member of public in the form of effective dose is 1 mSv/y, and on the article 41 also stated that the permit holder should specify dose limit to achieve optimization of radiation protection and safety. For that reason, a cover of disposal facility with a thickness of 0.6 m sample-1 clay could not be recommended.

When the thickness of disposal cover was added 2 times fold i.e. to 1.2 m, the effective dose equivalent received by workers at 1st year dispose will be much going down to 4.47 x 10^{-4} mSv/y or dropped 5,000 times lower than the previously received doses, see Figure 6. This is due to the presence of influences the shielding factor, depth of nuclide specific and thickness cover factor that play a sufficiently role to reduce large value of effective dose equivalent that will be received by the workers. At time of 100th years dispose, the dose becomes about 1.86 x 10^{-4} mSv/y. The values then were compared to the national regulation of radiation protection issued by regulation body indicated that the values are still below the dose limit, by 0.3 mSv/y [13]. By adding the cover thickness of disposal system, it can significantly reduce the magnitude of effective dose that will be accepted by the workers.

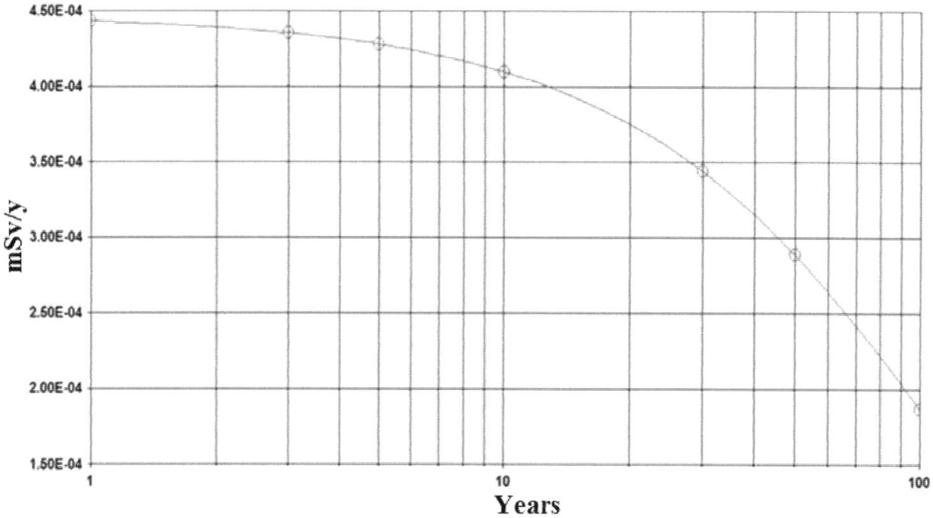

Figure 6. Total effective dose equivalent at a disposal facility at sample-1 clay host, with cover thickness 1.2 m.

In the analysis of radiological risk, the hazardous of ^{137}Cs to the human body is in associated with cell damage due to the result of ionizing radiation decay, and it will generate a potential cause of cancer. Estimation of the mortality risk by cancer assumes 6 of 100,000 peoples or 6 x 10^{-5} are exposed continuously to radiation from contaminated clay layers with 1 pCi/g of ^{137}Cs activity will suffer a fatal cancer [15].

The facility was assumed has clay cover with thickness 1.2 m. Figure 7 showed that FGR morbidity of radiological risk at 1th year operation of disposal facility with sample-1 as clay host which contaminated with ^{137}Cs. Probabilistic the individual to accept the radiation exposure that can result the radiological risk or cancer was 9.92 x 10^{-7} from all the possible causes of cancer to individual. While the probabilistic of radiological risk with cause of death to workers (mortality) was 6.85 x 10^{-7}. In the risk assessment of both samples, the result of radiological risk (morbidity and mortality) were not significant different. However, the area was represented by both results if compared with the ANL limitation has fulfil the safety requirement of individual effective dose equivalent received by workers and member of public. In the previous study stated that the location with higher Kd value the ^{137}Cs bound more strongly on a contaminated area, it was proved with the occurrence of small concentration of ^{137}Cs in the hypothesis well (with distance 30 m). Means that the ^{137}Cs less dispersed into environment than the location that has small Kd value [16].

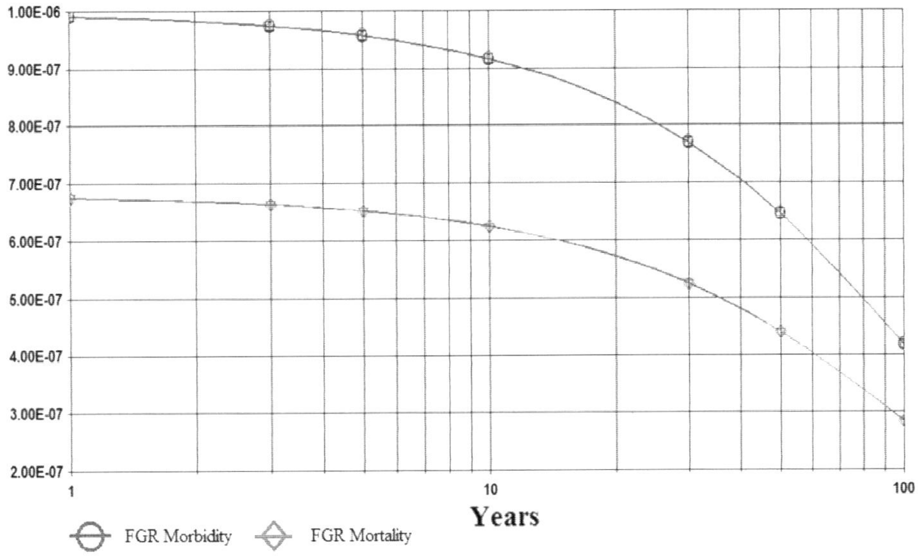

Figure 7. Estimation of radiological risk at sample-1

Summary of the study was compiled in Table 2, which some data came from the result of calculation (specially for sample-2).

Table 2. The results of RESRAD onsite calculation.

Parameters	Sample-1	Sample-2
Kd values	21,714 ml/g	4,035 ml/g
Dose at cover thickness 0.6 m	2.35 mSv/y	2.25 mSv/y
Dose at cover thickness 1.2 m	4.47×10^{-4} mSv/y	4.38×10^{-4} mSv/y
Estimation of cancer at cover thickness 1.2 m, FGR morbidity	9.92×10^{-7}	9.72×10^{-7}
Estimation of cancer at cover thickness 1.2 m, FGR mortality	6.85×10^{-7}	6.7×10^{-7}

5. Conclusion

The thickness changing of clay cover of disposal facility can reduce the quantities of effective dose equivalent was received by workers. At a higher Kd value will give the effective dose equivalent and radiological risk also higher, although the range of both values is not significant different. Thickness of clay cover that is 1.2 m has enough give the received dose values lower than the limit doses set by the regulatory body.

Acknowledgments

Many thanks are to the Head and staffs of Division of Predisposal and Disposal Technology-Center for Radioactive Waste Technology-BATAN for their supports and facilities.

References

1. International Atomic Energy Agency *2014 Near Surface Disposal Facilities for Radioactive Waste SSG-29* (Vienna, IAEA) p.4
2. Bergstrom U, Pers K, Almen Y 2011 *International Perspective on Repositories for Low Level Waste R-11-16* (Stockholm: SKB International AB) p.9
3. Arpansa 2015 *Radioactive Waste Safety in Australia Fact Sheet* (Victoria: ARPANSA) p.1

4. International Atomic Energy Agency 1999 *Near Surface Disposal Facilities of Radioactive Waste WS-R-1* (Vienna: IAEA) p.6
5. President of The Rep. Indonesia 2013 *Government Regulation No. 61/2013: Radioactive Waste Management* (Jakarta: Rep. of Indonesia) -in Indonesian.
6. Sucipta, Setiawan B, Pratomo B S, Suganda D 2010 *J. Waste Management Tech.**13** (1),* p.30-42
7. International Atomic Energy Agency 2009 *Classification of Radioactive Waste, GSG-1* (Vienna: IAEA) p.8
8. Setiawan B, Prihastuti S, Moersidik S S 2016 *Aceh Int. J. Sci. and Tech.* ***5** (3)*, p. 107-115
9. Yu C, Zielen A J, Cheng J J, LePoire D J, Gnanapragasam E, Kamboj S, Arnish J, Wallo III A, William W A and Peterson H 2001 *User's Manual for RESRAD Ver. 6 ANL/EAD-4* (Illinois: Argonne Nat., Lab) p.2-2
10. Yu C 2006 *Modelling Radionuclide Transport in the Environment and Assessing Risks to Human, Flora, and Fauna: The RESRAD Family of Code, Chapter 15, Applied Modelling and Computations in Nuclear Science*, ACS Symposium Series 945 (Washington D C: Am. Chem. Soc) p.14
11. Yu C 2007 *Tools for Assessing Radiological Doses to Human and Biota, in Radiological Assessment: Detection, Identification and Evaluation, Health Physic Soc. Professional Development School Textbook* (Virginia: HPS)
12. Smith B 2008 *Sci. for Democratic Action Vol 15 (4)* p. 1-16
13. Bapeten 2013 *Chairman Regulation No. 4/2013, Radiation Protection and Safety on Nuclear Energy Application* (Jakarta: BAPETEN) -in Indonesian.
14. Pandu D 2015 *Study of Scenario of Radionuclide Release on Water and Clay at Demonstration Plant of Near Surface Radioactive Waste Disposal Master Thesis* Depok: University of Indonesia) -in Indonesian.
15. Peterson J, MacDonell M, Haroun L, Monette F, Hildebrand R D, Taboas A 2007 *Radiological and Chemical Fact Sheets to Support Health Risk Analyses for Contaminated Areas* (Illinois: Argonne National Laboratory)
16. Setiawan B, Suganda D 2016 *Proc. Nat. Seminar of Waste Management Tech. XIV* (Jakarta: BATAN-UI) p. 93-104, – in Indonesian.

IOP Conf. Series: Journal of Physics: Conf. Series **962** (2018) 012054 doi:10.1088/1742-6596/962/1/012054

The Simulator Development for RDE Reactor

Muhammad Subekti, Syaiful Bakhri, Geni Rina Sunaryo

Center for Nuclear Reactor Technology and Safety, BATAN
Puspiptek Complex, Building No.80, Serpong, Tangerang Selatan 15310, Indonesia.

Email: subekti@batan.go.id

Abstract. BATAN is proposing the construction of experimental power reactor (RDE reactor) for increasing the public acceptance on NPP development plan, proofing the safety level of the most advanced reactor by performing safety demonstration on the accidents such as Chernobyl and Fukushima, and owning the generation fourth (G4) reactor technology. For owning the reactor technology, the one of research activities is RDE's simulator development that employing standard equation. The development utilizes standard point kinetic and thermal equation. The examination of the simulator carried out comparison in which the simulation's calculation result has good agreement with assumed parameters and ChemCAD calculation results. The transient simulation describes the characteristic of the simulator to respond the variation of power increase of 1.5%/min, 2.5%/min, and 3.5%/min.

1. Introduction

BATAN is proposing the experimental power reactor (RDE reactor) for pursuing the public acceptance on NPP development plan, proofing the safety level of the most advanced reactor by performing safety demonstration on the accidents such as Chernobyl and Fukushima, and owning the generation fourth (G4) reactor technology. RDE reactor is a High Temperature Gas Cooled Reactor (HTGR) with thermal power of 10 MWe that employed pebble bed fuel and helium coolant [1-2]. The reactor has planned to be constructed at Serpong, Tangerang Selatan, Indonesia, close to RSG-GAS reactor with thermal power of 30 MWt [3]. For owning the reactor technology, the research activities are the standard code verification and validation, transient calculation for safety analysis, as well as the simulator development.

The development of NPP simulator could be grouped in four classification level based on the simulator objective. The simulator objective is for basic study, classroom teaching, engineering and full scale for NPP staff operator training. The conventional full scale simulators based on analog/digital technology is dedicated for NPP operators. The engineering simulator is for initial training of the operator [4]. An example of engineering scale simulator has been developed by implementing a standard code of RELAP [5] and MELCOR [6]. The commercial NPP simulator such as PCTRAN is utilized for transient analysis [7]. Those developed simulators are mostly dedicated for understanding the accident process. However, the previous studies of method development [8-10] are a basic method for either normal and accident operation simulation. Therefore, the development of basic or classroom scale simulator with normal operation simulation is important to be done before carry out accident simulation.

The objective is to develop a RDE reactor simulator that employs neutronics, core thermal, and Nuclear Steam Supply System (NSSS) modules for simulating the transient reactor. The code development utilized LabVIEW software developer and Personal Computer for implementing the coupled kinetics-thermal calculation. The standard calculation of HTGR core used Germany standard of KTA [11]. The simulation demonstration followed the power increase arrangement by regulator and has a reactor operating pattern for avoiding shutdown by developed Reactor Protection System (RPS). Moreover, the reactor control employs PID and advanced technology to control the reactor, helium circulator, feedwater pump and steam valve in once time. Therefore, the normal and accident operation could be simulated with the developed RDE simulator including Reactor Core Cavity System (RCCS). The designed RCCS contains of three trains of cavity pipe, evaporation tank, and smaller cooling tower. The main cooling tower is utilized by the reactor for normal operation. For enhancing the user interaction, the simulator utilized three computer screens to show the reactor control, Balance of Plant (BOP), and RCCS.

2. Theory

2.1. RDE Reactor

The reactor main components consist of reactor vessel, blower, steam generator, turbine, secondary pump, condenser, cooling tower, and deaerator. The design of RDE reactor is assumed to approach the reactor HTR-10 [11] with several differences such as more number of small ball absorber, straight pipe for fuel discharge system, and unavailability of irradiation channel as showed in Fig. 1. Therefore, the number of control rod and small ball absorber holes are ten that arranged in side reflector. Consequently, the neutron balance will be different as well. The neutron flux is controlled symmetrically by 10 Control Rod (CRs) that could be inserted in CRs channel with speed of 10 mm/s in normal condition.

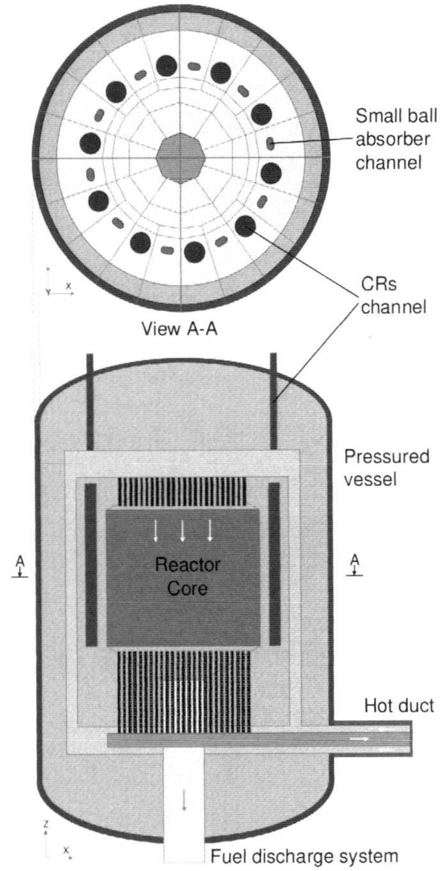

Figure 1. The design of RDE reactor is approaching the reactor HTR-10 with several differences such as more number of small ball absorber, straight pipe for fuel discharge system, and unavailability of irradiation channel.

2.2. Reactor equation

The reactor module utilizes point kinetics for neutron flux calculation and reactivity. The equation of point kinetics is described below [12]:

$$\frac{dn(t)}{dt} = \frac{\rho(t) - \beta}{\Lambda} n(t) + \sum_{i=1}^{6} \lambda_i C_i(t) + S(t) \tag{1}$$

$$\frac{dC_i(t)}{dt} = \frac{n(t)}{\Lambda} - \lambda_i C_i(t) \tag{2}$$

where :

$n(t)$	:	neutron population at time t (n/cm3.s)
$C_i(t)$	:	precursor for delayed neutron in energy group of i and time of t (n/cm3)
$S(t)$	:	neutron source production at time of t (n/s)
Λ	:	neutron generation time (s)
β_i	:	delayed neutron fraction for energy group of i (-)
λ_i	:	Decay constant of delayed neutron precursor for energy group of i (s^{-1})
$\rho(t)$	:	Reactivity at time t (pcm)

The reactivity value form control rod is calculated by using designed control rod data. Therefore, the total reactivity in the core could be summed from all contributed reactivity as below:

$$\rho_{Total} = \rho_{excess} + (\rho_{CR} + \rho_{Xe} + \rho_T) \tag{3}$$

$$\rho_{CR}=\sum_{i=1}^{10}\rho_{CR_i} \quad (4)$$

where :

ρ_{Total} : total reactivity (pcm)
ρ_{excess} : excess reactivity from fuel (pcm)
ρ_{CR} : total control rod reactivity (pcm)
ρ_{Xe} : poison reactivity from Xenon (pcm)
ρ_{Sm} : poison reactivity from Samarium (pcm)
ρ_T : negative temperature reactivity from Doppler effect (pcm)
ρ_{CR_i} : control rod reactivity from control rod number *i*=1-10 (pcm)

2.3. KTA standard equation

The simulator development for RDE reactor is based on empirical standard equation. The implemented equation that standardized in KTA [13] is heat transfer Q in a pebble bed fuel to surrounding below:

$$Q = \alpha A_k(T_k - T_g) \quad (5)$$

$$\alpha = \frac{Nu\lambda}{d} \quad (6)$$

$$\lambda = 2.682x10^{-3}T^{0.71(1.0-2.0x10^{-6}P)}(1.0 + 1.123x10^{-5}P) \quad (7)$$

where:

A_k : fuel wall area (m^2)
T_k : average temperature for pebble fuel (K)
T_g : temperature helium gas (K)
α : heat transfer coefficient for fuel wall (W/m^2K)
λ : thermal heat conductivity for helium gas (W/mK)
Nu : Nusselt number (·)
d : outer diameter for pebble bed fuel (m)
P : pressure (kPa)
T : temperature (K)

3. Methodology

The development for RDE simulator processes a modeling based on the assumed operation values of RDE reactor. The modeling includes the main component such as the reactor, steam generator, blower, turbine, deaerator, condenser, pump, and cooling tower. The next step of the development is coding development using LabVIEW developer tool with three main user interface screen. The main screens contains of reactor, NSSS and RCCS control. Therefore, the simulator development requires three screens to visualize the screen for reactor, NSSS, and RCCS controls. The comparison of steady state condition is required by utilizing the data taken from HTR-10 [13] and RDE assumed parameters [1] as well as ChemCAD calculation results. However, the engineering adjustment was done to reach the assumed outlet temperature of helium reactor coolant of 700°C and steam generator of 530°C. The adjustment considers the ChemCAD calculation results for variation of designed temperature inlet [14]. The maximum inlet steam temperature of 160°C is assumed to obtain good thermal performance.

The transient simulation starts from 0% to 100% power level with variance of power increase of 1.5%/min, 2.5%/min, and 3.5%/min. The examination of the RDE simulator carries out calculation result comparison and transient analysis in which the comparison includes the assumed parameter and ChemCAD calculation results. The transient analysis examines the control rod movement to follow the regulation of power increase and study the respond of control rod and blower controls.

4. Result and Discussion

4.1. Design Reference Comparison

Table 4 shows a comparison of the calculation result of the developing simulator and assumed parameters. The averaged value was calculated for 7200 s. The comparisons examine the difference of reactor outlet temperature at 10% and 100% power and include the calculated value from ChemCAD. The calculation method of simulator is similar with ChemCAD on mass and energy balance. Therefore, the difference of steam flowrate for assumed parameters could not be analyzed due to unknown explanation for assumed parameters in the conceptual document. However, the difference could be justified by comparing to the ChemCAD calculation results. The engineering adjustment based on ChemCAD calculation corrects the values for primary coolant and steam flowrate of 4.27 kg/s and 3.55 kg/s respectively. The consideration for the adjustment is to set the outlet temperature close to the assumed parameters.

Table 1. Comparison of the calculation result of the developing simulator and assumed parameters.

Parameter	HTR-10	RDE Parameters		
	Value[13]	Assumption[1]	ChemCAD	Simulator
Reactor Power (MWt)	10.00	10.00	10.00	10.00
Primary coolant inlet temperature (°C)	250.00	250.00	250.00	243.49
Primary coolant outlet temperature (°C)	700.00	700.00	700.00	701.43
Primary coolant pressure (bar)	30.00	30.00	30.00	30.00
Primary coolant flowrate (kg/s)	4.32	4.40	4.27	4.32
Steam-generator outlet temperature (°C)	450.00	530.00	530.00	529.89
Steam flowrate (kg/s)	4.00	4.00	3.55	3.55
Steam pressure (bar)	60.00	60.00	60.00	60.00
Feedwater temperature (°C)	145.00	145.00	145.00	145.50
Feedwater pressure (bar)	-	-	71.2	71.5

4.2. Benchmarking

The modeling of NSSS was done by using ChemCAD for steady state calculation that is shown in Fig. 2. Based on developed model, the similar model was developed for Simulator that the RDE reactor has main NSSS components such as steam generator, deaerator, turbine, condenser, and secondary pump. The transient simulation was started from reactor power 0% to 100%. The regulation of power increase was simulation in variance of 1.5%/min; 2.5%/min; and 3.5%/min.

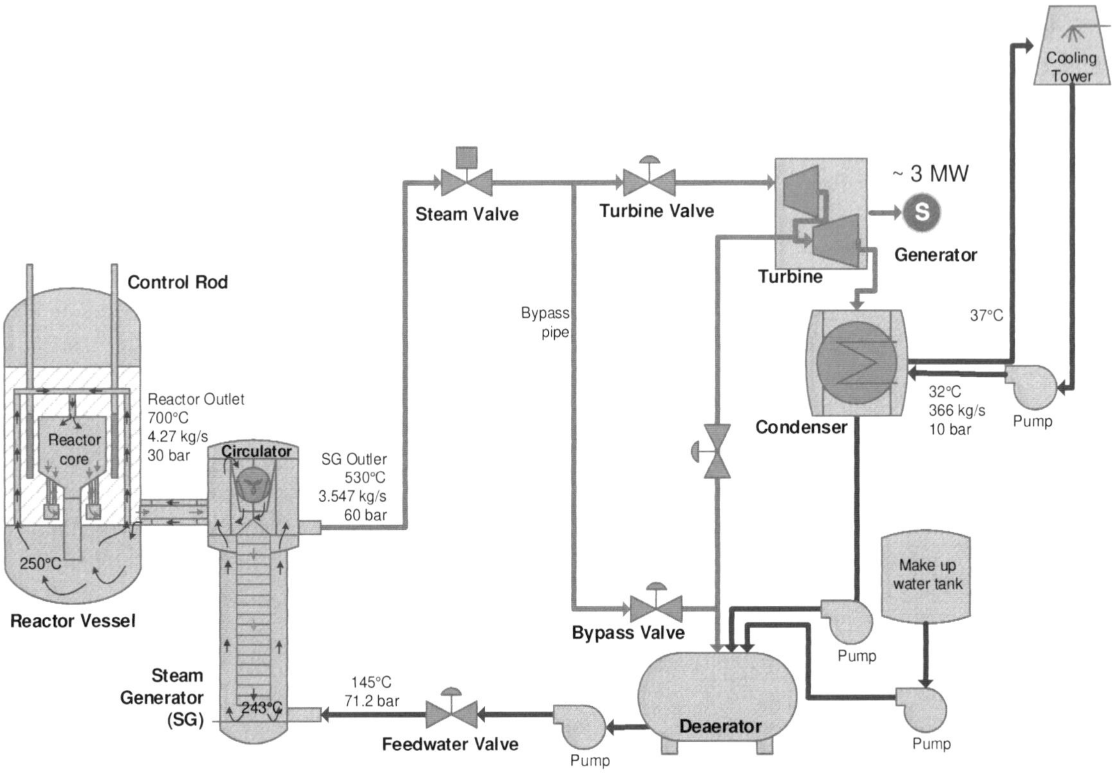

Figure 2. The steady state modelling of NSSS using ChemCAD.

Fig. 3 shows the transient simulation based on power increase of 1.5%/min. The reactor power increases from 0 MWt to 10 MWt by control rod withdrawal. The early simulation until t=300 s, control rod was withdrawal for reaching the maximum reactivity that is permitted by Reactor Protection System. The limit for minimum reactor period is 20 s. However, the warning in the simulator will actuated if the minimum reactor period of 40 s was exceeded. The figure cropping at t=4500s shows that the control rod has to adjust in reactivity insertion and correction to maintain the power increase surrounding 1.5%/min. The reactivity insertion and correction affect the increase and decrease of the reactor power in the same time frame. This power increase pattern continues until the power level reached 100% power with a small over shoot in Fig. 3.

The power increase rule that is stated by power plant regulation is commonly different between countries. Fig. 4 shows the transient simulation based on power increase variation of 1.5%/min, 2.5%/min, and 3.5%/min. The variation determines the reactor startup time from 0% power level to reach 100% power level. With power increase 2.5%/min or 3.5%/min, the 100% power level could be reached less than an hour. However, the 100% power level with power increase 1.5%/min could be reach in longer time, more than two hours.

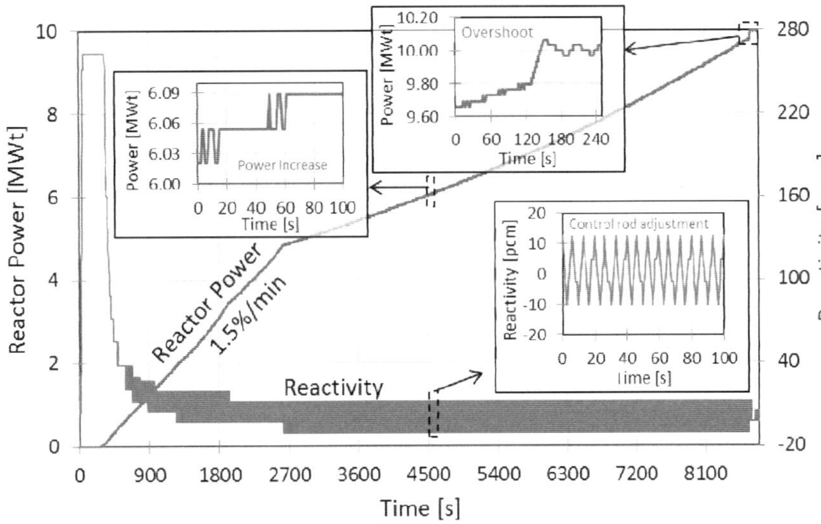

Figure 3. Transient simulation based on power increase of 1.5%/min.

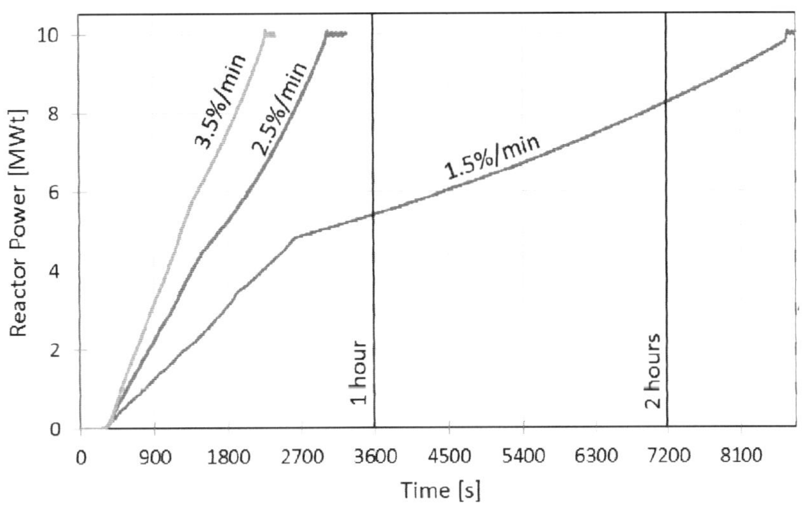

Figure 4. Transient simulation based on power increase variation.

The development also finished the simulator screen creation for (i) reactor, (ii) NSSS, and (iii) RCCS controls. Fig. 5 shows the developed simulator screen for reactor control. The screen is responsible for arranging the control rod position and Helium flow rate by blower. The reactor indicators that useful for reactor control is control rod positions, reactor period, reactivity, power level, etc. The brief operation of reactor startup in a manual mode, the early step of the operation is withdraw the control rod to increase the neutron flux in a safe condition showed by the parameters of total reactivity or neutron period. After reactor power increases significantly and affects to the Helium outlet temperature increase, the control rod position must be maintained so that the total reactivity or neutron period keep below the operation limit. The operation limit value considers the adequate margin to safety limit.

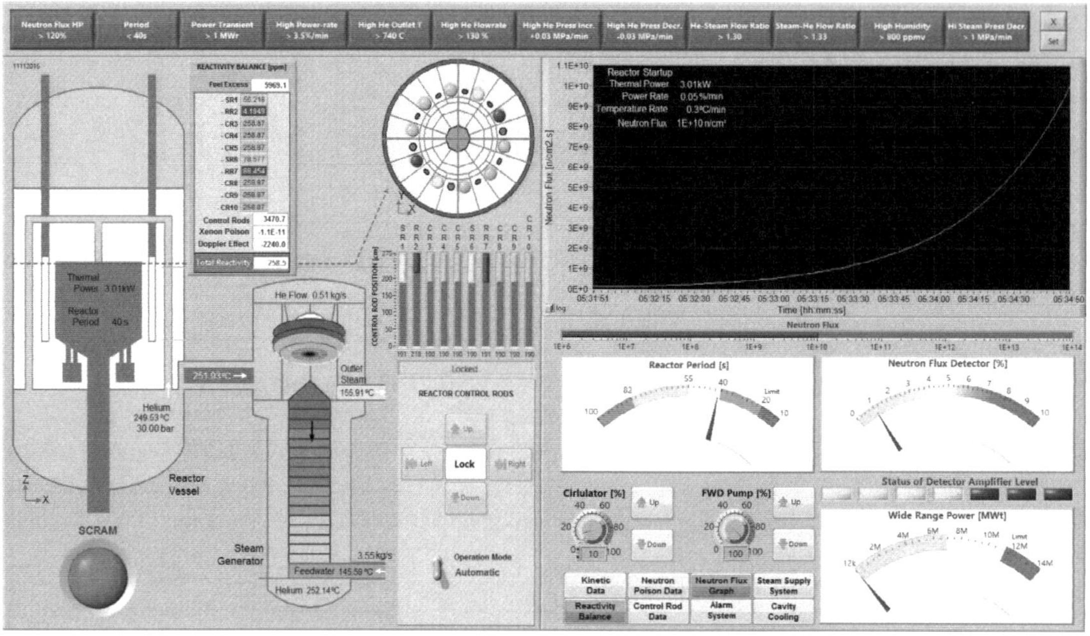

Figure 5. The developed simulator screen for reactor control.

5. Conclusion

The simulator development has finished the development of main code of reactor module that employed point kinetic method. The examination of the simulator carried out comparison in which the simulation's calculation result has good agreement with assumed parameters and ChemCAD calculation results. The variation of power increase was studied to determine the reactor startup time from 0% power level to reach 100% power level.

Acknowledgments

Author thanks the Center for Education and Training BATAN for the implementation of developed NPP simulator in the formal training in Nuclear Engineering I and II. The more advanced research for HTGR typed NPP was budgeted by Center for Nuclear Reactor Technology and Safety, BATAN 2016-2017.

References

[1] BATAN 2014 *User Requirement Document – Reaktor Daya Eksperimental* BATAN Jakarta.
[2] Subekti M Dibyo S and Gofar M A 2016 *Proc. of National Seminar on Nuclear Energy Technology* (BATAN).
[3] Subekti M Isnaini M D and Hastuti E P 2013 Tri Dasa Mega **15**(2):67-76.
[4] Zhe S Jun S Chunlin W and Yuanle M 2014 Nuclear Engineering and Design **271**:479-486.
[5] Sepielli M Negrenti E Parisi C and Cappelli M 2011 *20th International Conference Proceeding on Nuclear Energy for New Europe Slovenia* p 809.1.
[6] Gao Q Zhou K F Yang S G Mao H and Zhu J 2014 Applied Mechanics and Materials **1385**:496-500.
[7] Ibrahim S J Ewim D R and Edeoja O A 2013 Leonardo Electronic Journal of Practices and Technologies **22**:93.

IOP Conf. Series: Journal of Physics: Conf. Series **962** (2018) 012054 doi:10.1088/1742-6596/962/1/012054

[8] Subekti M Kristedjo Isnaini M D, and Endiah P H 2011 *Prosiding Seminar Nasional Sains dan Teknologi Nuklir* (BATAN) p 119.

[9] Subekti M Kristedjo and Isnaini M D *17th National Seminar Proceeding on Nuclear Power Plant Technology and Safety* (BATAN) p 374.

[10] Subekti M Isnaini M D and Hastuti E P 2012 *Prosiding Pertemuan dan Presentasi Ilmiah Penelitian Dasar Ilmu Pengetahuan dan Teknologi* Nuklir Book I (BATAN) p 76.

[11] Haipeng L Xiaojin H Liangju Z 2008 Annals of Nuclear Energy **35**:1642-1651.

[12] Wei C Sui Z Sun J and Shan W Z 2012 *Proceeding of 20th International Conference on Nuclear Engineering Collocated with the ASME 2012 Power Conference* (California, USA)

[13] 1984 "Reactor Core Design of High Temperature Gas-Cooled Reactors", Safety Standard of the Nuclear Safety Standard Commission (KTA), KTA 3102, Part 1-4.

[14] Dibyo S Irianto I D 2017 Tri Dasa Mega **19**(1):33-40.

Modelling the radiolysis of RSG-GAS primary cooling water

S L Butarbutar, R Kusumastuti, M. Subekti, G R Sunaryo

Center for Nuclear Reactor Technology and Safety, National Nuclear Energy of Agency (BATAN), Puspiptek Area Building 80, Banten, Indonesia

sofia@batan.go.id

Abstract. Water chemistry control for light water coolant reactor required a reliable understanding of radiolysis effect in mitigating corrosion and degradation of reactor structure material. It is known that oxidator products can promote the corrosion, cracking and hydrogen pickup both in the core and in the associated piping components of the reactor. The objective of this work is to provide the radiolysis model of RSG GAS cooling water and further more to predict the oxidator concentration which can lead to corrosion of reactor material. Direct observations or measurements of the chemistry in and around the high-flux core region of a nuclear reactor are difficult due to the extreme conditions of high temperature, pressure, and mixed radiation fields. For this reason, chemical models and computer simulations of the radiolysis of water under these conditions are an important route of investigation. FACSIMILE were used to calculate the concentration of O_2 formed at relatively long-time by the pure water γ and neutron irradiation (pH=7) at temperature between 25 and 50 ^oC. This simulation method is based on a complex chemical reaction kinetic. In this present work, 300 MeV-proton were used to mimic γ-rays radiolysis and 2 MeV fast neutrons. Concentration of O_2 were calculated at 10^{-6} - 10^6 s time scale.

Keywords: radiolysis, FACSIMILE, RSG-GAS, fast neutrons, γ-irradiation.

1. Introduction

RSG-GAS reactor is a multipurpose reactor that is utilized for research and scientific purposes. It is an open pool type reactor and use demineralized light (H_2O) water as coolant and neutron moderator at the same time. Primary heat transport system in this reactor operates at 40.5-50.57 ^oC [1]. Cooling water is circulating in the reactor core, where the radiation fields mix at the Linear Energy Tansfer (LET), energy deposited per unit length, typically vary ~0.3 sampai 40-60 keV/μm [2]. Irradiation process results in chemical decomposition of water molecule (radiolysis) which generates to the formation of free radical (e^-_{aq}, $H^\bullet$, $^\bullet OH$, dan $HO_2^\bullet/O_2^{\bullet -}$) and molecular product (H_2 dan H_2O_2), and O_2 as well. O_2 is a secondary product which can lead to corrosion of reactor component and associated reactor piping [2-5]. Corrosion can escalate operation and maintenance cost, radioactive contamination and radiation risk to the workers. Radiation mechanisms depend on several factors, while appropriate water chemistry plays an important role to suppress corrosionand its consequences to the material. RSG GAS have been operating since 30 years ago, when it attained the criticality at 1987, therefore maintening material structure is neccessary important to extend the life span of reactor.

Reliable understanding on the radiolysis process of reactor coolant is very important to control the reactor water chemistry [7]. There are two important key parameters to assess the chemical effect of ionizing radiation. Those key parameters are (1) yield or *G*-values for γ-ray and neutron (number of species formed per unit of absorbed energy) and (2) reaction rate constants of involved reactions between species. It is worth mentioning that *G*-values have important role to predict the oxidator concentration, in this case O_2, which can cause corrosion of material.

Direct observation or direct chemical measurement in and around high flux reactor core is difficult to conduct. Besides, there is no normal chemical measurement apparatus which is compatible towards the existing of high neutron flux. For these reasons, computer simulation of radiolysis under those conditions is a useful method. In this research work, FACSIMILE software was used to model the radiolysis mechanism then to calculate the concentration of generated O_2. The coolant radiolysis caused by the interaction of γ-ray and neutron is occured at the temperature range of 25 up to 50 °C. *G*-values for each ionizing radiation are obtained from the calculation results using Monte Carlo simulation.

This research were conducted to provide the radiolysis model of RSG GAS cooling water, since there is no available information in the literature currently, while plenty of information are available for nuclear power plants. It is considered that radiolysis model of RSG GAS cooling water using FACSIMILE can give the information of O_2 concentration which can lead to corrosion of reactor material.

2. Methodology

Radiolysis process causes formation of free radical species, molecular products and chemical reaction between formed species. In this research, batch system continuous irradiations were assumed. Radiolytic species was expected to reach the steady state concentration at longer time relatively to the time scale of homogeneous stage in the radiation interaction. To model this radiolysis, several parameters are needed, such as *G*-values, set of possible reactions, reaction rate constants and dose rate.

Simulation at room temperature (25 °C) should be done at first in order to validate reactions set feeded to the model. It should be done due to the abundant experimental data are exist at room temperature. Then, simulation for temperature above 25 °C were done. Radiolysis simulation is a simultaneously chemical reaction analysis and the effect of multivariable differential as a function of time. In this work, FACSIMILE software was used to integrate differential equations [8]. FACSIMILE software is a powerful modelling tool that is designed to efficiently solve differential equations, with the focus is on modelling the kinetics of physical and chemical systems. There are many benefits of using FACSIMILE such as to model transport and chemistry problems and to develop useful models rapidly with specific facilities for modelling chemical kinetics.

2.1. G-values

G-values of radiolytic species depend on several factors, such as temperature and LET. To gain the objective of this work, *G*-values of radiolytic species formed by the interaction of γ-ray and neutron with water at room and higher temperature are required. *G*-values which obtained from Monte Carlo simulation are shown at Table 1.

Table 1. *G*-values of radiolytic species for light water irradiation by γ-ray and neutron interaction

Species	γ-ray [9]		Neutron [10]	
	25 °C	50 °C	25 °C	50 °C
e^-_{aq}	2.64	2.62	0.43	0.81
$H^\bullet$	0.57	0.64	0.58	0.59
$^\bullet OH$	2.82	2.84	0.93	0.98
H_2	0.45	0.48	1.07	0.91
H_2O_2	0.65	0.63	1.11	1.05

2.2. Reaction rate constants

Reaction rate constants, reaction mechanisms from Atomic Energy of Canada Limited (AECL) compilation report of Elliot and Bartels were used in FACSIMILE [11]. This data base provides the best recommended value to be used in high temperature water radiolysis modelling, at temperature range 20-350 °C.

Table 2. Main reaction set and reaction rate constants for water radiolysis at room temperature (25 °C)[11]

Reaction	k (M^{-1} s^{-1})	Reaction	k (M^{-1} s^{-1})
$H^{\bullet} + H^{\bullet} \rightarrow H_2$	5.2×10^{9}	$e^{-}_{aq} + e^{-}_{aq} \rightarrow H_2 + 2\,OH^{-}$	7.3×10^{9}
$H^{\bullet} + {}^{\bullet}OH \rightarrow H_2O$	1.6×10^{10}	$e^{-}_{aq} + H^{+} \rightarrow H^{\bullet}$	2.1×10^{10}
$H^{\bullet} + H_2O_2 \rightarrow H_2O + {}^{\bullet}OH$	3.6×10^{7}	$e^{-}_{aq} + O_2^{\bullet-} \rightarrow H_2O_2 + 2\,OH^{-}$	1.3×10^{10}
$H^{\bullet} + e^{-}_{aq} \rightarrow H_2 + OH^{-}$	2.8×10^{10}	$e^{-}_{aq} + HO_2^{-} \rightarrow O^{\bullet-} + OH^{-}$	3.51×10^{9}
$H^{\bullet} + OH^{-} \rightarrow H_2O + e^{-}_{aq}$	2.4×10^{7}	$e^{-}_{aq} + O^{\bullet-} \rightarrow 2\,OH^{-}$	2.31×10^{10}
$H^{\bullet} + O_2 \rightarrow HO_2^{\bullet}$	1.3×10^{10}	$e^{-}_{aq} + H_2O \rightarrow H^{\bullet} + OH^{-}$	15.8
$H^{\bullet} + HO_2^{\bullet} \rightarrow H_2O_2$	1.1×10^{10}	$e^{-}_{aq} + O_2 \rightarrow O_2^{\bullet-}$	2.3×10^{10}
$H^{\bullet} + O_2^{\bullet-} \rightarrow HO_2^{-}$	1.1×10^{10}	$e^{-}_{aq} + HO_2^{\bullet} \rightarrow HO_2^{-}$	1.3×10^{10}
$H^{\bullet} + HO_2^{-} \rightarrow {}^{\bullet}OH + OH^{-}$	1.5×10^{9}	$e^{-}_{aq} + O(^3P) \rightarrow O^{\bullet-}$	2.0×10^{10}
$H^{\bullet} + O(^3P) \rightarrow {}^{\bullet}OH$	2.0×10^{10}	$e^{-}_{aq} + O_3 \rightarrow O_3^{\bullet-}$	3.6×10^{10}
$H^{\bullet} + O^{\bullet-} \rightarrow OH^{-}$	2.0×10^{10}	$H^{+} + O^{\bullet-} \rightarrow {}^{\bullet}OH$	5.0×10^{10}
$H^{\bullet} + O_3 \rightarrow O_2 + {}^{\bullet}OH$	3.7×10^{10}	$H^{+} + O_2^{\bullet-} \rightarrow HO_2^{\bullet}$	5.0×10^{10}
$H^{\bullet} + O_3^{\bullet-} \rightarrow OH^{-} + O_2$	1.0×10^{10}	$H^{+} + OH^{-} \rightarrow H_2O$	1.2×10^{11}
${}^{\bullet}OH + {}^{\bullet}OH \rightarrow H_2O_2$	6.3×10^{9}	$H^{+} + O_3^{\bullet-} \rightarrow {}^{\bullet}OH + O_2$	9.0×10^{10}
${}^{\bullet}OH + H_2O_2 \rightarrow HO_2^{\bullet} + H_2O$	2.9×10^{7}	$H^{+} + HO_2^{-} \rightarrow H_2O_2$	5.0×10^{10}
${}^{\bullet}OH + H_2 \rightarrow H^{\bullet} + H_2O$	4.0×10^{7}	$OH^{-} + O(^3P) \rightarrow HO_2^{-}$	4.2×10^{8}
${}^{\bullet}OH + e^{-}_{aq} \rightarrow OH^{-}$	3.6×10^{10}	$OH^{-} + HO_2^{\bullet} \rightarrow O_2^{\bullet-} + H_2O$	1.3×10^{10}
${}^{\bullet}OH + OH^{-} \rightarrow O^{\bullet-} + H_2O$	1.3×10^{10}	$O_2 + O^{\bullet-} \rightarrow O_3^{\bullet-}$	3.7×10^{9}
${}^{\bullet}OH + HO_2^{\bullet} \rightarrow O_2 + H_2O$	9.0×10^{9}	$O_2 + O(^3P) \rightarrow O_3$	4.0×10^{9}
${}^{\bullet}OH + O_2^{\bullet-} \rightarrow O_2 + OH^{-}$	1.1×10^{10}	$HO_2^{\bullet} + O_2^{\bullet-} \rightarrow HO_2^{-} + O_2$	9.7×10^{7}
${}^{\bullet}OH + HO_2^{-} \rightarrow HO_2^{\bullet} + OH^{-}$	8.3×10^{9}	$HO_2^{\bullet} + HO_2^{\bullet} \rightarrow H_2O_2 + O_2$	1.94×10^{8}
${}^{\bullet}OH + O(^3P) \rightarrow HO_2^{\bullet}$	2.02×10^{10}	$HO_2^{\bullet} + O(^3P) \rightarrow O_2 + {}^{\bullet}OH$	2.02×10^{10}
${}^{\bullet}OH + O^{\bullet-} \rightarrow HO_2^{-}$	1.0×10^{9}	$HO_2^{\bullet} + H_2O \rightarrow H^{+} + O_2^{\bullet-}$	1.4×10^{4}
${}^{\bullet}OH + O_3^{\bullet-} \rightarrow O_2^{\bullet-} + HO_2^{\bullet}$	8.5×10^{9}	$O_2^{\bullet-} + O^{\bullet-} \rightarrow O_2 + 2\,OH^{-}$	6.0×10^{8}
${}^{\bullet}OH + O_3 \rightarrow O_2 + HO_2^{\bullet}$	1.11×10^{8}	$O_2^{\bullet-} + H_2O \rightarrow HO_2^{\bullet} + OH^{-}$	0.155
$H_2O_2 + e^{-}_{aq} \rightarrow OH^{-} + {}^{\bullet}OH$	1.1×10^{10}	$O_2^{\bullet-} + O_3 \rightarrow O_3^{\bullet-} + O_2$	1.5×10^{9}
$H_2O_2 + OH^{-} \rightarrow HO_2^{-} + H_2O$	1.33×10^{10}	$HO_2^{-} + H_2O \rightarrow H_2O_2 + OH^{-}$	1.27×10^{6}
$H_2O_2 + O(^3P) \rightarrow HO_2^{\bullet} + {}^{\bullet}OH$	1.6×10^{9}	$HO_2^{-} + O^{\bullet-} \rightarrow O_2^{\bullet-} + OH^{-}$	8.02×10^{8}
$H_2O_2 + O^{\bullet-} \rightarrow HO_2^{\bullet} + OH^{-}$	5.55×10^{8}	$HO_2^{-} + O(^3P) \rightarrow O_2^{\bullet-} + {}^{\bullet}OH$	5.3×10^{9}
$H_2 + O(^3P) \rightarrow H^{\bullet} + {}^{\bullet}OH$	4.77×10^{3}	$O^{\bullet-} + O^{\bullet-} \rightarrow H_2O_2 + 2\,OH^{-}$	1.0×10^{8}
$H_2 + O^{\bullet-} \rightarrow H^{\bullet} + OH^{-}$	1.3×10^{8}	$O^{\bullet-} + O_3^{\bullet-} \rightarrow 2\,O_2^{\bullet-}$	7.0×10^{8}
$O(^3P) + O(^3P) \rightarrow O_2$	2.2×10^{10}	$O^{\bullet-} + H_2O \rightarrow {}^{\bullet}OH + OH^{-}$	1.3×10^{6}
$O(^3P) + H_2O \rightarrow 2\,{}^{\bullet}OH$	1.9×10^{3}	$O_3^{\bullet-} + H_2O \rightarrow O^{\bullet-} + O_2$	46.5

2.3. Dose rate
Dose rate is one of key parameters to model coolant radiolysis. The following is series of calculations to obtain average dose rate of each radiation type used in this work [12].

2.3.1. γ-ray dose rate
Total gamma sources of RSG-GAS (consist of radioactive capture, spontaneous fission and radioactive decay) in operating condition: 7.91E20 photon/s

Pool area of RSG-GAS:

2 x (60 cm x (8 x 7.71 cm)) + 2 x (60 cm x (8 x 8.1 cm)) + 2 x ((8 x 7.71 cm) (8 x 8.1 cm)) = 23,171 cm^2

γ-ray Flux = 7.91E20 photon/s x (1/23,171 cm^2)

Gamma (γ) flux conversion factor towards dose rate = 1.42E-4 (rem/hr)/(photon/cm^2 s)

Thus, γ-ray dose rate:

= 1.42E-4 (rem/hr)/(photon/cm^2 s) x 7.91E20 photon/s x (1/23,171 cm^2)

= 4.84E12 rem/hr ; if 1 rem = 0.01 Gy and 1 hr = 3600 s, so that

= 1.34E7 Gy/s

2.3.2. Neutron dose rate
Flux neutron = 3.34E14 neutron/ cm^2 s

neutron flux conversion factor towards dose rate = 1.04E-2 (rem/hr)/(neutron/cm^2 s)

Thus, neutron dose rate:

= 1.04E-2 (rem/hr)/(neutron/cm^2 s) x 3.34E14 neutron/ cm^2 s

= 3.48E12 rem/hr ; if 1 rem = 0.01 Gy and 1 hr = 3600 s, so that

= 9.67E6 Gy/s

3. Results and Discussion
Modelling the radiolysis of RSG GAS primary cooling water using FACSIMILE software has been conducted. This software has been validated with several calculations at both conditions, neutral and the existing of solution. In order to validate our feeded model, it is necessary to check mass equilibrium and the ratio of hydrogen atom number and oxygen atom number, principally, is about 2:1 [13]. Variation of O_2 concentrations towards time due to interaction of γ-ray and neutron with water are shown in red and blue solid line, respectively in the following figures.

3.1. Simulation at 25 ^oC
Figure 1 shows O_2 concentrations as a function of times that are obtained from our simulation at room temperature (25 ^oC) and aerated condition (pool type). The behaviour caused by these two radiation types are different. At early time, oxygen concentrations generated by γ-ray and neutron increase until

0.01 s. At this time scale, this phenomenon is due to the aeration condition,where the oxygen content over the pool could diffuse into the coolant. Another hypothetis that could be addressed to elucidate the escalation of O_2 concentrations is the contribution of chemical reactions as follows (the order is based on the high possibility to occur)

$$e^-_{aq} + O_2 \rightarrow O_2^{\bullet-} \quad (1)$$

$$H^\bullet + O_2 \rightarrow HO_2^\bullet \quad (2)$$

$$O_2 + O(^3P) \rightarrow O_3 \quad (3)$$

$$O_2 + O^{\bullet-} \rightarrow O_3^{\bullet-} \quad (4)$$

Those reactions consume O_2 in the water, while the reactions which form O_2 as follows (the order is based on the high possibility to occur).

$$H^\bullet + O_3 \rightarrow O_2 + {}^\bullet OH \quad (5)$$

$$HO_2^\bullet + O(^3P) \rightarrow O_2 + {}^\bullet OH \quad (6)$$

$$H^+ + O_3^{\bullet-} \rightarrow {}^\bullet OH + O_2 \quad (7)$$

$$O_2^{\bullet-} + O^{\bullet-} \rightarrow O_2 + 2\,OH^- \quad (8)$$

$$HO_2^\bullet + HO_2^\bullet \rightarrow H_2O_2 + O_2 \quad (9)$$

$$HO_2^\bullet + O_2^{\bullet-} \rightarrow HO_2^- + O_2 \quad (10)$$

The difference between these two ionizing radiations can be observed at time > 0.01 detik.

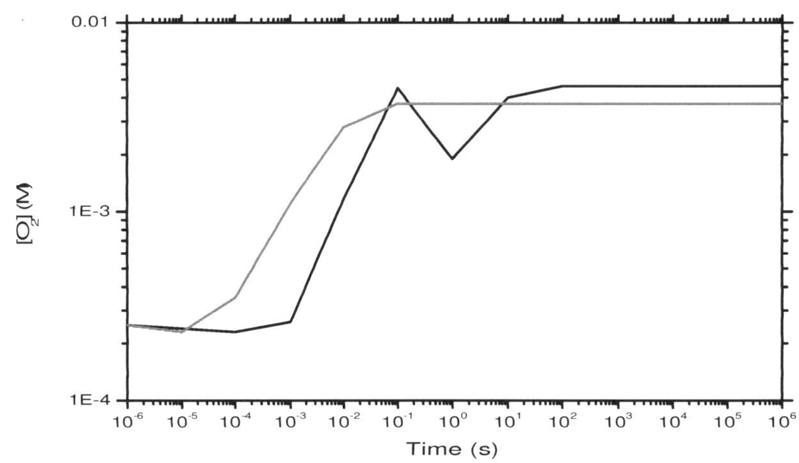

Figure 1. Variation of O_2 concentration towards time at 25 °C, red solid line for γ-ray interaction and black solid line for neutron interaction.

For γ-ray interaction, O_2 concentration reachs the steady state earlier (at time 0.1 s) rather than neutron interaction with water (which is at 100 s). Steady state condition is a condition when there are no more reactions that generate and consume O_2. In addition to that, O_2 concentration formed by γ-ray interaction with water is lower compared with neutron. It is known that O_2 is a secondary molecular product induced by radiolysis. Principally, radiolytic product, in this case free radicals, is found greater for radiolysis induced by γ-ray interaction (low-LET) rather than neutron interaction (high-LET), vice versa for molecular product formation such as O_2). This trend is signatured by the high-LET characteristic of neutron, itself, in which neutron deposit its energy at higher density as compared to γ-ray.

It is obvious that high LET deposits its energy at the vicinity thus spur/track which consists of free radical will overlap prior to spur/track lifetime. It results in the less free radical formation or in other words more molecular formation. Since there are more than one radiation types, thus oxygen formation will be accumulated.

At the time above 1E-5 s, both radiation types increase the number of oxygen concentration regardless the gap on neutron interaction at time 1s. The gap is induced by the reactions of oxygen consuming which play role at this time, having said that these reactions have relatively low reaction rate constant. Following 10 s later, there is a competition between reaction that consume oxygen and reaction that produce oxygen, from Fig 1, it looks like reactions to produce oxygen are more powerful prior to the steady state condition obtained.

3.2. Simulation at 50 °C

Figure 2 shows O_2 concentrations at aerated condition as a function of times as a result of γ-ray and neutron interaction with RSG-GAS cooling water. For γ-ray, O_2 concentration decline drastically after 0.01 s and even can be neglected. However for neutron, in contrary O_2 concentration increases significantly after 0.001 s after this radiation interacted with coolant. This phenomenon could be due to several reactions are sensitive to the temperature escalation, particularly reactions which form O_2 such as (5)-(10).

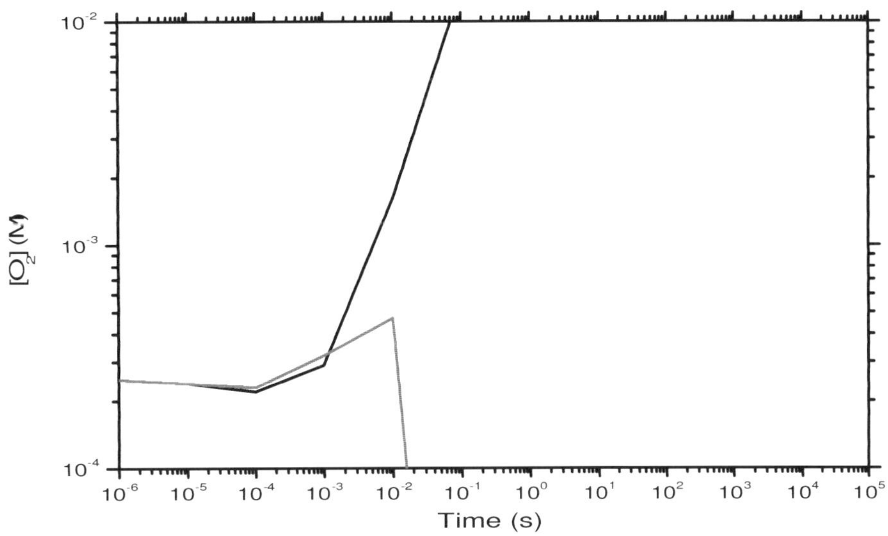

Figure 2. Variation of O_2 concentration towards time at 50 °C, red solid line for γ-ray interaction and black solid line for neutron interaction.

For γ-ray interaction, it looks like oxygen formation decrease significantly after 0.01 s. It can be confirmed that oxygen consuming reactions compete over the oxygen generating reactions. The opposite phenomenon is occured to the neutron interaction, where oxygen number increases as a function of time. However, a comprehensive investigation, especially at longer time, is required related to reactions behaviour towards the temperature above room temperature and radiation flux measurement. These important data will certainly give a clear model of coolant radiolysis. Moreover, the consequnces of the radiolysis of coolant towards reactor component can be predicted, take into account that this reactor has been operated for about 30 years.

4. Conclusion

In this research work, FACSIMILE was used to model radiolysis of RSG GAS cooling water in order to calculate O_2 concentration as a result of γ-ray and neutron interaction at temperature 25 and 50 ^{o}C. At 25 ^{o}C, O_2 concentration was calculated until it can reach the steady state condition at time between 10^{-6} and 10^{6} s. For γ-ray interaction, O_2 concentration reaches steady state at earlier time (at time 0.1 s) as compared to neutron interaction (at time 100 s). O_2 concentration formed by gamma is lower than neutron. At 50 ^{o}C, O_2 concentration as a result of γ-ray decreases drastically after 0.01 s and even can be neglected, however the contrary behaviour occured to neutron interaction. This phenomenon could be caused by several reactions which are sensitive with the escalation of temperature, therefore a comprehensive investigation are neccessary required. Moreover, to test our model and calculation, experimental data and measurement are needed.

Acknowledgment

The authors are wishing to thank national project fund (DIPA) of Center for Nuclear Reactor Technology and Safety, BATAN, year 2016 and Mr. Amir Hamzah for the valuable data and discussion regarding research reactor.

References

[1.] Laporan Analisis Keselamatan (LAK) RSG GAS , Rev 10.1, vol 2, Desember 2011

[2.] McCracken D R, Tsang K T, and Laughton, P J 1998 *Aspects of the physics and chemistry of water radiolysis by fast neutrons and fast electrons in nuclear reactors* Report AECL No.: 11895. Chalk River (Ontario): Atomic Energy of Canada Ltd.

[3.] Cohen P 1980 *Water coolant technology of power reactors* (La Grange Park (IL): American Nuclear Society).

[4.] Woods R J and Pikaev A K 1994 *Applied radiation chemistry: Radiation processing* (New York (NY): Wiley).

[5.] Research needs and opportunities in radiation chemistry workshop, Chesterton (IN), 19-22 April 1998, Report No.: DOE/SC-0003.Germantown (MD): U.S. Department of Energy,Office of Basic Energy Sciences; (1999) Available from http://www.science.doe.gov/production/bes/chm/Publications/RadRprt.pdf

[6.] Data processing technologies and diagnostics for water chemistry and corrosion control in nuclear power plants (DAWAC) 2006 Report No.: IAEA-TECDOC-1505.Vienna (Austria): International Atomic Energy Agency

[7.] Stuart C R 2000 *Radiation Research* vol 2 p 63

[8.] Curtis, AR *FACSIMILE/CHEKMAT User's Manual.*

[9.] Sanguanmith S, Meesungnoen J, Muroya Y, Lin M, Katsumura Y, Jay-Gerin J-P 2012 *Phys Chem Chem Phys* vol 14 p 16731

[10.] Butarbutar S L, Sanguanmith S, Meesungnoen J, Sunaryo G R, Jay-Gerin J-P 2014, *Radiat. Res., 181* 659

[11.] Elliot A J and Bartels D 2009 *The reaction set, rate constants and g-values for the simulation of the radiolysis of light water over the range 20 to 350 °C based on information available in*

2008, Report AECL No.:153-127160-450-001. Mississauga (Ontario): Atomic Energy of Canada Ltd.
[12.] Hamzah A 2015, personal communication
[13.] Takiguchi H, Ullberg M and Uchida S 2004 *J. Nucl. Sci. Technol.* vol 41 p 1881

Effect of Sol Concentration, Aging and Drying Process on Cerium Stabilization Zirconium Gel Produced by External Gelation

R Sukarsono, M Rachmawati, S R Susilowati, D Husnurrofiq, K. Nurwidyaningrum and A K Dewi

Center for Nuclear Fuel Technology, National Nuclear Energy Agency of Indonesia (Batan), Puspiptek Serpong, Tangerang, Indonesia

karsono_id@yahoo.com

Abstract. Cerium Stabilized Zirconium gel has been prepared using external gelation process. As the raw materials was used $ZrO(NO_3)_2$ and $Ce(NO_3)_4$ nitrate salt which was dissolved with water into Zr-Ce nitrate mixture. The concentration of the nitrate salt mixture in the sol solution was varied by varying the concentration of zirconium and cerium nitrate in the sol solution and the addition of PVA and THFA to produce a sol with a viscosity of 40-60 cP. The viscosity range of 40-60cP is the viscosity of the sol solution that was easy to produce a good gel in the gelation apparatus. Sol solution was casted in a gelation column equipped with following tools: a 1 mm diameter drip nozzle which was vibrated to adjust the best frequency and amplitude of vibration, a flow meter to measure the flow rate of sol, flowing of NH_3 gas to presolidification process. Gelation column was contained NH_4OH solution as gelation medium and gel container to collect gel product. Gel obtained from the gelation process than processed with ageing, washing, drying and calcinations to get round gel and not broken at calcinations up to 500°C. The parameters observed in this research are variation of Zr nitrate concentration, Ce nitrate concentration, ratio of Zr and Ce in the sol and ageing and drying process method which was appropriate to get a good gel. From the gelation processes that has been done, it can be seen that with the presolidification process can be obtained a round gel and without presolidification process, produce not round gel. In the process of ageing to get not broken gel, ageing was done on the rotary flask so that during the ageing, gels rotate in gelation media. Gels, then be washed by dilute ammonium nitrate, demireralized water and iso prophyl alcohol. The washed gel was then dried by vacuum drying to form pores on the gel which become the path for the gases resulting from decomposition of the gel to exit the gel. Vacuum drying can prevent cracking because the pores allow the gel to release the decomposition of the material during heating. Larger the concentration of nitric metal in sol solution, yields a gel with a larger diameter of gels. This research allows us to plan the diameter of the sintered particles to be made.

Keywords: high temperature reactor fuel, external gelation process, cerium stabilized zirconium,

I. Introduction

High temperature reactor (HTR) has a large prospect to be built in Indonesia. This reactor has a high level of safety because there will be no fuel melting in the accident state[1]. This reactor is also suitable to be built in Indonesia because it can be made modular in accordance with the needs so that it is suitable for Indonesia which consists of many islands.

Research on high temperature reactor fuel has been carried out in Batan since a few years ago. The preparation of the ADU Gel is a critic process in the HTR fuel making circuit because if the ADU gel produced from the gelation process is not good, it can not be further processed into HTR fuel. ADU gel making can be through 2 methods, namely the internal and external gelation methods. HTR fuel research in Indonesia has been decided to focus on gel preparation through external gelation methods. ADU gel preparation research aims to obtain a good gel, a gel that is round, uniform and not cracked on heating (drying, calcinations, reduction and sintering)[2]. This research needs to be done because many factors influence to get good gel result. This research will find the best method and operation condition in the gelation process preparation of ADU gel thorough surrogate materials of cerium stabilized zirconium (CSZ).

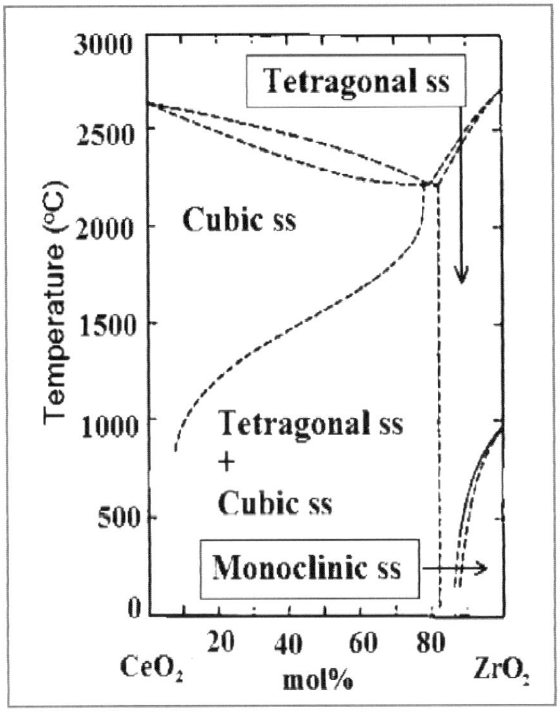

Figure 1. Phase Diagram of ZrO_2 and CeO_2

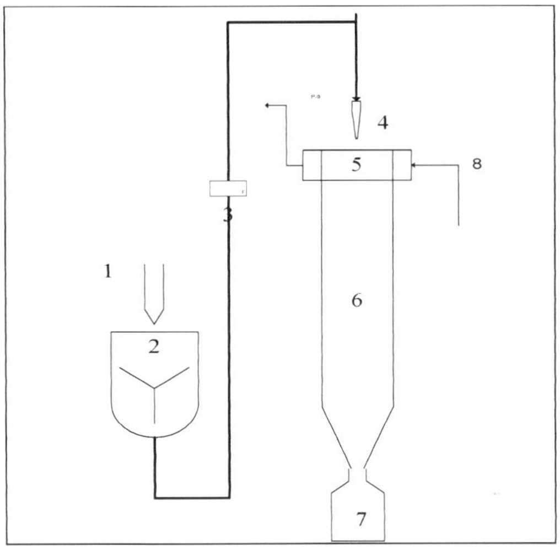

Explanation:
1. *Hoper*
2. Sol preparation tank
3. Flowmeter
4. Drip nozzle
5. NH_3 column
6. Casting column
7. Gel container
8. NH_3 gas flow

Figure 2. Gelation Column Diagram

Cerium stabilized zirconium (CSZ) is used as surrogate in studying the gelation process and futher processes of preparing uranium coated particle TRISO for core of HTR fuel[3]. The result of the gelation process using CSZ will be a reference in gelation process using uranium to gel ADU. The ADU gel is a gel particle that after further processing will be used to be the core of high temperature reactor fuel. The ADU gel particles are calcined into tri uranium octa oxide (U_3O_8) then reduced to UO_2 particles (uranium dioxide) and sintered into UO_2 sintered particles. The UO_2 sintered particles then are coated with pyrocarbon and silica carbide coating materials, into TRISO coated particles. Coated particles are the core of high temperature reactor fuel[4][5]. Some metals such as cerium stabilized zirconium, alumina, and thorium can be used as uranium surrogates because they have the same reaction in gel formation and similar change to metal oxides in calcinations, reduction and sintering process. This study uses CSZ as surrogate for uranium. Cerium stabilized zirconium can be made sol solution, using the same additive with is used in the uranium sol preparation. The CSZ sol is casted to gel that has properties similar to ADU gel.

The external gelation process is one of some method to prepare the metal nitrate to be gel that is used in industry. External gelation process use NH_4OH solution for media gelation. The CSZ sol can form a gel in the external gelation processes that can be treated by calcinations, reduction and sintering without the cracking. By using the CSZ gel as an ADU gel substitute, the process research can be carried out more freely and the operating conditions obtained can be used as reference in the manufacture of high temperature reactor (HTR) fuel using uranium.

The sol-gel process is used for the manufacture of metal oxide particles used in industry. The metals are made sol solution by converting them into nitrate salts and adding certain additives and then being gel particles. The gel particles then dried and calcined to form the metal oxide particles for various purposes in the industry. Making gel particles through the gelation process can be through 2 methods of internal and external gelation. In the internal gelation the polymerization process of gel occurs from operation condition change for example through temperature change and initiate the reaction of the compound that already present in the sol solution, while in the external gelation, reaction occurs by the material coming from outside of the sol for example from the medium of NH_4OH solution. The nitrate metal compounds used in this study were zirconium and cerium nitrate having high solubility of 693 g/L and 2,340 g/L respectively. The core of the gelation process is the change of pH in the sol solution that is dripped on the gelation medium. At the time the sol drop drops in the gelation medium there is a change of pH around 10, initiate the complex reaction and polymerization that solidify the droplet of the sol to gel. The mechanism of the gelation and polymerization reaction of metal nitrate can be explained as follows: The sol solution containing cerium nitrate, zirconium oxy nitrate and additives is dripped onto the NH_4OH solution medium and there is a following polymerization reaction[6][7]:

$$ZrO(NO_3)_3 + NH_4OH \rightleftarrows ZrO(OH)_2 + NH_4NO_3 \quad (1)$$
$$ZrO(OH)_2 + 9H_2O \rightleftarrows (Zr(H_2O)_8)^{4+} + 4OH^- \quad (2)$$
$$(Zr(H_2O)_8)^{4+} + H_2O \rightleftarrows (Zr(OH)(H_2O)_7)^{3+} + H^+ \quad (3)$$
$$(Zr(OH)(H_2O)_7)^{3+} + H_2O \rightleftarrows (Zr(OH)_2(H_2O)_6)^{2+} + H^+ \quad (4)$$
$$4(Zr(OH)_2(H_2O)_6)^{2+} \rightleftarrows (Zr(OH)_8(H_2O)_{16})^{8+} + 8\,H_2O \quad (5)$$

$(Zr_4(OH)_8(H_2O)_{16})^{8+}$ is called a zirconium tetrametric complex. The tetrametric zirconium complex joins one another and becomes a gel.

Reaction of cerium nitrate hydrolyzed in polymerization process

$$Ce(NO_3)_3 + H_2O \rightleftarrows [Ce(OH)_2](NO_3)_2 + HNO_3 \quad (6)$$
$$2[Ce(OH)_2](NO_3)_2 \rightleftarrows [Ce_2(OH)_4](NO_3)_4 \quad (7)$$
$$2[Ce_2(OH)_4](NO_3)_4 + H_{2O} \rightleftarrows [Ce_4(OH)_8](NO_3)_7 + H_2O \quad (8)$$

Cerium nitrate is hydrolyzed result polymerization of cerium nitrate to cerium gel.

Zirconium and cerium oxide mixtures form a solid solution with a crystalline structure that varies depending on the composition and temperature of the solid solution. According to Zhang, F.[8], the crystal structure formed from the solid mixtures ZrO_2 and CeO_2 can be seen in Figure 1. In the large mole ratio of Zr/mol Ce or the Ce mole % less than 10, the crystals are monoclinic and unchanged until heating is 500°C.

The CSZ gelation process is carried out in the gelation column unit that consists of a reactor for sol solution is made. This reactor also functions as a container for feeding of sol solution into gelation columns. Gelation column containing ammonium hydroxide liquid equipped with a nozzle system of 1 mm diameter which is vibrated at specific frequencies and amplitudes and equipped container at the bottom of the gelation column[3][8][9]. The gelation diagram can be seen in Figure 2 and the photo of the gelation apparatus can be seen in Figure 3.

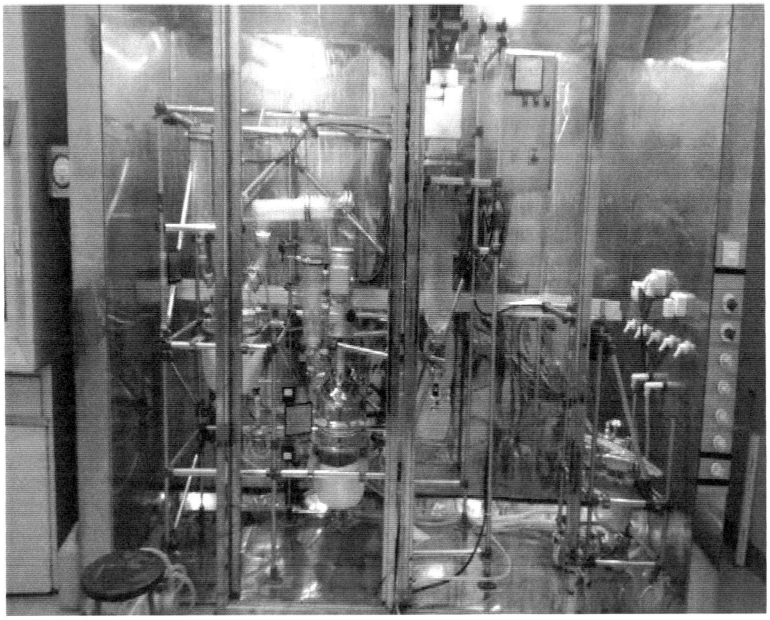

Figure 3. Gelation Column Unit

Gel resulted from gelation process, need to be aged to complete the polymerization reaction, continued by washing. The washing process use to remove residual materials and undesirable process results in the gels. Then the washed gel is dried to a gel free from reactant residues, byproducts reaction and washers [4][5][10]. The dried gel is then calcined to be converted to oxide and sintered into a sintered UO_2 particle[11][12][13].

2. Methodology

The materials used were cerium nitrate ($Ce(NO_3)_3.6H_2O$) and zirconium oxy nitrate $ZrO(NO_3)_2.2H_2O$ 99% purity. Both of materials were dissolved with mineral free water into nitrate salt mixtures. Parent solution of $ZrO(NO_3)_2$ was prepared by introducing $ZrO(NO_3)_2.2H_2O$ in a dissolution reactor than added by certain volume of pure water and heated to 70 °C. The concentration of $ZrO(NO_3)_2$ solutions was made to 277 g/L. A certain volume of $ZrO(NO_3)_2$ solution was added piecemeal powder $Ce(NO_3)_36H_2O$ with a mole weight ratio of Zr $(NO_3)_3$ and a mole of $Ce(NO_3)_3$ mole is varied between 8 to 13. Zirconium and cerium nitrate concentrations are varied to see the change in the resulting kernel diameter. Heating and stirring are continued until the soluble solids and then cooled. The solution (Zr-Ce) nitrate mixture was added by urea to the preneutralization process until pH 3 and stirred until homogeneous. Poly vinyl alcohol (PVA) solution of 220g/L concentration was prepared by dissolving PVA in pure water and heated to homogeneous. Sol solution was prepared by taking a mixture of Zr-Ce nitrate with a certain volume then added by a certain volume of 220g/L PVA solution and a certain volume of tetra hydro furfuryl alcohol (THFA), then stirred while heated to homogeneity.

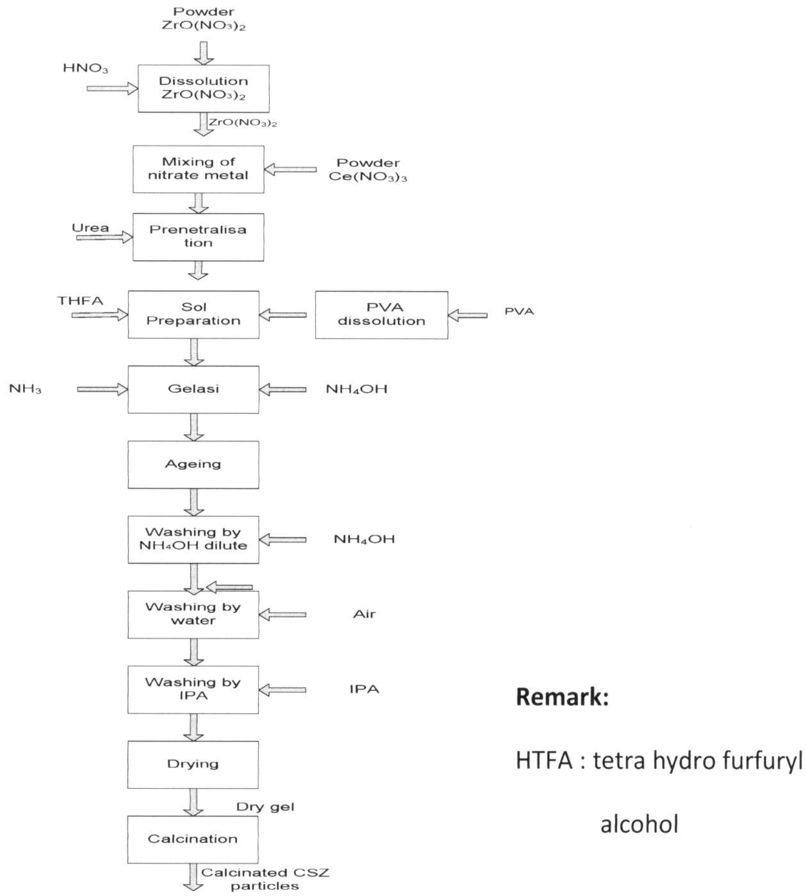

Figure 4. Flowchart of calcined CSZ particle Preparation

Sol solution after 1 night then is dripped (casting) in the gelation column by dripping the sol into a 14N ammonium hydroxide solution. The gelation apparatus can be seen in Figure 2 and 3. The gelation apparatus is prepared by filling the gelation column with NH_4OH solution and delivering NH_3 gas for gel presolidification. Sol solution in the feeder tank pressed by compressed air and fed to the dripping device. Sol solution flows through a nozzle of 1 mm diameter which is vibrated at a certain frequency and amplitude. The droplet drops before it falls onto the gelation medium through the NH_3 gas layer resulting in gel presolidification and surface hardening of the drops. The gelation process uses a specific vibration frequency, while the solvent flow rate and amplitude are set to produce a separate and uniform of drops. Gel resulted from gelation process is collected in the container and then after the gelation process, taken for next process.

The ageing process is a process to complete the gelation process by soaking the gel in a NH_4OH solution. The ageing process is done in a flask that is rotated so that NH_4OH contact evenly across the gel surface. The gel was introduced into a flask filled with a solution of NH_4OH 14 N gelation medium then rotated 20 rpm. The ageing process was done for 2 hours. Gel after ageing process needs to be washed. The washing process aims to remove the washer materials remaining on the surface gels, nitrate, residual additives and by-products of the gelation process from within the gel granules. Medium gelation and other ingredients still attached to the granular surface should be removed as it may interfere with the drying process, heating and causing the gel cracking. The first wash using a dilute NH_4OH to take the concentrated NH_4OH residue and the remaining organic material in the gel.

The 2nd washing was done using dematerialized water to take the remaining of NH_4OH. The third washing removes the remaining water in the gels with an organic solvent. Commonly used organic solvents include: CCl_4, iso prophyl alcohol (IPA), acetone, hexane and others which have low boiling point to volatilize. The gel was washed with dilute ammonia 3 times for 30 minutes, 6 times with demineralized water and last 3 times with isopropyl alcohol

Wet washing gel was dried in 2 ways: Drying Method A: Gel aerated 1 night then heated at 80°C for 1 hour, raised to 120°C held for 1 hour. Drying Method B: The wet gel is treated by vacuum heating at 60°C for 2 hours and 80°C for 1 hour, at a vacuum pressure of -0.3 atm. The calcinations process for converting the elements into oxides is carried out by gradual heating at temperatures of 300 ° C, 400 ° C and 500 ° C. The calcined CSZ gel preparation process diagram can be seen in Figure 4[3].

3. Results And Discussion

The successful formation of good gels in the gelation process using gelation column is determined by 2 factors, that are the tool factor and the sol solution properties factor. In terms of the tool factor, the variables that determine success include the dropout nozzle, the frequency and amplitude vibration setting, the diameter of nozzle, the adjustment of the sol flow rate, the nozzle height setting, the NH_3 gas stream for presolidification and facilities that facilitate the operation of tools such as gel container, feeder and expenditure materials, vacuum pump arrangement etc. In this research, the diameter of the drip nozzle is used 1 mm[3], the nozzle height which was determined from preliminary experiments was 22 cm, while the flow rate of solvent solution and amplitude is adjusted for certain vibration frequencies to produce the best droplets. Good droplets are determined from the separation of one drop with another and the stability of the droplets. The regulation of vacuum pressure and the NH_3 gas flow during the gelation process was also determined from a preliminary experiment with a NH_3 gas flow of 3 ml/min.

In terms of sol solution factor, the most important property is viscosity, because the viscosity determines whether the sol, when feed to the gelation process, can produce separate and homogeneous drops or not. The viscosity of the sol is also determined by the material and the type of additive, metal concentration, the ratio of the metals used when more than one metal is employed, the ratio of additive materials such as PVA and THFA, the surface tension of materials, the sol density and the pH of the sol. Variables in raw materials such as concentration, metal ratio in sol solution are usually have selected to produce a particular gel. To produce a sol solution with a certain viscosity which can be processed is by adjustment of the additive of PVA, THFA and the ratio between PVA and THFA[13]. In making sol solution, the first thing that must be considered is viscosity of sol. A sol solution that has a viscosity 50 cP (centi Poise) or that has a range of viscosity 40-60 cP, is easy to process in the gelation column. Changes in the sol composition may be performed but the latter should be followed by a change in the addition of the additive to allow viscosity to fall within the range [3].

Gels produced from gelation process, and then were washed with dilute NH_4OH, demineralized water and iso prophyl alcohol. Gels that were washed then was carried out by drying process. Drying is intended to remove residual IPA, and water in the gel. Drying has been carried out in 2 ways, the first by allowing the washed gel in open air for 1 night so that most of the IPA evaporates and the wet gel diameter shrinks, continued by drying in the furnace at 80°C and continuously drying the temperature of 120°C. Drying the 2nd way is by drying using a vacuum of -0.3 atm in the rotary flash at 60 ° C and 80 ° C.

Gel gelation process product, was performed by thermal gravimetric analysis (TGA)/differential thermal analysis (DTA) with an argon gas atmosphere to see gel characteristic at heated time up to 600°C. According to Kumar [9] for uranium and cerium gels at first low weight loss of 76°C-270°C due to dehydration characterized by endothermic reaction and followed by exothermic. The second weight reduction due to NH_3 evaporation at 270°C-520°C and the 3 rd weight reduction due to calcinations at temperature 520°C-550°C. In the CSZ gel the dehydration process occurs at the first weight reduction of temperature 70°C-100°C, followed by the removal of NH_3 gas at second weight reduction at temperatures of 100-260°C and calcinations from at temperature of 260°c - 310°C. The TGA/DTA analysis results for the washing gel can be seen in Figure 5 and for the heated gel at 120°C is shown in Figure. 6.

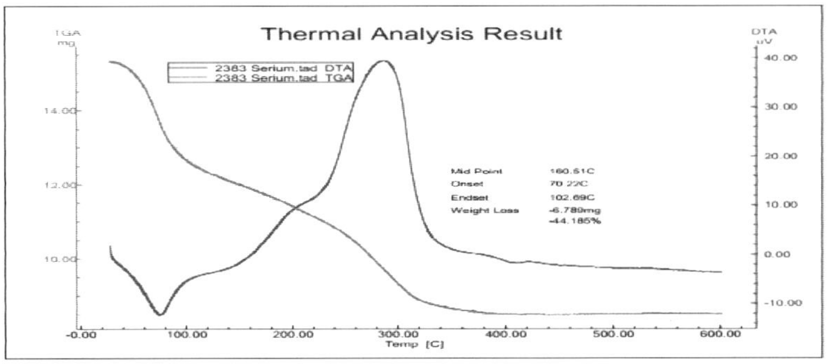

Figure 5. Thermal analysis result of washed (Zr-Ce) gel by heating rate of 2°C/men

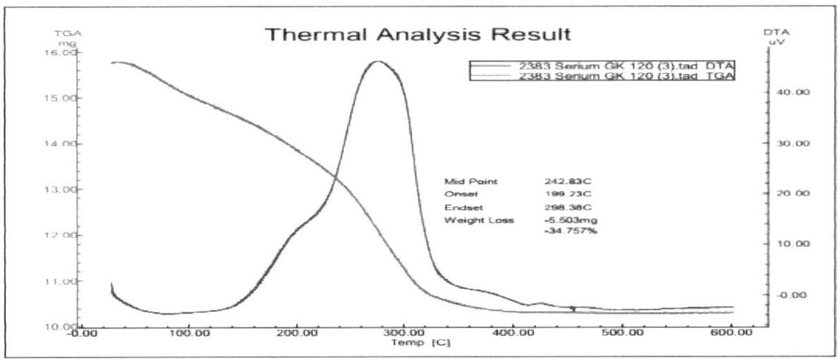

Figure 6. Thermal analysis result of 120°C heated (Zr-Ce) gel by heating rate of 2°C/men

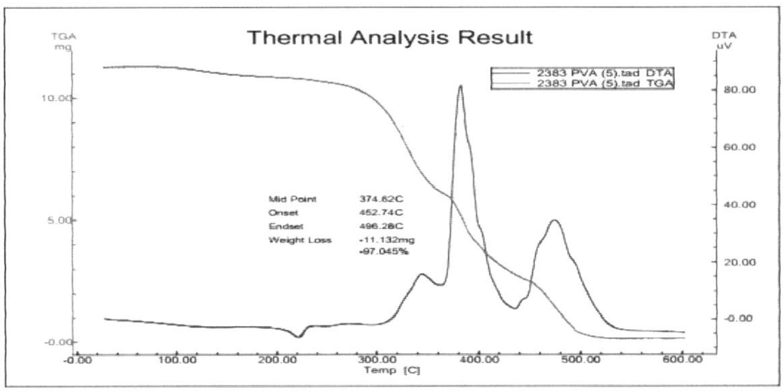

Figure 7. Thermal analysis result of PVA decomposition by heating rate of 2°C/men

Figure 5 is the result of TGA/DTA analysis for dried gel. In the Figure 5, the gel still contains residuals of washing process such as IPA and water which is indicated by weight reduction at a temperature of 60-120°C. Evaporation of IPA and water is a process of diffusion and for the diffusion process is required heat so that on the DTA looks endothermic process. After that there was a slight weight decrease at heating of 100-120°C and no increasing curve in DTA charts, when the water evaporated. At 120°C there is exothermic heat which indicates the decomposition of THFA. At a 265-280°C higher exothermic heat occurs due to decomposition of PVA and by-products such as ammonium nitrate. Temperatures of 300-400°C occur decomposition into zirconium oxide (ZrO_2) and cerium oxide (CeO_2). Figure 6 is the result of TGA/DTA analysis of the gel heated at 120°C. Gels after heated at 120°C, remain of water and IPA in the gels has been lost at the time of heating so that the graph is more gentle. The exothermic process occurs when the THFA decomposition is continued with PVA decomposition and ammonium nitrate byproducts.

Data of gel gelation and drying results can be seen in Figure 8 to 11. In Figure 8, gelation of zirconium oxy nitrate and cerium nitrate gel is shown. Concentration of zircon oxy nitrate 0.776 mol/L mol ratio Zr/mol Ce = 10.488, (Zr-Ce) concentration of 0.851 mol/L, PVA 4 gram/L of sol, THFA 15.4% volume, THFA/PVA = 3.85. The obtained sol has a viscosity of 54 cP so it can be casting in the gelation columns. The gelation result gel, then is ageing in the gel container at the bottom of the gelation column for 2 hours and then taken for washing. The washed gel is dried in open air for up to a minimum of 1 night and continued by drying in the furnace at 90°C and 112°C. If the Figure 8e and 8f, there are several gels of different color and are not uniform. These items seem more crack. This is possible because ageing with held in the container, there is no gel movement during the ageing process. Gel with different colors is caused because the process of ageing is not evenly distributed. Gel after washing and drying then is calcined at temperature of 200°C, 220°C, 320°C and 400°C.

Different temperatures of calcinations processes are to determine at what temperature the gel occurs cracking caused by weight reduction. If along the heating does not occur gel rupture, it is concluded that weight reduction due to dehydration, decomposition of organic material remnants or calcinations of gel, can be controlled, the gas out of the gel slowly and the diameter shrinkage does not cause gel rupture.

Figure 8a showed the washed gel. In the gel there is a material which is a mark of a gel line connecting one gel with another formed at the time of the gelation process. This happens because the gelation process between one drop and the other drops has not been completely separated so there is a line between drops. The gel color varies between heating results with each other because it is affected by the color of the dominant material that is still present in the gel. The color of the gel is initially yellow but after evaporation of the IPA material and water, the color becomes reddish from the red of Zr nitrate (Figure 8d). The red color will disappear at a warming temperature of 200°C and turn black (Figure 8e). At a temperature of 300-400°C is a whitish yellow oxide (Figure 8g, 8h, 8i). Calcinations at temperature of 200°C gels already occurred particles fracture. This causes gas to come out as a result of the decomposition of the material can not come out slowly and cause the particle crack.

Gel with open air drying method or gel left in open air for 1 night, can be seen in Figure 9. Gel is made from sol gelation process with zircon concentration of oxy nitrate 0.701 mol/L the ratio of mol Zr/mol Ce = 11.654, the concentration (Zr-Ce) nitrate 0.761mol/L, PVA 3.252 gram/L sol, THFA 12.19% volume, THFA/PVA = 3.75. The obtained sol has a viscosity of 59 cP so it can be gelation using the available gelation columns. Figure 9a shows a slightly transparent washing gel. The ageing process is carried out in a gel container at the bottom of the gelation column. Seen there are several different colored gels caused by ageing of the ageing process. The gel color varies between heating results with each other because it is affected by the color of the dominant material that is still present in the gel. The color of the gel is initially yellow but after the evaporation of the IPA and water materials, the color becomes reddish from red Zr nitrate (Figure 9c and 9d). The red color will disappear at 120°C heating temperature and turn black (Figure 9e). At a temperature of 300-400°C is a whitish yellow oxide (Figure 9i, 9j, 9k, 9l). Calcinations at temperature of 200°C, particles fracture already occurred in the gels. These causes by gas to come out as a result of the decomposition of the material can not come out slowly and cause the particle cracking.

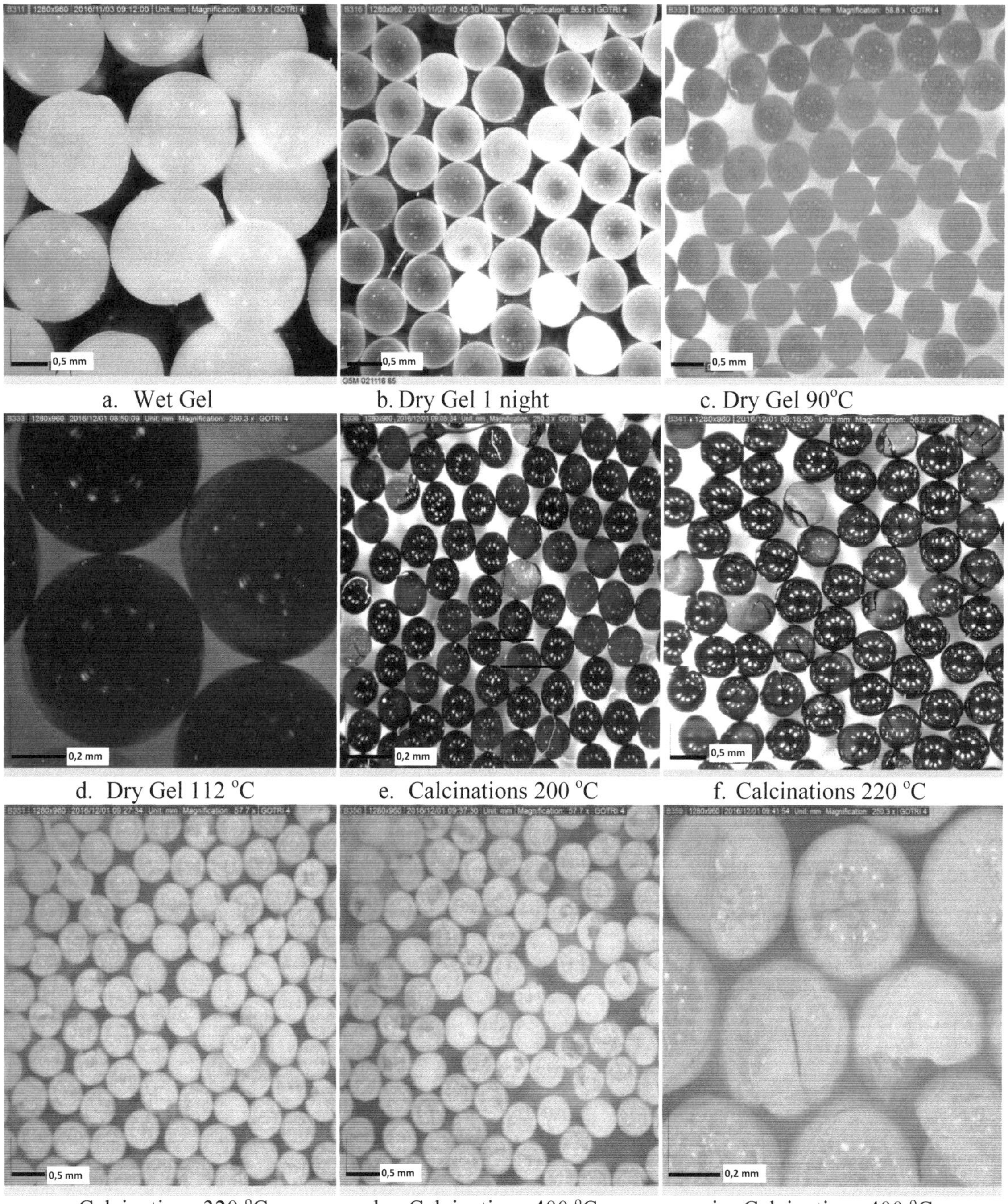

a. Wet Gel b. Dry Gel 1 night c. Dry Gel 90°C

d. Dry Gel 112 °C e. Calcinations 200 °C f. Calcinations 220 °C

g. Calcinations 320 °C h. Calcinations 400 °C i. Calcinations 400 °C

Figure 8. Gel processed with open air drying, drying in the furnaces at 90°c and 112°C and then calcined at 200°C, 220°C, 320°C and 400°C.

Gel with slightly different drying treatment with vacuum is shown in Figure 10. In Figure 10a, the gel was obtained from sol solution with $ZrO(NO_3)_2$ 1.056 mol/L concentration, mol ratio Zr/mol Ce = 13.447, (Zr-Ce) nitrate concentration 1.132 mol/L, PVA 3 Gram/L sol, THFA 12% volume and THFA/PVA ratio = 4 ml / g. The resulting sole has a viscosity of 54 cP so that the gelation process can be carried out, the same as the previous process. In the gelation process there is a note that the vacuum pump absorbing NH_3 flow above the gelation column is slightly too large and the NH_3 flow is too small. After gelation process was done using the sol solution, found that the gel was not round and slightly flat. According to preliminary experiments and previous studies [1] the gelation process without NH_3 flow will cause the gel to be flattened, then the non-round gel is the result, so the possibility of NH_3 gas is uneven over the gelation so that the process of presolidification is not perfect. The gel ageing process is done in a rotary flask so it is expected to occur evenly. In the washing process is the same as the previous process only in the drying process is carried out with a vacuum drying -0.3 atm at 60°C temperature followed by vacuum drying at 80°C. Gel that has been washed and dried looks more massive with smooth texture. The calcinations process at a temperature of 300°C and 500°C, result the gel that does not crack and just change the colors to more dark. The resulting gel is not round and slightly oval so the spericity is less. It is lack of spericity is due to the poor presolidification process.

Gel with other handling of vacuum drying is shown in Fig. 11. In Fig. 11a, the gel was obtained from solvent gelation with $ZrO(NO_3)_2$ 1.056 mol/L concentration, mol ratio Zr / mol Ce 8.447, (Zr-Ce) nitrate concentration 1.2 mol/L sol, PVA 3 gram/L solution, THFA 12% volume and THFA/PVA ratio = 4 ml/g. The resulting sol has a viscosity of 55 cP so that the gelation process can be carried out the same as the previous process. In the gelation process, NH_3 flow above the normal gelation columns, the vacuum adjustment was done so that NH_3 flow rate balanced, so the process of presolidification goes perfectly. The resulting gel ageing process, performed in a rotary flask, so that the ageing process evenly throughout the gel surface. In the washing process carried out drying with a vacuum of -0.3 atm at 60 ° C followed by vacuum drying at 80 ° C. The washed and dried gels are more transparent and the texture is smooth which does not break at heating from 300°C to 500°C. The resulting gel from the gelation process is round so that the spericity is high.

From the four ways of making the gel above there are some notes that can be written as follows:

1. Gelation with bad NH_3 gas flow will affect gel roundness. This is because without the NH_3 gas stream, the presolidification does not occur in the drops and when it falls on the gelation medium is formed deformation. The resulting droplet is not unanimous
2. The ageing process is done in the rotary flask, making the process of ageing evenly throughout the gel surface. The resulting gel does not have a different color as seen in the gel with ageing in gel container.
3. The drying process with vacuum - 0.3 atm in the rotary flask, prevents the occurrence of cracks in the gel. This is caused by increasing porosity of gel, and the release of the gas due to the warming and the smooth calcinations gently out of the gel and giving opportunity to the material to organize so as not to crack

IOP Conf. Series: Journal of Physics: Conf. Series **962** (2018) 012056 doi:10.1088/1742-6596/962/1/012056

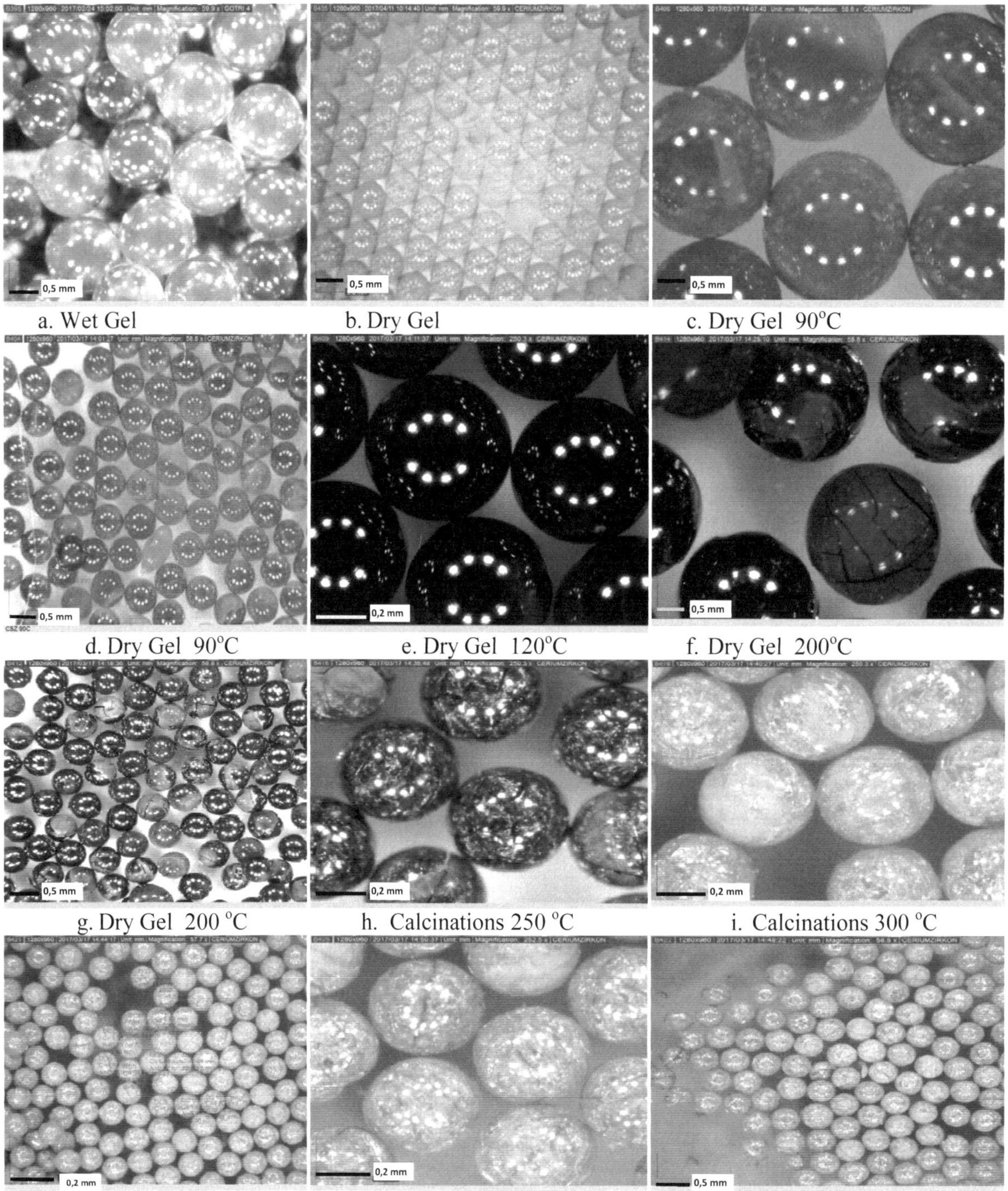

a. Wet Gel b. Dry Gel c. Dry Gel 90°C
d. Dry Gel 90°C e. Dry Gel 120°C f. Dry Gel 200°C
g. Dry Gel 200 °C h. Calcinations 250 °C i. Calcinations 300 °C
j. Calcinations 300 °C k. Calcinations 400 °C l. Calcinations 400 °C

Figure 9. Gel, processed with open air drying, drying in a furnace at 90°c and 120°C and then calcined at 200°C, 250°C, 300°C and 400°C.

a. Wet Gel b. Dry Gel Vacuum 80°C c. Calcinations 300 °C

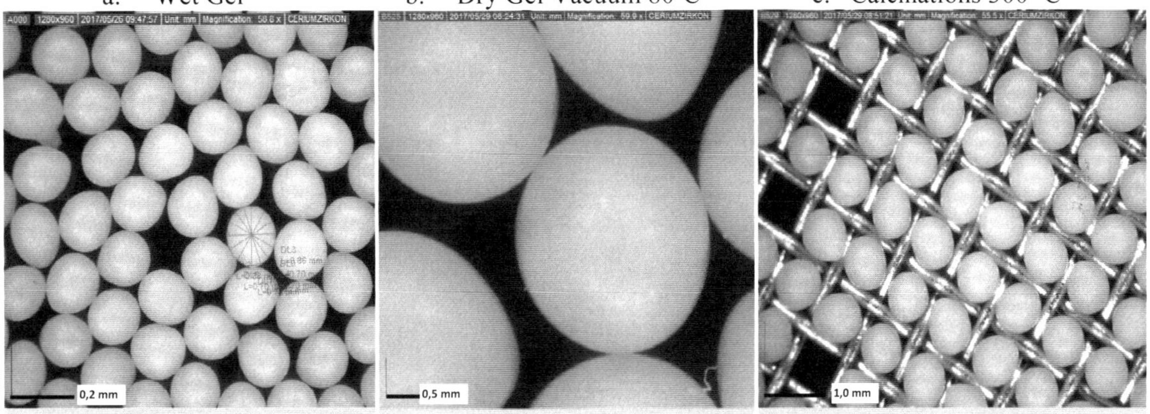

d. Calcinations 300 °C e. Calcinations 500 °C f. Calcinations 500 °C

Figure 10. Gel, processed in vacuum drying, and calcined in the furnace at 300°C and 500°C.

From the 4 kinds of gel are made, can be write a table of the different composition of sol and performance of gel produced. It can be seen at Table 1 and 2.

Table 1. Calcined Gel Result on Ratio of Mol Zr/Ce in Sol Solution at different heat treatment.

No	Ratio mol Zr/mol Ce	Percent mol Ce on mol (Zr-Ce)	Viscosity (cP)	Calcinations 300°C	Calcinations 500°C	Treatment
1	8.45	6.69	40	Not Crack	Not Crack	Vacuum Drying and Calcinations
2	10.49	5.46	59	Crack	Crack	Open air drying, heating and calcinations
3	11.66	4.94	54	Crack	Crack	Open air drying, heating and calcinations
4	13.88	4.18	55	Not Crack	Not Crack	Vacuum Drying and Calcinations

a. Wet Gel | b. Dry Gel Vacuum 80°C | c. Dry Gel

d. Calcinations 300°C | e. Calcinations 300°C | f. Calcinations 300°C

g. Calcinations 500°C | h. Calcinations 500°C | i. Calcinations 500°C

Figure 11. Gels processed in vacuum drying, and calcinations at 300°C and 500°C.

Table 2 shows the relationship of the calcinations state to the various sol solutions with the mole ratio of Zr/mol Ce and treated differently in the drying process. In the sol with a mole ratio of Zr/mol Ce = 8.45 and 10.88 treated with vacuum drying (Method 2) and sol with a mole ratio of Zr/mol Ce of 10.49 and 11.66 treated drying Method 1 without vacuum. It can be seen in Table 11 that by using vacuum in the drying process (no 1 and no 4); the resulting gel yield does not crack at the calcinations up to 500°C. According to literature no 2 it is said that by vacuuming at the time of drying, formed pores that will be a pathway of the decomposition of materials such as PVA, NH_4NO_3 byproducts etc. The presence of these pores facilitates gas expenditure and prevents the occurrence of cracking of gels on heating/calcinations.

Figure 1 shows the phase diagram between ZrO_2 and CeO_2. As seen in Table 2, the change in the ratio of Zr and Ce materials, giving% mol Ce to Zr-Ce ranges from 4-7. If heated to 500°C has not arrived at the point of change monoclinic ss to tetragonal ss, so there is no structural change at heating time up to 500°C. The change in the mol ratio of Zr/mol Ce does not cause the cracking as seen in the gels of Figures 8 and 9. Figure 8 and 9 shown that there is a particle crack; it must be caused by another cause. The cause is probably due to the difference in the drying process, i.e., the drying of air without vacuum on the gel yields in Figs. 8 and 9 and the vacuum of -0.3 atm in gels Figures 10 and 11. The vacuum on drying causes the formation of pores on Zr-Ce particles. Gel produced from the vacuum process, accommodates the decomposition and calcinations process and releases the gel outlet gases, without causing particle cracking.

Table 2. Relation of nitrate salt concentration in soles with gel diameter

No	Ratio Zr/mol Ce	Percent mol Ce in mol (Zr-Ce)	Metal nitrate concentration (mol/L)	Diameter (mm)				Treatment
				Wet Gel	Dry Gel	Calcinations 300°C	Calcinations 500°C	
1	8.450	6.690	1.200	1.585	0.883	0.663	0.624	Vacuum Drying and Calcinations
2	10.490	5.460	0.850	1.466	0.638	0.561	0.542	Open air drying, heating and calcinations
3	11.659	4.940	0.761	1.348	0.571	0.488	0.449	Open air drying, heating and calcinations
4	13.880	4.180	1.132	1.510	0.947	0.732	0.705	Vacuum Drying and Calcinations

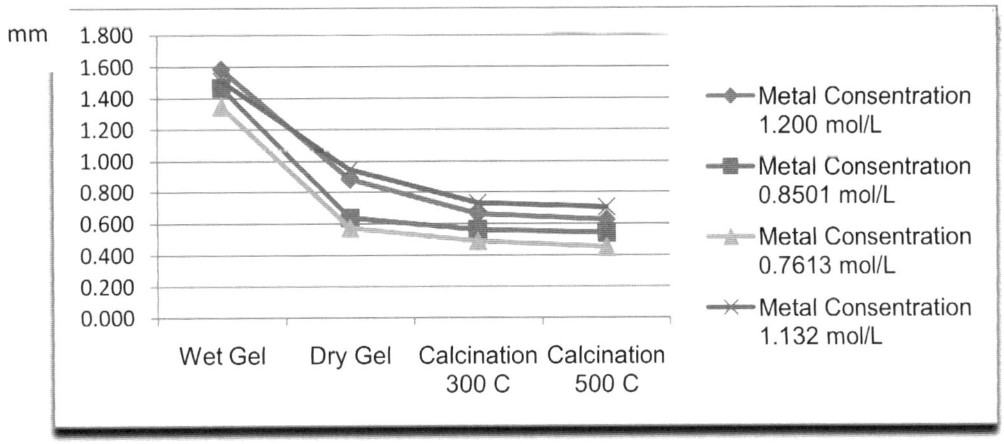

Figure 12. Relationship diameter of gel / particle versus nitrate metal concentration

Table 2 and Figure 12 show the relationship between the concentration of nitrate metal in the sol and the particle diameter from the wet gel to the calcined Zr-Cr oxide particles. It can be seen in Table 2 that there is a correlation of metal concentration in the sol with the resulting gel diameter. The greater the metal concentration in the sol solution, the larger the diameter of the gel/particle produced.

Given the relationship data of nitrate concentration with gel diameter, it can be planned to make gel with the final result of certain particle diameter.

4. Conclusion

From the research it can be concluded that there is correlation of metal concentration in sol with gel diameter. The greater the metal concentration the larger the diameter of the gel/ particle produced. In the gelation process, the NH3 gas stream will affect the gel roundness and yield of the round not gel. This is because with a good NH_3 gas flow, it causes the presolidification process of gels and will make drops harder when falls in gelation medium and does not occur drip form deformation. The resulting drops are rounded. The ageing process is done in a rotary flask, making the process of ageing evenly throughout the gel surface. The resulting gel does not have a different color as seen in the gel by ageing in the gel container during which the ageing gel is not moving. The drying process with a vacuum - 0.3 atm in the rotary flask prevents the occurrence of cracks in the gel. This is due to the possibility of pore formation so that the gas expenditure on heating and calcinations run smoothly. Preparation of gel (Zr-Ce) nitrate using good presolidification process, ageing on rotating flask and drying at vacuum temperature 60^oC followed by temperature 80^oC each for 1 hour, gel can be obtained round, and not cracked at calcinations of 500^oC temperature. The research data can be used to plan the gel-making process with the end result of certain particle diameter

References

[1] S. Ueta, J. Aihara, K. Sawa, A. Yasuda, M. Honda, and N. Furihata, “Development of high temperature gas-cooled reactor (HTGR) fuel in Japan,” *Prog. Nucl. Energy*, vol. 53, no. 7, pp. 788–793, 2011.

[2] Y. W. Lee *et al.*, “Development of HTGR-coated particle fuel technology in Korea,” *Nucl. Eng. Des.*, vol. 238, no. 11, pp. 2842–2853, 2008.

[3] R. Sukarsono; Meniek, “Pengaruh Viskositas Sol Dan Presolidifikasi Pada Gelasi Ekstrenal Dalam Pembuatan Gel Ammonium Diuranat Menggunakan Surrogate Cerium,” *J. Ilm. Daur Bahan Bakar Nukl. URANIA*, vol. 23, no. 2, pp. 69–86, 2017.

[4] G. Wang *et al.*, “Preparation of ceria-stabilized zirconia microspheres by external gelation: size control,” *J Sol-Gel Technol*, vol. 78, no. 3, pp. 514–522, 2016.

[5] Z. Xiangwen *et al.*, “Preparation of ammonium diuranate particles by external gelation process of uranium in INET,” *Nucl. Eng. Des.*, vol. 250, pp. 192–196, 2012.

[6] Z. Xiangwen *et al.*, “Preparation of spherical fuel elements for HTR-PM in INET,” *Nucl. Eng. Des.*, vol. 263, pp. 456–461, 2013.

[7] M. Li, “Making spherical zirconia particles from inorganic zirconium aqueous sols,” *Powder Technol.*, vol. 137, pp. 95–98, 2003.

[8] F. Zhang, C.-H. Chen, J. C. Hanson, R. D. Robinson, I. P. Herman, and S.-W. Chan, “Phases in Ceria-Zirconia Binary Oxide (1-x)CeO2-xZrO2 Nanoparticles: The Effect of Particle Size,” *J. Am. Ceram. Soc.*, vol. 89, no. 3, pp. 1028–1036, 2006.

[9] G. Wang *et al.*, “A comparative study of small-size ceria–zirconia microspheres fabricated by external and internal gelation Guogao,” *J Sol-Gel Sci Technol*, vol. 78, pp. 673–681, 2016.

[10] H. Shaochang, M. Jingtao, Z. Xingyu, W. Yang, and Z. Xiangwen, “Large-scale production of UO 2 kernels by sol – gel process at INET,” *Nucl. Eng. Des.*, vol. 271, pp. 158–161, 2014.

[11] K. Suresh Kumar, H. P. Nawada, and N. P. Bhat, “Comparative study of thermal decomposition of the sol-gel products of U and Ce by external and internal gelation processes,” *J. Nucl. Mater.*, vol. 321, no. 2–3, pp. 263–268, 2003.

[12] N. Kumar, Y. R. Bamankar, K. T. Pillai, and S. K. Mukerjee, “Effect of feed solution composition and heat treatment conditions on the morphology of uranium oxide microspheres prepared by sol – gel process,” *J. Nucl. Mater.*, vol. 359, pp. 80–92, 2006.

[13] Y.-W. L. Kyung-Chai Jeong, Sung-Chul Oh, Yeon-Ku Kim, “ADU compound particle Preparation.pdf,” *J. Ind. Eng. Chem*, vol. 13, no. 5, pp. 744–750, 2007.

Reactivity Coefficient Calculation for AP1000 Reactor Using the NODAL3 Code

Surian Pinem, Tagor Malem Sembiring, Tukiran, Deswandri, Geni Rina Sunaryo

Center for Nuclear Reactor Technology and Safety, National Nuclear Energy Agency of Indonesian (BATAN), Kawasan Puspiptek Building No. 80,Tangerang Selatan 15310, Banten, Indonesia

Email: pinem@batan.go.id

Abstract. The reactivity coefficient is a very important parameter for inherent safety and stability of nuclear reactors operation. To provide the safety analysis of the reactor, the calculation of changes in reactivity caused by temperature is necessary because it is related to the reactor operation. In this paper, the temperature reactivity coefficients of fuel and moderator of the AP1000 core are calculated, as well as the moderator density and boron concentration. All of these coefficients are calculated at the hot full power condition (HFP). All neutron diffusion constant as a function of temperature, water density and boron concentration were generated by the SRAC2006 code. The core calculations for determination of the reactivity coefficient parameter are done by using NODAL3 code. The calculation results show that the fuel temperature, moderator temperature and boron reactivity coefficients are in the range between -2.613 pcm/°C to -4.657pcm/°C, -1.00518 pcm/°C to 1.00649 pcm/°C and -9.11361 pcm/ppm to -8.0751 pcm/ppm, respectively. For the water density reactivity coefficients, the positive reactivity occurs at the water temperature less than 190 °C. The calculation results show that the reactivity coefficients are accurate because the results have a very good agreement with the design value.

Keywords: reactivity coefficient, inherent safety, NODAL3, SRAC2006, AP1000

1. Introduction

Westinghouse designed the advanced passive PWR (pressurized water reactor) reactor, AP1000 reactor, with a large electric power output of 1117 MW. AP1000 is designed with simple construction principles (simplicity) and modular to get cheap electrical charges, relatively [1]. The advantages of safety features of the AP1000 reactor have been evaluated in previous research works, such as criticality, peak power factor, reactivity, critical boron concentration time (critical boron concentration) and total outages reactivity (shutdown reactivity) [2-3].

The AP1000 designed has a better inherent safety since it has the negative total feedback reactivity coefficients. The negative reactivity coefficient ensures the reactor can stabilize the power when the reactor condition changes, such as fuel and moderator temperature increase when the power goes to the nominal level [4-7]. The fuel temperature coefficient (FTC) is reactivity change per fuel temperature change where the moderator temperature and density, as well as boron concentration, are maintained at constant condition. Another name applied to the FTC is the Doppler reactivity coefficient, often shortened to Doppler. Moderator temperature coefficient (MTC) is reactivity change per moderator temperature change where the fuel temperature, moderator density and boron concentration are maintained at a constant condition. The FTC is faster affects the power rise than the MTC because the fuel temperature immediately increases following a power increase, while the moderator temperature increase after several seconds.

This paper focuses on the evaluation of the FTC and MTC of the AP1000 for various water densities and boron concentrations because the temperature coefficient gives significantly to the inherent safety characteristics of the reactor. The MTC has been determined for the cold zero power (CZP) condition [8], however, the MTC for the hot and full power (HFP) condition is not carried out yet. Therefore, this paper shows the evaluation of the FTC and MTC for HFP condition. Besides the two main coefficient reactivities, the reactivity changes due to moderator densities and boron concentration are also carried out. The designed values of the AP1000 reactor are used as the references.

The reactivity coefficients are calculated by using the SRAC and the NODAL3 codes. The NODAL3 code solves steady-state as well as time dependent few-group nodal neutron diffusion equations in the 3-D Cartesian geometry and it is coupled with a simple thermal hydraulic model for typical PWRs. The verification has been conducted and their results showed that the NODAL3 code produces very satisfactory results for the steady-state and transient problems [9-11]. The SRAC code is used to generate the neutron diffusion group constants of fuel assemblies, control rod and reflector [12]. The core calculations using NODAL3 code are carried out to calculate the effective multiplication as a function of fuel and moderator temperatures, water densities and boron concentration.

2. AP1000 reactor core configuration

AP1000 reactor is a PWR reactor with two-loop which generates electrical energy of 1117 MW. The main design parameters for the AP1000 reactor resemble other PWR reactors. AP1000 reactor generates the thermal power of 3400 MW from 157 UO2 fuel assemblies with coolant and moderator are light water (H2O). The fuel has 3 uranium enrichments, 2.35%, 3.40% and 4.45%. Figure 1 shows the initial AP100 core configurations. Each of the AP1000 fuel assemblies comprises a 17 × 17 square lattice array of which 264 are fuel rods, 24 are guide tubes for reactor control and one central instrumentation tube.

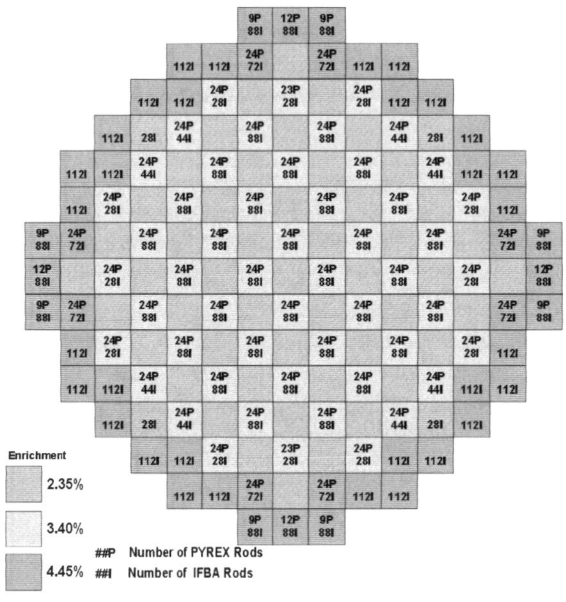

Figure 1. AP1000 core configuration [1]

Figure 2 shows the different arrangements of the Pyrex and IFBA rods in the AP1000 fuel assemblies (FA). According to Fig. 2, the Pyrex and IFBA rods are arranged in FA with three and five different configurations, respectively. The number of Pyrex in the FA are 9, 12 and 24. The number of IFBA in FA with enrichment of 4.45% is 28, 44, 72, 88 and 112, except the FA with enrichment of 2.35 % only has 28 IFBA.

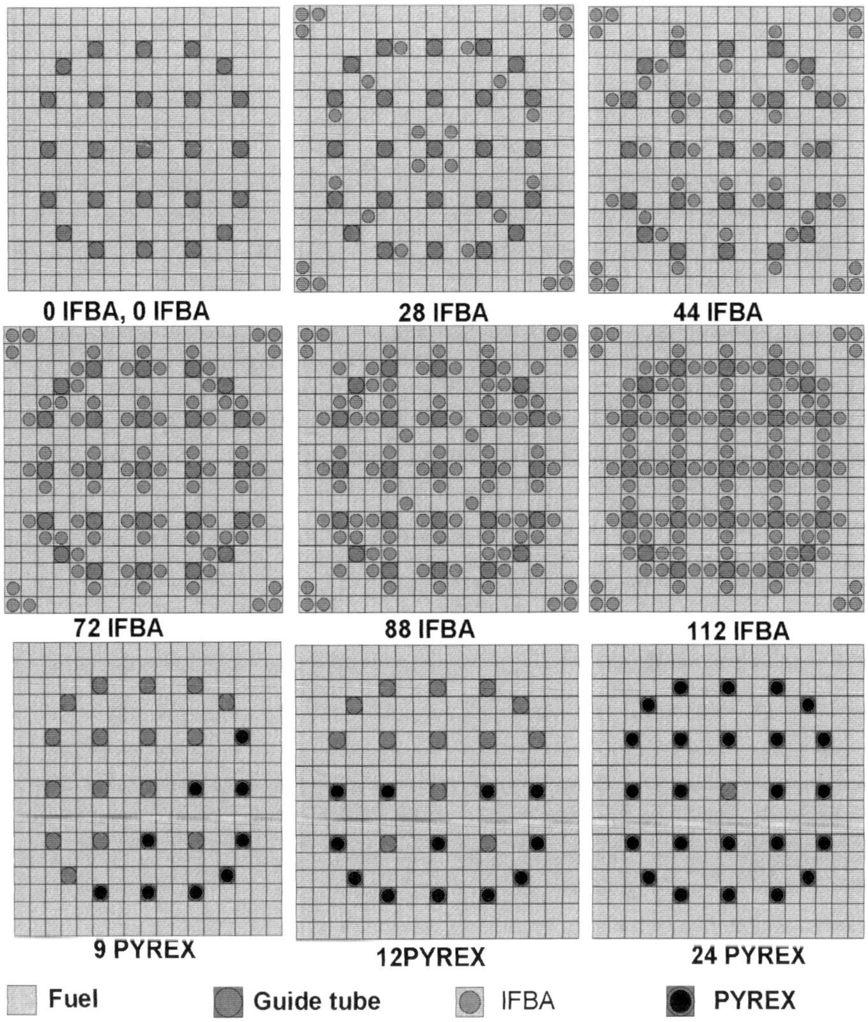

Figure 2. AP1000 fuel assemblies configuration [1]

3. Methodology

The NODAL3 code was used to calculate the criticality parameters, keff, by simulating the cores in the 3-D geometry of X-Y-Z model. All control rods were assumed to be fully up. The fuel and non-fuel zones of each standard and control fuel element were modeled separately. In the criticality calculation, some input data such as absorption macroscopic cross section (Σa), the ν-fission macroscopic cross section ($\nu\Sigma f$), the diffusion coefficient (D), the group scattering macroscopic cross section (Σs, $g \rightarrow g'$) and the fission spectrum for neutron energy groups, are needed in the NODAL3 code. The SRAC code was used for generating the diffusion group constants as a function of fuel temperatures and moderator temperatures as well as various boron concentration for all type of FA of the AP1000 reactor. The HFP condition is set with the fuel temperature of 900 K, moderator and cladding temperatures of 600 K, based on the document of AP1000.

The number of nodes in the direction of X, Y, and Z is 17, 17 and 21, respectively. Based on the previous work of sensitivity analysis [13], the all core calculations are carried out with the 2 × 2 radial nodes per assembly, 1×18 axial layers per assembly, the maximum time step of 10 ms, and 9 and 1 ring divisions for fuel pellet and cladding [13]. The core configuration is made in ¼ full core model and the axial area is divided into 19 layers with a size of 20.3 cm (2 layers), 8.9 cm (2 layers), 24.553333 cm (15 layers). The upper and lower reflector has a thickness of 24.5. The AP1000 core layout model according to FA type for radial and axial direction are shown in Figure 3 and Figure 4, respectively.

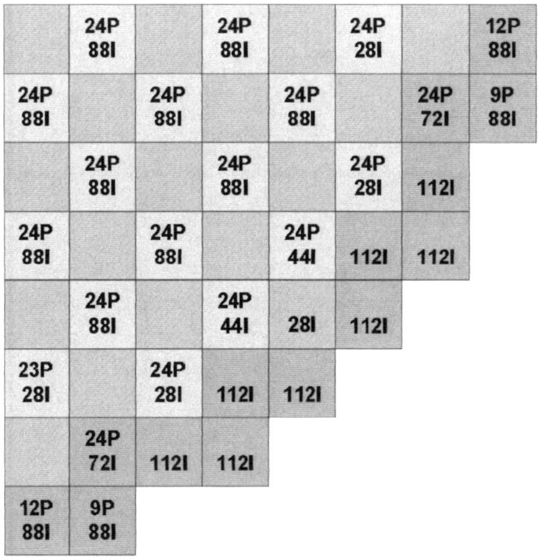

Figure 3. AP1000 quarter-core layout assemblies with number of Pyrex (P) and IFBA (I)

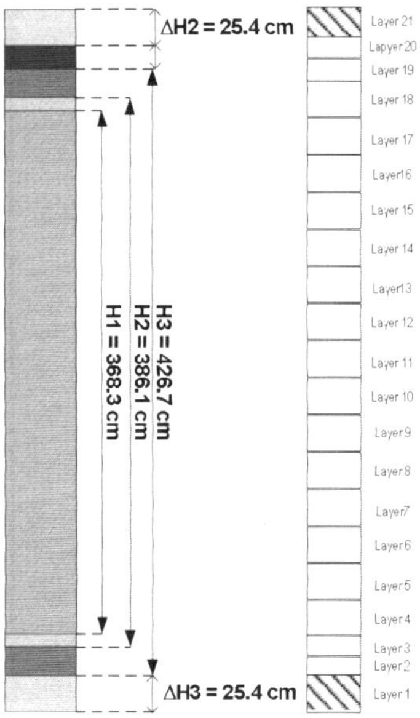

Figure 4. Axial zone of a AP1000 fuel assembly

In the calculation of FTC, the moderator and cladding were taken as 600 K and the fuel temperature was varied. The cross-sections were generated using SRAC code by varying the fuel temperature from 573 K - 1273 K (300 °C – 1000 °C). Meanwhile to calculate the MTC by varying the moderator temperature from 373 K - 573 K (100 °C – 300 °C, temperature range at the normal operation), where the fuel temperature is 900 K and the cladding temperature is 600 K.

In the calculation of moderator density coefficients (MDC), the water density was varied from 0.965422 g/cm^3 (corresponding to 300 K) to 0.659757 g/cm^3 (corresponding to 600 K) where the fuel temperature is 900 K, as well as the moderator and cladding temperatures are, 600 K. For the boron concentration coefficient (BCC), the concentration of boron in the moderator is varied in the range between 0 – 2500 ppm, (correspond to minimum and maximum boron concentration where the fuel temperature is 900 K and the moderator and cladding temperatures are 600 K, correspond to minimum and maximum boron concentration.

4. Results and discussions

Table 1 shows the accuracy of NODAL3 code (combination with SRAC2006 code) in the calculation of criticality parameters of the AP1000 reactor. The difference is about 0 % - 0.40% and the maximum difference occurs for full power without xenon condition. The NODAL3 calculation results showed a very good agreement with the designed value of the AP1000 reactor.

Table 1. Criticality parameters of the AP1000 reactor

Parameters	NODAL3 code	Reference (AP1000 design)[1]	Difference (%)
k_{inf} (cold, clean, unborated water)	1.330	1.330	0.00
k_{eff} (cold, zero power, boc, zero soluble boron)	1.207	1.205	-0.16
k_{eff} (full power, no xenon, hot RCCAs out, boron concentration 1184 ppm)	0.996	1.000	0.40

The distribution of FTC for the without boron and with boron (1184 ppm) in the moderator are shown in Figure 5. It showed that the FTCs are negative for all fuel temperature range (300 °C – 1000 °C) with the values in the range between -4.66 pcm/°C to -2.61 pcm/°C. If we compare to the design values, the range of -6.3 pcm/°C to -1.80 pcm/°C which is obtained by the NODAL3 code, it lies in the range of the design value. Figure 5 shows that at the higher temperature gives the lower negative FTC. This trend is same as the reference of the AP1000 reactor.

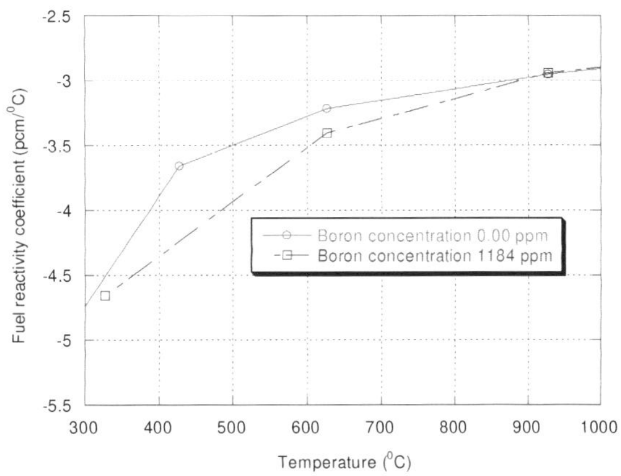

Figure 5. Fuel temperature coefficient for HFP condition

The FTC for with and without boron will overlap at the fuel temperature higher than 900 ^{o}C. This is because of the Doppler effect for without boron will decrease significantly at the temperature higher than 600 ^{0}C. On the other hand, for the with boron condition, the effect of boron is strong enough to keep the gradient of FTC.

The calculation results of the MTC are shown in Figure 6, where all values are negatives, with the range of -3.62 to -3.61 pcm/^{0}C and -1.006 to -1.005 pcm/^{0}C for without boron and with boron, respectively. In comparison to the design value of the AP1000 reactor, 0 pcm/^{0}C to -72 pcm/^{0}C [6], the calculation results are in the range of design event through the range is more narrow. It is noted that the effect of fission products are dominant in the MTC. The higher boron concentration affects the MTC value is a trend to be positive especially at the low temperature. This is due to the reduction of boron concentration in the moderator produces a large increase in thermal utilization (f) when the moderator is heated up. This increased contribution is sufficient to change the sign of the overall moderator coefficient from negative to positive.

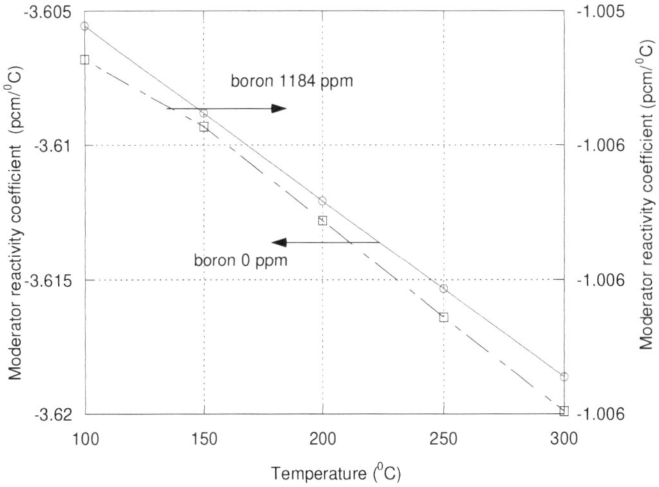

Figure 6. Moderator temperature coefficient for HFP condition

Before we discussed the MDC, it is understood when the temperature moderator increasing, moderator density decreases thereby increasing the possibility of leakage of neutrons and increase the probability of non-fission capture of these neutrons. Both effects cause a decrease in neutron population adding negative reactivity to the reactor core. Figure 7 shows the MDC results for the AP1000 reactor. It is found that MDC for the without and with boron (1184 ppm) are -18 pcm/g/cm^{3} to 10 pcm/g/cm^{3} and -21 pcm/g/cm^{3} to 45 pcm/g/cm^{3}, respectively. It is clear that the higher boron concentration give more positive MDC value since in the higher boron concentration makes the moderator density decreases there for it produces the lower leakage of the neutron. The same situation is found in the design value of the AP1000 reactor. To avoid the positive MDC, the power is started when the moderator temperature reach 100^{0}C (moderator density 0.95 g/cm^{3}) by water flow friction and the boron is added slowly. The negative MDC is beginning to decline from the temperature of 190 ^{0}C (moderator density 0.9 g/cm^{3}).

Figure 8 summarizes the calculation results of BCC (boron concentration coefficient) are -9.11361 to -8.0751 pcm/ppm, while the design value are -13.5 to -5 pcm/ppm. The calculation results are still within the range of design value. The BCC show slight decrease trend with increase boron concentration. Boron concentration is decreased (diluted) to compensate the feel depletion during one operational life cycle. The boron concentration (atoms/cm^{3}) decreases, resulting in a positive reactivity insertion and thermal utilization factor increases.

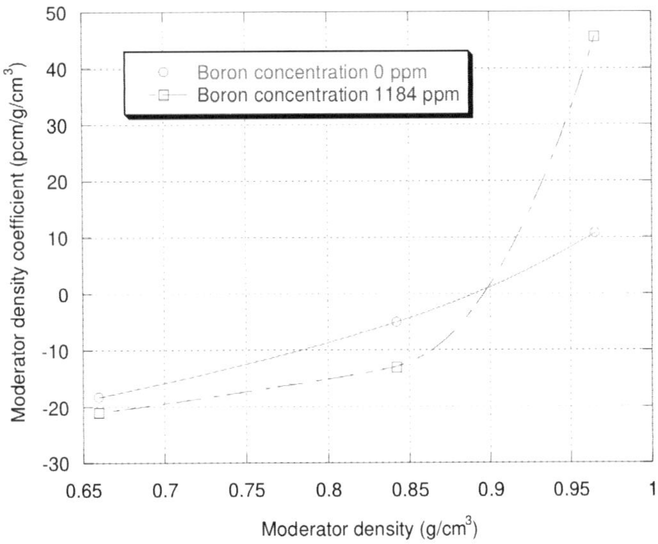

Figure 7. Moderator density coefficient for HFP condition

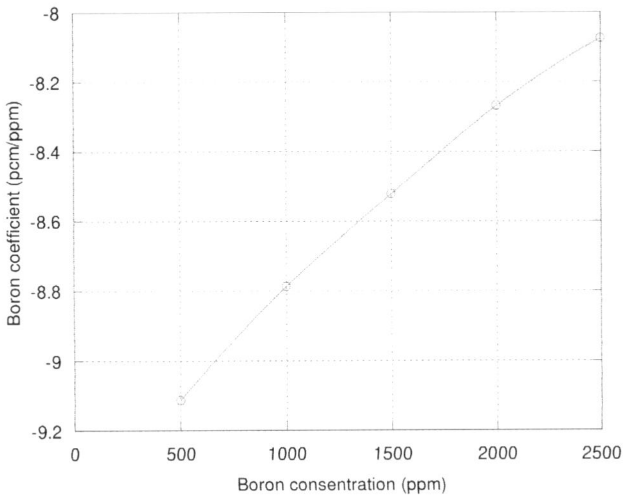

Figure 8. Boron concentration coefficient for HFP condition

5. Conclusions

In this study, the characteristic of the AP1000 reactivity coefficients, such as fuel temperature (FTC), moderator temperature (MTC), moderator density (MDC) as well as boron concentration (BCC) have been evaluated by using NODAL3 code. All reactivity coefficients of the AP1000 reactor are negative, except the MDC at the lower temperature (< 190 ^{0}C) is positive. However, total reactivity coefficient is negative. This fact shows that the transient analysis should concern the MTC in the range between 20 ^{0}C – 190 ^{0}C.

Acknowledgement

Our thanks to the Head of PTKRN and Dr Jupiter Sitorus Pane, M.Sc. as well as the staff of Reactor Physics and Technology Division of PTKRN-BATAN for their cooperation. This research is supported by using DIPA in the year of 2016.

References

[1] Schulz T L 2006 Westinghouse AP1000 advanced passive plant *Nucl. Eng. Des.* **236** 1547–57

[2] Sembiring T M 2011 Analysis of the 3-dimensional core model for evaluation of criticality parameters of the Advanced PWR 100 MW class *Tri Dasa Mega* **13** 78–95

[3] Pinem S and Surbakti T 2016 Analysis on neutronic parameters of the AP1000 Reactor Core *Seminar Nasional Teknologi Energi Nuklir 2016* pp 569–76

[4] Wang L, Chen B and Yao D 2013 Reactivity temperature coefficient evaluation of uranium zirconium hydride fuel element in power reactor *Nucl. Eng. Des.* **257** 61–6

[5] Motalab M A, Kim W and Kim Y 2015 Evaluation of CANDU6 PCR (power coefficient of reactivity) with a 3-D whole-core Monte Carlo Analysis *Nucl. Eng. Des.* **295** 127–37

[6] Vrban B, Lüley J, Farkas G, Haščík J, Hinca R, Petriska M, Slugeň V and Šimko J 2014 Temperature coefficients calculation for the first fuel loading of NPP Mochovce 3-4 *Ann. Nucl. Energy* **63** 646–52

[7] Stefani G L, Rossi P R, Maiorino J R and Santos T A 2015 Neutronic and Thermal-Hydraulic Calculations for the Ap-1000 Npp With the Mcnp6 and Serpent Codes *Proc. International Nuclear Atlantic Conference*

[8] Sembiring T . and Pinem S 2011 Evaluasi koefisien reaktivitas temperatur moderator reaktor daya PWR 1000 MWe *Pros. Semin. Nas. ke-17 Teknol. dan Keselam. PLTN Serta Fasilitas Nukl.* 164–74

[9] Sembiring T M and Pinem S 2012 Validasi paket program NODAL3 untuk kasus statis benchmark teras reaktor PWR *J. Iptek Nukl. Ganendra* **15** 82–92

[10] Pinem S, Sembiring T M and Liem P H 2014 The Verification of Coupled Neutronics Thermal-Hydraulics Code NODAL3 in the PWR Rod Ejection Benchmark *Sci. Technol. Nucl. Install.* **2014** 1–9

[11] Sembiring T M, Pinem S and Liem P H 2015 Validation of full core geometry model of the NODAL3 Code in the PWR transient benchmark problems *Tri Dasa Mega* **17** 141–8

[12] Okumura K, Kugo T, Kaneko K and Tsuchihashi K 2007 *SRAC2006 : A comprehensive neutronics calculation code system*

[13] Pinem S, Sembiring T M and Liem P H 2016 NODAL3 Sensitivity Analysis for NEACRP 3D LWR Core Transient Benchmark (PWR) *Sci. Technol. Nucl. Install.* **2016** 1–11

Study on Characteristic of Temperature Coefficient of Reactivity for Plutonium Core of Pebbled Bed Reactor

Zuhair, Suwoto, T. Setiadipura, S. Bakhri, G.R. Sunaryo

Center for Nuclear Reactor Technology and Safety – National Nuclear Energy Agency of Indonesia (BATAN), Puspiptek Complex, OB No. 80, Serpong, Tangerang Selatan 15310, Indonesia, Tel. (021)756-0912, Fax. (021)756-0913

zuhair@batan.go.id

Abstract. As a part of the solution searching for possibility to control the plutonium, a current effort is focused on mechanisms to maximize consumption of plutonium. Plutonium core solution is a unique case in the high temperature reactor which is intended to reduce the accumulation of plutonium. However, the safety performance of the plutonium core which tends to produce a positive temperature coefficient of reactivity should be examined. The pebble bed inherent safety features which are characterized by a negative temperature coefficient of reactivity must be maintained under any circumstances. The purpose of this study is to investigate the characteristic of temperature coefficient of reactivity for plutonium core of pebble bed reactor. A series of calculations with plutonium loading varied from 0.5 g to 1.5 g per fuel pebble were performed by the MCNPX code and ENDF/B-VII library. The calculation results show that the k_{eff} curve of 0.5 g Pu/pebble declines sharply with the increase in fuel burnup while the greater Pu loading per pebble yields k_{eff} curve declines slighter. The fuel with high Pu content per pebble may reach long burnup cycle. From the temperature coefficient point of view, it is concluded that the reactor containing 0.5 g-1.25 g Pu/pebble at high burnup has less favorable safety features if it is operated at high temperature. The use of fuel with Pu content of 1.5 g/pebble at high burnup should be considered carefully from core safety aspect because it could affect transient behavior into a fatal accident situation.

Keywords: *temperature coefficient of reactivity, plutonium core, pebble bed reactor, MCNPX, ENDF/B-VII*

1. Introduction

The superiority in quantity of resources and compatibility with environment makes nuclear energy becomes the most prospective energy among the various energy resources in overcoming the problem of increase in the world energy demand. Thirteen countries and institutions in the world that pursue the field of nuclear energy technology have been collaborating in the Generation IV international forum (GIF) with the aim of increasing the role of nuclear energy systems in helping meet the world energy needs in the future. Some Generation IV reactors with characteristic of using fuel more efficiently, reducing waste production, being economically competitive, and fulfilling stringent standards of safety and proliferation resistance will be demonstrated in the next decade, with commercial deployment beginning in 2030 [1,2].

The pebble bed reactor, a type of the HTGR which uses graphite as neutron moderator and helium as coolant, was selected as the most promising one of the six Generation IV nuclear reactors. This is because the reactor is designed to have a high operating temperature to produce high energy conversion efficiency supporting the application of industrial process heat such as hydrogen production, and others [3,4]. The pebble bed reactor is also designed to have the flexibility that offers the possibility of using fuel with a series of cycling to reduce the accumulation of plutonium resulted from a nuclear power plant (NPP) and military activities. The growing accumulation of plutonium clearly raises public concern about the radiotoxicity dangers and also opportunities to use plutonium not for peaceful purposes [5].

As a part of the solution searching that possible to control the plutonium, a current effort is focused on mechanisms to maximize consumption of plutonium. Plutonium core solution is a unique case in the high temperature reactor which is intended to reduce the accumulation of plutonium [6]. However, the performance of plutonium core which tends to produce a positive temperature coefficient of reactivity should be examined. The pebble bed inherent safety features which are characterized by a negative temperature coefficient of reactivity must be maintained under any circumstances.

The purpose of this study is to investigate the characteristic of temperature coefficient of reactivity for plutonium core of pebble bed reactor. A series of calculations were performed by the MCNPX [7] code and ENDF/B-VII library [8]. The calculation results of temperature coefficient with plutonium loading varied from 0.5 g to 1.5 g per fuel pebble were discussed to complete the analysis.

2. Pebble Bed Reactor

Pebble bed reactor has a cylindrical core with a diameter of 3 m and height of 9.43 m connected to a cone at the bottom of the reactor core with height of 0.61m. A power density of 3 MW/m^3 generates reactor thermal power of 200 MW. The reactor operates at helium inlet and outlet temperatures of 550 ºC and 950 ºC, respectively, with pressure of 4 MPa and the mass flow rate of 120 kg/s. The reactor parameters and core specifications of pebble bed reactor are given in table 1.

Table 1. Reactor parameter and core specification of pebble bed reactor [9].

Reactor parameter	
Thermal power (MW)	200
Core diameter/height (m/m)	3/9.43
Helium inlet/outlet temperature (^{o}C)	550/950
Helium pressure (MPa)	5.2
Helium flowrate (kg/s)	120
Helium density (g/cm^3)	1.78×10^{-4}
Core spesification	
Core specific power (MW/m^3)	3
Number of fuel pebble per m^3	5,394
Number of fuel pebble in core	359,548
Packing fraction	0.61
Natural boron impurity in graphite structure (ppm)	2

The reactor core is loaded by 359,548 fuel pebbles with a packing fraction of 0.61 while the cone is filled with 17,263 moderator pebbles with the same packing fraction. Packing fraction is defined as the ratio of total pebble to core volume. Helium coolant flows from top to bottom of the core through the space between pebbles arranged randomly. Fuel pebble and moderator pebble have the same diameter of 6 cm, but different material content. Moderator pebble is made of pure graphite with density of 1.84 g/cm^3. Fuel pebble consists of two regions: the first region, which is called the fueled-zone, 5 cm in diameter containing thousands TRISO particles dispersed in graphite matrix and the second region, which is called a graphite shell, surrounds the fueled-zone with 0.5 cm thick. Graphite matrix and graphite shell act as neutron moderator while moderator pebble as an extra moderator in the reactor core.

The TRISO particle is made of a plutonium oxide (PuO_2) kernel with a density of 10.4 g/cm^3 and it is wrapped by four coating layers: porous carbon buffer layer, inner pyrolitic carbon (iPyC) layer, a layer of silicon carbide (SiC) and outer pyrolitic carbon (oPyC). The fuel kernel has plutonium isotopic vector as shown in table 2. The four coating layers protect and preserve effectively the metallic and gas fission product from releasing to the environment for the temperatures up to 1,600 ºC. The overall diameter of TRISO particle is 660 cm. The pebble fuel specifications are summarized in table 3.

Table 2. Plutonium isotopic vector [10].

Isotope	Pu vector (%)
^{238}Pu	2.59
^{239}PU	53.85
^{240}Pu	23.66
^{241}Pu	13.13
^{242}Pu	6.77

Table 3. Fuel pebble specification of pebble bed reactor [10].

Fuel pebble			
Pebble diameter (cm)	6		
Fueled-zone diameter (cm)	5		
Graphite shell thickness (cm)	0.5		
Graphite shell density (g/cm^3)	1.75		
Graphite matrix density (g/cm^3)	1.75		
Natural boron impurity in grahite shell (ppm)	1.0		
Natural boron impurity in grahite matrix (ppm)	1.0		
TRISO coated particle			
Fuel kernel		iPyC layer	
Material	PuO_2	Thickness (μm)	40
Diameter (μm)	240	Density (g/cm^3)	1.90
Density (g/cm^3)	10.4	SiC layer	
Natural boron impurity in kernel (ppm)	1.0	Thickness (μm)	35
		Density (g/cm^3)	3.18
Buffer layer		oPyC layer	
Thickness (μm)	95	Thickness (μm)	40
Density (g/cm^3)	1.05	Density (g/cm^3)	1.90

During reactor operation, the fuel pebble can pass through the core to be recirculated several times in Multi-pass loading scheme, or only once in the once-through-then-out (OTTO) loading scheme. The mechanism of fuel loading in OTTO scheme is simple, so it is not necessary to have the equipment for fuel burnup measurement, fuel recirculation and extraction equipment to cancel the fuel to be reused. Multipass scheme requires all those mechanisms. The fuel which is not reused will be sent to the spent fuel storage tank for further processing.

3. Calculation Model

One of main factors in the pebble bed reactor modeling is feature of double heterogeneity characterized by randomness of TRISO coated particles distributed in the fuel pebble (first heterogeneity) and randomness of fuel pebbles distributed in the reactor core (second heterogeneity). MCNPX as a general purpose Monte Carlo radiation transport code has capability to model this double heterogeneity in detail and accurately. The modeling of pebble bed reactor consists of model of TRISO particles in fuel pebble and model of fuel pebbles in core region.

3.1. TRISO Particles Model

The randomness of TRISO particles in pebble was modeled by a simple cubic (SC) lattice. Each PuO_2 kernel with its four coating layers was placed in the center cubic lattice. This cubic lattice is defined as UNIVERSE. Fuel pebble was then modeled by applying LATTICE and FILL options with repetitive structures provided by MCNPX.

In this study, the content of plutonium per pebble was varied from 0.5 g to 1.5 g. Plutonium content (m_{Pu}) in fuel pebble is defined by the equation,

$$m_{Pu} = \frac{4\pi}{3} r_{kernel}^3 \times \rho_{Pu} \times N_{TRISO} \quad (1)$$

where,

ρ_{Pu} is plutonium density,

N_{TRISO} is number of TRISO particles per pebble, and

r_{kernel} is radius of the fuel kernel.

The calculation of the plutonium density (ρ_{Pu}) follows the equation,

$$\rho_{Pu} = \frac{A_{Pu}}{A_{Pu}+2A_O} \times \rho_{PuO2} \quad (2)$$

where,

ρ_{PuO2} is PuO_2 density,

A_{Pu} is plutonium atomic weight, and

A_O is oxygen atomic weight.

The calculation of the plutonium atomic weight (A_{Pu}) is undertaken by the equation,

$$A_{Pu} = \left[\frac{\alpha_{Pu238}}{A_{Pu238}} + \frac{\alpha_{Pu239}}{A_{Pu239}} + \frac{\alpha_{Pu240}}{A_{Pu240}} + \frac{\alpha_{Pu241}}{A_{Pu241}} + \frac{\alpha_{Pu242}}{A_{Pu242}}\right]^{-1} \quad (3)$$

where,

$A_{Pu238}, A_{Pu239}, A_{Pu240}, A_{Pu241}$ and A_{Pu242} are atomic weight of ^{238}Pu, ^{239}Pu, ^{240}Pu, ^{241}Pu and ^{242}Pu, respectively, and

$\alpha_{Pu238}, \alpha_{Pu239}, \alpha_{Pu240}, \alpha_{Pu241}$ and α_{Pu242} are isotopic vector of ^{238}Pu, ^{239}Pu, ^{240}Pu, ^{241}Pu and ^{242}Pu, respectively.

Plutonium content per pebble can be determined by changing the kernel radius where the number of TRISO particles is kept constant or by changing the number of TRISO particle in the fuel pebble where the kernel radius is kept constant. In this calculation, the kernel radius with standard size is preserved at fixed value. To get a different plutonium loading with specification from 0.5 to 1.50 g, the number of TRISO particles in the fuel pebble was changed from 7,526 to 22,577. The changed

TRISO particle number causes the size of SC lattice pitch (p) should be adjusted following the equation,

$$p = r_{zone} \sqrt[3]{\frac{4\pi}{3 \times N_{TRISO}}} \tag{4}$$

where,

r_{zone} is radius of the fueled-zone of pebble.

Table 4 shows the characteristics of plutonium content with TRISO number and SC lattice pitch used in the calculation. PuO_2 kernel concentration is calculated by a simple formula and presented in table 5. The concentration of the coating covered the kernel is calculated by more simple formula and presented in same table. The concentrations of graphite matrix and graphite shell which is identical are presented in table 6. Figure 1 illustrates the MCNPX model of TRISO particle in fuel pebble.

Table 4. Characteristics of plutonium content.

Pu content / pebble (g)	Number of TRISO	Lattice pitch SC (cm)
0.50	7526	0.20567
0.75	11,288	0.17967
1.00	15,051	0.16324
1.25	18,814	0.15154
1.50	22,577	0.14261

Table 5. Concentration of TRISO coated particle [10].

Kernel PuO_2			
^{238}Pu	6.01178×10^{-4}	^{242}Pu	1.54539×10^{-3}
^{239}Pu	1.24470×10^{-2}	^{16}O	4.60983×10^{-2}
^{240}Pu	5.44599×10^{-3}	^{10}B	1.14694×10^{-7}
^{241}Pu	3.00965×10^{-3}	^{11}B	4.64570×10^{-7}
Coating layers			
Buffer		iPyC/oPyC	
^{12}C	5.26449×10^{-2}	^{12}C	9.52621×10^{-2}
SiC			
^{28}Si	4.40158×10^{-2}	^{30}Si	1.47944×10^{-2}
^{29}Si	2.22871×10^{-2}	^{12}C	4.77240×10^{-2}

Table 6. Concentration of graphite matrix and graphite shell [10].

Graphite matrix		Graphite shell	
^{12}C	8.77414×10^{-2}	^{12}C	8.77414×10^{-2}
^{10}B	9.64977×10^{-9}	^{10}B	9.64977×10^{-9}
^{11}B	3.90864×10^{-8}	^{11}B	3.90864×10^{-8}

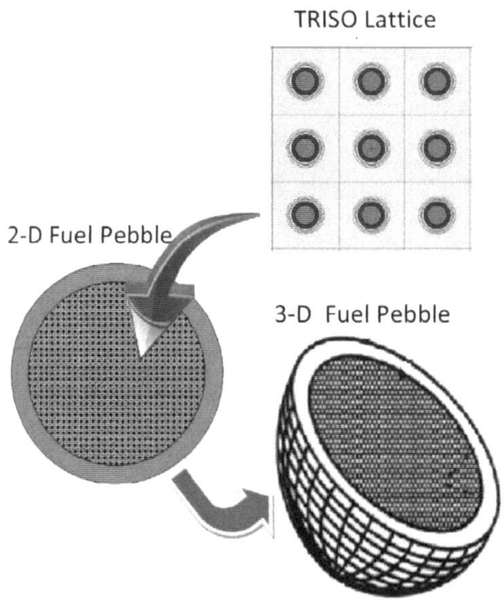

Figure 1. The MCNPX model for TRISO particle in fuel pebble.

3.2. Fuel Pebble Model

The randomness of fuel pebbles in the reactor core was modeled with body-centered cubic (BCC) lattice. There are two fuel pebbles in the BCC lattice: 1 is placed in the center and 1/8 in each corner of the cube. The BCC cubic lattice was defined as UNIVERSE. Similar to the modeling of fuel pebble, the reactor core were then modeled by applying LATTICE and FILL options with repetitive structures. The lattice pitch was calculated to be 7.185259 cm to adjust 359,548 fuel pebbles occupying the reactor core with packing fraction of 0.61.

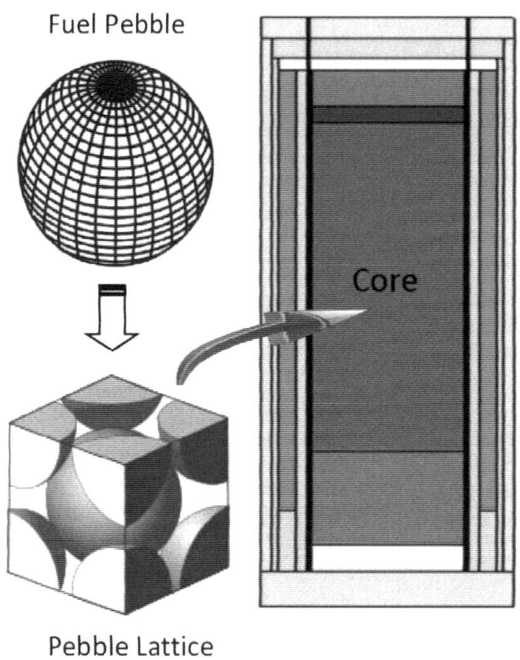

Figure 2. The MCNPX model of fuel pebble in the reactor core.

Moderator pebble in the conical region was also modeled with the same lattice and packing fraction with the fuel pebble. The edge effect of associated to the use of a repeated structure in fuel pebble modeling can be ignored because the packing fraction for the TRISO particles inside the fueled-zone of the pebble is very low, around 1.73-5.19 %. However, the same edge in core modeling effects can lead to non-negligible conditions in reactor calculation. Therefore, the packing fraction of 0.61 is reduced to 0.59 as a correction to eliminate the truncated pebble contribution in the surface of the core wall. Modeling of the reactor structure such as reflector, cooling channels, control rods and carbon layers around the reactor can be addressed in not difficult way. Modeling procedure of pebble bed reactor in detail and comprehensively with MCNP can also be found in our previous studies [11,12]. The CINDER90 modul integrated in the MCNPX was used to perform the nuclide inventory calculations as a function of fuel burnup. Figure 2 illustrates the MCNPX model of fuel pebble in the reactor core.

4. Result and Discussion

Temperature coefficient of reactivity is the response of the neutronic system to variations in temperature and defined as the ratio of reactivity change and temperature change. In this study, the calculated temperature coefficient of reactivity is a summation over the temperature coefficient of fuel, moderator and reflector. All calculations were done at various temperatures identified in ENDF/B-VII library. The problem was run using 110 cycles including 10 cycles skipped of 5000 particles per cycle to obtain exact results. The thermal scattering library S(α, ß) which depends on temperature was applied to consider the interaction of thermal neutrons with graphite moderator, graphite reflector, and graphite contained in the entire fuel pebble under the energy of ~4 eV. Initial neutron source was placed in the fuel kernel to lead fission reactions throughout the reactor system. Vacuum conditions was done on the outer boundary of pebble bed reactor.

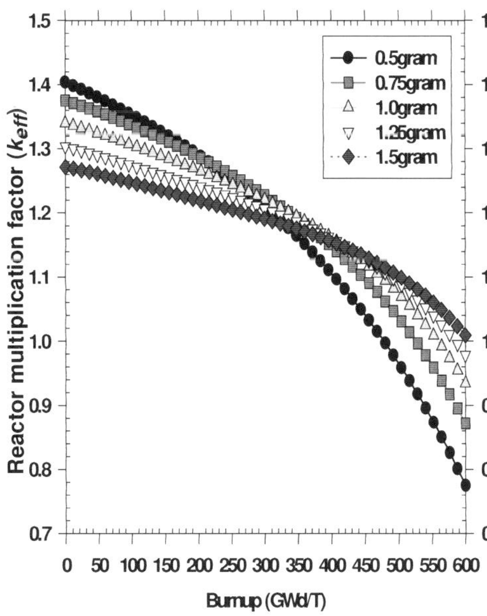

Figure 3. Reactor multiplication factor (k_{eff}) as a function of fuel burnup.

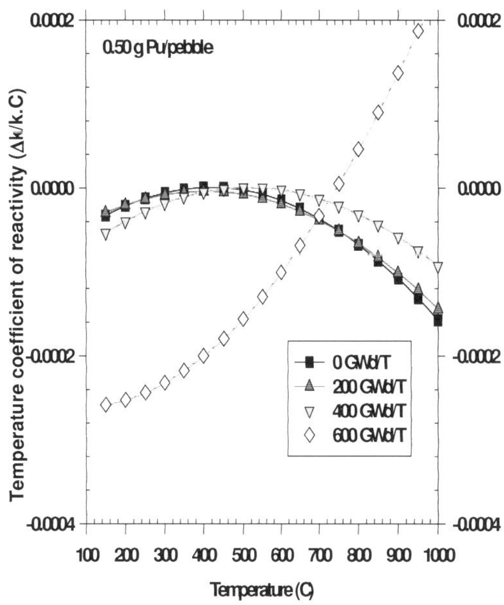

Figure 4. Temperature reactivity coefficient of 0.50 g Pu/pebble.

The calculation result of the reactor multiplication factor k_{eff} for the case of plutonium loading varied from 0.5 g to 1.5 g per fuel pebble is illustrated in figure 3. It can be observed here, the k_{eff} value of 1.25 g Pu/pebble at the beginning of the operating cycle is less than those of 1.50 g Pu/pebble. The k_{eff} initial value of 1.00 g Pu/pebble is smaller than those of 1.25 g Pu/pebble. And so on, the k_{eff} initial value of 0.50 g Pu/pebble is smaller than those of 0.75 g Pu/pebble. This phenomenon occurs because when the plutonium content per pebble is low, the presence of abundant graphite causes the better neutron thermalization. Consequently, the neutron flux will change to be higher and more fuel is burned. That is why the k_{eff} curve of 0.5 g Pu/pebble declines sharply with the increase in fuel burnup. The greater Pu loading per pebble yields k_{eff} curve declines slighter. The fuel with high Pu content per pebble may reach long burnup cycle. However, in general almost all the fuel can achieve burnup more than 450 GWd/T.

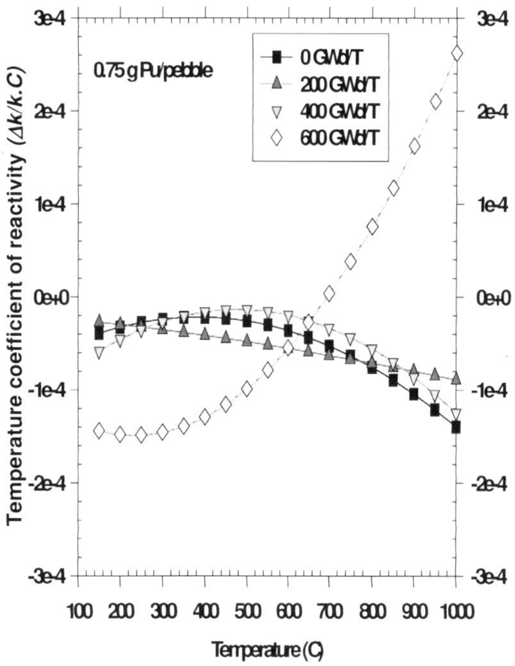

Figure 5. Temperature reactivity coefficient of 0.75 g Pu/pebble.

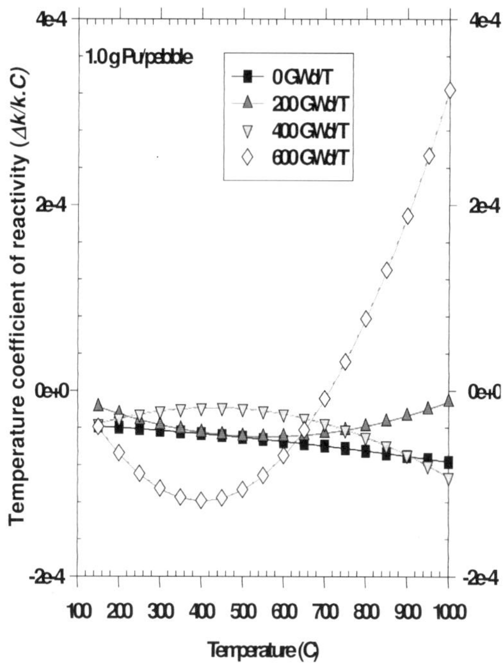

Figure 6. Temperature reactivity coefficient of 1.00 g Pu/pebble.

The calculation results of temperature reactivity coefficient are illustrated in figures 4 to 8. In the case of 0.5 g to 1.25 g plutonium loading per fuel pebble, each temperature coefficient of reactivity shows a similar tendency. These fuels become more negative with increasing temperature in fresh condition and burnup 200 and 400 GWd/T. Temperature reactivity coefficient values are in the range of -5.0×10^{-6} and -1.5×10^{-4}Δk/k/°Cfor the temperature range of 50 °C and 1000 °C. The fuel with burnup 600 GWd/T shows the most negative temperature coefficient of reactivity at low temperature, but then reducing drastically and become positive at temperature greater than 700 C. This means that the reactor containing 0.5 g-1.25 g Pu/pebble at high burnup has less favorable safety features if it operated at high temperature.

For the case of 1.5 g plutonium loading per fuel pebble, temperature coefficient reactivity shows a different trend. Fuel with burnup of 200 GWd/T shows more negative temperature coefficient of reactivity at temperatures of less than 600 ºC, but turn out to be less negative above 600 ºC. The temperature coefficient value at temperature of 1000 C is $-2.0\times10^{-5}\Delta k/k/^{o}C$ slightly more negative than $-1.5\times10^{-5}\Delta k/k/^{o}C$ at temperature of 150 ºC. The fuel with burnup of 400 GWd/T shows less negative temperature coefficient of reactivity than those of 200 GWd/T, but both have similar characteristics. At temperature of greater than 950 ºC, the temperature coefficient becomes positive. On the other hand, the fuel with burnup of 600 GWd/T shows the most negative temperature coefficient of reactivity at temperature of 150 ºC, but later reducing and turning positive after temperature of 300 ºC. This temperature coefficient returns to negative value at temperature above 900 ºC. These results give an impression that the use of fuel with high Pu content per pebble at high burnup should be considered carefully from the core safety aspect because it could affect transient behavior into a fatal accident situation.

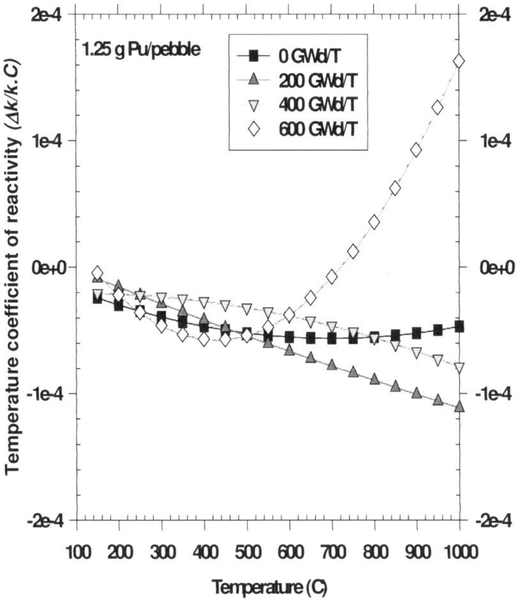

Figure 7. Temperature reactivity coefficient of 1.25 g Pu/pebble.

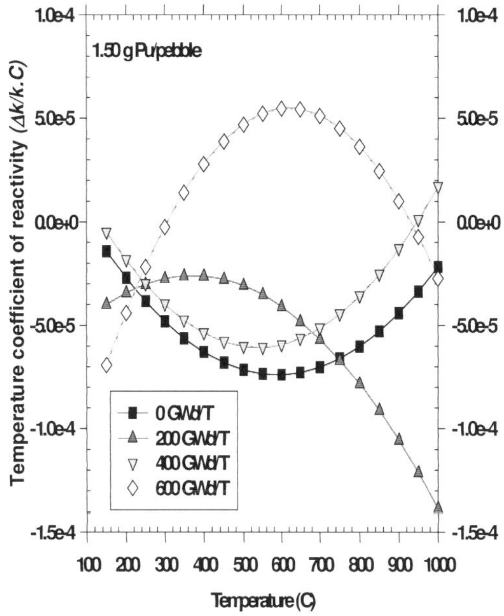

Figure 8. Temperature reactivity coefficient of 1.50 g Pu/pebble.

5. Conclusion

Study on characteristic of temperature coefficient of reactivity for plutonium core of pebble bed reactor has been conducted. A series of calculations with plutonium loading varied from 0.5 g to 1.5 g per fuel pebble were performed by the MCNPX code and ENDF/B-VII library. The calculation results show that the k_{eff} curve of 0.5 g Pu/pebble declines sharply with the increase in fuel burnup while the greater Pu loading per pebble yields k_{eff} curve declines slighter. The fuel with high Pu content per pebble may reach long burnup cycle. From the temperature coefficient point of view, it is concluded that the reactor containing 0.5 g-1.25 g Pu/pebble at high burnup has less favorable safety features if it is operated at high temperature. The use of fuel with Pu content of 1.5 g/pebble at high burnup should be considered carefully from core safety aspect because it could affect transient behavior into a fatal accident situation.

Acknowledgment
We would like to express our sincere gratitude to Dr. Geni Rina Sunaryo, M.Sc. for giving us motivation and support in conducting this research. We would also like to thank Dr. Syaiful Bakhri for his patience in taking the time to review and improve this paper. Moral encouragement from our colleagues in the Division of Reactor Physics and Technology is really appreciated.

References

[1] Generation IV International Forum (GIF) 2014 *Technology Roadmap Update for Generation IV Nuclear Energy Systems* (OECD Nuclear Energy Agency)
[2] Frima L L W 2013 *Burnup in a Molten Salt Fast Reactor* (Delft: Master Thesis Delft University of Technology)
[3] Graafland C 2014 *Modeling and Analysis of a Depressurized Loss of Forced Cooling Event in a Thorium Fueled High Temperature Reactor* (Delft: Bachelor Thesis Delft University of Technology)
[4] Bijman R 2014 *Radiotoxicity and Decay Heat Comparison of Nuclear Waste Produced by a Thorium or Uranium Fuelled Pebble Bed Reactor* (Delft: Bachelor Thesis Delft University of Technology)
[5] Meshesha A Z 2007 *Vialibility of Thorium-Based Reactors* (Guildford: Master of Science Thesis Department of Physics University of Surrey)
[6] Serfontein D E 2011 *Deep Burn Strategy for the Optimized Incineration of Reactor Waste Plutonium in Pebble Bed High Temperature Gas-Cooled Reactor* (Potchefstroom: Doctoral Thesis North-West University)
[7] J.S. Hendricks J S, McKinney G W, et al 2008 *MCNPX 2.6.0 Extensions*" (New Mexico: LA-UR-08-2216 Los Alamos National Laboratory)
[8] Chadwick M B, et al *ENDF/B-VII.1* 2011 *Nuclear Data for Science and Technology: Cross Sections, Covariances, Fission Product Yields and Decay Data* (Nuclear Data Sheets vol 112 issue 12) pp 2887-2996, December 2011
[9] Zuhair, Suwoto 2015 Analisis efek kecelakaan water ingress terhadap reaktivitas doppler teras rgtt200k *Jurnal Teknologi Reaktor Nuklir TRI DASA MEGA* **17** (1)
[10] Hosking G and Newton T D 2005 *Proposed Benchmark Specification for an HTR Fuelled with Reactor Grade Plutonium* (NEW/NSC/DOC(2003)22)
[11] Suwoto, Zuhair 2014 Analisis sensitivitas ketebalan reflektor grafit teras RGTT200K menggunakan perhitungan monte carlo *Jurnal Pengembangan Energi Nuklir* **16** (2)
[12] Zuhair, Suwoto, P Supriatna 2011 Studi efek fraksi packing triso dalam desain kritikalitas RGTT200K *Seminar Nasional ke-17 Teknologi dan Keselamatan PLTN serta Fasilitas Nuklir, Yogyakarta*

Effects of the Application of the New Nuclear Data Library ENDF/B to the Criticality Analysis of AP1000

Iman Kuntoro,T.M. Sembiring, Jati Susilo, Deswandri, G.R. Sunaryo

Center for Nuclear Reactor Technology and Safety, National Nuclear Energy Agency of Indonesia (BATAN), Kawasan Puspiptek Gd. No. 80 Serpong , Tangerang Selatan 15310 Indonesia

Email: imank@batan.go.id

Abstract. Calculations of criticality of the AP1000 core due to the use of new edition of nuclear data library namely ENDF/B-VII and ENDF/B-VII.1 have been done. This work is aimed to know the accuracy of ENDF/B-VII.1 compared to ENDF/B-VII and ENDF/B-VI.8. in determining the criticality parameter of AP1000. Analysis ws imposed to core at cold zero power (CZP) conditions. The calculations have been carried out by means of MCNP computer code for 3 dimension geometry. The results show that criticality parameter namely effective multiplication factor of the AP1000 core are higher than that ones resulted from ENDF/B-VI.8 with relative differences of 0.39% for application of ENDF/B-VII and of 0.34% for application of ENDF/B-VII.1.
Keywords: criticality analysis, AP1000, ENDF/B, MCNP

1. Introduction

For nuclear reactor, criticality is one of the most important parameter to be analyzed especially for reactor having high excess reactivity and a rather complicated of reactivity control system. PWR reactor core is loaded by big amount of fuel elements to produce high excess reactivity for providing a long period of operation with relatively efficient in operation cost. Besides that, the reactor is controlled by various modes namely, boron in moderator, burnable poison in fuel assembly such as Pyrex and IFBA.

AP1000 reactor is a PWR type of power reactor designed by Westinghouse Co. that can produce a nominal electric power of 1117 MWe or 3400 MW thermal generated by 157 fuel assemblies of UO2[1]. The fuel has 3 different enrichment 2.35 w/o, 3.40 w/o and 4.50 w/o. To control the reactivity, the AP1000 uses boric acid which is dissolved in moderator, Pyrex absorber rod and Integrated Fuel Burnable Absorber (IFBA). IFBA is a burnable absorber made from ZrB2 which is integrated in UO2 fuel, whereas Pyrex is an absorber rod made from B2O3 as control rod which is inserted into the guide tube.

Monte Carlo Method computer code, such as MCNP, performs very good analysis of criticality of AP1000 because it can make core model in detail in 3-dimension [2]. This MCNP code has been implemented to analyze criticality of AP1000 and gives results as compared to that of the design values[3]. Calculations were done by using ENDF/B-VI.8 library. Nowadays, MCNP version 6 uses newest evaluated nuclear data ENDF/B-VII.1[4-7]. It is reported in Ref. [4] that the ENDF/B-VII.1

produces better accurate values of reactivity or multiplication factor in comparison to that resulted from older version of nuclear data ENDF/B.

The use of ENDF/B-VII.1 were also widely applied in criticality analysis for some reactor types and even adopted in several computer codes [8-10]. The results showed that there are also better accuracy using this new nuclear data library. It is therefore, necessary to study on effect of newest nuclear data library ENDF/B in calculation of criticality parameter of AP1000 core.

The objective of the research is to obtain the accuracy of ENDF/B-VII.1 compared to ENDF/B-VII and ENDF/B-VI.8. in determining the criticality parameter (k_{eff}) of AP1000 core by using Monte Carlo method. The results are intended to give contribution on the accuracy of using the latest version of ENDF data library for AP1000 core.

2. Methodology

Criticality parameter or effective multiplication factors(k_{eff}) are calculated for 3-dimensional X-Y-Z geometry of AP1000 core. The calculation object is the AP1000 core at cold zero power (CZP) conditions what means at temperature of 200 ^{0}C, zero power and free xenon poisoning. The calculation is principally done in 2 (two) steps namely model preparation and then core calculation. The cross section view of the AP1000 core is described in the Fig.1.

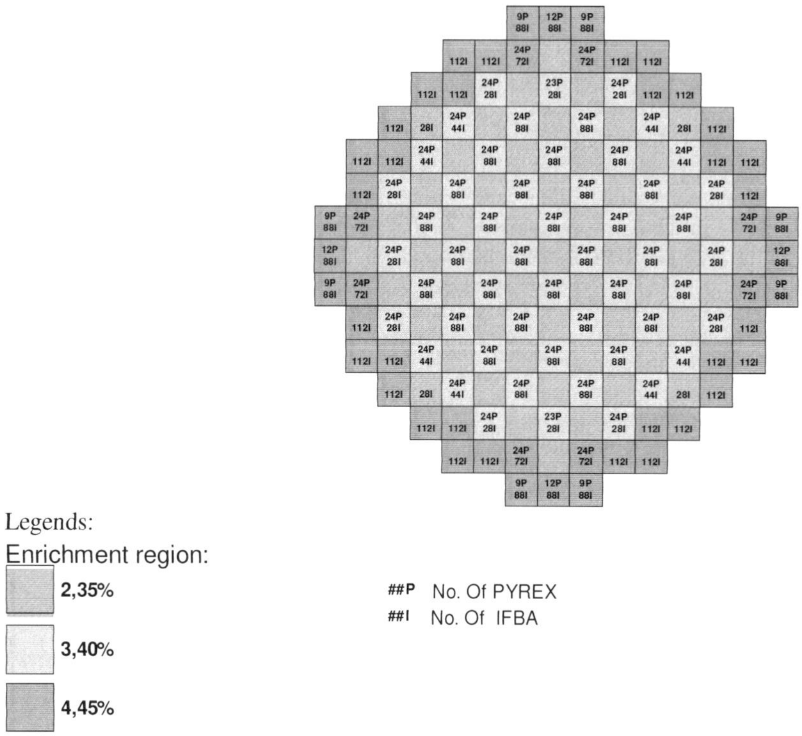

Figure1. Core configuration of the AP1000 with PYREX and IFBA [3]

The 3-dimensional core models of AP1000 used are the models resulted by T. Sembiring [3] that gave a good agreement to the design values of reactivity of AP 1000. The models are again shown in Fig. 2 and3 as 3-dimensional model of AP1000 core for radial and axial directions. They were generated byMCNP6 code for XY (radial) and XZ (axial) directions. The model are representing the core configuration of AP1000 as shown in Fig. 1.

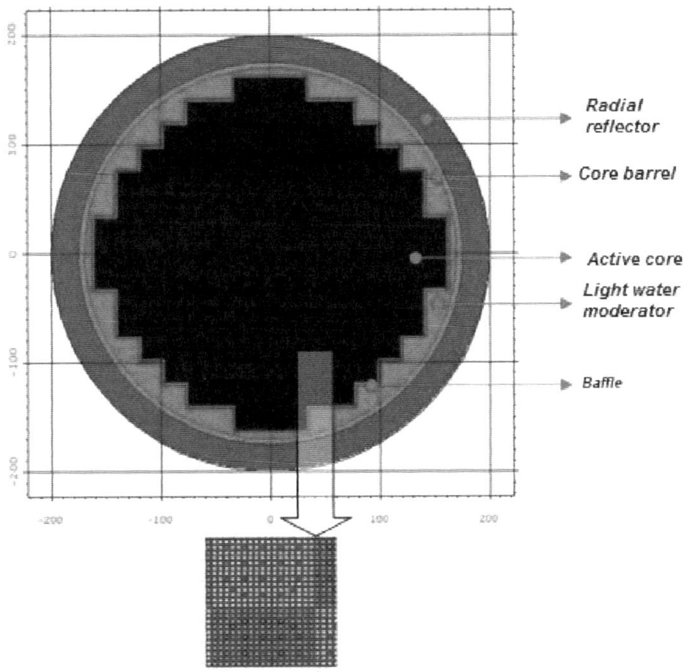

Figure 2. Three dimension model of AP1000 core in XY (radial) direction

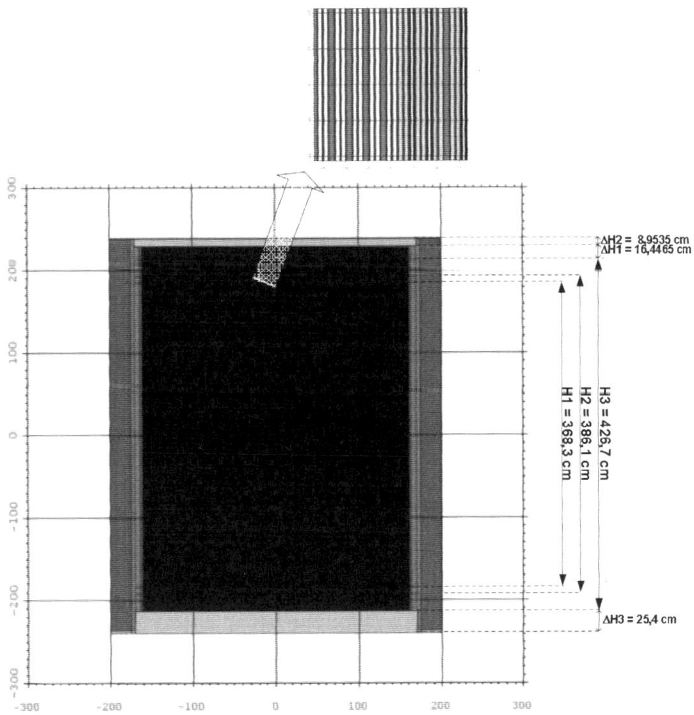

Figure 3. Three dimension model of AP1000 core in R-Z (axial) direction

Core calculations are then carried out by Monte Carlo computer code, MCNP6 version 1[2]. The code shows a very good accuracy for effective multiplication calculation for PWR reactor as validated by several research [5,11]. The calculations are done for two core conditions namely with and without boron.

3. Results and discussion

The calculation results of core reactivity of AP 1000 for conditions with and without boron are presented in Table 1 and Table 2. Table 1 shows that using new libraries either ENDF/B-VII or ENDF/B-VII.1 gives higher values of multiplication factor or reactivity of core. It means that reaction rate is higher as the libraries newer. The relative differences compared to those resulted by ENDF/B-VI are shown in Table2. For core without boron conditions, the relative differences of core multiplication factor are consecutively 0.39% and 0.35% for ENDF/B-VII and ENDF/B-VII.1.Whereas for core with boron conditions, those values are 0.39% and 0.33%.

Table 1. $k_{eff}$of the AP1000 core for with and with boron conditions

Core condition	ENDF/B-VI.8		ENDF/B-VII		ENDF/B-VII.1	
	k_{eff}	Standard deviation	k_{eff}	Standard deviation	k_{eff}	Standard deviation
without boron	1.20535	0.00010	1.21004	0.00010	1.21015	0.00010
with boron	0.97813	0.00010	0.98160	0.00010	0.98138	0.00010

Table 2. The relative difference of k_{eff} of the AP1000 core for with and with boron conditions

Core condition	ENDF/ B-VI.8	ENDF/ B-VII	Rel. diff	ENDF/B-VII.1	Rel. diff
	k_{eff}	k_{eff}	%	k_{eff}	%
without boron	1.20535	1.21004	0,39	1.21015	0.35
with boron	0.97813	0.98160	0,39	0.98138	0.33

The reasons of the increase those k_{eff} values as from ENDF/B-VII.1 can be followed from Fig. 4. This figure shows that there is differences in absorption rate in the energy region surrounding 0 – 0.02 eV and > 6.06 MeV. It tells us that the increase of that k_{eff} from ENDF/B-VII and VII.1 are caused by lower absorption rate in the energy region > 6.06 MeV in comparison to that of from ENDF/B-VI.8.

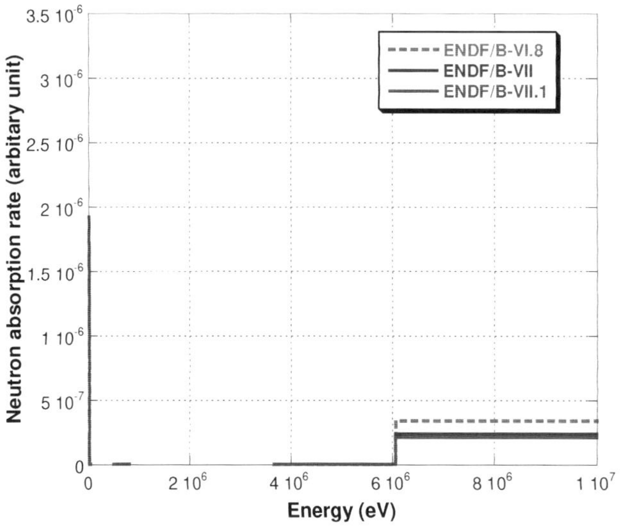

Figure 4. Neutron absorption rate as a function of energy for core without boron.

Fig. 5 shows in more detail the relative differences in using the ENDF/B-VII and ENDF/B-VII.1 compared to that of ENDF/B-VI.8. It is clearly shown that nuclear data ENDF/B-VII.1 has a significant differences in comparison with that of ENDF/B-VI.8 for the whole neutron energy. Whereas for ENDF/B-VII there is a relatively small difference in the region 0.0025 eV – 3.8 MeV. These relative differences for ENDF/B-VII and ENDF/B-VII.1 are in the range of -37% - 18% and -60% - 75% respectively. In comparison to Fig. 5, neutron absorption rate in the region > 6.06 MeV has relative difference of -30% and 60% for ENDF/B-VII and ENDF/B-VII.1. Figure 5 shows a proof that the differences of nuclear data in the energy region > 6.06 MeV cause the change of effective multiplication factor k_{eff} values while using ENDF/B-VII and ENDF/B-VII.1 nuclear data library.

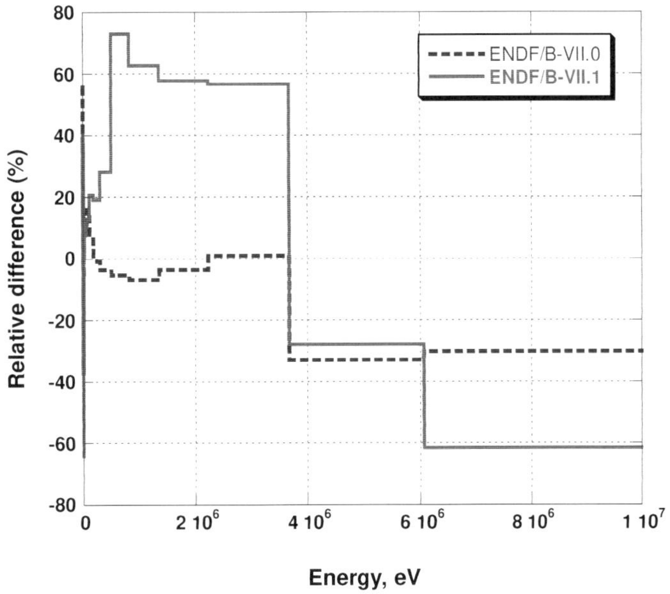

Figure 5. The relative difference of neutron absorption rate to the ENDF/B-VI.8 for without boron condition

Fig. 6 shows the neutron absorption rate for core with boron. It is similar to that one of core without boron that neutron energy in the region of 0 – 0.02 eV and > 6.06 MeV become the cause of producing difference k_{eff} from difference nuclear data library. For analyzing the neutron energy region that effects in producing the k_{eff} values, it is necessary to calculate the relative difference of neutron absorption rate and those are then compared to ENDF/B-VI.8 as presented in Fig. 7.

Fig.7 illustrates a shift of relative difference distribution of absorption rate especially for ENDF/B-VII.1 in the energy region of 3.68 – 6.06 MeV, where the relative difference drops to 0% compared to that of in Fig. 5. This case shows that boron has a role in the decrease of relative difference of the absorption rate. Similarly, for core without boron, the neutron energy region > 6.06 MeV effect the effective multiplication factor due to the difference of nuclear data library.

From the above discussions it can be summarized that the differences of calculated core reactivities are dependently on the status of nuclear data library. Nowadays, improvement of ENDF/B-VII is apparently of lower neutron absorption rate mainly in the energy region higher than 6.06 MeV. It cause higher value of core reactivity that give accuracy of about 0.39% and 0.34% for ENDF/B-VII and ENDF/B-VII. These numbers are proposed to be the corrrecton factors for multiplication values resulted from the former ENDF/BV-VI.8.

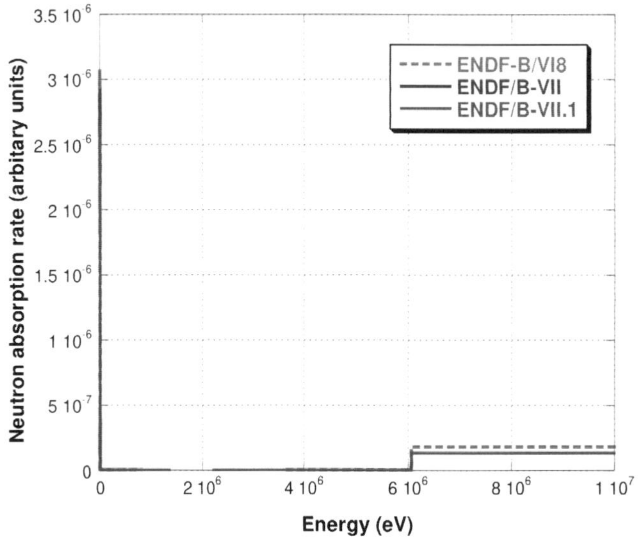

Figure 6. Neutron absorption rate as a function of energy for core with boron condition

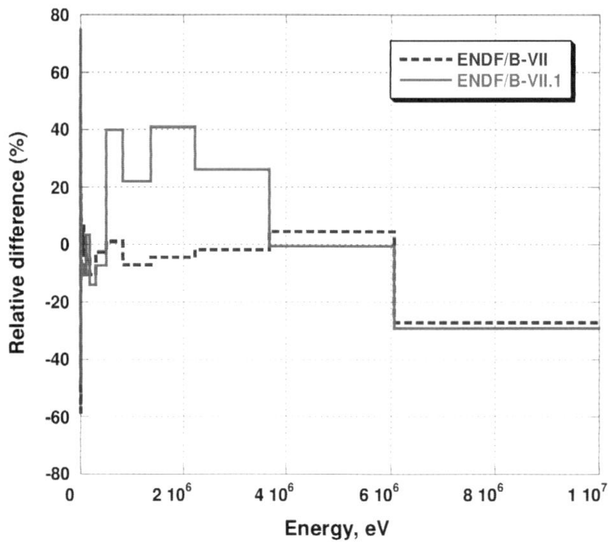

Figure 7. The relative difference of neutron absorption rate to the ENDF/B-VI.8 for with boron condition

4. Conclusions

Application of newer edition of nuclear data library ENDF/B-VII and ENDF/B-VII.1 in the calculation of core reactivity of AP 1000 reactor, produces higher value of multiplication factor compared to that of resulted from ENDF/B-VI.8. The relative difference values is of 0.39% for application of ENDF/B-VII and of about 0.34% for application of ENDF/B-VII.1.

Acknowledgments

Research work at Center for Nuclear Reactor Technology and Safety of BATAN was supported by the DIPA 2016.

References

[1] Schulz T L 2006 Westinghouse AP1000 Advanced Passive Plant *Nucl. Eng. Des.***236** 1547–57

[2] Goorley T, James M, Booth T, Brown F, Bull J, Cox L J, Durkee J, Elson J, Fensin M, Forster R A, Hendricks J, Hughes H G, Johns R, Kiedrowski B, Martz R, Mashnik S, Mckinney G, Pelowitz D, Prael R, Sweezy J, Waters L, Wilcox T and Zukaitis T 2016 Features of MCNP6 *Ann. Nucl. Energy* **87** 772–83

[3] Sembiring T M 2011 Analysis of the 3-Dimensional Core Model for Evaluation of Criticality Parameters of the Advanced PWR 100 MW Class *Tri Dasa Mega***13** 78-95

[4] Mosteller R 2014 Comparison of ENDF / B-VII . 1 and ENDF / B-VII . 0 Results for the Expanded Criticality Validation Suite for MCNP and for Selected Additional Criticality Benchmarks *Nucl. Data Sheets***118** 442–5

[5] Kahler A C, Macfarlane R E, Mosteller R D, Kiedrowski B C, Frankle S C, Chadwick M B, Mcknight R D, Lell R M, Palmiotti G, Hiruta H, Herman M, Arcilla R, Mughabghab S F, Sublet J C, Trkov A, Trumbull T H and Dunn M 2011 ENDF / B-VII . 1 Neutron Cross Section Data Testing with Critical Assembly Benchmarks and Reactor Experiments *Nucl. Data Sheets***112** 2997–3036

[6] van der Marck S C 2012 Calculations Shielding Benchmarks Calculations *Nucl. Data Sheets***113** 2935–3005

[7] Chadwick M B, Herman M, Obloˇ P, Pritychenko B, Arbanas G, Arcilla R, Brewer R, Brown D A, Capote R, Carlson A D, Cho Y S, Derrien H, Guber K, Hale G M, Hoblit S, Holloway S, Johnson T D, Kawano T, Kiedrowski B C, Kim H, Kunieda S, Larson N M, Leal L, Lestone J P, Little R C, Mccutchan E A, Macfarlane R E, Macinnes M, Mattoon C M, Mcknight R D, Mughabghab S F, Nobre G P A, Palmiotti G, Palumbo A, Pigni M T, Pronyaev V G, Vogt R L, Marck S C Van Der, Wallner A, White M C, Wiarda D and Young P G 2011 ENDF / B-VII . 1 Nuclear Data for Science and Technology : Cross Sections , Covariances , Fission Product Yields and Decay Data *Nucl. Data Sheets* **112** 2887–996

[8] Quan H, Honda Y, Goto M and Takada S 2017 Numerical investigation of the random arrangement effect of coated fuel particles on the criticality of HTTR fuel compact using MCNP6 *Ann. Nucl. Energy* **103** 114–21

[9] Zu T, Wan C, Cao L, Wu H and Shen W 2016 Total Uncertainty Analysis for PWR Assembly Based on the Statistical Sampling Method*Nuclear Science and Engineering***183** 371-86

[10] Jin H, Seog K, Gi S and Seung J 2017 An improved DeCART library generation procedure with explicit resonance interference using continuous energy Monte Carlo calculation *Ann. Nucl. Energy* **105** 95–105

[11] Douglass S, Rahnema F and Margulies J 2010 A stylized three dimensional PWR whole-core benchmark problem with Gadolinium *Ann. Nucl. Energy* **37** 1384–403

A Preliminary Design Of Application Of Wireless Identification And Sensing Platform On External Beam Radiotherapy

Heranudin[1,2] **and S. Bakhri**[3]

[1] Center for Radioisotope and Radiopharmaceutical Technology, National Nuclear Energy Agency of Indonesia, South Tangerang, Indonesia
[2] School of Physics, University of Wollongong, New South Wales, Australia
[3] Center for Nuclear Reactor Safety and Technology, National Nuclear Energy Agency of Indonesia, South Tangerang, Indonesia

heranudin@batan.go.id

Abstract. A linear accelerator (linac) is widely used as a means of radiotherapy by focusing high-energy photons in the targeted tumor of patient. Incorrectness of the shooting can lead normal tissue surrounding the tumor received unnecessary radiation and become damaged cells. A method is required to minimize the incorrectness that mostly caused by movement of the patient during radiotherapy process. In this paper, the Wireless Identification and Sensing Platform (WISP) architecture was employed to monitor in real time the movement of the patient's body during radiotherapy process. In general, the WISP is a wearable sensors device that can transmit measurement data wirelessly. In this design, the measurement devices consist of an accelerometer, a barometer and an ionizing radiation sensor. If any changes in the body position which resulted in incorrectness of the shooting, the accelerometer and the barometer will trigger a warning to the linac operator. In addition, the radiation sensor in the WISP will detect unwanted radiation and that can endanger the patient. A wireless feature in this device can ease in implementation. Initial analyses have been performed and showed that the WISP is feasible to be applied on external beam radiotherapy.

Keywords: Radiotherapy, Radiation dose, Accuracy, Wearable sensor.

1. Introduction

External beam radiotherapy is one of the most widely applied methods for modern cancer treatment [1][2][3][4]. Linear accelerator (linac) emits high energy photon output to patients who have a targeted malignant tumor. Unfortunately the normal tissue around the tumor also receives an abundance of external beam radiation from some shooting angles [1]. The intensity of radiotherapy and the accumulation of received doses have an impact on cell damage. Moreover, the normal tissue almost does not have a sufficient time to recover.

The above method of radiotherapy requires accurate calculation of depth-dose distribution between the target area and surrounding tissue to obtain an effective therapeutic process [2]. However, precise information about position and orientation of target is often not obtained during the treatment process. To deal with it, one of techniques is to increase the size of the target volume to accommodate the level of inaccuracy. As a result, the larger target size will extend more normal tissues receiving unnecessary radiation dose as well. Therefore, external beam radiotherapy requires the smallest addition of target volume assumptions but effectively to kill tumor cell.

The greatest cause of such inaccuracy is the uncertainty of the patient's movement during therapy. An illustration of beam path from some shooting angles as shown in figure 1(a) and the movement of the patient that being performed radiotherapy process is shown in figure 1(b) (c). In this paper we will introduce radiofrequency (RF) based monitoring system that can overcome the above problem. This design is able to monitor movement of patients in real time during the treatment process. In addition, the radiation sensor can detect unwanted radiation that can jeopardize the patient.

This system utilizes a platform of Wireless Identification and Sensing Platform (WISP) which has been developed by Intel Research Network [5]. In general, this system is a wearable sensor that attached to an area around targeted tumor. Then the device will send the measurement signal from the sensor in the form of both position and radiation dose level to RF receiver device. The RF receiver will then forward the information to the linac operator.

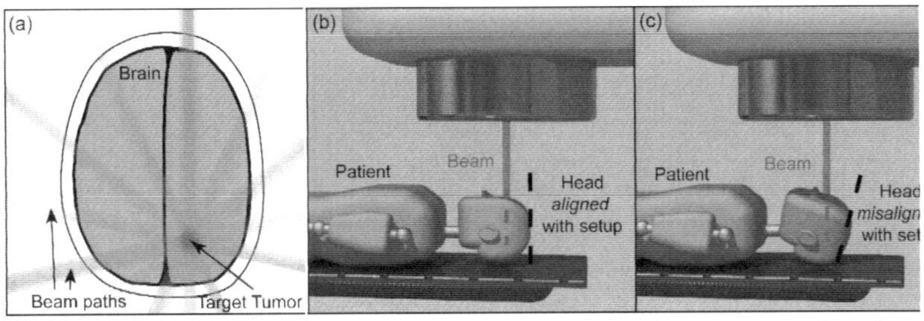

Figure 1. (a) External beam radiotherapy treatment for a targeted malignant tumor. **(b)** Initial position of the patient. **(c)** Uncertainty movement of the patient that can lead to receive unwanted radiation [1].

2. Methodology

WISP architecture harnesses Radio Frequency Identification (RFID) technology as a means of wireless communication. The RFID is an electronic device that uses a microchip and integrated antenna to transmit a unique identification data to the RFID reader or receiver. The advantage of WISP that conventional RFID does not have is a typical ability to program transmitted data [5][6][7][8]. The ability was used to transmit continuous measurement data from several sensors embedded in WISP [8]. In addition WISP also has an advantage to operate without battery that similar to conventional passive RFID tag.

A block diagram of the WISP architecture can be seen in figure 2. Antenna and impedance matching circuit are the initial and crucial part of receiving RF signals from RFID reader. The power harvester circuit converts the RF signal into a DC (Direct Current) voltage that used as a power supply for WISP. The demodulator circuit converts the RF signal into a data signal that can be read by a microcontroller. To perform data transmission, the microcontroller provides data to the modulator circuit by changing the antenna impedance. The WISP was designed to be compatible with standard commercial RFID device EPC class 1 generation 1 [5].

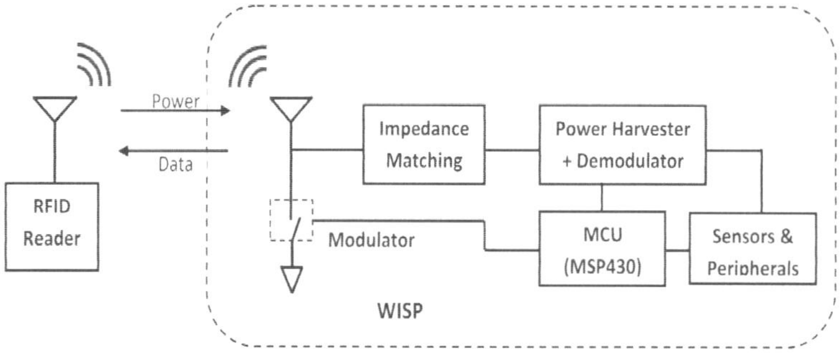

Figure 2. Blok diagram of WISP architecture

Due to limited power at the time of WISP operation, suitable circuit design and correct component selection are required. Ultra low power microcontroller MSP430F2132 is selected as the main controller of WISP system. This 16 bit microcontroller needs only a minimum voltage of 1.8 V and a current of about 600 μA for 4 MHz operation. This component can also perform flash memory operation and ADC (Analog to Digital Converter) task during operation. The microcontroller has several main functions in WISP such as sensor controller to perform measurements, power management and communication controller. The WISP hardware can be seen in figure 3.

Figure 3. The WISP hardware

Embedded sensors of accelerometer and barometer in WISP are able to monitor body movements in real time. Accelerometer uses X, Y and Z axis as references. Changes in WISP positions will alter the accelerometer results of Xg, Yg, and Zg. [5][6]. In figure 4, it can be seen that there is a change in Xg variable value caused by the movement of body position of the patient from sitting to lying down and vice versa. By correct setting of sensitivity of these sensors, the WISP will be able to detect small body movements such as a nod of the head. As a result, combination both accelerometer and barometer will provide more accurate information of the movement [8].

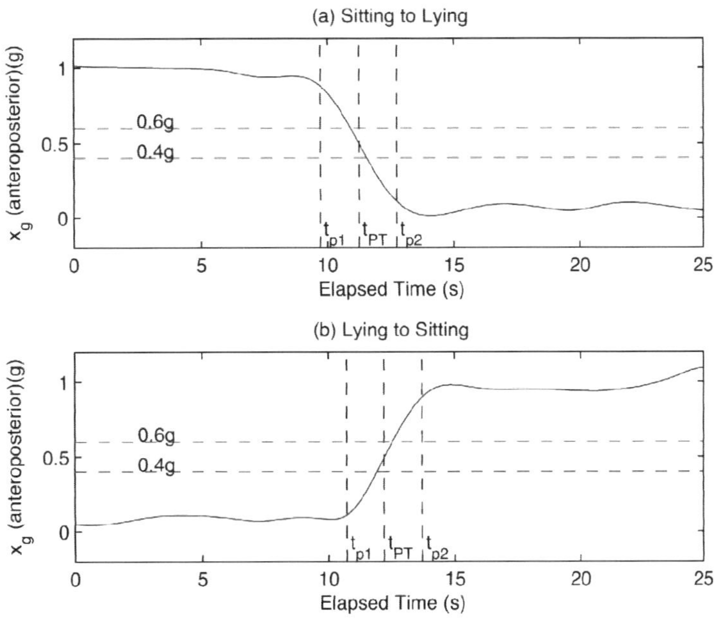

Figure 4. Detection of body movement using accelerometer [6]

In some circumstances, the microcontroller will deliver a warning signal to the operator in case of particular changes in body position that can endanger the patient. Afterwards, the shooting of linac can be adjusted precisely to the targeted tumor. Consequently, it can minimize damage to the surrounding normal tissue.

The WISP design also features a solid state sensor for ionizing radiation measurement. If ionizing radiation is detected by the sensor, it will generate changes in electrical charge. Difference in radiation dose level will also result in difference in the output voltage. The advantages of this sensor are the dimension is quite small and the power consumption is also lower than other types of radiation sensors [9][10][11].

The RD2014 sensor developed by Teviso Sensor Technology is able to detect gamma radiation with a dose rate of 0.1 μSv per hour to 100 mSv per hour as shown in figure 5 [12]. By operating the voltage between 3 volts to 5 volts and the current consumption of about 400 μA, this sensor will be compatible to be integrated into the WISP system. A threshold value of safe radiation is programmed into the WISP system. As a result, if any excess radiation is detected, the WISP will transmit warning information to the linac operator. However, a Monte Carlo simulation will be necessitated to analyze depth-dose distributions extensively [13].

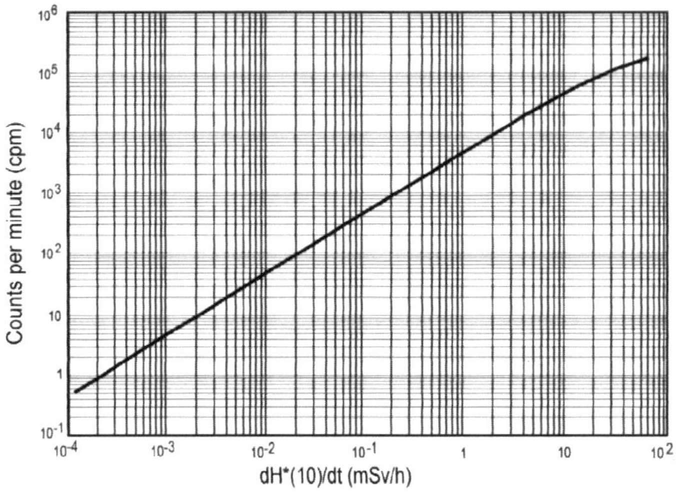

Figure 5. Radiation dose response of RD2014 [12]

3. Results and discussion

One of the most important things in this initial design is how to manage in such a way that the power harvested will be spent efficiently. For example MSP430F2132 working at 3 MHz frequency requires current consumption of 470 μA at 1.8 Volt, while the radiation sensor requires a current of about 400 μA and a working voltage of 3 volts. Eventually all components on WISP must be able to work with available power.

By using the minimum power input required for WISP activation, the working distance of WISP can be calculated by using Friis formula for path loss:

$$PRec = PTran - 20log\left(\frac{4\pi d}{\lambda}\right) + GTran + GRec - LPol \quad (1)$$

The power of the receiver device is PTran = 1 W = 30 dBm. With frequency EPC C1 G1 is 915 MHz, then the wavelength $\boldsymbol{\lambda}$ = 0.33 meters. Gain of transmitter antenna GTran = 6 dBi, which is an international standard. Gain of the receiver antenna GRec = 2 dBi, which is the standard gain for a dipole antenna. Loss of polarization, LPol is 3 dB.

By using a supply voltage of 3 Volts as shown in figure 6, then we get the working operation of WISP or Power received PRec = 6.1 dBm. The maximum working distance from WISP is about 3 meters. So it can be illustrated placement of WISP system on external file-based radiotherapy as in figure 7.

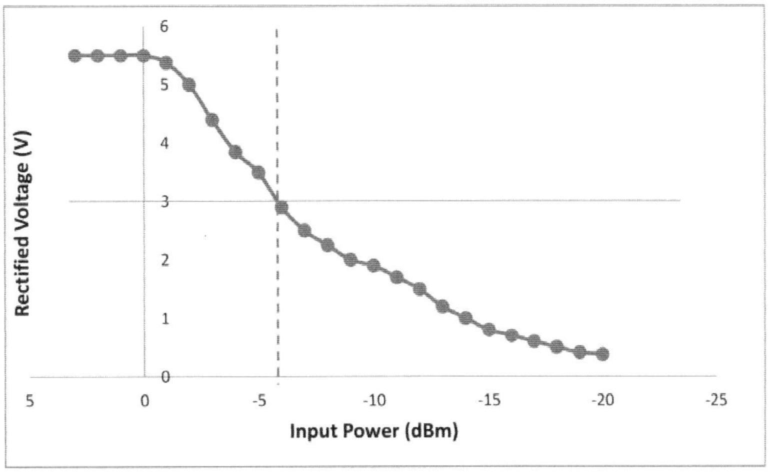

Figure 6. The WISP simulation for input power (dBm) vs rectified voltage (V)

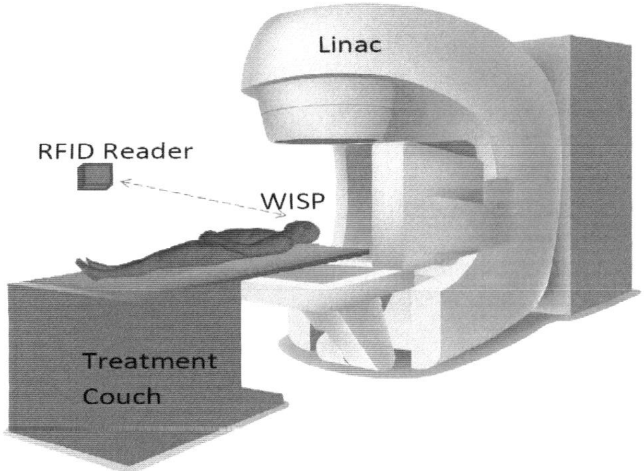

Figure 7. Design of WISP implementation for external beam radiotherapy treatment

4. Conclusion

In this initial design, it has been reported that WISP technology can support external beam radiotherapy treatment. Changes in body position of the patient can be monitored in real time with accelerometer and barometer embedded in WISP. Sensitivity of the accelerometer, which is measured in the variables Xg, Yg, dan Zg, can be set to obtain tolerable body motion thresholds. Radiation dose of the patient is also monitored by radiation sensors RD2014. The calculation with the Friis formula for path loss indicates that WISP can be placed near to the patient's body with the maximum distance of three meters from the receiver device or RFID reader.

The implementation of WISP to the radiotherapy treatment requires some further analysis. Among them are power management so that the WISP can run all the embedded sensors and components efficiently and also analysis of various positional changes of patients that affect the detection of the movement sensor. Furthermore, the Monte Carlo simulation will be required to simulate comprehensive depth-dose distributions. In addition, study of electronic component defect caused by received radiation should be considered.

References

[1] M. Ostyn, S. Kim, and W. H. Yeo 2016 A simulation study of a radiofrequency localization system for tracking patient motion in radiotherapy *Sensors (Switzerland)* **16** no.4

[2] A. Skalski, T. Zieliński, P. Kukołowicz, and P. Kédzierawski 2011 Computed tomography-based radiotherapy planning on the example of prostate cancer: application of level-set segmentation method guided by atlas-type knowledge *ACM Int. Conf. Proceeding Ser.* pp. 0–4

[3] A. Naseri and A. Mesbahi 2010 A review on photoneutrons characteristics in radiation therapy with high-energy photon beams *Reports Pract. Oncol. Radiother.* **15** 138–144

[4] F. d'Errico, M. Luszik-Bhadra, R. Nath, B. R. Siebert, and U. Wolf 2001 Depth dose-equivalent and effective energies of photoneutrons generated by 6-18 MV X-ray beams for radiotherapy *Health Phys.* **80**, 4-11

[5] A. P. Sample, D. J. Yeager, P. S. Powledge, A. V. Mamishev, and J. R. Smith 2008 Design of an RFID-based battery-free programmable sensing platform *IEEE Trans. Instrum. Meas.* **57** 2608–2615

[6] D. C. Ranasinghe, R. L. Shinmoto Torres, K. Hill, and R. Visvanathan 2014 Low cost and batteryless sensor-enabled radio frequency identification tag based approaches to identify patient bed entry and exit posture transitions *Gait Posture* **39** 118–123

[7] Y. Dong, A. Wickramasinghe, H. Xue, S. F. Al-Sarawi, and D. C. Ranasinghe 2015 A Novel Hybrid Powered RFID Sensor Tag *2015 IEEE Int. Conf. RFID* pp. 55–62

[8] Y. Su, A. Wickramasinghe, and D. C. Ranasinghe 2015 Investigating sensor data retrieval schemes for multi-sensor passive RFID tags *2015 IEEE Int. Conf. RFID* pp. 158–165

[9] T. Gupta, R. Sanket, R. Singh, A. Gehlot, E. Mehandiratta, A. Agarwal and S. Choudhury 2016 *Proceeding of International Conference on Intelligent Communication, Control and Devices* pp. 569-579

[10] B. Mukherjee, X. Wu, T. Maczka, T. Kwan, Y. Huang, and V. Mares 2016 Near space radiation dosimetry in Australian outback using a balloon borne energy compensated PIN diode detector *Radiat. Meas.* **94** 65–72

[11] M. Arsalan, A. Shamim, M. Shams, N. G. Tarr, and L. Roy 2012 Ultra low power CMOS-based sensor for on-body radiation dose measurements *IEEE J. Emerg. Sel. Top. Circuits Syst.* **2** 34–41

[12] Teviso 2015 *Product (Sensors) Instruction Manual of RD2014* Bern, Switzerland Teviso Sensor Technologies

[13] D. Sardari, R. Maleki, H. Samavat, and A. Esmaeeli 2010 Measurement of depth-dose of linear accelerator and simulation by use of Geant4 computer code *Reports of Practical Oncology and Radiotherapy* **15** 64-68.

Leaching kinetic of Nd. Y, Pr and Sm in rare earth hydroxide (REOH) use nitric acid

MV Purwani and Suyanti

Center for Accelerator Science and Technology
National Nuclear Energy Agency of Indonesia (BATAN)

Email : purwani@batan.go.id

Abstract. The purpose of this study were to determine the order of reaction, rate reaction constant and activation energy of reaction $Y(OH)_3$, $Nd(OH)_3$, $Pr(OH)_3$ and $Sm(OH)_3$ with HNO_3. The rate reaction constant is necessary to determine the residence time in the design of continuously stirred tank reactor (CSTR). The studied parameters were leaching temperature (60 – 90 °C) and leaching time (0-15 minutes). From the resulting data can be concluded that the leaching process were strongly influenced by the time and temperature process. Leaching rare earth hydroxide (REOH) using nitric acid follows second order. At leaching 10 grams of REOH using 40 ml HNO_3 0.0576 mol were obtained maximum conversion at 90 °C and leaching time 15 minutes for Y was 0.95 (leaching efficiency was 95%), for Nd was 0.97 (leaching efficiency was 97%), for Pr was 0.94 (leaching efficiency was 94%) and for Sm was 0.94 (leaching efficiency was 94%). The largest activation energy was Y of 23.34 kJ /mol followed by Pr of 20.00 kJ /mol, Sm of 17.94 kJ /mol and the smallest was Nd of 16.39 kJ /mol. The relationship between the rate constant of the reaction with T for Y was $k_Y = 338.26$ $e^{-23.34 / RT}$, for Nd was $k_{Nd} = 33.69$ $^{e -16.39 / RT}$, for Pr was $k_{Pr} = 102.04$ $e^{-20 / RT}$ and for Sm adalah was $k_{Sm} = 50.16$ $e^{-17.94/ RT}$
Keywords: REOH, HNO_3, leaching

1. Introduction

The rare earth elements (REEs) are a unique group of metals regarded as being among the most critical elements that are highly valued for their specialwased applications in many modern technologies. In the recent years, REEs have become a significant topic of interest in the metals industry due to a global supply shortage and strengthening demand[1] This has led many companies to develop processing techniques for extracting REEs from less common rare earth bearing deposits[2] and reclaiming REEs from end-of-life products by recycling (urban mining) spent batteries. phosphors and permanent magnets[3]. One of such examples was the apatite group of minerals (mainly fluorapatite) which are commonly mined for the production of phosphoric acid in the fertilwaser industry[4,5,6]. The monazite ores are known to contain significant quantities of REEs and therefore processing of monazite ores for REEs has become more prominent recently[7]. In the processing of monazite deposits however, often require a pre-leach stage prior to the acid bake stage at elevated temperature where the latter converts REE-phosphates to water soluble REE-sulphates[8].The purpose of the pre-leach stage of the ore / concentrate was to remove the disturbing calcium during the REE manufacturing stage.

Rare earths hydroxide (REOH) was a monazite process products containing REE elements namely Ce, La, Nd, Pr, Y, Sm and Gd[9]. For the separation of each of those elements. REOH should be leached, so that the solution was formed. The leaching reaction that occurs could be between the base with an acid and a base with an acid. The base will react with the acidic solution to form salt and water[10]. K. Stonea et al., who studied comparison of H_3PO_4, HCl, $HClO_4$ and HNO_3 usage leading to the leaching[11]. Leaching with HNO_3 also for laterite leaching[12]. Use of HNO_3, resulting in the greatest leaching efficiency. In general, leaching of REOH using HNO_3, because in addition of HNO_3 to easy and strong in leaching, it was also an oxidizer that will form Ce of III valence to Ce of IV valence[13]. Thus Ce will have different properties from other REE elements that have double valences of II and III, so Ce could be separated from other REE elements. Rare earth elements that are easily soluble or are leached are Y, Nd, Pr and Sm while Ce will dissolve later.

In this work, REOH was prepared from monazite sand. The leaching of REOH using HNO_3 0.0576 moles, the parameters were leaching temperature (60 – 90 °C) and leaching time (0-15 minutes). The solution product of leaching are analyzed by XRF.

The purpose of this study were to determine the order of reaction, rate reaction constant and activation energy of reaction $Y(OH)_3$, $Nd(OH)_3$, $Pr(OH)_3$ and $Sm(OH)_3$ with HNO_3. The rate reaction constant is necessary to determine the residence time in the design of continuously stirred tank reactor (CSTR).

2. Theory

The reaction rate represents the molarity of the solute in the reaction produced each reaction time[14].

$$A + B \rightarrow C$$

$$\text{The reaction rate } = r_A = -\frac{\Delta[A]}{\Delta t} = -\frac{\Delta[B]}{\Delta t} \quad (1)$$

The factors affecting the rate of reaction: temperature, catalyst, molarity and reactant concentration and factors affecting the constant or constant rate of reaction was temperature[15,16,17]. Determination of the constant and leaching reaction rate at the optimum condition of the use of the reactant ratio of REOH with HNO_3. The parameters studied were the variation of leaching time and temperature. The reactions that occurred between REOH and HNO_3 were as follows

$$RE(OH)_{3(s)} + 3HNO_{3(a)} \rightarrow RE(NO_3)_{3(a)} + 3H_2O_{(a)} \text{[18]} \quad (2)$$

s : solid, a : solution
RE : Y, Pr, Nd, Sm

$$-r_A = k_1.C_A^n.C_B = (k1\ C_B)\ C_A^n = k\ C_A^n \text{[19]} \quad (3)$$

n = reaction order
k = the overall reaction rate constant
C_A = concentration (mol) RE
C_B = concentration (mol) HNO_3

In the initial state or t_0, the number of reagents A or NA_0 and at time t of NA, the conversion of A (X_A) at a constant volume system was[14]:

$$X_A = \frac{N_{A0} - N_A}{N_{A0}} = 1 - \frac{C_A}{C_{A0}} \quad (4)$$

$$dX_A = \frac{dC_A}{C_{A0}} \tag{5}$$

Data analisys was by integral begins by selecting a certain rate equation to be tested in an integral way and comparing the approximate C-curve relationship with C data to t from the experiment. If they do not match, other estimates are then tested.

First Order Reaction

In general, the first order reaction can be represented by the following reaction equation:

A $\rightarrow$ product

If the first order reaction, then the equation of reaction rate was[20]:

$$-r_A = \frac{dC_A}{dt} = kC_A \tag{6}$$

The integration of equation (7) becomes:

$$-\int_{C_{A0}}^{C_A} \frac{dC_A}{C_A} = k\int_0^t dt \tag{7}$$

$$-ln\frac{C_A}{C_{A0}} = kt \tag{8}$$

CA_0 = RE concentration in the feed
C_A = RE concentration after reaction
If expressed in terms of conversion, the equation of rate reaction becomes (9):

$$\frac{dX_A}{dt} = k(1-X_A)\ [21] \tag{9}$$

$$-\ ln\ (1\text{-}X_A) = kt\ [22] \tag{10}$$

If the graph of the relationship ln (1-X_A) or ln (C_A / C_{A0}) to t was a linear line then the reaction which we observe was correct, following the first order reaction[23].

Second Order Reaction

For the second order reaction with the constant rate of reaction referring to the overall reaction rate constant, could use the following equation::

A + 3B $\rightarrow$ C+3D

$$-r_A = -C_{A0}\frac{dX_A}{dt} = -\frac{dC_B}{dt} = kC_AC_B^3 = k(C_{A0}-C_{A0}X_A)(C_{B0}-C_{A0}X_A) \tag{11}$$

$$-r_A = -C_{A0}\frac{dX_A}{dt} = kC_{A0}^2(1-X_A)(M-3X_A) \tag{12}$$

If it was integrated into [24,26,27]

$$\ln\frac{M-3X_A}{M(1-X_A)} = \ln\frac{C_BC_{A0}}{C_{B0}C_A} = \ln\frac{C_B}{MC_A} == C_{A0}(M-2)kt = (C_{B0}-C_{A0})kt \tag{13}$$

M = comparison (mole B / mol A) start = C_{B0} / C_{A0} *= mole of initial* HNO_3 */ mole RE*

The experimental variables in equation (13) are made in the form of a $\ln\frac{(M-3X_A)}{M(1-X_A)}$ vs-t relationship. If a linear line was obtained then the asummption of second order was correct and the CA_0 slope (M-2) k. The value of the reaction rate constant was the slope of the linear line divided by CA_0 (M-2).

The k relation as a function of temperature by refering to the Arrhenius equation,

$$k = e^{Ea / R.T} [28] \tag{14}$$

Han Wang[29], states the relationship between k and T was formulated in the equation

$$\ln k = \ln A - \left(\frac{Exp}{R}\right)\left(\frac{1}{T}\right) \tag{15}$$

A was the frequency factor or pre - exponential factor, R was the universal gas constant = 8.314J / gmol ^{0}K and T was the absolute temperature, ^{0}K, Ea = activation energy, J / gmole.

The equations for the determination of the reaction order may be used for the reaction of aqueous solids such as Na Thiosulfate with methyl iodide [14] so these equations may be used in the determination of REOH reaction with HNO_3.

3. Methodology

Materials

The materials used were REOH of the processed monazite sand which Ce and La was taken partly (2 times fractionation), technical HNO_3, water, filter paper, Nd_2O_3, Sm_2O_3, Pr_5O_{11}, Y_2O_3 Merck as analyswas standard materials.

Equipments

Glass equipments, analytic scales, Ikamag heater stirrers, thermometers, ovens, x-ray spectrometers (XRF)

Methods

Nitric acid (HNO_3) with the molarity of 14.4 M as much as 4 ml was put in the glass beaker added water up to 10 ml plus REOH of 10 grams, heated to a certain temperature of 60 °C while stirring. The heating time varied 3, 6, 9, 12 and 15 minutes. After each process, the residue was filtered. The leaching solution was taken as 1 ml diluted with water to 5 ml analyzed using XRF. The remaining solids or residues, dried and weighed. The work was repeated for temperatures of 70, 80 and 90 °C.

Measurable quantities

- The element content (%) in the REOH feed used XRF
- Concentrations of Y, Pr, Nd and Sm (ppm) elements in the leached solution

4. Results and discussion

The results of the composition analysis of REOH (Feed) using XRF were presented in Table 1. In general the largest element content in REOH was Nd, followed by Nd and La. In this research REOH used as the feed was REOH which was partly taken its Ce and La (fractination 2 times), so that the Nd content increases, La and Ce content decreases.

Table 1. The composition analysis of REOH (feed) using XRF

No	Component	Elements concentration %	Atomic Weight	Molecular Weight	Compound Concentration %
1	$Y(OH)_3$	5.36	89	140	8.43
2	$Ce(OH)_3$	19.07	140	191	26.01
3	$La(OH)_3$	6.36	139	190	8.69
4	$Nd(OH)_3$	20.68	144	195	28.00
5	$Sm(OH)_3$	13.18	150	201	17.66
6	$Pr(OH)_3$	4.26	141	192	5.80
7	$Dy(OH)_3$	0.94	162	213	1.24

IOP Conf. Series: Journal of Physics: Conf. Series **962** (2018) 012061 doi:10.1088/1742-6596/962/1/012061

8	$Gd(OH)_3$	3.14	157	208	4.16
	Total				100.00

Effects Temperature and Leaching Time on The Leaching Conversion

From the analysis result of REOH and leaching product solution could be determined or the leaching conversion (X) could be calculated at various time and temperature.

$$\text{Element conversion (X)} = \frac{weight\ of\ element\ in\ the\ feed - weight\ of\ element\ in\ the\ solution}{weight\ of\ element\ in\ the\ feed} \quad (16)$$

Weight of element in the feed = concentration of element x weight of REOH
Weight of element in the solution= concentration of element x volume of solution

The effect of time and leaching temperature on the conversion was shown in Figure 1a to Figure 1d. The longer time and the higher leaching temperature, the leaching conversion would be greater. In general reaction, reaction rate one of which was very influenced by temperature and the completely reaction was influenced by reaction time duration

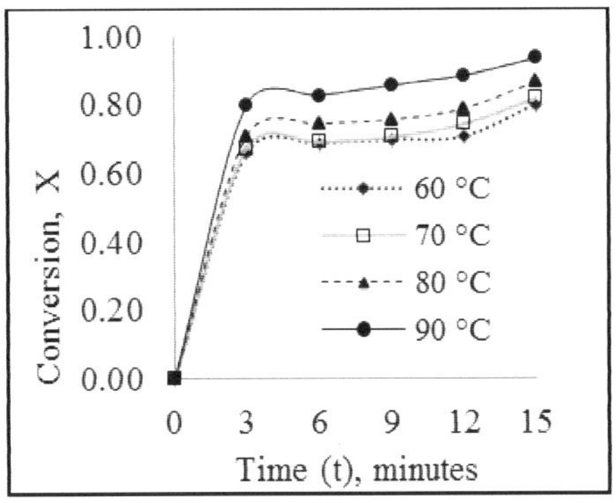

Figure 1a. Correlation between Y leaching time and conversion at various temperature

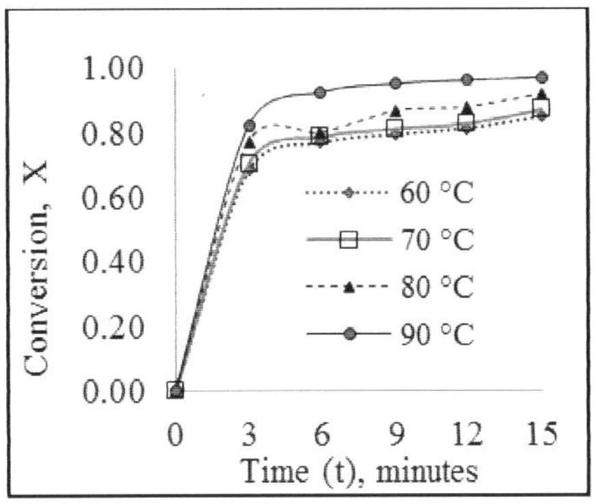

Figure 1b. Correlation between Nd leaching time and conversion at various temperature

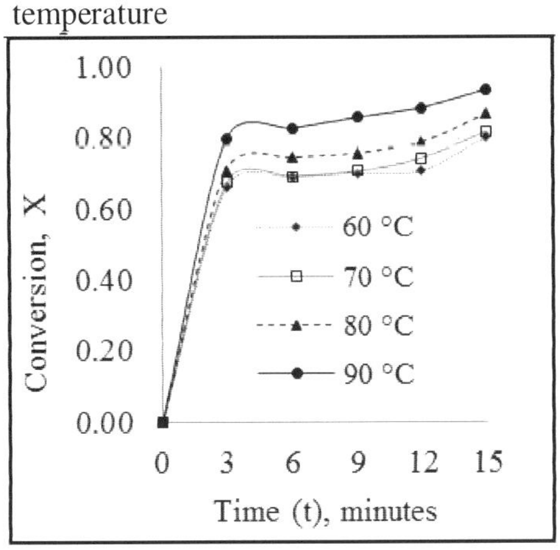

Figure 1c. Correlation between Pr leaching time and conversion at various temperature

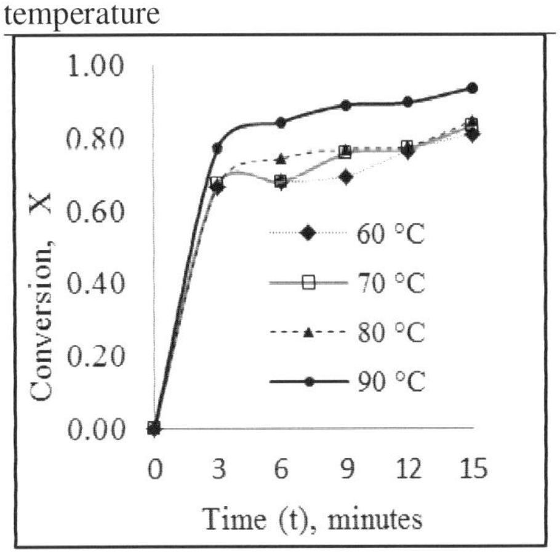

Figure 1d. Correlation between Sm leaching time and conversion at various temperature

The conversion value of Y was in the range of 0.00 - 0.95, the Nd conversion values were in the range of 0.00 - 0.97, Pr conversion value were in the range of 0.00 - 0.94 and Sm conversion value were in the range of 0.00 - 0.94.

Reaction order and rate reaction constant determination

First Order Reaction

In the calculation of reaction rate constant with first order reaction , it was needed the graph of time correlation with –ln (1-X) where X was conversion (mole). Based on the equation (13), first order reaction was experimented for finding the reaction constants. The reaction rate was the overall reaction rate constant with the assumption that the concentration of HNO_3 was constant because its amount tends to approach 1, therefore it could be mentioned that the occuring reaction was first order. The calculation results could be seen in Figure 2a to 2d and Table 2

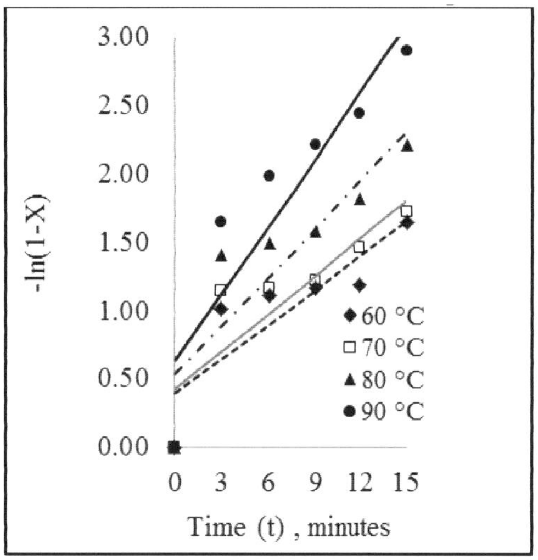

Figure 2a. Correlation between time and -ln(1-X) for determining of k value and linearity of first order reaction for Y element

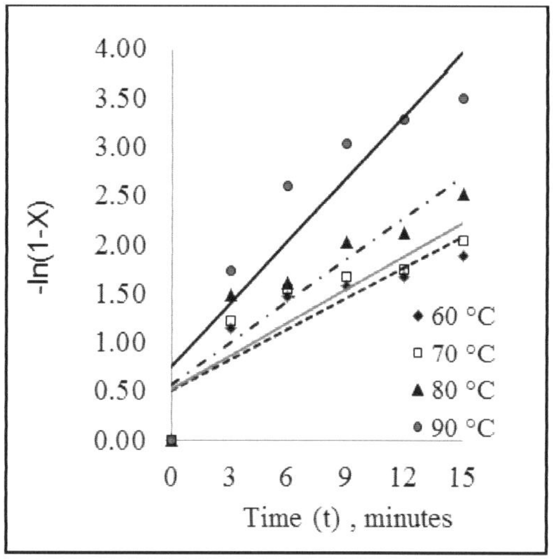

Figure 2b. Correlation between time and -ln(1-X) for determining k value and linearity of first order reaction for Nd element

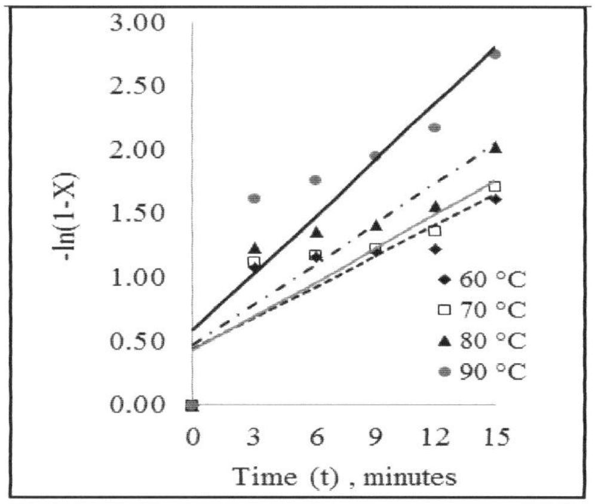

Figure 2c. Correlation between time and -ln(1-X) for determining k value and linearity of first order reaction for for Pr element

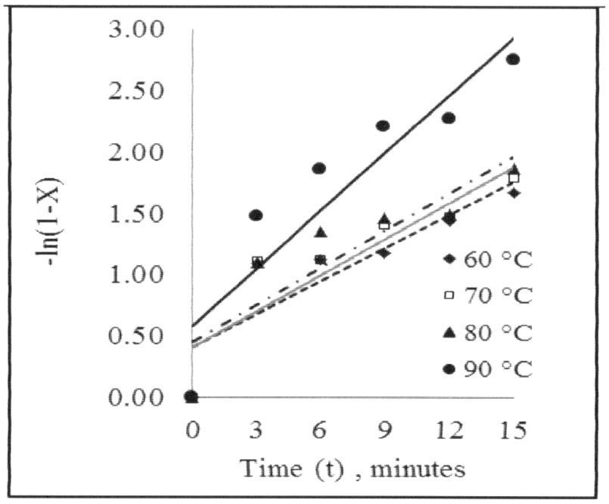

Figure 2d. Correlation between time and -ln(1-X) for determining k value and linearity of first order reaction for Sm element

Linear regression values were obtained from square root of R^2 of Y, Nd, Pr and Sm elements at various temperature as shown in Table 2.

Table 2. The equation and the value of linearity (R^2) in the first order reaction

	Temperature 60 °C	Temperature 70 °C	Temperature 80 °C	Temperature 90 °C
Y	y60 = 0.0835x + 0.394 R² = 0.7396 .	y70 = 0.0914x + 0.4322 R² = 0.7557	y80 = 0.1182x + 0.5317 R² = 0.7753	y90 = 0.1631x + 0.6384 R² = 0.8278
Nd	y60 = 0.1061x + 0.4993 R² = 0.7675	y70= 0.1139x + 0.5192 R² = 0.778	y80 = 0.1425x + 0.5606 R² = 0.8231	y90 = 0.2152x + 0.7458 R² = 0.8453
Pr	y60 = 0.0886x + 0.4336 R² = 0.7414	y70 = 0.0886x + 0.4336 R² = 0.7414	y80 = 0.1056x + 0.4739 R² = 0.7685	y90 = 0.1485x + 0.5911 R² = 0.8126
Sm	y60 = 0.0901x + 0.4123 R² = 0.7693	y70 = 0.0959x + 0.4728 R² = 0.7467	y80 = 0.1021x + 0.4366 R² = 0.7955	y90 = 0.1569x + 0.5837 R² = 0.8382

Second Order Reaction

Calculation of the second order reaction rate constant required the time-versus $\ln\frac{(M-3X_A)}{M(1-X_A)}$ correlation graph. To find the value of M, the number of Y, Nd, Pr and Sm moles in REOH and HNO_3 were initially calculated.

M = mol HNO_3 (C_{B0}) / mol C_{A0} (Y, Nd, Pr and Sm) hydroxide

Mol HNO_3 = 0.0576 and moles Y, Nd, Pr and Sm are seen in Table 3. Table 3 presents the M values for Y, Nd, Pr and Sm elements.

Table 3. M value for Y,Nd, Pr and Sm

M Y	M Pr	M Nd	M Sm
9.5624	19.056	4.01	6,5559

The results of second order reaction could be seen in Figures 3a– 3d and Table 4.

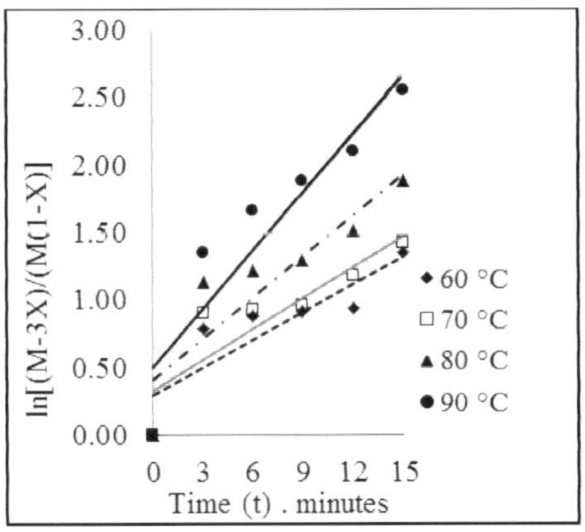

Figure 3a. Correlation between time and $\ln\frac{(M-3X_A)}{M(1-X_A)}$ for determining k value linearity of second order of Y element

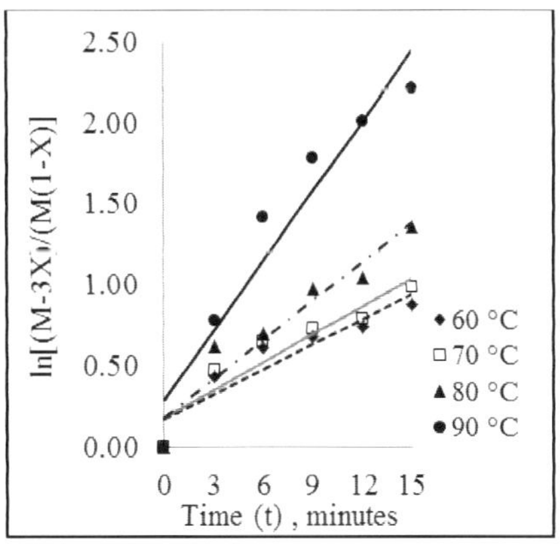

Figure 3b. Correlation between time and $\ln\frac{(M-3X_A)}{M(1-X_A)}$ for determining k value linearity of second order of Nd element

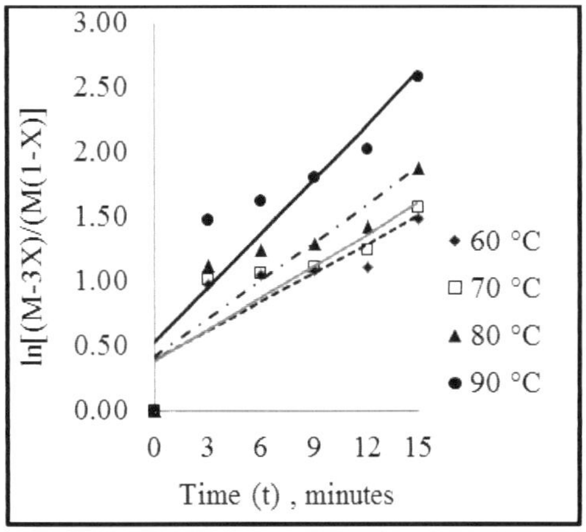

Figure 3c. Correlation between time and $\ln\frac{(M-3X_A)}{M(1-X_A)}$ for determining k value linearity of second order of Pr element

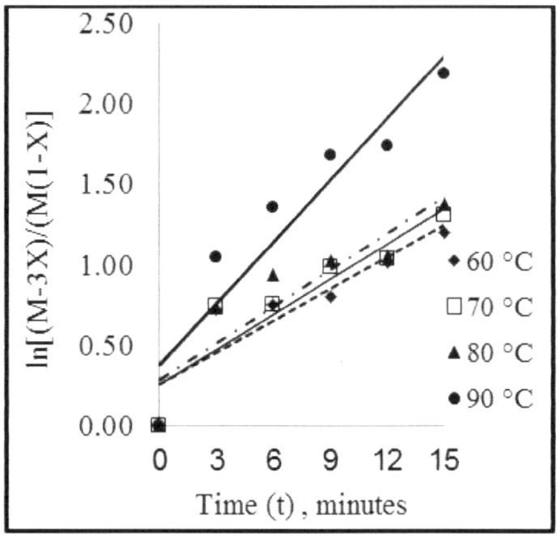

Figure 3d. Correlation between time and $\ln\frac{(M-3X_A)}{M(1-X_A)}$ for determining k value linearity of second order of Sm element

The linear regression value obtained from the square root of R^2 in the second order reaction of elements Y, Nd, Pr and Sm at various temperatures was presented in Table 4.

Table 4. The equation and the value of linearity (R^2) in the second order reaction

	Temperature 60 °C	Temperature 70 °C	Temperature 80 °C	Temperature 90 °C
Y	y60 = 0.0791x + 0.3636 R² = 0.7463	y70 = 0.0868x + 0.4 R² = 0.7637	y80 = 0.1132x + 0.495 R² = 0.7837	y90 = 0.1578x + 0.5976 R² = 0.8365
Nd	y60 = 0.0936x + 0.4124 R² = 0.7892	y70 = 0.1011x + 0.4296 R² = 0.8001	y80 = 0.1289x + 0.4641 R² = 0.8456	y90 = 0.2007x + 0.6357 R² = 0.8638
Pr	y60 = 0.0794x + 0.4217 R² = 0.7052	y70 = 0.0864x + 0.4178 R² = 0.7451	y80 = 0.1033x + 0.457 R² = 0.7723	y90 = 0.146x + 0.5717 R² = 0.8167
Sm	y60 = 0.0833x + 0.366 R² = 0.7814	y70 = 0.0889x + 0.423 R² = 0.7587	y80 = 0.0949x + 0.3883 R² = 0.808	y90 = 0.1489x + 0.5256 R² = 0.8509

Linearity values (R^2) of first order reaction (Table 2) were smaller compared to the linearity values(R^2) second order reaction (Table 4), so it could be stated that the reaction was second order.

In general, the determinant of the reaction rate was the phase with the smallest rate, especially for the chemical reactor design [30].

The list of k values of the 2nd order reaction at various temperatures can be arranged in Table 5.

Table 5. The k values of 2nd order reaction at various temperature

	Temperature, °C	T, °K	k , liter/mole.time	1/T	ln k	- ln k
Y	60	333	0.0791	0.0030	-2.537	2.537
	70	343	0.0868	0.0029	-2.444	2.444
	80	353	0.1132	0.0028	-2.179	2.179
	90	363	0.1578	0.0028	-1.846	1.846
Pr	60	333	0.0794	0.0030	-2.533	2.533

	70	343	0.0864	0.0029	-2.449	2.449
	80	353	0.1033	0.0028	-2.270	2.270
	90	363	0.0794	0.0028	-1.924	1.924
Nd	60	333	0.0936	0.0030	-2.369	2.369
	70	343	0.1011	0.0029	-2.292	2.292
	80	353	0.1289	0.0028	-2.049	2.049
	90	363	0.1489	0.0028	-1.904	1.904
Sm	60	333	0.0833	0.0030	-2.485	2.485
	70	343	0.0889	0.0029	-2.420	2.420
	80	353	0.0949	0.0028	-2.355	2.355
	90	363	0.1489	0.0028	-1.904	1.904

After obtaining the element's reaction rate constants of second order reaction at each temperature, the correlation of the reaction rate constant k as a function of temperature by referring to the Arrhenius equation, ln k = ln A - Ea /RT. (equation 15).

The calculation results are expressed in the linear line graph between ln k as ordinate and 1 / T as the abscissa for Y. Nd, Pr and Sm element could be observed in Figure 4. From Figure 4, the relationship between 1 / T and ln k, line alignment was expressed by R^2, the slope obtained was the activation energy and the intercept was the frequency factor.

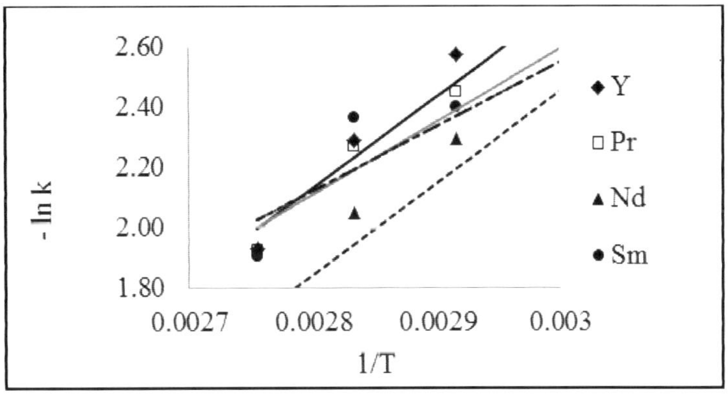

Figure 4a. Correlation between 1/T and -ln k for Y, Pr, Nd and Sm element

The linear equation of the correlation between 1/T (x) and -ln K (y) could be seen inTable 6. The activation energy was defined as the energy that must be exceeded so that chemical reactions can occur. Frequency factor (A) depends on the weight and molecular structure. frequency and position of collisions between molecules. The value of the frequency factor (A) could be determined from the intercept and the activation energy (E) was the value of the slope multiplied by R (gas constant 8.314 J/mole.°K). The calculation results of A and E values could be observed also in Table 6. From Table 6 it could be seen that the largest activation energy was Y of 23.34 kJ/mole followed by Pr of 20.00 kJ/mole. Sm of 17.94 kJ/mole and the smallest one was Nd of 16.39 kJ/mole

Table 6. Activation energy (E) and Frequency factors (A) values

Element	Y	Pr	Nd	Sm
Equation	y = 2807.3x - 5.8238	y = 1971.4x - 3.5173	y = 2157.6x - 3.9152	y = 2157.6x - 3.9152
Linearity	$R^2 = 0.9063$	$R^2 = 0.9063$	$R^2 = 0.9612$	$R^2 = 0.7663$
E/R	2807.3	2405.5	1971.4	2157.6
E, KJ	23.34	20.00	16.39	17.94
ln A	5.8238	4.6254	3.5173	3.9152
A	338.255	102.044	33.693	50.159
Arhenius Equation	$k_Y = 338.26\ e^{-23.34/RT}$	$k_{Pr} = 102.04\ e^{-20/RT}$	$k_{Nd} = 33.69\ e^{-16.39/RT}$	$k_{Sm} = 50.16\ e^{-17.94/RT}$

Ferdowsi et al, did and optimizing and kinetics of leaching process of cerium, lanthanum and neodymium elements of apatite using nitric acid. The maximum leaching efficiency of which were 66.1%, 56.8% and 51.7% respectively for Ce, La and Nd, achieved under the optimum leaching conditions with nitric acid concentration of 18%, 0.06 ratio of solid to liquid and leaching time of 38 minutes, the activation energy of Ce of 6.54 kJ / mole[20]

5. Conclusion

The leaching process was strongly influenced by the temperature and time of process. Leaching of REOH using nitric acid follows second order reaction. In leaching 10 g of REOH using 40 ml of HNO_3 0.0576 moles were obtained maximum conversion at 90 ^{0}C and leaching time of 15 minutes for Y was 0.95 (leaching efficiency of 95%). for Nd was 0.97 (leaching efficiency of 97%). for Pr was 0.94 (leaching efficiency of 94%) and for Sm was 0.94 (leaching efficiency of 94%). The largest activation energy was Y of 23.34 kJ / mole followed by Pr of 20.00 kJ/ mole, Sm of 17.94 kJ / mole and the smallest was Nd of 16.39 kJ / mole. The correlation between the reaction rate constant with T for Y was kY = 338.26 $e^{-23.34/RT}$. for Nd was kNd = 33.69 $e^{-16.39/RT}$, for Pr was kPr = 102.04 $e^{-20/RT}$ and for Sm was kSm = 50.16 $e^{-17.94/RT}$

Acknowledgments

On this occasion the authors would like to thank PT TIMAH who has given REOH and this research at the cost of INSINas in year 2017. Thanks also to Mrs. Suprihati who has helped me out in this research.

References

[1]. European Union 2014 Report on Critical Raw Materials for EU, http://ec.europa.eu/enterprise/policies/raw-materials/files/docs/crm-report-on-critical-raw-materials_en.pdf
[2]. Jorjani E, Bagherieh A, Chelgani S, 2011 Rare Earth Elements Leaching from Chadormalu Apatite Concentrate: Laboratory Studies and Regression Predictions, *Korean J. Chem. Eng.*, 28, pp. 557–562
[3]. Tunsu C, Petranikova M, Gergoric M, Ekberg C, Retegan T 2015 Reclaiming Rare Earth Elements from End-of-life Products: A Review of The Perspectives for Urban Mining Using Hydrometallurgical Unit Operations, *Hydrometallurgy*, 156 , pp. 239–258
[4]. Wang L. Long Z, Huang X, Yu Y, Cui D, Zhang G 2010 Recovery of Rare Earths from Wet-Process Phosphoric Acid, *Hydrometallurgy*, 101, pp. 41–47
[5]. Dippel S K 2012 Mineralogical and Geochemical Characterisation of Phosphogypsum Waste Material and Its Potential for Use as Backfill at WMC Fertilizers' Mine Site, (Phosphate Hill, N-W Queensland Thesis James Cook University, Townsville, QLD)
[6]. Amélie D, Benita P, Lukas P B, Anne S B, Torsten W V 2017 Evaluation of Potential Monazite Reference Materials for Oxygen Wasotope Analyses by SIMS and Laser Assisted Fluorination, *Chemical Geology*, Volume 450, pp 199–209

[7]. Farzaneh S, Fereshteh R, Aminib A 2017 Hydrometallurgical Digestion and Leaching of Iranian Monazite Concentrate Containing Rare Earth Elements Th, Ce, La and Nd *International Journal of Mineral Processing* Volume 159 pp 7–15

[8]. Gupta C K, Krishnamurthy N 2005 Extractive Metallurgy of Rare Earths (first ed.) (CRC Press, Boca Raton, Florida)

[9]. Agarwal V, Safarzadeh M S., 2017 Alkali Pug Bake Process for The Decomposition of Monazite Concentrates, *Minerals Engineering*, Volume 109, pp 32–41, L. Berry, J. Galvin,

[10]. Sasikumar C, Rao D S , Srikant S, Mukhopadhyay N K , 2007 Dissolution studies of Mechanically Activated Manavalakurichi Ilmenite with HCl and H_2SO_4., (Department of metallurgy Institute of Technology, (Banaras Hindu University, Varanasi, India)

[11]. Stonea K, Bandaraa AMLS, Senanayakea G, Jayasekeraa S 2016 Processing of Rare Earth Phosphate Concentrates: A comparative Study of Pre-*leaching* With Perchloric, Hydrochloric, Nitric and Phosphoric Acids and Deportment of Minor/major Elements, *Hydrometallurgy*, Volume 163 pp 137–147

[12]. Yonglu Z 2015 Pilot-scale Plant Study on The Innovative Nitric Acid Pressure Leaching Technology for Laterite Ores *Hydrometallurgy* volume 155 pp 88–94

[13]. Renata D A, Carlos AM 2010 Purification of Rare Earth Elements from Monazite Sulphuric Acid Leach Liquor and The Production of High-purity Ceric Oxide, *Minerals Engineering* vol 23 pp 538 – 540

[14]. Syukri S, 1999 Kimia Dasar 2, (ITB, Bandung)

[15]. Mugdha W, Connie K N, Gisele A, 2016 Process Investigation of The Acid Leaching of Rare Earth Elements from Phosphogypsum Using HCl, HNO_3, and H_2SO_4, *Hydrometallurgy*, Volume 166, pp 195–204.

[16]. Wenjuan Z, Jiangtao L, Zhongwei Z 2015 Leaching Kinetics of Scheelite with Nitric Acid and Phosphoric Acid, *International Journal of Refractory Metals and Hard Materials* Volume 52 pp 78–84

[17]. Krzysztof M 2013 Recovery of Vanadium, Potassium and Iron from a Spent Vanadium Catalyst by Oxalic Acid Solution Leaching, Precipitation and Ion Exchange Pocesses *Hydrometallurgy* Volumes 134–135

[18]. Kenneth S. 2017 Emeritus Professor of Biophysical Chemistry and Environmental Studies, Chapter 15 – Reaction Rates and Mechanisms, *Physical Chemistry, Concepts and Theory* (University of Missouri at Kansas City) pp 695–760

[19]. Sinclaira LK, Baekb DL, Thompsona J, Testera, JWRV. 2017 Rare Earth Elements Extraction from Pretreated Bastnäsite in Supercritical Carbon Dioxide *The Journal of Supercritical Fluids* Volume 124 pp 20–29

[20]. Ferdowsi A, Yoozbashizadeh H 2017 Process Optimization and Kinetics for Leaching of Cerium, Lanthanum and Neodymium Elements from Iron Ore Waste's Apatite by Nitric acid, *Transactions of Nonferrous Metals Society of China*, Volume 27 Issue 2, pp 420–428

[21]. Zárate GR, Lapidusb GT, Moralesa RD 2010 ca Depto. de Ing. Metalúrgica y Materiales, ESIQIE.IPN, Apdo. Postal 75-874, Col Lindavista, México D.F., C.P. 07-738, Mexico, Pressure Leaching of a Lead–zinc–silver Concentrate with Nitric Acid at Moderate Temperatures Between 130 and 170 °C, *Hydrometallurgy*, Volume 104, issue 1, pp 8–13

[22]. Chen G, Ilya K, Jiwchar G 2016 The Combined Effect of Temperature and pH on Albite Dissolution Rate Under Far-from-equilibrium Conditions *Geochimica et Cosmochimica Acta*, Volume 186 pp 154–167)

[23]. Ewelina K., Piotr B, Lori N 2017 The Redox Reaction Kinetics of Sinai Ore for Chemical Looping Combustion Applications *Applied Energy* Volume 190 pp 1258–1274

[24]. Levenspiel O 1972 *Chemical Reaction Engineering* Dept of Chem Engineering Oregon State University (Wiley Eastern Ltd New Delhi Bangalore Bombay Calcutta)

[25]. Petrucci RH Harwood WS and Herring FG 2002 *General Chemistry* 8th edn, (Prentice-Hall) p.666

[26]. Chang R 2007 *Chemwastry Ninth Edition* (New York Mc Graw Hill)
[27]. Juan R., Gregorio G, Raúl C, De Jesús MA , Abel HM, Alejandro CV 2016 Kinetics and Statistical Analysis of Nickel Leaching from Spent Catalyst in Nitric Acid Solution, *International Journal of Mineral Processing* Volume 148 pp 41–47
[28]. Ronghua Z, Xuetong Z, Shumin H 2015 Dissolution Kinetics of Quartz in Water at High Temperatures Across the Critical State of Water, The Journal of Supercritical Fluids, Volume 100 pp 58–69
[29]. Han W, Wewashu W, Qincheng B, Linchuan W 2015 Experimental Study of Heat Transfer and Flow Resistance of Supercritical Pressure Water in a SCWR Sub-channel, *The Journal of Supercritical Fluids* Volume 100 pp 15–25
[30]. Atkin PW 1999 Kimia Fisika (Penerbit Erlangga Bandung).

Solvent Selection for Extraction of Neodymium Concentrates of Monazite Sand Processed Product

Moch Setyadji. MV Purwani

Centre for Accelerator Science and Technology – BATAN Yogyakarta
Jalan Babarsari. Kotak Pos 6101 ykbb.Yogyakarta 55281

E-mail : mstyadji@batan.go.id

Abstract. The extraction of neodymium concentrates of monazite sand processed product has been done. The objective of this investigation was to determine the best solvent to separate Nd from Nd concentrate. As an aqueous phase was $Nd(OH)_3$ concentrated in HNO_3 and as solvent or the organic phase was trioctylamine (TOA). tryibuthyl phosphate (TBP). trioctylphosphine oxyde (TOPO) and di-ethyl hexyl phosphoric acid (D2EHPA) in kerosene. The investigated variables were HNO_3 concentration. feed concentration. solvent concentration or solvent in kerosene. time and stirring speeds. From the investigation on the selection of solvent for the extraction of $Nd(OH)_3$ concentrate with various solvents. it was concluded that the extraction of Nd could be carried out by using TBP or TOA. Extraction of Nd using TOA at the optimum HNO_3 concentration of 2M. feed concentration of 5 gram/10 mL. TOA in kerosene concentration of 6 %. stirring time of 15 minutes. stirring speed of 200 rpm was chosen if the Y concentration in Nd concentrate is small. In these condition D_{Nd} obtained was 0.65; extraction efficiency of Nd (E_{Nd})=37.10%. the concentrations of $Nd_2(C_2O_4)_3$ = 67.14%. $Ce_2(C_2O_4)_3$ = 1.79%. $La_2(C_2O_4)_3$ = 1.37% and $Y_2(C_2O_4)_3$ = 24.70%. Extraction of Nd using TBP at the optimum HNO_3 concentration of 1M. feed concentration of 5 gram/10 m. the TBP concentration in kerosene of 15%. stirring time of 15 minutes and stirring speed of 200 rpm was chosen if the Ce concentration in Nd concentrate is small. In these condition D_{Nd} obtained was 0.20. extraction efficiency of Nd (E_{Nd})=17%. concentration of $Nd_2(C_2O_4)_3$ =70.84%. $Ce_2(C_2O_4)_3$=15.53%. $La_2(C_2O_4)_3$ =0.00% and $Y_2(C_2O_4)_3$ = 8.63%.

Keywords: Solvent selection. neodymium concentrates. tri oxtyl amine (TOA). tributyl phosphate (TBP). diethyl hexyl phosphoric acid (D2EHPA) dan trioxthyl phosphone oxyde (TOPO)

1. Introduction

Neodymium is a chemical element in the periodic table having atomic number of 60. Neodymium was discovered by Welsbach in 1885. It has an atomic mass of 144.2 g mol-1. a density of 7.0 g cm-3. a melting point of 1024 ^{0}C and a point Boiling 3047^{0}C. Neodymium is including rare earth metals element whose abundant presence after cerium. belonging to the lanthanide group and can be found in monazite sands. Neodymium is one of the rare earth elements that can be used in home appliances such as color television. incandescent lamps. and energy saving lamps. A mixture of neodymium. iron. and boron metal is used in the manufacture of permanent magnets. This magnet is part of the vehicle component. also used on the loudspeaker and for data storage on the computer [1].

Monazite sand is a mineral which has a phosphate bond form containing thorium (Th) and rare earth metals such as yttrium (Y). cerium (Ce). lanthanum (La). gadolinium (Gd). dysporsium (Dy). neodymium (Nd) And samarium (Sm). The chemical formula of monazite sand is generally written as (RRE.Th) PO_4 [1,2]. The concentrate of Nd $(OH)_3$ of the processed monazite sand was obtained through several stages of the process from leaching of monasite by using H_2SO_4. Th separation by precipitation. Ce separation by oxidation and precipitation. the separation of La by fractional precipitation to obtain Nd $(OH)_3$ concentrate. This Nd $(OH)_3$ concentrate still contains other RRE elements[3].

The separation of Nd from the Nd $(OH)_3$ concentrate of the processed monasite sand product is necessary conducted considering the usefulness and the expensive price of Nd. The separation of Nd from Nd $(OH)_3$ concentrate was conducted by using solvent extraction process. The separation by using extraction is due to the greater advantages of this process. among of which is time saving/efficient. and that the equipment used is more simple. Based on the description. from this investigation it is expected to know the optimum condition of the extraction process to separate neodymium (Nd) from Nd $(OH)_3$ concentrate[4,5].

Solvent extraction involves the distribution of a solute between the two immiscible liquid phases[6,7]. Extraction techniques are useful for rapid and "clean" separations both for organic and inorganic substances. This way could be used for macro and micro analysis. Through the extraction process. the metal ion in the water solvent is pulled out by an organic solvent (organic phase). In general. extraction is the process of withdrawing a solute from its solution in water by another solvent which is immiscible with water (water phase).

The mass transfer mechanism on extraction[8]

-Mass transfer or metalic ion diffusion (A) from aqueous phase bodies through the film boundary layer to the interface between the water phase and the organic phase

-The metalic ion (A) reacts with the carrier or solvent (carrier = B) in the organic phase on the interface of water and organic phase.

-Mass transfer or diffusion of reaction products between metal and solvent in an organic phase from the water phase-organic phases interface to the organic phase body.

In this investigation it would be investigated the comparison of the efficiency of Nd extraction on various solvents. Extraction of Nd could be done by dissolving $Nd(OH)_3$ concentrate into nitric acid. This solvent selection is aimed to investigate the properties of each solvent and determine the best solvent for Nd extraction.

The reaction mechanism in the extraction process[9]:

• *Solvation*

Solvent extraction by solvation mechanism is used to extract uranium. plutonium. or thorium from acid solution. One of the solvents used for this purpose is the tributyl phosphate organophosphate (TBP) in the PUREX process commonly used in nuclear reprocessing. Shahryar Jafari Nejad[10] has studied the use of TBP for the extraction of La^{3+}, Ce^{3+} and Sm^{3+} ions. Siu Hua Chang et al[11]. studied the efficiency. stoichiometry and structure of Cu (II) recovering from the water phase using di-2-ethylhexylphosphoric (D2EHPA) and TBP. Dessouky et al[12] separated Zn (II). Fe (II). Fe (III) and Cd (II) by the solvent extraction using TBP and CYANEX 921 in kerosene from the chloride media. Mohsen Aliakbari et al[13] separated hafnium and zirconium by using TBP modified with ferromagnetic nano particles: the effect of acidity and metal concentration. Xingbin Li et al.[14] conducted the solvent extraction of vanadium from the acidic leached carbon stone solution using D2EHPA / TBP.

• *Ion Exchange*
Another known mechanism of extraction is the ion exchange mechanism. The acid of di -2-ethyl hexyl phosphate with the solute will occur an ion-exchange reaction [15]. Man Seung Lee et al. separated Hf (IV) -Zr (IV) in H_2SO_4 solution using D2EHPA or Cyanex 272. Diptendu et al[16] studied U (VI) extraction with a mineral acid from water phase using D2EHPA. PC88A and Cyanex 272. Torkaman et al[17] studied the solvent extraction kinetics of samarium in nitrate solution with D2EHPA and Cyanex 30. Mehdi Noori et al[18] recoveried and separated nickel and vanadium in sulfate solution using a mixture of D2EHPA and Cyanex 272.

• *Ion pair*
The solute will form ion pairs with the tri-n-octrylamine compound (TOA). Qi Wang et al[19] extracted 2-chloroethanol from the water phase using TOA. Mehdi Ghadiri et al[20] studied the extraction of Mo using TOA and stripping using ammonia solution.

• *Complex formation*
Solute with tri-n-octyl phosphine oxide (TOPO) forms a complex bond. Stas et al[21] studied the effects of uranium and iron co-extraction factors using DEHPA / TOPO- kerosen. Gadgil et al[22] studied the kinetics of uranium extraction in phosphoric acid using D2EHPA-TBP and D2EHPA-TOPO.

According to Nerst's distribution law. when into two immiscible solvents was entered the soluble solutes in both solvents there would be a solubility division. Both solvents are generally organic solvents and water solvents. In practice the solute would be distributed by itself into the two solvents after being shaken and left separately. The ratio of the solute concentration in both solvents remains constant. and it would be a constant at a fixed temperature. Those constants are called distribution constants or distribution coefficients.

The extraction process of Nd is very influenced by type and solvent concentration variations. concentration variation of HNO_3. feed weight variation of Nd $(OH)_3$ concentrate . time and stirring speed variations. so that in this investigation the extraction was performed with various variations as mentioned above.

2. Materials and methods

Materials used were:
Monazite sand. H_2SO_4 technical. NaOH technical. ammonia. $H_2C_2O_4$ technical grade. HNO_3 technical grade solution. D_2EHPA solution. TOA solution. TBP solution. TOPO crystalline. kerosene solution. aquadest. pure RRE materials for standard analysis.
Instrumentations used were:
X-ray or XRF (Ortec 7010) spectrometer. magnetic stirrer and heater (Ika® Werk). Analytical scales (Sartorius 2464). various size flasks. spray bottles. volumetric pipette. propipette. small bottle size of 10 mL vials. spex film. pH meter.

3. Experimental procedure

Preparation of Nd concentrate
Monazite sand of 100 grams was digested with 100 ml of concentrated H_2SO_4 technical grade at a temperature of 210^0C for 1 hour. The digested product was diluted with 2500 ml of water then it was filtered. The residue was disposed and the pH filtrate was adjusted by using 15% ammonia. The formed solid was the Th concentrate. then filtered. The filtrate was precipitated using 15% ammonia at a certain pH. then filtered and dried. The formed solids are RRE or RRE hydroxide (RREOH). Every 25 g of RREOH was dissolved in 70 mL of hot concentrated HNO_3 (165^0C) for 2 hours. After cooling . it was added by 15% ammonia by adjusting the pH. the precipitate formed Ce concentrate was filtered and dried . The filtrate was added by 15% ammonia by adjusting the pH. the precipitate

formed was filtered and dried. The precipitate formed was Nd $(OH)_3$ concentrate and La concentrate filtrate. All the processed products were analyzed using XRF.

Variation of HNO_3 concentration
The preparation of the feed solution was carried out by dissolving each of 20 g of Nd $(OH)_3$ concentrate into 100 mL of 1. 2. 3. 4. and 5 M HNO_3 solutions. Each 10 mL of feed solution (FA = water phase) with various HNO_3 concentrations . was put into five 50 mL chemical glasses. A 10 mL of a 6% TOA solution - kerosene (FO = organic phase) was added to FA. then it was stirred with a magnetic stirring bar for 15 minutes at the speed of 200 rpm. This process is called the extraction process. The water phase (FA) and the organic phase (FO) of the extraction products were separated. Each 5 mL of water phase was taken for being analyzed with XRF. The works was repeated for TBP - kerosen = 15%. D2EHPA - kerosen = 6% and TOPO - kerosen = 6%. From the variation of HNO_3 concentration having the highest D (optimum) used for optimization of feed concentration variation.

Variation of feed concentration
Amount of 3, 4, 6 and 7 g of Nd $(OH)_3$ concentrate were dissolved into 25 mL of the optimized product of HNO_3 solution as the feed (FA). Each 10 mL of feed was added by 10 mL of a 6%TOA - kerosene solution (FO) was then stirred with a magnetic stirer bar for 15 minutes at the stirring speed of 200 rpm. The aqueous phase (FA) and the extracted product FO were separated. Each 5 mL FA was taken for being analyzed using XRF. Work is repeated for TBP - kerosen = 15%. D2EHPA - kerosene = 6% and TOPO - kerosene = 6%. From the variation of HNO_3 concentration having the highest D(optimum) used for optimization of solvent concentration variation.

Variation of solvent concentration
Preparing a feed solution of the extraction optimized product of the feed concentration variation (FA). Each of 10 mL FA was added by 10 mL of TOA - kerosene solution with varying concentrations (2. 4. 8. and 10%). it was then stirred for 15 minutes at the stirring speed of 200 rpm. After extraction. FA and FO were separated. Each of 5 mL FA was taken for being analyzed using XRF. The works were repeated for TBP - kerosene = 5, 10, 20 and 25%. D2EHPA - kerosene = 2. 4. 8. and 10% and TOPO - kerosene = 2, 4, 8 and 10%. From the variation of HNO_3 concentration having the highest D(optimum) was used for the optimization of stirring speed variation.

Variation of stirring speed
Each 10 mL of feed solution from the optimized product of feed concentration (FA) was added by 10 mL of TOA - kerosene (FO) optimum solution. it was then stirred with a magnetic stirrer with a stirring time of 15 minutes. The stirring speed was varied by 100, 150, 250 and 300 rpm. Each 5 mL of water phase was taken for being analyzed using XRF. The work was repeated for % TBP - kerosene.% D2EHPA - kerosene and% TOPO - kerosene of the optimization products. From the variation of HNO_3 concentration having the highest D used for optimization of stirring time variation.

Variation of stirring time
Each 10 mL of feed solution of the optimized product of HNO3 concentration and the feed was extracted by 10 ml of TOA-kerosene. TBP-kerosene. D2EHPA-kerosene and TOPO-kerosene of the optimized product of solvent concentration with the optimized product of stirring speed. The stirring time or extraction was varied for 5, 10, 15, 20 and 25 minutes. After the extraction. FA and FO were separated. Each 5 mL of water phase was taken for being analyzed using XRF. From the analysis results can be known the optimum stirring time.

Stripping of the optimized products of extraction
The optimized products of extraction were stripped using 15% oxalic acid (stripping phase). the volume ratio of 1: 1 and it was restripped using water in order the remaining solute in organic phase could be wholly recovered in the stripping phase.

Precipitation of stripping products
The stripping products used 15%$H_2C_2O_4$ and water was mixed. added by 15% $H_2C_2O_4$ until no precipitate was formed. The formed precipitate was filtered and dried and then was analyzed by using XRF.
Determining of the distribution coefficient value (D)
The distribution coefficient is expressed by the following formula[23]:

$$D = \frac{C_2}{C_1} \quad or \quad D = \frac{C_o}{C_a} \tag{1}$$

Where D = distribution coefficient and C_1. C_2. C_o. and C_a respectively are solute concentrations in solvents 1. 2. organic. and water. In accordance with the agreement. the solute concentration in the organic solvent is written above and the solute concentration in the water solvent is written below. From the formula if the value of D is large. the solute will quantitatively tend to be more distributed into organic solvents. as well as the opposite.
The distribution coefficient value could be determined by comparing the organic phase with the water phase after the extraction.

$$D = \frac{element\ concentration\ in\ FO}{element\ concentration\ in\ FA} \tag{2}$$

The extraction efficiency could be determined by the following calculation:

$$Extraction\ efficiency\ (\%E) = \frac{solute\ concentration\ in\ organic\ phase}{solute\ concentration\ in\ the\ feed} x100\% \tag{3}$$

4. Research results and discussions

Concentrate of Nd $(OH)_3$
The result of analysis using XRF of monazite processing into the concentrate of $Nd(OH)_3$ is presented in Table 1. The concentrate of Nd $(OH)_3$ contains Ce. Y. La. water etc. (Sm. Pr)

Table 1. Element concentration and compound of $Nd(OH)_3$ concentrate of the processed product of monasite

		Element concentration, %		
Ce	Y	La	Nd	Water
2.87	14.94	3.61	39.42	2.87
		Compound concentration, %		
$Ce(OH)_3$	$Y(OH)_3$	$La(OH)_3$	$Nd(OH)_3$	Water
3.91	22.63	4.93	53.37	15.15

Variation of HNO_3 Concentrations

The graph of the correlation between the distribution coefficient (D) with the concentration of HNO_3. as well as the graph of the correlation of the extraction efficiency (% E) to the HNO_3 concentration. are shown in Figure 1 and Figure 2. In the extraction process. the acidity of the water phase is a very important factor. The addition of HNO_3 causes the increase of H^+ ions and NO_3^- ions complying with the following reactions[11]:

$$H^+ + NO_3^- \rightleftharpoons HNO_{3\,(a)} \quad (4)$$

$$HNO_{3\,(a)} \rightleftharpoons HNO_{3(o)} \quad (5)$$

$$Nd(OH)_3 + 3HNO_3 \longrightarrow Nd(NO_3)_3 + 3H_2O \quad (6)$$

$$Nd(NO_3)_3 \longrightarrow Nd^{+3} + 3NO_3^{-3} \quad (7)$$

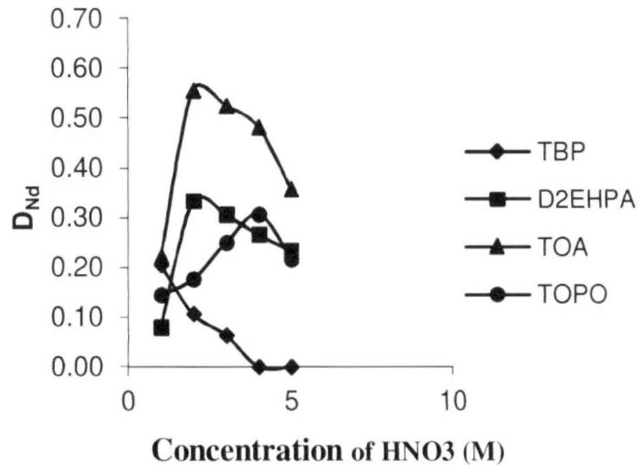

Figure 1. Graph of correlation of HNO_3 concentration with D_{Nd} using the solvents of TBP. D2EHPA. TOA. and TOPO

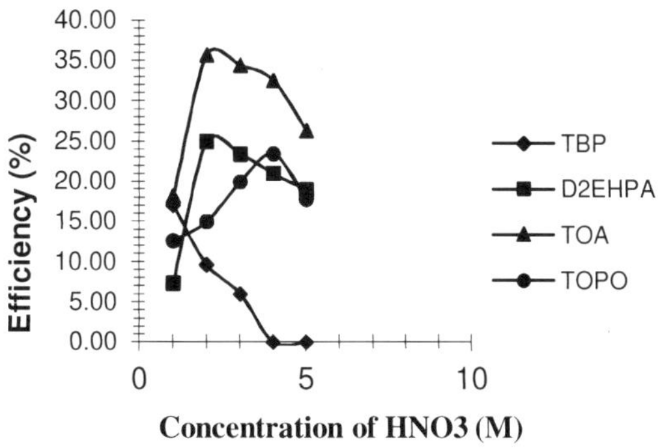

Figure 2. Graph of correlation of HNO_3 concentration with the extraction efficiency of Nd using the solvents of TBP. D2EHPA. TOA. and TOPO

Reaction with TBP[12-14]

$$HNO_{3(o)} + TBP_{(o)} \rightleftharpoons HNO_3.TBP_{(o)} \quad (8)$$

$$Nd^{3+} + 3(HNO_3.TBP) \rightleftharpoons Nd(NO_3)_3.3TBP_{(o)} + 3H^+ \quad (9)$$

Nitric acid was used to form (HNO_3.TBP) complex which functions as a complex intermediate-forming agent with metals as seen in the reactions (8) and (9). Excessive concentrations of HNO_3 could cause TBP degradation to Mono Butyl Phosphate (MBP) and Di Butyl Phosphate (DBP). thereby decreasing Kd. Mono Butyl Phosphate (MBP) and Di Butyl Phosphate (DBP) dissolve in the organic phase. Besides. if the DBP forms a complex compound with metal will become a strong bond so it will be difficult to be re-extracted. The highest value D= 0.2 at HNO_3concentration of 1 M.

The reaction with D_2EHPA could be seen in the following equation (14-16)

$$(H_2R_2) + 2HNO_3 \rightleftharpoons 2(HR.HNO_3) \quad (10)$$

$$Nd^{3+} + 2(HR.HNO_3) \rightleftharpoons Nd(NO_3)_3.2(HR) + 2H^+ \quad (11)$$

The existence of excessive HNO_3 would react with D2EHPA and cause a decrease in the amount of D2EHPA which reacts with the metal. Thereby reducing the reactivity of the element to the organic phase and causing a decrease in the value of D. In the use of HNO_3 concentrations above 2 M. the value of decreases more and the highest D= 0.35.

The reaction with TOA (R_3N) is as follows[18,19]:

$$HNO_{3(o)} + R_3N_{(o)} \rightleftharpoons HNO_3.R_3N_{(o)} \quad (12)$$

$$Nd^{3+} + 3(HNO_3.R_3N) \rightleftharpoons Nd(NO_3)_3.3R_3N_{(o)} + 3H^+ \quad (13)$$

The addition of HNO_3 as an intermediate for the formation of Nd $(NO_3)_3.3R_3N$ ions pair. For the addition of HNO_3 H^+ ions would subsequently increase in the system causing the equilibrium of the reaction shifts to the left. so that the formation of Nd $(NO_3)_3.3R_3N$ ions pair would decrease and the distribution coefficient would decrease as shown in Figure 1.
The reaction with TOPO as the following[20,21]

$$HNO_{3(o)} + R_3PO_{(o)} \rightleftarrows HNO_3.R_3PO_{(o)} \qquad (14)$$

$$Nd^{3+} + 3(HNO_3.R_3PO) \rightleftarrows Nd(NO_3)_3.3R_3PO_{(o)} + 3H^+ \qquad (15)$$

In Figure 3 and Figure 4 it could be seen that the distribution coefficient and Nd efficiency are greatest when using TOA solvents. compared to the use of TBP. TOPO and D2EHPA. At concentration of $HNO_3$2M. Nd has D= 0.56 and extraction efficiency = 35.70%.

Variation of Feed Concentration
The result of the extraction process of the weight variation of Nd $(OH)_3$ concentrate in the feed. hereinafter referred to as the feed concentration variation. the calculation result of D and% E are shown in Figure 3 and Figure 4

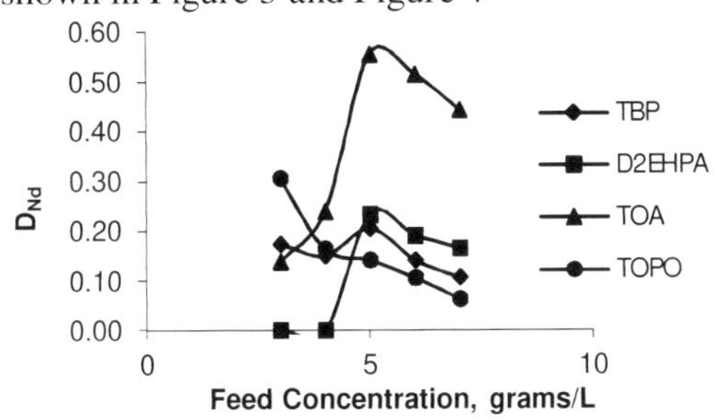

Figure 3. Graph of correlation of feed concentration with D_{Nd} using the solvents of TBP. D2EHPA. TOA. and TOPO.

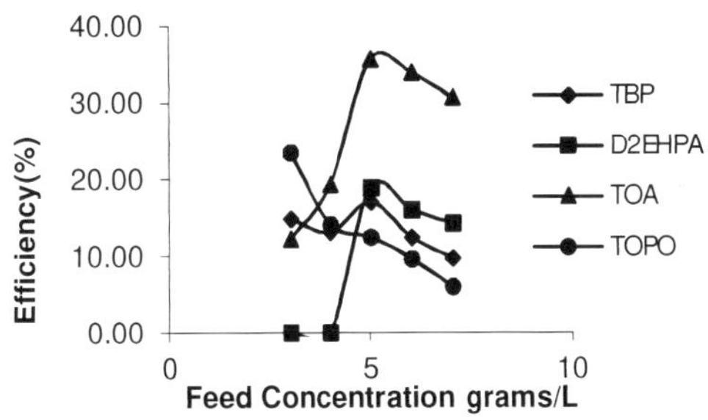

Figure 4. Graph of correlation of feed concentration with the extraction efficiency of Nd using the solvents of TBP. D2EHPA. TOA. and TOPO.

One of the most influential factors for the mass transfer rate from the water phase (FA) to the organic phase (FO) is the amount of solute concentration in the feed. The greater the solute concentration in the feed. the greater the mass transfer rate will be. In Figure 6 it can be seen that the extraction conditions will tend to increase as the feed increases. At a 6 g / 10 mL feed concentration. it is likely to give the highest distribution coefficient. but under these conditions Nd was not well separated from its impurities. hence based on the greatest its separation factor the concentration of the feed taken was 5 g / 10 mL. In Figures 3 and 4 it could be seen that the distribution coefficient and Nd efficiency are greatest when using TOA solvents. compared to the use of TBP and D2EHPA. At 5 g / 10 mL feed concentration. D_{Nd}= 0.56 and extraction efficiency (E_{Nd})= 35.70%.

Solvent Concentration Variation
The use of solvents greatly affects the separation of the elements in the feed in the form of Nd $(OH)_3$ concentrate. The greater the use of the solvent concentration so the better the extraction of elements in the feed. which is marked by the increase of the value of D. But after reaching a certain concentration. the value of D will more decrease because at the greater solvent concentration the solute transfer greater the transfer of the solute from the water phase to the organic phase will be more difficult. This can be explained by the Stokes-Einstein equation as follows[22]:

$$D_{AB} = \frac{k\ T}{6\pi\, r\mu_B} \qquad (16)$$

D_{AB} = diffusivity of A in dilute solution in B. k = Boltzmann constant. T = temperature. r = radius of solute particle. μ_B= = solvent viscosity. From the equation (16) it could be seen that the diffusivity was inversely proportional to the solvent viscosity. so the greater the solvent viscosity the more difficult it is to diffuse from the water phase to the organic phase. thereby lowering the value of D.

In Figure 5 it can be seen that Nd has the highest distribution coefficient at the solvent concentration of 20%. After passing the concentration of TBP 20% D value tends to decrease because the solvent is getting more viscous so it is difficult to diffuse into the organic phase and forms a complex with Ce, TBP, D2EHPA, TOA and TOPO solvents which gave a sufficiently good Nd extraction conditions can be seen in Figure 5.

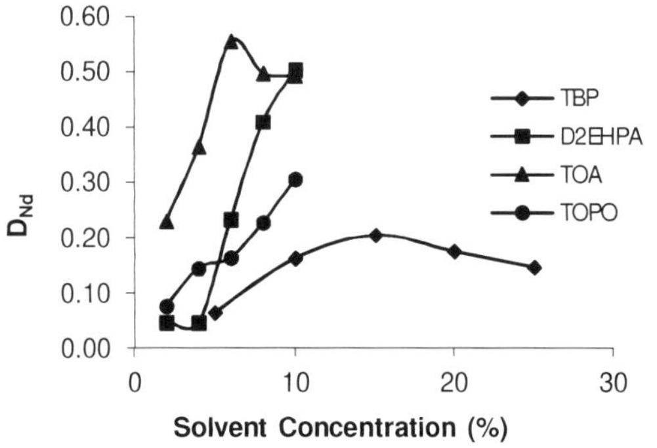

Fifure 5. Correlation of solvent concentration with D_{Nd} using the solvents of TBP. D2EHPA. TOA. and TOPO.

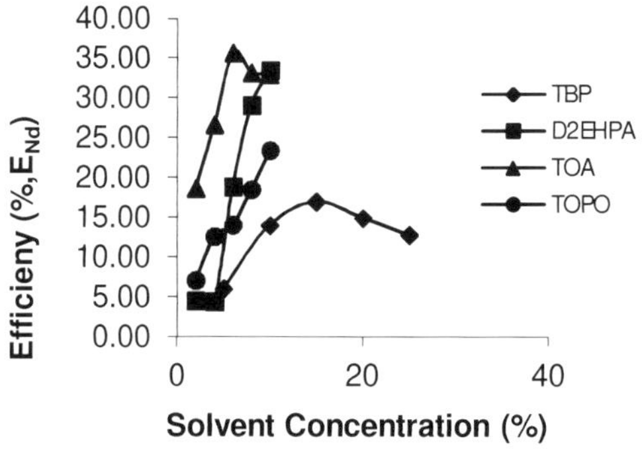

Figure 6. Correlation of solvent concentration with extraction efficiency of Nd using the solvents of TBP. D2EHPA. TOA. and TOPO.

Variation of Reaction Time

The stirring time is also related to the amount of stirred volume, so to vary the stirring time to the volume of 20 mL is reasonably sufficient to give an effective contact time. especially in the transfer of solutes from the water phase to the organic phase. Solvent TBP, D2EHPA and TOA gave the conditions of Nd extraction was good enough can be seen in Figure 7.

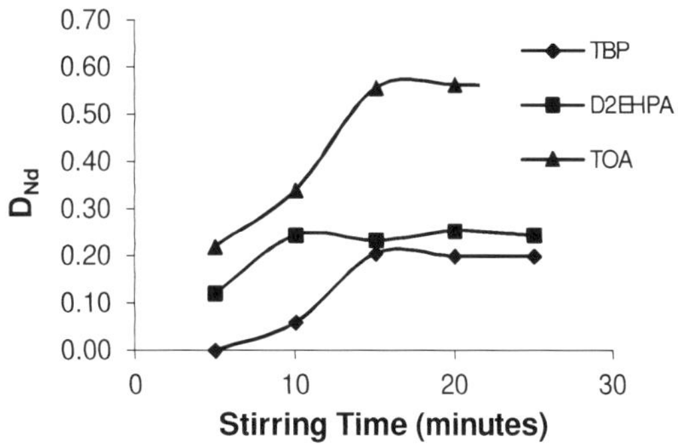

Figure 7. Correlation of stirring time with D_{Nd} with solvents of TBP. D2EHPA. and TOA.

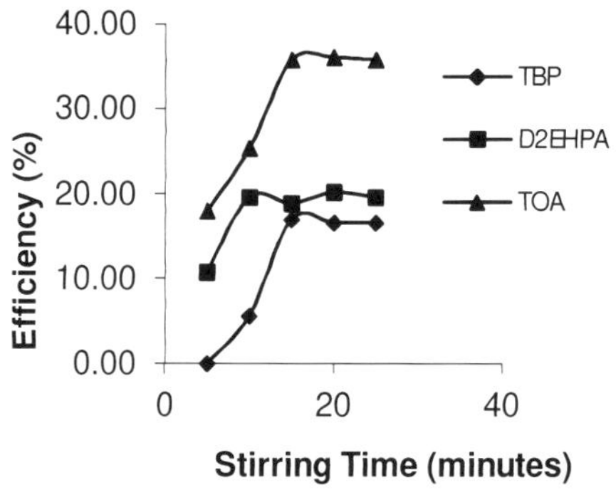

Figure 8. Correlation of extraction time with the extraction efficiency of Nd with solvents of TBP. D2EHPA. and TOA.

It was required a sufficient stirring time for the reaction to occur and the formation of the reaction products. so that the reaction and reaction products obtained was maximum. The reaction rate. the reduced reactants. or the increased of reaction products can be formulated mathematically as the following:

$$[A] = [A] o. E\text{-}kt \quad (17)$$

[A] = concentration of A after reacting. [A]o = initial concentration of A (before reacting).
k = reaction rate constant and t = reaction time.

When the reaction occured was first order it would get the equation $[A] = [A]_o . e^{-kt}$. This equation shows that in the first order reaction the concentration of the reactants will decrease exponentially with time while the product will increase.

In Figure 7 and Figure 8 it can be seen that the distribution coefficient and extraction efficiency of Nd were the greatest when TOA solvent was used. compared with that of TBP and D2EHPA. At the stirring time of 15 minutes. $D_{Nd} = 0.56$ and the extraction efficiency (E_{Nd}) = 35.70%.

Variation of Stirring Speed

The extraction process was also a phenomenon of mass transfer of two insoluble liquids, so that if there were not external forces as an assistance in the form of stirring. then the mass transfer both of the two liquids would be very slow. This stirring process would help the intermixture of water phase and organic phase where the stirring process would disperse the solute into the organic phase solution resulting in inter-phase contact. This phenomenon would increase the mass transfer of the solute from the feed into the organic phase solution. Profile of reactant concentration in the extraction with the chemical reaction is shown in Figure 8.

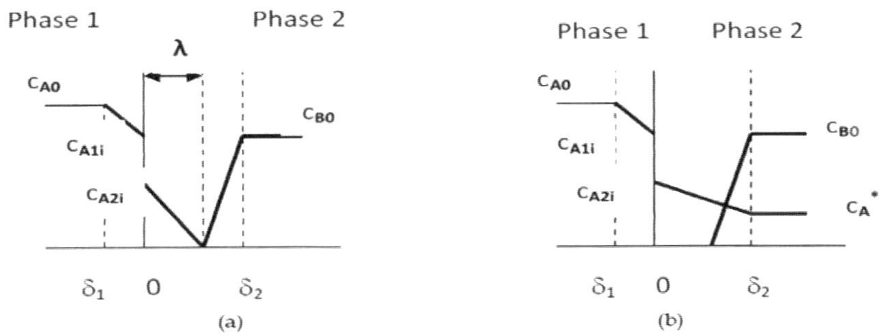

Figure 9. Profile of reactant concentration in the extraction with the chemical reaction
a) irreversible spontaneous reaction occured in the film layer of phase 2
b) slow reaction occured in the film layer of phase 2
Notes: Phase 1 : water phase. Phase 2 : organic phase . λ :film layer thickness

The occurrence of mass transfer from FA to FO is caused. due to chemical reactions and diffusion. Between FA and FO there was a layer of interface with certain imaginary thickness which is the resistance to the mass transfer rate from FA to FO or vice versa. The thickness of this thin layer of interfaces depends on the speed of stirring. The faster the stirring. the thinner is the layer thickness for the occuring of mass transfer. This layer thickness could be reduced by increasing the intensity of stirring. The D value will increase with the increase of stirring speed. because the intensity of the occuring inter-reactants collision is getting more extensive and faster.

With faster stirring. the layer thickness that inhibits the occuring of mass transfer due to smaller diffusion. but if the stirring intensity was enhanced. the layer thickness that inhibits the occuring of diffusion was already very thin and almost has not any more obstruction because it does not affect the mass transfer anymore. The optimum condition chosen was based on the largest separation factor value at 200 rpm. In Figure 10 shows the tendency of D value increases as the increase of stirring speed. The greater the stirring speed. the more extraction results will be obtained.

After experiencing a significantly increase. D value was slightly decline. this indicates that there was no increase of extracted solute to FO. Parameter of stirring speed is important carried out in order to know the optimum speed. In terms of diffusion. the diffusion rate = $D_A \partial^2 C_A / \partial z^2$. where z was the distance or the width traveled by elements or compounds that would diffuse from the water phase to the organic phase or vice versa. The wider z the diffusion will be slower. to shorten z the stirring was accelerated.

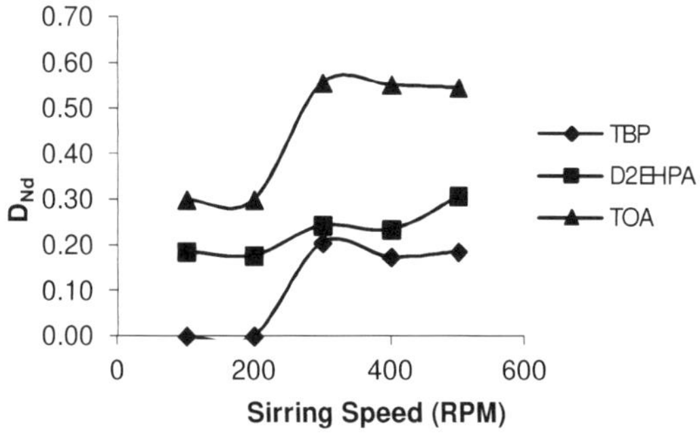

Figure 10. Correlation of stirring speed with D_{Nd} using the solvents of TBP. D2EHPA. and TOA.

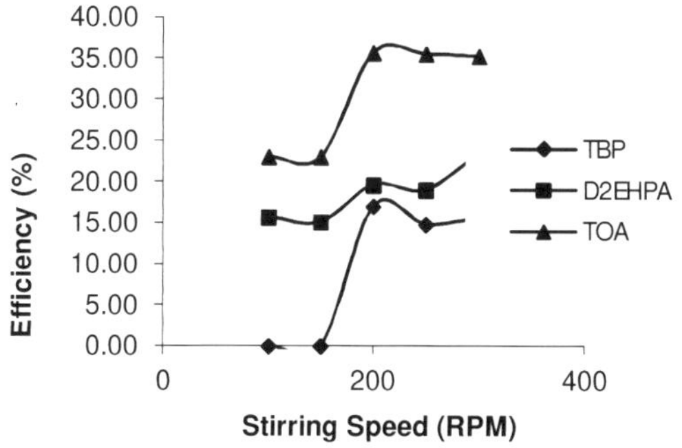

Figure 11. Correlation of stirring speed with the extraction efficiency of Nd using the solvents of TBP. D2EHPA. and TOA.

From the point of view of chemical reaction. k . [A] = - d $\frac{[A]}{dt}$. with k = the reaction rate constant. which is the value according to Arhenius. k = $Ae^{-E/RT}$, A = collision frequency, E = activation energy. T = temperature and R = ideal gas constants. The faster the stirring of the reaction will be more perfect because the area of impact would also be greater. In Figure 10 and Figure 11 it could be seen that the distribution coefficient and Nd efficiency were greatest when TOA solvents was used. compared with the use of TBP and D2EHPA. At a stirring speed of 200 rpm. D_{Nd} = 0.65 and extraction efficiency = 37.10%.

This result was competitive when compared with research results conducted by Kraikaew. J. et al[23]. In this investigation Nd extraction in rare earth using extractant 50% D2EHPA 1.5 M in kerosene. D_{Nd} obtained = 0.162 and extraction efficiency= 13.84%. For extractant 50% TBP 1.5 M in kerosene D_{Nd} obtained = 0.473 and extraction efficiency = 31.97%. While for TBP 1.8 M in kerosene D_{Nd} obtained = 0.681 and extraction efficiency = 40.47%. Meanwhile. in the same investigation by using an extractant mixture of TBP 1.2 M and D_2EHPA 0.3 M. D_{Nd} obtained = 0.331 and extraction efficiency=24.87%. While the researchers Panigrahia. M. et al[24] conducted an extraction of Nd ion using toluene extractant at various process conditions obtained the highest extraction efficiency of 33.56%. Another researcher. Gergoric. M.[25] carried out Nd extraction by using a 30% D2EHPA extractant in hexane with the highest efficiency obtained was 50%. The result of FO stripping using $H_2C_2O_4$15% was $RRE_2(C_2O_4)_3$ solid with the composition shown in Table 2.

Table 2. Products of FO stripping of extraction optimization results of various solvent and precipitation of stripping products

Solvent	Condi-tion					Extraction efficiency of elements. %			
	M	g/10mL	%	min	rpm				
						Ce	Y	La	Nd
TBP	1	5	15	15	200	50.5	4.49	0	17
D2EHPA	5	5	6	25	300	44.9	18	0	23.5
TOPO	4	3	10	15	200	28.1	18.1	1.3	23.45
TOA	2	5	6	15	200	13	28.7	7.89	36
						Element concentration, %			
						Ce	Y	La	Nd
TBP						7.99	3.70	0.00	36.96
D2EHPA						4.61	9.63	0.00	33.17
TOPO						2.98	10.00	0.17	34.19
TOA						0.92	10.58	0.70	35.03
						Compound concentration, %			
						$Ce_2(C_2O_4)_3$	$Y_2(C_2O_4)_3$	$La_2(C_2O_4)_3$	$Nd_2(C_2O_4)_3$
TBP						15.53	8.63	0.00	70.84
D2EHPA						8.96	22.47	0.00	63.57
TOPO						5.79	23.34	0.34	65.53
TOA						1.79	24.70	1.37	67.14

M: concentration of HNO_3, g / 10mL: Feed Concentration,%: solvent concentration, minute: Stirring time, rpm: Stirring speed.

Extraction efficiency of Nd with 15% TBP-kerosene of 17%, but the concentration of Nd_2 $(C_2O_4)_3$ obtained = 70.84% although the extraction efficiency of Ce was also great (50.5%) but because Ce concentration in Nd concentrate was insignificant, so the effect of Ce oxalate concentration in the products was insignificant (7.99%). Extraction efficiency of with 6% of TOA-kerosene was 36% , but the concentration of Nd_2 $(C_2O_4)_3$ obtained simply = 67.14%, , due to extraction efficiency of Y = 28.7% its effect was great on Y oxalate concentration in the product namely 24.70 %, since the number in Y in the Nd concentrate was large (the second largest after Nd). The largest impurities in extraction with TBP was Ce, while the largest impurities in extraction with TOA was Y. Thus if in the Nd concentrate containing Ce and Y it must be cleaned from Ce and Y.

5. Conclusion

The results of optimization of neodymium concentrate extraction process producing from processing of monazite sand using various solvents, it could be concluded that Nd extraction could be done by using TBP or TOA solvent. The optimum condition of Nd extraction using TOA solvent in HNO_3 concentration of 2 M, feed concentration of 5 g / 10 mL, TOA concentration in 6% kerosene, stirring time of 15 minutes, stirring speed of 200 rpm selected when Nd concentration has a small Y content. In this condition the distribution coefficient of Nd (D_{Nd})was 0.65, the extraction efficiency of Nd (E_{Nd})was 37.10%, the content of Nd_2 $(C_2O_4)_3$ was 67.14%, Ce_2 $(C_2O_4)_3$ was 1.79%, La_2 $(C_2O_4)_3$ was

1.37% and $Y_2(C_2O_4)_3$ was 24.70%. While Nd extraction using TBP at the optimum condition at HNO_3 concentration of 1 M, feed concentration of 5 g / 10 mL, TBP concentration in kerosene of 15%, stirring time of 15 minutes, stirring rate of 200 rpm selected if the concentrate of Nd has a insignificant Ce concentration. In this condition D_{Nd} obtained = 0.20, extraction efficiency of Nd (E_{Nd})= 17%, concentration of $Nd_2(C_2O_4)_3$ = 70.84%, $Ce_2(C_2O_4)_3$ = 15.53%, $La_2(C_2O_4)_3$ = 0,00 % And $Y_2(C_2O_4)_3$ = 8.63%.

Acknowledgements
On this occasion the authors would like to thanks Mrs. Suprihati for assisting this research and PSTA-BATAN and Kemenristekdikti with SINAS Insentive Programe for funding support.

References

[1]. Gupta C K and Krishnamurthy N 2005 Extractive Metallurgy of Rare Earths CRC Press ISBN 0-415-33340-7
[2]. Bambang Soetopo Lilik Subiantoro Priyo Sularto Dwi Haryanto 2012 Studi deposit monasit dan zirkon dalam batuan kuarter di daerah cerucuk belitung Pusat Pengembangan Geologi Nuklir – BATAN *Eksplorium* ISSN 0854 – 1418 Volume 33 No. 1, pp. 25- 40
[3]. Jacques Lucas 2015 Rare Earths Purification, Separation, Precipitation and Calcination, Science, Technology, Production and Use p 69–91
[4]. Syouhei N Hirai T and Komasana I 2003 Advanced Liquid-Liquid Extraction System for the Separation of Rare Earh ions by Combination of Conversion pof Metal Species With Chemical reaction, Journal of Solid State Chemistry p 171
[5]. Sánchez J.M Hidalgo M Valiente M and Salvadó V 1999 Neodymium This rare earth is extracted by di(2-ethyl-hexyl) phosphoric acid into hexane by an ion exchange mechanism.' *Solvent Extraction and Ion Exchange* p 455–474.
[6]. Vladimir S Kislik *2012* Recent Advances in Solvent Extraction Processes and Techniques Chapter 14 , *Solvent Extraction p 483-524,.*
[7]. Stevens GW TC Lo and M H I Baird 2014 Extraction, Liquid-Liquid", in Kirk-Othmer Encyclopedia of Chemical Technology
[8]. Koncsag C and Barbulescu A 2013 Liquid-liquid Extraction with and without Chemical Reaction in Elamin Mohan Mass transfer in multi System and its applications pp 207-232 ISBN 971913
[9]. Mohamed El-Amin 2011 Mass Transfer in Multiphase Systems and its Applications Mechanical Engineering »book edited by, ISBN 978-953-307-215-9
[10]. Shahryar Jafari Nejad, Hossein Abolghasemi, Mohammad A. Moosavian, Mohammad G. Maragheh 2011 Fractional factorial design for the optimization of supercritical carbon dioxide extraction of La^{3+} Ce^{3+} and Sm^{3+} ions from a solid matrix using bis (2,4,4-trimethylpentyl) dithiophosphinic acid + tributylphosphate *Chemical Engineering Research and Design, Vol. 89, Issue 6 p 827-835*
[11]. Siu Hua Chang Tjoon Tow Teng Norli Ismail *2011* Efficiency stoichiometry and structural studies of Cu(II) removal from aqueous solutions using di-2-ethylhexylphosphoric acid and tributylphosphate diluted in soybean oil *Chemical Engineering Journal, Volume 166 p 249-255*
[12]. Mohsen Aliakbari, Kamal Saberyan Mohammad Noaparast Hadi Abdollahi Ata Akcil *2014* Separation of Hafnium And Zirconium Using TBP Modified Ferromagnetic Nanoparticles: Effects of Acid And Metals Concentrations, *Hydrometallurgy, Volume 146 p 72-75*
[13]. Xingbin Li, Zhigan Deng, Chang Wei, Cunxiong Li, Minting Li, Gang Fan, Hui Huang *2014* Solvent extraction of vanadium from a stone coal acidic leach solution using D2EHPA/TBP: Continuous testing, *Hydrometallurgy In Press*

[14]. Man Seung Lee Raju Banda Soo Hwan Min *2015* Separation of Hf(IV)–Zr(IV) in H_2SO_4 solutions using solvent extraction with D2EHPA or Cyanex 272 at different reagent and metal ion concentrations *Hydrometallurgy p 84-90*

[15]. Diptendu Das VA Juvekar VH Rupawate K Ramprasad R Bhattacharya *2015* Effect of the nature of organophosphorous acid moiety on co-extraction of U(VI) and mineral acid from aqueous solutions using D2EHPA PC88A and Cyanex 272 *Hydrometallurgy, Volume 152 p 129-138*

[16]. R Torkaman J Safdari M Torab-Mostaedi MA Moosavian *2014* A kinetic study on solvent extraction of samarium from nitrate solution with D2EHPA and Cyanex 301 by the single drop technique *Hydrometallurgy Volume 150 p 123-129*

[17]. Mehdi Noori Fereshteh Rashchi Ataollah Babakhani, Ehsan Vahidi *2014* Selective recovery and separation of nickel and vanadium in sulfate media using mixtures of D2EHPA and Cyanex 272, *Separation and Purification Technology, Volume 136 p 265-273*

[18]. Qi Wang, Jiquan Liu, Lin Jiang *2012* Extraction of 2-chloroethanol from aqueous solutions by trioctylamine *Journal of Industrial and Engineering Chemistry Volume 18 p 1267-1271*

[19]. Mehdi Ghadiri Seyed Nezameddin Ashrafizadeh Mohammad Taghizadeh *2014* Study of molybdenum extraction by trioctylamine and tributylphosphate and stripping by ammonium solutions, *Hydrometallurgy, Volumes 144–145* p *151-155*

[20]. J Stas A Dahdouh H Shlewit S Khorfan *2002* Statistical study of factors affecting the co-extraction of uranium and iron in the second cycle of extraction with DEHPA/TOPO in kerosene *Hydrometallurgy Volume 65 p 23-30*

[21]. OD Gadgil V H Dalvi K T Shenoy H Rao SK Ghosh J B Joshi *2002* Kinetics of extraction of uranium from phosphoric acid by D2EHPA–TBP and D2EHPA–TOPO systems using constant interfacial area stirred cell *Chemical Engineering Science Volume 110* pages *169-184*

[22]. Welty RJ Wicks EC Wilson ER Rorrer G 2004 *Dasar-Dasar Fenomena Transport* Volum 3 Edisi Ke-4 Jakarta

[23]. Kraikaew J Srinuttrakul W and Chayavadhanakur C 2005 Solvent Extraction Study of Rare Earths from Nitrate Medium by the Mixtures of TBP and D2EHPA in Kerosene Journal of Metals Materials and Minerals Vol 15 No 2 pp.89-95

[24]. Panigrahia M Grabdaa M Kozaka D Doraia A Shibataa E Kawamuraa J 2017 Liquid-liquid extraction of neodymium ions from aqueous solutions of NdCl3 by phosphonium-based ionic liquids Imram Turkey Institute of Environmental Engineering of the Polish Academy of Science Zabrze Poland

[25]. Gergoric M Ekberg C Steenari B.M Retegan T 2017Separation of Heavy Rare-Earth Elements from Light Rare-Earth Elements Via Solvent Extraction from a Neodymium Magnet Leachate and the Effects of Diluents Nuclear Chemistry and Industrial Materials Recycling Chemistry and Chemical Engineering Chalmers University of Technology Gothenburg Sweden Journal of Sustainable Metallurgy

The Effects of Shielded Metal Arc Welding (Smaw) Welding On The Mechanical Characteristics With Heating Treatment inn S45c Steel

Munawar[1], Hammada Abbas[2], Ahmad Yusran Aminy[2]
[1] Mechanical Engineering Department Engineering Faculty University of Perjuangan Republik Indonesia
[2] Mechanical Engineering Department Engineering Faculty Hasanuddin University

email : munawarmustafa81@gmail.com

Abstract. Steel material has been used mainly for making tooling, automotive components, other household needs, power generators to frame buildings and bridges. This study aimed (1) to analyze the mechanical Characteristics of S45C steel with and without heating treatments, and (2) to analyze the temperature of heating treatment which could result in the maximal strength of S45C steel after the welding process. The research was conducted in the laboratory of mechanical engineering study program, Departement of mechanical Engineering, Christian university of indonesia paulus, makassar. The method used materials, instruments, and the dimensions determination of specimen based on the proposed testing standard, Next, was to determine the mechanical caracteristics of the S45C steel wich had been welded and heated.The tensile specimens, the hardness specimen, the impact specimen, and microstructures of which,each of the 3 specimens was the specimens was the specimen without treatment, the spesimen with the welding wthout heating and the specimen of 150°C, 250° C, 300° C. The research results indicated that the treatment process of 150^{0}C, 250°C and 300°C produced the changes of mechanic charateristics with the tensile strength of 42 kgf/mm2 when the temperature had reached 300°C, but at the temperature 300°C, the its toughness would decrease to Hi = 0.836 j/m2 and its hardness would increase to 40.83 at the temperature of 300°C. The value of the maximum strengs was reached at the heating temperature of 300°C for the tensile strength and the hardness, while at the temperature of 300°C its impact value would decrease.

Keywords: Steel S45C, SMAW, Heating treatment

1. Background

Steel material has been widely used for the manufacture of tools, automotive components, other household needs, generator power plants to the framework of buildings and bridges. In the application, the steel must have a strong structure, because all its structures will be affected by the frictional effect resulting in deformation or deformation. The effort to keep the metal more friction resistant is to do the heat, it plays an important role in efforts to increase steel hardness as needed. In addition, the selection of material types also needs to be considered in accordance with its function [1].

Hardening is a working heat process that is applied to produce hard workpieces. This treatment consists of heating the steel to its hardening temperature (austenitization temperature) and holding it at that temperature (holding time) for a certain period of time and then cooled to a very high quench rate to obtain the desired [2].

The process is hot up to temperature and temperatures. The heating is carried out at the required temperature, usually between 200°-600^oC depending on the need. The higher the heating temperature, the greater the decrease in the hardness of the elasticity increases. Hardened steel due to its heat and less suitable to be used with heat treatment, hardness and brittleness can be lowered to meet the use of the condition, which is being heated or hardened at temperatures below the critical temperature by cooling. This process is similar to an annealing process because here its properties can be carefully controlled, since there is no stability [3].

The electrode consists of two types of coated (flux) and non-coated parts which are the base for clamping the welding pliers. The function of the flux or electrode layer in the weld is to protect the liquid metal from the air environment to produce protective gas, stabilize the arc, the source of the alloying element. In welded steel wire, electrode wire is divided into electrodes for soft steel, high carbon steel, alloy steel, cast iron, and non-ferrous metals. The electrode material must have the same metallic properties [9].

Arifin [4] describes welding using electric arc welding requires a welding wire (electrode) consisting of coated metal lining of a chemical mixture. The function of the electrode as a generator and as an added material. Electric arc welding is the process of grafting metal by utilizing electric power as a heat source. Las arc welding is one type of electric weld where the heat source or the material pipe that is connected or welded comes from the electric arc.

Sonawan [7], explaining in the field of welding required an ideal relationship of open V connections and closed camp V connections. Connection V open to connect plate with thickness 6-15 mm with the angle between 600-800, root distance 2 mm, root height 1-2 mm.

Suharno [8], explaining about SMAW welding is a heat shielded arc welding generated from an electric arc between the end of the electrode and the metal to be welded. And another researcher [5] proposed a process of thermal cooling of mechanical properties and microstructure that is increased hardness, tensile strength and microstructure differences compared to non-heat treatments specimens and tempered specimens, the influence of oil quenching the formation of martensite and bainite structures so that the specimen becomes hard.

The S45C steel is of medium carbon steel, medium carbon steels containing carbon (C) between 0.3% to 0.6% and their carbon content allow the steel to be hardened by a suitable heating treatment [5].

Based on the above background, the important problem to be studied is how the difference of mechanical properties of Steel S45C with heating treatment and without heating effect and on mechanical condition of Steel S45C after welding. The purpose of this study was to analyze the mechanical properties of S45C Steel with heating and non-heating treatment and to analyze the temperature that can provide maximum power in S45C Steel after welding.

2. Materials and Methods

This research was conducted for six months at the Mechanical Technology Laboratory of Mechanical Engineering Program of the Christian University of Indonesia Paul Makassar. The equipment used in the research is Tensile Test Machine, Hardness Testing Equipment, Impact Test Equipment, Microscope, Polish Machine, Digital Camera. The materials used in this research are Steel S45C, Sulfuric Acid (H2SO4), Sandpaper.

The method used in this study was conducted using materials, tools, and the determination of the specimen dimensions based on the recommended testing standard then is to determine the mechanical properties of S45C steel that have been welded and subjected to heating treatment. Preparation of specimens consisting of tensile specimens, hardness specimens, impact test specimens and microstructures in which each of the three specimens was based on unheated specimens, specimens with non-heating welding and specimens with welding with heating 150°, 250°, and 300° celcius.

Statistical analysis was performed using ANOVA variance analysis on tensile and hardness test data. Varian Analysis (Analysis of variance, Annova) is a technique used to compare two or more population parameters, this technique is often used for research, especially in experimental research design [10].

3. Results

Calculation of Pull Test

Based on the tensile test on one of the normal specimens or without the welding process and the heating process, it is known that the material strength is based on the following data: Dimension of the specimen (Based on ASTM E8 test standard) is Length (Lo): 50 mm, 144 mm2. Data on tensile test results (Based on the maximum conditions) are yielding (Py) Expenses: 4952 kg, Added length in Pmax (ΔLmax): 20.8 mm. From the available equations, we can determine some components as information based on tensile test results as follows: Maximum Stress (σmax): 56.16 kgf / mm2, Strain (εmax): 0.4%, and Material elasticity (Emax): 136.97 kgf / mm2. Based on the results of calculations that have been done knowin that the material on the heating process the higher the heating process (0°C-300°C) the average tensile load until the maximum conditions increased. While the result between heating temperature to a tensile strength of material known as in Figure 1..

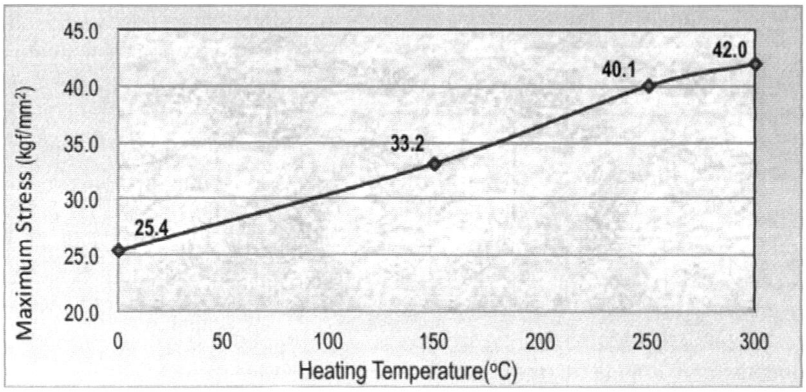

Figure 1. Graph of the relationship between heating temperature to strength Drag the material.

Calculation of impact test

From the calculation result, it is known that every increase of heat treatment temperature resulted in decreasing effort to break the specimen, as shown in Figure 2 graph. With the constant cross section for each test specimen, the decreasing of material price decreases as shown in Figure 3 The decline in the value of the effort to break the specimen and its impacted price is due to an increase in the brittle or brittle properties of the material. The condition can also be known by the easier the pendulum as a pounder load to break the specimen as shown in the graph of figure 4.

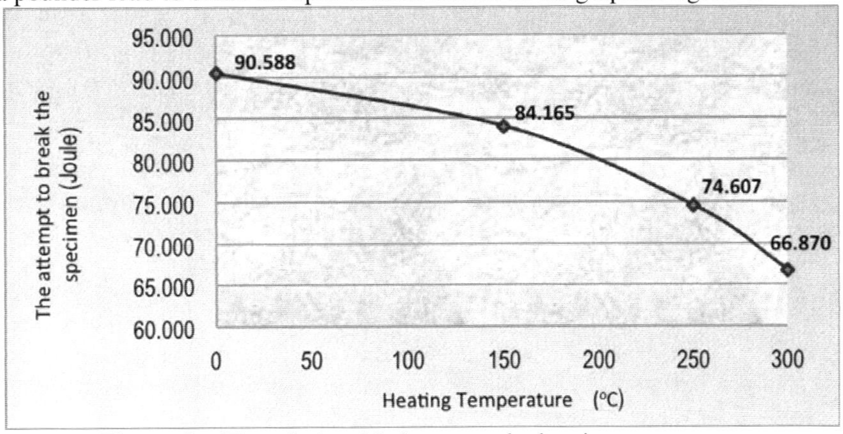

Figure 2. The relationship between the heating temperature of the business To break the specimen.

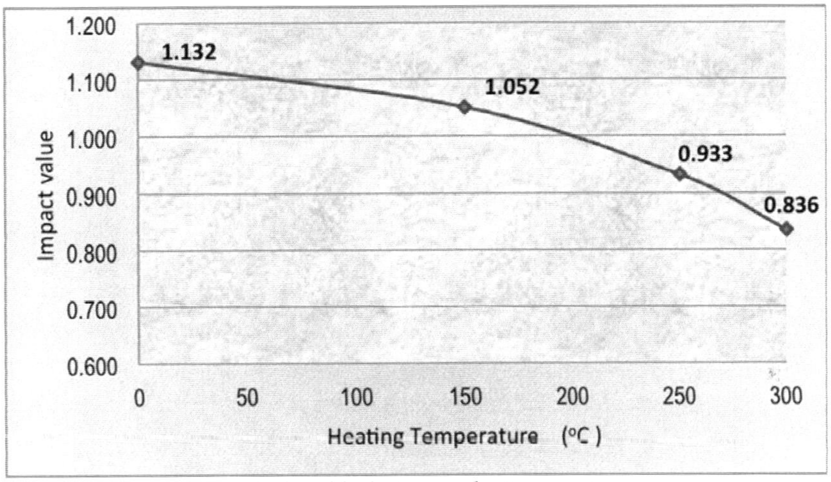

Figure 3. The relationship between the temperature of the heat treatment to Material impact value.

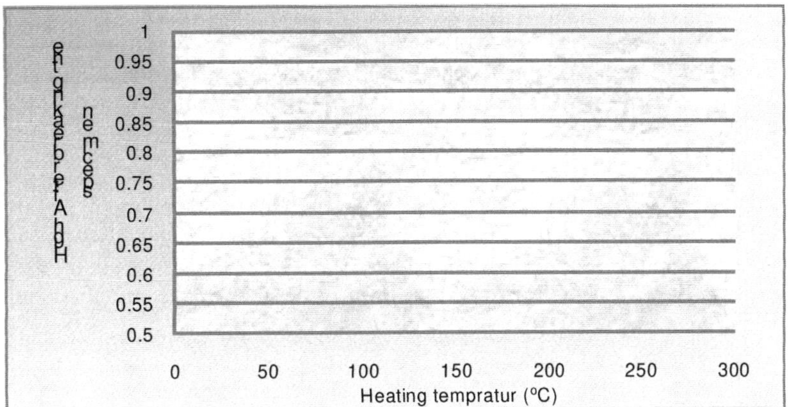

Figure 4. The Relationship of warming temperature relation to height of pendulum After breaking the specimen.

Calculation of hardness test

From the hardness test data, it is known that with the temperature change of heating, resulting in the change of hardness value of the material, the higher the given temperature, the higher the hardness value of the material. From a number of points of emphasis on the parent metal also experienced an increase in hardness but not significant where the minimum hardness value in the specimen without heating treatment can be seen in Table 1.

Table 1. Test result data and impact test calculation

No	Material	Speciment Spesification				Angle		h_1	h_2	h	U	Hi
		b(mm)	d(mm)	t(mm)	A(mm_2)	Alpha	beta					
1	Raw Material	10	10	2	80	130	67	1.429	0.531	0.898	92.529	1.157
2		10	10	2	80		65		0.679	0.750	77.295	0.966
3		10	10	2	80		67		0.531	0.898	92.529	1.157
			Average				66.33		0.580	0.846	87.451	1.093
4	without heating treatment	10	10	2	20	130	68	1.429	0.545	0.884	91.096	1.139
5		10	10	2	20		71		0.587	0.842	86.704	1.009
6		10	10	2	20		66		0.517	0.912	93.963	1.009
			Average				68.33		0.550	0.879	90.588	1.052
7	Heat Treatment 150C	10	10	2	20	130	68	1.429	0.545	0.884	91.096	1.139
8		10	10	2	20		75		0.646	0.783	80.700	1.009
9		10	10	2	20		75		0.646	0.783	80.700	1.009
			Average				72.667		0.612	0.817	84.165	1.052
10	Heat Treatment 250C	10	10	2	20	130	80	1.429	0.719	0.710	73.083	0.914
11		10	10	2	20		80		0.719	0.710	73.083	0.914
12		10	10	2	20		77		0.675	0.754	77.653	0.971
			Average				79.00		0.705	0.724	74.607	0.933
13	Heat Treatment 300C	10	10	2	20	130	83	1.429	0.765	0.664	68.423	0.855
14		10	10	2	20		83		0.765	0.664	68.423	0.855
15		10	10	2	20		86		0.81	0.619	63.763	0.797
			Average				84.00		0.780	0.649	66.87	0.836

Observation of Micro Structures

Microstructure testing was performed to determine the difference of metal structure in the specimen without heating treatment and a specimen of heating process 150°, 250°, and 300° Celcius. The microstructure was observed in the order of sand, polishing, etching with 5% H2SO4 solution for 3 sec, then examining micro structure with optical microscope Nikon Japan type 251565 and equipped with webcam camera with 50 to 500 times magnification.

Statistical Analysi

Statistical analysis was performed using annova variance analysis on tensile test and hardness test data. Based on the data of tensile test results known F smaller than the F crit (F <F crit) with 8.2699 difference. The analysis of hardness test is known to be smaller than F crit (F <F crit) by the difference of 8.235453 while the F test toughness is smaller than F crit (F <F crit) by the difference of 8.2699.

4. Discussion

This study shows that the effect of SMAW (Shielded Metal Arc Welding) welding effect on mechanical properties with heating treatment on S45C steel. The higher the heating temperature increases the strength of the material. The tensile test calculation is known that in the heating material in the heat treatment process the higher the heating temperature (0^oC - 300^oC) the average tensile load until the maximum condition increases, on the welding material without heating process Pmax = 3661.3 kgf, The temperature of 150oC average maximum tensile load of Pmax = 4777.0 kgf, subsequently increased and at a temperature of 300^oC the tensile load was at 6051.2 kgf. The condition also affects the tensile strength properties of material expressed by the maximum voltage, in the graph of figure 4.2. Based on the graph it is known at the temperature 0° C σmax = 25.4kgf mm2, at 150^oC σmax = 32.3 kgf / mm2, 250^oC σmax = 40.1 kgf / mm2 and at 300^oC σmax = 42.0kgf / mm2.

Impact Testing With a constant cross-sectional area for each test specimen, with the decreasing of the effort, the material impact price decreases. Due to the increase in heating temperature, there is a decrease of material impact price (Hi), on a unobserved specimen, Hi = 1,132 J / m2, and so on decreased to heating temperature 300^oC Hi = 0,836 J / m2. The impact price is due to increased brittle or brittle properties. The condition can also be known by the easier pendulum as a pounder load to break the specimen. The average of these changes is seen in the increasing angular degree of the impact test arm after breaking the specimen (β). Where on the specimen without heat treatment process h2 = 0,550 m and increased up to 300^oC heating temperature with h2 = 0,780 m.

Violence testing from a number of pressure points, on the parent metal also experienced an increase in hardness but not significant where the minimum hardness value in the specimen without heating treatment was 35.01 and the highest at 300^oC heating temperature of 35.52. In the weld metal part (Filer) also experienced an increase in hardness due to changes in warming temperature. In the material without heat treatment process, it is known that maximum hardness value is 37.25, at temperature 150^oC = 38,80, at temperature 250^oC = 40,49 and at temperature 300^oC = 40,83. The increase shows the ductile value of the material decreases and the hard properties increases.

Observation of microstructure based on microstructure photograph is known to start the process without heating treatment and process heating treatment 150^oC, 250^oC and 300^oC in micro structure there is a change of structure where the main composition of a micro structure consisting of ferrite, martensite and pearlite change either a percentage of amount or distribution. The change in the microstructure resulted in a change of strength of both tensile, impact and hardness. In specimens without heating treatment, the percentage of ferrite composition is more dominant than that of heat treatment up to 300^oC. This condition results in a material decrease in hardness and toughness (impact) and increased tensile strength at each heating temperature increase.

5. Testing And Advice

From the result of research and data analysis, it can be concluded that the heating process of 150^oC, 250^oC and 300^oC resulted in the change of mechanical properties where tensile strength will continue to increase until 300^oC at 42 kgf / mm2, while at temperature 300^oC its toughness will decrease with Hi = 0,836 J / M2 and its hardness increased to 300^oC = 40.83. The value of its strength at a heating temperature of 3000C for its attraction and its hardness at 300^oCt temperatures has a declining effect. Very highly recommended for a heating process with higher temperature.

References

[1] . Bambang. (2006). Comparison of Average Flow Control And Peak Flow In Power Factor Correction (PFC) Series AC-DC Boost Converter, Jurnal Sain dan Teknologi. Surabaya : Univ.Hang Tuah Surabaya. (in Bahasa, original title: Perbandingan Kendali Arus Rata-Rata Dan Arus Puncak Pada Rangkaian Koreksi Faktor Daya (PFC) AC-DC Boost Converter)

[2] . Anrinal. (2013). Physics Metallurgy. Yogyakarta. Andi. (in bahasa, original title: Metalurgi Fisik)

[3] . Alexander dkk. (1990). Dasar Metalurgi Untuk Rekayasa. Jakarta. PT Gramedia Pustaka Utama. (in bahasa, original title: Basic Metallurgy for Engineering)

[4] . Arifin. (1997). Karakterisasi Sifat Mekanis Dan Struktur Mikro Pipa Baja Jis Z 2201, Report Mechanical Engineering, Faculty of Engineering, Pandanaran Univ. (in bahasa, original title: Characterization of Mechanical Properties And Steel Structure of Jis Z 2201 Steel Pipe)

[5] . Daryanto. (2013). Material Science. Jakarta. Bumi Aksara, (in Bahasa, original title: Ilmu Bahan).

[6] . Kirono S. (2013). Analysis of Temperature Effect on Tempering Process on Mechanical Properties and Aisi 4360 Micro Steel Structure. Jakarta : Universitas Muhammadiyah. (in Bahasa, original title: Analisa Pengaruh Temperatur Pada Proses Tempering Terhadap Sifat Mekanis Dan Stuktur Mikro Baja Aisi 4340)

[7] . Sonawan. (2004). Introduction to Understanding Metal Welding ,. Bandung. Alfabeta. (in Bahasa original title: Pengantar Untuk Memahami Pengelasan Logam)

[8] . Suharno. (2008). The Great Influence of Bead Welding Temper Flow Against Toughness of SMAW Las Results on SS41 Steel,. Malang : Universitas Brawijaya. (In Bahasa, original title: Pengaruh Besar Arus Temper Bead Welding Terhadap ketangguhan Hasil Las SMAW Pada Baja SS41)

[9] . Suharto. (1991). Analysis of Effect of Cooling Media on ST Steel Pulling Strength After Welding Using Electrical Welding. Banjarbaru : Universitas Lambung Mangkurat. Kalimantan Selatan. (in Bahasa Original title: Analisis Pengaruh Media Pendingin Terhadap Kekuatan Tarik Baja ST 37 Pasca Pengelasan Menggunakan Las Listrik)

[10] . Sugiyono. (2012). Statistic for Research, (in Bahasa, Original Title Statistika Untuk Penelitian). Bandung. Publixher: Alfabeta.

Preliminary Development of Online Monitoring Acoustic Emission System for the Integrity of Research Reactor Components

S Bakhri[1], E Sumarno[1], R Himawan[1], T Y Akbar[2], M. Subekti[1], G. R. Sunaryo[1]

[1] Centre for Nuclear Reactor Technology and Safety, Puspiptek Complex, OB 80, Serpong, Setu, Tangerang Selatan, Indonesia
[2] Jurusan Fisika, Fakultas Matematika dan Ilmu Pengetahuan Alam, Universitas Sriwijaya

bakhrisy@batan.go.id

Abstract. Three research reactors owned by BATAN have been more than 25 years. Aging of (Structure, System and Component) SSC which is mainly related to mechanical causes become the most important issue for the sustainability and safety operation. Acoustic Emission (AE) is one of the appropriate and recommended methods by the IAEA for inspection as well as at the same time for the monitoring of mechanical SSC related. However, the advantages of AE method in detecting the acoustic emission both for the inspection and the online monitoring require a relatively complex measurement system including hardware software system for the signal detection and analysis purposes. Therefore, aim of this work was to develop an AE system based on an embedded system which capable for doing both the online monitoring and inspection of the research reactor's integrity structure. An embedded system was selected due to the possibility to install the equipment on the field in extreme environmental condition with capability to store, analyses, and send the required information for further maintenance and operation. The research was done by designing the embedded system based on the Field Programmable Gate Array (FPGA) platform, because of their execution speed and system reconfigurable opportunities. The AE embedded system is then tested to identify the AE source location and AE characteristic under tensile material testing. The developed system successfully acquire the AE elastic waveform and determine the parameter-based analysis such as the amplitude, peak, duration, rise time, counts and the average frequency both for the source location test and the tensile test.

Keywords : Acoustic Emission, monitoring, embedded system, research reactor

1. Introduction

Three research reactors in Indonesia, owned and operated by BATAN, i.e. the Triga Reactor Bandung, Yogyakarta and Multipurpose Research Reactor Siwabessy, Serpong, has been operated for more than 25 years. The aging condition requires in service inspection as well as a continuous condition monitoring of the systems, structures and components (SSC) for ensuring the safe operation of the reactor. In addition a comprehensive analysis for the integrity, the characteristics and performance of the SSC will also be very important to determine further regulatory permission for the life extension of the research reactors.

IAEA provides some guidelines for the integrity inspection and leakage monitoring of the SSC which are important for the safety of the reactor, such as the reactor vessel, isolation valve, cooling pumps and primary cooling pipes [1-3]. Several methods can be applied for the in-service inspection applications and at the same time for the condition monitoring of nuclear reactors such as ultrasonic testing, vibration monitoring, acoustic monitoring (acoustic emission –AE monitoring), and guided ultrasonic wave [4]. In case of component and structure However, the acoustic emission monitoring is promising due to the advantages such as relatively unaffected to the background mechanical vibration noise or resonance [5], real time monitoring [4, 6], early detection capability [7], covers a larger area of monitoring using a minimum number of sensors compare with other structural integrity detection method [8]. In addition, acoustic emission monitoring can also applied for the structural integrity inspection in nuclear facilities[9], leakage monitoring [10, 11], early crack detections [12-14], material testing [15, 16], rotating machine monitoring [5, 7, 17, 18], corrosion indication monitoring [19]. The example of typical positions and applications of the AE for the integrity and leaking monitoring of reactor and pipe and pump of Westinghouse PWR is given in [20]. The example for the pressure vessel of EPR is in [21] while the application of VVER NPP Steam Generator can be seen in [22].

However, the advantages of AE method in detecting the acoustic emission require a relatively complex measurement system including hardware software system for the signal detection and analysis purposes [23]. The AE measurement system requires high sampling rate data acquisition system, minimum two or three sensors with simultaneous data acquisition capability for better accuracy of source location, enough memory and data storage for the signal processing analysis and diagnosis. These requirements are necessary considering the challenges in nuclear reactor which requires better accuracy under the steady state as well as the early detection of emergency condition. In addition, the AE system is required to meet the purposes for the online monitoring as well as the inspection applications in research reactor, such as for coolant leakage monitoring, early crack detection in the material of safety related components, rotating machine monitoring and corrosion detection.

To address this challenge, therefore, the purpose of this work was to develop an AE system based on an embedded system which capable for doing the online monitoring of the research reactor's integrity structure. An embedded system was selected due to the possibility to install the equipment on the field in extreme environmental condition with capability to store, analyses, and send the required information for further maintenance and operation. The research was done by designing the embedded system based on the Field Programmable Field Array (FPGA), because of the execution speed and system reconfigurable opportunities. The AE embedded system was designed to accommodate parameter based approached and waveform-based approached. The AE embedded system is then tested by introducing simulated AE source events and tensile material testing.

2. Theory

Acoustic emission is produced when materials experience any deformations such as transient relaxation of stress and strain at a specific localized source in material [24]. This rapid and spontaneous phenomenon release several of energy, such as elastic, non-linear elastic, elastic-plastic, and elastic-viscoplastic. However, the elastic energy is mostly detected as the others are easily attenuated at short distance. Furthermore, among the elastic waves propagation modes, three of them are mostly used, i.e. Longitudinal (dilatational, P-) wave, Shear (or transverse, or S-) wave, and Rayleigh (or surface) wave. The P waves (the fastest wave), and S waves (slower than P waves), both represents the elastic waves in the sample material while the Rayleigh in on the surface.

This elastic wave energy, known also as AE event, are spread through the structures and eventually detected by AE sensors which are coupled with the sample material at the surface [25]. The AE transducer then identify this elastic energy as displacements which representation the nature of changes in the samples. The AE transducer produce electric signal based on the detected displacement under piezoelectric effect which are then amplified and processed into waveform signals for further parameter analysis. It should be noted that the ideal transducer can detect any changes of displacements both horizontally and vertically.

The basic known method to analyze the elastic waves are AE parameters-based approached and waveform-based approached [26]. The parameter-based approached covers amplitude, energy, hits, duration and the counts of ringing identification [27] as shown in Figure 1. The parameter based approached is relatively simple, with small storage data and fast recording. On the other hand the, waveform-based approached need to records all the waveform while providing more detail analysis various fracture analysis in fault-plane orientation, size, energy and mode. The waveform analysis employs various frequency analysis, such as FFT, SFFT, wavelet, invers and correlation.

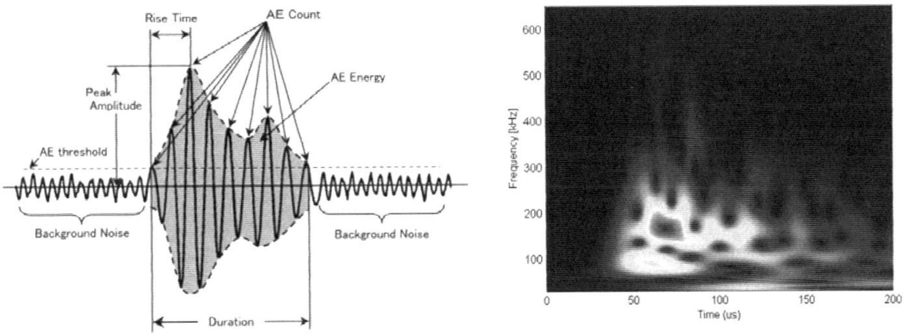

Figure 1. (a) The parameter based of AE [28] and (b) the waveform-based of AE using Continuous Wavelet Transform (CWT) [29]

3. Methodology

The research was done by designing the embedded system following several steps: performing conceptual design for the hardware and software of the AE, then followed by the development of the SE system based on the FPGA platform. The AE system was tested using generated acoustic events on a carbon steel plate with dimension 180 x 80 x 10 cm and with tensile testing. Some features of AE signal analysis, under time domain with parameter-based analysis, were also confirmed.

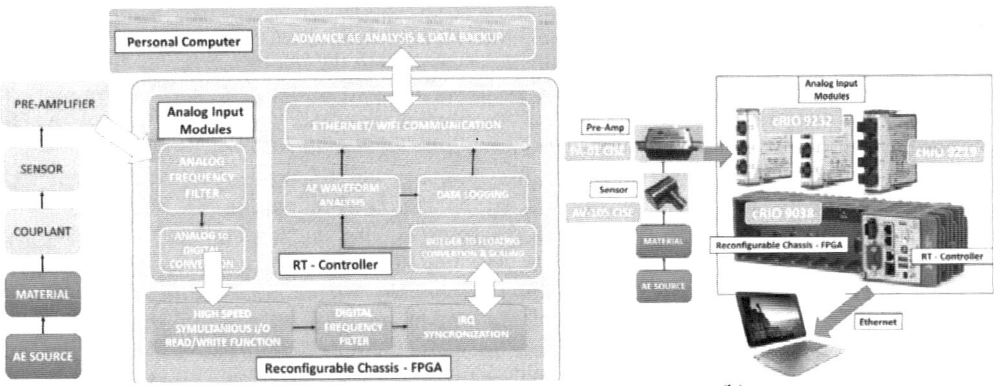

(a) (b)

Figure 2. (a) The conceptual design, and (b) hardware implementation of AE system

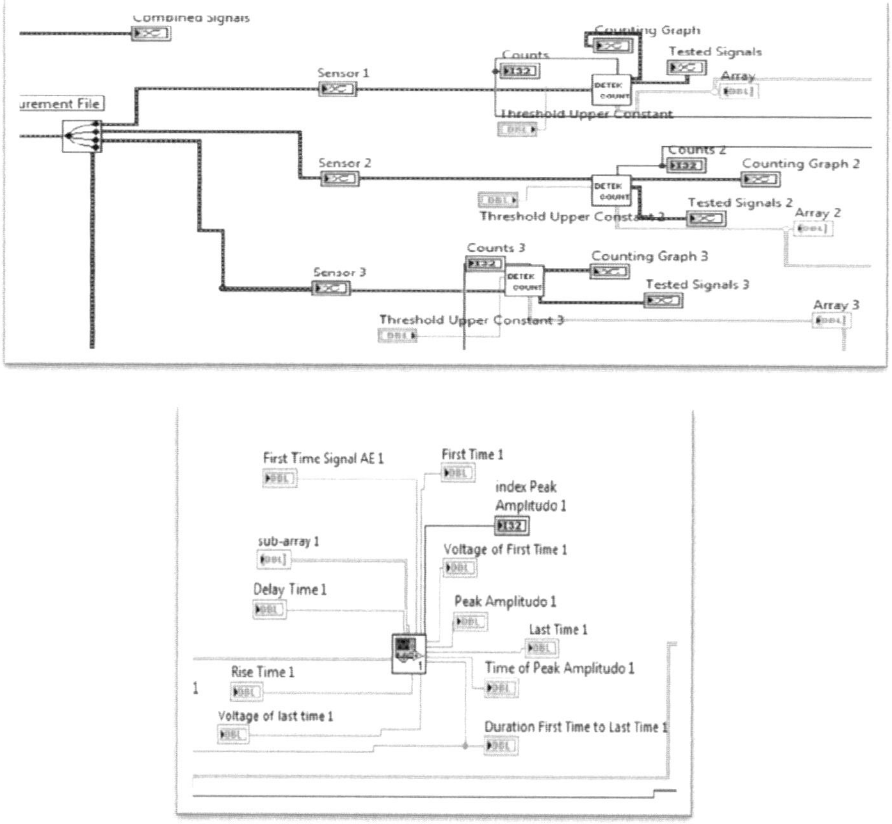

Figure 3. The developed software for the parameter-based AE analysis using LabVIEW

The conceptual design of the developed AE system is given in Figure 2 and Figure 3 showing the block diagram of the measurement chain system for the FPGA based embedded system and the developed parameter-based software detection system. Figure 2 shows that active sensors were employed in this paper, i.e. accelerometer, because of better AE wave conversion efficiency, better capacity in detecting any small displacements, and more immune from signal saturations even under high intensity source compare with other type of sensors. In this research, 4 AE sensors (AV-105 CISE) were employed. The preamplifiers (PA-01 CISE) were used to increase the output voltage of the signal, improve signal to noise ratio, and reduce the output impedance signals for long connector cables. Figure 2 also shows that analog and digital filter were applied in the simultaneous input modules and the FPGA module. A band-pass filter were developed to cut a low frequency of less than 1 KHz and high frequency for more than 300 KHz. Waveform analysis based on the basic feature extraction under time domain such as amplitude, the rise time duration, counts, RMS, energy and arrival time, were also determined. A special software for the detection of AE was made in this stage.

Following the AE system development, the final stage of this research are validation of the system by utilizing the statistical parameter-based test of simulated AE sources. Two type of experiments were done, i.e. fault location simulations and the tensile testing. The source fault locations may represent AE cracks, leaking or corrosion events. The AE sources ware created using pencil lead break method (2H, 0.5 mm)[4]. The generated AE were the then detected using 4 separated sensors as given in Figure 4. Furthermore, the tensile stress experiments was done as shown in Figure 5. The performance test of the AE system for the material testing is done at speed 200 mm/second using Aluminum-Magnesium alloys as one of the materials for the reactor tank and fuel cladding of RSG-GAS research reactor.

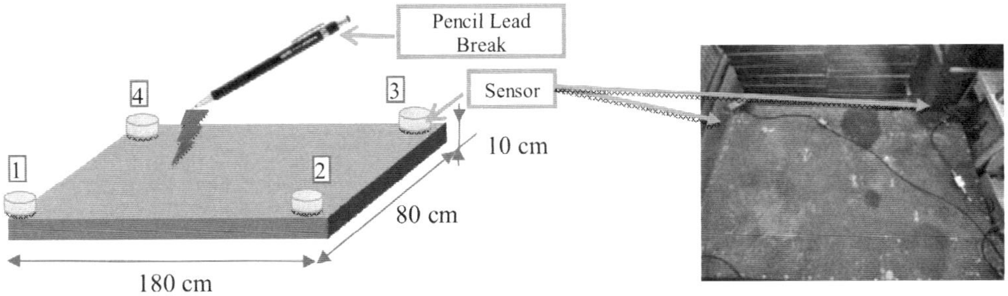

(a) (b)

Figure 4. Test arrangement for the pencil lead break acoustic emission test, (a) showing four sensors with the dimension of the iron-carbon plate testing (b) the actual photograph

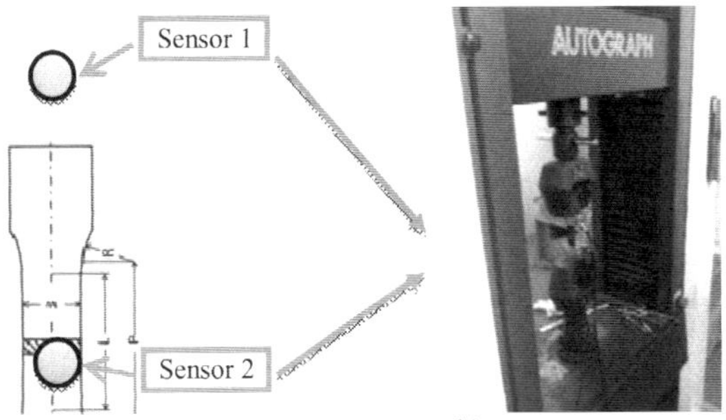

(a) (b)

Figure 5. Test arrangement for tensile test- testing showing the (a) sample with sensor position and (b) tensile-test machine

4. Results and Discussions

The purpose of this work is to demonstrate the preliminary development of embedded AE system for the online monitoring of research reactors. The elastic wave of the simulated AE is detected by four sensors. The generated AE events was done using pencil lead break, first event is generated at around (50,50) cm. The result show 4 signal burst of the sensor given in Figure 6.

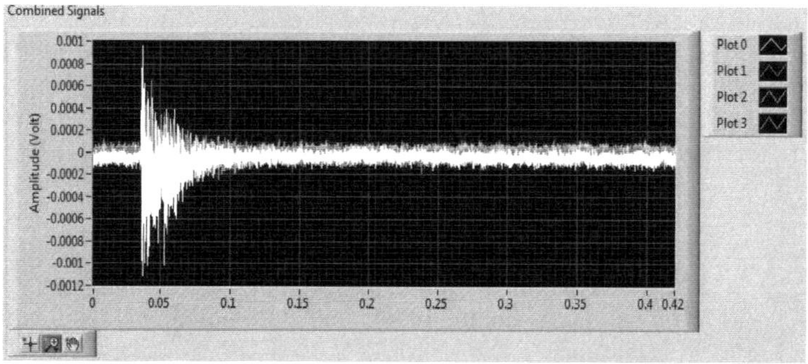

Figure 6. The simulated burst of the pencil lead break AE events of 4 sensors in LabVIEW.

Figure 6 confirms that the typical burst of the AE during the test is similar with the results of common AE signals. The AE signals indicate several parameters as the AE requirements, such as amplitude, peak time, rise time, duration, and noise. Detail of the parameter analysis requires zooming of the signals (Figure 7) where the interest parameters can be identified more easily. The applied threshold was applied slightly above the noise floor at around 0.000187 V where the highest peak is detected less than 0.01 V.

The example of detected AE parameters for the sensor 1 are given in Table 1, cover the parameter-based, both on the signal parameter level and the shape of the waveform. The table shows that the duration amplitude of the peak 959 μvolt, rise time 996.2 ms, and the duration of signals 29.886 ms. The duration of time represents the time signals from the start of signal crossing the threshold until the end. The crossing signals at the threshold at the start point and the end point are around 755 μvolt and 215 μvolt respectively. The produced software, listed the detected peaks crossing the positive threshold and arrange them into a 2D array as given in Figure 7 (b) where the detected signals above and below the threshold were given sign 1 and 0 respectively. Based on this array analysis, the number counts above the threshold are about 26 counts, with the average frequency (number of counts divided by the duration time) for about 869.97 kHz. The burst signal also shows that the longitudinal (P-) wave, and the Shear (or transverse, or S-) wave were not detected while the Rayleigh (or surface) wave was dominant. The P- (primary) and S- (secondary) wave which usually precede the Rayleigh wave might be around the noise floor and hard to be detected clearly. These detected parameters clearly indicate that the system is capable of performing the online monitoring for the source faults location based on the AE events for the crack, leaking and the corrosion.

Table 1. The AE parameters for source location.

No	Parameters	Values
1	The time of first crossing signal	35.9128 ms
	The amplitude of first crossing signal	755 μVolt
2	The time of last crossing signal	65.8078 ms
	The amplitude of first crossing signal	215 μVolt
3	Time of the peak	36.918 ms
	Amplitude of the peak	959 μVolt
4	Rise Time	996.2 ms
5	Duration	29.886 ms
6	Average frequency	869.97 kHz

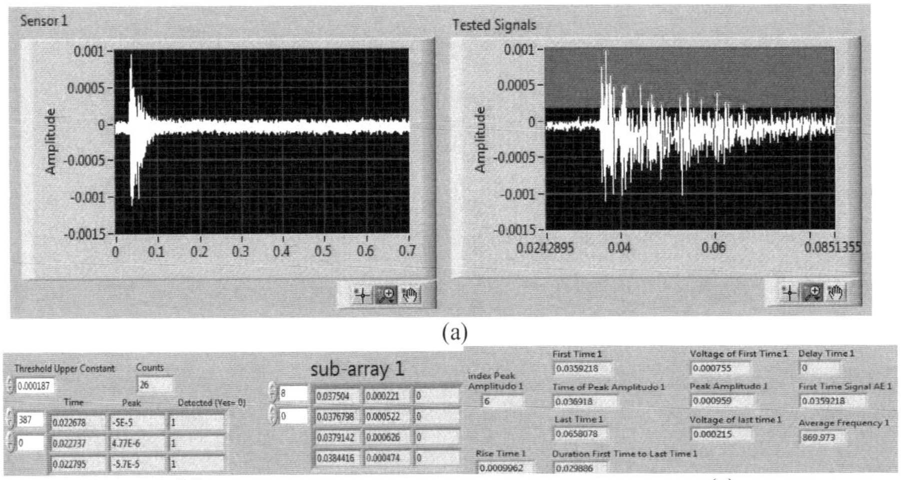

(a)

(b) (c)

Figure 7. The example of parameter analysis of a burst from sensor 1, presenting (a) detail of the signal (b) threshold, number of counts and the array of detected peaks, (c) the results of parameter analysis.

Further experiment is the tensile stress signals where the results are given in Figure 8. The detected signals in Figure 8 slightly different with the previous source fault location test. This test provides strength and characteristic information of the AlMg2 samples, as employed for the components of the RSG-GAS research reactor. The AE measurement is expected to detect the produced AE events mostly during the plastic transformation. The signal were acquired for more than 8 second to allow the material deformation and crack propagation can be identified. Figure 8 shows that the maximum burst at around 0.005 Volt is between 6 and 7 second.

The parameter-based approached analysis is given in Figure 9 demonstrating the detail analysis. Threshold was applied at around 0.5 mVolt producing 291 counts. The detail analysis shows that key parameters for the amplitude of the peak, rise time, duration, and the average of frequency are 3.22 mVolt, 28.0108 msecond, 0.100851 second, and 2885.46 kHz respectively as listed in Table 2. These recorded parameters show the Rayleigh wave during the plastic deformation before the crack occurs. Following this Rayleigh wave burst, several small burst indicate that the crack growth occur may attenuate the AE elastic wave to the sensors and it is hardly also detected by both sensors. The result shows that the AE embedded system able to meet the requirement for further tensile test experiment which also can be used to confirm any crack growth during the online monitoring of research reactor components.

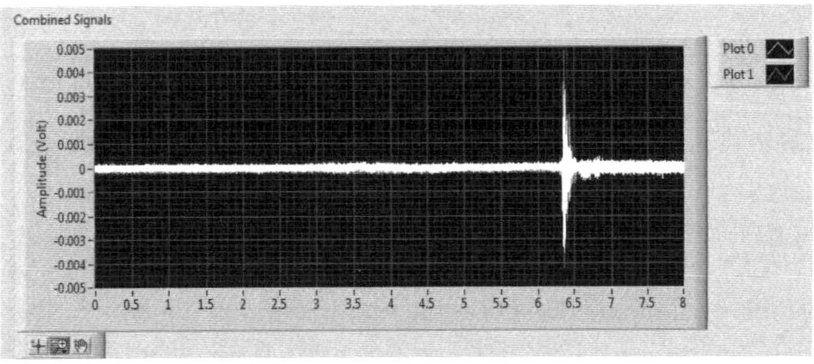

Figure 8. The simulated burst of the two sensors during the tensile test analysis.

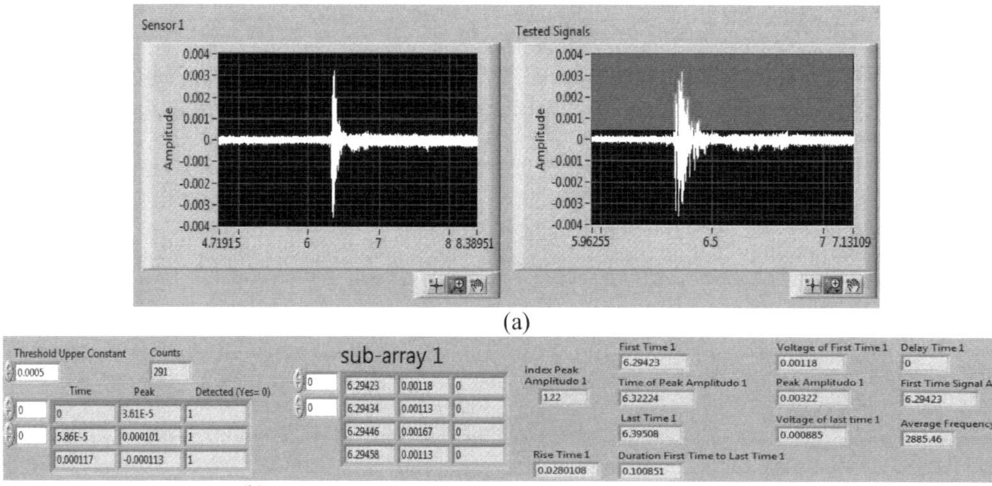

(a)

(b) (c)

Figure 9. The parameter-based analysis of a burst from sensor 1 in tensile stress, presenting (a) detail of the signal (b) threshold, number of counts and the array of detected peaks, (c) the results

Table 2. The AE parameters for the tensile test.

No	Parameters	Values
1	The time of first crossing signal	6.29423 second
	The amplitude of first crossing signal	1.18 mVolt
2	The time of last crossing signal	6.39508 sekon
	The amplitude of first crossing signal	0.885 mVolt
3	Time of the peak	6.32224 second
	Amplitude of the peak	3.22 mVolt
4	Rise Time	28.0108 msecond
5	Duration	0.100851 second
6	Average frequency	2885.46 kHz

The source location and tensile testing results prove that AE signals are identified clearly than the signal noises. Compare with the various typical AE source such as in [28], the signals can be clearly validated containing the required AE parameters which illustrates the AE elastic Rayleigh wave. The tensile test signal parameter even though has smaller amplitude, the AE embedded system is capable of sensing as well as defining the characteristic of signals.

5. Conclusion

It can be concluded that the development of an embedded AE system based on the FPGA has successfully detected the AE events for the online monitoring of components in research reactor. The performance was demonstrated through source location and tensile testing. Both of these tests denote the capacity of crack, leaking and corrosion detection for an online monitoring system. The AE embedded system successfully provide parameter-based analysis approached such as the peak of the amplitude, rise time, duration, counts and the average of frequency. These parameter-based results also indicates that this preliminary development system can be applied further to determine the coordinate of the AE events as well as to present the characteristic material of components during the crack growth before jeopardize the safety of the research reactor in general.

6. Acknowledgement

I would like to thank to the Centre for Nuclear Reactor Technology and Safety for the support of funding through DIPA 2015 and 2016 and the NDT facilities at the BPFKR division. I would to acknowledge the support of head of the research centre, Dr. Geni Rina Sunaryo, head of BPFKR, Drs Deswandri MEng, and the head of laboratory of NDT, Alim Mardi, ST, MSc.

References

[1] Agency I A E 2008 *On-Line Monitoring for Improving Performance of Nuclear Power Plants Part 2 : Process and Component Condition Monitoring and Diagnostics*: International Atomic Energy Agency)

[2] Koc R, Kazantzis N K, Nuttall W J and Ma Y H 2012 Economic assessment of inherently safe membrane reactor technology options integrated into IGCC power plants *Process Safety and Environmental Protection* **90** 436-50

[3] ________ 2013 *Advanced Surveilance, Diagnostic and Prognostic Techniques in Monitoring Structures, Systems, and Components in Nuclear Plants* (Vienna, Austria: INTERNATIONAL ATOMIC ENERGY AGENCY)

[4] Droubi M G, Faisal N H, Orr F, Steel J A and El-Shaib M 2017 Acoustic emission method for defect detection and identification in carbon steel welded joints *Journal of Constructional Steel Research* **134** 28-37

[5] Zhang Y, Lu W and Chu F 2017 Planet gear fault localization for wind turbine gearbox using acoustic emission signals *Renewable Energy* **109** 449-60

[6] Pomponi E and Vinogradov A 2013 A real-time approach to acoustic emission clustering *Mechanical Systems and Signal Processing* **40** 791-804

[7] Liu C, Wu X, Mao J and Liu X 2017 Acoustic emission signal processing for rolling bearing running state assessment using compressive sensing *Mechanical Systems and Signal Processing* **91** 395-406

[8] Xu C, Gong P, Xie J, Shi H, Chen G and Song G 2016 An acoustic emission based multi-level approach to buried gas pipeline leakage localization *Journal of Loss Prevention in the Process Industries* **44** 397-404

[9] Traore O I, Pantera L, Favretto-Cristini N, Cristini P, Viguier-Pla S and Vieu P 2017 Structure analysis and denoising using Singular Spectrum Analysis: Application to acoustic emission signals from nuclear safety experiments *Measurement* **104** 78-88

[10] Yu L and Li S Z 2017 Acoustic emission (AE) based small leak detection of galvanized steel pipe due to loosening of screw thread connection *Applied Acoustics* **120** 85-9

[11] 2013 Identifying leaking valves quickly and easily using acoustic emissions measurement technology *Sealing Technology* **2013** 11-2

[12] Barile C, Casavola C, Pappalettera G and Pappalettere C 2016 Analysis of crack propagation in stainless steel by comparing acoustic emissions and infrared thermography data *Engineering Failure Analysis* **69** 35-42

[13] Babu M N, Mukhopadhyay C K, Sasikala G, Albert S K, Bhaduri A K, Jayakumar T and Kumar R 2016 Study of fatigue crack growth in RAFM steel using acoustic emission technique *Journal of Constructional Steel Research* **126** 107-16

[14] Qin L, Ren H-W, Dong B-Q and Xing F 2014 Acoustic Emission Behavior of Early Age Concrete Monitored by Embedded Sensors *Materials* **7** 6908-18

[15] Dahmene F, Yaacoubi S, El Mountassir M, Bendaoud N, Langlois C and Bardoux O 2016 On the modal acoustic emission testing of composite structure *Composite Structures* **140** 446-52

[16] Chuluunbat T, Lu C, Kostryzhev A and Tieu K 2015 Investigation of X70 line pipe steel fracture during single edge-notched tensile testing using acoustic emission monitoring *Materials Science and Engineering: A* **640** 471-9

[17] Alfayez L, Mba D and Dyson G 2005 The application of acoustic emission for detecting incipient cavitation and the best efficiency point of a 60 kW centrifugal pump: case study *NDT & E International* **38** 354-8
[18] Dykas B and Harris J 2017 Acoustic emission characteristics of a single cylinder diesel generator at various loads and with a failing injector *Mechanical Systems and Signal Processing* **93** 397-414
[19] Zaki A, Chai H K, Behnia A, Aggelis D G, Tan J Y and Ibrahim Z 2017 Monitoring fracture of steel corroded reinforced concrete members under flexure by acoustic emission technique *Construction and Building Materials* **136** 609-18
[20] Jax P and Ruthrof K 1989 Acoustic emission inspections of nuclear components considering recent research programmes *Nuclear Engineering and Design* **113** 71-9
[21] Choi Y-C, Park J-H and Choi K-S 2011 An impact source localization technique for a nuclear power plant by using sensors of different types *ISA Transactions* **50** 111-8
[22] Bakirov M B, Povarov V P, Gromov A F and Levchuk V I 2015 Development of a technology for continuous acoustic emission monitoring of in-service damageability of metal in safety-related NPP equipment *Nuclear Energy and Technology* **1** 32-6
[23] Prateepasen A, Kaewwaewnoi W and Kaewtrakulpong P 2011 Smart portable noninvasive instrument for detection of internal air leakage of a valve using acoustic emission signals *Measurement* **44** 378-84
[24] Elfergani H A, Pullin R and Holford K M 2013 Damage assessment of corrosion in prestressed concrete by acoustic emission *Construction and Building Materials* **40** 925-33
[25] Mukhopadhyay C K, Jayakumar T, Haneef T K, Suresh Kumar S, Rao B P C, Goyal S, Gupta S K, Bhasin V, Vishnuvardhan S, Raghava G and Gandhi P 2014 Use of acoustic emission and ultrasonic techniques for monitoring crack initiation/growth during ratcheting studies on 304LN stainless steel straight pipe *International Journal of Pressure Vessels and Piping* **116** 27-36
[26] 2008 *Acoustic Emission Testing*: Springer Berlin Heidelberg)
[27] Geng J, Sun Q, Zhang W and Lü C 2016 Effect of high temperature on mechanical and acoustic emission properties of calcareous-aggregate concrete *Applied Thermal Engineering* **106** 1200-8
[28] 2016 *Practical Acoustic Emission Testing*: Springer Japan)
[29] Yaacoubi S, Dahmene F, Bouzenad A, El Mountassir M and Aouini M 2017 Modal acoustic emission for composite structures health monitoring: Issues to save computing time and algorithmic implementation *Composite Structures*

Power Peaking Effect of OTTO Fuel Scheme Pebble Bed Reactor

T. Setiadipura, Suwoto, Zuhair, S. Bakhri, G.R. Sunaryo

Center for Nuclear Reactor Technology and Safety – BATAN
Puspiptek Area, Office Building No. 80, Serpong, Tangerang Selatan 15310, Indonesia

tsdipura@batan.go.id

Abstract. Pebble Bed Reactor (PBR) type of Hight Temperature Gas-cooled Reactor (HTGR) is a very interesting nuclear reactor design to fulfill the growing electricity and heat demand with a superior passive safety features. Effort to introduce the PBR design to the market can be strengthen by simplifying its system with the Once-through-then-out (OTTO) cycle PBR in which the pebble fuel only pass the core once. Important challenge in the OTTO fuel scheme is the power peaking effect which limit the maximum nominal power or burnup of the design. Parametric survey is perform in this study to investigate the contribution of different design parameters to power peaking effect of OTTO cycle PBR. PEBBED code is utilized in this study to perform the equilibrium PBR core analysis for different design parameter and fuel scheme. The parameters include its core diameter, height-per-diameter (H/D), power density, and core nominal power. Results of this study show that diameter and H/D effectsare stronger compare to the power density and nominal core power. Results of this study might become an importance guidance for design optimization of OTTO fuel scheme PBR.

Keywords: pebble bed reactor, once-through-then-out, power peaking effect

1. Introduction

Pebble Bed Reactor (PBR) type of High Temperature Gas-cooled Reactor (HTGR) is one of Generation IV nuclear reactor design.It is a very interesting nuclear reactor design to fulfill the growing electricity and heat demand with a superior passive safety features. The PBR design release a very low radiation impact to environment in any operational condition, even in the worse accident scenario. In general, geometrical and material designs of the PBR core assure that the core will not goes beyond 1620°C, while in this temperature range the TRISO fuel particle assure a very low fission product release[1]. Both design characteristics guarantee a passive safety feature of PBR design. In addition, PBR cogeneration potential offers a possibility to optimally exploit the other mineral resources in a more environmently friendly such as coal liquefaction and gasification. In spite of many challenges [2], many efforts are in progress to developed and commercialize the PBR design [3]. Currently,Chinese's efforts with HTR-10, HTR-PM and the following HTR-600 lead this progress to commercialize PBR design [4]. PBR reactor physics design also flexible to accommodate wide fuel cycle options including thorium, plutonium, minor actinide, and rock-like oxide (ROX) based fuel beside the standard uranium oxide fuel [5,6,7].

Effort to introduce the PBR design to the market can be strengthen by –even more- simplifying its system with the once-through-then-out (OTTO) fuel scheme. Instead of circulating the pebble balls several times through the core, pebble balls are only have a single pass through the core. This fuel scheme omit the need for equipment of burnup measurement and pebble balls reloading. Reactor physics characteristic of OTTO fuel scheme also improve the cogeneration capability of the PBR core [8,9]. The nature of OTTO fuel scheme inherit a problem of power peaking effect due to only fresh fuel inserted from the core top. This is the main challenges which limit the efficiency, e.g. bunup and nominal power, of OTTO fuel scheme PBR. Additional burnable poisons to the fuel already studied to minimized this problem [10] although this might add some complexity to the PBR's fuel. Recent studies to applied the OTTO fuel scheme for different PBR design [11,12] show that there is a possibility that power peaking effect is not occurred for some design. As a follow up, purpose of this research is to investigate different core design parameters effect to the power peaking of OTTO fuel scheme design. Power peaking effect is represented by peak-to-average power density value of the equilibrium core with certain design parameters. Core diameter, geometrical height per diameter, average power density and nominal core power are the parameters which included in this study. Detail research methodology is provided in the next section, followed by the results and discussion also the conclusion of this study.

2. Research Methodology

General layout of the PBR core is given in Figure 1[13]. The cylindrical PBR's active core is surrounded by the graphite reflector in the radial, bottom, and upper side. Design of the pebble fuel is given in Table 1. A standard 6 cm pebble ball fuels are mixed randomly with around 61% fuel volume ratio in the active core bed.The pebble ball fuel design used in this study is based on the HTR-10 design with 17% U-235 enrichment. Each pebble ball fuel contains around 8335 TRISO coated fuel particles to make a 5 g uranium loading per pebble ball. In this study, the discharge cone at the bottom of the PBR core is omitted in the calculational model which physically acceptable due to low neutron flux in this area [13]. Diameter and height of the active core, average power density, and nominal core power are changing based on the parametric survey, while the thickness of the surrounding reflector are kept the same. The inlet and outlet helium temperature are kept to be 250°C and 700°C in this parametric survey by tuning the appropriate helium flow rate for each cases.

The first parametric survey is perform for a fixed average core power density of 4 W/cc, detailed parameters are shown in Table 2. While Table 3 shows a parameter used in the parametric survey with a fixed H/D of 1.5. In all the cases, fuel recirculation scheme is performed for 10, 7, 5, 2, 1 (OTTO) pass.All the results given in this study is the equilibrium core with k_{eff} slightly above one.

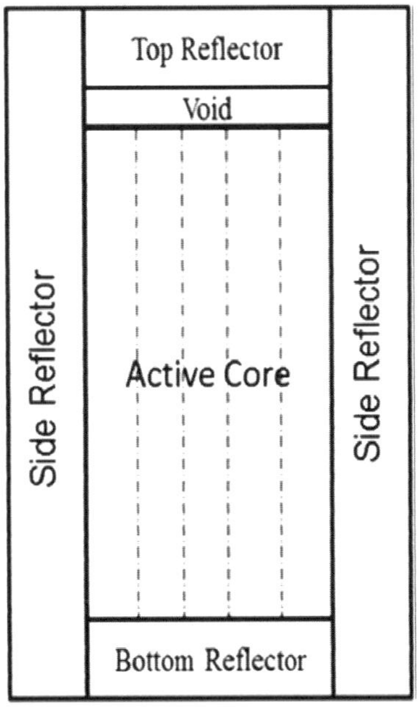

Figure 1. The schematic view of PBR core.

Table 1. Specification of fuel pebble and TRISO coated particle.

Fuel pebble	
Diameter of pebble (cm)	6
Diameter of fueled zone (cm)	5
Thickness of graphite shell (cm)	0.5
Density of graphite shell (g/cm^3)	1.75
Density of graphite matrix (g/cm^3)	1.75
TRISO coated particle	
Diameter of kernel (µm)	500
Density of kernel (g/cm^3)	10.4
Coating layers	
Material	C/iPyC/SiC/oPyC
Thickness (µm)	90/40/35/35
Density (g/cm^3)	1.04/1.88/3.15/1.88

Table 2. Parametric survey parameter for a fixed average power density of 4 W/cc

Core Diameter	Core heigth per diameter (H/D)			
	1.1	1.5	2	3
180cm	20 MWt	27.5MWt	36.6MWt	55MWt
240cm	47.2 MWt	65.2MWt	86.8MWt	130.3MWt
300cm	92.5 MWt	127.3MWt	169.5MWt	254.5MWt

Table 3. Parametric survey parameter for a fixed H/D=1.5

(height ; diameter)	Average power density		
	2 W/cc	3 W/cc	4 W/cc
(270cm ; 180cm)	13.7MWt	20.6MWt	27.5MWt
(360cm ; 240cm)	32.6MWt	48.9MWt	65.2MWt
(450cm ; 300cm)	63.7MWt	95.4MWt	127.2MWt

PEBBED code [14] is used in this research to perform an equilibrium core calculation of the PBR core. Neutron transport is calculated using the neutron diffusion approximation. Special burnup calculation and matrix recirculation method is applied in PEBBED to calculate the equilibrium core of the PBR's moving core. PEBBED solve simultaneously the neutron diffusion equation and burnup equation which accommodate the fuel movement as given in eq. 1. To achieve the equilibrium core PEBBED solved eq. 1 by omitting the time derivation term. Cross section generation calculation is performed based on ENDF/B-VII.0 nuclear data library by the COMBINE [15]code which is coupled with PEBBED. Homogenization of the double heterogeneity of the PBR's core are performed in the coated fuel particle level combining all the particle fuel layers, then the pebble fuel level combining the fuel zone and the graphite shell, then the active core level combining the fuel and helium.

$$\frac{\partial N_k(r,z,t)}{\partial t}+\frac{\partial N_k(r,z,t)}{\partial z}v_z(\mathrm{r},\mathrm{z})(1)$$
$$=\emptyset(r,z,t)\sum_{i=l}^{m}N_i(r,z,t)\sigma_{fi}y_{ik}+\emptyset(r,z,t)\sum_{s=r}^{q}N_s(r,z,t)\sigma_{as}y_{sk}+\sum_{i=l}^{m}N_i(r,z,t)\lambda_j\alpha_{jk}$$
$$-\lambda_k N_k(r,z,t)-\emptyset(\vec{r},t)N_k(r,z,t)\sigma_{ak}$$

where:

$N_k(r,z,t)$	: Nuclide density of isotope*k* at*(r,z) position*at time*t*
v_z	: Axial pebble fuel speed
$\emptyset(r,z,t)$	: Neutron flux at*(r,z) position*at time*t*
σ_{fi}	: Fission cross section of isotope*i*
y_{ik}	: Birth probability of isotope*k*from fission ofisotop*i*
σ_{as}	: Absorption cross section of isotope*s*
y_{sk}	: Birth probability of isotope*k*from absoption reaction of isotope*s*
λ_j	: Decay cross section of isotope*j*
α_{jk}	: Probability that isotope*j decay into* isotope *k*

3. Results and Discussions

Results of axial power density profile for different fuel scheme with a 4W/cc average power density, and diameter of 180cm dan height 360cmare given in Figure 2. It shows how the axial power density change for different fuel scheme. By reducing the number of pebble ball fuel pass through the core, the peak power density is higher and getting closer to the top part of the core, and finally it shows the power peaking effect for the OTTO fuel scheme. In the other way, Figure 2 shows how the peaking effect become smaller by increasing the pebble fuel pass, however the effectiveness of power peaking reduction become smaller as can be seen by comparing the case for 5,7 and 10 pass. Those results show that for each core design there will be its optimum fuel pass number giving a minimum power peaking with minimum fuel pass number. Lower fuel pass number might simplify the operational routine of the core, however if the pass is more than once (non-OTTO) the PBR system still need to be equipped with special fuel handling system and online burnup measurement which not meet the goal of system simplification of PBR core.

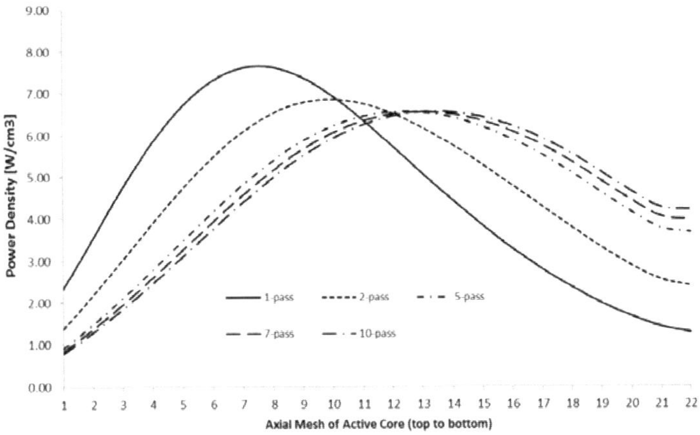

Figure 2. Axial power density profile for average power density 4 W/c m^3 ; D=180cm; H/D=2

Effect of different H/D for a fixed average power density are shown in Figures 3-5. In general, those figures show a common pattern where the OTTO fuel scheme will have a more significant power peaking in a thicker core diameter. For a thin core diameter, in particular with a low H/D, it is found that the power peaking effect of the OTTO fuel scheme can be omitted or insignificant. This area of design parameter or core design can be an interesting application area of the OTTO fuel scheme. For an area with low electrical demand and low backup frastructure this kind of reactor design will be appropriate. Figure 7 shows the axial power density for different H/D. Again it shows how the axial power density profile in low H/D become more homogeneous or flat. This phenomena might be due to the shorter leakage path in smaller active core.

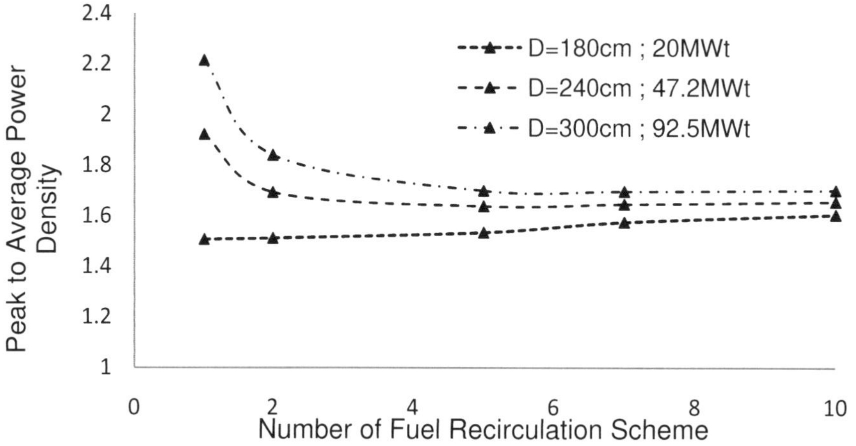

Figure 3. Parametric survey results for H/D=1.1, average power density 4W/cc

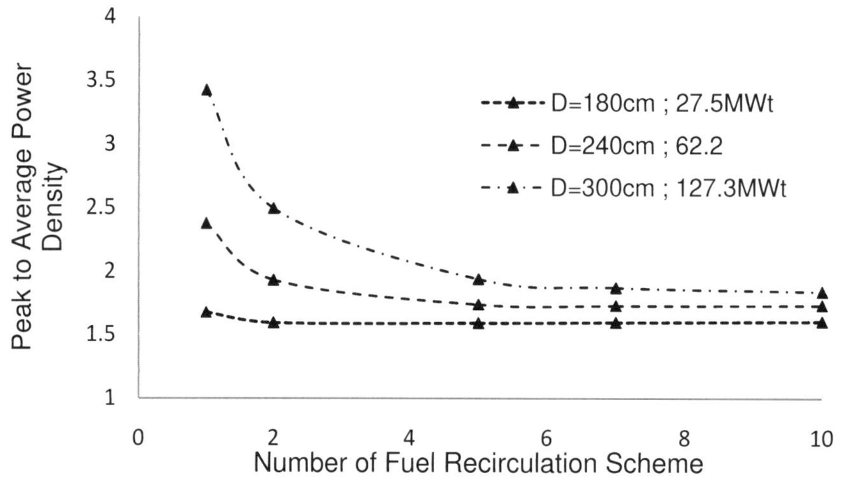

Figure 4. Parametric survey results for H/D=1.5, average power density 4W/cc

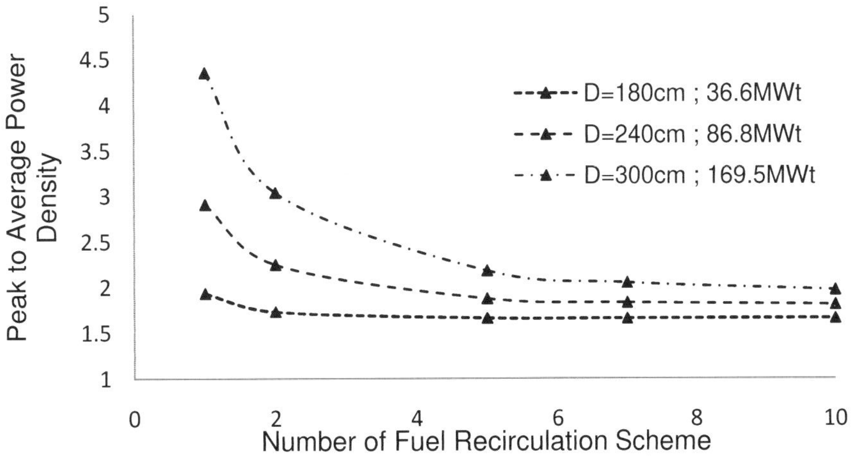

Figure 5. Parametric survey results for H/D=2, average power density 4W/cc

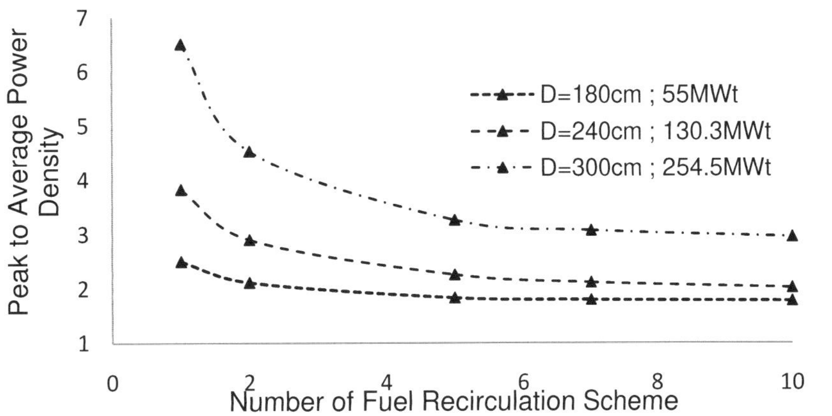

Figure 6. Parametric survey results for H/D=3, average power density 4W/cc

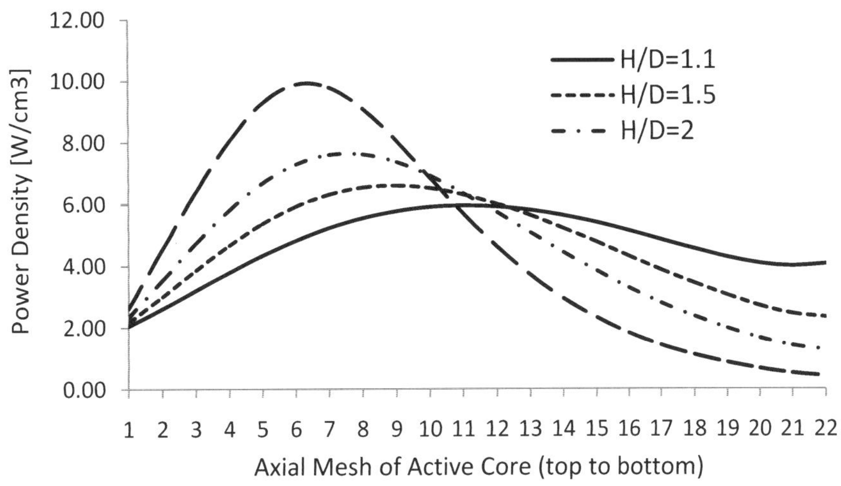

Figure 7. Axial power density profile for different H/D and average power density 4 W/cc

Effects of different average power density in several identic active core geometry to the power peaking of OTTO fuel scheme are shown in Figure 8. It shows clearly that different average power density of the core is insignificant to the power peaking effect of PBR core. Combining with the previous geometrical effect findings, the fact that increasing power density insignificant to the power peaking of OTTO fuel scheme PBR give an opportunity to further optimize the core efficiency, of a small core PBR, by increasing the nominal core power.

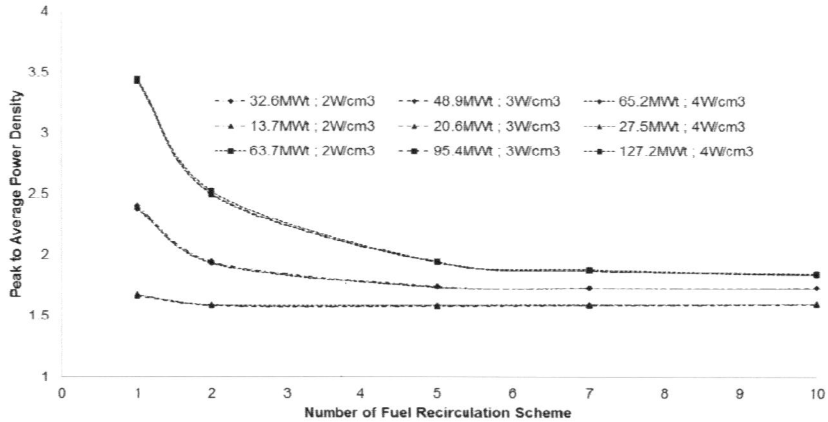

Figure 8. Parametric survey results for fixed H/D=1.5 for different average power density.

4. Conclusion
Parametric study on power peaking effect of OTTO fuel scheme PBR is already performed. Results of the study show that the active core diameter and core H/D parameter have a significant effect to the power peaking effect. However, this effect become smaller or negligible for certain small core design. Increasing power density is found to be insignificant to the power peaking effect. The finding of this study is an important guidance for optimum application of OTTO fuel scheme PBR.

Acknowledgment
We would like to express our gratitude to Dr. Hans Gougarat Idaho National Laboratory for the PEBBED code.

References
1. O. Seeger, M. Laurie, A. El Abjani, J. Ejton, D. Boudaud, D. Freis, P. Carbol, V.V. Rondinella, M. Futterer, H.J. Allelein, *Nucl. Eng. Des.* (2015)
2. M.V. Ramana, *Bulletin of the Atomic Scientist***72** (2016) 171
3. A.C. Kadak, *Engineering*, **2** (2016) 119
4. Z.Zhang, Y. Dong, F. Li, Z. Zhang, H. Wang, X. Huang, H. Li, B. Liu, X. Wu, *Engineering*, **2** (2016) 112
5. F.J. Wols, J.L. Kloosterman, D. Lathouwers, T.H.J.J van der Hagen, *Ann. Nucl. Energy***75** (2015) 542
6. C.Pohl, H.J. Rutten, *Nucl. Eng. Des.***248** (2012) 343
7. H.Q. Ho, T. Obara, *Ann. Nucl. Energy***87** (2016) 471
8. U.Hansen, R. Schulten, E. Teuchert, *Nucl. Sci. Eng.***47** (1972) 132
9. A. Bredimas, K. Kugeler, M.A. Futterer, *Nucl. Eng. Des.***271** (2014) 193

10. H.N. Tran, *Prog. Nucl. Energy***60** (2012) 47
11. T. Setiadipura, J.S. Pane, Zuhair, *JurnalPengembanganEnergi Nasional*, **18** (2016) 59 (in Indonesian)
12. T. Setiadipura, D. Irwanto, Zuhair, *Proceeding of Seminar Nasional on Nuclear Enegy Technology* (SENTEN) 2016
13. T. Setiadipura, T. Obara, *Ann. Nucl. Energy***71** (2014) 313
14. H.D Gougar, A.M. Ougouag, W.K. Terry, *Nucl. Sci. Eng.***165** (2010) 245
15. W.Y. Yoon, R.A. Grimesey, D.W. Nigg, R.L. Curtis, "COMBNE7.0 – A Portable ENDF/B-VII.0 Based Neutron Spectrum and Cross-Secction Generation Program", Idaho National Laboratory Report (2008)